W9-CPO-183

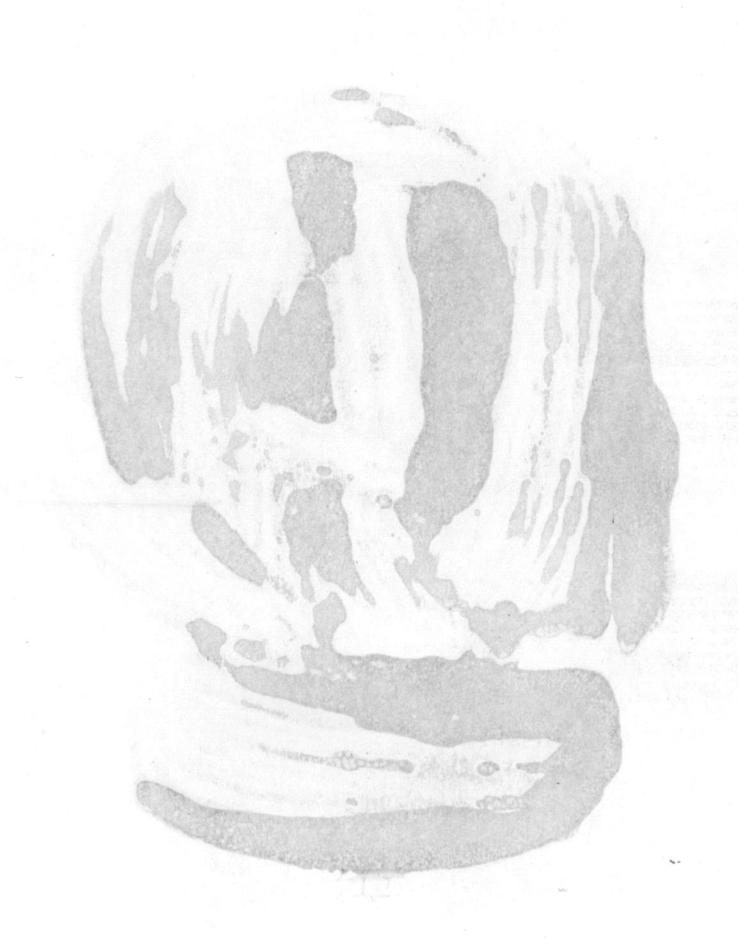

PROCEEDINGS OF THE
FIFTH BERKELEY SYMPOSIUM

VOLUME I

PROCEEDINGS *of the* FIFTH BERKELEY SYMPOSIUM ON MATHEMATICAL STATISTICS AND PROBABILITY

Held at the Statistical Laboratory
University of California
June 21–July 18, 1965
and
December 27, 1965–January 7, 1966

with the support of
University of California
National Science Foundation
National Institutes of Health
Air Force Office of Scientific Research
Army Research Office
Office of Naval Research

VOLUME I

STATISTICS

EDITED BY LUCIEN M. LE CAM
AND JERZY NEYMAN

UNIVERSITY OF CALIFORNIA PRESS
BERKELEY AND LOS ANGELES
1967

UNIVERSITY OF CALIFORNIA PRESS
BERKELEY AND LOS ANGELES
CALIFORNIA

CAMBRIDGE UNIVERSITY PRESS
LONDON, ENGLAND

COPYRIGHT © 1967, BY
THE REGENTS OF THE UNIVERSITY OF CALIFORNIA

The United States Government and its offices, agents, and employees, acting within the scope of their duties, may reproduce, publish, and use this material in whole or in part for governmental purposes without payment of royalties thereon or therefor. The publication or republication by the government either separately or in a public document of any material in which copyright subsists shall not be taken to cause any abridgment or annulment of the copyright or to authorize any use or appropriation of such copyright material without the consent of the copyright proprietor.

LIBRARY OF CONGRESS CATALOG CARD NUMBER: 49-8189

PRINTED IN THE UNITED STATES OF AMERICA

CONTENTS OF PROCEEDINGS VOLUMES I, II, III, IV, AND V

Volume I—Theory of Statistics

General Theory

Sequential Procedures

Information Theory

Nonparametric Procedures

Volume II—Part I—Theory of Probability

Probability on Algebraic Structures

Distributions in Functional Spaces

Stochastic Processes and Prediction

Martingales

Special Problems

Volume II, Part II—Theory of Probability

Markov Processes

Ergodic Theory

Volume III—Physical Sciences

Astronomy

Physics

Spectral Analysis

Control Processes

Reliability

Volume IV—Biology and Problems of Health

Information, Processing, and Cognition

Demography

Ecology

Epidemiology

Genetics

Chance Mechanisms in Living Organisms

Cellular Phenomena

Carcinogenesis

Experimentation

Decision Theory in Medical Diagnosis

Volume V—Weather Modification

Physical Background

Large Randomized Experiments

Nonrandomized Operations

Methodological Discussion

Observational Data

PREFACE

THE PURPOSE OF THE Berkeley Statistical Symposia, held every five years, is to stimulate research through lectures by carefully selected speakers and through prolonged personal contacts of scholars brought together from distant centers. Accordingly, particular Symposia last from four to seven weeks. On occasion, and this was the case with the Fifth Symposium, they are conducted in two parts, one in June–July, emphasizing theory, and the other in December–January, emphasizing applications. The winter part of the Fifth Symposium was held in conjunction with the 132nd Annual Meeting of the American Association for the Advancement of Science.

The Proceedings of the Symposia are intended to present a comprehensive cross-section of contemporary thinking on problems of probability and statistics and on selected fields of application. The rapid growth of research in statistics and especially in probability makes it increasingly difficult to achieve a complete coverage of the field, but sincere efforts are made to invite to the Symposia representatives of all the existing schools of thought, each individual having complete freedom of expression.

The organization of the theoretical part of the Fifth Berkeley Symposium was carried out, and the contributors were selected, with the participation of an Advisory Committee composed of Professors J. L. Doob, S. Karlin, and H. Robbins, delegated for this purpose by the American Mathematical Society and by the Institute of Mathematical Statistics. In addition, we had the assistance of Professor D. L. Burkholder, the Editor of the *Annals of Mathematical Statistics*. The interest of the American Mathematical Society and of the Institute of Mathematical Statistics and their help are deeply appreciated.

While a broad coverage of contemporary work in the theory of probability and statistics is difficult, the field of applications of these disciplines is currently so wide that the program of a single symposium can include no more than a few particular domains. The domains covered at the Fifth Symposium were selected on two principles. First, some applied problems appeared as subjects of studies by outstanding probabilists and statisticians invited to the Symposium on account of their work in theory. Second, an effort was made to delineate a few fields of substantive studies that appear particularly promising for probabilistic and statistical treatment. One of the most fruitful fields of this category is undoubtedly biology and problems of health. Here we profited greatly by the advice of Drs. LaMont Cole, Jerome Cornfield, F. N. David, Louis Hellman, Samuel Greenhouse, Hardin Jones, Samuel Karlin, David Krech, Lincoln Moses, Curt Stern, Michael B. Shimkin, and Cornelius Tobias. Quite a few of these colleagues are connected with the broad research activity of the National Institutes of Health and helped to bring to our attention many novel and important subfields of research.

In the field of astronomy we are deeply indebted to Drs. N. U. Mayall, Rudolph Minkowski, and Thornton L. Page.

For advice in the field of meteorology we are grateful to Drs. Earl Droessler, James Hughes, Dwight B. Kline, Morris Neiburger, Jerome Spar, Edward P. Todd, and P. H. Wyckoff. Special thanks are due to Dr. Kenneth B. Spengler, Secretary of the American Meteorological Society.

Following the established tradition, Volume I of the present *Proceedings* is given to the theory of statistics. Volume II is devoted to the theory of probability. Because of the large amount of material, about 1000 pages in print, this volume had to be divided into two parts formed through a somewhat arbitrary classification of papers. Volume III includes papers related to physical sciences: astronomy, theory of control, physics, and the theory of reliability. Volume IV, on biology and problems of health, includes papers on information and brain phenomena, on chance mechanisms in live organisms, on epidemiology, on genetics, on medical diagnosis, on clinical trials, on carcinogenesis and cellular phenomena, on demography, and on ecology. Some of these subdomains are already subjects of well developed statistical treatment. Others appear to offer interesting and important possibilities.

Compared to the *Proceedings* of the earlier Symposia, Volume V, being entirely given to the problem of artificial weather modification, is an innovation. With the classification adopted for the first four volumes, weather modification would fit Volume III. It is assigned a special volume because of the specificity of the domain and because of its separateness from all the other fields dealt with in Volume III. Also, the novelty of the problem of weather modification, considered by itself and as a field for statistical research, indicated the desirability of producing a comprehensive coverage of the more extensive experiments. Finally, it appears probable that the readership of the material being published in Volume V will be essentially different from that expected to be interested in Volume III.

The Fifth Symposium would not have been possible without very substantial financial support from various sources. Hearty thanks are due to Dr. Clark Kerr, President of the University of California, for a special grant made several years in advance of the Symposium. Without this grant, no planning and no initial steps for the organization of the Symposium would have been possible. This initial triggering grant of the University of California was later supplemented by the subsidy of the University Editorial Committee, without which the publication of the *Proceedings*, to be sold at a reasonable price, would have been a very difficult problem. To a very considerable extent, the theoretical part of the Symposium and the part concerned with physical sciences, were financed by The Program in Mathematics of the National Science Foundation, by the Air Force Office of Scientific Research, by the Army Research Office, and by the Office of Naval Research. The large program on biology and problems of health was made possible by a grant of the National Institutes of Health. The program on weather modification was organized using a grant of the Atmospheric Sciences Section of the National Science Foundation. Finally, we wish to record special help from the Office of Naval Research, in the form of air transportation for a number of foreign participants in the Symposium.

It is our pleasure to acknowledge gratefully the generosity of the governmental institutions enumerated. The vitality of our Symposia and the growth of the *Proceedings*, from 500 pages in 1945 to about 3,000 in 1965, seem to indicate that the funds provided are being spent to fill a real need.

The problems connected with the publication of such an amount of technical material are very substantial, especially since some of the material was originally written in languages other than English and required translation. All efforts were made toward speedy publication at a reasonable price, and we are pleased to acknowledge the excellent cooperation and assistance we received from the University of California Press.

For the translation of manuscripts, we are indebted to Drs. Amiel Feinstein, Morris Friedman, and Mrs. C. Stein. We are also indebted to several of our colleagues in the Department for work connected with the preparation of manuscripts for the printer. Special thanks are due to Professors E. L. Scott, M. Loève, E. W. Barankin, to Drs. Carlos-Barbosa Dantas, W. Bühler, Nora Smiriga, Grace Yang, to Mr. Steve Stigler, and Mrs. M. Darland. We are pleased to acknowledge the technical help of Mrs. Sharlee Guise and Mrs. Carol Rule Roth.

For taking care of the many complexities of editing technical manuscripts we are deeply indebted to Miss Susan Jenkins whose patience and skill deserve superlative praise. Thanks are also due to Mrs. Virginia Thompson for her greatly appreciated assistance in the same process.

To Mr. August Frugé, the Director of the University of California Press, we extend heartfelt thanks for financial, technical, and moral support in publishing so much difficult material. Special thanks are due also to Joel Walters, Editor of the University of California Press. In spite of all our efforts, we found ourselves unable to keep up with the schedule of publication proposed by the Press, but we must thank them for helping us to keep the delays at a minimum and for producing a publication in accordance with the usual excellent standards of the University of California Press.

Many thanks are due to our Administrative Assistant, Miss M. Genelly for taking care of many financial and organizational difficulties. For transportation, housing, and other logistic problems connected with the organization of the meeting itself, very valuable assistance was received from the staff of the Laboratory and in particular from Miss June Haynes and Mrs. J. Lovasich.

As was the case on many earlier simliar occasions, for supervising and taking care of the innumerable intricacies of local organization we are deeply indebted to our colleague Professor Elizabeth L. Scott. It is a pleasure to express here our deepest appreciation.

Last but not least we wish to thank the Department of Statistics of the University of California, Berkeley, and all our colleagues therein, for their sympathetic attitude and help. Particular thanks are due to David Blackwell.

During the winter part of the Fifth Symposium, the Statistical Laboratory lost one of its organizers as well as one of its most active members. Our colleague and cordial friend, Professor Evelyn Fix died of a heart attack on December 30, only a few hours after she acted as one of the hostesses at the banquet of the Symposium. *Sit ei terra levis!*

LUCIEN LE CAM

JERZY NEYMAN
Director, Statistical Laboratory

April, 1967

CONTENTS

General Theory

Sequential Procedures

Information Theory

SOME INEQUALITIES AMONG BINOMIAL AND POISSON PROBABILITIES

T. W. ANDERSON[1]
COLUMBIA UNIVERSITY
and
S. M. SAMUELS[2]
PURDUE UNIVERSITY

1. Introduction

The binomial probability function

$$(1.1)\qquad b(k; n, p) = \binom{n}{k} p^k (1-p)^{n-k}, \qquad k = 0, 1, \cdots, n,$$
$$= 0, \qquad k = n+1, \cdots,$$

can be approximated by the Poisson probability function

$$(1.2)\qquad p(k; \lambda) = e^{-\lambda}\frac{\lambda^k}{k!}, \qquad k = 0, 1, \cdots,$$

for $\lambda = np$ if n is sufficiently large relative to λ. Correspondingly, the binomial cumulative distribution function

$$(1.3)\qquad B(k; n, p) = \sum_{j=0}^{k} b(j; n, p), \qquad k = 0, 1, \cdots,$$

is approximated by the Poisson cumulative distribution function

$$(1.4)\qquad P(k; \lambda) = \sum_{j=0}^{k} p(j; \lambda), \qquad k = 0, 1, \cdots,$$

for $\lambda = np$. In this paper it is shown that the error of approximation of the binomial cumulative distribution function $P(k; np) - B(k; n, p)$ is positive if $k \leq np - np/(n+1)$ and is negative if $np \leq k$. In fact, $B(k; n, \lambda/n)$ is monotonically increasing for all n $(\geq \lambda)$ if $k \leq \lambda - 1$ and for all $n \geq k/(\lambda - k)$ if $\lambda - 1 < k < \lambda$, and is monotonically decreasing for all n $(\geq k)$ if $\lambda \leq k$. Thus

[1] Research supported by the Office of Naval Research under Contract Number Nonr-4259(08), Project Number NR 042-034. Reproduction in whole or in part permitted for any purpose of the United States Government.

[2] Research supported in part by National Science Foundation Grant NSF-GP-3694 at Columbia University, Department of Mathematical Statistics, and in part by Aerospace Research Laboratories Contract AF 33(657)11737 at Purdue University.

for most practical purposes the Poisson approximation overestimates tail probabilities, and the margin of overestimation decreases with n.

The probability function $b(k; n, \lambda/n)$ increases with n [to $p(k; \lambda)$] if $k \leq \lambda + \frac{1}{2} - (\lambda + \frac{1}{4})^{1/2}$ or if $\lambda + \frac{1}{2} + (\lambda + \frac{1}{4})^{1/2} \leq k$. These facts imply that for given n and λ $P(k; \lambda) - B(k; n, \lambda/n)$ increases with respect to k, to $\lambda + \frac{1}{2} - (\lambda + \frac{1}{4})^{1/2}$ and decreases with respect to k from $\lambda + \frac{1}{2} + (\lambda + \frac{1}{4})^{1/2}$.

When the Poisson distribution is used to approximate the binomial distribution for determining significance levels, in nearly all cases the actual significance level is less than the nominal significance level given by the Poisson, and the probability of Type I error is overstated. Similarly, the actual confidence level of confidence limits based on the Poisson approximation is greater than the nominal level. Section 4 gives precise statements of these properties.

In section 5 another approach is developed to the monotonicity of $B(k; n, \lambda/n)$ and $b(k; n, \lambda/n)$. Although the results are not as sharp as those in sections 2 and 3, the methods are interesting.

Throughout the paper we assume $\lambda > 0$ and $n \geq \lambda$ (or $1 \geq \lambda/(n-1)$ when used as a binomial parameter).

2. Inequalities among cumulative distribution functions

2.1. *A general inequality.* In this section we show that $B(k; n, \lambda/n)$ increases with n to $P(k; \lambda)$ if k is small relative to λ and decreases to $P(k; \lambda)$ if k is large.

For this purpose the following theorem of Hoeffding will be needed.

THEOREM (Hoeffding [1]). *Let $F(k)$ be the probability of not more than k successes in n independent trials where the i-th trial has probability p_i of success. Let $\lambda = p_1 + p_2 + \cdots + p_n$. Then*

$$(2.1)\qquad \begin{aligned} B(k; n, \lambda/n) &\geq F(k), \qquad & k \leq \lambda - 1,\\ B(k; n, \lambda/n) &\leq F(k), \qquad & k \geq \lambda. \end{aligned}$$

Equality holds only if $p_1 = \cdots = p_n = \lambda/n$.

It is possible to obtain from this the following result.

THEOREM 2.1.

$$(2.2)\qquad \begin{aligned} B(k; n, \lambda/n) &> B(k; n-1, \lambda/(n-1)), \qquad & k \leq \lambda - 1,\\ B(k; n, \lambda/n) &< B(k; n-1, \lambda/(n-1)), \qquad & k \geq \lambda. \end{aligned}$$

PROOF. If we choose $p_1 = 0$, $p_2 = \cdots = p_n = \lambda/(n-1)$, then $F(k) = B(k; n-1, \lambda/(n-1))$, and the result follows from Hoeffding's theorem.

COROLLARY 2.1.

$$(2.3)\qquad \begin{aligned} P(k; \lambda) &> B(k; n, \lambda/n), \qquad & k \leq \lambda - 1,\\ P(k; \lambda) &< B(k; n, \lambda/n), \qquad & k \geq \lambda. \end{aligned}$$

Alternatively, we may take $p_1 = 1$, $p_2 = \cdots = p_n = (\lambda - 1)/(n-1)$, in which case $F(k) = B(k-1; n-1, (\lambda-1)/(n-1))$. Hence we have the next theorem.

THEOREM 2.2.

$$B(k; n, \lambda/n) > B(k-1; n-1, (\lambda-1)/(n-1)), \quad k \leq \lambda - 1,$$
$$(2.4)\qquad B(k; n, \lambda/n) < B(k-1; n-1, (\lambda-1)/(n-1)), \quad k \geq \lambda.$$

The limits of (2.4) as $n \to \infty$ give the following corollary.

COROLLARY 2.2.

$$P(k; \lambda) \geq P(k-1; \lambda-1), \qquad k \leq \lambda - 1,$$
$$(2.5)\qquad P(k; \lambda) \leq P(k-1; \lambda-1), \qquad k \geq \lambda.$$

This corollary will be used in section 4. If λ increases by steps of 1, corollary 2.2 indicates that the probabilities in the tails $[0, \lambda - a]$ and $[\lambda + c, \infty]$ are increasing with λ if $a \geq 1$ and $c \geq 1$.

2.2. *Monotonicity near the mean.* If λ is an integer, theorem 2.1 includes all values of k. If λ is not an integer, what happens at the value of k which is between $\lambda - 1$ and λ? We shall show that $B(k; n, \lambda/n)$ increases with n if n is sufficiently large.

THEOREM 2.3.

$$B(k; n, \lambda/n) > B(k; n-1, \lambda/(n-1)), \quad k < \lambda, n \geq \lambda/(\lambda - k),$$
$$(2.6)\qquad B(k; n, \lambda/n) < B(k; n-1, \lambda/(n-1)), \quad k \geq \lambda.$$

PROOF. The second inequality is part of theorem 2.1. The first inequality is stated for $nk/(n-1) \leq \lambda$. It will follow from that inequality for $nk/(n-1) \leq \lambda \leq (n-1)k/(n-2)$ [lemma 2.2] and the following lemma.

LEMMA 2.1. *The function* $D(k; n, \lambda) = B(k; n, \lambda/n) - B(k; n-1, \lambda/(n-1))$ *has at most one sign change from negative to positive in the interval* $0 < \lambda \leq n - 1$.

PROOF. The case $k = n - 1$ is true since $B(n-1; n-1, \lambda/(n-1)) \equiv 1$. Hence we consider only $k < n - 1$. Since $D(k; n, \lambda)$ is 0 for $\lambda = 0$ and positive for $\lambda = n - 1$, it suffices to show that the derivative has at most two sign changes, from negative to positive to negative. Now

$$(2.7)\qquad \frac{d}{d\lambda} D(k; n, \lambda) = b(k; n-2, \lambda/(n-1)) - b(k; n-1, \lambda/n),$$

$$(2.8)\qquad \frac{b(k; n-2, \lambda/(n-1))}{b(k; n-1, \lambda/n)} = \frac{(n-1-k)(1+1/(n-1))^{n-1}}{(n-1-\lambda)(1+1/(n-1-\lambda))^{n-1-k}}.$$

Then (2.8) is increasing for $\lambda < k + 1$ and decreasing for $\lambda > k + 1$. Hence (2.7) has the desired sign change property and the lemma is proved.

LEMMA 2.2. *If* $nk/(n-1) \leq \lambda \leq (n-1)k/(n-2)$, *then*

$$(2.9)\qquad B(k; n, \lambda/n) > B(k; n-1, \lambda/(n-1)).$$

PROOF. In n independent trials with $n - 1$ of the probabilities of success equal to λ/n and the remaining probability equal to p, the probability of at most k successes is

$$(2.10)\qquad pB(k-1; n-1, \lambda/n) + (1-p)B(k; n-1, \lambda/n).$$

This is a decreasing function of p and is equal to $B(k; n, \lambda/n)$ when $p = \lambda/n$. The value of p which makes (2.10) equal to $B(k; n-1, \lambda/(n-1))$ is

$$p^* = \frac{B(k, n-1, \lambda/n) - B(k, n-1, \lambda/(n-1))}{b(k, n-1, \lambda/n)} \tag{2.11}$$

$$= \frac{n-1-k}{(\lambda/n)^k(1-\lambda/n)^{n-1-k}} \int_{\lambda/n}^{\lambda/(n-1)} u^k(1-u)^{n-2-k}\, du.$$

It suffices, then, to show that under the hypothesis $p^* > \lambda/n$. If

$$\lambda \leq (n-1)k/(n-2), \tag{2.12}$$

then the integrand in (2.11) is increasing in u over the range of integration. Hence for $\lambda \leq (n-1)k/(n-2)$

$$p^* > \frac{(n-1-k)(\lambda/(n-1) - \lambda/n)(\lambda/n)^k(1-\lambda/n)^{n-2-k}}{(\lambda/n)^k(1-\lambda/n)^{n-1-k}} \tag{2.13}$$

$$= \frac{(n-1-k)\lambda}{(n-1)(n-\lambda)},$$

which is at least λ/n if

$$\lambda \geq nk/(n-1), \tag{2.14}$$

which proves the lemma.

Theorem 2.2 follows from lemma 2.2, because lemma 2.1 indicates that if (2.9) holds for a given value of λ, it holds for all larger values.

COROLLARY 2.3.

$$\begin{aligned} P(k; \lambda) &> B(k; n, \lambda/n), \qquad k < \lambda, n \geq k/(\lambda - k), \\ P(k; \lambda) &< B(k; n, \lambda/n), \qquad k \geq \lambda. \end{aligned} \tag{2.15}$$

2.3. *Special cases.* Lemma 2.1 implies that for each n there is a number λ_n such that $B(k; n+1, \lambda/(n+1)) - B(k; n, \lambda/n)$ is negative if $\lambda < \lambda_n$ and is positive if $\lambda > \lambda_n$, and theorem 2.3 indicates that $k < \lambda_n < k + k/n$. There is no simple expression for λ_n. Is it true that $\lambda_n - k \sim k/n$? The answer is no, as can be seen in the case $k = 1$. The first derivative of $B(1; n, \lambda/n)$ with respect to n approaches 0 as $n \to \infty$, and the second derivative is negative for all $n > \lambda$ if $\lambda \geq (\frac{4}{3})^{1/2}$, while if $\lambda < (\frac{4}{3})^{1/2}$ it is negative if

$$n \geq [\lambda^2 + \lambda(4 - 3\lambda^2)^{1/2}/[2(\lambda^2 - 1)]. \tag{2.16}$$

Equivalently, the second derivative is negative if $\lambda > n/(n^2 - n + 1)^{1/2} \sim 1 + 1/(2n)$. Thus for $k = 1$, $\lambda_n < n/(n^2 - n + 1)^{1/2}$.

For the case $k = n - 1$, we can explicitly evaluate λ_n. We have $\lambda_n = n + 2 + 1/n - (1 + 1/n)^{n+1}$ so that $\lambda_n - k$ approaches $3 - e \sim .282$ as $n \to \infty$. Thus λ_n is smaller than $n - 1 + (n-1)/n$, as given by our general result. In particular, if $k = 1$ and $n = 2$, $\lambda_n = \frac{9}{8}$, which is smaller than $2/\sqrt{3}$ as given by the preceding paragraph.

3. Inequalities among probability functions

Let

$$r(k; n) = b(k; n, \lambda/n)/b(k; n-1, \lambda/(n-1)), \tag{3.1}$$

$$d(k; n) = b(k; n, \lambda/n) - b(k; n-1, \lambda/(n-1)). \tag{3.2}$$

Then

$$r(k; n)/r(k-1; n) = (n+1-k)(n-1-\lambda)/[(n-k)(n-\lambda)], \tag{3.3}$$

which is less or greater than one according to whether k is less or greater than $\lambda + 1$. Since $d(0; n) > 0$ (by theorem 2.3), $d(n; n) > 0$, and $\sum_{k=0}^{n} d(k; n) = 0$, we have the following proposition.

PROPOSITION 3.1. *For suitable nonnegative integers a_n and c_n, depending on λ,*

$$\begin{aligned} d(k; n) &> 0, \qquad 0 \le k \le a_n \quad \text{or} \quad c_n \le k \le n, \\ d(k; n) &\le 0, \qquad a_n < k < c_n, \end{aligned} \tag{3.4}$$

and

$$\begin{aligned} &\max_{0 \le k \le n} [B(k; n, \lambda/n) - B(k; n-1, \lambda/(n-1))], \\ &\min_{0 \le k \le n} [B(k; n, \lambda/n) - B(k; n-1, \lambda/(n-1))] \end{aligned} \tag{3.5}$$

occur at $k = a_n$ and c_n, respectively.

The following theorem gives a lower bound for a_n and an upper bound for c_n.

THEOREM 3.1. *If $0 \le k \le \lambda + \frac{1}{2} - (\lambda + \frac{1}{4})^{1/2}$ or if $\lambda + \frac{1}{2} + (\lambda + \frac{1}{4})^{1/2} \le k \le n$, then*

$$b(k; n, \lambda/n) > b(k; n-1, \lambda/(n-1)); \tag{3.6a}$$

if $\lambda + \frac{1}{2} - (\lambda + \frac{1}{4})^{1/2} < k < \lambda + \frac{1}{2} + (\lambda + \frac{1}{4})^{1/2}$ and n is sufficiently large,

$$b(k; n, \lambda/n) < b(k; n-1, \lambda/(n-1)). \tag{3.6b}$$

PROOF. The conditions on k in the first part of the theorem are equivalent to $\lambda \le k - k^{1/2}$ and $\lambda \ge k + k^{1/2}$. Since

$$r(k; n) = b(k; n, \lambda/n)/b(k; n-1, \lambda/(n-1)) \tag{3.7}$$

decreases for $\lambda < k$ and increases for $\lambda > k$, it suffices to consider $b(k; n, \lambda/n)$ at only the values $\lambda = k \pm k^{1/2}$. We prove (3.6a) by showing that $\log b(k; n, \lambda/n)$ is an increasing function of n at these two values of λ. Now

$$\begin{aligned} &\log b(k; n, \lambda/n) - \log \lambda^k + \log k! \\ &\quad = \sum_{j=1}^{k-1} \log (1 - j/n) + (n-k) \log (1 - \lambda/n) \\ &\quad = -\sum_{j=1}^{k-1} \sum_{r=1}^{\infty} (j/n)^r/r - (n-k) \sum_{r=1}^{\infty} (\lambda/n)^r/r \\ &\quad = -\lambda - \sum_{r=1}^{\infty} (1/n)^r \left[\sum_{j=1}^{k-1} j^r/r + \lambda^{r+1}/(r+1) - k\lambda^r/r \right]. \end{aligned} \tag{3.8}$$

Hence the first part of the theorem is proved if for $\lambda = k \pm k^{1/2}$, each bracketed expression is nonnegative. (Note that, if $k = 0$ or 1, the sums on j should be taken to be zero, and we see immediately that in these cases the theorem is true.)

For $r = 1$ the bracketed expression vanishes. In the general term, we replace the sum by a more manageable expression as follows: by convexity of z^r,

$$\int_x^{x+1} z^r\,dz \leq \tfrac{1}{2}[x^r + (x+1)^r]. \tag{3.9}$$

Hence,

$$\begin{aligned} \frac{(k-1)^{r+1}}{r+1} + \frac{(k-1)^r}{2} &= \int_0^{k-1} z^r\,dz + \tfrac{1}{2}(k-1)^r \\ &= \sum_{x=0}^{k-2} \int_x^{x+1} z^r\,dz + \tfrac{1}{2}(k-1)^r \\ &\leq \tfrac{1}{2} \sum_{x=0}^{k-2} [x^r + (x+1)^r] + \tfrac{1}{2}(k-1)^r \\ &= \sum_{j=0}^{k-1} j^r. \end{aligned} \tag{3.10}$$

Thus, each bracketed expression in (3.8) is at least

$$[(k-1)^{r+1} + \tfrac{1}{2}(r+1)(k-1)^r + r\lambda^{r+1} - (r+1)k\lambda^r]/[r(r+1)]. \tag{3.11}$$

Setting $\lambda = k + k^{1/2}$ and letting $u = k^{1/2}$, $v = k^{-1/2} \leq 1$, we can write (3.11) as

$$\begin{aligned} &[(u+1)^{r+1}(u-1)^{r+1} + \tfrac{1}{2}(r+1)(u+1)^r(u-1)^r + ru^{r+1}(u+1)^{r+1} \\ &\qquad - (r+1)u^{r+2}(u+1)^r]/[r(r+1)] \\ &= (u+1)^r\{(u-1)^r[u^2 + \tfrac{1}{2}(r-1)] - u^r(u^2 - ru)\}/[r(r+1)] \\ &= (u+1)^r u^{r+2}\{(1-v)^r[1 + \tfrac{1}{2}(r-1)v^2] - (1-rv)\}/[r(r+1)] \\ &\geq 0, \end{aligned} \tag{3.12}$$

since $(1-v)^r \geq 1 - rv$ for $v \leq 1$, $r \geq 1$. A similar argument establishes the result at $\lambda = k - k^{1/2}$, which completes the proof of the first part of the theorem. The second part of the theorem follows since the coefficient of $1/n$ in (3.8) is positive for $k - k^{1/2} < \lambda < k + k^{1/2}$.

COROLLARY 3.1. *If $m > n$ and if $0 \leq k \leq \lambda + \frac{1}{2} - (\lambda + \frac{1}{4})^{1/2}$ or $\lambda + \frac{1}{2} + (\lambda + \frac{1}{4})^{1/2} \leq k \leq m$, then*

$$p(k;\lambda) > b(k; m, \lambda/m) > b(k; n, \lambda/n). \tag{3.13}$$

Hence,

$$\max_{0 \leq k \leq \lambda + \frac{1}{2} - (\lambda + \frac{1}{4})^{1/2}} [P(k;\lambda) - B(k; n, \lambda/n)] \tag{3.14}$$

occurs at the largest integer which is not greater than $\lambda + \frac{1}{2} - (\lambda + \frac{1}{4})^{1/2}$, and

$$\min_{\lambda + \frac{1}{2} + (\lambda + \frac{1}{4})^{1/2} \leq k \leq n} [P(k;\lambda) - B(k; n, \lambda/n)] \tag{3.15}$$

occurs at the smallest integer which is at least $\lambda + \frac{1}{2} + (\lambda + \frac{1}{4})^{1/2}$.

4. Applications to statistical inference

4.1. *Testing hypotheses.* Suppose n independent trials are made, each trial with probability p of success. Consider testing the null hypothesis $p = p_0$, where p_0 is specified. Against alternatives $p < p_0$ a (uniformly most powerful) nonrandomized test is a rule to reject the null hypothesis if the observed number of successes is less than or equal to an integer $\underline{k}$. The significance level of the test is $B(\underline{k}; n, p_0)$. It may be approximated by $P(\underline{k}; np_0)$. If $\underline{k} \leq np_0 - 1$, then by corollary 2.1, $B(\underline{k}; n, p_0) < P(\underline{k}; np_0)$. (Corollary 2.3 allows us to raise the bound on k to $np_0 - np_0/(n + 1)$, but we shall use here the simpler bound.) The procedure is conservative in the sense that the actual significance level (defined by the binomial distribution) is less than the nominal significance level (given by the Poisson distribution). The condition $\underline{k} \leq np_0 - 1$, which can be verified by the statistician in defining the test procedure, enables him to say that the probability of rejecting the null hypothesis when it is true is less than the approximating probability.

Against alternatives $p > p_0$ a (uniformly most powerful) nonrandomized test consists in rejecting the null hypothesis if the number of successes is greater than or equal to an integer $\bar{k}$. The significance level of the test is

$$1 - B(\bar{k} - 1; n, p_0), \tag{4.1}$$

which may be approximated by $1 - P(\bar{k} - 1; np_0)$. If $np_0 \leq \bar{k} - 1$, then by corollary 2.1, $1 - B(\bar{k} - 1; n, p_0) < 1 - P(\bar{k} - 1; np_0)$ and the procedure is conservative. A nonrandomized test against two-sided alternatives $p \neq p_0$ consists in rejecting the null hypothesis if the number of successes is less than or equal to $\underline{k}$ or greater than or equal to $\bar{k}$. The significance level $B(\underline{k}; n, p_0) + 1 - B(\bar{k} - 1; n, p_0)$ may be approximated by $P(\underline{k}; np_0) + 1 - P(\bar{k} - 1; np_0)$. If $\underline{k} + 1 \leq np_0 \leq \bar{k} - 1$, the procedure is conservative.

A (uniformly most powerful) randomized test of the null hypothesis $p = p_0$ against alternatives $p < p_0$ consists of a rule to reject the null hypothesis if the observed number of successes is less than $\underline{k}$ and to reject the null hypothesis with probability $\underline{\pi}$ if the number of successes is $\underline{k}$. The significance level is $\underline{\pi}B(\underline{k}; n, p_0) + (1 - \underline{\pi})B(\underline{k} - 1; n, p_0)$, which may be approximated by $\underline{\pi}P(\underline{k}; np_0) + (1 - \underline{\pi})P(\underline{k} - 1; np_0)$. The Poisson approximation overestimates the significance level if $\underline{k} \leq np_0 - 1$, since both $P(k; np_0)$ and $P(k - 1; np_0)$ overestimate the corresponding binomial probabilities. A (uniformly most powerful) randomized test against alternatives $p > p_0$ consists of a rule to reject the null hypothesis if the observed number of successes is greater than $\bar{k}$ and to reject the null hypothesis with probability $\bar{\pi}$ if the number of successes is $\bar{k}$. The significance level is $1 - \bar{\pi}B(\bar{k} - 1; n, p_0) - (1 - \bar{\pi})B(\bar{k}; n, p_0)$, which may be approximated by $1 - \bar{\pi}P(\bar{k} - 1; np_0) - (1 - \bar{\pi})P(\bar{k}; np_0)$. The approximation is an overestimate if $np_0 \leq \bar{k} - 1$. A two-sided randomized test consists in rejecting the null hypothesis if the number of successes is less than $\underline{k}$ or greater than $\bar{k}$, rejecting the null hypothesis with probability $\underline{\pi}$ if the number of successes is $\underline{k}$ and rejecting the null hypothesis with probability $\bar{\pi}$ if the number of successes is $\bar{k}$. The significance level, $\underline{\pi}B(\underline{k}; n, p_0) + (1 - \underline{\pi})B(\underline{k} - 1; n, p_0) + 1 - \bar{\pi}B(\bar{k} - 1; n, p_0) - (1 - \bar{\pi})B(\bar{k}; n, p_0)$ is overestimated by $\underline{\pi}P(k; np_0) + (1 - \underline{\pi})P(\underline{k} - 1; np_0) + 1 - \bar{\pi}P(\bar{k} - 1; np_0) - (1 - \bar{\pi})P(\bar{k}; np_0)$ if $\underline{k} + 1 \leq np_0 \leq \bar{k} - 1$.

Some criteria for satisfying conditions for overestimation by the approximation will be derived from the following theorems.

THEOREM 4.1. *If* $P(\ell; \lambda) \leq P(h; m)$, *then* $m + \ell - h \leq \lambda$, *for* $\ell = h, h + 1, \cdots, m = h + 1, h + 2, \cdots, h = 0, 1, \cdots$.

PROOF. $P(\ell; \lambda) \leq P(h; m)$ implies $P(\ell; \lambda) \leq P(\ell; \ell + m - h)$ because $P(h; m) \leq P(\ell; \ell + m - h)$ by corollary 2.2, and $P(\ell; \lambda) \leq P(\ell; \ell + m - h)$ implies $\ell + m - h \leq \lambda$ because $P(\ell; \nu)$ is a decreasing function of ν. [In fact $dP(\ell; \nu)/d\nu = -p(\ell; \nu)$.]

Theorem 4.1 with $\ell = k$, $h = 0$, and $m = 1$ states that if $P(k; \lambda) \leq P(0; 1) = e^{-1} \sim .3679$, then $k \leq \lambda - 1$, which implies that $B(k; n, \lambda/n)$ is increasing in n to $P(k; \lambda)$. The theorem with $\ell = k - 1$, $h = 0$, and $m = 2$ states that if $P(k - 1; \lambda) \leq P(0, 2) = e^{-2} \sim .1353$ then $k \leq \lambda - 1$, which implies that $B(k; n, \lambda/n)$ as well as $B(k - 1; n, \lambda/n)$ are increasing in n. Hence

$$\underline{\pi}P(\underline{k}; np_0) + (1 - \underline{\pi})P(\underline{k} - 1; np_0) \leq P(0; 2) \tag{4.2}$$

implies $P(\underline{k} - 1; np_0) \leq P(0; 2)$, which implies

$$\underline{\pi}B(\underline{k}; n, p_0) + (1 - \underline{\pi})B(\underline{k} - 1; n, p_0) < \underline{\pi}P(\underline{k}; np_0) + (1 - \underline{\pi})P(\underline{k}; np_0). \tag{4.3}$$

THEOREM 4.2. *If* $P(h; m) \leq P(\ell; \lambda)$, *then* $\lambda \leq m + \ell - h$, *for* $\ell = h, h + 1, \cdots, h = m, m + 1, \cdots, m = 0, 1, \cdots$.

PROOF. $P(h; m) \leq P(\ell; \lambda)$ implies $P(\ell; \ell + m - h) \leq P(\ell; \lambda)$ because $P(\ell; \ell + m - h) \leq P(h; m)$ by corollary 2.2, and $P(\ell; \ell + m - h) \leq P(\ell; \lambda)$ implies $\lambda \leq m + \ell - h$ because $P(\ell; \nu)$ is a decreasing function of ν.

Theorem 4.2 with $\ell = k - 1$ and $h = m = 1$ states that if $P(k - 1; \lambda) \geq P(1; 1) = 2e^{-1} \sim .7358$, then $\lambda \leq k - 1$; hence $1 - B(k - 1; n, \lambda/n)$ is increasing in n. In this case, $k = 2, 3, \cdots$. The test which rejects the null hypothesis on the basis of one or more successes ($\bar{k} = 1$) is not covered; in fact, $1 - B(0; n, \lambda/n)$ is decreasing in n. The theorem with $\ell = k$, $h = 2$, and $m = 1$ states that if $P(k; \lambda) \geq P(2; 1) = 5e^{-1}/2 \sim .9197$, then $\lambda \leq k - 1$, which implies that $B(k - 1; n, \lambda/n)$ as well as $B(k; n, \lambda/n)$ are decreasing in n. Hence

$$1 - \bar{\pi}P(\bar{k} - 1; np_0) - (1 - \bar{\pi})P(\bar{k}; np_0) \leq 1 - P(2; 1) \sim .0803 \tag{4.4}$$

implies $1 - P(\bar{k}; np_0) \leq 1 - P(1; 2)$, which implies

$$1 - \bar{\pi}B(\bar{k} - 1; n, p_0) - (1 - \bar{\pi})B(\bar{k}; n, p_0) < 1 - \bar{\pi}P(\bar{k} - 1; np_0) - (1 - \bar{\pi})P(\bar{k}; np_0). \tag{4.5}$$

Since $k = 2, 3, \cdots$ here, the result does not cover tests for which $k = 1$ leads to rejection, with or without randomization.

These properties apply also to two-sided tests. If the nominal significance level is less than a given number, the nominal probability of rejection in each tail is less than that number, and we make the above deductions about $\lambda = np_0$.

Our conclusions can be simplified to the following rules:

RULE 1. *The actual significance level of a nonrandomized test of the parameter of a binomial distribution is less than the approximate significance level based on the Poisson distribution if the approximate significance level is less than or equal to .26, except for the test which accepts the null hypothesis for 0 success and rejects for every positive number of successes.*

RULE 2. *The actual significance level of a randomized test of the parameter of a binomial distribution is less than the approximate significance level based on the Poisson distribution if the approximate significance level is less than or equal to .08, except possibly for tests which accept the null hypothesis with some positive probability for 0 success, reject the null hypothesis with some positive probability for one success and always reject the null hypothesis for more than one success.*

Rule 2 could possibly be improved, because for any particular $\bar{\pi}$ (< 1), (4.5) may hold without $B(\bar{k} - 1; n, \lambda/n)$ decreasing in n. However, a study of conditions for every $\bar{\pi}$ would be very complicated; for example, lemma 2 of Samuels [2] shows that if $\bar{\pi} = \frac{1}{2}$, then $\frac{1}{2}B(k - 1; n, \lambda/n) + \frac{1}{2}B(k; n, \lambda/n)$ is decreasing if $\lambda \leq [k(k - 1)]^{1/2}(n - 1)/\{[k(k - 1)]^{1/2} + [(n - k)(n - k - 1)]^{1/2}\}$.

Rule 2 can be improved by omitting tests for which $\bar{k} = 1$ or 2. Then the Poisson probability exceeds the binomial probability if the Poisson probability is at least $P(0; 2) = e^{-2} \sim .1353$ from theorem 4.1; application of theorem 4.2 to randomized tests for $\bar{k} \geq 3$ gives the criterion of $1 - P(3; 2) = 1 - 19e^{-2}/3 \sim .1429$.

4.2. *Confidence limits.* The upper confidence limit $\bar{p}$ for the parameter p of a binomial distribution based on a sample of k successes in n independent trials at confidence level $1 - \epsilon$ is the solution in p of $B(k; n, p) = \epsilon$. The upper confidence limit $\bar{\lambda}$ for the parameter λ of a Poisson distribution based on k occurrences at confidence level $1 - \epsilon$ is the solution in λ of $P(k; \lambda) = \epsilon$. An approximation to $\bar{p}$ is $\bar{\lambda}/n$. If $k + 1 \leq \bar{\lambda}$, then $B(k; n, \bar{\lambda}/n) \leq P(k, \bar{\lambda})$; since $B(k; n, p)$ decreases in p $[dB(k; n, p)/dp = -b(k; n - 1, p)]$, $\bar{p} \leq \bar{\lambda}/n$. Thus $\bar{\lambda}/n$ is a conservative upper confidence limit for p in the sense that the actual confidence level $1 - B(k; n, \bar{\lambda}/n)$ is greater than the nominal confidence level $1 - P(k; \bar{\lambda}) = 1 - \epsilon$. This is true if $1 - \epsilon \geq 1 - e^{-1}$ (approximately .6321). In practice, if $\bar{\lambda}/n > 1$, the upper limit for p is taken as one.

The lower confidence limit $\underline{p}$ for the parameter p of a binomial distribution based on k successes in n trials at confidence level $1 - \delta$ is the solution in p of $1 - B(k - 1; n, p) = \delta$. The lower confidence limit $\underline{\lambda}$ of a Poisson distribution based on k occurrences at confidence level $1 - \delta$ is the solution in λ of $1 - P(k - 1; \lambda) = \delta$. Then $\underline{\lambda}/n$ is an approximation to $\underline{p}$. If $\underline{\lambda} \leq k - 1$, then $P(k - 1; \underline{\lambda}) < B(k - 1; n, \underline{\lambda}/n)$ and $\underline{\lambda}/n < \underline{p}$. Thus $\underline{\lambda}/n$ is conservative since $B(k - 1; n, \underline{\lambda}/n) > P(k - 1; \underline{\lambda}) = 1 - \delta$. This is true if $1 - \delta \geq 2e^{-1}$ (approximately .7358). The conservative procedure then is $\underline{p} = 0$ if $k = 0$, $\underline{p} = 1 - (1 - \delta)^{1/n}$ if $k = 1$ (the solution of $B(0; n, p) = (1 - p)^n = 1 - \delta$), and $\underline{\lambda}/n$ if $k > 1$.

A confidence interval of confidence $1 - \epsilon - \delta$ is $(\underline{p}, \overline{p})$, where $\underline{p}$ and $\overline{p}$ are defined above. An approximate procedure is to use $(\underline{\lambda}/n, \overline{\lambda}/n)$, except that $\underline{\lambda}/n$ is replaced by $\underline{p}$ if $k = 1$, and $\overline{\lambda}/n$ is replaced by 1 if $\overline{\lambda}/n > 1$.

RULE 3. *If the confidence level of the Poisson approximate confidence limits (with $\underline{p}$ if $k = 1$) is at least .74, the actual confidence level is greater than the nominal level.*

5. An alternative approach to inequalities of sections 2 and 3

The following theorem supplements Hoeffding's Theorem stated in section 2.1 and yields a corollary which is stronger than theorem 2.1 but slightly weaker than theorem 2.3.

THEOREM 5.1. *If $F(k)$ and λ are defined as in section 2.1 with*

$$\lambda > [(n - 1)/(n - 3)][k - 1/(n - k)], \tag{5.1}$$

and if $p_i \leq \lambda/(n - 1)$, $i = 1, \cdots, n$, then $F(k) \leq B(k; n, \lambda/n)$.

PROOF. The set of all vectors $(p_1, \cdots, p_n)$ which satisfy the hypothesis is compact, and $F(k)$ is a continuous function of p; hence, the supremum of $F(k)$ is attained. Suppose that the p_i's are not all equal; let $p_1 = \min_{i=1,\cdots,n} p_i$ and $p_2 = \max_{i=1,\cdots,n} p_i$. Let $f^*(j)$ and $F^*(j)$ be, respectively, the probabilities of j successes and of not more than j successes in trials 3 through n. Then

$$\begin{aligned} F(k) &= p_1p_2F^*(k - 2) + [p_1(1 - p_2) + p_2(1 - p_1)]F^*(k - 1) \\ &\qquad + (1 - p_1)(1 - p_2)F^*(k) \\ &= p_1p_2[f^*(k) - f^*(k - 1)] - (p_1 + p_2)f^*(k) + F^*(k). \end{aligned} \tag{5.2}$$

Since p_1 and p_2 are each at most $\lambda/(n - 1)$, the sum of the remaining p_i's is at least $(n - 3)\lambda/(n - 1)$. Theorem 2 of [2] states that, if the sum of these p_i's is greater than $k - 1/(n - k)$, then $f^*(k) > f^*(k - 1)$. Hence, if

$$\lambda > [(n - 1)/(n - 3)][k - 1/(n - k)], \tag{5.3}$$

we can increase $F(k)$ by replacing p_1 and p_2 by $(p_1 + p_2)/2$. Thus, under the hypothesis, the supremum of $F(k)$ is attained only when the p_i's are all equal, which gives the desired result.

If we take $p_1 = 0$, $p_2 = \cdots = p_n = \lambda/(n - 1)$, we have the following corollary.

COROLLARY 5.1. *If $k < \lambda$, then $B(k; n, \lambda/n) > B(k; n - 1, \lambda/(n - 1))$ for all n sufficiently large so that*

$$\lambda > [(n - 1)/(n - 3)][k - 1/(n - k)]. \tag{5.4}$$

Note that the right-hand side of (5.4) is greater than the right-hand side of (2.14).

It is possible to obtain a result almost as good as theorem 3.1 by a method analogous to that used in proving theorem 2.1. We begin with a theorem of Hoeffding [1] which is more general than that in section 2.1.

General theorem of Hoeffding. *Let $g(k)$ be any function of the number of successes k in n independent trials, and let λ be a number between 0 and n. Then the maximum and minimum values of Eg among all choices of $p_1, \cdots, p_n$ with $p_1 + \cdots + p_n = \lambda$ are attained for choices of the following form: r of the p_i's are 0, s of the p_i's are one, and the remaining $n - r - s$ of the p_i's are equal to $(\lambda - s)/(n - r - s)$.*

If $g(j)$ is one for $j \leq k$ and 0 for $j > k$, then $Eg = F(k)$. Evaluation of Eg for the possible values of r and s shows that if $\lambda \leq k + 1$, then $\min Eg$ is not obtained with $s > 0$, and if $\lambda \leq k$, $\min Eg$ is not attained with $r > 0$. This gives half of (2.2), and the other half is attained similarly.

We now take $g(j)$ to be one if $j = k$, and 0 otherwise. Then $Eg = f(k)$. We prove the following theorem.

Theorem 5.2. *If $\lambda \leq k - 1 - k^{1/2}$ or $\lambda \geq k + 1 + k^{1/2}$, then*

$$\max_{p_1+\cdots+p_n=\lambda} f(k) = b(k; n, \lambda/n). \tag{5.5}$$

Proof. From the general theorem of Hoeffding, we need only consider those choices of $p_1, \cdots, p_n$ with r of the p_i's equal to 0, s of them equal to one and the remaining $n - r - s$ equal to $(\lambda - s)/(n - r - s)$. Let us call such choices "candidates."

We shall show that if λ satisfies the hypothesis, and if $r > 0$ or $s > 0$, then there is another choice of the p_i's satisfying the constraint for which the probability of k successes is greater. To do this, we first note that

$$\begin{aligned} f(k) &= p_1p_2[f^*(k) - 2f^*(k-1) + f^*(k-2)] \\ &\quad + (p_1 + p_2)[f^*(k) - f^*(k-1)] + f^*(k), \end{aligned} \tag{5.6}$$

where p_1 and p_2 are the probabilities of success on any two specified trials, and $f^*(k)$ is the probability of k successes in the remaining $n - 2$ trials. If $p_1 < p_2$ and the coefficient of p_1p_2 is positive, then we can increase $f(k)$ without altering the sum $p_1 + p_2$ by replacing p_1 and p_2 by $(p_1 + p_2)/2$.

It can be shown that

$$b(k; n, p) - 2b(k - 1; n, p) + b(k - 2; n, p) \tag{5.7}$$

is negative if and only if

$$\begin{aligned} k - [k(n + 2 - k)/(n + 1)]^{1/2} &\leq (n + 2)p \\ &\leq k + [k(n + 2 - k)/(n + 1)]^{1/2}, \end{aligned} \tag{5.8}$$

and hence is positive if

$$p \leq (k - 1 - k^{1/2})/(n + 1) \quad \text{or} \quad p \geq (k + k^{1/2})/(n + 1). \tag{5.9}$$

For a candidate with $r > 0$, we take $p_1 = 0$, $p_2 = (\lambda - s)/(n - r - s)$, while for a candidate with $s > 0$, we take $p_1 = (\lambda - s)/(n - r - s)$, $p_2 = 1$. Then the coefficient of p_1p_2 is

$$
(5.10)\quad
\begin{aligned}
&b(k - s; n - r - s - 1, (\lambda - s)/(n - r - s)) \\
&\quad - 2b(k - s - 1; n - r - s - 1, (\lambda - s)/(n - r - s)) \\
&\quad + b(k - s - 2; n - r - s - 1, (\lambda - s)/(n - r - s)), \qquad \text{if} \quad r > 0; \\
&b(k - s + 1; n - r - s - 1, (\lambda - s)/(n - r - s)) \\
&\quad - 2b(k - s; n - r - s - 1, (\lambda - s)/(n - r - s)) \\
&\quad + b(k - s - 1; n - r - s - 1, (\lambda - s)/(n - r - s)), \qquad \text{if} \quad s > 0.
\end{aligned}
$$

From (5.9), the coefficient is positive and the theorem is proved.

Taking $r = 1$, $s = 0$, we have the following corollary.

COROLLARY 5.2. *If* $\lambda \leq k - 1 - k^{1/2}$ *or* $\lambda \geq k + 1 + k^{1/2}$, *then* $b(k; n, \lambda/n) \geq b(k - 1; n - 1, \lambda/(n - 1))$.

This, however, is weaker than theorem 3.1.

REFERENCES

[1] W. HOEFFDING, "On the distribution of the number of successes in independent trials," *Ann. Math. Statist.*, Vol. 27 (1956), pp. 713–721.

[2] STEPHEN M. SAMUELS, "On the number of successes in independent trials," *Ann. Math. Statist.*, Vol. 36 (1965), pp. 1272–1278.

AN OPTIMAL PROPERTY OF THE LIKELIHOOD RATIO STATISTIC

R. R. BAHADUR
UNIVERSITY OF CHICAGO

1. Introduction

Let $s = (x_1, x_2, \cdots, \text{ad inf})$ be a sequence of independent and identically distributed observations on a variable x with distribution depending on a parameter θ taking values in a set Θ. Let Θ_0 be a subset of Θ and consider the null hypothesis that θ is in Θ_0. For each n, let $T_n = T_n(x_1, \cdots, x_n)$ be a real-valued statistic such that, in testing the hypothesis, large values of T_n are significant. For any given s, let $L_n(s)$ be the level attained by T_n in the given case; that is, $L_n(s)$ is the maximum probability (consistent with θ in Θ_0) of obtaining a value of T_n as large or larger than $T_n(s)$. Then, in typical cases, L_n is asymptotically distributed uniformly over (0, 1) in the null case, and L_n tends to zero in probability, or perhaps even with probability one, in the nonnull case. The rate at which L_n tends to zero when a given nonnull θ obtains is a measure of the asymptotic efficiency of T_n against that θ. It is shown in this paper (under very mild restrictions on the family of possible distributions of x) that L_n cannot tend to zero at a rate faster than $[\rho(\theta)]^n$ when a nonnull θ obtains; here ρ is a parametric function defined in terms of the Kullback-Leibler information numbers such that, in typical cases, $0 < \rho < 1$ (theorem 1). It is also shown (under much more restrictive conditions on the distributions of x) that if $\hat{T}_n$ is (any strictly decreasing function of) the likelihood ratio statistic of Neyman and Pearson [1], and $\hat{L}_n$ is the level attained by $\hat{T}_n$, then $\hat{L}_n$ tends to zero at the rate $[\rho(\theta)]^n$ in the nonnull case (theorem 2). In short, the likelihood ratio statistic is an optimal sequence in terms of exact stochastic comparison as described and exemplified in [2], [3], and [4].

Theorems 1 and 2 are stated more precisely in section 2. Section 3 contains a discussion of these theorems. Proofs are given in sections 4 and 5.

2. Theorems

Let X be a space of points x, $\mathcal{B}$ a σ-field of sets of X, and for each point θ in a set Θ, let P_θ be a probability measure on $\mathcal{B}$. Let Θ_0 be a given subset of Θ.

This research was supported in part by Research Grant No. NSF-GP3707 from the Division of Mathematical, Physical and Engineering Sciences of the National Science Foundation and in part by the Statistics Branch of the Office of Naval Research. Reproduction in whole or in part is permitted for any purpose of the United States Government.

ASSUMPTION 1. *There exists a σ-finite measure λ on $\mathcal{B}$ such that each P_θ admits a probability density with respect to λ, say $dP_\theta = f(x, \theta)\, d\lambda$, $0 \leq f < \infty$.*

For any θ in Θ and θ_0 in Θ_0 let

$$K(\theta, \theta_0) = -\int_X \log\, [f(x, \theta_0)/f(x, \theta)]\, dP_\theta. \tag{1}$$

K is one of the information numbers introduced by Kullback and Leibler [5], [6]. It is easily seen that K is well-defined by (1); $0 \leq K \leq \infty$; $K = 0$ if and only if $P_\theta = P_{\theta_0}$ on $\mathcal{B}$; and $K < \infty$ implies that P_θ is absolutely continuous with respect to P_{θ_0}. Even if P_θ and P_{θ_0} are mutually absolutely continuous, K can be infinite.

ASSUMPTION 2. *For each θ in $\Theta - \Theta_0$ and θ_0 in Θ_0 such that $K(\theta, \theta_0) < \infty$, there exists a $t = t(\theta, \theta_0) > 0$ such that $\int_X [f(x, \theta)/f(x, \theta_0)]^t \cdot dP_\theta < \infty$.*

If $K(\theta, \theta_0) < \infty$, then $0 < f(x, \theta)/f(x, \theta_0) < \infty$ with probability one when θ obtains, so that the integral in the statement of assumption 2 is well defined for every t.

Let

$$J(\theta) = \inf\, \{K(\theta, \theta_0) : \theta_0 \in \Theta_0\}, \qquad \rho(\theta) = \exp\, [-J(\theta)]. \tag{2}$$

As stated in the introduction, in typical cases $0 < \rho < 1$ for θ in $\Theta - \Theta_0$, but we shall include the cases $J = 0$ and $J = \infty$ in the discussion because theorem 1 [theorem 2] is not entirely vacuous in case $J = 0$ [$J = \infty$].

Now let $s = (x_1, x_2, \cdots, \text{ad inf})$ be a sequence of independent and identically distributed observations on x. The probability distribution of s in its sample space when θ obtains is denoted by $P_\theta^{(\infty)}$, but we shall usually abbreviate $P_\theta^{(\infty}$ to P_θ.

For each $n = 1, 2, \cdots$, let $T_n(s)$ be an extended real-valued measurable function of s such that T_n depends on s only through $(x_1, \cdots, x_n)$. For each θ let $F_n(t, \theta)$ denote the left-continuous probability distribution function of T_n when θ obtains; that is,

$$F_n(t, \theta) = P_\theta(T_n(s) < t), \tag{3}$$

and let

$$G_n(t) = \inf\, \{F_n(t, \theta) : \theta \in \Theta_0\}, \qquad (-\infty \leq t \leq \infty). \tag{4}$$

Define

$$L_n(s) = 1 - G_n(T_n(s)). \tag{5}$$

For any ϵ with $0 < \epsilon < 1$ let $N(\epsilon, s) =$ the least positive integer m such that $L_n \leq \epsilon$ for all $n \geq m$, and let $N(\epsilon, s) = +\infty$ if no such m exists. As just defined, N is then the sample size required in order that the sequence $\{T_n\}$ of test statistics becomes (and remains) significant at the level ϵ.

The following theorem 1 is a generalization and extension of theorem 4.1 of [4] in the following respects: the null hypothesis is not necessarily simple, and no restrictions other than measurability are imposed on the sequence $\{T_n\}$.

THEOREM 1. *For each θ in $\Theta - \Theta_0$*

$$\liminf_{n\to\infty} \frac{1}{n} \log L_n(s) \geq -J(\theta), \tag{6}$$

and

$$\liminf_{\epsilon\to 0} \frac{N(\epsilon, s)}{\log\left(\frac{1}{\epsilon}\right)} \geq \frac{1}{J(\theta)} \tag{7}$$

with probability one when θ obtains.

It follows from (6) that, for each nonnull θ,

$$\liminf_{n\to\infty} \frac{1}{n} \log E_\theta(L_n) \geq -J(\theta) \tag{8}$$

and

$$\lim_{n\to\infty} P_\theta(L_n > r^n) = 1 \qquad \text{if} \quad 0 < r < \rho(\theta). \tag{9}$$

The conclusions (8) and (9) are more useful than (6) or (7) in case L_n does not necessarily tend to 0 with probability one in the nonnull case.

For each n, let λ_n be the likelihood ratio statistic; that is,

$$\lambda_n(s) = \frac{\sup\left\{\prod_{i=1}^{n} f(x_i, \theta_0) : \theta_0 \in \Theta_0\right\}}{\sup\left\{\prod_{i=1}^{n} f(x_i, \theta) : \theta \in \Theta\right\}}. \tag{10}$$

In case the numerator and denominator in (10) are both 0, or both ∞, let $\lambda_n = 1$. Then λ_n is well-defined, with $0 \leq \lambda_n \leq 1$. It is assumed that λ_n is measurable for each n.

Since small values of λ_n are significant, we consider instead an equivalent statistic, $\hat{T}_n$ say, such that $\hat{T}_n$ is a strictly decreasing function of λ_n for each n. The particular choice of $\hat{T}_n$ is immaterial since only the exact levels attained are being considered, and we choose

$$\hat{T}_n(s) = -n^{-1} \log \lambda_n(s) \tag{11}$$

mainly because this choice facilitates some of the writing. Let $\hat{F}_n$, $\hat{G}_n$, and $\hat{L}_n$ be defined by (3), (4), and (5) by taking T_n to be $\hat{T}_n$, and let $\hat{N}$ be determined as above by the sequence $\{\hat{L}_n\}$.

Suppose now that, in addition to assumptions 1 and 2, assumptions 3–6 of section 5 are also satisfied.

THEOREM 2. *For each θ in $\Theta - \Theta_0$*

$$\lim_{n\to\infty} \frac{1}{n} \log \hat{L}_n(s) = -J(\theta), \tag{12}$$

and

$$\lim_{\epsilon\to 0} \frac{\hat{N}(\epsilon, s)}{\log\left(\frac{1}{\epsilon}\right)} = \frac{1}{J(\theta)} \tag{13}$$

with probability one when θ obtains.

It follows from (7) and (13) that for any given sequence $\{T_n\}$, the resulting sample size N required to attain the level ϵ satisfies

$$\liminf_{\epsilon\to 0} \frac{N(\epsilon, s)}{\hat{N}(\epsilon, s)} \geq 1 \tag{14}$$

with probability one whenever a nonnull θ with $0 < J(\theta) < \infty$ obtains.

It follows from (12) that for each nonnull θ,

$$\lim_{n\to\infty} \frac{1}{n} \log E_\theta(\hat{L}_n) = -J(\theta) \tag{15}$$

and

$$\lim_{n\to\infty} P_\theta(r_1^n < \hat{L}_n < r_2^n) = 1 \qquad \text{if} \quad r_1 < \rho(\theta) < r_2. \tag{16}$$

The likelihood ratio statistic is sometimes defined to be the right-hand side of (10) but with Θ replaced with $\Theta - \Theta_0$ in the denominator. This modified definition of λ_n is usually, but not always, equivalent to the definition (10). It can be seen from section 5 that under the same assumptions 1–6, theorem 2 holds also for the modified $\hat{T}_n$.

3. Remarks

(a) Let us say that a sequence $\{T_n\}$ is optimal when a given $\theta \in \Theta - \Theta_0$ obtains if, with L_n the level attained by T_n, $n^{-1} \log L_n \to -J(\theta)$ with probability one. According to theorems 1 and 2, this definition of optimality is plausible and $\{\hat{T}_n\}$ is an optimal sequence for every nonnull θ. Optimality in the present sense is, however, a rather weak property and is enjoyed, presumably, by a fairly wide class of statistics. An example of an optimal sequence other than $\{\hat{T}_n\}$ has already been mentioned at the end of section 2, and other examples are described in the following remarks (b) and (c). Further comparison of two optimal sequences requires, in general, an analysis very much deeper than is available at present. A similar difficulty arises in a theory of estimation closely related to the stochastic comparison of tests (cf. [4], section 6).

(b) The optimal exponential rate of convergence of levels, namely ρ^n, depends on the null set Θ_0 and on the particular alternative θ in $\Theta - \Theta_0$ under consideration, but not on the entire set of alternatives $\Theta - \Theta_0$. It follows, in particular, that if Δ is a subset of $\Theta - \Theta_0$, and if $T_n^* = -n^{-1} \log \lambda_n^*$, where λ_n^* is the likelihood ratio statistic for testing Θ_0 against Δ, then $\{T_n^*\}$ and $\{\hat{T}_n\}$ are both optimal sequences whenever a θ in Δ obtains. To consider the matter from another viewpoint, suppose that the initial nonnull set $\Theta - \Theta_0$ is enlarged to a set Σ by admitting certain additional nonnull distributions, and suppose that assumptions 1–6 are satisfied in the enlarged framework. Let $T_n^0 = -n^{-1} \log \lambda_n^0$, where λ_n^0 is the likelihood ratio statistic for testing Θ_0 against Σ. Then $\{T_n^0\}$ is an optimal sequence everywhere on Σ and hence also on $\Theta - \Theta_0$. Presumably, however, closer analysis will show that when a θ in $\Theta - \Theta_0$ obtains, T_n^0 is distinctly inferior to

$\hat{T}_n$ in the sense that $L_n^0 > \hat{L}_n$ with probability one for all sufficiently large n. This last is the case, for example, if X is the real line, P_θ denotes the normal distribution with mean θ and variance 1, $\Theta = [0, \infty)$, $\Theta_0 = \{0\}$, and $\Sigma = (-\infty, \infty) - \{0\}$; in this example, $\hat{L}_n = \frac{1}{2}L_n^0$ for all sufficiently large n when a positive θ obtains.

(c) Suppose that the maximum likelihoods in (10) are replaced by average likelihoods over Θ_0 and Θ with respect to appropriate averaging distributions. Then, under certain conditions, the resulting statistic remains optimal against each Θ in $\Theta - \Theta_0$. This important remark was suggested by Dr. P. J. Bickel at the reading of this paper at the Symposium. Dr. Bickel and the author hope to present an adequate treatment of the remark elsewhere, but it may be worthwhile to state here the following. Suppose that assumptions 1–6 and some additional assumptions are satisfied. Then $\Theta - \Theta_0$ and Θ_0 are metric spaces. Let ξ be a fixed prior probability distribution such that each neighborhood of each point in either space has positive probability. For each n let $\pi_n(s)$ be the posterior probability of Θ_0 given $(x_1, \cdots, x_n)$, and let $\overline{T}_n(s) = n^{-1} \log [(1 - \pi_n)/\pi_n]$. Then the relevant asymptotic properties (cf. (19) and (20) below) of $\overline{T}_n$ are exactly the same as those of $\hat{T}_n$.

(d) For given n and s let $L_n(s)$ defined by (3), (4), and (5) be written temporarily as $L_n(s, T_n)$ to indicate its dependence on T_n. Let $M_n(s) = \inf \{L_n(s, T_n)\}$, the infimum being taken over the class of all measurable statistics T_n which depend on s only through $(x_1, \cdots, x_n)$. Although M_n is not the level attained by any statistic (that is, there exists no T_n such that $M_n(s) = L_n(s, T_n)$ for all s), it is of some theoretical interest to study the behavior of M_n. We consider two special cases.

Suppose first that for each x in X the set $\{x\}$ is $\mathcal{B}$-measurable and $P_\theta(\{x\}) = 0$ for all θ in Θ_0. In this case $M_n(s) = 0$ for all s and all n.

Suppose next that X is a finite set and that $\mathcal{B}$ is the class of all subsets of X. In this case $M_n(s) = \sup \{\prod_{i=1}^n f(x_i, \theta) : \theta \in \Theta_0\}$ where $f(x, \theta) = P_\theta(\{x\})$. It follows hence by lemma 4 of section 5 that

$$\lim_{n \to \infty} \frac{1}{n} \log M_n = -J(\theta) - H(\theta) \tag{17}$$

with probability one when θ obtains, where

$$H(\theta) = -\sum_{x \in X} f(x, \theta) \log f(x, \theta) \tag{18}$$

is the Shannon information number. It follows that with $N'(\epsilon, s)$ the sample size required to make $M_n \leq \epsilon$, we have $N' \leq \hat{N}$ and $\lim_{\epsilon \to 0} \{N'/\hat{N}\} = J(\theta)/[J(\theta) + H(\theta)]$ with probability one in the nonnull case. If X contains k points, $H(\theta) \leq \log k$, so that $J/[J + H] \leq J/[J + \log k]$ for all θ.

To consider a simple example, suppose that X consists of the two points 0 and 1, $\Theta = (0, 1)$, $P_\theta(\{1\}) = 1 - P_\theta(\{0\}) = \theta$, and $\Theta_0 = \{\frac{1}{2}\}$. In this example, J and H are functions of $|\theta - \frac{1}{2}|$, and the values of $J/[J + H]$ for $|\theta - \frac{1}{2}| = .0(.1).5$ are .00, .03, .12, .27, .53, and 1.00.

(e) Assumptions 1 and 2 of theorem 1 are very weak and even these can be dispensed with to a certain extent (cf. the last paragraph of section 4). Unfortunately, some of the additional assumptions 3–6 required by the present proof of theorem 2 are quite restrictive, and what is perhaps worse, it is often difficult to determine whether they hold in a given case. The troublesome assumptions include versions of the compactifiability and integrability conditions introduced by Wald [7] in his proof of the consistency of maximum likelihood estimates. As is pointed out in [8], it is often difficult and sometimes impossible to verify such conditions, even in certain apparently simple cases where the estimates themselves are visibly consistent, and the likelihood function behaves as it should. It may be added here that at least some of the conditions embodied in assumptions 3–6 are indispensable to a general proof of theorem 2; this may be seen from [9].

In many examples it is a relatively simple matter to show directly that (12) and (13) are satisfied, as follows. First it is shown that

$$\hat{T}_n \to J(\theta) \tag{19}$$

with probability one when θ obtains. Next it is shown that the distribution function $\hat{G}_n$ satisfies the following condition: for each positive t in some neighborhood of J,

$$\frac{1}{n} \log [1 - \hat{G}_n(t)] \to -t \quad \text{as} \quad n \to \infty. \tag{20}$$

It is then immediate from (19) and (20) that (12) holds, and (12) implies (13). Of course, (20) is not quite necessary for (12); in fact, there are simple examples where even assumptions 1–6 hold but (20) as stated does not.

The proof of (19) is troublesome in the general case (cf. section 5) but quite trivial in many examples. Proofs of (20), or of versions thereof, are always nontrivial since (20) is an assertion about very small tail probabilities of the exact null distribution of $\hat{T}_n$.

The present regularity assumptions give little or no trouble in certain fairly general circumstances. Assumptions 1–6 are satisfied in case X is a finite set (that is, the multinomial case) no matter what Θ and Θ_0 may be, provided that $\theta_1 \neq \theta_2$ implies $P_{\theta_1} \neq P_{\theta_2}$ for θ_1 and θ_2 in Θ. (This last proviso is harmless in the present context.) Only assumption 2 requires verification in case Θ is a finite set, no matter what X may be. Assumptions 2–6 are usually satisfied but require verification in case Θ is an interval on the real line, assumption 1 holds, and $f(x, \theta)$ is continuous in θ over Θ for each x.

It is worthwhile to note that the regularity conditions under discussion do not include conditions required by the asymptotic null distribution theory of maximum likelihood and likelihood ratios; consequently, the present conditions are satisfied in many so-called irregular cases. For example, if X is the real line, $\Theta = (-\infty, +\infty)$, $\Theta_0 = \{0\}$, and P_θ represents the uniform distribution over $(\theta - \frac{1}{2}, \theta + \frac{1}{2})$, then assumptions 1–6 hold with $J(\theta) = \infty$ for each nonnull θ. In this example there exists a random variable $m = m(s)$ with $1 \leq m \leq \infty$ such that $P_\theta(m < \infty) = 1$ for each nonnull θ, and such that $\hat{L}_n = 1$ for $n < m$ and $\hat{L}_n = 0$ for $n \geq m$ for every s; hence, $\hat{N} = m$ for every ϵ and s.

(f) As pointed out in [2], [3], and [4], stochastic comparison has several connections with power function considerations. In particular, theorems 1 and 2 can be shown to yield the following conclusions concerning the asymptotic properties of critical regions. Consider a particular nonnull θ. For each n, let W_n be a critical region in the sample space of $(x_1, \cdots, x_n)$ such that $P_\theta(W_n) \to p$ as $n \to \infty$, where $0 < p < 1$. Let $\alpha_n = \sup \{P_{\theta_0}(W_n): \theta_0 \in \Theta_0\}$ be the size of W_n. Then $\liminf_{n\to\infty} n^{-1} \log \alpha_n \geq -J(\theta)$. Next, let $\hat{W}_n$ be a critical region of the form $\{s: \hat{T}_n \geq \hat{k}_n\}$, with the constants $\hat{k}_n$ chosen so that $P_\theta(\hat{W}_n) \to \hat{p}$ where $0 < \hat{p} < 1$; then $n^{-1} \log \hat{\alpha}_n \to -J(\theta)$. In other words, if the power of the critical region against a given alternative is held fixed, the rate of convergence to zero of the resulting size is optimal for regions based on $\hat{T}_n$. Related but much deeper optimality conclusions concerning critical regions based on $\hat{T}_n$ have been obtained previously by Hoeffding [10] in the case when X is a finite set.

4. Proof of theorem 1

The following lemma 1 is required in the proofs of theorems 1 and 2. Let z be an extended real-valued random variable such that $P(-\infty \leq z < \infty) = 1$, and let $\varphi(t) = E(e^{tz})$ be the moment generating function (m.g.f.) of z, $0 \leq \varphi \leq \infty$.

LEMMA 1. *Let n be a positive integer, and let $z_1, \cdots, z_n$ be mutually independent replicates of z. Then $P(z_1 +, \cdots, + z_n \geq 0) \leq [\varphi(t)]^n$ for $t > 0$.*

PROOF. The lemma (and much more) is well known (cf. [11], [12], [13]), but for the sake of completeness we include here the proof given in [11]. Let $Z_n = \sum_{i=1}^n z_i$. Then $P(Z_n \geq 0) = P(\exp (tZ_n) \geq 1) \leq E(\exp (tZ_n)) = [\varphi(t)]^n$.

Now choose and fix a θ in $\Theta - \Theta_0$, a θ_0 in Θ_0, and an $\epsilon > 0$. Let $r_1 = \exp [-K(\theta, \theta_0) - \epsilon]$, $0 \leq r_1 < 1$. Let W_n denote an event which depends on s only through $x_1, \cdots, x_n$. The following lemma is closely related to a theorem of C. Stein (cf. [6], pp. 76–77).

LEMMA 2. *There exists $r_2 = r_2(\theta, \theta_0, \epsilon)$, $0 < r_2 < 1$, such that for each n and W_n,*

$$P_{\theta_0}(W_n) \geq r_1^n[P_\theta(W_n) - r_2^n]. \tag{21}$$

PROOF. Consider a fixed n. If $K = \infty$, then $r_1 = 0$ and (21) holds trivially with $r_2 = \frac{1}{2}$ (say). Suppose then that $K < \infty$. Let

$$A_n = \left\{s: \prod_{i=1}^n f(x_i, \theta_0) \geq r_1^n \prod_{i=1}^n f(x_i, \theta)\right\}. \tag{22}$$

Then

$$\begin{aligned} P_{\theta_0}(W_n) &\geq P_{\theta_0}(A_n \cap W_n) \\ &\geq r_1^n P_\theta(A_n \cap W_n) \qquad \text{by (22)} \\ &\geq r_1^n[P_\theta(W_n) - [1 - P_\theta(A_n)]]. \end{aligned} \tag{23}$$

Now consider the random variable $y = \log [f(x, \theta)/f(x, \theta_0)]$ when θ obtains; y is well-defined and $P_\theta(-\infty < y < \infty) = 1$. The m.g.f. of y is ≤ 1 at $t = -1$, and is finite for a positive t by assumption 2. Thus the m.g.f. of y is finite in a neighborhood of $t = 0$. Let $z = y - K - \epsilon$. Then the m.g.f. of z, $\varphi(t)$ say, is finite in a neighborhood of $t = 0$ and $\varphi'(0) = E_\theta(z) = -\epsilon < 0$ by (1). Since $\varphi(0) = 1$, there exists a $t_2 > 0$ such that with $r_2 = \varphi(t_2)$ we have $0 < r_2 < 1$. It follows from (22) that, in an obvious notation, $1 - P_\theta(A_n) = P_\theta\left(\sum_{i=1}^n z_i > 0\right)$. Hence

$$1 - P_\theta(A_n) \leq r_2^n \tag{24}$$

by lemma 1. It is plain from (23) and (24) that (21) holds.

Let there be given a sequence of measurable statistics T_n as in section 2. By putting $W_n = \{s: T_n \geq t\}$ in (21) it follows from (3) that

$$1 - F_n(t, \theta_0) \geq r_1^n[1 - F_n(t, \theta) - r_2^n] \tag{25}$$

for all t and all n.

Lemma 3. *With probability one when θ obtains,*

$$1 - F_n(T_n(s), \theta) \geq n^{-2} \tag{26}$$

for all sufficiently large n.

Proof. It is easily verified that if T is an extended real-valued random variable, and $F(t) = P(T < t)$, then $P(1 - F(T) < r) \leq r$ for all r in $[0, 1]$. It follows hence that $\sum_{n=1}^\infty P_\theta(1 - F_n(T_n, \theta) < n^{-2}) \leq \sum_{n=1}^\infty n^{-2} < \infty$.

Proof of theorem 1. Choose and fix a θ in $\Theta - \Theta_0$. Let $B = B(\theta)$ be the set of all s such that (26) holds for all sufficiently large n. Then B is a measurable set with $P_\theta^{(\infty)}(B) = 1$. We shall show that (6) and (7) hold for each s in B. Choose and fix an s in B, and let $m = m(s)$ be an integer such that (26) holds for all $n \geq m$.

Let θ_0 be a point in Θ_0 and let ϵ be a positive constant. For each n let $t = T_n(s)$ in (25). It then follows from (25) and (26) that $1 - F_n(T_n(s), \theta_0) \geq r_1^n[n^{-2} - r_2^n]$ for $n \geq m$. Since $L_n(s)$ defined by (3), (4), and (5) cannot be less than $1 - F_n(T_n(s), \theta_0)$, it follows that $L_n(s) \geq r_1^n[n^{-2} - r_2^n]$ for $n \geq m$. Hence

$$\liminf_{n\to\infty} \frac{1}{n} \log L_n(s) \geq -K(\theta, \theta_0) - \epsilon \tag{27}$$

by the definition of r_1. Since θ_0 and ϵ in (27) are arbitrary, (6) holds.

Since (7) holds trivially if $J = \infty$, suppose that $0 \leq J < \infty$. It then follows from (6) that $L_n > 0$ for all sufficiently large n. If $\limsup_{n\to\infty} L_n(s) > 0$, then $N = \infty$ for all sufficiently small ϵ and (7) again holds trivially. Suppose then that $\lim_{n\to\infty} L_n = 0$. In this case $1 \leq N < \infty$ for all ϵ; $N \to \infty$ through a subsequence of the integers as $\epsilon \to 0$; and $L_N \leq \epsilon$ for all ϵ. It follows hence that

$$\overline{\lim_{\epsilon\to 0}}\, \{N^{-1} \log (1/\epsilon)\} \leq \overline{\lim_{\epsilon\to 0}}\, \{-N^{-1} \log L_N\} \leq \overline{\lim_{n\to\infty}}\, \{-n^{-1} \log L_n\} \leq J(\theta) \tag{28}$$

by (6), and this establishes (7). This completes the proof of theorem 1.

It is plain from the preceding proof that assumptions 1 and 2 can be weakened considerably. Indeed, there is a version of theorem 1 which holds without any regularity assumptions whatsoever. To describe this version, for any θ and θ_0 in Θ let $K(\theta, \theta_0) = \int_X [\log (dP_\theta/dP_{\theta_0})]\, dP_\theta$ if P_θ is absolutely continuous with respect to P_{θ_0}, and let $K = \infty$ otherwise. Let J be defined by (2). It then follows by a slight modification of the preceding proof (using the law of large numbers instead of lemma 1, and $E_\theta[1 - F_n(T_n, \theta)] \geq \frac{1}{2}$ instead of lemma 3) that (8) holds for each nonnull θ. It follows from (8) that (6) and (7) are satisfied with both inferior limits replaced by superior limits. It would be interesting to know whether theorem 1 as stated holds (with the present definition of J) without any assumptions whatsoever.

5. Proof of theorem 2

We shall first state the additional assumptions required of the given framework X, $\mathcal{B}$, $\{P_\theta : \theta \in \Theta\}$, $\Theta_0 \subset \Theta$, and $f(x, \theta) = dP_\theta/d\lambda$. In order to avoid needless loss of generality, most of these assumptions are stated below in more or less the forms required by the proof itself. Certain stronger but more readily verifiable conditions are also given.

Let $\bar{\Theta}$ be a metric space of points θ, and let δ denote the given metric on $\bar{\Theta}$. We shall say that $\bar{\Theta}$ is a suitable compactification of Θ if the following conditions (i)–(iv) are satisfied: (i) $\bar{\Theta}$ is compact; (ii) $\Theta \subset \bar{\Theta}$, and Θ is everywhere dense in $\bar{\Theta}$; (iii) for each $\theta \in \bar{\Theta}$ there exists $d_1 = d_1(\theta) > 0$ such that, for each d in $(0, d_1)$,

$$g(x, \theta, d) = \sup \{f(x, \theta_1) : \theta_1 \in \Theta, \delta(\theta, \theta_1) < d\} \tag{29}$$

is $\mathcal{B}$-measurable, $0 \leq g \leq \infty$; and (iv) for each $\theta \in \bar{\Theta}$,

$$\int_X g(x, \theta, 0)\, d\lambda \leq 1, \tag{30}$$

where $g(x, \theta, 0) = \lim_{d \to 0} g(x, \theta, d)$. In typical cases $g(x, \theta, 0) = f(x, \theta)$ for $\theta \in \Theta$, so that g is an extension of the given function f on $X \times \Theta$ to a function on $X \times \bar{\Theta}$.

A slightly different formulation of the notion of suitable compactification, and many nontrivial examples, are given in [8].

Assumption 3. *There exists a suitable compactification of* Θ, *say* $\bar{\Theta}$. *With* $\bar{\Theta}_0$ *the closure of* Θ_0 *in* $\bar{\Theta}$, $\bar{\Theta}_0$ *is a suitable compactification of* Θ_0.

The second part of this assumption is to the effect that, for each $\theta_0 \in \bar{\Theta}_0$,

$$g_0(x, \theta_0, d) = \sup \{f(x, \theta_1) : \theta_1 \in \Theta_0, \delta(\theta_0, \theta_1) < d\} \tag{31}$$

is $\mathcal{B}$-measurable for all sufficiently small $d > 0$. With $g_0(x, \theta_0, 0) = \lim_{d \to 0} g_0(x, \theta_0, d)$, it is plain from (29) and (31) that $g_0(x, \theta_0, 0) \leq g(x, \theta_0, 0)$ for all x and $\theta_0 \in \bar{\Theta}_0$; consequently, in view of (30), the required condition

$$\int_X g_0(x, \theta_0, 0)\, d\lambda \leq 1 \tag{32}$$

is automatically satisfied.

For any $\theta \in \Theta$ and $\theta_0 \in \bar{\Theta}_0$, let $\bar{K}(\theta, \theta_0)$ be defined by (1) with $f(x, \theta_0)$ replaced by $g_0(x, \theta_0, 0)$. It follows from (32) that $\bar{K}$ is well-defined and $0 \leq \bar{K} \leq \infty$. Since g_0 may be thought of as an extension of the function $f(x, \theta_0)$ on $X \times \Theta_0$ to $X \times \bar{\Theta}_0$, $\bar{K}$ is to be thought of as an extension of K on $\Theta \times \Theta_0$ to $\Theta \times \bar{\Theta}_0$. An alternative method of extending K is to use g instead of g_0, but the present approach is preferable in that the following assumptions 4 and 5 are weaker than the corresponding assumptions in terms of g.

ASSUMPTION 4. *For each θ in $\Theta - \Theta_0$,*

$$J(\theta) = \inf \{\bar{K}(\theta, \theta_0): \theta_0 \in \bar{\Theta}_0\}. \tag{33}$$

It is plain from (2) that (33) holds if $\bar{K}$ is indeed an extension of K and if, for the given θ, $\bar{K}(\theta, \theta_0)$ is either continuous in θ_0 over $\bar{\Theta}_0$, or $= \infty$ for θ_0 in $\bar{\Theta}_0 - \Theta_0$.

ASSUMPTION 5. *For given θ in $\Theta - \Theta_0$ and θ_0 in $\bar{\Theta}_0$, there exists $d = d(\theta, \theta_0) > 0$ such that*

$$\int_X \log^+ [g_0(x, \theta_0, d)/f(x, \theta)] \, dP_\theta < \infty. \tag{34}$$

Assumptions 4 and 5 are automatically satisfied if Θ_0 is a finite set, and in particular, if the null hypothesis is simple.

It is convenient to restate assumption 5 here as follows. For given $\theta \in \Theta - \Theta_0$ and $\theta_0 \in \bar{\Theta}_0$, let d be restricted to sufficiently small values so that $g_0(x, \theta_0, d)$ is measurable. Consider

$$y_0 = y_0(x, \theta_0, d:\theta) = \log [g_0(x, \theta_0, d)/f(x, \theta)] \tag{35}$$

when θ obtains. Then y_0 is well-defined and $-\infty \leq y_0 < \infty$ with probability one. The condition (34) is that $E_\theta(y_0)$ exists and $-\infty \leq E_\theta(y_0) < \infty$. Since $g_0(x, \theta_0, d)$ decreases to $g_0(x, \theta_0, 0)$ as d decreases to zero, and since $-\bar{K}$ is by definition the expected value of $y_0(x, \theta_0, 0:\theta)$ when θ obtains, (34) implies (and is implied by)

$$\lim_{d \to 0} E_\theta(y_0(x, \theta_0, d:\theta)) = -\bar{K}(\theta, \theta_0), \tag{36}$$

even if $\bar{K} = \infty$. (Cf. [8], section 2.)

ASSUMPTION 6. *Given τ, $0 < \tau < 1$, $\epsilon > 0$, and θ in $\bar{\Theta}$, there exists $d = d(\tau, \epsilon, \theta) > 0$ such that*

$$\int_X [g(x, \theta, d)/f(x, \theta_0)]^\tau \, dP_{\theta_0} < 1 + \epsilon \tag{37}$$

for all θ_0 in Θ_0.

In order to discuss this assumption, consider a particular $\theta_0 \in \Theta_0$ and sufficiently small $d > 0$. Consider

$$y = y(x, \theta, d:\theta_0) = \log [g(x, \theta, d)/f(x, \theta_0)] \tag{38}$$

when θ_0 obtains. Then y is well-defined and $-\infty \leq y < \infty$ with probability one. Let the integral in (37) be denoted by $\psi(\tau|\theta, d, \theta_0)$; ψ is the m.g.f. of y. It follows from the convexity of m.g.f.'s that for $0 < \tau < 1$, $\psi(\tau)$ cannot exceed

$\max \{P_{\theta_0}(y > -\infty), \int_X g(x, \theta, d)\, d\lambda\}$. Hence $\psi(\tau) \leq \max \{1, \int_X g(x, \theta, d)\, d\lambda\}$; this bound does not depend on θ_0 (or on τ). We observe next that

$$\int_X g(x, \theta, d)\, d\lambda \to \int_X g(x, \theta, 0)\, d\lambda \leq 1 \quad \text{as} \quad d \to 0,$$

provided that

$$\int_X g(x, \theta, d_1)\, d\lambda < \infty \qquad \text{for some} \quad d_1 = d_1(\theta) > 0 \tag{39}$$

It follows that (39) is a sufficient condition for the validity of assumption 6 at the given $\theta \in \bar{\Theta}$, no matter what the null set Θ_0 may be. Condition (39) is satisfied if, for example, X is a countable set, $\mathcal{B}$ is the class of all subsets of X, and there exists $h(x)$ such that $P_\theta(\{x\}) \leq h(x)$ for all θ and all x, and $\sum_x h(x) < \infty$. It is plain that condition (39) is satisfied whenever Θ is a finite set.

Condition (39) is, however, much stronger than is generally necessary. To obtain weaker or different sufficient conditions, suppose that $\psi(\tau|\theta, d, \theta_0) < \infty$ for some $d > 0$. It then follows that

$$\lim_{d \to 0} \psi(\tau|\theta, d, \theta_0) = \psi(\tau|\theta, 0, \theta_0). \tag{40}$$

The right-hand side in (40) is the m.g.f. of $y(x, \theta, 0{:}\theta_0)$. Since this last m.g.f. does not exceed one for $0 < \tau < 1$, assumption 6 will hold at the given θ if (40) holds uniformly for θ_0 in Θ_0. Uniformity is guaranteed by Dini's theorem if for each d in some interval $[0, d_1)$ with $d_1 > 0$, $\psi(\tau|\theta, d, \theta_0)$ is continuous in θ_0 over Θ_0 and has a continuous extension to $\bar{\Theta}_0$, and (40) holds for the extended functions for each θ_0 in $\bar{\Theta}_0$. This last condition is satisfied, in particular, if there exists a $\mathcal{B}$-measurable $h(x)$ such that $f(x, \theta_0) \leq h(x)$ for all x and all $\theta_0 \in \Theta_0$ and such that $\int_X [g(x, \theta, d_1)]^\tau [h(x)]^{1-\tau}\, d\lambda < \infty$, and if $g_0(x, \theta_0, 0)$ is an extension of $f(x, \theta_0)$ and is continuous over $\bar{\Theta}_0$ for each x.

We proceed to establish theorem 2. Assumptions 3, 4, and 5 are used to obtain lemma 4 below, and assumptions 3 and 6 to obtain lemma 5. Theorem 2 is a straightforward consequence of theorem 1 and lemmas 4 and 5.

For any set $\Gamma \subset \bar{\Theta}$ such that $\Gamma \cap \Theta$ is nonempty and any $\theta \in \Theta$, let

$$\begin{aligned} R_n(\Gamma, \theta) &= R_n(s{:}\Gamma, \theta) \\ &= n^{-1} \log \frac{\sup \left\{ \prod_{i=1}^{n} f(x_i, \theta_1){:}\theta_1 \in \Gamma \cap \Theta \right\}}{\left\{ \prod_{i=1}^{n} f(x_i, \theta) \right\}}. \end{aligned} \tag{41}$$

R_n is well-defined (with $-\infty \leq R_n \leq \infty$) with probability one when θ obtains. It is not required, however, that R_n be a measurable function of s.

LEMMA 4. *For each* $\theta \in \Theta - \Theta_0$,

$$R_n(\Theta_0, \theta) \to -J(\theta) \tag{42}$$

with probability one when θ *obtains.*

PROOF. Choose and fix $\theta \in \Theta - \Theta_0$ and suppose θ obtains. Let $a > 0$, $b > 0$ be constants, and let $H = \max\{-J(\theta) + a, -b\}$. Let θ_0 be a point in $\bar{\Theta}_0$. According to (36), there exists $d = d(\theta_0) > 0$ such that, with $y_0(x)$ defined by (35), $E_\theta(y_0) < \max\{-\bar{K}(\theta, \theta_0) + a, -b\}$. Hence $E_\theta(y_0) \leq H$, by assumption 4. Let Γ be the open sphere in $\bar{\Theta}_0$ with center θ_0 and radius d, and let $\Gamma^0 = \Gamma \cap \Theta_0$. It is then plain from (31), (35), and (41) that $R_n(s:\Gamma^0, \theta) \leq n^{-1} \sum_{i=1}^n y_0(x_i)$ for every s and n. Hence $\limsup_{n\to\infty} R_n(\Gamma^0, \theta) \leq H$ with probability one.

Since $\bar{\Theta}_0$ is compact, we can find a finite number of spheres $\Gamma_1, \cdots, \Gamma_k$ such that $\bigcup_j \Gamma_j = \bar{\Theta}_0$, and such that the conclusion of the preceding paragraph holds for each $\Gamma_j^0 = \Gamma_j \cap \Theta_0$. Since $R_n(\Theta_0, \theta) = \max\{R_n(\Gamma_j^0, \theta): j = 1, \cdots, k\}$, it follows that $\limsup_{n\to\infty} R_n(\Theta_0, \theta) \leq H$ with probability one. Since a and b are arbitrary, we conclude that $\limsup_{n\to\infty} R_n(\Theta_0, \theta) \leq -J$ with probability one.

With θ_0 a point in Θ_0, $R_n(\Theta_0, \theta) \geq R_n(\{\theta_0\}, \theta)$ by (41); hence,

$$\liminf_{n\to\infty} R_n(\Theta_0, \theta) \geq -K(\theta, \theta_0)$$

with probability one, by (1) and (41). Since θ_0 is arbitrary, we see from (2) that $\liminf_{n\to\infty} R_n(\Theta_0, \theta) \geq -J(\theta)$ with probability one.

LEMMA 5. *Given* $\epsilon > 0$ *and* τ, $0 < \tau < 1$, *there exists a positive integer* $k = k(\epsilon, \tau)$ *such that*

$$1 - \hat{G}_n(t) \leq k \cdot (1 + \epsilon)^n \cdot e^{-n\tau t} \tag{43}$$

for all n *and* t.

PROOF. Let θ be a point in $\bar{\Theta}$, and $d = d(\theta) > 0$ be such that, with g defined by (29), (37) holds for all θ_0 in Θ_0. Let Γ denote the open sphere in $\bar{\Theta}$ with center θ and radius d.

Consider a particular $\theta_0 \in \Theta_0$ and suppose that θ_0 obtains. Let $y(x)$ be given by (38), and let ψ be the m.g.f. of y. According to (37), $\psi(\tau) < 1 + \epsilon$. It is plain from (29), (38), and (41) that $R_n(\Gamma, \theta_0) \leq n^{-1} \sum_{i=1}^n y(x_i) = S_n$, say. An application of lemma 1 (with $z = y - t$, $t = \tau$, and $\varphi(\tau) = \psi(\tau) \exp(-t\tau)$) shows that $P_{\theta_0}(S_n \geq t) \leq (1 + \epsilon)^n \exp(-n\tau t) = b_n(t)$, say, for all n and t.

Since $\bar{\Theta}$ is compact, we can find a finite number of open spheres $\Gamma_1, \cdots, \Gamma_k$ such that $\bigcup_j \Gamma_j = \bar{\Theta}$, and such that, for each $\theta_0 \in \Theta_0$ and j, there exists a random variable $S_{nj} = S_{nj}(\theta_0)$ with $R_n(\Gamma_j, \theta_0) \leq S_{nj}$ and $P_{\theta_0}(S_{nj} \geq t) \leq b_n(t)$. Now, it is clear from (10), (11), and (41) that, when a given θ_0 obtains,

$$\begin{aligned} \hat{T}_n \leq R_n(\Theta, \theta_0) &= \max\{R_m(\Gamma_j, \theta_0): j = 1, \cdots, k\} \\ &\leq \max\{S_{nj}(\theta_0): j = 1, \cdots, k\}. \end{aligned} \tag{44}$$

Since $\hat{T}_n$ is measurable by assumption, it follows from (44) that

$$P_{\theta_0}(\hat{T}_n \geq t) \leq \sum_j P_{\theta_0}(S_{nj} \geq t) \leq \sum_j b_n(t) = k \cdot b_n(t). \tag{45}$$

Thus

$$1 - \hat{F}_n(t, \theta_0) \leq k \cdot (1 + \epsilon)^n \cdot \exp(-n\tau t). \tag{46}$$

Since (46) holds for every finite t, it follows by letting $t \to \infty$ that $P_{\theta_0}(\hat{T}_n = \infty) = 0$, that is, (46) holds for $t = \infty$ also. Since θ_0 in (46) is arbitrary, we see from (3) and (4) that (43) holds.

PROOF OF THEOREM 2. Suppose that a given θ in $\Theta - \Theta_0$ obtains. Since $\hat{T}_n \geq -R_n(\Theta_0, \theta)$ by (10), (11), and (41), it follows from (42) that

$$\liminf_{n\to\infty} \hat{T}_n \geq J(\theta) \tag{47}$$

with probability one. It will be shown later that in fact (19) holds.

Choose ϵ and τ as in lemma 5. Since $\hat{L}_n \equiv 1 - \hat{G}_n(\hat{T}_n)$, we see from (43) that

$$n^{-1} \log \hat{L}_n \leq -\tau \hat{T}_n + n^{-1} \log k + \log (1 + \epsilon) \tag{48}$$

for every s and n. It follows from (47) and (48) that $\limsup_{n\to\infty} \{n^{-1} \log \hat{L}_n\} \leq -\tau J(\theta) + \log (1 + \epsilon)$ with probability one. Since ϵ and τ are arbitrary, $\limsup \{n^{-1} \log \hat{L}_n\} \leq -J(\theta)$ with probability one. Theorem 1 applied to $\hat{T}_n$ now shows that (12) holds with probability one.

If $J = 0$ for the given θ, theorem 1 applied to $\hat{T}_n$ shows that (13) holds with probability one. Suppose then that $0 < J \leq \infty$, and choose and fix an s such that (12) is satisfied. Suppose first that $\hat{L}_n = 0$ for all sufficiently large n. Then $\hat{N}$ is a bounded function of ϵ, and $J = \infty$ by (12), so (13) holds. Suppose now that $\hat{L}_n > 0$ for infinitely many n. It is plain from (12) and $J > 0$ that $\hat{L}_n \to 0$. Consequently, $1 \leq \hat{N}(\epsilon, s) < \infty$ for every ϵ; $\hat{N}$ increases to ∞ through a subsequence of the integers as ϵ decreases to zero; and

$$\hat{L}_{\hat{N}-1} > \epsilon \geq \hat{L}_{\hat{N}} \tag{49}$$

for all ϵ such that $\hat{N} \geq 2$. It follows easily from (49) by using (12) that $\lim_{\epsilon\to 0} \{\hat{N}^{-1} \log (1/\epsilon)\} = J$. This completes the proof of theorem 2.

It may be worthwhile to note that the present assumptions imply that (19) holds with probability one. Choose ϵ and τ as in lemma 5. It follows from (48) by theorem 1 applied to $\hat{T}_n$ that

$$\liminf_{n\to\infty} (-\tau \hat{T}_n) + \log (1 + \epsilon) \geq -J(\theta) \tag{50}$$

with probability one. Since ϵ and τ are arbitrary, (50) implies that $\limsup \hat{T}_n \leq J(\theta)$ with probability one, and (19) now follows from (47).

In view of (42), (19) is equivalent to

$$R_n(\Theta, \theta) \to 0 \tag{51}$$

in the case when $J(\theta) < \infty$. Condition (51) is of the same formal structure as (42), since J vanishes when Θ_0 is replaced by Θ on the right-hand side of (2). It follows that a direct proof of (51) (and thereby of (19)) can be given along the lines of the proof of lemma 4. This direct proof requires, however, that the integrability condition of assumption 5 hold for each θ_0 in $\bar{\Theta}$ and with g_0 replaced by g.

REFERENCES

[1] J. Neyman and E. S. Pearson, "On the use and interpretation of certain test criteria for purposes of statistical inference," *Biometrika*, Vol. 20-A (1928), pp. 175–240 and 264–299.

[2] R. R. Bahadur, "Simultaneous comparison of the optimum and sign tests of a normal mean," *Contributions to Probability and Statistics*, Stanford, Stanford University Press, 1960, pp. 79–88.

[3] ———, "Stochastic comparison of tests," *Ann. Math. Statist.*, Vol. 31 (1960), pp. 276–295.

[4] ———, "Asymptotic efficiency of tests and estimates," *Sankhyā*, Vol. 22 (1960), pp. 229–252.

[5] S. Kullback and R. A. Leibler, "On information and sufficiency," *Ann. Math. Statist.*, Vol. 22 (1951), pp. 79–86.

[6] S. Kullback, *Information Theory and Statistics*, New York, Wiley, 1959.

[7] A. Wald, "Note on the consistency of the maximum likelihood estimate," *Ann. Math. Statist.*, Vol. 20 (1949), pp. 595–601.

[8] J. Kiefer and J. Wolfowitz, "Consistency of the maximum likelihood estimator in the presence of infinitely many incidental parameters," *Ann. Math. Statist.*, Vol. 27 (1956), pp. 887–906.

[9] R. R. Bahadur, "Examples of inconsistency of maximum likelihood estimates," *Sankhyā*, Vol. 20 (1958), pp. 207–210.

[10] W. Hoeffding, "Asymptotically optimal tests for multinominal distributions," *Ann. Math. Statist.*, Vol. 36 (1965), pp. 369–408.

[11] H. Chernoff, "A measure of asymptotic efficiency for tests of a hypothesis based on the sum of observations," *Ann. Math. Statist.*, Vol. 23 (1952), pp. 493–507.

[12] R. R. Bahadur and R. Ranga Rao, "On deviations of the sample mean," *Ann. Math. Statist.*, Vol. 31 (1960), pp. 1015–1027.

[13] W. Hoeffding, "Probability inequalities for sums of bounded random variables," *J. Amer. Statist. Assoc.*, Vol. 58 (1963), pp. 13–30.

THE USE OF THE LIKELIHOOD FUNCTION IN STATISTICAL PRACTICE

GEORGE A. BARNARD
IMPERIAL COLLEGE

1. Introduction

The title I have chosen deliberately echoes that of the paper by L. J. Savage, [1] because it is written with an objective which closely corresponds to Savage's —to encourage practical statisticians to explore the ways in which the study of the likelihood function generated by a set of data can help in its interpretation. At the same time I hope theoretical statisticians will be encouraged to study the theory of likelihood with a view to explaining in detail how the likelihood function can be used, and what its limitations are. It appears to be high time we did this, because for some years now it has been common for geneticists to express themselves in terms of likelihood, and the following quotation indicates that high energy physicists are following suit: "How then can an experimenter present the results of his work in an 'objective' fashion, that is, without introducing his own prior beliefs? One way, *often used by physicists* (my italics, G.B.), is to present $L(x_i^{obs}; \alpha)$ as a function of α for his particular observations $x_i^{obs}; \ldots$" [2].

Whereas in some fields it is still possible to attribute failure to use "orthodox" statistical methods to mere ignorance, such an explanation is untenable in the case of statistically sophisticated areas such as these two. It must be here that the "orthodox" methods have been tried and found wanting.

Lest my allusion to Savage be misinterpreted to mean that I accept the subjective Bayesian position, let me hasten to specify some of the ways in which we differ. Whereas Savage, if I understand him aright, would regard a specification of the likelihood function as *always* providing, at least in principle, the solution to problems of statistical inference, I conceive of likelihood methods as rigorously applicable only to those situations where the distribution of the observations x over the sample space S can be taken as known to belong to a (usually continuously) parametrized family of distributions with probability functions $f(x, \theta)$, the parameter θ ranging over a well-defined parameter space Ω. The primary problem in such a case is often how to express the order of preference among the different values of θ which may be said to be rationally induced

This paper was written while the author was a Visiting Professor at Stanford University, attached to the Computing Group, Stanford Linear Accelerator Center, under the auspices of the United States Atomic Energy Commission.

when the observations x are known. It is this problem which, it seems to me, is answered by giving the likelihood function

$$L(\theta) = f(x, \theta)/\sup_{\theta} f(x, \theta). \tag{1}$$

This problem is different from that of establishing any particular one of the θ's as "true" in some absolute sense, or of rejecting any such particular value as incredible. Being equipped with a likelihood function is like being equipped with a balance, but no weights, and being confronted with several specimens, $A_1, A_2, A_3, \cdots; B_1, B_2, B_3, \cdots; C_1, C_2, C_3, \cdots$ of each of several denominations A, B, C of coins. With such a balance we could establish, for example, that a coin of denomination A is more than twice, but less than three times as heavy as one of denomination B. But we could not express the weight of any of the coins in grams. And the analogy with the balance would be misleading, unless we imagined that the coins have the peculiarity that different denominations react chemically to dissolve into thin air when put together on the same scale pan. It would then have no verifiable meaning to say that the weight of a coin of denomination A is equal to the combined weight of a coin of denomination B put together with a coin of denomination C. Correspondingly, if the likelihood of θ_1 against θ_2 is $\frac{4}{1}$, while that of θ_1 against θ_3 is $\frac{4}{2}$, it will have no meaning to say that the likelihood of θ_1 against "θ_2 or θ_3" is $4/(1 + 2) = \frac{4}{3}$, because the likelihood of "θ_2 or θ_3" is not well defined on the data x.

If each of θ_2 and θ_3 had a definite prior probability, and these prior probabilities stood in the ratio $p:(1 - p)$, then we could interpret "θ_2 or θ_3," in relation to x, as meaning "θ_2, with probability p, or θ_3, with probability $1 - p$." Then the likelihood for "θ_2 or θ_3," given x, would be proportional to $pf(x, \theta_2) + (1 - p)f(x, \theta_3)$. My unwillingness to suppose that such combinations of likelihoods always have meaning is another respect in which my position differs from that of the subjective Bayesians.

But perhaps the major difference lies in my view that there are inference problems not of this comparative form, where it *cannot* be taken as known that the distribution of x belongs to a well-defined family. In relation to an hypothesis H, it may well be appropriate to define a (real-valued) measure of discrepancy $T(x)$ and, given an observed result x_0, to calculate the probability $\Pr\{T(x) \geq T(x_0)|H\} = \alpha(x_0)$. If this probability is small, we shall be faced with the disjunction: either an event of small probability has occurred, or H is false. Thus with sufficiently small α, we shall be disposed to think up some alternative to H which fits the observations better.

When, in such a case, we say that H is rejected on the data at the α level of significance, we are using α as a "measure" of credibility of H, on the data, in the simple but useful sense, that as α ranges from 1 down to 0 our disposition to think of some alternative will increase. On the other hand, of course, if H is rejected by data x on the α level of significance, while another hypothesis H' is rejected by other data y on the α level of significance, it will not at all follow that our disposition to think of an alternative to H' must be the same as our disposition to think of an alternative to H. Only if H' is rejected by the same data, using the same criterion T, on a higher level of significance than H will it be illogical, in the absence of other evidence, to accept H' while rejecting H.

It is an elementary error, of course, to think of Pr $\{T(x) \geq T(x_0)|H\}$, when it is known that $T(x) \geq T(x_0)$, as being the *probability* of H. It is a measure, or ranking, of credibility of a different, cruder, but nonetheless useful kind. Likelihood may, in a loose sense, be thought of as standing intermediate between this crude measure of credibility, and the very precise measure provided by a statement of probability.

I should, perhaps, make clear that I am making no serious claim to originality in this paper. It might well be regarded as a sermon on the text, from an early edition of Fisher's *Statistical Methods:* "The mathematical concept of probability is inadequate to express our mental confidence or diffidence in making such inferences (from sample to population—G.B.), and . . . the mathematical quantity which appears to be appropriate for measuring our order of preference among different possible populations does not in fact obey the laws of probability. To distinguish it from probability, I have used the term 'Likelihood' to designate this quantity*; since both the words 'likelihood' and 'probability' are loosely used in common speech to cover both kinds of relationship. . . ."

"*A more specialized application of the likelihood is its use, under the name of 'power function', for comparing the sensitiveness, in some chosen respect, of different possible tests of significance."

I shall try to expound what I take this to mean.

2. The likelihood function as an operating characteristic

My first point is a gloss, or extension, on Fisher's footnote. One of the great services that have been rendered by Neyman and Pearson to mathematical statistics is the emphasis they have placed on the operating characteristic of a statistical procedure. First applied by them in connection with hypothesis testing, or two-way decision procedures, the concept has more recently been generalized to cover multiple decision procedures. If the sample space S is finite (as we shall later argue is really *always* the case) and there are k possible decisions D_i $(i = 1, 2, \cdots, k)$, and if C_i is the region in S where we take decision D_i, then the operating characteristic of our decision procedure specifies Pr $\{D_i|\theta\}$ as a function of θ; that is, it gives

$$P_i(\theta) = \sum_{x \in C_i} f(x, \theta) = \sum_{x \in C_i} w(x) L_x(\theta), \tag{2}$$

where $L_x(\theta)$ is the likelihood function, given the observation x, and

$$w(x) = \sup_{\theta} L_x(\theta), \tag{3}$$

the normalizing factor of the likelihood function, may be regarded here as a (nonnegative) weighting factor. In other words, *the i-th component of the operating characteristic $P_i(\theta)$ is a weighted sum of the likelihood functions for the points leading to the i-th decision.*

(Here we take the operating characteristic to consist of the set of functions which gives the probability of taking the i-th decision as a function of the parameter value θ. This is a more direct generalization of the original idea of the power curve than the (single) function which gives the mean value of the loss as a function of θ. The former becomes a special case of the latter, of course, if we allow vector-valued loss functions. But the former, though less general, has the advantage that it retains a precise meaning in situations, often arising in practice, where the exact evaluation of losses is impossible.)

We can go further if we restrict consideration to decision procedures which are admissible. For, under wide conditions, we know that such procedures must be Bayes with respect to some prior on Ω, and hence have the property that two points x, x' in S for which the likelihoods $L_x(\theta)$, $L_{x'}(\theta)$ are the same, must lead to the same decision.

Thus, in such conditions we can say that any component of the operating characteristic of any admissible procedure is obtainable by forming weighted sums of possible likelihood functions.

Now let us introduce a new notion—that of a *nonrestricting* class of possible decisions. The idea behind this is that when the number k of possible decisions is small we often, in practice, feel that we should be able to have a wider choice; and it often happens that when we review the possibilities open to us, we are in fact able to choose from among a wider class of decisions than we thought. For example, we often find the simple two-way decision alternatives in an hypothesis testing problem restrictive—to declare rejection at the 5% level, or nonrejection, say—and we may give ourselves three alternatives: rejection at 5%, rejection at 1%, or nonrejection. Or, again, we may take as a third alternative, in such a case, to continue sampling.

The artificiality of a restricted class of decisions is perhaps most marked in connection with procedures such as those which have been developed to select the k largest from a set of m means. It can easily happen that we find ourselves with clear evidence that $k - 1$ of the means are larger than the rest, but we really have no substantial grounds for selecting the k-th one from the remaining $(m - k + 1)$. Then we may well regret having said we would choose k, and would prefer to be allowed to choose only $k - 1$. Or again, we may find that $k + 1$ of the means are clearly larger than the rest, but there are no appreciable differences between these. Here we would wish to be allowed to choose $k + 1$ instead of k; and this may, on re-examination of the real situation, prove to be allowable in such a case. To represent the practical situation fully, we would need to set up the problem as one of selecting *approximately* k means, with a loss function for departure from k, as well as a loss function for selecting a smaller rather than a larger mean. If we do not do this, it is only because the theory of such a problem could well become unmanageable.

On examination, it appears that the difficulties we experience, in situations such as these, arise from the fact that limitations on the alternatives allowed to us, force us to make the same decision for two distinct observations x, x', even though the information which x provides about θ is different from that which x' provides, in the sense that the value of any minimal sufficient statistic for θ is different at x from what it is at x'. A nonrestrictive class of decisions may then be associated with an admissible decision rule which is such that the class C_i of observations x which lead to decision D_i consists of those observations, and only those observations, for which a minimal sufficient statistic takes a given value. But in such a case, the class C_i will consist of those observations, and only those, for which the likelihood function is a fixed $L_i(\theta)$. And then $P_i(\theta)$, which we have seen is a weighted sum of likelihood functions, will evidently be proportional to this $L_i(\theta)$. Thus, in this sense, a decision function which makes full use of the information provided by the observations will have an operating characteristic whose components are proportional to the set of possible likelihood functions.

It is important to notice that this result is "robust under approximation," in the following sense: we may say that a set of possible decisions is not *seriously* restrictive if it does not force us to make the same decision for two observations x, x' for which the information about θ is *very* different; for this last will mean that the likelihood functions $L_x(\theta)$ and $L_{x'}(\theta)$ will be approximately equal, and thus their weighted sum will be approximately equal to either one of them. Thus, when the set of possible decisions is not seriously restrictive, the i-th component $P_i(\theta)$ of the operating characteristic of any admissible decision rule will be approximately proportional to the likelihood function $L_x(\theta)$ for any x in C_i.

Therefore, one reason for looking at the set of possible likelihood functions which can arise from a given experiment is that it gives us a conspectus of the set of operating characteristics of admissible decision rules. Hence, it may well lead to economy of thought to look at these likelihood functions, especially in the many practical situations where it is not at all easy to obtain a true conspectus of the set of all possible decisions, together with their associated loss functions.

To conclude this section we may note that presumably what Fisher had in mind in writing his footnote was that the choice of a critical region C in S, together with the decision rule to reject the hypothesis tested if x fell in C, and not to reject it if x fell in $S - C$, was tantamount to introducing a new observable y, taking the value 1 if x fell in C, and the value 0 if x fell in $S - C$. The sample space S would then be mapped onto a two-point sample space $\{0, 1\}$. The likelihood function of θ, given $y = 1$, would then be proportional to the power curve associated with C; and indeed, if, as would often be the case, $P(\theta)$ had supremum 1, the likelihood function and the power curve would be identical.

3. Frequency interpretations

One of the red herrings most frequently drawn across discussions on the foundations of statistics is based on the false idea that some measures of uncertainty can be given frequency interpretations, but others cannot. An associated fallacy, against which Charles Stein has spoken out clearly, but few others have, is the idea that such frequency interpretations are unique. It is worthwhile stressing the universality, and the nonuniqueness of frequency interpretations, because this helps to get rid of the idea that the *meaning* of measures of uncertainty is to be found in a particular set of frequency interpretations. The meaning of ideas like probability, significance level, and likelihood, is really to be found only in their use; frequency interpretations are expository only.

Thus the commonest frequency interpretation of a level of significance asks us to imagine a long series of cases, in each of which the hypothesis tested is true. Then if, whenever significance level α is attained, we say that the hypothesis tested is false, the long run frequency of errors is α. More precisely, and a little more generally, if we use the same rule for assertion in a long series of cases, in which the hypothesis tested is sometimes true and sometimes false, the long run frequency of errors cannot exceed α. But such an interpretation of 'significance level α' runs into difficulty when we carry out multiple tests on a single set of data. For this reason, among others, I personally prefer an account which runs as follows.

Imagine that, at any particular time, one's view of the world is represented by the tentative acceptance of a number of hypotheses H_i, each relating to a set of data x_i, the sets and the hypotheses being reasonably independent of each other. If we have a measure of discrepancy (or test criterion) T defined for each set of data, in relation to its associated hypothesis, the level of significance associated with x_i, in regard to H_i, is

$$\alpha_i = \Pr\{T \geq T(x_i)|H_i\}. \tag{4}$$

Now the α_i, if all the H_i were true, would be distributed (nearly) uniformly over the interval [0, 1]. For example, roughly 1% of the α_i should be less than 0.01, and if the actual proportion is greatly in excess of this, we should consider modifying some of the associated H_i. It would also call for some consideration if far fewer than 1% of the α_i were less than 0.01, though attention would in this case first be directed at the formation of the criterion T, or the calculation of its distribution. Hypotheses giving low values to α should thus be regarded much as luxuries by a person of moderate means—not to be avoided entirely, but not to be indulged in to excess.

But neither this nor the more common frequency interpretation of significance level is to be regarded as more than expository. The meaning of "H is rejected at significance level α" is "Either an event of probability α has occurred, or H is false," and our disposition to disbelieve H arises from our disposition to disbelieve in events of small probability.

In the same way, to say that, on data x, θ is k times as likely as θ' *means* that $L(\theta)/L(\theta') = k$, simply; but we can give such a statement a frequency interpretation as follows: imagine a long series of experiments, in each of which we have to decide between two alternatives such as θ and θ'. If, in such a series, we assert the truth of the more likely hypothesis only when the likelihood ratio (such as $L(\theta)/L(\theta')$) exceeds k, then the "odds against error" are bounded below by k. Here we define "odds against error" as the ratio

$$\frac{\text{Long run number of times we rightly decide}}{\text{Long run number of times we wrongly decide}}. \tag{5}$$

(As was first noted by C. A. B. Smith, the truth of this interpretation follows from a simple application of Markov's inequality to the nonnegative random variable L/L', which has mean value 1 on θ'.) We can thus attach an appropriate form of "inductive behavior" to likelihood ratios, if we wish to. But in truth, as with significance levels, our disposition to prefer θ to θ' (assuming k greater than 1) arises from our disposition to prefer that hypothesis which makes more probable what we know to be true. In this case, the disposition to prefer is measured in the rather precise sense, that further independent data having a likelihood ratio less than $1/k$ would need to be forthcoming before our preference between θ and θ' was reversed.

Another form of frequency interpretation for $L(\theta)$ can be given, and is useful, in cases where Ω can be regarded as a separable topological space, in which $L(\theta)$ is always continuous, with a nonnegative measure m defined on open sets. It is then possible to define a sequence $T = \{\theta_r\}$, $r = 1, 2, 3, \cdots$ of points in Ω such that, for any pair of open sets A, B in Ω,

$$\lim_{n\to\infty} \frac{\text{Number of times } T \text{ visits } A\text{, in first } n \text{ terms}}{\text{Number of times } T \text{ visits } B\text{, in first } n \text{ terms}} = \frac{m(A)}{m(B)}. \tag{6}$$

If, now, we imagine a long run of cases, in which the true value of θ on the r-th occasion is θ_r, and select from such a long run all those cases in which the likelihood function was $L(\theta)$, then in the subsequence so selected, for any pair of open sets A, B,

$$\lim_{n\to\infty} \frac{\text{Number of times the subsequence visits } A\text{, in first } n \text{ terms}}{\text{Number of times the subsequence visits } B \text{ in the first } n \text{ terms}} = \left(\int_A L(\theta)\, dm\right)\Big/\left(\int_B L(\theta)\, dm\right) \tag{7}$$

and if, in particular, $\int_\Omega L(\theta)\, dm$ is finite, then $\int_A L(\theta)\, dm / \int_\Omega L(\theta)\, dm$ is the relative frequency with which the subsequence visits A. (These statements are true, with probability 1, in the infinite product space $S \times S \times S \times \cdots$, in which the measure in the r-th component is defined by $f(x, \theta_r)$. They follow from a countable number of applications of the strong law of large numbers.)

This last form of frequency interpretation may be contrasted with that commonly given to a set of confidence intervals $C(x)$, for a parameter θ, with confidence coefficient $1 - \alpha$. Here we imagine the rule $C(x)$ applied to every one of a sequence of cases, in which θ varies in any manner, and we can say that the relative frequency with which $C(x)$ contains the true value of θ is $1 - \alpha$ with probability 1. Now, whereas this frequency interpretation is more general than that just previously discussed, it is open to the objection that it may require us to group together cases where $C(x)$ consists of the whole of Ω, when the true value is certainly in $C(x)$, and cases where $C(x)$ is quite small, and it appears most implausible that θ should be in $C(x)$. The truth of the statement of long run frequency cannot be doubted, but it is hard to justify its relevance to individual cases, when they fall into such discrepant sets. (That cases where this happens are by no means pathological is indicated, for example, by the problem of the ratio of two normal means.) It would seem that lumping such cases together is just as 'arbitrary' as choosing a particular measure m, especially if the value of the ratio

$$\int_A L(\theta)\, dm \Big/ \int_\Omega L(\theta)\, dm \tag{8}$$

does not depend critically on the choice of m.

Perhaps it may be well to spell this out somewhat. By 'arbitrary,' we mean 'dependent on an act of will.' Now the distinguished author of the theory of confidence intervals has emphasized that his notion of 'inductive behavior' involves an act of will. And I am now suggesting that this act of will is one which not all of us would in fact make, since it may involve us in grouping together statements about parameters, some of which are obviously true (since they assert merely $\theta \in \Omega$), while others are obviously doubtful (since they assert '$\theta \in A$,' where A is a comparatively small subset of Ω). The act of will involved in collecting together statements of the form '$\theta \in A$' and statements of the form '$\theta \in \Omega$,' and labeling them all with the same confidence coefficient is not one which all of us will wish to make. In the same way, if we refer a statement of the form '$\theta \in A$,' on the basis of a given sample, to a sequence of cases satisfying the conditions of the measure m above, and so attach to this statement a 'measure of credibility' given by the above formula, we are also making an act of will (in choosing m), but one which, in many cases, could be less objectionable than that involved with the confidence interval.

The objection here leveled against confidence intervals is much weakened, if not removed, by inverting the usual form of statement. If, instead of asserting '$\theta \in A$,' with confidence coefficient $1 - \alpha$, we say "any value of θ in the complement $-A$ of A is rejected at significance level α," those cases where the confidence set covers the whole parameter space lose their absurd appearance; since we are then merely saying that the data are insufficient to enable us to reject any of the possible values of θ. And even if, as many of us feel, it is inappropriate to try to inflate the 'significance' of precise experiments, by adding to the denominator of the frequency a number of experiments which were insufficiently precise, we still have recourse to the simple disjunction, "either θ is not in $-A$, or an event of probability $\leq \alpha$ has occurred."

4. An example from particle physics

It is fundamental to the point of view of the present paper that the theory of statistical inference exists to serve those who are collecting and using empirical data, either to further natural knowledge or to improve the making of decisions; and it is more especially (though not exclusively) with the furtherance of natural knowledge that we are here concerned. It follows that our judgment of the value of any proposed statistical technique will be based on the way in which the technique in question appears to meet the purposes of those engaged in this pursuit. It seems worthwhile, therefore, to consider a specific scientific situation in which these ideas come into play. The broad nature of the problems facing workers in particle physics is sufficiently well known to provide a useful example.

If a beam of polarized Λ particles is observed to decay into a proton and a pion in a bubble chamber, it is possible for each such particle to measure the cosine x of the angle between the track of one of the decay particles and the direction of polarization of the decaying particle. It is then known that observations on distinct particles are independent of each other, and each x follows a distribution with density

$$f(x, \theta) = \tfrac{1}{2}(1 + \theta x), \qquad -1 \leq x \leq +1, \tag{9}$$

where θ is the product of the degree of polarization of the decaying particles, and the parity nonconservation parameter.

A set of n observations $x_1, \cdots, x_i, \cdots, x_n$ then gives a likelihood function $L(\theta)$ proportional to

$$\prod_i (1 + \theta x_i) = 1 + s_1\theta + s_2\theta^2 + \cdots + s_n\theta^n \tag{10}$$

where the s_r are the elementary symmetric functions formed by taking the sum of products of the x_i, r at a time. Since the likelihood is a polynomial in θ with real roots, all of them outside the interval $-1 \leq \theta \leq +1$, it is easy to see that unless all the x_i have the same sign (an event whose probability depends on θ, but which cannot happen with n as small as 10 in more than about 5% of cases), there will be a unique maximum of $L(\theta)$ lying between the largest negative root of $L(\theta)$ and the smallest positive root. It will turn out more convenient to deal with the logarithmic derivative of the likelihood,

$$g(\theta) = \partial(\log L(\theta))/\partial\theta = \sum_i (x_i/(1 + \theta x_i)), \tag{11}$$

which is obviously monotone decreasing, with probability 1, in all cases. We shall argue that for most purposes the best summary that can be provided of the data consists in specifying those values of θ for which $g(\theta)/g'(\hat{\theta})$ takes given values—if one is restricted to giving three numbers, then the roots of $g(\theta) = 0, \pm 2g'(\hat{\theta})$ will probably be most useful, provided that they all lie between -1 and $+1$; while if they do not, instead it will be better to quote the value of $g(1)$, or of $g(-1)$, or both, as the case may be. If more than three numbers can be given, then the roots of $g(\theta)/g'(\hat{\theta}) = 0, \pm 2, \pm 3$, with corresponding provisos, will be best. We shall suggest that ideally, of course, it would be best to specify all values of $g(\theta)$ in the interval $(-1, +1)$ if this were practicable. When the root $\hat{\theta}$ of $g(\theta) = 0$ lies in the interval, it is the value which maximizes the likelihood, and $L(\theta)$ can be recovered as

$$L(\theta) = \exp \int_{\hat{\theta}}^{\theta} g(\theta)\, d\theta. \tag{12}$$

It is easy to see how $L(\theta)$ could be recovered in other cases also.

Now why should we suggest that an approximate specification of $g(\theta)$ is the 'best' way of summarizing the data? Such a judgment must involve an assessment of the purposes for which the data are being collected. In this case the main object is to collect numerical data on the properties of 'elementary' particles, with a view to finding some underlying regularity. The situation is very similar to that obtained in chemistry around the time when Newlands, Mendeleyev, and Lothar Meyer worked towards establishing the periodic classification of the chemical elements, when poor estimates of some atomic weights as well as the fact that some elements had not yet been found, gave rise to difficulties in establishing the classification—indeed a case could be made out for the thesis that the classification was established just as soon as the numerical properties of the elements were established with sufficient precision. With the 'elementary' particles of today it may well be that some underlying regularities are also being obscured by inaccuracies in the data.

The function of the data analysis, then, is to *point towards* some values for the masses, decay cross-sections, and so on, of the various particles, as being more plausible than others. At the same time, the analysis should indicate how far a theoretical value may differ from an experimental value, without having an overwhelming weight of evidence against it. An important feature is, that we should be able in some way to associate a 'weight of evidence' with each deviation of theory from experiment, in such a way that the combined deviations can be assessed. In practice this is often beset with computational difficulties, which may mean, for example, that only first-order perturbation values are available from theory; but should the extent to which a theory fits other facts appear to justify it, further calculations are usually undertaken. Thus there will usually be an element of judgment involved in our assessment of how well a theory fits observed facts; but that part of this assessment which can be quantified surely should be, and the likelihood of the theoretical values seem very well suited for this purpose.

If one theory gives predicted values $t_1, t_2, t_3, \cdots$, for a series of parameters $\theta_1, \theta_2, \theta_3, \cdots$, independently determined, while another theory predicts values $t_1', t_2', t_3', \cdots$, then the first theory gives the better fit if

$$L(t_1)L(t_2)L(t_3) \cdots > L(t_1')L(t_2')L(t_3') \cdots, \tag{13}$$

and vice versa if the inequality is reversed. For reasons just indicated, in addition to other complications which may arise, it may well be difficult, and not altogether necessary, to actually carry through such a comparison with full numerical precision. But the principle is there, and rough estimates can be made. Low values of L for particular parameters will indicate in what respects a proposed theory needs modification.

It is in a specific scientific situation such as we are contemplating that the difficulties of the subjective Bayesian position appear most strongly. For, quite apart from the fact that anyone who could make a plausible survey of the prior distributions involved in particle physics would have achieved a major feat of scientific imagination, there is the question, what would a posterior distribution do for us which the likelihood function will not? We could, of course, place bets rationally; but this is not a serious aim—even though it may perhaps be indulged in for amusement. And if we found a theory T to be more probable, a posteriori, than a theory T', while being less likely on the data, this could only be because T was considered a priori much more plausible than T'.

But on what grounds could we go against the evidence in this way? On examination, any case where such a thing appeared to be happening would surely turn out to be one in which we were taking account of further relevant information, in the guise of the prior distribution; however, then the likelihood computation, taking this further information into account, would agree with the posterior probability. In introducing a prior distribution, we thus are abandoning the important claim to objectivity in return for an illusory gain.

It is an important advantage of the likelihood method that, when the number of observations grows large enough, we can appeal to asymptotic theory to derive arguments of the significance test type, which enable us to rule out certain theories as too much at variance with the facts to be worth serious consideration. But the probability levels we would invoke would normally be much smaller than those commonly used, for example, with confidence intervals. For instance, the probability of decay of a Λ particle into a proton, a muon, and a neutrino, has been estimated to be about 1.5×10^{-4}, so that we might, by over-simple application of statistical arguments, be led to suppose that a particle seen to decay in this way was not a Λ particle.

Among the set of 95% confidence intervals we might obtain for the (perhaps) 1000 quantities independently measured in particle physics, there should be some 50 which would fail to contain the true values. We would have no indication of which 50 these might be; and it could easily happen that one theory gave values inside all the confidence intervals, whereas another gave several values outside their confidence intervals. Yet the latter theory was more plausible, on the data, than the former.

The practical advantage of the procedure suggested—to specify the values at which $g(\theta)/g'(\hat{\theta})$ equals 0, ± 2, and ± 4—arises from the fact that, in combining the results from two independent experiments we merely have to add the corresponding g's—easily done graphically—while in going out as far as $g(\theta) = \pm 4$, we would be taking care of any departures from linearity in $g(\theta)$ which would be likely to arise in practice.

5. Singularities in the likelihood function

Some writers have felt that the use of likelihood by itself cannot be justified in general, because infinite values may be encountered in connection with hypotheses which are not at all plausible. For example, if we have a single observation x from a normal distribution with unknown mean and unknown variance, the likelihood function is proportional to

$$L(\mu, \sigma) = (1/\sqrt{2\pi}\sigma) \exp -\tfrac{1}{2}((x - \mu)/\sigma)^2, \tag{14}$$

which becomes infinite when $\mu = x$ and $\sigma = 0$. Or again, as Bruce Hill has pointed out, when we fit the three-parameter log-normal distribution, $y = \log (x - \tau)$, with y normal with mean μ and standard deviation σ, there is a singularity as τ approaches the smallest of the observations x_i. And I am indebted to Dr. A. W. F. Edwards for drawing my attention to a most interesting example, arising in the theory of evolutionary genetics, where we suppose we are observing a Brownian movement in x with variance increasing at unit rate per unit time t. Given two pairs of observations (x_1, t_1), (x_2, t_2), with $t_1 < t_2$, we want to estimate the epoch t_0, and the value x_0 of x, at which the process began. We find a singularity of the likelihood at $(x_0, t_0) = (x_1, t_1)$.

It is trivial that all these paradoxes would be disposed of if it were accepted that the distributions involved were never really continuous, but always discrete, and that the continuous expressions for them were approximations introduced for mathematical convenience. For then the likelihood function, like the discrete probabilities, would necessarily always be finite. And it is an important fact of nature that the resolving power of any measuring instrument is always finite, so that the distribution of any quantity really observable is always discrete. But it may be felt that the quantitative effect of this discretization will be insufficient—although the likelihood will cease to be infinite, it will remain larger than intuition would lead us to expect. Therefore, some further discussion is called for.

Let us take first the (obviously imaginary) case of the single observation from a normal population. Bearing in mind the discretization, together with the important point that likelihood measures *relative* plausibility, this says that if we have an arbitrarily precise observation x, this provides arbitrarily strong evidence in favor of an hypothesis H which says that x will certainly have this value, as against any hypothesis H' which gives x a nonzero variance. Now let us imagine an experimental situation which might fancifully be thought to correspond—one in which a space probe is sent to Mars to measure its magnetic field, if any. The hypothesis H asserts that Mars has no magnetic field whatsoever, and H' asserts that there is a weak magnetic field which, like that of the earth, fluctuates with time, continuously. We suppose the fluctuation to be such that if measured at a randomly chosen epoch, such as that of the arrival of the probe, the magnetic field will be normally distributed, with mean 0.01 and standard deviation 0.02. The probe takes one reading at its nearest approach to Mars, and then begins transmitting the binary digits of the magnetic field strength. These are all zero. To begin with, we shall feel that there is little evidence in favor of either H or H', but surely, as the successive digits (assumed to be utterly reliable) turn out to be zero, we shall feel more and more disposed to favor H as against H'. In fact, as the digits accumulate, without limit, so our disposition to favor H against H' will increase without limit, just as the likelihood function suggests it should.

To turn now to the case of the log-normal distribution on analysis, it turns out that the distribution of x, which gives rise to the singularity here, is one which has a very high but narrow peak at the lowest observed value of x, x_0, say, together with a long, low tail covering larger values of x. Thus the hypothesis to which the likelihood is pointing is one which says, in effect, that a moderate fraction (perhaps 20%) of the values of x will lie very near to x_0, that no value of x will be observed less than x_0, while the remaining (say) 80% of values will be spread out rather thinly over the range above x_0.

Now the data discussed by Hill related to the date of appearance of symptoms after inoculation for smallpox, and were in fact recorded to the nearest day. On carrying through the arithmetic, it appears that the paradox could only arise in this case if we supposed the times of appearance of symptoms to be recorded to the nearest exp $\{-150\}$-th part of a day, that is, correct to about 10^{-59} seconds. It may perhaps bring home the absurdity involved to point out that merely to avoid the uncertainty due to the finite velocity of light signals, it would be necessary, for this sort of accuracy, for the distance from the observer to the subject to be controlled to less than the nuclear radius. But now suppose we had a medical theory which predicted the time of occurrence of symptoms with this sort of precision; would not observations in accordance with it be taken as strong confirmation?

6. Likelihood sets and confidence intervals

If the view is accepted that the likelihood serves to rank possible parameter values in a rational order of plausibility on given data, then those subsets $A(\lambda; x)$ of the parameter space defined by

$$A(\lambda; x) = \{\theta: L(\theta, x) \geq \lambda\} \tag{21}$$

acquire some importance. For a given x, those points belonging, for example, to $A(\frac{1}{2}; x)$ will be such that no alternative value of θ can be specified for which the given data will give a likelihood ratio more than 2 to 1 in favor. In this sense, these are the values of θ which this experiment will not by itself contradict. Or, we may convert the values of λ, ranging from 0 to 1, to another scale of value of C, also ranging from 0 to 1, by the condition

$$\int_{A(\lambda,x)} L(\theta, x)\, dm \Big/ \int_{\Omega} L(\theta, x)\, dm = C, \tag{22}$$

and so obtain a measure, C, of the assurance with which we might assert that θ was in $A(\lambda; x)$. This measure of assurance would have the frequency interpretation referred to in the above section; it would not, in general, be a "confidence coefficient," in the sense of confidence interval theory.

Finally, we might establish λ as a function of x, with parameter α, in such a way that the resulting $A(\lambda; x)$ would be a confidence set for θ, with confidence coefficient α. In this way we can obtain three ways of parametrizing the sets A, for a given x, with the parameters λ, C, and α, all of which will range between 0 and 1. The parametrization with λ will involve nothing which is not given in the original specification; the parametrization with C will involve an arbitrary choice of the measure m for its interpretation; the parametrization with α will also involve an arbitrary classification of the observed result along with others not observed. In the asymptotic situation to which the classical maximum likelihood theory applies, natural choices in the two latter cases will lead to an equation of C with α, and a unique functional relationship of each with λ. In small samples, this will not be so; and perhaps the more liberal minded of us will feel inclined to quote all three parametrizations, since all of them convey useful information.

REFERENCES

[1] L. J. SAVAGE, "Subjective probability and statistical practice," *Foundations of Statistical Inference*, London, Methuen, 1962.

[2] FRANK SOLMITZ, "Analysis of experiments in particle physics," *Ann. Rev. Nuclear Science*, Vol. 14 (1964), pp. 375–402.

[3] R. A. FISHER, *Statistical Methods for Research Workers*, London, Oliver and Boyd, p. 10, 1958 (13th ed.).

PROBLEMS RELATING TO THE EXISTENCE OF MAXIMAL AND MINIMAL ELEMENTS IN SOME FAMILIES OF STATISTICS (SUBFIELDS)

D. BASU
UNIVERSITY OF NORTH CAROLINA
and
INDIAN STATISTICAL INSTITUTE

1. Summary

In statistical theory one comes across various families of statistics (subfields). For each such family, it is of some interest to ask oneself as to whether the family has maximal and/or minimal elements. The author proves here the existence of such elements in a number of cases and leaves the question unsolved in a number of other cases. A number of problems of an allied nature are also discussed.

2. Introduction

Let $(\mathfrak{X}, \mathfrak{A}, \mathcal{P})$ be a given probability structure (or statistical model). A statistic is a measurable transformation of $(\mathfrak{X}, \mathfrak{A})$ to some other measurable space. Each such statistic induces, in a natural manner, a subfield (abbreviation for sub-σ-field) of $\mathfrak{A}$ and is, indeed, identifiable with the induced subfield.

Between subfields of $\mathfrak{A}$ there exists the following natural partial ordering.

DEFINITION 1. *The subfield $\mathfrak{A}_1$ is said to be larger than the subfield $\mathfrak{A}_2$ if every member of $\mathfrak{A}_2$ is also a member of $\mathfrak{A}_1$.*

A slightly weaker version of the above partial order is the following.

DEFINITION 2. *The subfield $\mathfrak{A}_1$ is said to be essentially larger than the subfield $\mathfrak{A}_2$ if every member of $\mathfrak{A}_2$ is $\mathcal{P}$-equivalent to some member of $\mathfrak{A}_1$.*

As usual, two measurable sets A and B are said to be $\mathcal{P}$-equivalent if their symmetric difference $A \Delta B$ is P-null for each $P \in \mathcal{P}$.

Given a family $\mathfrak{F}$ of subfields (statistics), one naturally inquires as to whether $\mathfrak{F}$ has a largest and/or least element in the sense of definition 1. In the absence of such elements in $\mathfrak{F}$, one may inquire about the possible existence of maximal and/or minimal elements. An element $\mathfrak{A}_0$ of $\mathfrak{F}$ is a maximal (minimal) element of $\mathfrak{F}$, if there exists no other element $\mathfrak{A}_1$ in $\mathfrak{F}$ such that $\mathfrak{A}_1$ is larger (smaller) than $\mathfrak{A}_0$. In the absence of maximal (minimal) elements in $\mathfrak{F}$, one may look for elements that are essentially largest (least) or are essentially maximal (minimal) in the sense of the weaker partial order of definition 2.

The particular case in which $\mathcal{F}$ is the family of all sufficient subfields has received considerable attention. The largest element of $\mathcal{F}$ is clearly the total subfield $\mathcal{A}$ itself. If $\mathcal{P}$ is a dominated family of measures, then it is well known that $\mathcal{F}$ has an essentially least element in terms of the weaker partial order of definition 2. In general, $\mathcal{F}$ does not have even essentially minimal elements. If, however, an essentially minimal element exists, then it must be essentially unique, and thus, the essentially least element of $\mathcal{F}$ (see corollary 3 to theorem 4 in [3]).

In [1] the author considers the family $\mathcal{F}$ of ancillary subfields. A subfield $\mathcal{A}_0$ is said to be ancillary if the restriction to $\mathcal{A}_0$ of the class $\mathcal{P}$ of probability measures shrinks the class down to a single probability measure. The least ancillary subfield is clearly the trivial subfield, consisting of only the empty set $\varnothing$ and the whole space $\mathcal{X}$. The existence of maximal elements in the family of ancillary subfields is demonstrated in [1]. In general, there exists a multiplicity of maximal ancillary subfields.

In sections 3 to 6 we list four problems that are similar to the problem of ancillary subfields. In section 7 we develop a general method to demonstrate the existence of maximal elements in these four cases. In section 8 we discuss some related questions, and in section 9 we list a number of other problems.

3. The family $\mathcal{F}_1$ of $\mathcal{B}$-independent subfields

Let $\mathcal{B}$ be a fixed subfield. A subfield is said to be $\mathcal{B}$-independent (independent of $\mathcal{B}$) if $P(BC) \equiv P(B)P(C)$ for all $B \in \mathcal{B}$, $C \in \mathcal{C}$ and $P \in \mathcal{P}$.

Let $\mathcal{F}_1$ be the family of all $\mathcal{B}$-independent subfields. Clearly, the least element of $\mathcal{F}_1$ is the trivial subfield. Even in very simple situations, $\mathcal{F}_1$ has no largest, or essentially largest, element. In section 7 we shall show that $\mathcal{F}_1$ always has maximal elements. Consider the two examples.

EXAMPLE 1(a). Let $\mathcal{X}$ consist of the four points a, b, c, and d, and let $\mathcal{P}$ consist of only one probability measure—the one that allots equal probabilities to the four points. Let $\mathcal{B}$ consist of the four sets $\varnothing$, $\mathcal{X}$, $[a, b]$, and $[c, d]$. Then the two subfields $\mathcal{C}_1$ and $\mathcal{C}_2$, consisting respectively of

$$\begin{aligned} &\mathcal{C}_1\colon \varnothing, \mathcal{X}, [a, c] \quad \text{and} \quad [b, d],\\ &\mathcal{C}_2\colon \varnothing, \mathcal{X}, [a, d] \quad \text{and} \quad [b, c], \end{aligned} \tag{3.1}$$

are both maximal $\mathcal{B}$-independent subfields. Incidentally, in this case, $\mathcal{C}_1$ and $\mathcal{C}_2$ happen to be independent of each other.

EXAMPLE 1(b). Let $x_1, x_2, \cdots, x_n$ be n independent normal variables with equal unknown means φ and equal unknown standard deviations θ. Let $\mathcal{B}$ be the subfield induced by the statistic

$$\bar{x} = (x_1 + x_2 + \cdots + x_n)/n, \tag{3.2}$$

and let $\mathcal{C}$ be induced by the set of differences

$$D = (x_1 - x_n, x_2 - x_n, \cdots, x_{n-1} - x_n). \tag{3.3}$$

Here $\mathcal{C}$ is $\mathcal{B}$-independent, but it is not the largest $\mathcal{B}$-independent subfield. Indeed, in this situation there are infinitely many maximal elements in $\mathfrak{F}_1$ (see example 1 in [1]). However, it is possible to show that $\mathcal{C}$ is an essentially maximal element in $\mathfrak{F}_1$. In the above example, one may reverse the role of $\bar{x}$ and D and ask oneself as to whether $\bar{x}$ is a maximal D-independent statistic. It is of some interest to speculate about the truth or falsity of the following general proposition.

PROPOSITION 1. *If $\mathcal{C}$ is a maximal (or essentially maximal) $\mathcal{B}$-independent subfield, then $\mathcal{B}$ is a maximal (or essentially maximal) $\mathcal{C}$-independent subfield.*

4. The family $\mathfrak{F}_2$ of φ-free subfields

Let us suppose that the members of the class $\mathcal{P}$ are indexed by two independent parameters θ and φ; that is,

$$\mathcal{P} = \{P_{\theta,\varphi} | \theta \in \Theta, \varphi \in \Phi\}, \tag{4.1}$$

the parameter space being the Cartesian product $\Theta \times \Phi$.

A subfield $\mathcal{C}$ is called φ-free if the restriction of $\mathcal{P}$ to $\mathcal{C}$ leads to a class of probability measures that may be indexed by θ alone; that is for all $C \in \mathcal{C}$ the probability $P_{\theta,\varphi}(C)$ is a function of θ only. Let $\mathfrak{F}_2$ be the family of all φ-free subfields. Evidently, the concept of φ-free subfields is a direct generalization of the concept of ancillary subfields.

The trivial subfield is again the least element of $\mathfrak{F}_2$. That $\mathfrak{F}_2$ always has maximal elements will be demonstrated later. In general, $\mathfrak{F}_2$ has a plurality of maximal elements.

EXAMPLE 2(a). Let $\mathfrak{X}$ consist of the five points a, b, c, d, and e, and let $\mathcal{P} = \{\mathcal{P}_{\theta,\varphi}\}$ consist of the probability measures

(4.2)

x	a	b	c	d	e
$P_{\theta,\varphi}(x)$	$1-\theta$	$\theta\varphi$	$\theta\varphi$	$\theta(\frac{1}{2}-\varphi)$	$\theta(\frac{1}{2}-\varphi)$

where $0 < \theta < 1$ and $0 < \varphi < \frac{1}{2}$.

There are exactly 12 subsets of $\mathfrak{X}$ whose probability measure is φ-free, and they are $\mathfrak{X}$, $[a]$, $[b, d]$, $[b, e]$, $[c, d]$, $[c, e]$, and their complements. As these 12 sets do not constitute a subfield, it is clear that there cannot exist a largest element in $\mathfrak{F}_2$. The two subfields $\mathcal{C}_1$ and $\mathcal{C}_2$ consisting respectively of

$$\begin{aligned} &\mathcal{C}_1: \mathfrak{X}, [a], [b, d], [c, e] \quad \text{and their complements,} \\ &\mathcal{C}_2: \mathfrak{X}, [a], [b, e], [c, d] \quad \text{and their complements,} \end{aligned} \tag{4.3}$$

are the two maximal elements of $\mathfrak{F}_2$.

EXAMPLE 2(b). Let $x_1, x_2, \cdots, x_n$ be n independent and identically distributed variables with a cumulative distribution function (cdf) of the type $F(x - \varphi/\theta)$, $-\infty < \varphi < \infty$, $0 < \theta < \infty$, where the function F is known and φ, θ are the so-called location and scale parameters.

The subfield $\mathcal{C}$ generated by the $n - 1$ dimensional statistic,

$$D = (x_1 - x_n, x_2 - x_n, \cdots, x_{n-1} - x_n), \tag{4.4}$$

is φ-free in the sense defined before. In general, it is not true that $\mathcal{C}$ is the largest element of the family $\mathfrak{F}_2$ of φ-free subfields. In the particular case where F is the cdf of a normal variable, the subfield $\mathcal{C}$ may be shown to be an essentially maximal element of $\mathfrak{F}_2$. Let us observe that in this particular case, $\mathfrak{F}_2$ is the same as $\mathfrak{F}_1$ of example 1(b). The following proposition may well be true.

PROPOSITION 2. *Whatever may be F, the subfield $\mathcal{C}$ (as defined above) is an essentially maximal element of the family $\mathfrak{F}_2$ of φ-free (φ being the location parameter) subfields.*

Suppose in example 2(b) we reverse the role of φ and θ and concern ourselves with the family $\mathfrak{F}_2^*$ of θ-free subfields, that is, with subfields every member of which has a probability measure that does not involve the scale parameter θ. The author believes that the following proposition is generally true.

PROPOSITION 3. *Every θ-free subfield is also φ-free, that is, $\mathfrak{F}_2^* \subset \mathfrak{F}_2$.*

In the particular case where F is the *cdf* of a normal variable, the truth of proposition 3 has been established in [4].

5. The family $\mathfrak{F}_3$ of $\mathcal{G}$-similar subfields

Let $\mathcal{G} = \{g\}$ be an arbitrary but fixed class of measurable transformations of $(\mathcal{X}, \mathcal{A})$ into itself. For each $P \in \mathcal{P}$, the transformation $g \in \mathcal{G}$ induces a probability measure Pg^{-1} on $(\mathcal{X}, \mathcal{A})$. A subfield $\mathcal{C}$ will be called $\mathcal{G}$-similar if, for each $g \in \mathcal{G}$ and $P \in \mathcal{P}$, the restriction of the two measures P and Pg^{-1} to $\mathcal{C}$ are identical. In other words, $\mathcal{C}$ is $\mathcal{G}$-similar if for all $C \in \mathcal{C}$,

$$Pg^{-1}(C) \equiv P(C) \qquad \text{for all} \quad P \in \mathcal{P} \quad \text{and} \quad g \in \mathcal{G}. \tag{5.1}$$

Let $\mathfrak{F}_3$ be the family of all $\mathcal{G}$-similar subfields. One may look upon $\mathfrak{F}_3$ as the family of subfields that are induced by statistics $T(x)$ such that $T(x)$ and $T(gx)$ are identically distributed for each $P \in \mathcal{P}$ and $g \in \mathcal{G}$. The least element of $\mathfrak{F}_3$ is, of course, the trivial subfield. As we shall see later, $\mathfrak{F}_3$ always has maximal elements and, in general, a plurality of them.

EXAMPLE 3(a). Let $\mathcal{X}$ be the real line and $\mathcal{P} = \{P_\theta | -\infty < \theta < \infty\}$, where P_θ is the uniform distribution over the interval $(\theta, \theta + 1)$. Let $\mathcal{G}$ consist of the single transformation g defined as $gx =$ the fractional part of x. It is easy to check that for all θ in $(-\infty, \infty)$, $P_\theta g^{-1} = P_0$.

In this example, the subfield $\mathcal{C}$ is $\mathcal{G}$-similar if and only if each member of $\mathcal{C}$ has a probability that is θ-free. Thus, the family $\mathfrak{F}_3$ of $\mathcal{G}$-similar subfields is the same as the family of ancillary subfields. Here, $\mathfrak{F}_3$ has a largest element, and that is the subfield of all Borel sets A such that the two sets A and $A + 1$ are essentially equal with respect to the Lebesgue measure.

EXAMPLE 3(b). Let $(\mathcal{X}, \mathcal{A}, \mathcal{P})$ be as in example 2(b) where F is known and φ, θ are the location and scale parameters. Define the shift transformation g_a as

(5.2) $$g_a(x_1, x_2, \cdots, x_n) = (x_1 + a, x_2 + a, \cdots, x_n + a),$$

where a is a fixed real number. Let $\mathcal{G} = \{g_a | -\infty < a < \infty\}$ be the class of all shift transformations.

Denoting the joint distribution of $(x_1, x_2, \cdots, x_n)$ by $P_{\varphi,\theta}$, we note at once that

(5.3) $$P_{\varphi,\theta}\, g_a^{-1} = P_{\varphi+a,\theta}.$$

In this example, the family $\mathfrak{F}_3$ of $\mathcal{G}$-similar subfields is the same as the family $\mathfrak{F}_2$ of φ-free subfields.

Let us call the set A $\mathcal{G}$-invariant if $A \in \mathcal{A}$ and $g^{-1}A = A$ for all $g \in \mathcal{G}$. Likewise, let us call A almost $\mathcal{G}$-invariant if the two sets $g^{-1}A$ and A are $\mathcal{P}$-equivalent for all $g \in \mathcal{G}$. Let $\mathcal{B}_i$ and $\mathcal{B}_a$ be respectively the class of $\mathcal{G}$-invariant and almost $\mathcal{G}$-invariant sets. It is easy to check that $\mathcal{B}_i$ and $\mathcal{B}_a$ are members of the family $\mathfrak{F}_3$ of $\mathcal{G}$-similar subfields. The following proposition should be provable under some conditions.

PROPOSITION 4. *The subfield $\mathcal{B}_a$ of almost $\mathcal{G}$-invariant sets is a maximal $\mathcal{G}$-similar subfield.*

Under some general conditions it should also be true that the subfield $\mathcal{B}_i$ of $\mathcal{G}$-invariant sets is an essentially maximal element of $\mathfrak{F}_3$. This is so in the case of example 3(b) where F is the cdf of a normal variable.

6. The family $\mathfrak{F}_4$ of $\mathcal{B}$-linked subfields

Let $\mathcal{B}$ be a fixed subfield of $\mathcal{A}$. A subfield $\mathcal{C}$ will be called $\mathcal{B}$-linked if $\mathcal{B}$ is sufficient for $(\mathcal{C}, \mathcal{P})$; that is, for every $C \in \mathcal{C}$, there exists a $\mathcal{B}$-measurable mapping $Q(C, \cdot)$ of $\mathcal{X}$ into the unit interval such that, for all $B \in \mathcal{B}$ and $P \in \mathcal{P}$,

(6.1) $$P(BC) = \int_B Q(C, \cdot)\, dP(\cdot).$$

Let $\mathfrak{F}_4$ be the family of all $\mathcal{B}$-linked subfields. The trivial subfield is again the least element of $\mathfrak{F}_4$. We shall presently see that $\mathfrak{F}_4$ always has maximal elements.

EXAMPLE 4(a). (i) Let $\mathcal{B}$ be the trivial subfield. It is easy to see, in this instance, that $\mathfrak{F}_4$ is the same as the family of all ancillary subfields.

(ii) Let us suppose that $\mathcal{P}$ is indexed by the parameters φ and θ. Let $\mathcal{B}$ be a fixed φ-free subfield, that is, a member of $\mathfrak{F}_2$ as defined in section 4. In this instance, every $\mathcal{B}$-linked subfield is also φ-free.

(iii) Let $\mathcal{B}$ be a sufficient subfield. In this case $\mathfrak{F}_4$ is the family of all subfields.

EXAMPLE 4(b). Let $(\mathcal{X}, \mathcal{A}, \mathcal{P})$ be as in example 1(b), and let $\mathcal{B}_0$ be the subfield induced by the sample variance $\Sigma(x_i - \bar{x})^2/n$. If $\mathcal{C}$ is the subfield induced by

(6.2) $$D = (x_1 - x_n, x_2 - x_n, \cdots, x_{n-1} - x_n),$$

then it is easy to check that $\mathcal{C}$ is $\mathcal{B}_0$-linked. Since $\mathcal{B}_0$ is φ-free, it follows that every $\mathcal{B}_0$-linked subfield is also φ-free. It is possible to show that $\mathcal{C}$ is an essentially maximal $\mathcal{B}_0$-linked subfield. The truth of the following proposition is worth investigating.

PROPOSITION 5. *If $\mathcal{B}_0$ and $\mathcal{C}$ are as in example* 4(b), *then $\mathcal{C}$ is an essentially largest element of the family $\mathcal{F}_4$ of the $\mathcal{B}_0$-linked subfields.*

7. Existence of maximal elements

In this section we develop some general methods to prove the existence of maximal elements in the families $\mathcal{F}_1$, $\mathcal{F}_2$, $\mathcal{F}_3$, and $\mathcal{F}_4$. Let us first note a common feature of the four families of subfields. Each $\mathcal{F}_i$ $(i = 1, 2, 3, 4)$ is the totality of all subfields that can be embedded in a certain class $\mathcal{E}_i$ of measurable sets. This will be clear once we define the four classes $\mathcal{E}_1$, $\mathcal{E}_2$, $\mathcal{E}_3$, and $\mathcal{E}_4$ of measurable sets.

DEFINITIONS. (i) *Let $\mathcal{E}_1$ be the class of all $\mathcal{B}$-independent (see section 3) sets;* $\mathcal{E}_1 = \{A|P(AB) = P(A)P(B), \textit{for all } P \in \mathcal{P}, B \in \mathcal{B}\}$.

(ii) *Let $\mathcal{E}_2$ be the class of all φ-free (see section 4) sets;* $\mathcal{E}_2 = \{A|P_{\varphi,\theta}(A) \textit{ does not involve } \varphi\}$.

(iii) *Let $\mathcal{E}_3$ be the class of all $\mathcal{G}$-similar (see section 5) sets;* $\mathcal{E}_3 = \{A|P(g^{-1}A) = P(A) \textit{ for all } P \in \mathcal{P}, g \in \mathcal{G}\}$.

(iv) *Let $\mathcal{E}_4$ be the class of all $\mathcal{B}$-linked (see section 6) sets; $A \in \mathcal{E}_4$ if and only if there exists a $\mathcal{B}$-measurable mapping $Q(A, \cdot)$ of $\mathcal{X}$ into the unit interval such that* $P(AB) = \int_B Q(A, \cdot)\, dP(\cdot)$ *for all $P \in \mathcal{P}$ and $B \in \mathcal{B}$.*

It is now clear that, for $i = 1, 2, 3, 4$,

$$\mathcal{F}_i = \{\mathcal{C}|\mathcal{C} \text{ is a subfield and } \mathcal{C} \subset \mathcal{E}_i\}, \tag{7.1}$$

that is, $\mathcal{F}_i$ is the family of all subfields that can be embedded in the class $\mathcal{E}_i$ of measurable sets.

Our first general result is the following.

THEOREM 1. *Each $\mathcal{E}_i$, $(i = 1, 2, 3, 4)$ has the following properties:*

(a) $\varnothing \in \mathcal{E}_i$, $\mathcal{X} \in \mathcal{E}_i$;

(b) $A \in \mathcal{E}_i$, $B \in \mathcal{E}_i$, $A \subset B \Rightarrow B - A \in \mathcal{E}_i$;

(c) *$\mathcal{E}_i$ is closed for countable disjoint unions.*

The proof of theorem 1 is routine and hence omitted. An immediate consequence of theorem 1 is the following.

COROLLARY. *Each $\mathcal{E}_i$, $(i = 1, 2, 3, 4)$ is a monotone class of sets.*

The following are our fundamental existence theorems.

THEOREM 2. *If $\mathcal{E}$ is a given monotone class of sets, and $\mathcal{F}$ is the family of all Borel fields that could be embedded in $\mathcal{E}$, then corresponding to each element $\mathcal{C}$ of $\mathcal{F}$, there exists a maximal element $\tilde{\mathcal{C}}$ of $\mathcal{F}$ such that $\mathcal{C} \subset \tilde{\mathcal{C}}$.*

PROOF. Let $\{\mathcal{C}_t|t \in T\}$ be an arbitrary subfamily of $\mathcal{F}$, which is linearly ordered with respect to the partial order of inclusion relationship, and let $\mathcal{C}_0 = \bigcup_{t \in T} \mathcal{C}_t$.

Since $\{\mathcal{C}_t\}$ is linearly ordered, it follows that $\mathcal{C}_0$ is a field of sets. The monotone extension of $\mathcal{C}_0$ is then the same as the Borel extension $\mathcal{C}_1$ of $\mathcal{C}_0$. Since $\mathcal{E}$ is monotone and $\mathcal{C}_0 \subset \mathcal{E}$, it follows that $\mathcal{C}_1 \subset \mathcal{E}$ and hence $\mathcal{C}_1 \in \mathcal{F}$. Thus, every linearly ordered subfamily of $\mathcal{F}$ has an upper bound in $\mathcal{F}$. Theorem 2 is then a consequence of Zorn's Lemma.

An immediate consequence of theorems 1 and 2 is theorem 3.

THEOREM 3. *For each* $\mathcal{C} \in \mathfrak{F}_i$ *there exists a maximal element* $\tilde{\mathcal{C}}$ *in* $\mathfrak{F}_i$ *such that* $\mathcal{C} \subset \tilde{\mathcal{C}}$, $(i = 1, 2, 3, 4)$.

8. Some general results

Let $\mathcal{E}$ be a class of measurable sets having the same characteristics as those of the classes $\mathcal{E}_i$ in theorem 1. That is,

(a) $\varnothing \in \mathcal{E}$, $\mathfrak{X} \in \mathcal{E}$;

(b) $A \in \mathcal{E}, B \in \mathcal{E}, A \subset B \Rightarrow B - A \in \mathcal{E}$;

(c) $\mathcal{E}$ is closed for countable disjoint unions.

Let $\mathfrak{F}$ be the family of all the subfields that may be embedded in $\mathcal{E}$, and let $\mathfrak{F}_0$ be the subfamily of all the maximal elements in $\mathfrak{F}$. That $\mathfrak{F}_0$ is not vacuous has been established in theorem 2.

Two members A and B of $\mathcal{E}$ are said to 'conform' if $AB \in \mathcal{E}$. The set $A \in \mathcal{E}$ is said to be 'conforming' if $AB \in \mathcal{E}$ for all $B \in \mathcal{E}$. If every member of $\mathcal{E}$ is conforming, then $\mathcal{E}$ must itself be a Borel field; hence, there is no problem since $\mathfrak{F}_0$ consists of a single member, namely $\mathcal{E}$ itself. A subfield is 'conforming' if every one of its members is so.

THEOREM 4. *Let* $\mathfrak{D}$ *be the class of all the conforming sets in* $\mathcal{E}$, *that is,*

$$A \in \mathfrak{D} \Leftrightarrow A \in \mathcal{E} \quad \text{and} \quad AB \in \mathcal{E} \quad \text{for all} \quad B \in \mathcal{E}. \tag{8.1}$$

Let $\mathfrak{M}$ *stand for a typical element of* $\mathfrak{F}_0$; *that is,* $\mathfrak{M}$ *is a maximal element of* $\mathfrak{F}$:

(i) $\mathfrak{M}$ *is a maximal element of* $\mathfrak{F}$ *if and only if* $A \in \mathcal{E} - \mathfrak{M}$ *implies that* A *does not conform to at least one member of* $\mathfrak{M}$;

(ii) $\mathfrak{D}$ *is a subfield and is equal to the intersection of all the maximal elements in* $\mathfrak{F}$. *It is the largest conforming subfield;*

(iii) $\mathcal{C}$ *is a conforming subfield if and only if for* $\mathcal{B} \in \mathfrak{F}$ *it is true that* $\mathcal{C} \vee \mathcal{B} \in \mathfrak{F}$, *where* $\mathcal{C} \vee \mathcal{B}$ *stands for the least subfield containing both* $\mathcal{C}$ *and* $\mathcal{B}$.

PROOF. Let $\mathfrak{M} \in \mathfrak{F}_0$, and let A be a fixed member of $\mathcal{E} - \mathfrak{M}$. If possible, let A conform to all the members of $\mathfrak{M}$. Consider the class $\mathfrak{M}^*$ of sets of the type $AM_1 \cup A'M_2$, where M_1 and M_2 are arbitrary members of $\mathfrak{M}$. It is easy to check that $\mathfrak{M}^* \in \mathfrak{F}$ and that $A \in \mathfrak{M}^*$ and $\mathfrak{M} \subset \mathfrak{M}^*$. This violates the supposition that $\mathfrak{M}$ is a maximal element of $\mathfrak{F}$. Thus, the 'only if' part of (i) is proved. To prove the 'if' part we have only to observe that if $\mathfrak{M}$ is not maximal, then there exists a larger subfield $\mathfrak{M}^* \subset \mathcal{E}$ and this implies the existence of an $A \in \mathcal{E} - \mathfrak{M}$ that conforms to every member of $\mathfrak{M}$.

Since every member of $\mathfrak{D}$ conforms by definition to every member of $\mathcal{E}$, it is an immediate consequence of (i) that $\mathfrak{D} \subset \mathfrak{M}$ for each $\mathfrak{M} \in \mathfrak{F}_0$, that is, $\mathfrak{D} \subset \bigcap \mathfrak{M}$.

Now let M and E be typical members of $\bigcap \mathfrak{M}$ and $\mathcal{E}$ respectively. From theorem 2 there exists a maximal element $\mathfrak{M}_0$ in $\mathfrak{F}$ which contains the subfield consisting of $\varnothing$, E, E', and $\mathfrak{X}$. Thus, M and E are together in the subfield $\mathfrak{M}_0$, and hence they must conform. Since E is arbitrary, it follows that $M \in \mathfrak{D}$. We have thus proved the equality of $\mathfrak{D}$ and $\bigcap \mathfrak{M}$, and have incidentally proved the equality of $\mathcal{E}$ and $\bigcup \mathfrak{M}$. Since each $\mathfrak{M}$ is a subfield, it is now clear that $\mathfrak{D} = \bigcap \mathfrak{M}$

is also a subfield. That it is the largest conforming subfield follows from its definition.

Now let $\mathcal{C}$ be an arbitrary conforming subfield; that is, let $\mathcal{C}$ be a subfield of $\mathfrak{D}$. For each $\mathcal{B} \in \mathfrak{F}$ there exists (theorem 2) a maximal element $\mathfrak{M}$ of $\mathfrak{F}$ such that $\mathcal{B} \subset \mathfrak{M}$. But $\mathcal{C} \subset \mathfrak{D} \subset \mathfrak{M}$. Therefore, $\mathcal{C} \vee \mathcal{B} \subset \mathfrak{M} \subset \mathcal{E}$, that is $\mathcal{C} \vee \mathcal{B} \in \mathfrak{F}$. This proves the 'only if' part of (iii). The 'if' part is trivial.

For example, let $\mathcal{E}$ be the class of all $\mathcal{B}$-linked sets (see sections 6 and 7) in the probability structure $(\mathfrak{X}, \mathcal{A}, \mathcal{P})$, where $\mathcal{B}$ is a fixed subfield of $\mathcal{A}$. If the set A is $\mathcal{B}$-linked, that is, if there exists a $\mathcal{B}$-measurable function $Q(A, \cdot)$ satisfying definition (iv) of section 7, then it is easily seen that AB is $\mathcal{B}$-linked for every $B \in \mathcal{B}$. We have only to define $Q(AB, \cdot)$ as $Q(A, \cdot)\, I(B, \cdot)$, where $I(B, \cdot)$ is the indicator of B.

In this case, $\mathcal{B}$ is a conforming subfield. Theorem 4(iii) then asserts that for every $\mathcal{B}$-linked subfield $\mathcal{C}$, the subfield $\mathcal{B} \vee \mathcal{C}$ is also $\mathcal{B}$-linked. It will be of some interest to find out conditions under which $\mathcal{B}$ is the largest conforming subfield, that is, $\mathcal{B} = \mathfrak{D}$.

9. Some further problems

In this section we list four problems that are mostly unsolved.

(A) *Separating subfields.* Let $\mathcal{P}$ be a class of 'distinct' probability measures on a measurable space $(\mathfrak{X}, \mathcal{A})$. That is, for each pair P_1, P_2 of members of $\mathcal{P}$ there exists a measurable set $A \in \mathcal{A}$ such that $P_1(A) \neq P_2(A)$. A subfield $\mathcal{B}$ will be called 'separating' if the restriction of $\mathcal{P}$ to $\mathcal{B}$ gives rise to a class of distinct measures. For example, every sufficient subfield is separating. No ancillary or φ-free (see section 4) subfield is separating.

Let $\mathfrak{F}_5$ be the family of all separating subfields. By definition, $\mathcal{A}$ is the largest element of $\mathfrak{F}_5$. What can we say about the existence of minimal elements in $\mathfrak{F}_5$? A variant of this problem has recently received some attention in the USSR ([6], [8]). A partition Π of $\mathfrak{X}$ into a class of disjoint measurable sets $\{A_t\}$ will be called 'separating' if, for each pair P_1, P_2 of member of $\mathcal{P}$, there exists a member A_t of the partition Π such that $P_1(A_t) \neq P_2(A_t)$. A separating partition is called minimal if there exists no other separating partition with a smaller number of parts. Let $\nu(\mathcal{P})$ stand for the number, possibly infinite, of parts in a minimal separating partition. What can we say about $\nu(\mathcal{P})$?

EXAMPLE 5(a). Consider the class $\mathcal{P}$ of all normal distributions on the real line with unit variances. Here $\nu(\mathcal{P}) = 2$. Any partition of the real line into two half lines is clearly separating and, of course, minimal. The corresponding subfield is a minimal element of $\mathfrak{F}_5$.

EXAMPLE 5(b). Let $\mathcal{P}$ be the family of uniform distributions on $[0, \theta]$, $0 < \theta < 1$. In this case $\nu(\mathcal{P}) = 3$ (see [6]).

EXAMPLE 5(c). If $\mathcal{P}$ consists of a finite number of measures $P_1, P_2, \cdots, P_n$, then $\nu(\mathcal{P}) \leq n$. If $\mathcal{P}$ consists of a countable number of continuous measures, then $\nu(\mathcal{P}) = 2$ (see [6], [8]).

(B) *Partially sufficient subfields.* The notion of partial sufficiency, as introduced by Fraser [5], is as follows.

Let $\mathcal{P} = \{P_{\varphi,\theta}\}$, $\varphi \in \Phi$, $\theta \in \Theta$, be a family of probability measures indexed by the two independent parameters φ and θ. A subfield $\mathcal{B} \subset \mathcal{A}$ will be called θ-sufficient for $\mathcal{A}$ (or simply θ-sufficient) if

(i) $\mathcal{B}$ is φ-free in the sense of section 4, and

(ii) for each $A \in \mathcal{A}$ there exists a choice of the conditional probability (function) of A given $\mathcal{B}$ that does not depend on θ; that is, for each $\varphi \in \Phi$, there exists a $\mathcal{B}$-measurable function $Q_\varphi(A, \cdot)$ that maps $\mathfrak{X}$ to the unit interval in such a manner that

$$P_{\varphi,\theta}(AB) \equiv \int_B Q_\varphi(A, \cdot)\, dP_{\varphi,\theta}(\cdot) \tag{9.1}$$

for all $B \in \mathcal{B}$ and $\theta \in \Theta$.

Let $\mathfrak{F}_6$ be the family of all θ-sufficient subfields. Under what conditions can we prove that $\mathfrak{F}_6$ is not vacuous? What about the minimal and maximal elements in $\mathfrak{F}_6$?

(C) *Complete subfields.* Given a probability structure $(\mathfrak{X}, \mathcal{A}, \mathcal{P})$, we call a subfield $\mathcal{B}$ 'complete' if for a $\mathcal{B}$-measurable, $\mathcal{P}$-integrable function f, the integral $\int_{\mathfrak{X}} f\, dP \equiv 0$, for all $P \in \mathcal{P}$, when, and only when, f is $\mathcal{P}$-equivalent to zero. Let $\mathfrak{F}_7$ be the family of all complete subfields. What can we say about the existence of maximal and minimal elements in $\mathfrak{F}_7$?

Let us terminate this list of problems with a final one.

(D) *Complementary subfield.* Let $(\mathfrak{X}, \mathcal{A})$ be a given measurable space and let $\mathcal{B}$ be a fixed subfield of $\mathcal{A}$. A subfield $\mathcal{C}$ will be called a complement to $\mathcal{B}$ if $\mathcal{B} \vee \mathcal{C} = \mathcal{A}$, that is, if $\mathcal{A}$ is the least Borel field that contains both $\mathcal{B}$ and $\mathcal{C}$.

Let $\mathfrak{F}_8$ be the family of all subfields that are complements to $\mathcal{B}$. For example, if $\mathcal{B}$ is the trivial subfield, then $\mathfrak{F}_8$ consists of a single element, namely $\mathcal{A}$ itself. If $\mathcal{B} = \mathcal{A}$, then $\mathfrak{F}_8$ consists of all subfields of $\mathcal{A}$.

Of course, $\mathcal{A}$ is the largest element of $\mathfrak{F}_8$. It is easy to construct examples where $\mathfrak{F}_8$ has a multiplicity of minimal elements. Whether $\mathfrak{F}_8$ always has a minimal element is not known.

10. An addendum

Of the several speculatory statements made (and listed as propositions) in this paper, E. L. Lehmann has recently proved proposition 3 under some conditions on F. Counterexamples to propositions 1 and 2 have been obtained by J. K. Ghosh.

REFERENCES

[1] D. Basu, "The family of ancillary statistics," *Sankhyā*, Vol. 21 (1959), pp. 247–256.

[2] ———, "On maximal and minimal sub-fields of certain types," *Institute of Statistics Mimeo Series, No. 422*, University of North Carolina, 1965.

[3] D. L. Burkholder, "Sufficiency in the undominated case," *Ann. Math. Statist.*, Vol. 32 (1961), pp. 1191–1200.

[4] G. B. Dantzig, "On the non-existence of tests of Student's hypothesis having power functions independent of σ," *Ann. Math. Statist.*, Vol. 11 (1941), pp. 186–191.

[5] D. A. S. Fraser, "Sufficient statistics with nuisance parameters," *Ann. Math. Statist.*, Vol. 27 (1956), pp. 838–842.

[6] A. M. Kagan and V. N. Sudakov, "Separating partitions of certain families of measures," Vestnik Leningrad University, No. 13 (1964), pp. 147–150. (In Russian.)

[7] T. S. Pitcher, "Sets of measures not admitting necessary and sufficiency statistics or sub-fields," *Ann. Math. Statist.*, Vol. 28 (1957), pp. 267–268.

[8] S. M. Visik, A. A. Cobrinksii, and A. L. Rosenthal, "A separating partition for a finite family of measures," *Teor. Verojatnost. i Primenen.*, Vol. 9 (1964), pp. 165–167. (In Russian.)

ON CONFIDENCE INTERVALS AND SETS FOR VARIOUS STATISTICAL MODELS

YU. K. BELYAEV
UNIVERSITY OF MOSCOW

1. Introduction

In addition to obtaining point estimates, one of the central problems of statistical inference is the construction of confidence intervals. In most works, as a rule, considerations are limited to independent sampling, which restricts the range of application without justification. The present paper intends to show that the problems of constructing confidence sets and intervals may be solved for diverse models of mathematical statistics. Underlying the methods is the concept of systems of confidence sets (see [1], [2], [3]). The material expounded below is part of the lectures in a course in mathematical statistics read to students in the Mathematics-Mechanics Faculty of Moscow University in the Fall semester of 1965.

2. Construction of confidence intervals

We consider the statistical model $[X, \mathcal{B}_X, \Theta, P_\theta]$ where $X = \{x\}$ is the set of possible results x of the experiment, and $\mathcal{B}_X$ is a σ-algebra of events. A family of stochastic measures P_θ governing the outcome of the experiment is given on $\mathcal{B}_X$. The object Θ is a set of unknown parameters. Let us consider the subset H of points (θ, x) in the direct product $\Theta \times X$. The sets $H_\theta = \{x\colon (\theta, x) \in H\}$ are called θ-sections of H. The sets $H_x = \{\theta\colon (\theta, x) \in H\} \subseteq \Theta$ are called x-sections of H. The subsets $\{H_x\}$ of the set Θ are called confident with a coefficient of confidence not less than (equal to) γ if the set $\{\theta \in H_x\} \in \mathcal{B}_X$ and

$$\inf_{\theta \in \Theta} P_\theta\{\theta \in H_x\} \geq (=)\gamma. \tag{1}$$

THEOREM 1. (See [1].) *If the θ-sections H_θ of the set H are measurable, and if for every $\theta \in \Theta$ they satisfy the condition*

$$\inf_{\theta \in \Theta} P_\theta\{x \in H_\theta\} \geq (=)\gamma, \tag{2}$$

then the x-sections of the set $H \subseteq \Theta \times X$ form a system of confidence sets with a coefficient of confidence not less than (equal to) γ, or briefly, a γ system.

The proof is a direct consequence of the equivalence of the events

$$\{(\theta, x) \in H\} \equiv \{x \in H_\theta\} \equiv \{\theta \in H_x\}. \tag{3}$$

Let us consider the following example. Let us assume that the space X is formed by sets of integers $x = (d_1, \cdots, d_m)$, $d_i = 0, 1, \cdots$, whose coordinates are mutually independent Poisson random variables with unknown parameters forming the vector $\theta = (\lambda_1, \cdots, \lambda_m)$, that is, $M_\theta d_i = \lambda_i$. Here Θ is a positive quadrant of m-dimensional Euclidean space. Let us construct θ-sections of the set H by means of the formula

$$H_\theta = \{x: d_1 + \cdots + d_m \geq d_\gamma\}, \tag{4}$$

where d_γ is the greatest integer for which

$$P_\theta\{d_1 + \cdots + d_m \geq d_\gamma\} = \sum_{k=d_\gamma}^{\infty} \frac{(\lambda_1 + \cdots + \lambda_m)^k}{k!} e^{-(\lambda_1 + \cdots + \lambda_m)} \geq \gamma. \tag{5}$$

It is easy to verify (see [2]) that the x-sections of such a set H are defined by the formula

$$H_x = \{(\lambda_1, \cdots, \lambda_m): \lambda_1 + \cdots + \lambda_m \leq \Delta_{1-\gamma}(d_1 + \cdots + d_m), \lambda_i \geq 0\}, \tag{6}$$

where $\Delta_\alpha(k)$ is the solution of the transcendental equation $\sum_{k=0}^{d} (z^k/k!)e^{-z} = \alpha$. By virtue of (4), condition 2 of theorem 1 is satisfied and the sets H_x given by formula (6) form a γ system.

In statistical models to which multidimensional spaces of unknown parameters Θ correspond, the need often arises for a construction of the confidence interval for the function $f(x, \theta)$. It is assumed that $f(x, \theta)$ is $\mathcal{B}_X$ measurable for each $\theta \in \Theta$. The interval $[\underline{f}(x), \bar{f}(x)]$ is called confident for $f(x, \theta)$ with a coefficient of confidence not less than (equal to) γ, if $\underline{f}(x), \bar{f}(x)$ are $\mathcal{B}_X$ measurable, and

$$\inf_{\theta \in \Theta} P_\theta\{\underline{f}(x) \leq f(x, \theta) \leq \bar{f}(x)\} \geq (=)\gamma. \tag{7}$$

Let us designate such stochastic intervals briefly as γ intervals. The need to extend the problem to functions dependent on x as well as on θ arises in a natural way in problems of statistical acceptance testing, say [2].

If all the constructions of confidence intervals in mathematical statistics are analyzed, we then shall see that they either explicitly or implicitly follow the plan of the following theorem.

THEOREM 2. *If $\{H_x\}$ is a γ system, and*

$$\underline{f}(x) = \inf_{\theta \in H_x} f(x, \theta) \quad \text{and} \quad \bar{f}(x) = \sup_{\theta \in H_x} f(x, \theta) \tag{8}$$

are $\mathcal{B}_X$ measurable functions, then the interval $[\underline{f}(x), \bar{f}(x)]$ is a γ interval for $f(x, \theta)$.

The proof is a consequence of the relation $\{\theta \in H_x\} \subseteq \{\underline{f}(x) \leq f(x, \theta) \leq \bar{f}(x)\}$ from which we obtain for any $\theta \in \Theta$

$$\gamma \leq P_\theta\{\theta \in H_x\} \leq P_\theta\{\underline{f}(x) \leq f(x, \theta) \leq \bar{f}(x)\}. \tag{9}$$

It is sometimes useful to keep the following in mind.

COROLLARY. *If one is given a priori the supplementary information that* $\theta \in \Theta_0 \subset \Theta$, *then a narrower* γ *interval may be constructed by means of the formulas*

$$\underline{f}'(x) = \inf_{\theta \in H_x \cap \Theta_0} f(x, \theta) \quad \text{and} \quad \bar{f}'(x) = \sup_{\theta \in H_x \cap \Theta_0} f(x, \theta). \tag{10}$$

The following trivial remark may also turn out to be useful sometimes.

THEOREM 3. *If* $\{H_x^i\}$ *is a* γ_i *system,* $i = 1, 2$, *then the system of sets* $\{H_x = H_x^1 \cap H_x^2\}$ *is a* $(\gamma_1 + \gamma_2 - 1)$ *system.*

It follows from theorems 2 and 3 that in certain cases of statistical models with a space of large dimensionality Θ the problem of seeking the upper bound of a one-sided γ interval $[0, \bar{f}(x)]$ takes the specific form of a concave nonlinear programming problem. More specifically, in some cases it is required to find the upper confidence level for the concave function $f(\theta)$ when the γ system is formed by random convex polyhedra. In such cases the $\max f(\theta)$ is sought at the vertices of the confidence polyhedra. For example, for the γ system described by sets of the form (6) and the function $f(\theta) = \sum_{i=1}^m f_i(\lambda_i)$, where the f_i are concave functions, the upper bound of a one-sided γ interval equals

$$\bar{f}(x) = \max_{i=1,m} \left\{ f_i\left(\Delta_{1-\gamma}\left(\sum_{j=1}^{m} d_i \right)\right)\right\}, \tag{11}$$

with $f_i(0) = 0$.

The construction of a γ system is done in a simpler way by using the assignment of stochastic variables dependent on both the outcome of the experiment $x \in X$ and on the parameter $\theta \in \Theta$. Generally, let $g(x, \theta)$ be a vector stochastic variable. In the space G of values of the stochastic variable $g(x, \theta)$ we select for each $\theta \in \Theta$ a subset $G_\theta(\gamma)$ such that

$$P_\theta\{g(x, \theta) \in G_\theta(\gamma)\} \geq \gamma. \tag{12}$$

It follows from condition (12) that the set $H_\theta = \{x\colon g(x, \theta) \in G_\theta(\gamma)\}$ may be considered as a θ-section of some set H in the product space $\Theta \times X$. We obtain the following assertion from theorem 1.

THEOREM 4. *The system of sets*

$$H_x = \{\theta\colon g(x, \theta) \in G_\theta(\gamma)\}, \tag{13}$$

where the $G_\theta(\gamma)$ *satisfy relation* (12), *is a* γ *system.*

The method of constructing γ systems is particularly simple in those cases when $g(x, \theta)$ has a distribution independent of the unknown parameters θ. Here $G_\theta(\gamma) = G(\gamma)$, that is, independent of the unknown parameter θ.

Let us illustrate the method of constructing γ systems by two examples.

Let the test outcomes be $x = (t_{r_1}^{(1)}, \cdots, t_{r_m}^{(m)})$, where $t_{r_i}^{(i)}$ is the time of the appearance of the r_i-th event in a Poisson process with index i and unknown intensity λ_i. It is assumed that the $t_{r_i}^{(i)}$ are stochastic variables which are mutually independent with respect to i. The sets $\theta = (\lambda_1, \cdots, \lambda_m)$, $\lambda_i \geq 0$ play the part of the unknown parameter θ. It is easy to check that $2\lambda_i t_{r_i}^{(i)}$ has a χ^2 distribution with $2r_i$ degrees of freedom. Correspondingly, the stochastic variable $g(x, \theta) = 2 \cdot \sum_{i=1}^m \lambda_i t_{r_i}^{(i)}$ has a χ^2 distribution with $2\sum_{i=1}^m r_i$ degrees of freedom.

If we select the interval $[0, \chi^2_\gamma(2 \sum_{i=1}^m r_i)]$ as the set $G_\theta(\gamma)$, where χ^2_γ is the quantile level γ for the χ^2 distribution with $2 \sum_{i=1}^m r_i$ degrees of freedom, then in conformity with (13), the sets

$$H_x = \left\{(\lambda_1, \cdots, \lambda_m): 2 \sum_{i=1}^m \lambda_i t_{r_i}^{(i)} \leq \chi^2_\gamma \left(2 \sum_{i=1}^m r_i\right)\right\} \tag{14}$$

generate a γ system. Thus the confidence sets are formed by points $(\lambda_1, \cdots, \lambda_m)$, cut out of the first quadrant by the hyperplane $2 \sum_{i=1}^m \lambda_i t_{r_i}^{(i)} = \chi^2_\gamma(2 \sum_{i=1}^m r_i)$.

Let us use the second example to illustrate the methods of constructing confidence intervals for one of the components of the unknown parameter, when the other parameters may be considered as nuisances.

Let β_t be a Wiener process with $M\beta_t = \mu \cdot t$ and variance $M(\beta_t - \mu t)^2 = t$. The process β_t is observed up to the cut-off time t^*. It is assumed that $P\{t^* > t\} = e^{-\lambda t}$ and that t^* is independent of the value of β_t. Hence, the trial outcome is a piece of the trajectory of the Wiener process β_s, $0 \leq s \leq t^*$. The unknown parameters are $\theta = (\mu, \lambda)$, that is, the local drift coefficient μ and the cut-off intensity λ. If we start from sufficient statistics [3], then we may limit ourselves to the space of values $X = \{x\}$, where $x = (t, y)$ are the coordinates of the Wiener process $\beta_t = y$ at the cut-off time $t^* = t$. It follows from the conditions of the problem that the probability density is

$$p_\theta(x) = \lambda \cdot e^{-\lambda t} \cdot \frac{1}{\sqrt{2\pi t}} \exp \{(y - \mu t)^2/2t\}. \tag{15}$$

Let us consider the two-dimensional stochastic variable

$$g(x, \theta) = (s_1, s_2); \qquad s_1 = \lambda t, \ s_2 = \sqrt{\lambda}\, (\beta_t - \mu t). \tag{16}$$

It is easy to verify that its distribution is independent of the unknown values of the parameter θ and is given by the density

$$p(s_1, s_2) = \frac{1}{\sqrt{2\pi s_1}} \exp \left\{-\frac{s_2^2}{2s_1} - s_1\right\}. \tag{17}$$

In conformity with theorem 4, we may select any domain $G(\gamma)$ in the plane of the points (s_1, s_2) such that

$$\int_{G(\gamma)} p(s_1, s_2)\, ds_1\, ds_2 = \gamma. \tag{18}$$

In conformity with (13), the confidence sets generating the γ system are

$$H_x = \{(\mu, \lambda): (\lambda t^*, \sqrt{\lambda}(\beta_{t^*} - \mu t^*)) \in G(\gamma)\}. \tag{19}$$

It is possible to formulate and solve the problem of selecting the best domain $G(\gamma)$, which would minimize the chosen "width" index of the confidence interval. In this example we limited ourselves to a rectangular domain of the form $G(\gamma) = \{s_1 \geq a, -b \leq s_2 \leq b\}$ where a and b are connected by means of (18). For such a domain $G(\gamma)$ the confidence domain is

$$H_x = \{(\mu, \lambda): \lambda t^* \geq a, -b \leq \sqrt{\lambda}\, (\beta_{t^*} - \mu t^*) \leq b\}, \tag{20}$$

or finally,

$$H_x = \left\{(\mu, \lambda): \lambda \geq \frac{a}{t^*}, \frac{\beta_{t^*} - b\lambda^{-1/2}}{t^*} \leq \mu \leq \frac{\beta_{t^*} + b\lambda^{-1/2}}{t^*}\right\}. \tag{21}$$

Let us now assume that we are required to construct a γ interval, for the parameter μ when λ plays the part of the nuisance parameter. To apply theorem 2, let us note that $f(\theta) = \mu$. From (8) and (21) we find

$$\underline{\mu} = \inf_{\theta \in H_x} \mu = \frac{\beta_{t^*} - bt^{*1/2}a^{-1/2}}{t^*}, \qquad \bar{\mu} = \sup_{\theta \in H_x} \mu = \frac{\beta_{t^*} + bt^{*1/2}a^{-1/2}}{t^*}. \tag{22}$$

The selection of the optimum a and b may be carried out by using the method of Lagrange multipliers.

Let us now consider the generalization of an example of Fisher [4], [5] to elementary models of Markov processes with a countable set of states which are observed up to a certain stopping time. The material expounded below is an extension of results of L. N. Bol'shev [6], [7], and [2], which are associated with observations of a Poisson process.

Let us assume that the space of trial outcomes X can be made into a completely ordered set by introducing the relation $x \prec y$ meaning "x to the left of y." The reader might at once imagine the points of the stopping limit of a stochastic process as the generalization of a line. Furthermore, let us assume that for any $z \in X$ the probability $\mathfrak{F}(z, \theta) = P_\theta\{x \preceq z\}$ is a nonincreasing (nondecreasing) function of the parameter θ, which we consider to be a real number in this case. Let us use the notation $\mathcal{G}(z, \theta) = P_\theta\{z \succeq x\}$. It is convenient to consider that the family of probabilistic measures P_θ assigned on X is consistent in the sense that for each interval $[a, b]$, $P_\theta\{a \preceq x \preceq b\} > 0$ is continuous with respect to θ, with the possible exception of the critical value of θ only. It is thereby implicitly assumed that those points x which cannot be observed as a result of the experiment are excluded from consideration. For each value of θ in the space X let us prescribe an interval $[\underline{x}(\theta), \bar{x}(\theta)]$, as narrow as possible, for which

$$P_\theta\{\underline{x}(\theta) \preceq x \preceq \bar{x}(\theta)\} \geq (=)\, \gamma. \tag{23}$$

Since $\mathfrak{F}(x, \theta)$ is nonincreasing (nondecreasing) in θ, the bounds of $\underline{x}(\theta)$, $\bar{x}(\theta)$ may be chosen as nondecreasing (nonincreasing) functions of θ. It is traditional to select such values of the trial outcomes for which

$$\begin{cases} \mathcal{G}(\underline{x}(\theta), \theta) \geq 1 - \epsilon_1 \geq \sup\limits_{x_1 \succ \underline{x}(\theta)} \mathcal{G}(x_1, \theta) \\ \mathfrak{F}(\bar{x}(\theta), \theta) \geq 1 - \epsilon_2 \geq \sup\limits_{x_2 \prec \bar{x}(\theta)} \mathfrak{F}(x_2, \theta) \end{cases} \tag{24}$$

as $H_\theta = [\underline{x}(\theta), \bar{x}(\theta)]$. Analogous relationships are written down for nondecreasing $\mathfrak{F}(x, \theta)$. If ϵ_1 and ϵ_2 satisfy the condition $\gamma = 1 - (\epsilon_1 + \epsilon_2)$, then (23) follows from (24). Hence, the interval $[\underline{x}(\theta), \bar{x}(\theta)]$ may be considered as a θ-section of the set $H \subset \Theta \times X$; the x-sections generate a γ system. From the fact that the functions $\underline{x}(\theta)$, $\bar{x}(\theta)$ are nondecreasing (nonincreasing) it follows that the

x-sections are the intervals $[\underline{\theta}(x), \bar{\theta}(x)]$. The boundaries of the confidence intervals for the observation of the outcome x^* are found from the relations

$$\text{(25)} \qquad \begin{cases} \bar{\theta}(x) = \sup \{\theta: \mathcal{G}(x^*, \theta) \geq 1 - \epsilon_1 \geq \sup\limits_{x_1 > x^*} \mathcal{G}(x_1, \theta)\}, \\ \underline{\theta}(x) = \inf \{\theta: \mathcal{F}(x^*, \theta) \geq 1 - \epsilon_2 \geq \sup\limits_{x_2 < x^*} \mathcal{F}(x_2, \theta)\}, \end{cases}$$

when $\mathcal{F}(x, \theta)$ is nonincreasing, and from the relations

$$\text{(26)} \qquad \begin{cases} \bar{\theta}(x) = \sup \{\theta: \mathcal{F}(x^*, \theta) \geq 1 - \epsilon_2 \geq \sup\limits_{x_2 < x^*} \mathcal{F}(x_2, \theta)\}, \\ \underline{\theta}(x) = \inf \{\theta: \mathcal{G}(x^*, \theta) \geq 1 - \epsilon_1 \geq \sup\limits_{x_1 > x^*} \mathcal{G}(x_1, \theta)\}, \end{cases}$$

when the function $\mathcal{F}(x, \theta)$ is nondecreasing in θ. The following theorem is therefore proved.

THEOREM 5. *If the set of trial outcomes can be completely ordered in such a way that the function* $\mathcal{F}(z, \theta) = P_\theta\{x^* \leq z\}$ *is nonincreasing (nondecreasing) in* θ, *then the boundaries of the* γ *intervals are found from formulas* (25), ((26)).

Let us note that in those cases where $\mathcal{F}(x, \theta)$ is a continuous function of x, equations (25) and (26) take a simple form. Here $\bar{\theta} = \bar{\theta}(x^*)$, $\underline{\theta} = \underline{\theta}(x^*)$ are solutions of the equations

$$\text{(25')} \qquad \mathcal{F}(x^*, \bar{\theta}) = \epsilon_1, \qquad \mathcal{F}(x^*, \underline{\theta}) = 1 - \epsilon_2,$$

$$\text{(26')} \qquad \mathcal{F}(x^*, \underline{\theta}) = \epsilon_1, \qquad \mathcal{F}(x^*, \bar{\theta}) = 1 - \epsilon_2,$$

which is, however, a simple consequence of the fact that under this assumption the stochastic variable $\mathcal{F}(x^*, \theta)$ has a uniform distribution in the interval [0, 1] (see [4]).

As a nontrivial application to life-testing of the theorem proved above, let us consider the following statistical model. A Markov process ξ_s of the pure birth type is observed whose possible values are the integers and $\xi_0 = 0$. Let the intensity of the transitions from the state k to the state $k + 1$ be $\mu_k(\theta)$. Here $\theta > 0$ is an unknown value of the parameter which affects this intensity. A set S of stopping points is given in the plane of (t, k) points, where t is the time coordinate and k is the value of the process ξ_t. As a result of testing, the trajectory of the process ξ_s is observed up to the time of first hitting one of the points of the set S. We make the following assumptions relative to the set of stopping points S and the process ξ_s: (a) for any $\theta > 0$ the trajectory of ξ_s reaches one of the points of S with probability 1; (b) the set S can be completely ordered; hence, for an arbitrary nondecreasing function $k(s)$ taking integer values, any of the boundary points x not lying below the graph $(s, k(s))$ will be "to the left" of any of the boundary points lying below this graph, if the graph $(s, k(s))$ is drawn up to the first hit of the set S; (c) $\mu_k(\theta)$ is a nondecreasing function of θ for every k; $\min_k \mu_k(\theta) \to \infty$, $\theta \to \infty$.

It is useful to employ the following rule in establishing the order relation $x_1 < x_2$ or "x_1 is to the left of x_2" for $x \in S$. If x_1 and x_2 belong to one segment in the (t, k) plane formed by stopping points of the form (s, k), $s' \leq s \leq s''$,

$k = \text{const.}$, then we assert that $x_1 = (t_1, k_1) \prec x_2 = (t_2, k_2)$, when $k_1 = k_2 = k$, $s' \leq t_1 < t_2 \leq s''$. If the values $t_1 = t_2$ coincide at the points x_1, x_2, we assert $x_1 \prec x_2$ when $k_1 > k_2$. Furthermore, the order relation is established in such a manner as to satisfy (b). This is possible for a broad class of sets S.

THEOREM 6. *Assuming conditions* (a)–(c), *the confidence* γ *intervals for* θ *are defined by* (26), *where* $\mathfrak{F}(x, \theta)$ *is the probability of not stopping "to the right," and* $\mathcal{G}(x, \theta)$ *is the probability of not stopping "to the left" of the point* $x \in S$.

The absorption probabilities $\mathfrak{F}(x, \theta)$, $\mathcal{G}(x, \theta)$ may be calculated by using conventional Markov process techniques (see details in [8]). When the points of S have the form $(t, k(t))$, where $k(t)$ is a nonincreasing function of t,

$$\mathfrak{F}(x, \theta) = \sum_{l=0}^{k} p_l(t, \theta), \qquad \mathcal{G}(x, \theta) = \sum_{l \geq k} p_l(t, \theta). \tag{27}$$

Here $x = (t, k)$ and $p_l(t, \theta)$ is the probability that the value of the process is $\xi_t = l$ at time t for free motion without stopping points. For example, for the Poisson process $\mu_k(\theta) = \theta$, $p_l(t, \theta) = ((\theta_t)^l)/l!)e^{-\theta t}$, from which one of the L. N. Bol'shev results [7] easily follows.

The proof of theorem 6 may be obtained by using a stochastic transformation of the time along the trajectory of the process ξ_s. If $\mu_k(\theta'') > \mu_k(\theta')$ for $\theta'' > \theta'$, then the transition to the value θ'' corresponds to a decrease of the sojourn time in the state k, which also increases the probability of stopping the transformed trajectory at the points $y \in S$ "to the left" of x.

In conclusion, let us make several remarks on the construction of the shortest system of confidence intervals. For the case of one-parameter exponential families, the shortest confidence intervals are connected in a definite manner with the most powerful tests of hypotheses (see [3]). If series of statistical models $[X_t, \mathcal{B}_{X_t}, \Theta, P_{\theta,t}]$ dependent on the "time" t of data accumulation are considered, it is recommended to start from the effective estimates $g(x)$ in constructing confidence intervals for $f(\theta)$. The γ interval is constructed by means of (13). Unfortunately, such a procedure is difficult to carry out in practice for statistical models with spaces of large dimensionality, because of the complexity of the algorithms giving $\underline{f}(x)$, $\bar{f}(x)$ by means of (8). It is necessary to compromise and to construct simpler γ systems which take into account in some way the specific properties of the function $f(\theta)$, and at the same time use comparatively simple algorithms to achieve (8).

For $f(\theta) = \sum_{i=1}^{m} f_i(\lambda_i)$, where the $f_i(\lambda_i)$ are concave functions satisfying conditions leading to the systems (6), better results than those of (11) are obtainable for small values of d_i by using the following system of sets:

(28)

$$H_x = \left\{(\lambda_1, \cdots, \lambda_m) \colon \sum_{i=1}^{m} \lambda_i \leq \Delta_{1-\gamma_1}\left(\sum_{i=1}^{m} d_i\right), 0 \leq \lambda_i \leq \Delta_{1-\gamma_2}(d_i), i = 1, \cdots, m\right\}.$$

In conformity with theorem 3 the sets (28) generate $(\gamma_1 + \gamma_2^m - 1)$ systems.

The H_x are convex polyhedra. The upper confidence bound is $\max f(\theta)$ taken over the vertices of the polyhedron H_x. It is easy to show that the coordinates of those vertices of the polyhedron H_x at which $\max f(\theta)$ is achieved are the following. The set $S \cup i_0$, $S \subseteq (1, \cdots, m)$, for which

$$\sum_{i \in S} \Delta_{1-\gamma_2}(d_i) \leq \Delta_{1-\gamma_1}\left(\sum_{i=1}^{m} d_i\right) < \sum_{i \in S} \Delta_{1-\gamma_2}(d_i) + \max_{j \notin S} \Delta_{1-\gamma_2}(d_j) \tag{29}$$

corresponds to each vertex. The values of the vertex coordinates λ_i, $i \in S$ are assumed to equal $\Delta_{1-\gamma_2}(d_i)$. The remaining are $\lambda_i = 0$ with the exception of $\lambda_{i_0} = x$, $i_0 \notin S$, which is the solution of the equation

$$\sum_{i \in S} \Delta_{1-\gamma_2}(d_i) + x = \Delta_{1-\gamma_1}\left(\sum_{i=1}^{m} d_i\right), \quad x < \Delta_{1-\gamma_2}(d_{i_0}). \tag{30}$$

In practice it is impossible to sort out all such vertices since they are many and to find the $\max f(\theta)$. However, it is possible to mention a completely realizable algorithm whose complexity depends not so much on m as on the number of different values of d_i.

The algorithmic character of the problems on confidence intervals for complicated spaces Θ is apparently typical.

REFERENCES

[1] H. Cramér, *Mathematical Methods of Statistics,* Princeton, Princeton University Press, Vol. IL, 1946.
[2] B. V. Gnedenko, Yu. K. Belyaev, and A. D. Soloviev, *Mathematical Methods in Reliability Theory,* Moscow, Nauka, 1965.
[3] E. Lehmann, *Testing Statistical Hypotheses,* New York, Wiley, 1964 (3d ed.).
[4] R. A. Fisher, "The fiducial argument in statistical inference," *Ann. Eugenics,* Vol. 5 (1935), pp. 391–398.
[5] J. Neyman, "Outline of the theory of statistical estimation based on the classical theory of probability," *Philos. Trans. Roy. Soc., London, Ser. A,* Vol. 236 (1937), pp. 233–380.
[6] L. N. Bol'shev, "On the construction of confidence limits," *Teor. Verojatnost. i Primenen.,* Vol. 10 (1965), pp. 187–192.
[7] ———, "Comparison of the intensities of the simplest flows," *Teor. Verojatnost. i Primenen.,* Vol. 7 (1962), pp. 353–355.
[8] Kai Lai Chung, *Markov Chains with Stationary Transition Probabilities,* Vol. X, Berlin, Springer, 1960.

LIMIT THEOREMS FOR REGRESSIONS WITH UNEQUAL AND DEPENDENT ERRORS

FRIEDHELM EICKER
UNIVERSITY OF FREIBURG IM BREISGAU
and
COLUMBIA UNIVERSITY, NEW YORK

1. Summary

This paper deals with the asymptotic distribution of the vectorial least squares estimators (LSE) for the parameters in multiple linear regression systems. The regression constants are assumed to be known; the errors are assumed (a) to be independent but not necessarily identically or normally distributed (section 3), or (b) to constitute a generalized linear discrete stochastic process (section 4). The latter part includes the case of regression for time series. Conditions are studied under which the joint distribution functions (d.f.'s) of the vectorial LSE's tend to a multivariate normal d.f. as the sample size increases. In the proof a central limit theorem (CLT) for weighted averages of independent random variables is used. In case (a), a theorem for large classes of linear regressions is proved (theorem 3.2), whose conditions are in a certain sense also necessary. The theorem simultaneously permits consistent estimation of the limiting covariance matrix of the LSE's. The results in case (b) are contained in theorems 4.2, 4.3, 4.4, 4.5, 4.6. They are not naturally of as closed a form as those pertinent to case (a) because of the more complicated nature of the problem. Some use of spectral theory is made. Several examples are discussed (section 3.3). The assumptions made in this paper are weaker than those of results published earlier in the literature. (For a more recent survey, compare [6].) Their structure is quite simple so that they ought to be useful in applications. Section 4.4 contains some remarks on multivariate regression equations.

2. Introduction (notations)

There exists a considerable number of publications dealing with the asymptotic normality of parameter estimates for linear regressions, many of which deal with specific cases, however, or are unnecessarily narrow in the assumptions made. The most general paper among these, and the one closest to

Research supported in part by National Science Foundation Grant NSF-GP-3694 at Columbia University.

the present note, seems to be [9]. In that paper vectorial regression equations are considered while we are predominantly interested in scalar ones. For that case, however, the assumptions of [9] are more restrictive than those of the present note.

For the individual components of the vectorial LSE the asymptotic normality was already proved under general assumptions in [1] for the case of independent errors.

The system of (scalar) linear regression equations is denoted by

$$y_t = x_{t1}\beta_1 + \cdots + x_{tq}\beta_q + \epsilon_t, \qquad t = 1, \cdots, n, \tag{2.1}$$

$n \geq q$ being the sample size. In matrix notation this becomes

$$y(n) = X_n\beta + \epsilon(n), \tag{2.2}$$

where $y(n)$ is the (column) vector of observations (n-dimensional), $X_n = (x_{tj})$, the $(n \times q)$-matrix of known regression constants assumed to be of full rank throughout, $\beta = (\beta_1, \cdots, \beta_q)'$ is the vector of unknown regression parameters ($'$ denotes the transpose), and $\epsilon(n) = (\epsilon_1, \cdots, \epsilon_n)'$ is the n-vector of error random variables (r.v.'s) about which we assume throughout that

$$E\epsilon_t = 0, \qquad 0 < E\epsilon_t^2 < \infty, \qquad \text{for all } t. \tag{2.3}$$

All quantities are real.

Let $P_n = X_n'X_n$. Then the vectorial LSE for β, denoted by $b(n) = (b_1(n), \cdots, b_q(n))'$, become

$$b(n) = P_n^{-1}X_n'y(n) = \beta + P_n^{-1}X_n'\epsilon(n). \tag{2.4}$$

The row vectors of X_n will be denoted by $r_1', \cdots, r_n'$, and the column vectors by $x_1(n), \cdots, x_q(n)$. By F, we denote a (nonempty) set of d.f.'s whose elements G have the properties

$$\int x\, dG(x) = 0, \qquad 0 < \int x^2\, dG(x) < \infty. \tag{2.5}$$

3. Independent nonidentically distributed errors

3.1. *The asymptotic normality of the* $b(n)$. In order to find a limiting d.f., the vectors

$$b(n) - \beta = P_n^{-1}X_n'\epsilon(n) \tag{3.1}$$

have to be normalized by premultiplication by certain matrices B_n. Let

$$\Sigma_n = \operatorname{cov} \epsilon(n)\epsilon'(n) = \operatorname{diag}(\sigma_1^2, \cdots, \sigma_n^2), \qquad \sigma_k^2 = \operatorname{var} \epsilon_k, \tag{3.2}$$

be the covariance matrix of the error vector, and write

$$B_n^2 = P_n^{-1}X_n'\Sigma_n X_n P_n^{-1}. \tag{3.3}$$

If B_n is the unique positive definite square root of B_n^2, the q-vectors

$$B_n^{-1}P_n^{-1}X_n'\epsilon(n) \tag{3.4}$$

all have expectation zero and covariance matrix I_q (= q-dimensional identity matrix).

Let $\mathfrak{F}(F)$ be the set of all sequences $\epsilon \equiv \{\epsilon_1, \epsilon_2, \cdots\}$ of independent error r.v.'s ϵ_t (independent within each sequence) whose d.f.'s belong to some set F subject to (2.5). For a given sequence ϵ, the vectors $\epsilon(n)$ have as components the first n members of ϵ, $n = 1, 2, \cdots$.

Theorem 1 of [4] then applies without further ado and yields the following theorem.

THEOREM 3.1. *The d.f.'s of $B_n^{-1}P_n^{-1}X_n'\epsilon(n)$ tend to the q-dimensional normal d.f. $N(0, I_q)$ and the summands of $B_n^{-1}P_n^{-1}X_n'\epsilon(n)$, are infinitesimal both uniformly for all sequences $\epsilon \equiv \{\epsilon_1, \epsilon_2, \cdots\} \subset \mathfrak{F}(F)$, if and only if the following three conditions are satisfied:*

$$\text{(I)} \qquad \max_{k=1,\cdots,n} r_k'P_n^{-1}r_k \to 0,$$

$$\text{(II)} \qquad \sup_{G \in F} \int_{|x|>c} x^2\, dG(x) \to 0, \qquad \text{as } c \to \infty,$$

$$\text{(III)} \qquad \inf_{G \in F} \int x^2\, dG(x) > 0.$$

(All limits throughout the paper hold for $n \to \infty$ unless otherwise stated.)

The fact that the assertion of the theorem holds for all sequences $\{\epsilon_t\} \in \mathfrak{F}(F)$ makes it particularly useful in practice, since one usually does not know the error d.f.'s if they are not identical. It may also be pointed out that condition (I) on the regression matrices does not necessitate any knowledge about the error sequence present in a particular regression. Analogously, (II) and (III) concern only the set F of admissible error d.f.'s. If the ordinary CLT were applied, one would obtain conditions concerning, simultaneously, the error sequence and the regression sequences. It is interesting that the consideration of the whole class $\mathfrak{F}(F)$ implies the necessity of the conditions (I)–(III). Condition (II) means *uniform integrability* of the variance integrals with respect to the class F.

As it stands, theorem 3.1 is still of limited practical use since the normalizing matrices require the knowledge of the usually unknown error variances σ_k^2. Applying a law of large numbers for nonnegative random variables ([5], p. 143), this defect can be removed by replacing σ_k^2 by the square of the k-th residual

$$e_k(n) = y_k - r_k'b(n) = \epsilon_k - r_k'P_n^{-1}X_n'\epsilon(n), \qquad k = 1, \cdots, n. \tag{3.5}$$

The matrix $D_n^2 \equiv P_n^{1/2}B_n^2P_n^{1/2}$ is then replaced by

$$C_n^2 = P_n^{-1/2}X_n'S_nX_nP_n^{-1/2} \tag{3.6}$$

with $S_n = \operatorname{diag}(e_1^2(n), \cdots, e_n^2(n))$. This replacement amounts to an estimation of the matrix D_n^2 in the sense of (3.9), as will be shown below. After this substitution the estimator

$$C_n^{-1/2}P_n^{-1}X_n'y(n) \tag{3.7}$$

no more contains any unknown quantity. Without any new assumptions we then obtain the next theorem.

THEOREM 3.2. *Under the assumptions (I), (II), (III) of the preceding theorem, the d.f.'s of*

$$C_n^{-1} P_n^{-1/2}(b(n) - \beta) \tag{3.8}$$

tend to $N(0, I_q)$ uniformly for all error sequences $\epsilon \in \mathfrak{F}(F)$.

For the proof we need a lemma on matrices whose entries are random variables.

LEMMA 3.1. *A sequence of symmetrical random $q \times q$-matrices $A_n \rightarrow I_q$, i.p. if and only if $c'A_nc \rightarrow 1$, i.p. for all unimodular constant q-vectors c.*

PROOF. The "only if" part follows from Slutsky's theorem. To show the converse, first take $c = v_k$ [$=$ k-th unit vector], $k = 1, \cdots, q$, and then c proportional to $v_k + v_j$, any pair $k \neq j$.

PROOF OF THEOREM 3.2. We show first

$$D_n^{-1}C_n^2D_n^{-1} \rightarrow I_q, \qquad \text{i.p.} \tag{3.9}$$

We introduce the vectors

$$(u_1(n), \cdots, u_n(n))' = c'D_n^{-1}P_n^{-1/2}X_n', \qquad n = q, q + 1, \cdots, \tag{3.10}$$

with some unimodular constant q-vector c. Then

$$c'D_n^{-1}C_n^2D_n^{-1}c = \sum_{k=1}^{n} u_k^2(n)e_k^2(n). \tag{3.11}$$

Since $\sum_k u_k(n)\epsilon_k = c'D_n^{-1}P_n^{-1/2}(b(n) - \beta)$ is a sum of independent infinitesimal r.v.'s whose d.f. for $n \rightarrow \infty$ tends to $N(0, 1)$ as a consequence of theorem 3.1, we have

$$\sum_k u_k^2(n)\epsilon_k^2 \rightarrow 1, \qquad \text{i.p.} \tag{3.12}$$

by theorem 4 of ([5], p. 143).

Taking account of the second terms of $e_k(n)$ as given by (3.5), we have

$$E(r_k'P_n^{-1}X_n'\epsilon(n))^2 \leq Mr_k'P_n^{-1}r_k \tag{3.13}$$

where the existence of $M = \sup_G \int x^2\, dG(x) < \infty$ is implied by (II). Putting $m = \inf_G \int x^2\, dG(x) [> 0]$ and denoting by $\|\cdot\|$ the Euclidean norm, we have

$$\sum_k u_k^2(n) = c'D_n^{-1}c \leq \frac{1}{m}. \tag{3.14}$$

Hence, by (I),

$$E(\sum_k u_k^2(n)(r_k'P_n^{-1}X_n'\epsilon(n))^2) \leq (M/m) \max_k r_k'P_k^{-1}r_k \rightarrow 0, \tag{3.15}$$

and consequently, $\sum_k u_k^2(n)(r_k'P_n^{-1}X_n'\epsilon(n))^2 \rightarrow 0$, i.p. Finally, $\sum_k u_k^2(n)e_k^2(n) \rightarrow 1$, i.p. for all unimodular vectors c. Because of lemma 3.1, this proves (3.9).

We now prove

$$C_n - D_n \rightarrow 0, \qquad \text{i.p.} \tag{3.16}$$

Put $E_n = C_n^2 - D_n^2$. By (3.9) there exists a sequence of events Ω_n with $P\Omega_n \rightarrow 1$ such that $\sup_{\Omega_n} \|E_n\| \rightarrow 0$ where $\|E_n\| \equiv \max_{i,j=1,\cdots,q} |(E_n)_{ij}|$. In the following all quantities, and equations in quantities, showing the index n are considered only on the event Ω_n, $n = q, q + 1, \cdots$.

Putting $(D_n^2 + E_n)^{1/2} = D_n + E_n^*$, we then have to show $\|E_n^*\| \to 0$, that is, according to our convention, $\sup_{\Omega_n} \|E_n^*\| \to 0$. Now $E_n = D_nE_n^* + E_n^*D_n + E_n^{*2}$; E_n^*, being a real symmetric matrix, has only real characteristic values (c.r.), to be denoted by $\lambda_{1n}(\omega) \geq \lambda_{2n}(\omega) \geq \cdots \geq \lambda_{qn}(\omega)$, $\omega \in \Omega_n$, and possesses q orthogonal real characteristic vectors (c.v.). Suppose $\sup_{\omega \in \Omega_n} \lambda_{1n}(\omega) \equiv \Lambda_n > 0$ for an infinite set Γ of naturals. We assume for simplicity that there exists a matrix $E_n^*(\omega_n)$, $\omega_n \in \Omega_n$, that actually possesses Λ_n as a c.r. (otherwise we can always find an E_n^* whose maximum c.r. differs arbitrarily little from Λ_n). Let $v_n \in S_q$ (the unit sphere $\subset R_q$) be a c.v. of $E_n^*(\omega_n)$ associated with Λ_n. Then at $\omega = \omega_n$ respectively,

$$v_nE_nv_n = 2\Lambda_nv_n'D_nv_n + \Lambda_n^2 \to 0, \qquad n \in \Gamma, \tag{3.17}$$

since $\|E_n\| \to 0$ and the c.v. of D_n are bounded between the finite positive constants m and M.

On the other hand, suppose $\inf_{\omega \in \Omega_n} \lambda_{qn}(\omega) = \lambda_n < 0$ on some infinite set Γ' of integers, and let $u_n \in S_q$ be a c.v. associated with λ_n and a suitable matrix E_n^*. Then again $u_n'E_nu_n = \lambda_n(\lambda_n + 2u_n'D_nu_n) \to 0$, $n \in \Gamma'$, and hence either $\lambda_n \to 0$ or $\lambda_n + 2u_n'D_nu_n \to 0$ for some infinite sequence $\Gamma'' \subset \Gamma'$. But $0 < u_n'(D_n^2 + E_n)^{1/2}u_n = u_n'D_nu_n(1 - 2) + o(n)$ for $n \in \Gamma''$ which is impossible. Thus all c.r.'s of E_n^* tend to zero, which implies (3.16).

Finally, (3.16) implies $C_n^{-1}D_n \to I_q$ i.p., and premultiplication of

$$D_n^{-1}P_n^{-1/2}X_n'\epsilon(n)$$

yields the assertion.

Uniformity in $\epsilon \in \mathfrak{F}(F)$ of (3.8) follows from the fact that the preceding proof remains valid if instead of one and the same ϵ for each n, we take for each n an arbitrary $\epsilon(n) \in \mathfrak{F}(F)$. Thus, (3.8) holds for all sequences of sequences $\epsilon(n)$, and this is equivalent with uniformity in ϵ.

3.2. *Remarks.* (1) In practice, for finite n, one uses theorem 3.2 in the form

$$\text{d.f. } (b(n)) \sim N(\beta, C_n^2). \tag{3.18}$$

In certain situations this relation may save the trouble of computing the inverse square root of C_n^2.

(2) Theorem 2 of [1] states the asymptotic normality of the single components $b_j(n)$, after suitable normalization. We remark without proof that this theorem also remains valid under unchanged assumptions if, as in section 3.1, the unknown variances σ_k^2 in the normalizing factor are replaced by the squares of the residuals (3.5). Thus, the additional assumptions given in theorem 3 of [1] are in fact superfluous.

The progress of the present paper over [1] lies essentially in the determination of the joint asymptotic d.f. of the vectorial LSE $b(n)$ which was not possible by the method used in [1]. Besides that, condition (I) of the present paper is simpler than the corresponding condition in [1].

(3) If F contains only one element, say G, then (II) and (III) reduce to $0 < \int x^2 \, dG(x) < \infty$. The errors are, in this case, identically distributed.

(4) The assumptions (I), (II), and (III) are no longer necessary in theorem 3.2. However, since they are necessary in theorem 3.1, the necessary and sufficient assumptions of theorem 3.2 presumably do not differ very much from (I)–(III).

(5) Concerning the admissible sequences of regression matrices we prove the following lemma.

LEMMA 3.2. *Condition* (I) *implies*

$$\lambda_{\min}(P_n) \to \infty \tag{3.19}$$

and

$$\max_{k=1,\cdots,n} |x_{k,j}|/||x_j(n)|| \to 0, \qquad j = 1, \cdots, q. \tag{3.20}$$

Here $\lambda_{\min}$ denotes the minimum characteristic value and $||\cdot||$ the Euclidean norm. Regression vectors $x_j(n)$ satisfying (3.20) are called *slowly increasing* (compare [7], p. 233).

PROOF. We introduce the $q \times q$ diagonal matrices

$$D_n = \operatorname{diag}\,(||x_1(n)||, \cdots, ||x_q(n)||), \qquad n = 1, 2, \cdots. \tag{3.21}$$

Since $\operatorname{tr}(D_n^{-1}P_nD_n^{-1}) = q$, we have

$$\sup_n \lambda_{\max}(D_n^{-1}P_nD_n^{-1}) \le q. \tag{3.22}$$

Now

$$r_k'P_n^{-1}r_k = r_k'D_n^{-1}(D_n^{-1}P_nD_n^{-1})^{-1}D_n^{-1}r_k \ge q^{-1}||D_n^{-1}r_k||^2. \tag{3.23}$$

By (I), $\max_k ||D_n^{-1}r_k|| \to 0$ and thus (3.20) follows. Equation (3.20) implies $||x_j(n)|| \to \infty$, for all j, which in turn implies (3.19).

(6) Sufficient for (I) is the relation (3.20) together with

$$\inf_n \lambda_{\min}(D_n^{-1}P_nD_n^{-1}) > 0, \tag{3.24}$$

as can be seen from an inequality similar to (3.23). However, conditions (3.24) plus (3.20) are, in general, not necessary. In particular, (3.24) is satisfied if

$$D_n^{-1}P_nD_n^{-1} \to R \tag{3.25}$$

where R is some positive definite $q \times q$ matrix.

3.3. *Examples.* We now discuss some examples of regression matrices and check whether they possess property (I) or not.

(1) *Polynomial regression.* Let $x_{k,j} = k^{c_j}$, $c_1 > \cdots > c_q > -\frac{1}{2}; j = 1, \cdots, q$, $k = 1, 2, \cdots$. The c_j need not be integers. Then (compare [2], p. 469) the $x_j(n)$ are slowly increasing and $D_n^{-1}P_nD_n^{-1} \to H$ where H is a positive definite submatrix of the Hilbert matrix. Hence, (3.25), and consequently (I), is satisfied.

(2) *Trigonometric regression.* Let

$$x_{k,2j-1} = \cos w_jk, \qquad x_{k,2j} = \sin w_jk, \qquad j = 1, \cdots, q, \qquad k = 1, 2, \cdots \tag{3.26}$$

where the w_j are such that rank $X_n = 2q$. Then (see [2], p. 477) the $x_j(n)$ are slowly increasing and $n^{-1}P_n \to I_q^*$ where I_q^* is a diagonal matrix with diagonal elements 1 or $\frac{1}{2}$. Again (3.25), and thus (I), is satisfied.

(3) *Mixed trigonometric and polynomial regression.* Let

$$\begin{cases} x_{kj} = k^{c_j}, & j = 1, \cdots, q \qquad \text{with} \quad c_1 > \cdots > c_q > -\tfrac{1}{2}, \\ x_{kj} = e^{ikw_j}, & j = q+1, \cdots, Q \\ & \text{with} \quad 0 < w_j < 2\pi, \qquad w_j \neq w_k \quad \text{for} \quad j \neq k. \end{cases} \tag{3.27}$$

Again (3.25) is satisfied with a matrix R of the form

$$R = \begin{pmatrix} H & 0 \\ 0 & I_{Q-q} \end{pmatrix} \tag{3.28}$$

where H is as in example 1. In order to prove this, we observe first that

$$n^{-c-1} \sum_{k=1}^{n} k^c e^{ikw} \to 0 \qquad \text{for} \quad n \to \infty \tag{3.29}$$

if $c = 0, 1, \cdots$ and $0 < w < 2\pi$. This can be seen by deriving $\sum_{k=1}^{n} e^{ikw} = e^{iw}(1 - e^{inw})/(1 - e^{iw})$ repeatedly with respect to iw. In order to prove (3.29) for nonintegers c, we derive the left-hand side with respect to c and obtain

$$n^{-c-1} \sum_k k^c e^{ikw} \ln (k/n), \tag{3.30}$$

which remains bounded for $c > -\frac{1}{2}$ as $n \to \infty$, since

$$\int_0^1 \left(\frac{t}{n}\right)^c \ln \frac{t}{n}\, d\left(\frac{t}{n}\right) = -(c+1)^{-2}. \tag{3.31}$$

and since we have proved (3.29) already for integers c, it holds also for nonintegers $c > -\frac{1}{2}$.

Finally, because the matrix R is positive definite, property (I) holds.

(4) *Analysis of variance case.* Consider a one-way classification with q classes having $N_1, \cdots, N_q$ observations respectively. The regression matrix is given by

(3.32)

$$X_n' = \begin{pmatrix} 1, \cdots, 1, & 0, \cdots, 0, & & \cdots & 0 \\ 0, \cdots, 0, & 1, \cdots, 1, & 0, \cdots, & & 0 \\ & \cdots & & & \\ 0, & \cdots & & 0, & 1, \cdots, 1 \end{pmatrix}_{q \times n}, \qquad n = N_1 + \cdots + N_q.$$

Then $P_n = \operatorname{diag}(N_1, \cdots, N_q)$. Condition (I) is satisfied if $\min_j N_j \to \infty$. Hence, in this case the LSE of the effects are asymptotically normally distributed for every error sequence $\epsilon \in \mathfrak{F}(F)$.

(5) *Exponential regression.* The regression vectors $x_j(n)$ with $x_{kj} = c_j^k$ where $c_1 > c_2 > \cdots > c_q > 1$, are not slowly increasing since

$$c_j^{2n} \Big/ \sum_{k=1}^{n} c_j^{2k} > \frac{c_j}{c_j^2} > 0. \tag{3.33}$$

Therefore, by lemma 3.2, (I) cannot be satisfied.

(6) *Mixed polynomial-exponential regression*, $x_{kj} = k^{c_j} d_j^k$, $d_j > 1$, $c_j > 0$. Again the regression vectors are not slowly increasing, since $||x_j(n)||^2 = O(n^{2c_j} d_j^{2n})$.

4. Dependent errors (regression for time series)

4.1. *The case of one regression vector.* Since for dependent errors the results are not of as closed a form as for independent errors, we consider first the case of only one regression vector ($q = 1$). This case already shows the main deviations from the previous results.

System (2.1) reduces now to

$$y_t = x_t\beta + \epsilon_t, \qquad t = 1, 2, \cdots, n, \tag{4.1}$$

in vectorial notation $y(n) = x(n)\beta + \epsilon(n)$. We assume, as is typical also for the analysis of time series,

$$\epsilon_t = \sum_{j=-\infty}^{\infty} c_j \eta_{t+j} = \sum_{j=-\infty}^{\infty} c_{j-t}\eta_j, \qquad t = 1, 2, \cdots, \tag{4.2}$$

where the sequence $c \equiv \{c_j\}$ of real constants is square summable, that is, $c \in l^2$ (the Hilbert space of all square summable sequences of real numbers). In all of the previous publications the stronger assumption $\sum_{j=-\infty}^{\infty} |c_j| < \infty$ has been made (see, for example, [9]).

As is well known, the condition $c \in l^2$ is necessary in order that (4.2) holds as a limit in quadratic mean. The r.v.'s η_j are assumed to be independent with expectations zero, but they need not be identically distributed. Let their d.f.'s, as in section 3, lie in a set F where F satisfies (2.5) and conditions (II) and (III) of section 3.1. Then each ϵ_t is, in fact, defined as a limit in the mean of the sums $\sum_{j=-n}^{n} c_j\eta_{t+j}$ for every sequence $\{\eta_i\} \in \mathfrak{F}(F)$ (see section 3.1). Random sequences $\{\epsilon_t\}$ of this type have been called *generalized linear processes* ([3]). If the η_t are identically distributed, the ϵ_t form a strictly stationary linear stochastic process.

With $P_n = ||x(n)||^2$, the (scalar) LSE's of β are

$$b(n) = ||x(n)||^{-2}x'(n)y(n), \qquad n = 1, 2, \cdots. \tag{4.3}$$

In order to investigate the asymptotic normality of the sequence $\{b(n)\}$, put

$$\begin{aligned} \zeta_n &= (\operatorname{var} b(n))^{-1/2}(b(n) - \beta) \\ &= (\operatorname{var} b(n))^{-1/2}||x(n)||^{-2}x'(n)\epsilon(n) \\ &= (\operatorname{var} b(n))^{-1/2}||x(n)||^{-2} \sum_{j=-\infty}^{\infty} \left(\sum_{t=1}^{n} x_t c_{j-t}\right) \eta_j. \end{aligned} \tag{4.4}$$

We have $E\zeta_n = 0$, $\operatorname{var} \zeta_n = 1$ for all n. Put

$$\begin{aligned} A_{nj} &= \sum_{t=1}^{n} x_j c_{j-t}, \\ S_n &= \sum_{j=-\infty}^{\infty} A_{nj}^2, \end{aligned} \tag{4.5}$$

Clearly, always $S_n < \infty$. In ([3], p. 319) the following proposition, which also holds in a more general context, has been proved.

THEOREM 4.1. *Let $S_n > 0$ for all n. In order that* (A): d.f. $(\zeta_n) \to N(0, 1)$, *and* (B): *the contributions of the summands of ζ_n in the last expression of* (4.4) *are infinitesimal, both for every $\{\eta_j\} \in \mathfrak{F}(F)$, the conditions* (II) *and* (III) *of theorem* 3.1 *and*

$$\sup_{j=-\infty,\cdots,\infty} A_{nj}^2/S_n \to 0 \tag{4.6}$$

are jointly necessary and sufficient.

We do not emphasize the validity of the theorem for the whole class $\mathfrak{F}(F)$. We rather consider a sequence $\{\eta_i\} \in \mathfrak{F}(F)$ to be given and shall now analyze in detail the remaining condition (4.6).

Condition (4.6) may be verified directly for a given sequence $c \in l^2$ and a given sequence $x \equiv \{x_1, x_2, \cdots\}$ of regression constants. It would be more convenient, however, if for each c the class $\mathfrak{X}_c$ of all x's satisfying (4.6), or for each x the class $\mathfrak{C}_x$ of all c's satisfying (4.6) were known. It then remains only to be checked whether a given x belongs to $\mathfrak{X}_c$, or a given c belongs to $\mathfrak{C}_x$. Since in a regression problem x is known but c usually is not, the classes $\mathfrak{C}_x$ are of greater interest. We shall, therefore, direct our attention mainly on $\mathfrak{C}_x$. If we are unable to determine a class $\mathfrak{C}_x$ completely, we shall try to find as large a subclass as possible.

Let $c(\lambda) \in L^2$ (the space of the complex valued functions over $\Lambda = \{\lambda: -\frac{1}{2} \leq \lambda \leq \frac{1}{2}\}$ whose moduli are Lebesgue square integrable) be such that

$$c_j = \int_\Lambda e^{-2\pi i\lambda j} c(\lambda)\, d\lambda, \qquad c(\lambda) \sim \sum_j c_j e^{2\pi i j\lambda}, \tag{4.7}$$

and put

$$x_n(\lambda) = \sum_{t=1}^{n} x_t e^{2\pi i t\lambda}. \tag{4.8}$$

Then for sufficiently large n,

$$S_n = \int_\Lambda |x_n(\lambda)c(\lambda)|^2\, d\lambda > 0 \tag{4.9}$$

as is required in theorem 4.1; null sequences x and c are of course excluded.

Because of their importance in practical applications, one will not want to exclude all finite sequences c from any $\mathfrak{C}_x$. But if $\mathfrak{C}_x$ contains any finite nonnull sequence, then (4.6) implies

$$\max_{k=1,\cdots,n} |x_k|/\|x(n)\| \to 0, \tag{4.10}$$

as will be seen from lemma 4.3 (x is slowly increasing). In order to investigate the behavior of the left-hand side of (4.6) under this additional assumption, we prove the following lemma.

LEMMA 4.1. *Equation* (4.10) *implies*

$$\sup_j |A_{n,j}|/||x(n)|| \to 0 \tag{4.11}$$

for all $c \in l^2$.

Presumably, (4.11) (for fixed c) also implies (4.10), but we do not prove it here, except for finite c-sequences (lemma 4.3).

PROOF. Choose $m_n < n$, $m_n \to \infty$ such that

$$m_n \max_k |x_k|/||x(n)|| \to 0. \tag{4.12}$$

There exists always an integer j_n such that $|A_{n,j_n}| = \sup_j |A_{n,j}|$. Put $J_n = \{j_n - n, \cdots, j_n - 1\}$ and split the sum

$$A_{n,j_n} = \sum_{t=1}^{n} x_t c_{j_n - t} = \sum_{t \in J_n} x_{j_n - t} c_t \tag{4.13}$$

into the sum $\alpha_n = \sum_{t \in J_n \cap K_n} x_{j_n - t} c_t$ and into the remainder β_n; here K_n is the index set $\{t: t| < [m_n/2]\}$.
Now

$$|A_{n,j_n}| \leq |\alpha_n| + |\beta_n| \leq m_n \sup_j |c_j| \max_k |x_k| + ||x(n)|| (\sum_{|t| \geq [m_n/2]} c_t^2)^{1/2}. \tag{4.14}$$

After division by $||x(n)||$, this tends to zero for $n \to \infty$.

THEOREM 4.2. *Let* (4.10) *be true and*

$$\operatorname*{ess\,inf}_{\lambda \in \Lambda} |c(\lambda)| > 0. \tag{4.15}$$

Then (4.6) *holds, and consequently, statement* (A) *is valid.*

PROOF. The proof follows from the preceding lemma and

$$S_n \geq ||x(n)||^2 \operatorname*{ess\,inf}_{\lambda} |c(\lambda)|^2. \tag{4.16}$$

For a large class of slowly increasing regression vectors, condition (4.15) is not necessary. Assume, besides (4.10), that

$$\lim_{n \to \infty} \sum_{t=1}^{n-h} x_{t+h} x_t / ||x(n)||^2 = \tilde{R}_h, \qquad h = 1, 2, \cdots, \tag{4.17}$$

exists. Put $\tilde{R}_{-h} = \tilde{R}_h$. Then $\{\tilde{R}_h\}$ is a positive definite sequence, and there exists a d.f. $M(\lambda)$, $\lambda \in \Lambda$, of finite variation such that

$$\tilde{R}_h = \int_{\Lambda} e^{2\pi i h \lambda} \, dM(\lambda) \tag{4.18}$$

([7], p. 233). We have $M(\frac{1}{2}) - M(-\frac{1}{2}) = 1$ and

$$||x(n)||^{-2} \int_{-1/2}^{\lambda} |x_n(\mu)|^2 \, d\mu \equiv M_n(\lambda) \to M(\lambda) \tag{4.19}$$

at continuity points of $M(\lambda)$.

Now let $\{I_1, \cdots, I_K\}$ be any partition of Λ into disjoint intervals, whose end points are not jump points of $M(\lambda)$. Let

$$\text{(4.20)} \qquad \operatorname*{ess\,inf}_{I_k} |c(\lambda)|^2$$

denote the essential infimum of $|c(\lambda)|^2$ over I_k. Then

$$\text{(4.21)} \qquad S_n/\|x(n)\|^2 \geq \sum_{k=1}^{K} \operatorname*{ess\,inf}_{I_k} |c(\lambda)|^2 \Delta M_n(I_k)$$

where $\Delta M_n(I_k)$ is the variation of $M_n(\lambda)$ over I_k. Because of (4.19),

$$\text{(4.22)} \qquad \liminf_n S_n/\|x(n)\|^2 \geq \sum_{k=1}^{K} \operatorname*{ess\,inf}_{I_k} |c(\lambda)|^2 \Delta M(I_k).$$

The relation remains valid if we take on the right-hand side the supremum with respect to all admissible partitions with arbitrary K.

DEFINITION. *The function $|c|$ on Λ is called essentially positive at λ if*

$$\text{(4.23)} \qquad \sup_{I \ni \lambda} \operatorname*{ess\,inf}_{I} |c| > 0.$$

where I denotes on interval $\subset \Lambda$.

We now deduce theorem 4.3.

THEOREM 4.3. *Let* (4.10) *and* (4.17) *be true, and let $|c(\lambda)|$ be essentially positive on at least one point of the spectrum of $M(\lambda)$. Then the right-hand side of* (4.22) *is positive, and consequently* (4.6) *and statement* (A) *are valid.*

For some sequences of regression constants it is possible to obtain a complete characterization of the class $\mathfrak{C}_x$.

EXAMPLE. Let $x_t = 1$, for all t. Then $\mathfrak{C}_{\{1\}}$ consists of all $c \in l^2$ with only a small exceptional class characterized by $\lim_{n\to\infty} \sum_{j=-n}^{n} c_j = 0$ and convergence of $\sum_{j=0}^{n} c_j$ to T, say. The c's of this subclass satisfy (4.6) if and only if $\sum_{j=-n}^{n} c_j$ does not converge too fast to zero, namely, if and only if

$$\text{(4.24)} \qquad \sum_{n=1}^{N} \left(\left(\sum_{j=0}^{n} c_j - T \right)^2 + \left(\sum_{j=-n}^{-1} c_j + T \right)^2 \right) \to \infty$$

for $N \to \infty$ ([3], p. 325). If $c_{-1} = c_{-2} = \cdots = 0$, (4.24) reduces to

$$\text{(4.25)} \qquad \sum_{n=1}^{N} \left(\sum_{j=0}^{n} c_j \right)^2 \to \infty .$$

We conclude this section with a remark concerning the convergence properties of the sequence of functions

$$\text{(4.26)} \qquad x_n(\lambda)c(\lambda) \sim \sum_{j=-\infty}^{\infty} A_{n,j} e^{2\pi i j \lambda}.$$

Let $d_n(\lambda)$, $n = 1, 2, \cdots, -\frac{1}{2} \leq \lambda \leq \frac{1}{2}$, be any sequence of functions with $d_n(\lambda) \in L^2$,

$$\text{(4.27)} \qquad d_n(\lambda) \sim \sum_{j=-\infty}^{\infty} d_{nj} e^{2\pi i j \lambda}.$$

Then

$$\text{(4.28)} \qquad \int_\Lambda \overline{d_n(\lambda)} x_n(\lambda) c(\lambda)\, d\lambda = \sum_j \overline{d_{n,j}} A_{n,j}.$$

LEMMA 4.2. *Let* $\sum_j |d_{n,j}|^2 = 1$, $\sum_j |d_{n,j}|^2$ *converge uniformly in* n. *Then* (4.6) *implies*

$$S_n^{-1/2} \int_\Lambda \overline{d_n(\lambda)} x_n(\lambda) c(\lambda)\, d\lambda \to 0 \tag{4.29}$$

uniformly with respect to the functions $d_n(\lambda)$ *out of the considered class.*

PROOF. Choose $m_n < n$, $m_n \to \infty$ such that

$$m_n \sup_j |A_{nj}|/S_n^{1/2} \to 0. \tag{4.30}$$

Then

$$S_n^{-1/2} |\sum_j \overline{d_{n,j}} A_{n,j}| \leq 2m_n \sup_j |A_{n,j}| S_n^{-1/2} + (\sum_{|j| \geq m_n} |d_{n,j}|^2)^{1/2} \to 0. \tag{4.31}$$

Let us now put $\gamma_n(\lambda) = S_n^{-1/2} x_n(\lambda) c(\lambda)$, and assume $\gamma_n(\lambda)$ converges boundedly in measure to an integrable limiting function $\gamma(\lambda)$ on Λ. Let $\mathrm{Re}\gamma(\lambda)$ be of one sign in some interval $\subset \Lambda$ and take all $d_n(\lambda) = d(\lambda)$, the characteristic function of this interval. Then by the bounded convergence theorem,

$$\int_\Lambda d(\lambda) \gamma_n(\lambda)\, d\lambda \to \int_\Lambda d(\lambda) \gamma(\lambda)\, d\lambda. \tag{4.32}$$

By the above lemma the integrals on the left tend to zero under (4.6), so that $\mathrm{Re}\gamma(\lambda) = 0$ [a.e.] on the considered interval. Repeating the argument for $\mathrm{Im}\gamma(\lambda)$, we obtain $\gamma(\lambda) = 0$ [a.e.]. This, however, is in contradiction with the fact that

$$\int_\Lambda |\gamma_n(\lambda)|^2\, d\lambda = 1 \qquad \text{for all } n, \tag{4.33}$$

which implies

$$\int_\Lambda |\gamma(\lambda)|^2\, d\lambda = 1. \tag{4.34}$$

The same argument holds if any infinite subsequence of $\{\gamma_n(\lambda)\}$ is taken or any measurable subset of Λ is considered instead of Λ.

Thus we have the following: *under* (4.6), *no subsequence of* $\{\gamma_n(\lambda)\}$ *converges boundedly in measure to an integrable function.*

Let $c(\lambda) \equiv 1$. Then (4.6) is equivalent with (4.10). In addition, let (4.17) be true, so that (4.19) holds. Assume $M(\lambda)$ possesses a bounded derivative $M'(\lambda)$. Then the preceding proposition is somewhat surprising in view of the fact that

$$\int_{-1/2}^{\lambda} |\gamma_n(\mu)|^2\, d\mu = \|x(n)\|^{-2} \int_{-1/2}^{\lambda} |x_n(\mu)|^2\, d\mu \to \int_{-1/2}^{\lambda} M'(\mu)\, d\mu \tag{4.35}$$

for all λ. One may guess that $\gamma_n(\lambda)$ must be increasingly oscillatory for $n \to \infty$, which may be due to the fact that $\gamma_n(\lambda)$ is complex-valued.

Here the remark may be of interest that always

$$\sum_{t=1}^{n} |x_t| \Big/ \left(\sum_{t=1}^{n} x_t^2 \right)^{1/2} \to \infty \tag{4.36}$$

if (4.10) holds so that, at least sometimes, the boundedness condition of the convergence will be violated.

4.2. *One regression vector (errors are finite moving averages).* For further study of condition (4.6), we now restrict ourselves to finite c-sequences. Let $k_1 < k_2$, $c_{k_1}, c_{k_2} \neq 0$, $c_j = 0$ for $j < k_1$, $j > k_2$. Then

$$A_{n,j} = 0 \qquad \text{for} \quad j \leq k_1 \quad \text{and} \quad j > k_2 + n, \qquad n = 1, 2, \cdots, \tag{4.37}$$

$$A_{n,j} \equiv A_j = \sum_{k=k_1}^{k_2} c_k x_{j-k} \qquad \text{if} \quad k_2 < j \leq k_1 + n, \qquad n > k_2 - k_1, \tag{4.38}$$

$$S_n = \sum_{j=k_1+1}^{k_1+n} A_j^2 + \sum_{j=k_1+n+1}^{k_2+n} A_{n,j}^2, \qquad n > k_2 - k_1, \tag{4.39}$$

with $A_j = x_1 c_{j-1} + \cdots + x_{j-k_1} c_{k_1}$ for $k_1 < j \leq k_2$. Since

$$\inf_n \sup_j |A_{n,j}| > 0 \tag{4.40}$$

(4.6) implies $\lim S_n = +\infty$, we have by (4.6)

$$\sum_{j=k_1+n+1}^{k_2+n} A_{n,j}^2 / S_n \to 0. \tag{4.41}$$

Thus (4.6) also implies

$$\sum_{j=k_1+1}^{n} A_j^2 \to \infty, \tag{4.42}$$

and it is equivalent to

$$\sup_j A_{n,j}^2 \Big/ \sum_{j=k_1+1}^{n} A_j^2 \to 0. \tag{4.43}$$

We have, moreover, the following lemma.

LEMMA 4.3. *If c is a finite nonnull sequence, then* (4.6) *implies* (4.10), *and* (4.10) *is equivalent to* (4.11).

PROOF. We have $\max_j A_{n,j}^2 \geq c_{k_2}^2 x_n^2$. Since $c(\lambda)$ is now continuous, we have $\sup_\lambda |c(\lambda)|^2 = \gamma < \infty$. Hence by (4.9),

$$S_n \leq \gamma \|x(n)\|^2. \tag{4.44}$$

Thus

$$\sup_j A_{n,j}^2 / S_n \geq \gamma' x_n^2 / \|x(n)\|^2, \qquad \gamma' > 0, \tag{4.45}$$

and (4.6) implies

$$|x_n| / \|x(n)\| \to 0. \tag{4.46}$$

As seen above, (4.6) implies $S_n \to \infty$; hence,

$$\|x(n)\| \to \infty. \tag{4.47}$$

However, this together with (4.46) is equivalent to (4.10), since otherwise, with k_n chosen such that

$$|x_{k_n}| = \max_{k=1,\cdots,n} |x_k|, \tag{4.48}$$

$|x_{k_n}| / \|x(k_n)\| \nrightarrow 0$, in contradiction to (4.46).

The proof of the equivalence of (4.10) and (4.11) is similar. (End of proof.)

If $c(\lambda)$, which is now a Fourier polynomial, has no zeros, and if (4.10) is true, then we have asymptotic normality by theorem 4.2.

Condition (4.42) obviously is not satisfied (and thus we have no asymptotic normality) if the sequence $\{x_t\}$ satisfies the recursive system

$$\sum_{k=k_1}^{k_2} c_k x_{j-k} = A_j = 0, \qquad j > k_2. \tag{4.49}$$

If the equation

$$c_{k_1} t^{k_2-k_1} + \cdots + c_{k_2-1} t + c_{k_2} = 0 \tag{4.50}$$

possesses a root of modulus one, but no larger ones ($c(\lambda)$ then has a zero in $[-\frac{1}{2}, \frac{1}{2}]$), then $\{x_t\}$ satisfies (4.10), but (4.6) does not hold if (4.49) is true. Thus the converse of the first statement of lemma 4.3 does not hold. A criterion for the validity of (4.49) is that all the determinants

$$\begin{vmatrix} x_1, & \cdots, & x_n \\ & \cdots & \\ x_n, & \cdots, & x_{2n-1} \end{vmatrix} \tag{4.51}$$

vanish for $n > k_2 - k_1$.

There are many cases of pairs (x, c) of sequences incompatible with (4.6) that are of practical interest. For example, let $x_t = 1$, for all t, and $c_0 = -c_m = 1$, all other $c_j = 0$. Then $x_t - x_{t-m} = 0$, and $t = m + 1, m + 2, \cdots$ is a recursive system.

From this example we see that a theorem of the type "If x is any element of a class of regression sequences independent of c, and if c is any element of a class independent of x, then asymptotic normality holds," is not desirable because it excludes too many important cases. It is for this reason that we entered into a sharper analysis and tried to obtain the classes $\mathcal{C}_x$.

For the case that $c(\lambda)$ has zeros, we can state the following: if (4.10) is true, and if the sequence of functions

$$\Big(\max_{j=1,\cdots,n} x_j^2\Big)^{-1} \int_{-1/2}^{\lambda} |x_n(\mu)|\, d\mu, \qquad -\tfrac{1}{2} \leq \lambda \leq \tfrac{1}{2}, \tag{4.52}$$

or any subsequence thereof does not tend to a pure jump function whose jump points are all zeros of $c(\lambda)$, then asymptotic normality holds.

About the estimation of the normalizing factors used in (4.4) some remarks are made in section 4.4 (3).

4.3. *Multiple regression (dependent errors).* We use the notation of sections 2 and 3.1 with $q > 1$ and errors given by (4.2). Let $B_n^2 = \operatorname{cov}(b(n)b'(n))$. This matrix has no longer the simple structure (3.3) because $\sum_n$ in general is not diagonal. In order to derive the next theorem, we introduce the q-vectors

$$\zeta(n) = B_n^{-1} P_n^{-1} X_n' \epsilon(n) = B_n^{-1} P_n^{-1} \sum_{j=-\infty}^{\infty} \left(\sum_{t=1}^{n} r_t c_{j-t}\right) \eta_j, \tag{4.53}$$

which all have expectation zero and covariance matrix I_q. (The vector r_k was defined as the k-th column of X_n'. The matrices $P_n (n \geq q)$ are assumed non-singular.)

Now

$$\text{d.f.}\ (\zeta(n)) \to N(0, I_q) \tag{4.54}$$

if and only if

$$\text{d.f.}\ (d'(n)\zeta(n)) \to N(0, 1) \tag{4.55}$$

for all sequences of constant q-vectors $d(n)$ with $d(n) \in S_q$ (the unit sphere in R_q). For

$$d(n) = B_n P_n f(n) \| B_n P_n f(n) \|^{-1}, \qquad f(n) \in S_q, \tag{4.56}$$

we have

$$d'(n)\zeta(n) = \| B_n P_n f(n) \|^{-1} \sum_j \left(\sum_{t=1}^{n} f'(n) r_t c_{j-t} \right) \eta_j. \tag{4.57}$$

Here $\{\eta_j\} \in \mathfrak{F}(F)$, F subject to conditions (II) and (III) of theorem 3.1. We now have reduced the problem of finding conditions asserting (4.54) to the one-dimensional case. By theorem (4.1), (4.55) holds if

$$\sup_j A_{n,j}^2 / S_n \to 0, \qquad S_n = \sum_{j=-\infty}^{\infty} A_{n,j}^2, \tag{4.58}$$

where now

$$A_{n,j} = \sum_{t=1}^{n} f'(n) r_t c_{j-t} = f'(n) X_n' \begin{pmatrix} c_{j-1} \\ \cdot \\ \cdot \\ \cdot \\ c_{j-n} \end{pmatrix}. \tag{4.59}$$

Relation (4.54) holds if (4.58) holds for every sequence $\{f(n)\}$, $f(n) \in S_q$. That is the case if and only if

$$\sup_{f \in S_q} \left[\sup_j (f' X_n' c(j, n))^2 \Big/ \sum_{k=-\infty}^{\infty} (f' X_n' c(k, n))^2 \right] \to 0; \tag{4.60}$$

here $c(j, n) = (c_{j-1}, \cdots, c_{j-n})'$. Putting

$$\begin{gathered} \sum_{j=-\infty}^{\infty} c_{j+k} c_j = R_k = R_{-k}, \\ R(n) = \begin{pmatrix} R_0, & \cdots, & R_{n-1} \\ & \cdots & \\ R_{n-1}, & \cdots, & R_0 \end{pmatrix}, \end{gathered} \tag{4.61}$$

the denominator in (4.60) becomes

$$S_n = f' X_n' R(n) X_n f. \tag{4.62}$$

In analogy to (4.60), we also have, with

$$x_n(\lambda) = \sum_{t=1}^{n} f' r_t e^{2\pi i t \lambda}, \tag{4.63}$$

$$S_n = \int_\Lambda |x_n(\lambda) c(\lambda)|^2 \, d\lambda. \tag{4.64}$$

Suppose

$$\text{ess}\inf_{\lambda} |c(\lambda)|^2 \equiv \gamma_0 > 0. \tag{4.65}$$

Then

$$S_n \geq \gamma_0 f' P_n f, \qquad P_n = X_n' X_n, \tag{4.66}$$

and (4.60) holds, according to lemma 4.1, if

$$\max_t \sup_{f \in S_q} (f' r_t)^2 / f' P_n f \to 0. \tag{4.67}$$

This is equivalent to

$$\max_t \sup_{f \in S_q} (f' P_n^{-1/2} r_t)^2 = \max_t r_t' P_n^{-1} r_t \to 0. \tag{4.68}$$

Hence, the following theorem holds.

THEOREM 4.4. *If*

$$\max_t r_t' P_n^{-1} r_t \to 0 \tag{I}$$

and ess $\inf_\lambda |c(\lambda)| > 0$, *then* (4.54) *holds.*

As in section 4.1, assumption (4.65) may be weakened by assuming that the regression matrices X_n allow for a harmonic analysis. Generalizing (4.17), we assume

$$\max_{t=1,\cdots,n} x_{tj}^2 / ||x_j(n)||^2 \to 0 \qquad \text{for } j = 1, \cdots, q, \tag{4.69}$$

$$\sum_{t=1}^{n} x_{t+k,r} x_{t,s} / ||x_r(n)|| \, ||x_s(n)|| \to R_h^{(r,s)}, \qquad r, s = 1, \cdots, q, \quad h = 0, \pm 1, \cdots. \tag{4.70}$$

Then $R_{-h}^{(s,r)} = R_h^{(r,s)}$, and the $(q \times q)$-matrices $\tilde{R}_h = \{R_h^{(r,s)}\}$ admit the spectral representation

$$\tilde{R}_h = \int_\Lambda e^{2\pi i h \lambda} \, dM(\lambda), \qquad h = 0, \pm 1, \cdots, \tag{4.71}$$

where the elements of the $(q \times q)$-matrix $M(\lambda)$ are functions of bounded variation and $M(\lambda_2) - M(\lambda_1)$ is positive semidefinite for every $\lambda_2, \lambda_1, -\frac{1}{2} \leq \lambda_1 < \lambda_2 \leq +\frac{1}{2}$ (see [7], p. 233). The set of all points λ with $M(\lambda_2) - M(\lambda_1)$ positive definite for all $\lambda_1 < \lambda < \lambda_2$ is called the *spectrum* of $M(\lambda)$.

We assume

$$\tilde{R}_0 = \int_\Lambda dM(\lambda) = M(\tfrac{1}{2}) - M(-\tfrac{1}{2}) \equiv M \tag{4.72}$$

to be nonsingular and put

$$D_n = \text{diag}\,(||x_1(n)||, \cdots, ||x_q(n)||). \tag{4.73}$$

Then ([7], p. 238)

$$D_n P_n^{-1} X_n' R(n) X_n P_n^{-1} D_n \to M^{-1} \int_\Lambda |c(\lambda)|^2 \, dM(\lambda) M^{-1} \equiv \tilde{M}. \tag{4.74}$$

Now from (4.62), for $n \to \infty$,

$$S_n = f' X_n' R(n) X_n f = f' P_n D_n^{-1} \tilde{M} D_n^{-1} P_n f \geq \lambda_q ||D_n^{-1} P_n f||^2 \tag{4.75}$$

where we have put $\lambda_q = \lambda_{\min}(\tilde{M})$. Since by (4.70) $D_n^{-1}P_nD_n^{-1} \rightarrow \tilde{R}_0$, we have in the limit $S_n \geq \gamma \|D_n f\|^2$ with some new constant γ that is positive if $\lambda_q > 0$.

We now assume $\lambda_q > 0$ and have (4.60) if

$$(4.76) \qquad \sup_j \sup_{f \in S_q} (f' D_n D_n^{-1} X_n' c(j, n))^2 / \|D_n f\|^2 = \sup_j \sup_{f \in S_q} (f' D_n^{-1} X_n' c(j, n))^2 \rightarrow 0.$$

Choosing $f = D_n^{-1}X_n'c(j, n)/\|D_n^{-1}X_n'c(j, n)\|$, we see that

$$(4.77) \qquad \sup_j \|D_n^{-1}X_n'c(j, n)\| \rightarrow 0,$$

or that

$$(4.78) \qquad \sup_j \sum_{k=1}^{n} x_{tk} c_{j-k}| / \|x_k(n)\| \rightarrow 0 \qquad \text{for all} \quad k = 1, \cdots, q$$

is sufficient for (4.60). The latter, however, is true because of (4.69) and lemma (4.1).

A condition implying $\lambda_q > 0$ is that $\int_\Lambda |c(\lambda)|^2 \, dM(\lambda)$ be nonsingular in case this integral is defined and is the limit of the corresponding finite approximations. It is the higher dimensional generalization of (4.22).

In summary, the following holds.

THEOREM 4.5. *Let* (4.69) *and* (4.70) *be true, and let* $\tilde{R}_0$ *be nonsingular. Let* $c(\lambda)$ *be essentially positive on at least one point of the spectrum of* $M(\lambda)$. *Then* (4.54) *holds.*

Concerning the estimation of the normalizing matrices B_n, the reader is referred to section 4.4 (3).

Finally, it may be noticed that ess inf$_{\lambda \in \Lambda} |c(\lambda)| > 0$ can always be achieved by simply adding to the y_t's independent random variables of an artificial sequence $\{\rho_t\}$, which are also independent of the η_j. The function $c^*(\lambda)$, associated with the new combined error sequence $\rho_t + \epsilon_t$, always satisfies ess inf$_{\lambda \in \Lambda} |c^*(\lambda)| > 0$. This method may be considered as a particular type of prewhitening. It always implies, however, an increase in variance of the LSE $b(n)$. A similar proposal has been made by Hannan [9].

4.4. *Concluding remarks.* (1) As is well known, the Gauss-Markov estimators

$$(4.79) \qquad b_G(n) = (X_n' \Sigma_n^{-1} X_n)^{-1} X_n' \Sigma_n^{-1} y(n)$$

are the minimum variance linear unbiased estimators for β, whether the ϵ_t are correlated or not. One may, therefore, try to use $b_G(n)$ instead of the above considered LSE $b(n)$. However, in the first place, the covariance matrix Σ_n usually is unknown, and a useful estimate for Σ_n^{-1} (or for the functions of Σ_n^{-1} that occur in (4.79)) cannot be obtained from a single sequence of observations $\{y_t\}$. Instead, it appears to be more appropriate to use a distribution free method. In the second place, there is not much point in preferring $b_G(n)$ to $b(n)$, because it is known that both are equally efficient asymptotically for a rather large class of error sequences $\{\epsilon_t\}$ and of sequences of regression matrices $\{X_n\}$ [7].

(2) The results of the preceding sections can be extended to the case of *vectorial or multivariate regression equations*

$$y_t = x_{t1}\beta_1 + \cdots + x_{tq}\beta_q + \epsilon_t \tag{4.80}$$

where now $y_t = (y_t^1, \cdots, y_t^k)'$, $x_{tj} = (x_{tj}^1, \cdots, x_{tj}^k)'$, $\epsilon_t = (\epsilon_t^1, \cdots, \epsilon_t^k)'$ are k-vectors and the β_j are scalars as before ([9], for a review see [11]). Sometimes vectorial regression equations are expressed in the form

$$y_t = B\phi_t + \epsilon_t \tag{4.81}$$

where the vector $\phi_t = (\phi_{t1}, \cdots, \phi_{tp})'$ contains the known regression constants, and the $(k \times p)$-matrix B contains the unknown regression constants. However, (4.81) is a special case of (4.80), as is seen by putting

$$(x_{t1}, \cdots, x_{tq}) = \begin{pmatrix} \phi_t', & 0, & \cdots 0 \\ 0, & \phi_t', & \cdots 0 \\ \cdot & \cdot & \cdot \\ \cdot & \cdot & \cdot \\ \cdot & \cdot & \cdot \\ 0, & 0, & \phi_t' \end{pmatrix} \tag{4.82}$$

where the zeros represent zero row vectors of p dimensions. With $q = kp$, both sides are $(k \times kp)$-matrices. We get the row vector β by placing the rows of B one behind the other.

If the k-vectors ϵ_t are independent for different t with possibly dependent components for each fixed t, and assuming $E\epsilon_t = 0$, we have almost the case considered in section 3, except that the error sequence $(\epsilon_1^1, \cdots, \epsilon_1^k, \epsilon_2^1, \cdots, \epsilon_2^k, \cdots)$ now is k-dependent. We might apply a CLT for k-dependent r.v.'s. However, it appears to be almost as easy to appeal directly to the CLT for independent r.v.'s. For this purpose, let

(i) R be a nonempty set of strictly positive definite $(k \times k)$-covariance matrices,

(ii) F be the set of d.f.'s defined in section 2,

(iii) $\mathcal{G}(F, R)$ be the set of all sequences $\{\epsilon_t\}$ of independent k-vectors ϵ_t with cov $(\epsilon_t\epsilon_t') \in R$ and d.f. $(\epsilon_t^j) \in F$ for all t, j,

(iv) $b(n)$ be the vectorial LSE for β,

$$b(n) = P_n^{-1}X_n'y(n). \tag{4.83}$$

Here now $X_n = (x_{tj}; t = 1, \cdots, n, j = 1, \cdots, q)$ is a $(kn \times q)$-matrix, $P_n = X_n', X_n$, and $y(n) = (y_1^1, \cdots, y_1^k, y_2^1, \cdots, y_2^k, \cdots, y_n^k)'$. Similarly, $\epsilon(n)$ is defined. Furthermore,

$$\begin{aligned} B_n^2 &= \text{cov}\,(b(n)b'(n)) = P_n^{-1}X_n'\textstyle\sum_n X_n P_n^{-1} \\ &= P_n^{-1} \sum_{j=1}^{n} x_j'\rho_j x_j P_n^{-1} \end{aligned} \tag{4.84}$$

with

$$\begin{aligned} x_j &= (x_{j1}, \cdots, x_{jq})_{k\times q}, \\ \rho_j &= E(\epsilon_j\epsilon_j')_{k\times k}. \end{aligned} \tag{4.85}$$

Let r_m be the m-th column vector of X'_n, $m = 1, \cdots, kn$. Finally, let $\lambda_k(\rho)$ be the smallest characteristic root of $\rho \in R$. Then we have

THEOREM 4.6. *The d.f.* $(B_n^{-1}(b(n) - \beta) \rightarrow N(0, I_q)$ *for all* $\{\epsilon_t\} \in \mathcal{G}(F, R)$ *if*

(I) $$\max_{m=1,\cdots,kn} r'_m P_n^{-1} r_m \rightarrow 0,$$

(II) $$\sup_{G \in F} \int_{|x|>c} x^2 \, dG(x) \rightarrow 0 \quad \textit{for} \quad c \rightarrow \infty,$$

(III) $$\inf_{\rho \in R} \lambda_k(\rho) > 0.$$

We indicate the *proof* only for $q = 1$. By (4.83) we have

$$B_n^{-1}(b(n) - \beta) = B_n^{-1} P_n^{-1} \sum_{j=1}^{n} x'_j \epsilon_j \tag{4.86}$$

where $x'_j \epsilon_j = \sum_{s=1}^{k} x_{j1}^s \epsilon_j^s$ are independent r.v.'s. Putting $\inf_{\rho \in R} \lambda_k(\rho) = \lambda$ and $\max_{s=1,\cdots,k} (x_{j1}^s)^2 P_n^{-1} = \kappa_{n,j}$, we obtain

$$B_n^2 \geq P_n^{-2} \lambda \sum_{j=1}^{n} \|x_j\|^2 = \lambda P_n^{-1}. \tag{4.87}$$

Now for any $\delta > 0$,

$$\begin{aligned} P(|B_n^{-1} P_n^{-1} x'_j \epsilon_j| > \delta) &\leq P(B_n^{-1} P_n^{-1} \|x_j\| \, \|\epsilon_j\| > \delta) \\ &\leq P(\|\epsilon_j\|^2 > \lambda \delta^2 / (k \kappa_{n,j})) \\ &\leq \sum_{j=1}^{k} P((\epsilon_j^s)^2 > \lambda \delta^2 / (k^2 \kappa_{n,j})) \\ &\leq \sum_{s=1}^{k} \frac{k^2 \kappa_{n,j}}{\lambda \delta^2} \int_{x^2 > \lambda \delta^2 / (k^2 \kappa_{n,j})} x^2 \, dG_j^s(x) \\ &\leq k^3 \kappa_{n,j} \lambda^{-1} \delta^{-2} \phi \left(\left(\frac{\lambda \delta^2}{k^2 \kappa_{n,j}} \right)^{1/2} \right) \end{aligned} \tag{4.88}$$

where $G_j^s = \text{d.f.}\ (\epsilon_j^s)$ and $\phi(c) = \sup_{G \in F} \int_{|x|>c} x^2 \, dG(x)$. Now $\sum_{j=1}^{n} \kappa_{n,j} \leq P_n / P_n = 1$. Putting $\kappa_n = \max_{j=1,\cdots,n} \kappa_{n,j}$, we finally have

$$\sum_{j=1}^{n} P(|B_n^{-1} P_n^{-1} x'_j \epsilon_j| > \delta) \leq k^3 \lambda^{-1} \delta^{-1} \phi \left(\left(\frac{\lambda \delta^2}{k^2 \kappa_n} \right)^{1/2} \right) \rightarrow 0 \tag{4.89}$$

by (II). Hence, the CLT holds.

The case where the random k-vectors ϵ_t are generated by a moving average process $\epsilon_t = \sum_{j=-\infty}^{\infty} A_j \eta_{t-j}$ with independent and identically distributed random vectors η_j of r components and with constant $(k \times r)$-matrices A_j has been considered by Hannan [9]. There it is assumed that

$$\sum_{j=-\alpha}^{\infty} (\lambda_{\max}(A'_j A_j))^{1/2} < \infty, \tag{4.90}$$

and that the regression matrices allow for a generalized harmonic analysis in order to derive the asymptotic normal distribution of the LSE. Although the results are likely to be true under more general conditions similar to those of sections 4.1 to 4.3, we do not discuss this possibility here.

(3) An estimation of the normalizing constants $B_n^2 = \operatorname{var} b(n)$, or of the normalizing matrices B_n^2 in multiple regression, for an unknown sequence of dependent errors ϵ_t with unequally distributed residuals η_j (see (4.2)) seems to be impossible if only a single sequence $\{y_t\}$ of observations is given. By estimation of B_n^2 we mean a relation like (3.9) with a suitably adapted statistic C_n^2. The reason for the difficulty lies, first of all, in the admission of nonidentically distributed residuals η_j in (4.2) which, together with the unknown c_j's, introduces to many unknown parameters.

Therefore, we restrict ourselves in the following to strictly stationary error sequences $\{\epsilon_t\}$, that is, *we assume the η_j to be independently identically distributed with variance one* (besides $E\eta_j = 0$). In fact, it suffices to have only identical variances of the η_j not necessarily equal to one. We also restrict ourselves to simple regression ($q = 1$). Then

$$B_n^2 = ||x(n)||^{-4}E(x'(n)\epsilon(n))^2 = ||x(n)||^{-4}x'(n)R(n)x(n) \tag{4.91}$$

where

$$R(n) = E(\epsilon(n)\epsilon'(n)) = \begin{pmatrix} R_0, & R_1, & \cdots, & R_{n-1} \\ R_1, & R_0, & \cdots, & R_{n-2} \\ & & \cdots, & \\ R_{n-1}, & R_{n-2}, & \cdots, & R_0 \end{pmatrix} \tag{4.92}$$

is the covariance matrix of the ϵ_t with

$$R_k = R_{-k} = E(\epsilon_t\epsilon_{t+k}) = \sum_{j=-\infty}^{\infty} c_jc_{j+k}, \qquad k = 0, 1, \cdots. \tag{4.93}$$

Putting

$$\tilde{R}_{k,n} = \tilde{R}_{-k,n} = ||x(n)||^{-2}\sum_{t=1}^{n-k} x_tx_{t+k}, \qquad k = 0, 1, \cdots, n-1, \tag{4.94}$$

we obtain

$$B_n^2 = ||x(n)||^{-2}\sum_{k=-n+1}^{n-1} R_k\tilde{R}_{k,n} = ||x(n)||^{-4}S_n \tag{4.95}$$

where S_n was defined in section 4.1.

In order to estimate B_n^2, we introduce first the sample covariances of $\{\epsilon_t\}$,

$$R_{k,n} = \frac{1}{n-|k|}\sum_{h=1}^{n-|k|} \epsilon_h\epsilon_{h+|k|}, \qquad k = 0, \pm 1, \cdots, \tag{4.96}$$

and, as an estimator for B_n^2,

$$B_{n,\alpha}^2 = ||x(n)||^{-2}\sum_{k=-n+1}^{n-1} R_{k,n}\tilde{R}_{k,n} \tag{4.97}$$

(later we shall replace the ϵ_h by the known residuals $e_h(n)$). Since $ER_{k,n} = R_k$, then $EB_{n,\alpha}^2 = B_n^2$.

We now introduce the *additional assumption*

$$\sum_{k=-\infty}^{\infty} R_k^2 < \infty. \tag{4.98}$$

Then ([10], p. 16) with some $\gamma < \infty$,

$$\text{var } R_{k,n} < \gamma/(n - |k|). \tag{4.99}$$

By Schwarz's inequality,

$$\begin{aligned} \text{var } B^2_{n,\alpha} &\leq \|x(n)\|^{-4} \sum_{k=-n+1}^{n-1} \tilde{R}^2_{k,n} \text{ var } R_{k,n} \\ &\leq \frac{\gamma}{\|x(n)\|^4} \sum_{k=-n+1}^{n-1} \tilde{R}^2_{k,n}/(n - |k|). \end{aligned} \tag{4.100}$$

We have to divide var $B^2_{n\alpha}$ by B^4_n and need the ratio tending to zero. We assume

$$S_n > \gamma' \|x(n)\|^2, \qquad \text{for some} \quad \gamma' > 0. \tag{4.101}$$

This is, for instance, the case under the assumptions of theorems 4.2 and 4.3 where the asymptotic normality of the LSE $b(n)$ for the regression parameters is proved.

Now with (4.95), (4.100), and (4.101),

$$\begin{aligned} B_n^{-4} \text{ var } B^2_{n,\alpha} &= \|x(n)\|^8 S_n^{-2} \text{ var } B^2_{n,\alpha} \\ &= O\left(\sum_{k=-n+1}^{n-1} \tilde{R}^2_{k,n}/(n - |k|)\right). \end{aligned} \tag{4.102}$$

If a regression sequence $\{x_t\}$ has the property that the last expression tends to zero, then $B_n^{-2} B^2_{n,\alpha}$ is a (strongly) consistent sequence of estimators of one. This is certainly true if

$$\sup_n \sum_{k=-n+1}^{n-1} \tilde{R}^4_{k,n} < \infty, \tag{4.103}$$

since because of the slowly increasing character of $\{x_t\}$ there exists a sequence of integers $m_n \to \infty$ such that

$$\sum_{k=m_n}^{n-1} \tilde{R}^2_{k,n}/(n - k) \to 0, \tag{4.104}$$

and also

$$\sum_{k=n-m_n}^{n} k^{-2} \to 0. \tag{4.105}$$

Now estimate the central part of

$$\sum_{k=-n+1}^{n-1} \tilde{R}^2_{k,n}/(n - |k|) \tag{4.106}$$

for $k = -m_n, \cdots, m_n$ by Schwarz's inequality. Because of (4.103) and (4.105), it tends to zero. The rest of the sum tends to zero by (4.104).

As pointed out before $B^2_{n,\alpha}$ is not yet an estimate of B^2_n, since $B^2_{n,\alpha}$ contains the error r.v.'s ϵ_t. We now proceed to replace them by the residuals

$$e_k(n) = y_k - x_k b(n) = \epsilon_k - \|x(n)\|^{-2} x_k x'(n) \epsilon(n). \tag{4.107}$$

Let

$$\hat{R}_{k,n} = \frac{1}{n - |k|} \sum_{h=1}^{n-|k|} e_h(n) e_{h+|k|}(n). \tag{4.108}$$

Putting

$$R'_{k,n} = \frac{(x'(n)\epsilon(n))^2}{(n-|k|)||x(n)||^4} \sum_{h=1}^{n-|k|} x_h x_{h+|k|} = \frac{(x'(n)\epsilon(n))^2}{(n-|k|)||x(n)||^2} \tilde{R}_{k,n}, \tag{4.109}$$

$$R''_{k,n} = \frac{x'(n)\epsilon(n)}{(n-|k|)||x(n)||^2} \sum_{h=1}^{n-|k|} (\epsilon_h x_{h+|k|} + \epsilon_{h+|k|} x_h), \tag{4.110}$$

we obtain

$$\hat{R}_{k,n} = R_{k,n} + R'_{k,n} - R''_{k,n}. \tag{4.111}$$

Consider now,

$$\hat{B}_n^2 = ||x(n)||^{-2} \sum_{k=-n+1}^{n-1} \hat{R}_{k,n}\tilde{R}_{k,n} = B_{n,\alpha}^2 + B_{n,\beta}^2 - B_{n,\gamma} \tag{4.112}$$

where

$$B_{n,\beta}^2 = ||x(n)||^{-2} \sum_{k=-n+1}^{n-1} R'_{k,n}\tilde{R}_{k,n} = ||x(n)||^{-4}(x'(n)\epsilon(n))^2 \sum_{k=-n+1}^{n-1} (n-|k|)^{-1}\tilde{R}_{k,n}^2, \tag{4.113}$$

$$B_{n,\gamma} = ||x(n)||^{-2} \sum_{k=-n+1}^{n-1} R''_{k,n}\tilde{R}_{k,n}. \tag{4.114}$$

Now

$$B_n^{-2}EB_{n,\beta}^2 = \sum_{k=-n+1}^{n-1} (n-|k|)^{-1}\tilde{R}_{k,n}^2 \to 0, \tag{4.115}$$

making use of our assumption concerning (4.102). Since $B_{n,\beta}^2 \geq 0$, this implies

$$B_n^{-2}B_{n,\beta}^2 \to 0, \qquad \text{i.p.} \tag{4.116}$$

Concerning $B_{n,\gamma}$, we proceed as follows. First,

$$R''_{k,n} = (n-|k|)^{-1}||x(n)||^{-2} \sum_{j,\ell=-\infty}^{\infty} A_{nj}(A_{n-|k|,\ell-|k|} + A_{n\ell} - A_{|k|\ell})\eta_j\eta_\ell. \tag{4.117}$$

After some straightforward computations, we obtain

$$\mathrm{E}(\sum_{j,\ell} A_{n,j}A_{n-|k|,1-|k|}\eta_j\eta_\ell)^2 < \text{const}\, S_n S_{n-|k|}, \tag{4.118}$$

and similar relations hold for the other terms of $E(R''_{k,n})^2$. Hence,

$$\begin{aligned} EB_{n,\gamma}^2 &< \text{const}\, ||x(n)||^{-8}S_n \sum_{|k|<n} \tilde{R}_{k,n}^2 \frac{1}{n-|k|} \sum_{k=1}^{n} (S_k + S_n - S_{n-k})k^{-1}, \\ B_n^{-4}EB_{n,\gamma}^2 &< \text{const} \sum_{|k|<n} R_{k,n}^2 \frac{1}{n-|k|} ||x(n)||^{-2} \sum_{k=1}^{n} k^{-1}(||x(k)||^2 + ||x(n)||^2 - ||x(n-k)||^2). \end{aligned} \tag{4.119}$$

This tends to zero, and hence $B_n^{-2}B_{n,\gamma} \to 0$, i.p., if we assume

$$\|x(n)\|^{-2} \max_{t=1,\cdots,n} x_t^2 = O(n), \tag{4.120}$$

besides what we have assumed previously. In summary we have the following theorem.

THEOREM 4.7. *The d.f.'s of the statistics* $(\hat{B}_n^2)^{-1}(b(n) - \beta)$ *tend to* $N(0, 1)$ *if* $\{\epsilon_t\}$ *is a strictly stationary linear process, if* ess $\inf_\lambda |c(\lambda)| > 0$ *and* (4.98) *holds, and if the regression sequence satisfies* (4.120) *and*

$$\sum_{k=-n+1}^{n-1} \tilde{R}_{k,n}^2 (n - |k|)^{-1} \to 0. \tag{4.121}$$

Condition (4.120) is satisfied, for example, for polynomial regression sequences $x_t = t^c$, $x > -\frac{1}{2}$. However, (4.121) is not satisfied at least for some polynomial sequences. A slightly weaker assumption than (4.120) that is sufficient is

$$\|x(n)\|^{-2} \sum_{t=1}^{n} x_t^2 \ln (n/t) = O(1). \tag{4.122}$$

In this subsection it was not our aim to achieve the utmost in generality, we rather wanted to indicate one possibility of replacing the normalizing constant B_n^2 by a statistic. The assumptions on the regression sequence, in particular (4.121), can be weakened considerably, if stronger assumptions are imposed on the admissible error sequences $\{\epsilon_t\}$, such as

$$\sum_{k=-\infty}^{\infty} |R_k| < \infty, \tag{4.123}$$

and if instead of $\hat{B}_n^2$, a different estimator is used, for instance,

$$\hat{\hat{B}}_n^2 = \|x(n)\|^{-2} \sum_{|k|<\sqrt{n}} R_{k,n} \tilde{R}_{k,n}. \tag{4.124}$$

Clearly, (4.123) is satisfied, for example, for finite c-sequences.

REFERENCES

[1] F. EICKER, "Asymptotic normality and consistency of the least squares estimators for families of linear regressions," *Ann. Math. Statist.*, Vol. 34 (1963), pp. 447–456.

[2] ———, "Uber die Konsistenz von Parameterschätzfunktionen für ein gemischtes Zeitreihen-Regressionsmodell," *Z. Wahrscheinlichkeitstheorie*, Vol. 1 (1963), pp. 456–477.

[3] ———, "Ein Zentraler Grenzwertsatz für Summen von Variablen einer verallgemeinerten linearen Zufallsfolge," *Z. Wahrscheinlichkeitstheorie*, Vol. 3 (1965), pp. 317–327.

[4] ———, "A central limit theorem for linear vector forms with random coefficients," to be published.

[5] B. V. GNEDENKO and A. N. KOLMOGOROV, *Limit Distributions for Sums of Independent Random Variables*, Cambridge, Addison-Wesley, 1954.

[6] A. J. GOLDMAN and M. ZELEN, "Weak generalized inverses and minimum variance linear unbiased estimation," *J. of Research*, National Bureau of Standards, Washington, D.C., Vol. 68B (1964), pp. 151–172.

[7] U. GRENANDER and M. ROSENBLATT, *Statistical Analysis of Stationary Time Series*, New York, Wiley, 1957.

[8] E. J. HANNAN, *Time Series Analysis*, London, Methuen, 1960.
[9] ———, "A central limit theorem for systems of regressions," *Proc. Cambridge Philos. Soc.*, Vol. 57 (1961), pp. 583–588.
[10] Z. A. LOMNICKI and S. K. ZAREMBA, "On estimating the spectral density function of a stochastic process," *J. Roy. Statist. Soc. Ser. B*, Vol. 18 (1956), pp. 13–37.
[11] M. ROSENBLATT, "Statistical analysis of stochastic processes with stationary residuals," *Probability and Statistics* (Cramér volume), edited by U. Grenander, New York, Wiley, 1959.

WEAK LIMITS OF SEQUENCES OF BAYES PROCEDURES IN ESTIMATION THEORY

R. H. FARRELL
CORNELL UNIVERSITY

1. Introduction

Let $(X, \mathcal{B}, \mu)$ be a totally σ-finite measure space and $\{f(\cdot, \omega), \omega \in \Omega\}$ a family of (generalized) density functions relative to $(X, \mathcal{B}, \mu)$. If $a \in \Omega$ and b is a (randomized) decision procedure for the decision space $\mathfrak{D}$, we borrow Stein's [5] notation and write

$$K(a, b) = \int W(a, t)b(x, dt)f(x, a)\mu(dx), \tag{1.1}$$

where $W(a, t)$ is the measure of loss if $t \in \mathfrak{D}$ is decided and $a \in \Omega$ is the case.

In the sequel we will always suppose that Ω and $\mathfrak{D}$ are locally compact metric spaces and will make suitable measurability assumptions about W, b, f. As is known from the work of Wald [7], under fairly liberal assumptions an admissible procedure b is Bayes in the wide sense. That is, we may find sequences $\{b_n, n \geq 1\}$ and $\{\lambda_n, n \geq 1\}$ such that if $n \geq 1$, b_n is Bayes relative to λ_n, $K(a, b_n) \leq K(a, b)$ for all a in the support of λ_n, and $\lim_{n\to\infty} \int (K(a, b) - K(a, b_n))\lambda_n(da) = 0$. Under convexity assumptions on W one may suppose $b = \text{weak} \lim_{n\to\infty} b_n$, as is explained in the appendix.

If $\mathfrak{D} = (-\infty, \infty)$ and $(\partial W/\partial t)$ is well-defined, then with suitable hypotheses the statement, b_n is Bayes relative to λ_n, is equivalent to the statement, for almost all x, for all t, in the support of $b_n(x, \cdot)$,

$$0 = \int \left(\frac{\partial W}{\partial t}\right)(\omega, t)f(x, \omega)\lambda_n(d\omega). \tag{1.2}$$

If t is vector-valued, (1.2) may be replaced by a system of equations.

Logically, given that b is Bayes in the wide sense relative to $\{b_n, n \geq 1\}$ and $\{\lambda_n, n \geq 1\}$, one would hope to determine a measure $\lambda(\cdot)$ such that for almost all x, for all t in the support of $b(x, \cdot)$,

$$0 = \int \left(\frac{\partial W}{\partial t}\right)(\omega, t)f(x, \omega)\lambda(d\omega). \tag{1.3}$$

Research sponsored in part by the Office of Naval Research under Contract Number Nonr 401(50).

Since not every admissible procedure is Bayes, (1.3) cannot always be solved using probability measures $\lambda(\cdot)$.

Various people have observed that if λ is allowed to be σ-finite on Ω, then many decision procedures of interest are solutions of equations like (1.3). This provided the basis for admissibility proofs in Karlin [3]. And it is strongly suggested that if $b = \text{weak}\lim_{n\to\infty} b_n$, b_n Bayes relative to λ_n, then one might always be able to choose constants $\{k_n, n \geq 1\}$ such that a nonzero σ-finite $\lambda = \text{weak}\lim_{n\to\infty} k_n^{-1}\lambda_n$ was defined for which (1.3) holds. Sacks [4] tried to prove such a result but discovered that it is false.

The arguments used by Sacks required $\{f(\cdot, \omega), \omega \in \Omega\}$ to be an exponential family of density functions. In particular, if $x, y \in X$ and $x \neq y$, then $f(x, \omega)/f(y, \omega)$ is a finite but unbounded function of ω. It was observed by Sacks that if one weakened the restrictions to allow $f(x, \omega) = 0$ for some $(x, \omega) \in X \times \Omega$, then one could give counter examples to show no σ-finite λ could exist.

In this paper we reformulate the problem somewhat and thereby obtain a theory including many examples not covered by Sacks [4]. In particular, one can suppose $f(x, \omega) > 0$ for all $x \in X$, $\omega \in \Omega$ and make the ratios $f(x, \omega)/f(y, \omega)$ very smooth. Yet the result remains false.

In order to consider weak convergence of sequences of measures, one needs to compactify Ω to Ω^*. Given a suitable compactification, one then easily finds examples where mass escapes to the boundary $\Omega^* - \Omega$. The reformulation of the problem and discussion of such details as escape of mass to the boundary constitute section 2.

Section 2 is an exposition designed to give a reformulation of the problem, a compactification of Ω, the "right" renormalization $k_n^{-1}\lambda_n$ of the measures λ_n, $n \geq 1$. Late in section 2 the results are formulated in the theorem of section 2. Lemmas 2.2 and 2.3 are intended as observations useful in various applications of the theorem.

Section 3 gives a necessary and sufficient condition for admissibility in certain estimation problems where strictly convex loss is used. This condition is similar to a condition used by Blyth [1] and Stein [6] to prove admissibility. Suppose b is admissible. To obtain a necessary condition, the main problem is to show that one can pick a compact set E with the following property. Suppose $\{b_n, n \geq 1\}$, $\{\lambda_n, n \geq 1\}$ are any sequences such that $b = \text{weak}\lim_{n\to\infty} b_n$, b_n Bayes relative to λ_n. There exists an integer N such that if $n \geq N$, then $\lambda_n(E) > 0$, and, for every compact set F such that $E \subset F$,

$$\limsup_{n\to\infty} \lambda_n(F)/\lambda_n(E) < \infty. \tag{1.4}$$

In lemma 3.2 we state sufficient conditions for such a result to be true. The hypotheses of lemma 3.2 are satisfied by the examples studied in sections 4 and 5.

Related is the idea of a procedure being admissible outside every compact parameter set. That is, if $E \subset \Omega$, E compact, then the procedure is admissible relative to the parameter space $\Omega - E$. Kiefer and Schwartz (to appear) have studied examples of Bayes tests for the independence of sets of normal variates. These tests have the property that given any compact subset $E \subset \Omega$ there is a probability measure supported on $\Omega - E$, relative to which the given procedure is Bayes.

In estimation problems using strictly convex loss, suppose $E \subset \Omega$, E is an open set, and the procedure δ is not admissible for the parameter space $\Omega - E$. Suppose $f(x, \omega) > 0$ for all $(x, \omega) \in X \times \Omega$, and for each $x \in X$, $f(x, \cdot)$ is a continuous function on Ω. Then there is another procedure δ^* such that $K(\omega, \delta^*) < K(\omega, \delta)$ for all $\omega \in \Omega - E$, and if $F \subset \Omega - E$ is any compact set, $\inf_{\omega \in F} (K(\omega, \delta) - K(\omega, \delta^*)) > 0$ (see lemma 3.1). If δ is admissible, then we may find sequences $\{\delta_n, n \geq 1\}$ and $\{\lambda_n, n \geq 1\}$ such that δ_n is Bayes relative to λ_n, $n \geq 1$, and

$$\lim_{n\to\infty} (\lambda_n(E))^{-1} \int (K(\omega, \delta) - K(\omega, \delta_n))\lambda_n(d\omega) = 0. \tag{1.5}$$

This is a consequence of results of Stein [5]. Let F be a compact set, $E \subset F$; then $F - E$ is a compact set. Since δ^* is not Bayes, we find

$$\begin{aligned}(\lambda_n(E))^{-1} \int (K(\omega, \delta) - K(\omega, \delta_n))\lambda_n(d\omega) \\ \geq (\lambda_n(E))^{-1} \int_E (K(\omega, \delta) - K(\omega, \delta^*))\lambda_n(d\omega) \\ + (\lambda_n(E))^{-1} \int_{\Omega - E} (K(\omega, \delta) - K(\omega, \delta^*))\lambda_n(d\omega).\end{aligned} \tag{1.6}$$

The first integral is bounded, provided risk functions are bounded on compact sets. The integrand of the second integral is strictly positive, and on $F - E$, is bounded away from zero. We find

$$\limsup_{n\to\infty} \lambda_n(F)/\lambda_n(E) < \infty. \tag{1.7}$$

In the sequel we will see that a nonzero σ-finite measure

$$\lambda(\cdot) = \operatorname{weak\,lim}_{n\to\infty} (\lambda_n(E))^{-1}\lambda_n(\cdot) \tag{1.8}$$

is defined, and from the above it follows that

$$\int_{\Omega - E} (K(\omega, \delta) - K(\omega, \delta^*))\lambda(d\omega) < \infty. \tag{1.9}$$

This shows that the "growth" properties of the measure λ at infinity are limited by the finiteness of these integrals.

The development in sections 2 and 3 depend on the general results of decision theory. The results as formulated in the appendix have been developed by Wald [7], Le Cam (unpublished), and others, and are widely known. But in this case, the literature seems to be lagging badly behind the development of the subject. It was therefore decided to put a few needed results in an appendix, along with proofs.

Details of a number of examples have been worked out when $\Omega = \mathfrak{D} = (-\infty, \infty)$. In section 4 we classify these examples into three groups, and we work out the details for two groups of examples.

Details for the third group of examples are given in section 5. Here it will be seen that in order to obtain the "right" result, the methods of section 2 apply, but a different functional equation and a different type of normalization are required.

Section 6 gives the construction of an example of an admissible procedure b, sequences $\{b_n, n \geq 1\}$ and $\{\lambda_n, n \geq 1\}$, such that $b = \text{weak} \lim_{n\to\infty} b_n$, and for the renormalized sequence of measures some mass does escape to the boundary.

In case $W(\omega, \cdot)$ is a strictly convex function, $\omega \in \Omega$, all Bayes procedures and all admissible procedures are nonrandomized. Further, every admissible procedure will be a weak limit of Bayes procedures. One can then interpret the results of section 4 as saying every admissible procedure solves a nondegenerate functional equation; and the results of section 5 as saying every nonlinear admissible procedure solves a nondegenerate functional equation.

Using such functional equations and additional smoothness assumptions, one can infer things about the continuity and differentiability of admissible estimators. We do not pursue this subject here.

2. Formulation of the problem

It is not the primary purpose of this paper to prove complete class theorems. Consequently, instead of starting by saying "Suppose δ is admissible," we start by saying "Suppose $\delta = \text{weak} \lim_{n\to\infty} \delta_n$." Throughout we will suppose $\{\delta_n, n \geq 1\}$ is a sequence of Bayes decision procedures, δ_n Bayes relative to $\{\lambda_n, n \geq 1\}$, with λ_n supported on Ω. And we shall restrict the study to those δ's which are weak limits of such sequences. The meaning of weak limit, as explained in the appendix, is as a sequence of bilinear forms acting on elements of a Banach space.

Nonetheless, in section 3 we will use the results of this section to obtain a complete class theorem. The result is about estimation of a vector parameter using a strictly convex loss function.

The discussion of this section represents as much as anything an exposition of a method or concept for treating a certain problem. This makes it difficult to formally state results as theorems. Nonetheless, towards the end of the section a theorem is stated. The statement assumes the preceding discussion as understood.

Throughout we suppose that $\mathfrak{D}$ is a locally compact metric space and that $\{\delta_n(\cdot, \cdot), n \geq 1\}$ is a sequence of decision procedures such that for each $x \in X$, $n \geq 1$, $\delta_n(x, \cdot)$ is a regular Borel probability measure on $\mathfrak{D}$.

By saying δ_n is Bayes relative to λ_n, we mean the following. We suppose given k functions $V_1(\cdot, \cdot), \cdots, V_k(\cdot, \cdot)$, and that each one is a continuous real-valued function on $\Omega \times \mathfrak{D}$ in the product topology. Thus, in terms of the introduction, $k = 1$ and $V_1(\omega, t) = (\partial W/\partial t)(\omega, t)$. We suppose for all $x \in X$, $n \geq 1$, that if t is in the support of the measure $\delta_n(x, \cdot)$, then

$$0 = \int V_i(\omega, t) f(x, \omega) \lambda_n(d\omega), \qquad i = 1, \cdots, k. \tag{2.1}$$

Basically we wish to prove that if $\{\delta_n, n \geq 1\}$ are Bayes and if

$$\operatorname*{weak\,lim}_{n \to \infty} \delta_n = \delta, \tag{2.2}$$

then the probability measures λ_n can be renormalized so that the renormalized sequence converges weakly on compact subsets of Ω. This description will prove to be inadequate, and much of the sequel is about compactification of Ω.

In order to compactify Ω, we assume that there is a positive continuous real valued function $V(\cdot)$ defined on Ω, which we call the normalizing function, satisfying the following.

(i) If E is a compact subset of $\mathfrak{D}$, then

$$\sup_{\omega \in \Omega} \sup_{t \in E} |V_i(\omega, t)|/V(\omega) < \infty, \qquad 1 \leq i \leq k. \tag{2.3}$$

(ii) If E is a compact subset of $\mathfrak{D}$, then $V_i(\omega, t)/V(\omega)$ is a uniformly continuous function of $(\omega, t) \in \Omega \times E$.

It may be helpful to consider the example $\Omega = \mathfrak{D} = (-\infty, \infty)$, $W(\omega, t) = (\omega - t)^\alpha$, $k = 1$, $V_1(\omega - t) = (\partial W/\partial t)(\omega - t) = -\alpha(\omega - t)^{\alpha - 1}$. A suitable choice of $V(\cdot)$ would be $V(\cdot) = |\omega|^{\alpha-1} + 1$.

In addition we assume

(iii) Ω is a locally compact metric space, and

(iv) the topology on Ω is such that for all $x \in X$, $f(x, \cdot)$ is a continuous function of its second variable. Since each $f(\cdot, \omega)$ is a density function, (iv) implies

$$\lim_{\omega \to \omega_0} \int |f(x, \omega) - f(x, \omega_0)| \mu(dx) = 0. \tag{2.4}$$

We find it necessary to assume

(v) for all $x \in X$, $\omega \in \Omega$ that $f(x, \omega) > 0$.

The main reason for this assumption is to ensure that

(vi) for all $x, y \in X$, $f(x, \omega)/f(y, \omega)$ is a bounded continuous function on compact ω sets.

The basic set of equations studied is

$$\begin{aligned} 0 &= \int V_i(\omega, t) f(x, \omega) \lambda_n(d\omega) \\ &= \int [V_i(\omega, t)/V(\omega)][V(\omega) f(x, \omega)/k_n(x)] \lambda_n(d\omega), \qquad 1 \leq i \leq k. \end{aligned} \tag{2.5}$$

We introduce the normalization

$$k_n(x) = \int V(\omega) f(x, \omega) \lambda_n(d\omega), \qquad x \in X, \quad n \geq 1, \tag{2.6}$$

and define a sequence of probability measures on Ω by

$$k_n(x) \nu_n(x, E) = \int_E V(\omega) f(x, \omega) \lambda_n(d\omega), \qquad x \in X, \quad n \geq 1. \tag{2.7}$$

In common examples, to each $t \in \mathfrak{D}$ we may find a compact subset $C_t \subset \Omega$ such that $\inf_{\omega \notin C_t} W(\omega, t)/V(\omega) > 0$. When this is the case, for any Bayes procedure δ, Bayes relative to λ, with finite Bayes risk, it follows by Fubini's theorem that

$$\int V(\omega)f(x, \omega)\mu(d\omega) < \infty \qquad \text{for almost all} \quad x \in X. \tag{2.8}$$

Throughout the remainder of this paper, we suppose that these integrals are finite, this assumption being the analogue of the supposition that a Bayes procedure has finite Bayes risk.

Then (2.5) may be written as

$$0 = \int [V_i(\omega, t)/V(\omega)]\nu_n(x, d\omega), \qquad 1 \leq i \leq k, \quad n \geq 1. \tag{2.9}$$

The problem is now phrased in terms of integration of bounded continuous functions by probability measures. And we wish to establish results about the weak convergence of $\{\nu_n(x, \cdot), n \geq 1\}$. To do this we need to have these act on a compact space. Below we compactify Ω to Ω^*. Then we may compute weak $\lim_{n\to\infty} \nu_n(x, \cdot) = \nu(x, \cdot)$, $x \in X$. The embedding used is such that Ω is a Borel subset of Ω^*. In the general case $\nu(x, \Omega^* - \Omega) > 0$ is possible, and one can construct examples where $\nu(x, \Omega^* - \Omega) = 1$ for all $x \in X$. In the special case considered by Sacks [4], the only possibility is $\nu(x, \Omega^* - \Omega) = 0$ for all x.

We will show below that with suitable restrictions,

(2.10)
if t is in the support of $\delta(x, \cdot)$, then there are t_n,
t_n is in the support of $\delta_n(x, \cdot)$, and
$t = \lim_{n\to\infty} t_n$.

Further,

$$\begin{aligned} 0 &= \lim_{n\to\infty} \int [V_i(\omega, t_n)/V(\omega)]\nu_n(x, d\omega) \\ &= \int [V_i(\omega, t)/V(\omega)]\nu(x, d\omega), \qquad 1 \leq i \leq k. \end{aligned} \tag{2.11}$$

To obtain the compactification of Ω, we map Ω into a product space as follows. Let $\rho(\cdot, \cdot)$ be a metric on Ω which gives the topology of Ω and which satisfies $\rho(\omega_1, \omega_2) \leq 1$ for all $\omega_1 \in \Omega$, $\omega_2 \in \Omega$. Take countable dense subsets $\{\omega_i, i \geq 1\}$ of Ω and $\{t_i, i \geq 1\}$ of $\mathfrak{D}$. Then for each $\omega \in \Omega$, we associate the value $\phi(\omega)$ given by

$$\phi(\omega) = \{\rho(\omega, \omega_i), V_j(\omega, t_p)/V(\omega), 1 \leq i, 1 \leq p, 1 \leq j \leq k\}. \tag{2.12}$$

This is a one-to-one continuous mapping into a product space (with a countable number of coordinates). Therefore, the set $\phi(\Omega)$ is a Borel subset of the product space (in the product topology), and the closure Ω^* is a compact subset of the product space (see Hausdorff [2]). We may, and do in the sequel, identify Ω with $\phi(\Omega)$. The functions $V_j(\cdot, t_p)/V(\cdot)$, $1 \leq p$, have continuous extensions to Ω^* which are coordinate mappings (projections). The assumption (ii) allows us to approximate each $V_j(\cdot, t)/V(\cdot)$, $1 \leq j \leq k$, $t \in \mathfrak{D}$, uniformly by coordinate mappings, so these functions also have unique continuous extensions. The sequences $\{\lambda_n, n \geq 1\}$ and $\{\nu_n(x, \cdot), n \geq 1, x \in X\}$ are extended to the Borel sets of Ω^* by $\lambda_n(E) = \lambda_n(E \cap \Omega)$, $\nu_n(x, E) = \nu_n(x, E \cap \Omega)$, $n \geq 1$, $x \in X$. To study the question of convergence, the following lemma is basic.

LEMMA 2.1. *Let $X - F \subset X$ be the set of x such that for* every *t in the support of $\delta(x, \cdot)$ there is an integer sequence $\{n_i(x, t), i \geq 1\}$ and a real number sequence $\{t(x, t, n_i(x, t)), i \geq 1\}$ such that $t(x, t, n_i(x, t))$ is in the support of $\delta_{n_i(x,t)}(x, \cdot)$, $\lim_{i\to\infty} n_i(x, t) = \infty$, and $\lim_{i\to\infty} t(x, t, n_i(x, t)) = t$. Then F is a $\mathfrak{B}$-measurable set and $\mu(F) = 0$.*

PROOF. Let $\{U_i, i \geq 1\}$ be a countable base for the open sets of $\mathfrak{D}$. Let

$$F(i, m) = \{x | \delta(x, U_i) > 0, \delta_n(x, U_i) = 0, n \geq m\}. \tag{2.13}$$

We will show that $F = \bigcup_{i,m=1}^{\infty} F(i, m)$.

Let $x \in F$. Then for some t in the support of $\delta(x, \cdot)$, t is bounded away from the support of $\delta_n(x, \cdot)$ for n sufficiently large. That is, there is a U_i in the countable base, and an integer m, such that $t \in U_i$, and if $n \geq m$, the support of $\delta_n(x, \cdot)$ is disjoint from U_i. Therefore, if $n \geq m$, $\delta_n(x, U_i) = 0$ and $\delta(x, U_i) > 0$. Thus $x \in F(i, m)$.

Conversely, if $x \in \bigcup_{i,m=1}^{\infty} F(i, m)$, then $x \in F(i, m)$ for some i, m. Since $\delta(x, U_i) > 0$, there is a number $t \in U_i$, t in the support of $\delta(x, \cdot)$. If $n \geq m$, then $\delta_n(x, U_i) = 0$ so the support of $\delta_n(x, \cdot)$ is bounded away from t by U_i. Hence, t cannot be a limit of a subsequence as described, and $x \in F$ follows.

Let U_i be given and $g_i(\cdot)$ a real valued continuous function on $\mathfrak{D}$ such that $g_i(t) = 0$ if $t \notin U_i$, $g_i(t) > 0$ if $t \in U_i$, $1 \leq i$. Then, by the meaning of weak limits (see appendix),

$$0 = \lim_{n\to\infty} \iint_{F(i,m)} g_i(t)\delta_n(x, dt)\mu(dx) \tag{2.14}$$

$$= \iint_{F(i,m)} g_i(t)\delta(x, dt)\mu(dx).$$

Since $\delta(x, U_i) > 0$ for all $x \in F(i, m)$, it follows that $\mu(F(i, m)) = 0$. Since this holds for $i \geq 1$, $m \geq 1$, it follows that $\mu(F) = 0$. The proof of lemma 2.1 is complete.

THEOREM. *Let $\delta =$ weak $\lim_{n\to\infty} \delta_n$. Let $\{\lambda_n, n \geq 1\}$ be a sequence of regular Borel probability measures defined on the subsets of Ω. Suppose that if $n \geq 1$, $x \in X$, and if t is in the support of $\delta_n(x, \cdot)$; then*

$$0 = \int V_i(\omega, t)f(x, \omega)\lambda_n(d\omega), \qquad 1 \leq i \leq k. \tag{2.15}$$

Let the compactification Ω^* and the probability measures $\{\nu_n, n \geq 1\}$ be as above. Let $\{n_i, i \geq 1\}$ be an integer sequence such that for almost all $x[\mu]$, $\nu(x, \cdot) =$ weak $\lim_{i\to\infty} \nu_{n_i}(x, \cdot)$. Then there exists a $\mathfrak{B}$-measurable set F such that $\mu(F) = 0$. If $x \notin F$ and t is in the support of $\delta(x, \cdot)$, then

$$0 = \int [V_i(\omega, t)/V(\omega)]\nu(x, dt), \qquad 1 \leq i \leq k. \tag{2.16}$$

PROOF. We will suppose that the subsequence $\{n_i, i \geq 1\}$ is the entire sequence. This involves no loss of generality. Let F_1 be a $\mathcal{B}$-measurable set such that if $x \notin F_1$ then $\{\nu_n(x, \cdot), n \geq 1\}$ is weakly convergent. Here $\mu(F_1) = 0$ and we write $\nu(x, \cdot) = \text{weak} \lim_{n\to\infty} \nu_n(x, \cdot)$. Choose $x \notin F_1$ and let t be in the support of $\delta(x, \cdot)$. Let F_2 be chosen in accordance with lemma 2.1 relative to the sequence $\{\delta_n, n \geq 1\}$. Then if $F = F_1 \cup F_2$ and $x \notin F$, there are sequences $\{n_i, i \geq 1\}$ and $\{t_{n_i}, i \geq 1\}$ such that $\lim_{i\to\infty} n_i = \infty$, $\lim_{i\to\infty} t_{n_i} = t$ and t_{n_i} is in the support of $\delta_{n_i}(x, \cdot)$, $i \geq 1$. Therefore,

$$0 = \int [V_i(\omega, t_{n_i})/V(\omega)]\nu_{n_i}(x, d\omega), \qquad 1 \leq j \leq k. \tag{2.17}$$

Since by (ii)

$$\lim_{i\to\infty} V_j(\omega, t_{n_i})/V(\omega) = V_j(\omega, t)/V(\omega) \tag{2.18}$$

uniformly in $\omega \in \Omega$, and since $\text{weak} \lim_{i\to\infty} \nu_{n_i}(x, \cdot) = \nu(x, \cdot)$, it follows that

$$0 = \int [V_j(\omega, t)/V(\omega)]\nu(x, d\omega), \qquad 1 \leq j \leq k. \tag{2.19}$$

That completes the proof of the theorem.

The discussion so far has not used (v) or (vi). The above theorem is therefore valid quite generally. In practice the hypothesis that there be a single integer sequence $n \geq 1$ on which $\{\nu_n(x, \cdot), n \geq 1\}$ is weakly convergent for almost all x is difficult to verify. In the examples we explore, (v) and (vi) are used to verify this hypothesis. It will also appear in the exposition of examples that one will want the ratios $f(x, \omega)/f(y, \omega)$ as functions on $\Omega \to (0, \infty)$ to have continuous extensions to functions on $\Omega^* \to [0, \infty]$. In order to achieve this, one may have to modify the construction of Ω^* to have the form

$$\phi(\omega) = \{\rho(\omega, \omega_i), V_j(\omega, t_p)/V(\omega), f(x_j, \omega)/f(y_k, \omega), \quad 1 \leq i, 1 \leq p, 1 \leq j, 1 \leq k\} \tag{2.20}$$

taken over suitable countable dense subsets (see (2.12)).

We now prove several lemmas which are useful in verifying the hypotheses of the theorem.

LEMMA 2.2. *Suppose for all $x, y \in X$ that $\sup_{\omega\in\Omega} f(y, \omega)/f(x, \omega) < \infty$ and that $f(y, \cdot)/f(x, \cdot)$ has a unique continuous extension to Ω^*. Let $x_0 \in X$, and let $\{n_i, i \geq 1\}$ be a sequence of integers such that $\lim_{i\to\infty} n_i = \infty$ and $\nu(x_0, \cdot) = \text{weak} \lim_{i\to\infty} \nu_{n_i}(x_0, \cdot)$ exists. Then $\nu(y, \cdot) = \text{weak} \lim_{i\to\infty} \nu_{n_i}(y, \cdot)$ exists for all $y \in X$. For every Borel set E in Ω^*,*

$$\nu(y, E) = \int_E [f(y, \omega)/f(x_0, \omega)]\nu(x_0, d\omega) \bigg/ \int [f(y, \omega)/f(x_0, \omega)]\nu(x_0, d\omega). \tag{2.21}$$

If for some $x_0 \in X$ and for some sequence $\{n_i, i \geq 1\}$,

$$\nu(x_0, \cdot) = \text{weak} \lim_{i\to\infty} \nu_{n_i}(x_0, \cdot) \text{ and } \nu(x_0, \Omega) = 0, \tag{2.22}$$

then

$$\nu(y, \cdot) = \text{weak} \lim_{i\to\infty} \nu_{n_i}(y, \cdot) \tag{2.23}$$

exists for all $y \in X$ *and* $\nu(y, \Omega) = 0$. *If* t *is in the support of* $\delta(y, \cdot)$, *then*

$$0 = \int [V_j(\omega, t)/V(\omega)]\nu(y, d\omega) \tag{2.24}$$
$$= \int [V_j(\omega, t)f(y, \omega)/V(\omega)f(x_0, \omega)]\nu(x_0, d\omega), \qquad 1 \leq j \leq k.$$

PROOF. We may consider $f(y, \cdot)/f(x, \cdot)$ as being a continuous function on Ω^*. Then

$$\lim_{i\to\infty} k_{n_i}(y)/k_{n_i}(x_0) = \lim_{i\to\infty} \int [f(y, \omega)/f(x_0, \omega)]\nu_{n_i}(x_0, d\omega) \tag{2.25}$$
$$= \int [f(y, \omega)/f(x_0, \omega)]\nu(x_0, d\omega).$$

If $g(\cdot)$ is a bounded continuous function on Ω^*,

$$\lim_{i\to\infty} \int g(\omega)\nu_{n_i}(y, d\omega) \tag{2.26}$$
$$= \lim_{i\to\infty} [k_{n_i}(x_0)/k_{n_i}(y)] \int [g(\omega)f(y, \omega)/f(x_0, \omega)]\nu_{n_i}(x_0, d\omega)$$
$$= \int [g(\omega)f(y, \omega)/f(x_0, \omega)]\nu(x_0, d\omega) \Big/ \int [f(y, \omega)/f(x_0, \omega)]\nu(x_0, d\omega).$$

A standard approximation argument proves (2.21). The proof of the remainder of the lemma is obvious.

LEMMA 2.3. *Given the hypotheses of lemma 2.2, suppose that*

$$\nu(x, \cdot) = \text{weak} \lim_{i\to\infty} \nu_{n_i}(x, \cdot) \qquad \text{for all} \quad x \in X. \tag{2.27}$$

Suppose $\nu(x, \Omega) > 0$ *for all* x. *If* $E \subset \Omega$ *is an open set having closure* $\overline{E}$, *and if* $\overline{E} \subset \Omega$ *and* $\nu(x, \overline{E} - E) = 0$ *for all* $x \in X$, *then for each* $x \in X$ *the sequence of measures* $\lambda_{n_i}(\cdot)/k_{n_i}(x)$ *converges weakly when restricted to* $\overline{E}$, *and if* $\lambda(x, \cdot)$ *is the limiting* σ*-finite measure on* Ω,

$$\int_{\overline{E}} V(\omega)f(x, \omega)\lambda(x, d\omega) = \nu(x, \overline{E}). \tag{2.28}$$

PROOF. If $g(\cdot)$ is a continuous real-valued function defined on $\overline{E}$, we let $g^*(\cdot)$ be a continuous extension of $g(\cdot)$ to all of Ω^*, such that $g^*(\cdot)$ has compact support. (Note that by lemma 2.2, the measures $\{\nu(x, \cdot), x \in X\}$ are mutually absolutely continuous with respect to each other.) We have

$$\lim_{i\to\infty} \int_{\overline{E}} g(\omega)\lambda_{n_i}(d\omega)/k_{n_i}(x) = \lim_{i\to\infty} \int_{\overline{E}} [g^*(\omega)/V(\omega)f(x, \omega)]\nu_{n_i}(x, d\omega) \tag{2.29}$$
$$= \int_{\overline{E}} [g^*(\omega)/V(\omega)f(x, \omega)]\nu(x, d\omega)$$
$$= \int_{\overline{E}} [g(\omega)/V(\omega)f(x, \omega)]\nu(x, d\omega).$$

This shows that the limit exists for each such $g(\cdot)$, establishing weak convergence. Choose $g(\omega) = V(\omega)f(x, \omega)$ for $\omega \in \overline{E}$. The conclusion of the lemma follows.

3. A necessary and sufficient condition for admissibility

In this section we will suppose $\Omega = \mathfrak{D}$ is Euclidean n-space and that for each $\omega \in \Omega$, the measure of loss $W(\omega, \cdot)$ is a strictly convex function, for each $t \in \mathfrak{D}$, $W(\cdot, t)$ is a continuous function. We suppose $W(\omega, t) \geq 0$ for all $(\omega, t) \in \Omega \times \mathfrak{D}$. To apply decision theory results we need to assume that if C is a compact subset of Ω, then

$$\lim_{t\to\infty} \inf_{\omega\in C} W(\omega, t) = \infty. \tag{3.1}$$

We consider only procedures having bounded risk on compact Ω sets.

We shall suppose the hypotheses (i)–(vi) and (2.4)–(2.8) of section 2 hold, where the V_j are the partial derivatives of W, and in particular, we will make heavy use of (v), that $f(x, \omega) > 0$ for all $(x, \omega) \in X \times \Omega$. We consider a fixed procedure δ and suppose the following can be proven.

Let $\{\delta_n, n \geq 1\}$ and $\{\lambda_n, n \geq 1\}$ be any sequences such that δ_n is Bayes relative to λ_n, $n \geq 1$, such that $\delta = \text{weak} \lim_{n\to\infty} \delta_n$, and $\lambda_n(\Omega) = 1$, $n \geq 1$. For every such sequence it must follow that there exists an integer sequence $\{n_i, i \geq 1\}$ such that for almost all $x, y[\mu]$,

$$0 < \lim_{i\to\infty} k_{n_i}(x)/k_{n_i}(y) < \infty, \tag{3.2}$$

and

$$\nu(x, \cdot) = \text{weak} \lim_{i\to\infty} \nu_{n_i}(x, \cdot) \quad \text{exists.} \tag{3.3}$$

We suppose $\nu(x, \Omega) > 0$ for almost all $x[\mu]$ necessarily follows.

The examples considered by Sacks [4] satisfy these hypotheses. Several later sections of this paper consider examples where these hypotheses are satisfied. In the sequel we write $K(\omega, \delta^*)$ for the risk of δ^* evaluated at ω.

THEOREM 3.1. *Suppose the hypotheses made above hold. A necessary and sufficient condition that a non-Bayes procedure δ shoula be admissible is that there exist an open set E with compact closure $\overline{E}$ and sequences $\{\delta_n, n \geq 1\}$, $\{\lambda_n, n \geq 1\}$ such that*

$$\delta = \text{weak} \lim_{n\to\infty} \delta_n; \tag{3.4}$$

(3.5) *δ_n is Bayes relative to the probability measure $\lambda_n(\cdot)$, $n \geq 1$. If C_n is the support of $\lambda_n(\cdot)$, then $\overline{E} \subset C_n \subset C_{n+1} \subset \Omega$, $n \geq 1$. The sets C_n are compact, $n \geq 1$;*

(3.6) *for almost all $x \in X$, $\lim_{n\to\infty} \lambda_n(E)/k_n(x)$ exists, finite and positive;*

$$\lim_{n\to\infty} (\lambda_n(\overline{E}))^{-1} \int (K(\omega, \delta) - K(\omega, \delta_n))\lambda_n(d\omega) = 0. \tag{3.7}$$

We will need the following lemma in the proof of sufficiency. We consider only nonrandomized procedures.

LEMMA 3.1. *Suppose δ^* is as good as δ. Let $A_1 = \{x|\delta(x) \neq \delta^*(x)\}$, and suppose $\mu(A_1) > 0$. Let C be a compact parameter set. Then there exists a real number $\gamma > 0$ such that if $\omega \in C$, then*

$$\gamma + K(\omega, (\delta + \delta^*)/2) \leq K(\omega, \delta). \tag{3.8}$$

PROOF. By hypothesis of this section, if $B \in \mathcal{B}$, then $\mu(B) > 0$ if and only if $\int_B f(x, \omega)\mu(dx) > 0$ for all $\omega \in \Omega$. We set

$$A_2(\alpha) = \{x| \text{ if } \omega \in C, \text{ then } \alpha + W(\omega, (\delta(x) + \delta^*(x))/2) \leq (\tfrac{1}{2})\, (W(\omega, \delta(x)) + W(\omega, \delta^*(x))\}. \tag{3.9}$$

Since $W(\omega, t)$ is strictly convex in t and continuous in ω, if $\delta(x) \neq \delta^*(x)$, then

$$0 < \inf_{\omega \in C} \{\tfrac{1}{2}(W(\omega, \delta(x)) + W(\omega, \delta^*(x))) - W(\omega, (\delta(x) + \delta^*(x))/2)\}. \tag{3.10}$$

Therefore, $A_1 = \bigcup_{n=1}^{\infty} A_2(1/n)$. If $n \geq 1$, then $A_2(1/n) \subset A_2(1/(n + 1))$, so it follows that

$$\int_{A_1} f(x, \omega)\mu(dx) = \lim_{n \to \infty} \int_{A_2(1/n)} f(x, \omega)\mu(dx). \tag{3.11}$$

By Dini's theorem this limit is uniform in $\omega \in C$, since all functions involved are continuous functions of ω. Further, since $0 < \inf_{\omega \in C} \int_{A_1} f(x, \omega)\mu(dx)$, it follows that we may find an integer n such that if $\omega \in C$, then

$$\int_{A_2(1/n)} f(x, \omega)\mu(dx) \geq 1/n. \tag{3.12}$$

If we take $\gamma = (1/n)^2$, the lemma now follows.

PROOF OF SUFFICIENCY. If δ^* is as good as δ, then by lemma 3.1 we may suppose $K(\omega, \delta^*) + \gamma \leq K(\omega, \delta)$ for all $\omega \in \overline{E}$ where $\gamma > 0$. Then we must have

$$\gamma\lambda_n(\overline{E}) \leq \int_{\overline{E}} (K(\omega, \delta) - K(\omega, \delta^*))\lambda_n(d\omega) \leq \int (K(\omega, \delta) - K(\omega, \delta_n))\lambda_n(d\omega). \tag{3.13}$$

Application of (3.7) leads to the contradiction that $\gamma = 0$.

We will need the following lemma in the proof of necessity.

LEMMA 3.2. *There exists an open subset U of Ω having compact closure (in Ω) with the following property. Suppose $\{\delta_n, n \geq 1\}$ is a sequence of decision procedures, δ_n is Bayes relative to λ_n, $n \geq 1$, and $\lambda_n(\Omega) = 1$, $n \geq 1$. Suppose $\delta =$ weak $\lim_{n \to \infty} \delta_n$, and $k_n(x)$, $n \geq 1$, is defined as in (2.6) for the sequence $\{\lambda_n, n \geq 1\}$. Then*

$$0 < \liminf_{n \to \infty} \lambda_n(U)/k_n(x) < \limsup_{n \to \infty} \lambda_n(U)/k_n(x) < \infty. \tag{3.14}$$

(The given open set U is to have this property for all choices of sequences $\{\delta_n, n \geq 1\}$, $\{\lambda_n, n \geq 1\}$.)

PROOF. If the lemma is false, let $\{U_n, n \geq 1\}$ be an increasing sequence of open sets with compact closure, $\Omega = \bigcup_{n=1}^{\infty} U_n$. Corresponding to each U_n there are sequences $\{\delta_{m,n}, m \geq 1\}$ and $\{\lambda_{m,n}, m \geq 1\}$ such that the hypotheses of the lemma are satisfied, yet $\lim_{m \to \infty} \lambda_{m,n}(U_n)/k_{m,n}(x) = 0$. On the basis of the assumptions made in section 2, $\lim_{m \to \infty} \lambda_{m,n}(U_n)/k_{m,n}(x) = \infty$ is impossible.

We take countable dense subsets $\{f_i, i \geq 1\}$ of $L_1(X, \mathcal{B}, \mu)$ and $\{g_i, i \geq 1\}$

of $C(\mathfrak{D}^*)$. Choose $x_0 \in X$. For each $n \geq 1$ we may choose an integer N_n such that if $m \geq N_n$, then for all $1 \leq i, j \leq n$,

$$\left| \int f_i(x)g_j(t)\delta(x, dt)\mu(dx) - \int f_i(x)g_j(t)\delta_{m,n}(x, dt)\mu(dx) \right| \leq 1/n, \tag{3.15}$$

and

$$\lambda_{m,n}(U_n)/k_{m,n}(x_0) \leq 1/n. \tag{3.16}$$

If we interpret $\delta_{m,n}$ as a bilinear form $(\cdot, \cdot)_{m,n}$, then $|(f, g)_{m,n}| \leq \|f\| \|g\|$ for all $f \in L_1(X, \mathfrak{B}, \mu)$, $g \in C(\mathfrak{D}^*)$. Using this it is easy to show

$$\delta = \text{weak} \lim_{n \to \infty} \delta_{N_n, n}. \tag{3.17}$$

If we choose a subsequence $\{n_i, i \geq 1\}$ on which for almost all $x, y \in X$,

$$0 < \lim_{i \to \infty} k_{N_{n_i}, n_i}(x)/k_{N_{n_i}, n_i}(y) < \infty, \tag{3.18}$$

which is possible by hypothesis, and such that for almost all x,

$$\nu(x, \cdot) = \text{weak} \lim_{i \to \infty} \nu_{N_{n_i}, n_i}(x, \cdot), \tag{3.19}$$

then for every $g(\cdot)$ continuous on Ω with compact support,

$$\lim_{i \to \infty} \int g(\omega)\nu_{N_{n_i}, n_i}(y, d\omega) = \int g(\omega)\nu(y, d\omega) \tag{3.20}$$

exists for all y. Our construction is such that if $N_n \geq m$, then,

$$\lambda_{N_n, n}(U_m)/k_{N_n, n}(x_0) \leq 1/N_n. \tag{3.21}$$

Consequently, if $g(\omega) = 0$ outside of U_m, it follows that $\int g(\omega)\nu(y, d\omega) = 0$ for almost all $y \in X$. Since we may take m large enough that the support of $g(\cdot)$ is contained in U_m, it follows that $\int g(\omega)\nu(y, d\omega) = 0$ for all $g(\cdot)$ having compact support contained in Ω. Since we suppose Ω is open in Ω^*, $\Omega = \bigcup_{n=1}^{\infty} U_n$, it follows that $\nu(y, \Omega) = 0$ for almost all $y \in X$. This contradicts the basic hypothesis of section 3 that $\nu(y, \Omega) > 0$ for almost all $y \in X$. This contradiction shows that the lemma must be correct.

PROOF OF NECESSITY. In view of the lemma, we may suppose an open set U is given having compact closure such that U has the property stated in lemma 3.2 relative to the admissible procedure δ. We now apply a theorem on admissibility due to Stein [5]. To adapt Stein's notation, if $a \in \Omega$ and b is a decision procedure, we set

$$K(a, b) = \int W(a, t)b(x, dt)f(x, a)\mu(dx). \tag{3.22}$$

With a minor modification of Stein's proofs and results, we may prove the following. Let ξ be a probability measure supported on U. Relative to $\gamma > 0$ and the risk function

$$\int (K(a, b) - K(a, \delta))\xi(da) + \gamma(K(a, b) - K(a, \delta)), \tag{3.23}$$

let b_γ be a minimax procedure. Let

$$-\epsilon_\gamma = \sup_a \left\{\int (K(a, b_\gamma) - K(a, \delta))\xi(da) + \gamma(K(a, b_\gamma) - K(a, \delta))\right\} \tag{3.24}$$

be the minimax risk of b_γ. Then Stein shows $\lim_{\gamma\to 0} \epsilon_\gamma = 0$. The weak compactness condition used by Stein is satisfied in our problem. See theorem 2A of the appendix.

Since b_γ is minimax, for all $a \in \Omega$,

$$\int K(a, \delta)\xi(da) + \gamma K(a, \delta) \geq \int K(a, b_\gamma)\xi(da) + \gamma K(a, b_\gamma). \tag{3.25}$$

If we divide by γ and let $\gamma \to \infty$, we find

$$\limsup_{\gamma\to\infty} K(a, b_\gamma) \leq K(a, \delta). \tag{3.26}$$

For each γ we choose a compact parameter set C_γ, satisfying the following. If $\gamma_1 < \gamma_2$, then $C_{\gamma_1} \subset C_{\gamma_2}$, and $\bigcup_{\gamma>0} C_\gamma = \Omega$. If b_γ^* is minimax for $a \in C_\gamma$ relative to the risk

$$\int (K(a, b) - K(a, \delta))\xi(da) + \gamma(K(a, b) - K(a, \delta)), \tag{3.27}$$

and Bayes relative to ξ_γ^*, then the minimax risk of b_γ^* is $\geq -2\epsilon_\gamma$. We are using here theorem A4 of the appendix. We find that if $a \in C_\gamma$, then

$$0 \geq \int (K(a, b_\gamma^*) - K(a, \delta))\xi(da) + \gamma(K(a, b_\gamma^*) - K(a, \delta)), \tag{3.28}$$

so that as above, if $a \in \Omega$,

$$\limsup_{\gamma\to\infty} K(a, b_\gamma^*) \leq K(a, \delta). \tag{3.29}$$

If we write the above as follows,

$$\begin{aligned} 0 &\geq (1+\gamma)\left\{\int K(a, b_\gamma^*)\xi(da)/(1+\gamma) + \gamma \int K(a, b_\gamma^*)\xi_\gamma^*(da)/(1+\gamma)\right. \\ &\qquad \left. - \int K(a, \delta)\xi(da)/(1+\gamma) - \gamma \int K(a, \delta)\xi_\gamma^*(da)/(1+\gamma)\right\} \\ &\geq -2\epsilon_\gamma, \end{aligned} \tag{3.30}$$

then we find b_γ^* to be Bayes relative to the risk $K(a, b)$ and the probability $\xi_1/(1+\gamma) + \gamma\xi_\gamma^*/(1+\gamma) = \xi_\gamma$.

Since we assume strictly convex loss, and since we suppose δ is admissible, it follows from (3.29) and theorem A3 of the appendix that

$$\delta = \operatorname{weak}\lim_{\gamma\to\infty} b_\gamma^*. \tag{3.31}$$

Thus (3.4) and (3.5) hold. Equation (3.7) follows from (3.30). If $\overline{E}$ is the closure of an open subset E of Ω, and if $U \subset E$, then

$$\xi_\gamma(\overline{E}) \geq \xi_\gamma(U) \geq 1/(1+\gamma). \tag{3.32}$$

Therefore, by (3.30),

(3.33) $$\lim_{\gamma\to\infty} (\xi_\gamma(\overline{E}))^{-1} \int (K(a, b_\gamma^*) - K(a, \delta))\xi_\gamma(da) = 0.$$

We will complete the proof of the theorem by verifying (3.6). Take $\gamma = n_i$ where $\{n_i, i \geq 1\}$ is an integer sequence on which for almost all x,

(3.34) $$\nu(x, \cdot) = \text{weak} \lim_{i\to\infty} \nu_{n_i}(x, \cdot).$$

We take E to be an open subset of Ω with closure $\overline{E} \subset \Omega$, such that $U \subset \overline{E}$ and such that $\nu(x_0, \overline{E} - E) = 0$. This implies $\nu(x, \overline{E} - E) = 0$ for almost all $x \in X$. Then it follows that for almost all x,

(3.35) $$\lim_{i\to\infty} \lambda_{n_i}(\overline{E})/k_{n_i}(x)$$

exists. We use here (2.26). By lemma 3.2, this limit must be positive. It follows from the construction of section 2 that the limit is finite.

The proof of the theorem is complete.

4. Estimation of a real parameter

We suppose $\Omega = \mathfrak{D} = (-\infty, \infty)$ and examine some of the common examples. We will restrict the discussion to functions $V_1(\cdot, \cdot)$ and $V(\cdot)$ satisfying

(4.1) $$\lim_{\omega\to-\infty} V_1(\omega, t)/V(\omega) = 1, \qquad \lim_{\omega\to\infty} V_1(\omega, t)/V(\omega) = -1.$$

Many typical loss functions $W(\omega, t) = w(\omega - t)$ have this property, where $V_1(\omega, t) = -(dw/dx)_{x=\omega-t}$.

As was suggested in section 2, the analysis depends on the ratios

(4.2) $$\{f(x, \cdot)/f(y, \cdot), x \in X, y \in X\}.$$

We define sets

(4.3) $$A_{\alpha,\beta}^y = \{x | \lim_{\omega\to-\infty} f(x, \omega)/f(y, \omega) = \alpha, \lim_{\omega\to\infty} f(x, \omega)/f(y, \omega) = \beta\}.$$

We consider in this section the following cases.

Case Ia. For all y, $\mu(X - A_{1,1}^y) = 0$.

Case Ib. For all y,

(4.4) $$X = \bigcup_{\alpha>0,\beta>0} A_{\alpha,\beta}^y,$$

and

(4.5) $$\mu(\{x | \lim_{\omega\to-\infty} f(x, \omega)/f(y, \omega) \neq \lim_{\omega\to+\infty} f(x, \omega)/f(y, \omega)\}) > 0.$$

Case II. For all y, $X = A_{0,\infty}^y \cup A_{\infty,0}^y$ and

(4.6) $$\mu(A_{0,\infty}^y) > 0, \qquad \mu(A_{\infty,0}^y) > 0.$$

Case II includes the examples considered by Sacks [4], as well as including many examples of monotone likelihood ratios. The family of Cauchy densities falls in case Ia.

In terms of the construction of section 2, the compactification $\Omega^* = [-\infty, \infty]$.

Consequently, if a limit measure ν puts mass on the boundary, we have only $P_+(x) = \nu(x, \{\infty\})$ and $P_-(x) = \nu(x, \{-\infty\})$ to consider.

Suppose δ is given, $\{\delta_n, n \geq 1\}$ and $\{\lambda_n, n \geq 1\}$ are given, and

$$\delta = \text{weak} \lim_{n \to \infty} \delta_n. \tag{4.7}$$

In case I, we may use lemma 2.2. According to this lemma, we may pick an integer sequence $\{n_i, i \geq 1\}$ such that for all $x \in X$, weak $\lim_{i\to\infty} \nu_{n_i}(x, \cdot) = \nu(x, \cdot)$ exists. Consequently, we may apply the theorem of section 2. If t is in the support of $\delta(x, \cdot)$ (except for $x \in F$, an exceptional set, $\mu(F) = 0$),

$$\begin{aligned} 0 &= \int_{\Omega^*} [V_1(\omega, t)/V(\omega)]\nu(x, d\omega) \\ &= P_-(x) - P_+(x) + \int_{\Omega} [V_1(\omega, t)/V(\omega)]\nu(x, d\omega). \end{aligned} \tag{4.8}$$

By lemma 2.2,

(4.9)

$$\begin{aligned} P_+(y)/P_-(y) &= \int_{\{\infty\}} [f(y, \omega)/f(x, \omega)]\nu(x, d\omega) \Big/ \int_{\{-\infty\}} [f(y, \omega)/f(x, \omega)]\nu(x, d\omega) \\ &= \frac{\lim_{\omega\to\infty} f(y, \omega)/f(x, \omega)}{\lim_{\omega\to-\infty} f(y, \omega)/f(x, \omega)} (P_+(x)/P_-(x)). \end{aligned}$$

Therefore, in case Ia, $P_+(y)/P_-(y) = P_+(x)/P_-(x)$, except for a set of measure zero, whereas in case Ib, $P_+(y)/P_-(y) \neq P_+(x)/P_-(x)$ on a set of positive measure.

If (4.8) holds and $\nu(x, \Omega) = 0$, then we find $P_+(x) = P_-(x) = \frac{1}{2}$. By lemma 2.2, if $\nu(x, \Omega) = 0$ for a single x, then $\nu(x, \Omega) = 0$ for all x, implying $P_+(x) = P_-(x) = \frac{1}{2}$ for all x.

We shall show in section 5 that in case Ia this is indeed possible. We have already seen that in case Ib this is not possible.

LEMMA 4.1. *In case Ib, if* $\delta =$ weak $\lim_{n\to\infty} \delta_n$, δ_n *is Bayes relative to* λ_n, $n \geq 1$, *then there exists an integer sequence* $\{n_i, i \geq 1\}$ *such that*

$$\nu(x, \cdot) = \text{weak} \lim_{i\to\infty} \nu_{n_i}(x, \cdot) \qquad \text{for all } x \text{ and } \nu(x, \Omega) > 0 \text{ for all } x. \tag{4.10}$$

Using lemma 4.1 and lemma 2.3 one can prove at once the following theorem.

THEOREM 4.1. *In case Ib, if* $\delta =$ weak $\lim_{n\to\infty} \delta_n$, δ_n *Bayes relative to* λ_n, $n \geq 1$, *then there exists a nonzero* σ*-finite regular Borel measure* $\lambda(\cdot)$ *defined on the real line and a real-valued function* $\rho(\cdot)$ *such that for almost all* $x[\mu]$, *if* t *is in the support of* $\delta(x, \cdot)$, *then*

$$0 = \rho(x) + \int V_1(\omega, t) f(x, \omega)\lambda(d\omega). \tag{4.11}$$

In section 6 we give an example of an admissible estimator δ for which (4.10) holds and $\rho(x) \neq 0$.

The main result in case II is as follows.

THEOREM 4.2. *Suppose in case II that* $\delta = \text{weak} \lim_{n\to\infty} \delta_n$, δ_n *Bayes relative to* λ_n. *Then for all* x, y, $\liminf_{n\to\infty} k_n(x)/k_n(y) > 0$. *There exists an integer sequence* $\{n_i, i \geq 1\}$ *such that for all* $x \in X$,

$$\nu(x, \cdot) = \text{weak} \lim_{i\to\infty} \nu_{n_i}(x, \cdot) \tag{4.12}$$

exists. $\nu(x, \Omega) = 1$ *for all* x. *There exists a nonzero* σ*-finite regular Borel measure* $\lambda(\cdot)$ *such that for almost all* $x[\mu]$, *if* t *is in the support of* $\delta(x, \cdot)$, *then*

$$0 = \int_\Omega V_1(\omega, t) f(x, \omega) \lambda(d\omega). \tag{4.13}$$

The proof of this theorem involves considerable detail. This is broken down into several lemmas.

LEMMA 4.2. *Given the hypotheses of theorem* 4.2, *for all* x, y,

$$\liminf_{n\to\infty} k_n(x)/k_n(y) > 0. \tag{4.14}$$

PROOF. We suppose to the contrary that for some x_0, y_0, and sequence $\{n_i, i \geq 1\}$, that $\lim_{i\to\infty} k_{n_i}(x_0)/k_{n_i}(y_0) = 0$. That is,

$$0 = \lim_{i\to\infty} \int [f(x_0, \omega)/f(y_0, \omega)] \nu_{n_i}(y_0, d\omega). \tag{4.15}$$

Under the conditions of case II we must have either subcase A:

$$\lim_{\omega\to\infty} f(x_0, \omega)/f(y_0, \omega) = 0, \tag{4.16}$$

or subcase B:

$$\lim_{\omega\to-\infty} f(x_0, \omega)/f(y_0, \omega) = 0. \tag{4.17}$$

Subcase A. Let x' be such that $\lim_{\omega\to\infty} f(x', \omega)/f(y_0, \omega) = \infty$. We will show that $\nu_{n_i}(x', \cdot) \to \nu(x', \cdot)$ satisfying $\nu(x', \{\infty\}) = 1$.

In subcase A, $\lim_{\omega\to-\infty} f(x_0, \omega)/f(y_0, \omega) = \infty$. Therefore, for every integer N, $1 = \lim_{i\to\infty} \nu_{n_i}(y_0, [N, \infty])$. We use this in the calculation.

Let g_1, g_2 be nonnegative real-valued continuous functions on Ω^* satisfying the following. For some integer n, if $\omega \geq n$, then $g_2(\omega) = 0$, and if $\omega \leq n$ $g_1(\omega) = 0$. If $\omega \geq n$, then $g_1(\omega) > 0$ and $g_1(\infty) = 1$.

Let $\{m_i, i \geq 1\}$ be a subsequence of $\{n_i, i \geq 1\}$ on which

$$\text{weak} \lim_{i\to\infty} \nu_{m_i}(x', \cdot) = \nu(x', \cdot) \tag{4.18}$$

exists. Then

$$\begin{aligned} \int g_1(\omega)\nu(x', d\omega) \Big/ \int g_2(\omega)\nu(x', d\omega) &= \lim_{i\to\infty} \frac{[k_{m_i}(x')]^{-1} \int g_1(\omega) V(\omega) f(x', \omega) \lambda_{m_i}(d\omega)}{[k_{m_i}(x')]^{-1} \int g_2(\omega) V(\omega) f(x', \omega) \lambda_{m_i}(d\omega)} \\ &= \lim_{i\to\infty} \frac{\int [f(x', \omega)/f(y_0, \omega)] g_1(\omega) \nu_{m_i}(y_0, d\omega)}{\int [f(x', \omega)/f(y_0, \omega)] g_2(\omega) \nu_{m_i}(y_0, d\omega)} = \infty. \end{aligned} \tag{4.19}$$

Since $\int g_1(\omega)\nu(x', d\omega) < \infty$, this can happen only if $\int g_2(\omega)\nu(x', d\omega) = 0$. Since the choice of $g_2(\cdot)$ is arbitrary subject to $g_2(\omega) = 0$ if $\omega \geq n$, n arbitrary, $\nu(x', \{\infty\}) = 1$ follows. This argument shows that every limit point of the sequence $\{\nu_{n_i}(x', \cdot)\}$ is the same. Therefore, weak $\lim_{i\to\infty} \nu_{n_i}(x', \cdot) = \nu(x', \cdot)$ where $\nu(x', \{\infty\}) = 1$.

By definition of case II,

$$A_{0,\infty}^{y_0} = \{x | \lim_{\omega\to\infty} f(x, \omega)/f(y_0, \omega) = \infty\} \tag{4.20}$$

has positive μ measure. Suppose $x \in A_{0,\infty}^{y_0}$ and t is in the support of $\delta(x, \cdot)$. By lemma 2.1, for almost all such x we may find a subsequence (depending on x) $\{m_i, i \geq 1\}$ of $\{n_i, i \geq 1\}$ and real numbers t_{m_i} in the support of $\delta_{m_i}(x, \cdot)$, $i \geq 1$, such that $t = \lim_{i\to\infty} t_{m_i}$. Then we obtain

$$\begin{aligned} 0 &= \lim_{i\to\infty} \int [V_1(\omega, t_{m_i})/V(\omega)]\nu_{m_i}(x, d\omega) \\ &= \int [V_1(\omega, t)/V(\omega)]\nu(x, d\omega) = -1. \end{aligned} \tag{4.21}$$

This contradiction shows that subcase A cannot happen.

Subcase B is dual to subcase A, and similar arguments lead to a similar contradiction. That proves lemma 4.2.

LEMMA 4.3. *Suppose for some $x \in X$ and integer sequence $\{n_i, i \geq 1\}$ that* weak $\lim_{i\to\infty} \nu_{n_i}(x, \cdot) = \nu(x, \cdot)$. *Then $\nu(x, \{-\infty\} \cup \{\infty\}) = 0$.*

PROOF. Suppose $\nu(x, \{\infty\}) > 0$. Choose y such that $\lim_{\omega\to\infty} f(y, \omega)/f(x, \omega) = \infty$. Then

$$\lim_{i\to\infty} k_{n_i}(y)/k_{n_i}(x) = \lim_{i\to\infty} \int [f(y, \omega)/f(x, \omega)]\nu_{n_i}(x, d\omega) = \infty. \tag{4.22}$$

Contradiction. The supposition $\nu(x, \{-\infty\}) > 0$ leads to a similar contradiction of lemma 4.2.

LEMMA 4.4. *Choose $x_0 \in X$ and an integer sequence $\{n_i, i \geq 1\}$ such that $\lim_{i\to\infty} n_i = \infty$ and for every continuous function g having compact support,*

$$\lim_{i\to\infty} \int g(\omega)\lambda_{n_i}(d\omega)/k_{n_i}(x_0) \tag{4.23}$$

exists. Then the weak $\lim_{i\to\infty} \lambda_{n_i}(\cdot)/k_{n_i}(x_0) = \lambda(\cdot)$ *is a nonzero σ-finite measure. Further,* weak $\lim_{i\to\infty} \nu_{n_i}(x_0, \cdot) = \nu(x_0, \cdot)$ *exists, and for every integrable real-valued function $g(\cdot)$,*

$$\int g(\omega)\nu(x_0, d\omega) = \int g(\omega)V(\omega)f(x_0, \omega)\lambda(d\omega). \tag{4.24}$$

PROOF. Let $g(\cdot)$ be continuous with compact support. Then if E is the support of $g(\cdot)$, $\inf_{\omega\in E} V(\omega)f(x_0, \omega) > 0$. This implies

$$\sup_{n\geq 1} \int g(\omega)\lambda_n(d\omega)/k_n(x_0) < \infty. \tag{4.25}$$

By considering a countable subset of g's (dense in the continuous functions vanishing at $\pm\infty$) with compact support, we may choose an integer sequence

$\{n_i, i \geq 1\}$ such that for every continuous function with compact support,

$$\lim_{i\to\infty} \int g(\omega)\lambda_{n_i}(d\omega)/k_{n_i}(x_0) \tag{4.26}$$

exists. Use a diagonalization argument. Let $\lambda(\cdot)$ be the limiting σ-finite measure determined by these limits.

If $g(\cdot)$ is a continuous function with compact support, then so is

$$g(\cdot)V(\cdot)f(x_0, \cdot). \tag{4.27}$$

We find

$$\begin{aligned}\lim_{i\to\infty} \int g(\omega)\nu_{n_i}(x_0, d\omega) &= \lim_{i\to\infty} \int g(\omega)V(\omega)f(x_0, \omega)\lambda_{n_i}(d\omega)/k_{n_i}(x_0) \\ &= \int g(\omega)V(\omega)f(x_0, \omega)\lambda(d\omega).\end{aligned} \tag{4.28}$$

If we choose from $\{n_i, i \geq 1\}$ a subsequence $\{m_i, i \geq 1\}$ such that $\nu(x_0, \cdot) = \text{weak}\lim_{i\to\infty} \nu_{m_i}(x_0, \cdot)$ exists, then for every compact set $E \subset \Omega$, (4.28) uniquely determines $\nu(x_0, E)$. By lemma 4.3, $\nu(x_0, \{\infty\} \cup \{-\infty\}) = 0$. Therefore, $\nu(x_0, \cdot)$ is uniquely determined by (4.28), which proves that $\text{weak}\lim_{i\to\infty} \nu_{n_i}(x_0, \cdot) = \nu(x_0, \cdot)$ exists. By lemma 4.3, (4.28), and a standard approximation argument, (4.24) follows.

LEMMA 4.5. *Let x_0, $\{n_i, i \geq 1\}$ and $\lambda(\cdot)$ be as in lemma 4.4. Then for every $y \in X$,*

$$\nu(y, \cdot) = \text{weak}\lim_{i\to\infty} \nu_{n_i}(y, \cdot) \text{ exists.} \tag{4.29}$$

In addition,

$$\lim_{i\to\infty} k_{n_i}(y)/k_{n_i}(x_0) = \int_\Omega f(y, \omega)V(\omega)\lambda(d\omega), \tag{4.30}$$

and for every ν integrable Ω^* measurable g,

$$\int_{\Omega^*} g(\omega)\nu(y, d\omega) = \int_\Omega g(\omega)f(y, \omega)V(\omega)\lambda(d\omega) \Big/ \int_\Omega f(y, \omega)V(\omega)\lambda(d\omega). \tag{4.31}$$

PROOF. Let $\{m_i, i \geq 1\}$ be a subsequence of $\{n_i, i \geq 1\}$ such that $\nu(\cdot) = \text{weak}\lim_{i\to\infty} \nu_{m_i}(y, \cdot)$ exists. If $g(\cdot)$ is a function having compact support, then

(4.32)

$$\begin{aligned}\int g(\omega)\nu(d\omega) &= \lim_{i\to\infty} \int g(\omega)\nu_{m_i}(y, d\omega) \\ &= \lim_{i\to\infty} \left[\int g(\omega)[f(y, \omega)/f(x_0, \omega)]\nu_{m_i}(x_0, d\omega)\right] [k_{m_i}(x_0)/k_{m_i}(y)]. \\ &= \left[\int g(\omega)f(y, \omega)V(\omega)\lambda(d\omega)\right] [\lim_{i\to\infty} k_{m_i}(x_0)/k_{m_i}(y)].\end{aligned}$$

By lemma 4.3, $\nu(\Omega) = 1$. Using a standard approximation argument, we obtain

$$\lim_{i\to\infty} k_{m_i}(y)/k_{m_i}(x_0) = \int_\Omega f(y, \omega)V(\omega)\lambda(d\omega). \tag{4.33}$$

Since this holds on every subsequence $\{m_i, i \geq 1\}$ for which $\{\nu_{m_i}, i \geq 1\}$ has a weak limit, (4.30) follows.

Since the value of the limit in (4.32) is independent of the subsequence used, (4.31) follows. For continuous $g(\cdot)$ having compact support, since $\nu(y, \{-\infty\} \cup \{\infty\}) = 0$, these functions are dense in the L_1 space of $\nu(y, \cdot)$, which proves (4.31).

PROOF OF THEOREM 4.2. We may pick an integer sequence $\{n_i, i \geq 1\}$ with $\lim_{i\to\infty} n_i = \infty$ such that for every $x \in X$, weak $\lim_{i\to\infty} \nu_{n_i}(x, \cdot)$ exists. Therefore, the theorem of section 2 applies. For almost all x, if t is in the support of $\delta(x, \cdot)$, then

$$0 = \int [V_1(\omega, t)/V(\omega)]\nu(x, d\omega) \tag{4.34}$$

$$= \int V_1(\omega, t) f(x, \omega)\lambda(d\omega).$$

Thus (4.13) is proven and the proof is complete.

5. Estimation in case Ia

We will suppose that for all x, y in X,

$$\lim_{\omega\to-\infty} f(x, \omega)/f(y, \omega) = \lim_{\omega\to\infty} f(x, \omega)/f(y, \omega) = 1. \tag{5.1}$$

In keeping with section 4, we suppose $\Omega = \mathfrak{D} = (-\infty, \infty)$. The results of this section depend on assuming X is Euclidean p space.

In case Ia one can give examples of sequences $\{\delta_n, n \geq 1\}$ which converge weakly, δ_n Bayes relative to λ_n, $n \geq 1$, and such that weak $\lim_{n\to\infty} \nu_n(x, \cdot) = \nu(x, \cdot)$ exists for all x, yet $\nu(x, \Omega) = 0$ for all x. We give such an example, where $X = (-\infty, \infty)$.

Let $W(\omega, t) = (\omega - t)^2$, $-\infty < t, \omega < \infty$, and $f(x, \omega) = c/(1 + (x - \omega)^4)$, the constant c being appropriately chosen. Let $\lambda_n(\cdot)$ put mass $\frac{1}{2}$ at n and mass $\frac{1}{2}$ at $-n$. A direct calculation shows that the (nonrandomized) Bayes estimator is

$$\delta_n(x) = \frac{8x^3n + 8xn^4}{2 + 2x^4 + 12x^2n^2 + 2n^4}. \tag{5.2}$$

For given x, $\lim_{n\to\infty} \delta_n(x) = 4x$, which implies weak $\lim_{n\to\infty} \delta_n(x) = 4x$. For given n, $\lim_{x\to\infty} \delta_n(x) = 0$ and $\lim_{x\to-\infty} \delta_n(x) = 0$.

The parameter ω is a location parameter. An easy direct calculation shows that estimators $\delta(x) = \alpha x$ are inadmissible estimators of a location parameter if $\alpha > 1$. In particular, $\delta(x) = 4x$ is not admissible.

The limiting procedure in case Ia depends on the asymptotic behavior of the first and second partial derivatives of $W(\cdot, \cdot)$ on its second variable and upon the asymptotic behavior of the first partial derivatives of $f(\cdot, \cdot)$ with respect to its first variable x. We suppose for $\omega \in X$ that $W(\omega, \cdot)$ is a strictly convex function. In the sequel we write W_2 and W_{22} for the first and second partial derivatives of W on its second variable. Further, if $x = (x_1, \cdots, x_p)$, then we write $f_i(x, \omega)$ for the partial derivative of f with respect to x_i evaluated at (x, ω), $1 \leq i \leq p$.

The "right" normalization in case Ia is not the normalization discussed in section 2. Instead, one wants a normalizing function $V(\cdot)$ satisfying the following.

If E is a compact $\mathfrak{D}$ set, then

$$\begin{aligned} &\sup_{t \in E, \omega \in \Omega} |W_{22}(\omega, t)|/V(\omega) < \infty; \\ (5.3)\qquad &\lim_{|\omega| \to \infty} W_{22}(\omega, t)/V(\omega) = \beta_1 > 0 \quad \text{uniformly in} \quad t \in E; \\ &W_{22}(\omega, t)/V(\omega) \quad \text{is uniformly continuous in} \quad (\omega, t) \in \Omega \times E. \end{aligned}$$

We then define

$$(5.4)\qquad k_n'(x) = \int V(\omega) f(x, \omega) \lambda_n(d\omega), \qquad x \in X, \quad n \geq 1,$$

and

$$(5.5)\qquad \nu_n'(x, F) = \int_F V(\omega) f(x, \omega) \lambda_n(d\omega)/k_n'(x)$$

for Borel subsets of Ω.

THEOREM 5.1. *Given the regularity conditions stated below there exist constants* $\beta_2^{(1)}, \cdots, \beta_2^{(p)}, \beta_3$ *with the following property. Let* $\{n_i, i \geq 1\}$ *be an integer sequence for which* $\nu'(x, \cdot) = \text{weak} \lim_{i \to \infty} \nu_{n_i}'(x, \cdot)$ *exists for all* $x \in X$ (see lemma 2.2). *Then* $\nu'(x, \Omega) = 0$ *for all* x *or* $\nu'(x, \Omega) > 0$ *for all* $x \in X$.

If $\nu'(x, \Omega) = 0$ *for all* $x \in X$, *then* $\delta(x) = (-\beta_3/\beta_1) \sum_{i=1}^p \beta_2^{(i)} x_i$, *where* $x = (x_1, \cdots, x_p)$.

If $\nu'(x, \Omega) > 0$ *for all* $x \in X$, *let* $\alpha = (\alpha_1, \cdots, \alpha_p)$ *be any* p *dimensional row vector of Euclidean* p*-space. The pair* $\delta(x), \delta(x + \alpha)$ *satisfies the functional equation*

$$\begin{aligned} (5.6)\qquad 0 = &\int W_2(\omega, \delta(x + \alpha))[f(x + \alpha, \omega) - f(x, \omega)]/[V(\omega) f(x, \omega)]\nu'(x, d\omega) \\ &+ \int [(W_2(\omega, \delta(x + \alpha)) - W_2(\omega, \delta(x)))/V(\omega)]\nu'(x, d\omega). \end{aligned}$$

We first define the constants $\beta_2^{(i)}$, $1 \leq i \leq p$, β_3, and give the regularity conditions needed. We suppose

$$(5.7)\qquad \lim_{\omega \to -\infty} \frac{\omega f_i(t, \omega)}{f(t, \omega)} = \lim_{\omega \to \infty} \frac{\omega f_i(t, \omega)}{f(t, \omega)} = \beta_2^{(i)}, \qquad 1 \leq i \leq p,$$

uniformly in t in compact subsets of X, and

$$(5.8)\qquad \sup_{\omega \in \Omega, t \in K} \left| \frac{\omega f_i(t, \omega)}{f(t, \omega)} \right| < \infty, \qquad 1 \leq i \leq p,$$

for every compact subset K of X. We suppose for compact subsets K of $\mathfrak{D}$ that

$$(5.9)\qquad \lim_{|\omega| \to \infty} (\operatorname{sgn} \omega) W_2(\omega, t)/(1 + |\omega| V(\omega)) = \beta_3 > 0,$$

uniformly in $t \in K$, and

$$(5.10)\qquad \sup_{\omega \in \Omega, t \in K} |W_2(\omega, t)|/(1 + |\omega| V(\omega)) < \infty.$$

We suppose for compact subsets K of $\mathfrak{D}$ that

$$\sup_{\substack{t_1, t_2 \in K \\ \omega \in \Omega}} f(t_1, \omega)/f(t_2, \omega) < \infty \tag{5.11}$$

and

$$\lim_{|\omega| \to \infty} f(t_1, \omega)/f(t_2, \omega) = 1 \qquad \text{uniformly in } \quad t_1, t_2 \in K. \tag{5.12}$$

The theorem is proven from the relation, for all x, α in p-space,

$$\begin{aligned} 0 &= \int W_2(\omega, \delta_n(x + \alpha)) f(x + \alpha, \omega) \lambda_n(d\omega), \\ &= \int W_2(\omega, \delta_n(x + \alpha))[f(x + \alpha, \omega) - f(x, \omega)]\lambda_n(d\omega) \\ &+ \int [W_2(\omega, \delta_n(x + \alpha)) - W_2(\omega, \delta_n(x))] f(x, \omega)\lambda_n(d\omega). \end{aligned} \tag{5.13}$$

In case $\nu'(x, \Omega) > 0$ for all x, the final part of the theorem is simply the result of taking weak limits. We note that if weak $\lim_{n\to\infty} \int_{(\cdot)} V(\omega) f(x, \omega)\lambda_n(d\omega)/k_n'(x)$ does not exist for all x, then by lemma 2.2, we may find a subsequence on which this limit does exist for all x. We therefore suppose, without loss of generality, that this is the entire integer sequence. Using lemma 2.1, given x and α, we choose a subsequence on which

$$\lim_{i \to \infty} \delta_{n_i}(x) = \delta(x), \qquad \lim_{i \to \infty} \delta_{n_i}(x + \alpha) = \delta(x + \alpha). \tag{5.14}$$

The argument will not be affected by supposing this is the entire sequence.

On the assumption that $\nu'(x, \Omega) = 0$ for all $x \in X$, using (5.4) and (5.5), the mean value theorem, and (5.3), we find for the third integral in (5.13), after normalization by $(k_n'(x))^{-1}$, and after letting $n \to \infty$, the limiting value $\beta_1(\delta(x + \alpha) - \delta(x))$.

To evaluate the second integral in (5.13), write

$$\begin{aligned} (k_n'(x))^{-1} W_2(\omega, \delta_n(x + \alpha))[f(x + \alpha, \omega) - f(x, \omega)]\lambda_n(d\omega) \\ = (k_n'(x))^{-1} \int \frac{W_2(\omega, \delta_n(x + \alpha))}{1 + |\omega| V(\omega)} (1 + |\omega| V(\omega)) \\ \left(\sum_{i=1}^{p} \alpha_i \frac{f_i(\eta_i(x, \omega), \omega) f(\eta_i(x, \omega), \omega)}{f(\eta_i(x, \omega), \omega) f(x, \omega)} \right) f(x, \omega)\lambda_n(d\omega). \end{aligned} \tag{5.15}$$

The numbers $\eta_1(x, \omega), \cdots, \eta_p(x, \omega)$ are determined by the use of the mean value theorem, and $\eta_i(x, \omega)$ lies in value between x_i and $x_i + \alpha_i$, $1 \le i \le p$. Use of the given regularity conditions and passage to the limit gives for the limiting value $\beta_3 \sum_{i=1}^{p} \alpha_i \beta_2^{(i)}$.

From the two limiting results obtained, the first part of the theorem follows.

In order to get some feeling about sizes, we consider location parameters, and $p = 1$. Suppose

$$f(x - \omega) = c/(1 + |x - \omega|^{-\beta_2}), \qquad \beta_2 < 0. \tag{5.16}$$

Then $(\partial f/\partial x)(x - \omega)/f(x - \omega) \sim \beta_2$ as $\omega \to \infty$. We take $W(\omega, t) = |\omega - t|^\alpha$ and $V(\omega) = 1 + |\omega - t|^{\alpha-2}$. Then $\beta_1 = \alpha(\alpha - 1)$ and $\beta_3 = \alpha$.

In order that $\delta(x) = [(-\beta_2)/(\alpha - 1)]x$ have finite risk, we require $1 > \alpha + \beta_2$. It is therefore not possible to attain $-\beta_2\beta_3/\beta_1 = 1$ and have an estimator with finite risk. As previously observed, estimators of a location parameter with $-\beta_2/\alpha - 1 > 1$ are not admissible.

6. An example in which some mass escapes to the boundary

By taking examples in which X, Ω, $\mathfrak{D}$ are compact, and $W(\cdot, \cdot)$, $f(\cdot, \cdot)$ are jointly continuous, it is easy to construct examples, by removing a point from Ω, in which a sequence of probability measures on Ω in the limit puts mass on the boundary, and yet the Bayes procedures converge to an admissible procedure.

We now consider an example for which $W(\cdot, \cdot)$ is unbounded, $\Omega = (0, \infty)$, and in the limit mass is placed at $+\infty$ in the sense of section 2. We suppose X is compact, $\mu(\cdot)$ is a probability measure defined on the Borel subsets of X, $\{f(\cdot, \omega), \omega \in \Omega\}$ is a family of generalized probability measures relative to $\mu(\cdot)$. If $x_0 \in X$, we suppose for all $x \in X$ that $f(x, \omega)/f(x_0, \omega)$ is a bounded uniformly continuous function of $x \in X$, $\omega \in \Omega$, and $\lim_{\omega\to\infty} f(x, \omega)/f(x_0, \omega) = 1$, uniformly in x.

We suppose $W(\cdot, \cdot)$ is a strictly convex function of its second variable and write $W_2(\cdot, \cdot)$ for the partial derivative of $W(\cdot, \cdot)$ on its second variable. We want $W(t, t) = 0$ for all $t \geq 0$, $\lim_{t\to\infty} W_2(\omega, t) > 0$, and $-1 = \lim_{\omega\to\infty} W_2(\omega, t)$, $t \geq 0$, $\omega \geq 0$. For example, $W_2(\omega, t) = \phi(t - \omega)$ for suitable ϕ. We take the normalizing function $V(\omega) = 1$, $0 \leq \omega < \infty$.

In order that mass move to $+\infty$, let $\{\alpha_n, n \geq 1\}$ be a nonnegative real number sequence such that $\lim_{n\to\infty} \alpha_n f(x_0, n) = 1$. In view of our assumption about $f(\cdot, \cdot)$, $\lim_{n\to\infty} \alpha_n f(x, n) = 1$ for all $x \in X$.

Let $\lambda(\cdot)$ be a probability measure on the Borel sets of Ω. We will need to assume λ, W satisfy (6.5) given below. Define $\{\lambda_n(\cdot), n \geq 1\}$ by $\lambda_n(E) = \lambda(E)$ if $n \notin E$, $\lambda_n(E) = \lambda(E) + \alpha_n$ if $n \in E$, $n \geq 1$. Let $\delta_n(\cdot)$ solve the equation

$$(6.1) \qquad 0 = \int W_2(\omega, \delta_n(x))f(x, \omega)\lambda(d\omega) + W_2(n, \delta_n(x))f(x, n)\alpha_n.$$

In the sequel we prove that $\lim_{n\to\infty} \delta_n(x) = \delta(x)$, where $\delta(x)$ solves (6.6), uniformly in $x \in X$ and that $\delta(\cdot)$ is admissible.

First, we show (6.1) is solvable. Since $W_2(\omega, \cdot)$ is strictly increasing, by the monotone convergence theorem,

$$(6.2) \qquad \lim_{t\to 0} \left[\int W_2(\omega, t)f(x, \omega)\lambda(d\omega) + W_2(n, t)f(x, n)\alpha_n\right]$$
$$= \int W_2(\omega, 0)f(x, \omega)\lambda(d\omega) + W_2(n, 0)f(x, n)\alpha_n < 0,$$

whereas

$$(6.3)\qquad \lim_{t\to\infty}\left[\int W_2(\omega, t)f(x, \omega)\lambda(d\omega) + W_2(n, t)f(x, n)\alpha_n\right]$$
$$= \int W_2(\omega, \infty)f(x, \omega)\lambda(d\omega) + W_2(n, \infty)f(x, n)\alpha_n > 0.$$

Therefore, (6.1) is solvable for each $x \in X$, $n \geq 1$.

To prove $\lim_{n\to\infty} \delta_n(x) = \delta(x)$ uniformly in x, suppose $\{x_n, n \geq 1\}$ and $\{t_n, n \geq 1\}$ are sequences such that if $n \geq 1$, then $x_n \in X$, $t_n \geq 0$, $t_n = \delta_n(x_n)$, $\lim_{n\to\infty} x_n = x$ (recall that X is compact) and $\lim_{n\to\infty} t_n = t$ (we allow $t = +\infty$).

Case I. $\text{Limit}_{n\to\infty}\, t_n = \infty$ and $t_n > n$ infinitely often, say on the sequence $\{t_{n_i}, i \geq 1\}$. Since $W_2(n_i, t_{n_i}) > 0$, and since (using the bounded convergence theorem)

$$(6.4)\qquad \lim_{i\to\infty} \int W_2(\omega, t_{n_i})f(x_{n_i}, \omega)\lambda(d\omega)$$
$$= \lim_{i\to\infty} \int W_2(\omega, t_{n_i})[f(x_{n_i}, \omega)/f(x_0, \omega)]f(x_0, \omega)\lambda(d\omega)$$
$$= \int W_2(\omega, \infty)f(x, \omega)\lambda(d\omega) > 0,$$

equation (6.1) is not solvable for large values of n_i.

Case II. $\text{Limit}_{n\to\infty}\, t_n = \infty$ and $t_n \leq n$ except for a finite number of values of n. Note that $0 \geq W_2(n, t_n) \geq W_2(n, 0)$. Our hypothesis was that $\lim_{n\to\infty} W_2(n, 0) = -1$. Therefore, $\liminf_{n\to\infty} W_2(n, t_n)f(x_n, \omega)\alpha_n \geq -1$. Since $\lim_{n\to\infty} \int W_2(\omega, t_n)f(x_n, \omega)\lambda(d\omega) = \int W_2(\omega, \infty)f(x, \omega)\lambda(d\omega)$, (6.1) will be unsolvable for large values of n if we assume

$$(6.5)\qquad \int W_2(\omega, \infty)f(x, \omega)\lambda(d\omega) > 1, \qquad x \in X.$$

Case III ($t < \infty$). Using the bounded convergence theorem

$$(6.6)\qquad 0 = \int W_2(\omega, t)f(x_n, \omega)\lambda(d\omega) + \lim_{n\to\infty} W_2(n, t_n)f(x_n, n)\alpha_n$$
$$= \int W_2(\omega, t)f(x, \omega)\lambda(d\omega) - 1.$$

This equation has a unique solution. Therefore $\delta(x) = \lim_{n\to\infty} \delta_n(x_n)$ as was to be shown.

We now prove that the risks converge. That is,

$$(6.7)\qquad 0 = \lim_{n\to\infty} \int [W(\omega, \delta(x)) - W(\omega, \delta_n(x))]f(x, \omega)\mu(dx)\lambda(d\omega)$$
$$+ \lim_{n\to\infty} \alpha_n \int [W(n, \delta(x)) - W(n, \delta_n(x))]f(x, n)\mu(d\omega).$$

The limit $\delta(x)$ solves (6.6). From this it follows at once that $\sup_x |\delta(x)| < \infty$. Since $\lim_{n\to\infty} \delta_n(x) = \delta(x)$ uniformly in x, there is a $K > 0$ such that $\sup_{x\in X, n\geq 1} |\delta_n(x)| \leq K$. Since $\sup_{\omega,t} |W_2(\omega, t)| < \infty$, it follows that

$$(6.8)\qquad \sup_{\omega,x} |W(\omega, \delta(x)) - W(\omega, \delta_n(x))| < \infty$$

and

$$\lim_{n\to\infty} |W(\omega, \delta(x)) - W(\omega, \delta_n(x))| = 0. \tag{6.9}$$

Therefore, using the bounded convergence theorem,

$$0 = \lim_{n\to\infty} \int [W(\omega, \delta(x)) - W(\omega, \delta_n(x))]f(x, \omega)\lambda(d\omega). \tag{6.10}$$

Since $\sup_{x,\omega} f(x, \omega)/f(x_0, \omega) < \infty$, and $\lim_{n\to\infty} \alpha_n f(x_0, n) < \infty$, it follows that $\sup_{n\geq 1, x\in X} \alpha_n f(x, n) < \infty$. Since $\lim_{n\to\infty} |W(n, \delta(x)) - W(n, \delta_n(x))| = 0$ uniformly in x, and since $\mu(X) < \infty$,

$$0 = \lim_{n\to\infty} \alpha_n \int [W(n, \delta(x)) - W(n, \delta_n(x))]f(x, n)\mu(dx) = 0. \tag{6.11}$$

That proves (6.7).

To prove that $\delta(\cdot)$ is admissible, suppose $\delta'(\cdot)$ is as good as $\delta(\cdot)$. Let $K(\omega, \delta)$, $K(\omega, \delta')$, and $K(\omega, \delta_n)$ be the risks of δ, δ', and δ_n evaluated at ω. Since δ_n is a Bayes procedure,

$$\begin{aligned} 0 \leq \int (K(\omega, \delta) - K(\omega, \delta'))\lambda_n(d\omega) &\leq \int (K(\omega, \delta) - K(\omega, \delta_n))\lambda(d\omega) \\ &+ \alpha_n \int (K(\omega, \delta) - K(\omega, \delta_n)). \end{aligned} \tag{6.12}$$

As $n \to \infty$, the right-hand side of (6.12) tends to zero. Since the loss function is strictly convex, using lemma 3.1 of section 3, the admissibility of δ follows.

APPENDIX. DECISION THEORY

Parts of this paper lean heavily on the interpretation of statistical decision procedures as continuous bilinear forms on certain pairs of Banach spaces. This interpretation is well known, but necessary details do not seem to be available anywhere. We will first discuss bilinear forms abstractly and then discuss statistical procedures.

Let F_1, F_2 be Banach spaces. Given are norms $\|x\|_1$ of $x \in F_1$ and $\|y\|_2$ of $y \in F_2$. A bilinear form $(\cdot, \cdot)$ on $F_1 \times F_2$ is a real-valued function of two variables such that to each $x \in F_1$, $(x, \cdot)$ is a linear functional on F_2, to each $y \in F_2$, $(\cdot, y)$ is a linear functional on F_1.

If a bilinear form is continuous in each variable, then using the uniform boundedness theorem one easily shows there is a constant K satisfying

$$K = \sup_{x\in F_1, y\in F_2} |(x, y)|/\|x\|_1\|y\|_2. \tag{A.1}$$

The constant K is the norm of the bilinear form. Conversely, if a bilinear form satisfies an inequality of the form

$$|(x, y)| \leq K\|x\|_1\|y\|_2 \tag{A.2}$$

for all $x \in F_1$, $y \in F_2$, then the bilinear form is a continuous linear functional in each variable.

The space F_{12} of continuous bilinear forms on $F_1 \times F_2$ is a Banach space under the norm defined above. The weak topology in F_{12} is the weakest topology such that for each pair $(x, y) \in F_1 \times F_2$, the mapping $(\cdot, \cdot) \to (x, y)$ defined on F_{12} to the real line is a continuous mapping. A standard argument of embedding F_{12} in a product space will show the unit ball of F_{12} is compact in the weak topology.

Generally, in most decision theory discussions, the assertion that a set of decision procedures is compact is the assertion that a set of bilinear forms is weakly compact. We develop this idea here.

In the context of this paper, $\mathfrak{D}$ is the set of decisions, and we assume that $\mathfrak{D}$ is a locally compact Hausdorff space and that $\mathfrak{D}^*$ is the one point compactification of $\mathfrak{D}$. We suppose $F_2 = C(\mathfrak{D}^*)$, the set of bounded continuous real-valued functions on $\mathfrak{D}^*$. We take $F_1 = L_1(X, \mathfrak{B}, \mu)$. If $\delta(\cdot, \cdot)$ is a statistical decision procedure, then for every Borel subset A of $\mathfrak{D}^*$, $\delta(\cdot, A)$ is a bounded measurable function, and, if $x \in X$, then $\delta(x, \cdot)$ is a probability measure on the Borel subsets of $\mathfrak{D}^*$. We now state and prove a converse to this.

THEOREM A1. *Suppose* $F_1 = L_1(X, \mathfrak{B}, \mu)$ *and* $(X, \mathfrak{B}, \mu)$ *is a totally* σ-*finite measure space. Suppose* $\mathfrak{D}^*$ *is a compact metric space. Let* $F_2 = C(\mathfrak{D}^*)$. *If* $(\cdot, \cdot)$ *is a continuous bilinear form on* $F_1 \times F_2$ *of norm* K, *then there exists* $\delta(\cdot, \cdot)$ *satisfying the following:*

(i) *to each* $x \in X$, $\delta(x, \cdot)$ *is a countably additive finite measure on the Borel subsets of* $\mathfrak{D}^*$;

(ii) *to each Borel set* $E \subset \mathfrak{D}^*$, $\delta(\cdot, E)$ *is a bounded* $\mathfrak{B}$-*measurable function;*

(iii) *if* $f(\cdot) \in L_1(X, \mathfrak{B}, \mu)$ *and* $g(\cdot) \in C(\mathfrak{D}^*)$, *then*

$$(f, g) = \iint f(x)g(t)\delta(x, dt)\mu(dx);$$

(iv) *for all* $x \in X$ *and Borel subsets* E *of* $\mathfrak{D}^*$, $|\delta(x, E)| \leq K$.

This theorem is known. An unpublished proof, different from the proof given below, has been given by Le Cam.

PROOF. The space $C(\mathfrak{D}^*)$ is a separable metric space. We take a countable dense subset $\{g_n^*(\cdot), n \geq 1\}$ of $C(\mathfrak{D}^*)$. By discarding some of these functions, we may find a subset $\{g_n(\cdot), n \geq 1\}$ which are linearly independent over the rational numbers, and such that every $g_n^*(\cdot)$ is a linear combination of functions in $\{g_n(\cdot), n \geq 1\}$, $n \geq 1$.

Since $(\cdot, g_n)$ is a continuous linear functional on $L_1(X, \mathfrak{B}, \mu)$, and since $(X, \mathfrak{B}, \mu)$ is a totally σ-finite measure space, we may find a bounded $\mathfrak{B}$-measurable function $\delta(\cdot, g_n)$ satisfying

$$(f, g_n) = \int f(x)\delta(x, g_n)\mu(dx) \tag{A.3}$$

for all $f \in L_1(X, \mathfrak{B}, \mu)$, and

$$\sup_x |\delta(x, g_n)| \leq K\|g_n\|_2. \tag{A.4}$$

Let $C_R(\mathfrak{D}^*)$ be the linear span of $\{g_n, n \geq 1\}$ over the rational numbers. I $g \in C_R(\mathfrak{D}^*)$, we represent $g(\cdot)$ uniquely as a finite sum

$$g(\cdot) = \sum r_i g_i(\cdot), \tag{A.5}$$

and define for all $x \in X$,

$$\delta(x, g) = \sum r_i \delta(x, g_i). \tag{A.6}$$

Then if $f \in L_1(X, \mathcal{B}, \mu)$,

$$\int f(x)\delta(x, g)\mu(dx) = \sum r_i(f, g_i) = (f, g) \leq K\|f\|_1\|g\|_2. \tag{A.7}$$

Since this holds for all $f \in L_1(X, \mathcal{B}, \mu)$,

$$\operatorname*{ess\,sup}_x |\delta(x, g)| \leq K\|g\|_2, \qquad \text{for all} \quad g \in C_R(\mathfrak{D}^*). \tag{A.8}$$

We may then find a set $N \in \mathcal{B}$, $\mu(N) = 0$, such that if $g \in C_R(\mathfrak{D}^*)$, then $\sup_{x \notin N} |\delta(x, g)| \leq K\|g\|_2$. This is possible since $C_R(\mathfrak{D}^*)$ is countable.

If $x \notin N$, then $\delta(x, \cdot)$ is a continuous linear functional defined on $C_R(\mathfrak{D}^*)$. It has a unique continuous extension to $C(\mathfrak{D}^*)$. So, if $x \notin N$, $g \in C(\mathfrak{D}^*)$, we write $\delta(x, g)$ for the extension and have $|\delta(x, g)| \leq K\|g\|_2$.

Now if $g \in C(\mathfrak{D}^*)$, we may find a subsequence $\{g_{n_i}, i \geq 1\}$ in $C_R(\mathfrak{D}^*)$ such that $\lim_{i\to\infty} \sup_{t\in\mathfrak{D}^*} |g_{n_i}(t) - g(t)| = 0$. Then if $x \notin N$, $\delta(x, g) = \lim_{i\to\infty} \delta(x, g_{n_i})$. Therefore, $\delta(\cdot, g)$ is a $\mathcal{B}$-measurable function.

By the Riesz representation theorem, the continuous linear functional $\delta(x, \cdot)$ is representable by a countably additive measure. For Borel sets E of $\mathfrak{D}^*$, we write $\delta(x, E)$ for the value of the measure. Then if $x \notin N$, for all Borel subsets E of $\mathfrak{D}^*$, $|\delta(x, E)| \leq K$.

We now show that for each Borel set E, $\delta(\cdot, E)$ is a $\mathcal{B}$-measurable function. Let $\mathcal{C}$ be the set of all Borel subsets of $\mathfrak{D}^*$ for which this is true. The set $\mathcal{C}$ is clearly a monotone class, and by considering monotone sequences of continuous functions, one easily shows $\mathcal{C}$ to contain all compact sets. Therefore, $\mathcal{C}$ contains all Borel sets. (This type of argument works even if the measures are signed measures.)

One may show at once, using the bounded convergence theorem, that in extending δ from $C_R(\mathfrak{D}^*)$ to $C(\mathfrak{D}^*)$, we have for all $f \in L_1(X, \mathcal{B}, \mu)$, $g \in C(\mathfrak{D}^*)$, $(f, g) = \int f(x)\delta(x, g)\mu(dx)$.

From the form of the Riesz representation,

$$(f, g) = \iint f(x)g(t)\delta(x, dt)\mu(dx). \tag{A.9}$$

The integral is absolutely convergent. That completes the proof.

A statistical decision procedure is a bilinear form satisfying (v) if $f \in L_1(X, \mathcal{B}, \mu)$, $f(x) \geq 0$ for all $x \in X$, $g \in C(\mathfrak{D}^*)$, $g(t) \geq 0$ for all $t \in \mathfrak{D}^*$, then $(f, g) \geq 0$; (vi) if $f \in L_1(X, \mathcal{B}, \mu)$, then $(f, 1) = \int f(x)\mu(dx)$. It is easily checked that the set of bilinear forms satisfying these conditions is a weakly closed subset of the unit ball of F_{12}.

COROLLARY. *Let* $(X, \mathcal{B}, \mu)$ *be totally* σ*-finite, let* $L_1(X, \mathcal{B}, \mu)$ *be a separable*

Banach space, and suppose $\mathfrak{D}^*$ is a compact metric space. Then the set of statistical decision procedures is sequentially compact.

In the usual statistical problem a set Ω of density functions is given, if $f \in \Omega$ then $f \in L_1(X, \mathcal{B}, \mu)$. The set of decisions $\mathfrak{D}$ need not be compact, but we suppose $\mathfrak{D}$ has a compactification $\mathfrak{D}^*$ containing $\mathfrak{D}$ as a Borel set. For each $f \in \Omega$, $t \in \mathfrak{D}$ we suppose a measure of loss $W(f, t) \geq 0$ is given. We assume $W(f, \cdot)$ has an extension to $\mathfrak{D}^*$ such that for each $f \in \Omega$, the extended function is lower semicontinuous.

THEOREM A2. *Suppose $\{\delta_n, n \geq 1\}$ is a sequence of statistical decision procedures. There exists a subsequence $\{\delta_{n_i}, i \geq 1\}$ and a procedure δ such that for all $f \in \Omega$,*

$$\int W(f, t)\delta(x, t)f(x)\mu(dx) \leq \liminf_{i\to\infty} \int W(f, t)\delta_{n_i}(x, dt)f(x)\mu(dx). \tag{A.10}$$

It may be that $\delta(x, \mathfrak{D}^* - \mathfrak{D}) > 0$ for some x.

PROOF. Extend δ_n to $\mathfrak{D}^*$ by $\delta_n(x, \mathfrak{D}^* - \mathfrak{D}) = 0$ for all $x \in X$, $n \geq 1$. Choose a subsequence such that $\delta = \text{weak}\lim_{i\to\infty} \delta_{n_i}$. Let $W_N(f, \cdot)$ be an increasing sequence of continuous functions on $\mathfrak{D}^*$ satisfying $\lim_{N\to\infty} W_N(f, t) = W(f, t)$ for all $f \in \Omega$, $t \in \mathfrak{D}^*$. Then

$$\begin{aligned}\liminf_{i\to\infty} \int W(f, t)\delta_{n_i}(x, dt)f(x)\mu(dx) &\geq \lim_{t\to\infty} \int W_N(f, t)\delta_{n_i}(x, dt)f(x)\mu(dx) \\ &= \int W_N(f, t)\delta(x, dt)f(x)\mu(dx).\end{aligned} \tag{A.11}$$

Let $N \to \infty$ and apply the monotone convergence theorem. The result follows.

In some applications $\mathfrak{D}$ is a finite dimensional space, and for each $f \in \Omega$, $W(f, \cdot)$ is a strictly convex function. If $\lim_{t\to\infty} W(f, t) = \infty$, then we obtain the following result.

THEOREM A3. *Let $\{\delta_n, n \geq 1\}$ be a sequence of decision procedures; let δ be an admissible procedure, and*

$$\limsup_{n\to\infty} \int W(f, t)\delta_n(x, dt)f(x)\mu(dx) \leq \int W(f, t)\delta(x, dt)f(x)\mu(dx) \tag{A.12}$$

for all $f \in \Omega$. Let δ^* be a weak limit point of $\{\delta_n, n \geq 1\}$. Then for all x, $\delta^*(x, \mathfrak{D}^* - \mathfrak{D}) = 0$, and if $A = \{x | \delta(x, \cdot) \neq \delta^*(x, \cdot)\}$, then $\int_A f(x)\mu(dx) = 0$ for all $f \in \Omega$.

PROOF. Let $\delta^* = \text{weak}\lim_{i\to\infty} \delta_{n_i}$ for some subsequence. By theorem A2,

$$\int W(f, t)\delta^*(x, dt)f(x)\mu(dx) \leq \int W(f, t)\delta(x, t)f(x)\mu(dx). \tag{A.13}$$

Since δ is admissible, δ must be nonrandomized; therefore, we write

$$\int W(f, t)\delta(x, t)f(x)\mu(dx) = \int W(f, \delta(x))f(x)\mu(dx). \tag{A.14}$$

Further, we set $\delta^*(x) = \int t\delta^*(x, dt)$ and obtain by Jensen's inequality that

$$\int W(f, \delta^*(x))f(x)\mu(dx) \leq \int W(f, \delta(x))f(x)\mu(dx). \tag{A.15}$$

Since $\delta(\cdot)$ is admissible, if $f \in \Omega$, this must be equality. Let

$$A_1 = \{x | \delta^*(x, \{\delta^*(x)\}) \neq 1\}. \tag{A.16}$$

Since W is a strictly convex function, $\int_{A_1} f(x)\mu(dx) = 0$ for all $f \in \Omega$. Again, since W is a strictly convex function, if $A = \{x | \delta(x) \neq \delta^*(x)\}$, then

$$\int_A f(x)\mu(dx) = 0 \tag{A.17}$$

for all $f \in \Omega$. That completes the proof.

We consider here the minimax theorem in the context needed for section 3. We suppose $W(\ , \cdot)$ is bounded on compact $\Omega \times \mathfrak{D}$ subsets, $W(\omega, \cdot)$ is lower semicontinuous for each $\omega \in \Omega$, and $W(\cdot, t)$ is continuous for each $t \in \mathfrak{D}$. We assume that $W(\omega, t) \geq 0$ for all $(\omega, t) \in \Omega \times \mathfrak{D}$, and if $C \subset \Omega$ is a compact set, then

$$\lim_{t \to \infty} \inf_{\omega \in C} W(\omega, t) = \infty. \tag{A.18}$$

We suppose that we are given a family $\{f(\cdot, \omega), \omega \in \Omega\}$ of density functions relative to the σ-finite measure space $(X, \mathfrak{B}, \mu)$, and to each $x \in X$, $f(x, \cdot)$ is a continuous function on Ω.

THEOREM A4. *Let $r(\cdot)$ be a nonnegative lower semicontinuous function on Ω, and C a compact parameter set. Assume $\sup_{\omega \in C} r(\omega) < \infty$. Relative to the measure of loss $W(\omega, t) - r(\omega)$ there exists a minimax procedure δ (ω is restricted to C) which is Bayes relative to λ supported on C and*

$$\text{minimax risk} = \iint (W(\omega, t) - r(\omega))\delta(x, dt)f(x, \omega)\mu(dx)\lambda(d\omega). \tag{A.19}$$

PROOF. If δ is an admissible procedure for $\omega \in C$, then the values of δ lie in a compact subset of $\mathfrak{D}$. Indeed, take $a_0 \in \mathfrak{D}$. By hypothesis we can find a compact subset E of $\mathfrak{D}$ such that $a_0 \in E$, and if $a \notin E$, then $\sup_{\omega \in C} W(\omega, a_0) < \inf_{\omega \in C} W(\omega, a)$. Therefore, one always does better to decide a_0 than a.

Consequently, we may suppose there is a constant K such that for all $\omega \in C$, all $x \in X$, and all t in the support of $\delta(x, \cdot)$, $W(\omega, t) \leq K$. It follows that

$$K(\omega, \delta) = \iint W(\omega, t)\delta(x, dt)f(x, \omega)\mu(dx) \tag{A.20}$$

is continuous in ω and that $K(\cdot, \delta) - \delta(\cdot)$ is upper semicontinuous.

We let R_1 be the set of real-valued upper semicontinuous functions on C such that if $g \in R_1$, then for some δ, $g(\omega) = K(\omega, \delta) - r(\omega)$ for all $\omega \in C$. Then R_1 is a convex set of functions. We let R_2 be the set of real-valued continuous functions such that if $g_2 \in R_2$, there is a $g_1 \in R_1$ such that for all $\omega \in C$, $g_1(\omega) \leq g_2(\omega)$. Then R_2 is a convex set of continuous functions, and each $g_1 \in R_1$ is the limit of a monotone decreasing sequence of functions in R_2.

We apply the now classical construction. Let $R(\epsilon)$ be the set of all real-valued continuous functions on C such that if $g \in R(\epsilon)$, then $\sup_{\omega \in C} g(\omega) < \epsilon$. Then for each ϵ, $R(\epsilon)$ is a convex subset of the continuous functions on C, and $R(\epsilon)$ has an interior point in the sup topology. Further, if $\sup_{\omega \in C} r(\omega) < \infty$, there exist ϵ such that $R(\epsilon)$ and R_2 are disjoint. We consider the functions in R_2 as restricted to C. Since $R(\epsilon_0) = \bigcup_{\epsilon<\epsilon_0} R(\epsilon)$, there is a largest ϵ such that $R(\epsilon)$ and R_2 are disjoint.

The convex sets $R(\epsilon)$ and R_2 may be separated by a hyperplane. By the Riesz representation theorem there is a finite countably additive measure ξ on the Borel subsets of C and a number α such that if $g \in R(\epsilon)$, then $\int g(\omega)\xi(d\omega) \leq \alpha$; if $g \in R_2$, then $\int g(\omega)\xi(d\omega) \geq \alpha$. For integer $N \geq 1$, if $C' \subset C$ is a compact subset of C, we may approximate $-N\chi_{C'}$ by monotone limits of functions in $|\epsilon| + R(\epsilon)$. This implies $-N\xi(C') \leq \alpha + |\epsilon|\xi(\Omega)$. Let $N \to \infty$ and obtain $\xi(C') \geq 0$ for every compact subset of C. Since $\xi \neq 0$, we may suppose ξ is normalized to be a probability measure.

Since functions in R_1 may be approximated from above by functions in R_2, it follows that the hyperplane determined by ξ, α separates R_1 and $R(\epsilon)$. Further if $\beta > 0$ and $(\epsilon + \beta)1$ is the constant function of value $\epsilon + \beta$, then R_1 contains a function g satisfying $g(\omega) \leq \epsilon + \beta$ for $\omega \in C$.

Since R_1 has the weak compactness property of theorem A2, R_1 contains the risk function of a minimax procedure satisfying

$$\text{(A.21)} \qquad \sup_{\omega \in C} g(\omega) \leq \epsilon; \qquad \int g(\omega)\xi(d\omega) = \epsilon.$$

Since every procedure δ which is Bayes with respect to ξ gives rise to a risk function in R_1, and since every such procedure must therefore have Bayes risk $\geq \epsilon$ relative to ξ, it follows that g is Bayes relative to ξ and the class of *all* procedures δ.

That completes the proof.

REFERENCES

[1] C. R. Blyth, "On minimax procedures and their admissibility," *Ann. Math. Statist.*, Vol. 22 (1951), pp. 22–42.
[2] F. Hausdorff, *Mengenlehre*, New York, Dover, 1944 (3d rev. ed.).
[3] S. Karlin, "Admissibility for estimation with quadratic loss," *Ann. Math. Statist.*, Vol. 29 (1958), pp. 406–463.
[4] J. Sacks, "Generalized Bayes solutions in estimation problems," *Ann. Math. Statist.*, Vol. 34 (1963), pp. 751–768.
[5] C. Stein, "A necessary and sufficient condition for admissibility," *Ann. Math. Statist.*, Vol. 26 (1955), pp. 518–522.
[6] ———, "The admissibility of Pitman's estimator of a single location parameter," *Ann. Math. Statist.*, Vol. 30 (1959), pp. 970–979.
[7] A. Wald, *Statistical Decision Functions*, New York, Wiley, 1950.

OPTIMUM MULTIVARIATE DESIGNS

R. H. FARRELL,[1] J. KIEFER,[1] and A. WALBRAN
CORNELL UNIVERSITY

1. Introduction

1.1. *Notation and preliminaries.* This paper is concerned with the computation of optimum designs in certain multivariate polynomial regression settings.

Let $f = (f_1, \cdots, f_k)$ be a vector of k real-valued continuous linearly independent functions on a compact set X. We shall work in the realm of the approximate theory discussed in many of the references, wherein a design is a probability measure ξ (which can be taken to be discrete) on X. The information matrix $M(\xi)$ of the design ξ for problems where the regression function is $\sum_1^k \theta_i f_i(x)$ (with $\theta = (\theta_1, \cdots, \theta_k)$ unknown and with uncorrelated homoscedastic observations and quadratic loss considerations of best linear unbiased estimators) has elements $m_{ij}(\xi) = \int f_i f_j \, d\xi$. Thus, $\det M^{-1}(\xi)$ is proportional to the generalized variance of the best linear estimators of all θ_i. We denote by Γ the space of all $M(\xi)$. We shall have occasion to consider the set of all *distinct* functions of the form $f_i f_j$, $i \geq j$, and shall write them as $\{\phi_t, 1 \leq t \leq p\}$. We then write $\mu_t(\xi) = \int \phi_t \, d\xi$. Whether or not some ϕ_t is a nonzero constant (as it is in our polynomial examples), we define $\phi_0(x) \equiv 1$ and $\mu_0 = 1$.

The main results of this paper characterize, for certain X and f, some designs ξ^* which are D-optimum; that is, for which

$$\det M(\xi^*) = \max_{\xi} \det M(\xi). \tag{1.1}$$

Define, for $M(\xi)$ nonsingular,

$$\begin{aligned} d(x, \xi) &= f(x) M^{-1}(\xi) f(x)', \\ \bar{d}(\xi) &= \max_{x \in X} d(x, \xi). \end{aligned} \tag{1.2}$$

The quantity $d(x, \xi)$ is proportional to the variance of the best linear estimator of the regression $f(x)\theta'$ at x. A result of Kiefer and Wolfowitz [8] is that ξ^* satisfies (1.1) if and only if it satisfies the G-(global-) optimality criterion

$$\bar{d}(\xi^*) = \min_{\xi} \bar{d}(\xi), \tag{1.3}$$

and that (1.1) and (1.3) are satisfied if and only if

$$\bar{d}(\xi^*) = k. \tag{1.4}$$

If the support of an optimum design is exactly k points, then ξ is uniform on those points. Our main way of finding D- and G-optimum (hereafter simply

[1] Research supported by ONR Contract Nonr-401(03).

called "optimum") designs and of verifying their optimality is thus to guess a ξ^* (perhaps by minimizing det $M(\xi)$ over some subset of designs depending on only a few parameters) and then to verify (1.4). We also record here the fact that all optimum ξ^* have the same $M(\xi^*)$, and that they all satisfy

$$\xi^*(\{x : d(x, \xi^*) = k\}) = 1. \tag{1.5}$$

It is often the case that there is a compact group $G = \{g\}$ of transformations on X, with associated transformations $\{\bar{g}\}$ on $\{\xi\}$, and such that $d(gx, \xi) = d(x, \bar{g}\xi)$. In such a case (see Kiefer [6]), there is G-invariant optimum design ξ^* (that is, such that $\xi^*(gA) = \xi^*(A)$ for all g and A), and the function $d(\cdot, \xi^*)$ and set of (1.5) are G-invariant.

Whereas some of our discussion refers to general X and f, our detailed examples of sections 2, 3, and 4 all treat problems of *polynomial regression in q variables, of degree $\leq m$.* Here X is a compact q-dimensional Euclidean set whose points we usually denote by $x = (x_1, \cdots, x_q)$, and the $f_i(x)$ are of the form $\Pi_{j=1}^{q} x_j^{m_j}$ where the m_j are nonnegative integers with sum $\leq m$. It is well known in this case that

$$k = \binom{m+q}{q}. \tag{1.6}$$

Moreover, since the $f_i f_j$ are all the monomials of degree $\leq 2m$, we have

$$p = \binom{2m+q}{q}. \tag{1.7}$$

The three examples we shall treat in detail are (in section 4) the unit q-ball

$$\left\{x : \sum_1^q x_i^2 \leq 1\right\}; \tag{1.8}$$

(in section 3) the q-cube

$$\{x : \max_{1 \leq i \leq q} |x_i| \leq 1\}; \tag{1.9}$$

and (in section 2) the unit simplex, which it is more convenient to represent in barycentric coordinates $x = (x_0, x_1, \cdots, x_q)$ as

$$\left\{x : \min_{0 \leq i \leq q} x_i \geq 0, \sum_0^q x_i = 1\right\}. \tag{1.10}$$

These are perhaps the three generalizations which are simplest, most natural, and of greatest practical importance, of the unit interval ($q = 1$), which is discussed in section 2. Unfortunately, the simple structure which is present when $q = 1$ and which is reflected in the elegant results of Guest [3] and Hoel [4] does not carry over to $q > 1$, and the results depend strongly on the shape of X; even in the case of the simplex where at least some analogous results seem to hold, they cannot be obtained by the same methods. We now indicate how this is reflected in the geometry of Γ.

1.2. *The geometry of* Γ. The set Γ can clearly be regarded as a convex body in p-dimensional Euclidean space with coordinates μ_t, $1 \leq t \leq p$; of course,

$p \leq k(k+1)/2$. Write $a = (a_0, a_1, \cdots, a_p)$. Let $\sum_1^p a_t\mu_t + a_0 = 0$ be a supporting hyperplane of Γ with $\sum_1^p a_t\mu_t + a_0 \geq 0$ in Γ. Clearly, the supporting polynomial $T(x; a) = \sum_1^p a_t\phi_t(x) + a_0$ is nonnegative on X.

(For future reference, the reader should note in connection with the previous and next paragraphs that, if ξ^* is optimum and $\gamma^* = M(\xi^*)$, then ξ^* is admissible and hence γ^* is a boundary point, and $k - d(x, \xi^*)$ supports Γ at γ^*.)

Let $\gamma_0 = M(\xi_0)$ be a boundary point of Γ. A supporting polynomial $T(\cdot\,; a^0)$ which supports Γ at γ_0 is then ≥ 0 on X and is 0 on the support of ξ_0. Thus, an analysis of what the set of zeros of a T of the above form can be can yield information about the boundary points of Γ (while the extreme points are clearly a subset of the points corresponding to ξ's with one-point support). For example, in the well-known univariate polynomial case $X = [0, 1]$, $k = m + 1$, $f_i(x) = x^{i-1}$, any such T is a nonnegative polynomial on X of degree $\leq 2m$, which (if not identically zero) therefore has at most $m + 1$ zeros, at most m of which are in the interior of X. In this example, moreover, if $\gamma = M(\xi)$ is an arbitrary point of Γ and $\xi^{(0)}(0) = 1$, the line from the boundary point $M(\xi^{(0)})$ through γ passes through another boundary point $\gamma' = M(\xi')$, so that $\gamma = M(\lambda\xi^{(0)} + (1 - \lambda)\xi')$ with $0 \leq \lambda \leq 1$; thus one concludes that any point of Γ can be represented as $M(\xi'')$ for a ξ'' supported by at most $m + 1$ points. One can also characterize the admissible ξ easily in this example as the boundary points with at most $m - 1$ points of support in the interior of X (Kiefer [5]).

Unfortunately, the examples studied in the present paper (as well as non-polynomial, and especially non-Chebyshev systems in one dimension) do not yield such simple analyses. This is clear when one considers the more complex sets on which a T can now vanish. For example, in the case of linear regression ($m = 1$) on the square (1.9) with $q = 2$, any supporting T which is not identically zero, being quadratic, vanishes either on a subset of the corners of X, or at a single point of X, or on a line of X. In the latter case we invoke the one-dimensional result to conclude that at most two points are needed to support a ξ yielding this $M(\xi)$; thus, every boundary point of Γ is obtainable from a ξ supported either by a subset of the corners or else by at most two other points. Replacing $\xi^{(0)}$ in the argument of the previous paragraph by the measure which assigns all probability to the point $(-1, -1)$, we conclude that every point of Γ can be obtained from a ξ which is supported either by a subset of the corners or else by at most three points of X. If we replace (1.9) by (1.10) with $q = 2$, we obtain that at most 3 points rather than 4 are needed. The admissible points can be characterized similarly, but it is clear that the difficulty of obtaining such characterizations will be much greater for larger q and m. As for the optimum design, it is the uniform distribution on the 3 corners in the case (1.10), on the 4 corners in the case (1.9), and, for another example, on the 5 corners if X is a symmetric pentagon. The uniqueness in all three cases can be proved by the method given in the next paragraph, but in other cases, such as (1.8), there is no uniqueness. Section 3.3 of [6] characterizes optimum designs for linear regression on general compact X in q dimensions.

The increased complexity in higher dimensions is also present in the uniqueness question: given $\gamma = M(\xi^*)$, when is there no other ξ with $M(\xi) = \gamma$? This can sometimes be answered as follows. Suppose γ is a boundary point and that a supporting polynomial T at γ has exactly L zeros $x^{(1)}, x^{(2)}, \cdots, x^{(L)}$ on X. Any ξ with $M(\xi) = \gamma$ must be supported by a subset of $\{x^{(1)}, \cdots, x^{(L)}\}$, and must satisfy $\sum_j \phi_t(x^{(j)})[\xi^*(x^{(j)}) - \xi(x^{(j)})] = 0$ for $0 \leq t \leq p$. Hence, if rank $\{\phi_t(x^{(j)}), 0 \leq t \leq p, 1 \leq j \leq L\} = L$, then ξ^* is the unique design yielding γ. In the univariate polynomial example of the second paragraph above, each boundary point γ can be proved by this device to be yielded by a unique ξ^*.

The prescription outlined just below (1.4) for verifying optimality, and which has worked well when $q = 1$ or $m \leq 2$, is difficult to apply in other cases. This is because $k - d(x, \xi^*)$ can no longer be written as a sum of a small number of obviously nonnegative simple polynomials, but may instead require a large number of rational functions for such a representation. The decision procedures (Tarski, Henkin, and others) for representing or verifying nonnegativity of such polynomials are unwieldy to implement in these problems. The example of the simplex (1.10) with $q = 2$, $m = 3$, treated in section 2 by direct analysis, illustrates the increased complexity. In other cases we have been unable to obtain analytical verifications of optimality and have used machine methods to obtain results which are satisfactory from a practical point of view but which, theoretically, only yield statements of results which hold to within a certain accuracy, rather than complete proofs of the exact results.

We end this subsection with a simple observation which is often useful in optimum design theory for polynomial regression on a q-dimensional set X. If B is a subset of X such that for some $q \times q$ orthogonal matrix A and some scalar b with $|b| > 1$, the set $bAB = \{x: b^{-1}A^{-1}x \in B\}$ is also a subset of X, then no design ξ supported by B can be optimum. This follows at once upon defining ξ' by $\xi'(C) = \xi(b^{-1}A^{-1}C)$ for $C \subset X$ and computing $\det M(\xi') = b^k \det M(\xi)$. In particular, if X is such that $x \in X \Rightarrow ax \in X$ for $0 \leq a < 1$, then the support of any optimum design must contain at least one point of the boundary of X. The considerations of this paragraph can be modified in an obvious way for admissibility questions.

1.3. *Number of points needed for an optimum design.* An aspect of the geometry of Γ which is of particular practical importance is the minimum number N of points such that there is an optimum design supported by N points. (It will be clear how to modify much of the discussion which follows to treat this question for points of Γ other than those corresponding to D-optimum designs, but for brevity we will treat only the latter.) An optimum design will be called *minimal* if no proper subset of its support is the support of an optimum design. We shall see that this property is broader than that of being an optimum design on N points; the latter will be called *absolutely minimal.*

Clearly $N \geq k$. On the other hand, if there is a matrix B of rank b such that $\sum_{j=1}^{p} b_{ij}\phi_j(x)$ is a constant function of x for each i, then Γ has dimension $\leq p - b$.

Since the extreme points of Γ can be obtained from ξ's with one-point support, we obtain the trivial bounds

$$k \le N \le \min (p - b + 1, k(k + 1)/2), \tag{1.11}$$

where the well-known bound $k(k + 1)/2$, which is relevant only when $p = k(k + 1)/2$ and $b = 0$, is a consequence of the fact that optimum designs correspond to certain boundary points of Γ. In the polynomial case we have $b = 1$ (since 1 is a ϕ_t) and thus, from (1.6) and (1.7),

$$\binom{m + q}{q} \le N \le \binom{2m + q}{q}. \tag{1.12}$$

Of greater use is the upper bound one can obtain once one knows some optimum design ξ^*. Let $V = \{x: d(x, \xi^*) = k\}$ (see (1.5)), and let W be the support of ξ^*. We denote the number of points in these sets by v and w. (When v or w is infinite, as in the example of the ball (1.8) for $q \ge 2$ in section 4, it is easy to see that (1.14) below still holds, but we shall usually treat the finite case.) Let $U = V$ or W (and $u = v$ or w). The $p + 1$ linear equations

$$\sum_{x \in U} \phi_t(x)\xi(x) = \mu_t(\xi^*), \qquad 0 \le t \le p \tag{1.13}$$

in the unknowns $\xi(x)$ are consistent (since $\{\xi^*(x)\}$ is a solution), so that the dimensionality of the linear set H (say) of solutions of (1.13) is $u - h$ where $h = \text{rank } \{\phi_t(x), 0 \le t \le p, x \in U\}$.

Considering H as a set in the u-dimensional space with coordinates $\xi(x)$, $x \in U$, we know that ξ^*, with all coordinates nonnegative, is in H, and conclude easily that H contains a point with all coordinates nonnegative and with at least h zero coordinates. Hence,

$$N \le \text{rank } \{\phi_t(x), 0 \le t \le p, x \in U\}. \tag{1.14}$$

(Of course, if U is replaced by X, this becomes the $p - b + 1$ of (1.11).) We will illustrate the use of this in the polynomial case (where $1 \in \{\phi_t\}$ so that the domain of t in (1.14) can be taken to be $1 \le t \le p$) in section 3.3, in the case of quadratic regression on the q-cube (1.9). We can think of such polynomial applications in terms of finding a matrix C of rank c such that $\sum_{j=1}^{p} c_{ij}\phi_j(x) \equiv 0$ on W. We can then conclude that, if N_W is the minimum number of points in W supporting an optimum design, then (paralleling (1.11))

$$N_W \le p - c. \tag{1.15}$$

On the other hand, it is obvious from (1.13) that, for $U = V$ or W,

$$N_U = u \quad \text{if rank} \quad \{\phi_t(x), 0 \le t \le p, x \in U\} = u. \tag{1.16}$$

It is easy to give examples which illustrate the fact that we can have $N_W > N$; that is, that minimality and absolute minimality do not coincide. For example, in the case $m = 1$, $q = 2$ of linear regression on the unit disc (1.8), the discussion of the fourth paragraph of section 1.2 shows that the set of j equally spaced points on the boundary is minimal if $j = 3$, 4, or 5, but is absolutely minimal only when $j = 3$.

Also, the designs whose optimality is easiest to verify are often ones which are symmetric, that is, invariant in the sense described below (1.5), and there is no reason why minimal designs should be of this form. For example, in the case of the q-cube (1.9) with $m = 1$, the uniform distribution on the 2^q corners is the only optimum design invariant under the symmetries of the q-cube, but if q is such that there exists a $(q + 1) \times (q + 1)$ Hadamard matrix (for example, if $q + 1$ is a power of 2), then it is well known that there is an optimum design on $k = q + 1$ corners (namely, the corners of an inscribed regular q-simplex). This example illustrates another technique for reducing an upper bound on N or N_W; in sections 3 and 4 we shall see how the use of various known results on orthogonal arrays and rotatable configurations can be used similarly.

The search for absolutely minimal designs can be described as a programming problem, of finding a nonnegative solution of (1.13) with $U = V$, which has a minimum number of nonzero elements. Analytical or machine methods for solving this problem would seem important.

2. The simplex

We have mentioned in section 1 that this case (1.10) evidences the most regular mathematical behavior among q-dimensional sets X. In the linear and quadratic cases it has been known for some time that the simplex exhibits a behavior (described precisely, below) very much like that present when $q = 1$. This phenomenon appears to carry over to cubic and perhaps higher degree regression, although we have as yet proved only one small fragment of the conjectured general result, and have machine computations in only two other cases. To describe these results, let E_m be the set of $m + 1$ points supporting the Guest-Hoel design when $q = 1$. (Thus, $E_1 = \{x_0 = 0, 1\}$; $E_2 = \{x_0 = 0, \frac{1}{2}, 1\}$; $E_3 = \{x_0 = 0, (1 \pm 5^{-1/2})/2, 1\}$; $E_4 = \{x_0 = 0, \frac{1}{2}, 1, (1 \pm (\frac{3}{7})^{1/2})/2\}$; and so on.)

The results in the linear and quadratic cases for general dimension q (Kiefer [7]) can be summarized by stating that, for degrees $m = 1$ and 2, the unique optimum design assigns equal measure to each of the points which is in the E_m of some edge of the q-simplex (when that edge is considered as a 1-simplex). We cannot hope for this pattern for $m > 2$, since the E_m points on all edges will be fewer in number than the k of (1.6). However, one can still conjecture that one or all of the following are true: (1) there is an optimum design whose support includes the E_m points on all edges (and no other points on edges); (2) there is an optimum design which assigns equal measure to each of k points; (3) the optimum design is unique; (4) generalizing the vertex- and edge-stationarity of (1), for fixed m there are optimum designs for dimension q which have the same support on the r-dimensional faces of X for $q \geq r$; (5) the design of (4) has points of support only on faces of dimension $\leq \min(m - 1, q)$.

What we have succeeded in treating analytically is the case $m = 3$, $q = 2$, and the details will be found at the end of this section. We have also observed, by machine search, that (a) the optimum design in the case $m = 3$, $q = 3$ appears to give equal weights to the E_3-points on edges (including vertices) and the midpoints of 2-dimensional faces, just as in the cases $m = 3$, $q = 1$ or 2; (b) the optimum design in the case $m = 4$, $q = 2$ appears to give equal weights to the E_4-points on edges and to the three points of the form $\{x_h = 0.567, x_i = x_j\}$. One can also prove analytically for $m = 3$ and general k that, among all designs which assign equal weights to the vertices, midpoints of 2-dimensional faces, and points on edges satisfying $x_i = b = 1 - x_j$, the choice $b = (1 \pm 5^{-1/2})/2$ minimizes the generalized variance for each k. (The generalized variance for such designs for $q \geq 2$ is proportional to $[v(1 - 2b)^{1/2}]^{-2q(q+1)}$, where $v = b(1 - b)$.) All of the above results conform with the conjectures of the previous paragraph.

If any of the general conjectures are true, they would constitute a deep new result in the area of multidimensional moment and approximation theory. Evidently a new approach is needed, perhaps even to verify analytically (b) and (c) of the previous paragraph. The technique employed for low dimensions and/or degrees, for example, by Kiefer [7] and Uranisi [11], has been that described at the end of section 1.2, and the difficulties encountered for larger q or m are as described there. Even in the case $m = 3$, $q = 2$ which we now consider, a much more brutal approach is used, and it does not suffice when $q = 3$.

THEOREM 2.1. *For $m = 3$, $q = 2$, the unique optimum design ξ^* assigns measure $\frac{1}{10}$ to each of the three vertices, the point $x_0 = x_1 = x_2 = \frac{1}{3}$, and the six points $\{x_h = 0, x_i = 1 - x_j = (1 + 5^{-1/2})/2.\}$.*

PROOF. We shall show that $10 - d(x, \xi^*) \geq 0$ on X, with equality only at the ten points supporting ξ^*. Since the function d is the same for all optimum designs, any optimum design must have this same support, and the weights are unique since there are 10 points and 10 functions. This yields uniqueness.

It is convenient to consider, in place of the coordinates x_i, the coordinates $\beta, t (-\frac{1}{2} \leq \beta \leq 1, 0 \leq t \leq 1)$ satisfying $3x_0 = 1 - t, 3x_1 = 1 - \beta t, 3x_2 = 1 + (\beta + 1)t$ on the portion $0 \leq x_0 \leq x_1 \leq x_2$ of X which, because of the symmetry of ξ^*, is all we need consider. For fixed β, variation of t from 0 to 1 yields a segment from center to edge of X. Write $L = \beta^2 + \beta + 1$ (so that $\frac{3}{4} \leq L \leq 3$). A simple computation yields

$$9 \sum_{i<j} x_i x_j = 3 - Lt^2,$$

$$81 \sum_{i<j} x_i^2 x_j^2 = 3 - 6(L - 1)t^3 + L^2 t^4,$$

$$729 \sum_{i<j} x_i^3 x_j^3 = 3 + 3Lt^2 - 21(L - 1)t^3 + 3L^2 t^4 + 3L(L - 1)t^5 + (-L^3 + 3L^2 - 6L + 3)t^6, \tag{2.1}$$

$$27 \prod_i x_i = 1 - Lt^2 + (L - 1)t^3.$$

A straightforward computation of $M(\xi^*)$ and $d(x, \xi^*)$ (for example, in terms of the functions x_i, $x_i x_j$, $x_i x_j (x_i - x_j)$, and $\prod_i x_i$, with $i < j$) yields

$$(2.2)\qquad 1 - d(x, \xi^*)/10 = 12 \sum_{i<j} x_i x_j - 120 \sum_{i<j} x_i^2 x_j^2 + 300 \sum_{i<j} x_i^3 x_j^3 + \prod_i x_i \Big[-102 + 410 \sum_{i<j} x_i x_j - 1512 \prod_i x_i\Big].$$

From (2.1) and (2.2) we have, writing $g(t, L) = 729[1 - d(x, \xi^*)/10]/6t^2$,

$$(2.3)\qquad g(t, L) = 131L - 318(L-1)t - 77L^2t^2 + 449L(L-1)t^3 - [50L^3 + 102(L-1)^2]t^4.$$

We must show $g(t, L) \geq 0$ for $0 < t \leq 1$, $\frac{3}{4} \leq L \leq 3$. We note that $g(1, L) = 2(3 - L)(5L - 6)^2$, so that the zeros of g on the boundary of X are precisely the vertex $(L = 3)$ and E_3-point $L = \frac{6}{5}$.

Writing $f(t, L) = [g(t, L) - g(1, L)]/(1 - t)$, we obtain

$$\begin{aligned}(2.4)\qquad f(t, L) &= [50L^3 - 270L^2 + 563L - 216] \\ &\quad + [50L^3 - 270L^2 + 245L + 102]t \\ &\quad + [50L^3 - 347L^2 + 245L + 102]t^2 + [50L^3 + 102(L-1)^2]t^3 \\ &= D(L) + C(L)t + B(L)t^2 + A(L)t^3 \quad \text{(say)}.\end{aligned}$$

We shall show that $f > 0$ for $0 \leq t \leq 1$, $\frac{3}{4} \leq L \leq 3$, and this will complete the proof.

We have $D(\frac{3}{4}) > 0$ and $D'(L) = 150L^2 - 540L + 563 > 0$. Thus,

$$(2.5)\qquad A(L) > 0, \qquad D(L) > 0, \qquad \tfrac{3}{4} \leq L \leq 3.$$

Also, one sees easily that, for $\frac{3}{4} \leq L \leq 1$, we have $B(L) \geq 50L^3$ and $C(L) \geq 0$. We conclude that $f(t, L) > 0$ for $0 \leq t \leq 1$, $\frac{3}{4} \leq L \leq 1$. We divide the region $1 \leq L \leq 3$ into two parts, the division λ being the zero of $C(L)$ in $1 \leq L \leq 3$ $(1.6 < \lambda < 1.7)$.

For $\lambda \leq L \leq 3$, we have $C(L) \leq 0$. We shall show the positivity of something $\leq f$, namely,

$$\begin{aligned}(2.6)\qquad f(t, L) + C(L)(1-t)^2(1+t) &= [100L^3 - 540L^2 + 808L - 114] \\ &\quad - 77L^2t^2 + [100L^3 - 168L^2 + 41L + 204]t^3 \\ &= E(L) - 77L^2t^2 + F(L)t^3 \text{ (say)}.\end{aligned}$$

We first note that $-23L^2 + 132L - 114$ is positive at $L = 1.6$ and $L = 3$ and, hence, for $\lambda \leq L \leq 3$. Therefore,

$$\begin{aligned}(2.7)\qquad 0 &< 100L(L - 2.6)^2 - 23L^2 + 132L - 114 = E(L) - 3L^2 \\ &< E(L) - \frac{4(77)^3}{27(84)^2}L^2 = E(L) + L^2 \min_{0 \leq t \leq 1} t^2(84t - 77) \\ &\leq E(L) - 77L^2t^2 + 84L^2t^3.\end{aligned}$$

Thus, the expression (2.6) will be proved positive if we show that $0 < F(L) - 84L^2 = h(L)$ (say). But an easy computation shows that $h(1.6) > 0$, $h'(1.6) > 0$, and $h''(L) > 0$ for $L \geq 1.6$.

In the region $1 \leq L \leq \lambda$, we have $C(L) \geq 0$, and thus need only show that

$D(L) + B(L)t^2 + A(L)t^3 > 0$. Because of (2.5), this is immediate if $B(L) \geq 0$, so we need only consider the possibility $B(L) < 0$, in which case

$$(2.8) \quad \begin{aligned} &D(L) + B(L)t^2 + A(L)t^3 \\ &\geq [D(L) + B(L) \max_{0 \leq t \leq 1} (t^2 - t^3/2)] + t^3[A(L) + B(L)/2] \\ &= [D(L) + B(L)/2] + t^3[A(L) + B(L)/2] = R(L) + S(L)t^3 \text{ (say).} \end{aligned}$$

One sees easily that $R(L) > 0$ for $1 \leq L \leq 2$ and $S(L) \geq 0$ for $L \geq 1$, completing the proof.

3. Symmetric regions; the cube

The case of the q-cube (1.9) exhibits less regularity than either the simplex or ball. This is seen even in the linear case described in section 1.3, where more than k points of support may be required (as when $q = 2$ and the unique optimum design is uniform on the 4 corners); and, even more, in the quadratic case, where optimum designs can be written down explicitly almost immediately for the ball and simplex, but require at least some consideration for the cube, regarding weights assigned to the points of the 3^q array J of points with coordinates 0, 1, -1. We shall now study this quadratic case in considerable detail. We begin by characterizing some properties of optimum quadratic designs for more general symmetric regions. (For general linear regression see Kiefer [6].)

3.1. *Quadratic regression on symmetric regions.* We introduce some of the ideas by considering, in the present paragraph, the q-cube. The fact that, when $m = 2$, the support of every optimum ξ^* is a subset of the 3^q array J, is easily seen as follows: $d(x, \xi^*)$ for any optimum ξ^* is symmetric under the group of symmetries of the cube (see discussion just below (1.5)), goes to $+\infty$ with $|x|$, and is a positive quartic on Euclidean q-space. Writing B for the subset of X where $d(x, \xi^*) = k$ (so that the support of ξ^* is contained in B), we will show that the existence of points in $B - J$ leads to a contradiction. Calling vertices, edges, and so on, the 0-, 1-, $\cdots$, skeleton of X, suppose that x' in $B - J$ lies in the r-skeleton, and hence in the relative interior of some r-cube G of that skeleton. By symmetry of B, there is another point x'' of $B - J$ which is also in the relative interior of G. The function d attains its maximum on G at x' and x'', and hence cannot be a positive quartic on q-space unless it is a constant, in which case it does not go to $+\infty$ with $|x|$.

We turn now to more general symmetric regions X to which we can apply some similar arguments.

We consider quadratic regression in q variables $x_1, \cdots, x_q$ on a symmetric region X of Euclidean q-space. The meaning of saying X is symmetric is that X is invariant under permutations $(x_1, \cdots, x_q) \rightarrow (x_{\sigma_1}, \cdots, x_{\sigma_q})$ and is invariant under sign changes $(x_1, \cdots, x_q) \rightarrow (\epsilon_1 x_1, \cdots, \epsilon_q x_q)$, $\epsilon_1 = \pm 1, \cdots, \epsilon_q = \pm 1$. The discussion just below (1.5) states that there are optimum designs which are symmetric; also, it implies that the function $d(\cdot, \xi^*)$ for any optimum design ξ^*

is a symmetric polynomial in the variables $x_1^2, \cdots, x_q^2$, of degree 2 in these variables. If we write $s = x_1^2 + \cdots + x_q^2$ and $t = x_1^4 + \cdots + x_q^4$, then the general polynomial of this type is

$$P(s, t) = as^2 + bs + c + dt. \tag{3.1}$$

The map $h: (x_1, \cdots, x_q) \to (x_1^2 + \cdots + x_q^2, x_1^4 + \cdots + x_q^4)$ maps the region X to a region X^* in the s, t plane. Clearly, $d(h^{-1}(s, t), \xi) = d^*((s, t), \xi)$ (say) is well-defined for any symmetric ξ and any (s, t) in X^*. We will be concerned with an examination of the values of $d^*(\cdot, \xi^*)$ at points of X^*, for an optimum ξ^*. We know from (1.4) and (1.6) that, throughout X,

$$d(x, \xi^*) - (q + 1)(q + 2)/2 \leq 0, \tag{3.2}$$

the equality holding at points including the support of ξ. We now show that there are two possibilities:

(i) the zeros of $d^*(\cdot, \xi) - (q + 1)(q + 2)/2$ lie entirely on the boundary of X^*;

(ii) the coefficient $d = 0$ in (3.1), so that the polynomial has the form $as^2 + bs + c$. (In this case the design is a rotatable design which can be shown to be optimum for the problem wherein X is replaced by the smallest ball centered at the origin and containing X, minus the largest open ball contained in its complement (which subtraction is vacuous if X contains the origin).

To see the validity of this assertion, suppose (s_0, t_0) is an interior point of X^* at which $P(s_0, t_0) - (q + 1)(q + 2)/2 = 0$. In view of (3.2) and continuity of the map h, (s_0, t_0) is a local maximum of the polynomial P. Therefore, the first partial derivatives vanish at (s_0, t_0), so that $d = 0$ follows. That proves the assertion.

An analysis of which of (i) and (ii) holds requires more precise knowledge of X, as we see by contrasting the cases (1.8) and (1.9). Although we already know from the first paragraph of this subsection that (i) holds for the q-cube (1.9), we shall continue our analysis for that example along the present lines, both to illustrate this method which can be applied to other symmetric regions similarly, and also because we will then use the method for cubic regression on the q-cube.

Thus, we now suppose X is given by (1.9). We will see that X^* is a closed bounded set which may be described in terms of an upper and lower boundary curve. The upper curve consists of q pieces:

$$\{t = (s - i)^2 + i, \quad i \leq s \leq i + 1\}, \qquad i = 0, 1, \cdots, q - 1. \tag{3.3}$$

The lower curve may be described by the single equation

$$\{qt = s^2, 0 \leq s \leq q\}. \tag{3.4}$$

This last assertion follows at once by the Cauchy-Schwarz inequality since

$$s^2 = (x_1^2 + \cdots + x_q^2)^2 \leq q(x_1^4 + \cdots + x_q^4) = qt. \tag{3.5}$$

We observe that equality holds in (3.5) if and only if $x_1^2 = x_2^2 = \cdots = x_q^2$.

To obtain the upper boundary we suppose the value of $t = x_1^4 + \cdots + x_q^4$ is fixed and seek to minimize s. We may suppose at the start that $x_1 \geq x_2 \geq \cdots \geq x_q \geq 0$.

Consider x_q as a function of x_1 and take partial derivatives with $x_2, \cdots, x_{q-1}$ fixed. This gives $\partial x_q/\partial x_1 = -x_1^3/x_q^3$ and $\partial s/\partial x_1 = 2x_1(x_q^2 - x_1^2)/x_q^2$. We suppose here $x_q > 0$. Since the derivative is negative, we decrease s by increasing x_1 and decreasing x_q, and this preserves the ordering $x_1 \geq x_2 \geq \cdots \geq x_q \geq 0$.

Using this it may be seen that if $t = i + \delta^4$, $0 \leq \delta < 1$, is the fixed value of t, then the minimum for s is obtained by taking $x_1 = x_2 = \cdots = x_i = 1$, $x_{i+1} = \delta$, $x_{i+2} = \cdots x_q = 0$. Thus $s = i + \delta^2$ and $t = i + (s - i)^2$, as asserted in (3.3).

This argument shows even more, that the minimum value of s can be obtained from $x_1, \cdots, x_q$ if and only if $x_1 = 1, \cdots, x_i = 1$ (when $t = i + \delta^4$).

We now show, using an argument like that of the first paragraph of this subsection, that the only possible location of a zero of $d^*(\cdot, \xi^*)$ (for ξ^* optimum) on the boundary segment $\{t = i + (s - i)^2, i \leq s \leq i + 1\}$, is at an end point. The polynomial $d^*((s, i + (s - i)^2), \xi^*)$ is a quadratic in s for $0 \leq s < \infty$ which, being equal to $f(x)M^{-1}(\xi^*)f(x)'$ with $x_1 = x_2 = \cdots = x_i = 1$, $x_{i+1} = s^{1/2}$, $x_{i+2} = \cdots = x_q = 0$, is nonnegative for $0 \leq s < \infty$ and goes to infinity with s, so that it cannot have a local maximum over the interval $i \leq s \leq i + 1$ at an interior point of the latter.

Finally, using the same type of argument, we show that no zero of $d^*(\cdot, \xi^*) - k$ can occur interior to the segment $\{qt = s^2, 0 \leq s \leq q\}$. This is so because $d^*((s, s^2/q), \xi^*)$ is a quadratic in s for $0 < s < \infty$ which, being equal to $f(x)M^{-1}(\xi^*)f(x)'$ with $x_1 = x_2 = \cdots = x_q = (s/q)^{1/2}$, goes to infinity with s, and can thus not attain its maximum over $0 \leq s \leq q$ at an interior point of the latter.

Our discussion has not yet excluded the possibility that the optimal design is rotatable, that is, that $d^*((s, t), \xi^*)$ has the form $as^2 + bs + c$ for an optimum ξ^*. Were this the case, then the design would be optimum for X replaced by the ball $K = \{x: \sum_1^q x_i^2 \leq q\}$, since the argument of the previous paragraph shows that if the optimum ξ^* is rotatable, then $d(x, \xi^*)$ takes on its maximum value $(q + 1)(q + 2)/2$ at the point $(1, 1, \cdots, 1)$ satisfying $\sum_1^q x_i^2 = q$. The moment matrix $M(\xi)$ for an optimal design ξ on K is uniquely determined and is known (Kiefer [6]) to put mass at $s = 0$ and on $s = q$, the moments of the conditional distribution on $s = q$ being those of the uniform measure on this surface.

But, in our problem, the design ξ must be concentrated in the cube (1.9), and the only points in common between the cube and the shell $s = q$ are the corners $(\pm 1, \cdots, \pm 1)$. It is easily seen that every symmetric ξ which is concentrated on the corners and at the origin makes $\int x_1^4 \xi(dx) = \int x_1^2 x_2^2 \xi(dx)$, which is not the case for the optimum rotatable design on the ball. Hence the optimal design for quadratic regression on the q-cube cannot be rotatable.

The discussion substantiates what was already known from simpler calculations in this special case, as indicated earlier. We now bring these ideas to bear on the problem of cubic regression on the cube. We shall return in sections 3.3–3.5 to quadratic regression and shall consider at length there the possible supporting sets for optimum designs.

3.2. *Cubic regression on the q-cube.* We first treat the case $q = 2$, for the sake of simplicity and explicitness of numerical results. The function $d(\cdot, \xi^*)$ for an optimum ξ^* is now a nonnegative polynomial of degree 6 in the variables x_1, x_2, having the following properties. From (1.4) and (1.6),

$$\bar{d}(\xi^*) \leq 10; \tag{3.6}$$

and d is symmetric in x_1, x_2 and invariant under sign changes.

It follows that if we write $t = x_1^4 + x_2^4$ and $s = x_1^2 + x_2^2$ as before, then we can define $d^*(\cdot, \xi)$ for symmetric ξ as in the paragraph below (3.1). This will now be a polynomial of degree 3 in s, t, of the form

$$P(s, t) = as^3 + bs^2 + cs + d + est + ft. \tag{3.7}$$

The domain X^*, from (3.3) and (3.4) with $q = 2$, is the closed bounded set whose boundary consists of the three curves

$$\begin{aligned} &\{s^2 = t, 0 \leq s \leq 1\}, \\ &\{(s-1)^2 + 1 = t, 1 \leq s \leq 2\}, \\ &\{s^2 = st, 0 \leq s \leq 2\}. \end{aligned} \tag{3.8}$$

We consider first the implication of assuming that for some (s_0, t_0) interior to this region $d((s_0, t_0), \xi^*) = 10$. This would be a local maximum interior to the domain X^*, so that $P(\sigma + s_0, \tau + t_0) = a\sigma^3 + b'\sigma^2 + c'\sigma + d' + e\sigma\tau + f\tau$ (say), defined in a neighborhood of $(\sigma, \tau) = (0, 0)$, would have a local maximum at $(0, 0)$. We would therefore have $e = f = 0$, and $P(s, t)$ would be a function only of s, which is the definition of ξ^* being rotatable. We also note that in this case we would have

$$P(s, t) = (as + b'')(s - s_0)^2 + 10 \tag{3.9}$$

with $as + b'' \leq 0$ for $0 \leq s \leq 2$ and with $a > 0$ (since $P \to \infty$ as $s \to \infty$).

Thus, if the design is not rotatable, then the polynomial $d^*(\cdot, \xi^*)$ can vanish only on the boundary curves of (3.8). Substitution of any one of the three relations of (3.8) for t into $P(s, t) - 10$ gives a cubic in s which does not change sign. Hence, any root s interior to the interval determined by the substitution would require the root to be a double root. Therefore, each of the three sections of boundary in the s, t plane can have at most two points at which the polynomial $P - 10$ vanishes. Since the boundary of the square X is mapped into the curve $\{(s-1)^2 + 1 = t, 1 \leq s \leq 2\}$, it follows from the final paragraph of section 1.2 that this curve contains at least one point at which $P - 10$ vanishes.

We now eliminate the possibility of a rotatable design being optimum. The theory of optimum designs for polynomial regression on the ball has been developed by Kiefer [6] and in section 4 of the present paper. An optimum design ξ^* for the square, if rotatable, would have a $d^*(\cdot, \xi^*)$, given by (3.9), attaining its maximum on the square at the corners ($s = t = 2$) and satisfying $d^*((s, t), \xi^*) \leq 10$ on the image under h of the ball K of radius $2^{1/2}$. (This image is bounded by the curves $\{s^2 = 2t, 0 \leq s \leq 2\}$, $\{s^2 = t, 0 \leq s \leq 2\}$, and

$\{s = 2, 2 \le t \le 4\}$.) Hence ξ^* would be optimum for the problem of cubic regression on $K = \{x: x_1^2 + x_2^2 \le 2\}$, with $d(x, \xi^*)$ attaining its maximum over K on the two circles $s = 2$ and $s = s_0$. Thus, $s_0 = 2\rho^2$, where ρ and 1 are the radii of circles where $d(x, \xi') = 10$, where ξ' is optimum for cubic regression when X is the unit ball of (1.8), to be discussed further in section 4. Although this ξ' is not unique, its moments up to those of order 4 (that is, the elements of $M(\xi')$), and the total masses β and $1 - \beta$ which it assigns to the circles of radii ρ and 1, respectively, are unique. (For the first of these facts, see just above (1.5); for the second, replace ξ' by the optimum design ξ'' defined by $\xi''(A) = \int \xi'(gA)\nu(dg)$, where ν is the invariant probability measure on the orthogonal group in two dimensions as in [6], and use the uniqueness of the masses in such a ξ'', proved in [6].) Since $\int_0^{2\pi} (2\pi)^{-1} \cos^4 \theta \, d\theta = \frac{3}{8}$, we would thus obtain

$$\int x_1^4 \xi^*(dx) = 4 \int x_1^4 \xi'(dx) = 4 \left(\tfrac{3}{8}\right) [(1 - \beta) + \rho^4 \beta]. \tag{3.10}$$

Since ξ^* is supported within the square (1.9), we also have $\int x_1^4 \xi^*(dx) \le 1$. Hence, (3.10) yields

$$\frac{3(1 - \beta)}{2} \le \left(\tfrac{3}{2}\right) [(1 - \beta) + \rho^4 \beta] \le 1, \tag{3.11}$$

or $\beta \ge \frac{1}{3}$. But (from table 4.1 of section 4) $\beta = .3077$. We conclude that the optimum design cannot be rotatable.

The following maximization problem was solved numerically. Put mass $p_1/4$ at each of $(1, 1)$, $(1, -1)$, $(-1, 1)$ and $(-1, -1)$. Put mass $p_2/8$ at each of the eight points $(\pm 1, \pm a)$ and $(\pm a, \pm 1)$, $0 < a < 1$. Put mass $(1 - p_1 - p_2)/4$ at each of the four points $(\pm b, \pm b)$, $0 < b < 1$. (In each of the above, the two $\pm$ signs act independently.) Our earlier discussion shows a design of this form to be a candidate for being optimum (although we did not yet eliminate certain other forms). The determinant of $M(\xi)$ was maximized on the Cornell CDC 1604 as a function of the four parameters involved, giving

$$\begin{aligned} a &= 0.35880, \\ b &= 0.48000, \\ p_1 &= 0.36770, \\ p_2 &= 0.46100. \end{aligned} \tag{3.12}$$

For this design ξ, the quantity $\bar{d}(\xi)$ was computed numerically and was found to be ≤ 10 to five decimal places.

We now leave the case $q = 2$ to discuss cubic regression on the cube (1.9) for general $q \ge 3$, where an analysis similar to the one just given for $q = 2$ may again be carried out. If ξ is symmetric, $d(x, \xi)$ is now a sixth degree symmetric polynomial in $x_1, \cdots, x_q$, in which any monomial term involving an odd exponent has zero coefficient. We now need three symmetric functions,

$$\begin{aligned} s &= x_1^2 + \cdots + x_q^2, \\ t &= x_1^4 + \cdots + x_q^4, \\ u &= x_1^6 + \cdots + x_q^6, \end{aligned} \tag{3.13}$$

and define $h: X \to X^*$ by $h(x) = (s, t, u)$. (When $q = 2$, we do not need u since $2u = 3st - s^3$ in that case.) Again $d^*(\cdot, \xi)$ on X^* is well-defined for symmetric ξ by $d^*(h(x), \xi) = d(x, \xi)$, and now has the form

$$P(s, t, u) = as^3 + bs^2 + cs + d + est + ft + gu. \tag{3.14}$$

If this polynomial has a local maximum in the interior of X^*, then one may show $e = f = g = 0$ as before, and therefore one may conclude that the design is rotatable. (In fact, this conclusion clearly holds if (1.9) is replaced by an arbitrary compact symmetric q-dimensional set.)

We now extend the argument we used when $q = 2$, to conclude again that an optimum design ξ^* cannot be rotatable. Such a ξ^* would, by the same argument as before, be optimum for cubic regression on $K = \{x: x_1^2 + \cdots + x_q^2 \leq q\}$. Let β and $1 - \beta$ again denote the total mass assigned to the spheres $\{\sum_1^q x_i^2 = \rho\}$ and $\{\sum_1^q x_i^2 = 1\}$, by each optimum design for cubic regression on the unit ball (1.8). The integral of x_1^4, with respect to the uniform probability measure on $\{\sum_1^q x_i^2 = 1\}$, is now $3/q(q + 2)$. Also, just as before, $\int x^4\xi^*(dx) \leq 1$. Thus, the analogue of (3.10) and the second inequality of (3.11) is that, if ξ^* is an optimum design for K, then

$$q^2[3/q(q + 2)][(1 - \beta) + \beta\rho^4] = \int x_1^4\xi^*(dx) \leq 1. \tag{3.15}$$

If one writes down equations from which the parameters of an optimum rotatable design on (1.8) may be calculated, then one obtains (see section 4, equation (4.5))

$$\begin{aligned} &\frac{(q + 3)(q + 2)(q + 1)}{6} \\ &\qquad = \frac{q+1}{1 - \beta} + \frac{(q - 1)(q + 2)}{2((1 - \beta) + \beta\rho^4)} + \frac{(q + 4)q(q - 1)}{6((1 - \beta) + \beta\rho^6)}. \end{aligned} \tag{3.16}$$

Since $\rho < 1$, we may replace ρ^6 by ρ^4 and also drop the first term on the right in (3.16), and may then divide both sides by $(q + 1)/6$, obtaining

$$q^2 + 5q + 6 > [q^2 + 5q - 6]/[(1 - \beta) + \beta\rho^4]. \tag{3.17}$$

Substituting this inequality for $1 - \beta + \beta\rho^4$ into (3.15) yields

$$\begin{aligned} 0 &> 3q[q^2 + 5q - 6] - (q + 2)[q^2 + 5q + 6] \\ &= 2q^3 + 8q^2 - 34q - 12. \end{aligned} \tag{3.18}$$

The last polynomial is easily seen to be positive for $q \geq 3$. We have thus proved theorem 3.1.

THEOREM 3.1 *For cubic regression on the q-cube* (1.9) *with $q \geq 2$, an optimum design cannot be rotatable.*

Thus, the problem reduces to a study of the nature of the boundary of X^* in the (s, t, u)-space and of the solution of the appropriate maximization problem. We do not attempt to do this in the present paper.

3.3. *Optimum symmetric designs for quadratic regression on the q-cube.* We have already given a short proof in the first paragraph of section 3.1, that all optimum designs for $m = 2$ on the q-cube (1.9) are supported by a subset of the 3^q-array J. For $j = 0, 1, \cdots, q$, let J^j be the subset of J consisting of those $2^{q-j}\binom{q}{j}$ points with exactly j coordinates equal to zero. (Thus, J^j consists of the midpoints of all j-dimensional faces of X.) Kiefer [7] obtained optimum designs supported by $J^0 \cup J^1 \cup J^2$ when $q \leq 5$, showed that this set could not support an optimum design when $q \geq 6$, and described ([7], footnote 5) the method, obtained with Farrell, for obtaining optimum designs for each q on the union of three J^j's and certain subsets of such a union by solving (3.22) below. This joint work is the subject of the present subsection. Subsequently, Kono [9], citing this description in [7], also showed that optimum symmetric designs for each q can only be supported by a subset of J, and carried out detailed calculations for an optimum design on the union of three J^j's when $q \leq 9$, obtaining optimum symmetric designs on $J^0 \cup J^1 \cup J^q$, again by solving equations (3.22) below, for that choice of the J^j's.

We shall first characterize those sets of the form

$$J(j_1, j_2, j_3) = \bigcup_{i=1}^{3} J^{j_i} \tag{3.19}$$

which can support a symmetric optimum design. We shall take $j_1 < j_2 < j_3$. (It can be seen that two J^j's cannot suffice when $q \geq 2$, by noting that no α_{j_i} can be 0 in the demonstration which follows.) Such a design assigns probability $\alpha_{j_i}/2^{q-j_i}\binom{q}{j_i} > 0$ to each point of $J^{j_i} (i = 1, 2, 3)$, where $\sum_1^3 \alpha_{j_i} = 1$. The pertinent moments of such a design are computed, as in (4.1) of [7], to be

$$\begin{aligned} u(\xi^*) &= \int x_1^2 \xi^*(dx) = \int x_1^4 \xi^*(dx) = \sum_1^3 \alpha_{j_i}(q - j_i)/q, \\ v(\xi^*) &= \int x_1^2 x_2^2 \xi^*(dx) = \sum_1^3 \alpha_{j_i}(q - j_i)(q - j_i - 1)/q(q - 1). \end{aligned} \tag{3.20}$$

Write as in (4.5) of [7],

$$\begin{aligned} U_q &= \frac{(q + 3)}{4(q + 1)(q + 2)^2} \{(2q^2 + 3q + 7) \\ &\qquad + (q - 1)[4q^2 + 12q + 17]^{1/2}\}, \\ V_q &= \frac{(q + 3)}{8(q + 2)^3(q + 1)} \{(4q^3 + 8q^2 + 11q - 5) \\ &\qquad + (2q^2 + q + 3)[4q^2 + 12q + 17]^{1/2}\}. \end{aligned} \tag{3.21}$$

One computes $M^{-1}(\xi^*)$, as in (4.3) of [7], and then observes, exactly as in [7], that $\bar{d}(\xi^*) = k$ if and only if

$$u(\xi^*) = U_q, \qquad v(\xi^*) = V_q. \tag{3.22}$$

To solve these equations for the j_i and α_{j_i}, we may think of plotting in the (u, v)-plane, for fixed $q \geq 2$, the $q + 1$ points

$$(u_j, v_j) = ((q - j)/q, (q - j)(q - j - 1)/q(q - 1)), \qquad 0 \leq j \leq q, \tag{3.23}$$

and the point (U_q, V_q). Then clearly (3.22) is satisfied for nonnegative j_i if and only if (U_q, V_q) lies in the triangle with vertices (u_{j_i}, v_{j_i}). Even though we shall not use any such geometric considerations in the demonstration which follows, they help in understanding the results, and also in understanding what is involved in considering unions of more than three J^i's, which we shall forego. (We also remark to the reader that the idea of the demonstration which follows, without such tedious computational details, can be obtained by replacing U_q and V_q by their asymptotic values $2 - q^{-1} + 3q^{-2}/2$ and $1 - 2q^{-1} + 5q^{-2}$, and following through the argument for "large q".)

We first note, replacing $[4q^2 + 12q + 17]^{1/2}$ by the smaller value $(2q + 3)$, that it is easy to verify that $U_q > (q - 1)/q \geq u_j, j > 0$, from which we conclude that $j_1 = 0$. Substituting this fact into (3.22) (and using $\sum_1^3 \alpha_{j_i} = 1$), we obtain

$$\begin{aligned} \alpha_{j_2} &= [q/j_2(j_3 - j_2)]\{-(q - 1)V_q + (2q - j_3 - 1)U_q - (q - j_3)\}, \\ \alpha_{j_3} &= [q/j_3(j_3 - j_2)]\{(q - 1)V_q - (2q - j_2 - 1)U_q + (q - j_2)\}. \end{aligned} \tag{3.24}$$

For fixed q, the expression $\{(q - 1)V_q - (2q - j - 1)U_q + (q - j)\} = F_q(j)$ (say) is linear and (since $U_q < 1$) decreasing in j. We shall show in the next two paragraphs that

$$F_q(2) > 0 > F_q(3) \qquad \text{for} \quad q > 5. \tag{3.25}$$

It follows then from (3.24) that (3.22) can be satisfied for $q > 5$ (with positive α_{j_i}'s) if and only if

$$0 < j_2 < 3 \leq j_3. \tag{3.26}$$

The first inequality of (3.25) can be written as

$$q - 2 > (2q - 3)U_q - (q - 1)V_q, \tag{3.27}$$

or

$$\begin{aligned} 8(q - 2)(q + 2)^3(q + 1)/(q + 3) > 4q^4 + 12q^3 + 7q^2 - 6q& \\ - 89 + (2q^3 - q^2 - 16q + 15)[4q^2 + 12q + 17]^{1/2}&. \end{aligned} \tag{3.28}$$

A direct computation shows that the left side of (3.28) equals

$$\begin{aligned} 8\{q^4 + 2q^3 - 2q^2 - 10q - 4 + (2q - 4)/(q + 3)\}& \\ > 8q^4 + 16q^3 - 16q^2 - 80q - 32&. \end{aligned} \tag{3.29}$$

Using the fact that $[4q^2 + 12q + 17]^{1/2} < 2q + 3 + 4/(2q + 3)$ and that $4(2q^3 - q^2 - 16q + 15)/(2q + 3) < 2q(2q - 3)$, we obtain that the right side of (3.28) is less than $8q^4 + 16q^3 - 24q^2 - 30q - 44$. This last is less than the right side of (3.29) by $(q-6)(8q+2)$, proving (3.28) (and thus (3.27)) for $q \geq 6$.

The second inequality of (3.25) can be written as

$$q - 3 < (2q-4)U_q - (q-1)V_q, \tag{3.30}$$

or

$$8(q-3)(q+2)^3(q+1)/(q+3) < 4q^4 + 8q^3 - 7q^2 - 32q \\ - 117 + (2q^3 - 3q^2 - 18q + 19)[4q^2 + 12q + 17]^{1/2}. \tag{3.31}$$

The left side of (3.31) is

$$8\{q^4 + q^3 - 6q^2 - 16q - 4 - 12/(q+3)\} \\ < 8q^4 + 8q^3 - 48q^2 - 128q - 32. \tag{3.32}$$

Using the fact that $[4q^2 + 12q + 17] > 2q + 3 + 4/(2q+3) - 16/(2q+3)^3$ and that $[4/(2q+3) - 16/(2q+3)^3](2q^3 - 3q^2 - 18q + 19) > 4q^2 - 12q - 22$, we obtain that the right side of (3.31) is greater than $8q^4 + 8q^3 - 48q^2 - 60q - 82$. The latter is clearly greater than the right side of (3.32), proving (3.31) and thus (3.30).

In the same manner that (3.26) was proved (or by direct calculation in the few cases $q < 6$), one can show that (3.26) is replaced by $0 < j_1 < 2 \leq j_2$ for $2 \leq q \leq 6$. To summarize, then, we have the following theorem.

THEOREM 3.2. *The set* $J(j_1, j_2, j_3)$ *of* (3.19) *supports a symmetric optimum design for quadratic regression on the q-cube, if and only if*

$$\begin{array}{lllll} j_1 = 0, & j_2 = 1, & 2 \leq j_3 \leq q, & \text{when} & 2 \leq q \leq 5, \\ j_1 = 0, & j_2 = 1 \text{ or } 2, & 3 \leq j_3 \leq q, & \text{when} & q \geq 6. \end{array} \tag{3.33}$$

We remark that, among the sets $J(0, j_2, j_3)$ permitted by (3.33), the set $J(0, 1, q)$ consisting of vertices, midpoints of edges, and center, has the smallest number of points $(2^q + q2^{q-1} + 1)$ of any optimum symmetric design. In view of (1.12), such designs are quite unsatisfactory for large q, and in the remainder of section 3 we shall therefore seek asymmetric designs on fewer points.

The weights α_{j_i} for any optimum symmetric design on a set (3.19) permitted by (3.33) may be obtained from (3.24) and $\alpha_0 = 1 - \alpha_{j_2} - \alpha_{j_3}$. For $j_2 = 1$, $j_3 = 2$, and $q \leq 5$, $\alpha_{j_i}/2^{q-j_i}\binom{q}{j_i}$ is tabled in [7]; for $j_2 = 1$, $j_3 = q \leq 7$, the α_j are tabled in [9].

3.4. *Bounds on N for quadratic regression on the q-cube.* We now apply the considerations of section 1.3 regarding the minimum number N of points needed to support an optimum design on the q-cube when $m = 2$. We first note

THEOREM 3.3. *The optimum design for quadratic regression on the q-cube is unique if and only if* $q \leq 2$, *in which case the support is* J.

PROOF. The lack of uniqueness when $q \geq 3$ follows from (3.33). The uniqueness when $q = 1$ or 2 (the former of which is well known) can be proved by using (1.16) with $U = J$; the matrix $\{\phi_t(x), x \in J, 1 \leq t \leq p\}$ is easily seen to have rank 3 or 9 in these two cases. (We recall that the subscript value $t = 0$ need not be included in polynomial regression.)

We next improve the upper bound $\binom{q+4}{4}$ of (1.12) by use of (1.14). We shall take $U = W = J(0, 1, q)$ in the calculation which follows. For $x \in J(0, 1, q)$ (in fact, for $x \in J$), the relations

$$\begin{aligned} x_i &= x_i^3, \\ x_i^2 &= x_i^4, \\ x_i x_j &= x_i x_j^3, \qquad i \neq j \end{aligned} \tag{3.34}$$

for $1 \leq i, j \leq q$ are satisfied. These are $q^2 + q$ linearly independent relations among the $\phi_t (1 \leq t \leq p)$. Among the set of ϕ_t which remain after deleting those on the right side of (3.34), the relations

$$\begin{aligned} (1 - x_j^2)(x_r^2 - x_s^2) &= 0, \\ (x_i^2 - x_j^2)(x_r^2 - x_s^2) &= 0, \end{aligned} \tag{3.35}$$

are satisfied on $J(0, 1, q)$ with all subscripts unequal and between 1 and q, inclusive. (Equalities (3.35) are vacuous if $q < 3$.) This is so because either all $x_i^2 = 0$, or else at most one $x_i^2 = 0$. The relations (3.35) among the ϕ_t are not linearly independent when $q \geq 3$, so we must find the dimension of the vector space spanned by the ϕ_t. To this end, we write $L = q + 1$, $y_i = x_i^2$ for $1 \leq i \leq q$, and $y_{L+1} = 1$.

For $L \geq 4$, let Q be the vector space over the reals of all linear combinations of the polynomials $(y_i - y_j)(y_r - y_s)$ with i, j, r, s distinct integers between 1 and L, inclusive (a subspace of the quadratic polynomials in L variables). We shall show the following lemma.

LEMMA 3.4. *For $L \geq 4$, we have*

$$\dim Q = L(L - 3)/2. \tag{3.36}$$

PROOF. All subscripts in the proof which follows are to be reduced mod L. We first show that the $L(L - 3)/2$ special polynomials

$$(y_j - y_{j+1})(y_{j+i+1} - y_{j+i+2}), \tag{3.37}$$

with all subscripts distinct, span Q. (Note, for example, that $j = L$ is permitted.) We must show that any polynomial $(y_i - y_j)(y_r - y_s)$ is a linear combination of these special polynomials. There are two cases to which any other can be reduced by symmetry.

Case 1 $(i < j < r < s)$. Then

$$(y_i - y_j)(y_r - y_s) = \sum_{0 \leq u < j-i, 0 \leq v < s-r} (y_{i+u} - y_{i+u+1})(y_{r+v} - y_{r+v+1}). \tag{3.38}$$

Case 2 $(i < r < j < s)$. Use the identity

$$(y_i - y_j)(y_r - y_s) = (y_i - y_r)(y_j - y_s) - (y_r - y_j)(y_s - y_i) \tag{3.39}$$

to reduce to case 1.

To conclude the proof of (3.36), we need only show that the special polynomials (3.37) are linearly independent. To this end, we obtain an appropriate ordering of the special polynomials, say $\{g_\alpha, 1 \le \alpha \le L(L-3)/2\}$, and of the functions $y_i y_j (i \ne j)$, say $\{h_\beta, 1 \le \beta \le L(L-1)/2\}$; then writing $g_\alpha = \sum_\beta c_{\alpha\beta} h_\beta$, we show that $c_{\alpha\alpha} \ne 0$ and $c_{\alpha\beta} = 0$ for $\alpha > \beta$, which proves the desired result. The g_α are, in order, the special polynomials of (3.37) with $i = 1$ and $j = 0, 1, \cdots, L-1$; then, with $i = 2$ and $j = 0, 1, \cdots, L-1; \cdots, i =$ {greatest integer $\le (L-3)/2$} and $j = 0, 1, \cdots, L-1$; if L is even, there are then $L/2$ additional special polynomials with $i = (L-2)/2$ and $j = 0, 1, \cdots, (L-2)/2$. Note that those functions in the ith collection of L polynomials (or of $L/2$ if L is even and $i = (L-2)/2$) have i as the minimum distance between subscripts γ, δ which appear in any term $y_\gamma y_\delta$ entering with nonzero coefficient in the special polynomial. Moreover, for fixed i, these $y_\gamma y_{\gamma+i}$ appear in the order $\gamma = 1, 2, \cdots, L$. Thus, when we order the h_β as $y_j y_{j+1}$ for $j = 1, 2, \cdots, L$, and then $y_j y_{j+2}$ for $j = 1, \cdots, L$, and so on, we see at once that the $c_{\alpha\beta}$ have the desired property. This completes the proof of the lemma.

Putting $L = q + 1$ in (3.36), we conclude that the ϕ_t on the left side of (3.35) span a real vector space of dimension $(q+1)(q-2)/2$ (which is also correct if $q = 2$). Adding this to the number $q^2 + q$ of restrictions (3.34), which are independent of (3.36), and subtracting the result from $p = \binom{q+4}{4}$ and using (1.14) with $\{0 \le t \le p\}$ replaced by $\{1 \le t \le p\}$, we obtain theorem 3.5.

THEOREM 3.5. *For quadratic regression on the q-cube,*

$$N \le (q+1)(q^3 + 9q^2 - 10q + 48)/24. \tag{3.40}$$

3.5. *The use of orthogonal arrays to reduce the number of points of support.* We have already mentioned in section 1 how orthogonal arrays of strength 2 are used classically to reduce the number of points of support for an optimum design for linear regression on the q-cube. Similar techniques can be employed in other settings, as we now illustrate for quadratic regression on the q-cube. We shall consider the following particular scheme of application.

Suppose, for each positive integer r, that we can find a subset A_r of the 2^r corners of the r-cube ((1.9) with $q = r$), such that the uniform probability measure on A_r has the same moments of order ≤ 4 as the uniform probability measure on the 2^r corners. Suppose A_r has n_r points. Then, suppose we replace J^0 in $J(0, 1, q)$ by A_q (with α_0/n_q probability per point); replace the 2^{q-1} points of J^1 with $x_i = 0$ by the n_{q-1} points of the form

$$x_i = 0, \qquad (x_i, \cdots, x_{i-1}, x_{i+1}, \cdots, x_q) \in A_{q-1},$$

for $1 \le i \le q$ (with probability α_1/qn_{q-1} per point); retain J^q, with probability α_q. Since only moments of order ≤ 4 are present in $M(\xi)$, and because of the way in which zero coordinate values enter into the replacement of J^1, we obtain a design with the same M as the optimum symmetric design on $J(0, 1, q)$, and which is therefore also optimum. It is supported by

$$n_q + qn_{q-1} + 1 \tag{3.41}$$

points.

The classical construction of such an A_r is in terms of an *orthogonal array of strength 4 with 2 levels*, that is, an $n_r \times r$ matrix T_r with entries ± 1 such that every 4-row submatrix has the property that each of the 16 possible 4-vectors with entries ± 1 appears equally often in the submatrix. We then consider the n_r columns of the matrix T_r as the points of A_r and clearly obtain the required moment properties. The reader is referred to such references as [1] for detailed discussion of orthogonal arrays.

Orthogonal arrays of strength ≥ 3 have been considered extensively by Rao, Bose, Bush, Seiden, and others. The principal method of construction is geometric. A set C of r points in the finite projective space $PG(d-1, 2)$ of dimension $d-1$, no 4 of which lie on the same 2-dimensional flat, yields a T_r with $n_r = 2^d$; for, writing B for the $r \times d$ matrix whose rows are the points of C (each of the d coordinates of such a point being an element of the Galois field $GF(2)$), and writing D for the $d \times 2^d$ matrix whose columns are the different d-vectors with coordinate values in $GF(2)$, one sees easily that BD has the required properties of T_r, except that the elements ± 1 of T_r are replaced by 0, 1 (of $GF(2)$) in BD. (It is not always known when this geometric construction yields the maximum r for given d.)

For fixed r, Rao's lower bound on n_r, usually given in geometric terms, can be obtained for general orthogonal arrays $T_n = \{t_{ij}, 1 \leq i \leq r, 1 \leq j \leq n_r\}$ of strength 4 with elements ± 1, as follows. Let τ_0 be the row vector of n_r 1's; let τ_i be the i-th row of T_r, and for $1 \leq i < i' \leq r$ let $\tau_{i,i'} = (t_{i1}t_{i'1}, \cdots, t_{in_r}t_{i'n_r})$. The $1 + r + \binom{r}{2}$ vectors $\tau_0, \tau_1, \cdots, \tau_r, \tau_{12}, \cdots, \tau_{(r-1)r}$ are easily shown to be orthogonal, because of the properties of T_r. Hence,

$$n_r \geq (r^2 + r + 2)/2. \tag{3.42}$$

In the other direction, it is simple to give a geometric construction which yields an orthogonal array of strength 4 satisfying

$$n_r = \text{largest number } 2^d \text{ which is } \leq 1 + r + \binom{r}{2} + \binom{r}{3}. \tag{3.43}$$

For, in $PG(d-1, 2)$, if we have chosen j points, no 4 of which are coplanar, there are $\binom{j}{2}$ pairs of points each of which determines a line with one point outside the pair, and $\binom{j}{3}$ triples of points, each of which determines a plane with one point not on the lines just mentioned. Thus, as long as $j + \binom{j}{2} + \binom{j}{3} < 2^d - 1$ (equal to the number of points in $PG(d-1, 2)$), there remains a $(j+1)$st point which can be chosen without destroying noncoplanarity. Continuing in this way, we can obtain r points, where r is the smallest integer for which $r + \binom{r}{2} + \binom{r}{3} \geq 2^d - 1$. This yields (3.43).

The reader should have no trouble in writing down analogues of (3.42) and (3.43) for other Galois fields, and, in fact, for arrays of different strength.

When q is large, the use of (3.41) with (3.43) yields an optimum design on $\leq q^4(1 + o(1))/3$ points. This is $0(q^4)$ like p or the bound (3.40), but these last are both $q^4(1 + o(1))/24$. Thus, we do not know whether or not the orthogonal array approach can, for large q, yield a design with no more points than (3.40), let alone whether the order r^3 of (3.43) rather than the order r^2 of (3.42) (or neither) is the best possible as $r \to \infty$. We do know that there are some small values of q for which the method of using orthogonal arrays cannot yield a value of (3.41) which is less than (3.40) or even p. This is a consequence of the fact ([10], [12]) that the minimum possible values of n_r for $r = 4, 5, 6, 7, 8$ are known to be 16, 32, 32, 64, 64, so that the numbers listed in the last column of table I below, and which were obtained by using these values in (3.41),

TABLE I

q	k	p	(3.40)	Points in $J(0, 1, q)$	Achievable Using (3.41)
2	6	15	9	9	9
3	10	35	21	21	21
4	15	70	45	49	49
5	21	126	87	113	113
6	28	210	154	257	225
7	36	330	254	577	289
8	45	495	396	1281	577
9	55	715	590	2817	705
10	66	1001	847	6145	1409
11	78	1365	1179	13313	1537
12	91	1820	1599	28673	1793
13	105	2380	2121	61441	3585
16	153	4845	4454	589825	4353
17	171	5985	5544	1245185	4608
Asymptotic Value	$q^2/2$	$q^4/24$	$q^4/24$	$q2^{q-1}$	$\leq q^4/3$

cannot be improved upon by using orthogonal arrays for $q \leq 8$. We have also used the values $n_r = 128$ for $9 \leq r \leq 11$ and $n_r = 256$ for $12 \leq r \leq 17$ in this table. These are the best values obtainable geometrically [12], but it is not yet known whether a nongeometric construction can yield better orthogonal arrays in these cases. (For values like $q = 10$ or 13, where $n_{q-1} > n_{q-2}$, the number obtained from (3.41) is at its worst compared with p or (3.40); similarly, for $q = 8$, 9, 11, 12, and 17, the comparison is more favorable.)

We are indebted to Professor Esther Seiden for several communications concerning the construction of these orthogonal arrays of strength 4.

In view of the unattainability of p or (3.40) for some values q by using the method of this subsection, it is clear that further study is needed of designs which have less symmetry. For example, by considering nonuniform measures on smaller sets than A_r, and perhaps subsets of more than three J^i's, one should be able to do considerably better. Perhaps one can even reduce the number of points required from $0(q^4)$ to a smaller order such as $0(q^3)$. One other obvious attempt to obtain $0(q^3)$ is to seek an optimum design with equal mass on each point of J, plus additional masses on J^0 and J^q, and thus to replace the q arrays used with J^1 in (3.41) by an orthogonal array of strength 4 with three levels, which is used in place of J (the n_q and 1 being present in (3.41) as before). The analogue of (3.43) for $GF(3)$ shows that this three-level array again requires only $0(q^3)$ points, so that (3.41) would yield $0(q^3)$. Unfortunately, one cannot choose positive probabilities on J^0, J, and J^q so as to satisfy the analogue of (3.22) for large q.

4. The q-ball

We now suppose X to be the unit q-ball (1.8). For regression of degree m, a rough characterization of optimum designs was given by Kiefer [6]): every optimum ξ assigns measure one to $(m + 1)/2$ spherical shells centered at 0, where one of these shells is the boundary of X and where 0 counts as $\frac{1}{2}$ shell. Some weighted mixture of uniform measures on these shells is optimum (although other measures with the same first $2m$ moments are also optimum). The weights and radii of shells are hard to compute for $m > 2$; when $m = 2$, measure $2/(q + 1)(q + 2)$ is assigned to the origin and the remainder is assigned to the boundary of X.

Two problems of interest here are (1) to obtain at least approximate information on the radii and weights when $m > 2$, and (2) to obtain discrete measures on the shells supported by as few points as possible. In most of the remaining paragraphs of this section we shall indicate the type of treatment of problem (1) which is possible for $m > 2$, considering here the example $m = 3$. Problem (2) entails considerations related to those of section 3 and also to the extensive literature on the construction of rotatable designs. It differs from the latter in its specification of the radii and weights and in its allowing of unequal masses on points which may not be symmetrically spaced. The implementation of the resulting optimum designs of the approximate theory for specified sample sizes by discrete designs which approximate them, will yield nonrotatable designs which can be expected to involve fewer distinct points and to have better performance characteristics than the rotatable designs which are usually used. The payment for this in the form of a design matrix which is harder to invert may be worthwhile with modern computing equipment.

An optimum design ξ^* in the case $m = 3$ on the q-ball can be described in terms of two parameters: measure β is spread uniformly on a sphere of radius $\rho < 1$, and measure $1 - \beta$ is assigned to the unit sphere (equal to the boundary of X). For such a design, $d(x, \xi^*)$ depends only on $r^2 = \sum x_i^2$, say $d(x, \xi^*) = d^*(r, \xi^*)$. The optimum ρ and β can be found either by solving the two equations $d^*(\rho, \xi^*) = \binom{q+1}{3}$ $(=k)$ and $\partial d^*(r, \xi^*)/\partial r|_{r=\rho} = 0$ (see [6]), or else by maximizing $\det M(\xi)$ with respect to ρ and β. We shall exhibit the second method.

Grouping the functions into four sets as $\{1, x_1^2, \cdots, x_k^2\}$, $\{x_i, x_i^3, x_i x_j^2; i \neq j\}$,

$\{x_i x_j; i < j\}$, $\{x_h x_i x_j; h < i < j\}$, one sees that the product of two elements from different sets has zero integral. Thus, for ξ of the specified form, det $M(\xi)$ can be evaluated as the product of four determinants; one obtains, with C_q denoting a constant depending on q,

$$\begin{aligned}\log \det M(\xi) = C_q + 2q \log \rho + (q+1) \log [\beta(1-\beta)(1-\rho^2)^2] \\ + \frac{(q+2)(q-1)}{2} \log [(1-\beta) + \beta\rho^4] \\ + \frac{(q+4)q(q-1)}{6} \log [(1-\beta) + \beta\rho^6].\end{aligned} \tag{4.1}$$

The two equations obtained by setting equal to zero the derivatives with respect to each of ρ and β, are not very manageable analytically. (This is also true of the equations obtained by the other approach mentioned in the previous paragraph.) These equations, however, can be solved easily by machine, and the results of this computation on the Cornell Computing Center CDC 1604 are given in table II below. We note here that the behavior of the maximizing values of ρ_q and β_q (say) as $q \to \infty$ are easily discernible from (4.1). A routine analysis shows that $\beta_q = hq^{-2} + o(q^{-2})$ and $\rho_q = \rho^* + o(1)$ where $0 < q < \infty$ and $0 < \rho^* < 1$ and where h and ρ^* maximize the coefficient of q in (4.1):

$$\begin{aligned} h^{-1} + (1-\rho^{*6})/6 = 0, \\ 2/\rho^* - 4\rho^*/(1-\rho^{*2}) + h\rho^{*5} = 0. \end{aligned} \tag{4.2}$$

Thus, as $q \to \infty$,

$$\begin{aligned} \rho_q^2 \sim \rho^{*2} = (3^{1/2} - 1)/2 = .3660254, \\ \beta_q \sim hq^{-2} = 4q^{-2}(1 + 3^{-1/2}) = 6.309401q^{-2}. \end{aligned} \tag{4.3}$$

A finer analysis can be used to produce further terms in an asymptotic expansion.

We digress in this paragraph to derive a result which was used in section 3.2. The matrix $M(\xi)$ may be inverted explicitly for ξ of the form we have been considering, the answer being expressed in terms of ρ_q and β_q. This allows one to write an expression for

$$\begin{aligned} d^*(r, \xi^*) = \frac{(1-\beta_q)(1-r^2)^2 + \beta_q(r^2-\rho_q^2)^2}{(1-\beta_q)\beta_q(1-\rho_q^2)^2} \\ + q\,\frac{(1-\beta_q)r^2(1-r^2)^2 + \beta_q r^2\rho_q^2(r^2-\rho_q^2)^2}{(1-\beta_q)\beta_q\rho_q^2(1-r_q^2)^2} \\ + \frac{(q+2)(q-1)}{2}\,\frac{r^4}{(1-\beta_q)+\beta_q\rho_q^4} \\ + \frac{(q+4)q(q-1)}{6}\,\frac{r^6}{(1-\beta_q)+\beta_q\rho_q^6}. \end{aligned} \tag{4.4}$$

Taking $r = 1$ gives (since $d^*(1, \xi^*) = k$)

$$\begin{aligned} \frac{(q+1)(q+2)(q+3)}{6} = \frac{q+1}{1-\beta_q} + \frac{(q+2)(q-1)}{2(1-\beta_q)+\beta_q\rho_q^4} \\ + \frac{(q+4)q(q-1)}{6(1-\beta_q)+\beta_q\rho_q^6}. \end{aligned} \tag{4.5}$$

Equation (4.5) was used as (3.16) in section 3.2 to show that optimum designs for cubic regression on the q-cube could not be rotatable.

The following table of numbers for β_q and ρ_q were computed as described above. (Of course, for $q = 1$ we have the Guest-Hoel design.)

TABLE II

q	β_q	$q^2\beta_q$	ρ_q^2
1	0.5000	0.500	0.2000
2	0.3077	1.231	0.2657
3	0.2455	2.210	0.2970
4	0.1695	2.712	0.3142
5	0.1241	3.102	0.3249
6	0.09483	3.414	0.3321
7	0.07490	3.670	0.3373
8	0.06068	3.884	0.3412
9	0.05019	4.065	0.3442
10	0.04221	4.221	0.3465
10^2	0.6020×10^{-3}	6.020	0.364381
10^3	0.6279×10^{-5}	6.279	0.365866
10^4	0.6306×10^{-7}	6.306	0.366010
10^5	0.6309×10^{-9}	6.309	0.366024
∞		6.309401	0.3660254

From a practical point of view, what is important for other examples (for instance, larger m on the ball) is the indication that the use of the limiting values h and ρ^* for fairly small values of q leads to a value of $\max_x d(x, \xi)$ which is not too large; this aspect deserves further machine study in other contexts.

For $m \geq 4$, the same approach can be used, but of course the larger number of parameters makes the analysis messier, especially if $q > 2$. We remark that when $m = 4$, $q = 2$, the optimum weights are $\frac{1}{15}$, 0.343912, and 0.589422, at $r^2 = 0$, 0.460249, and 1, respectively.

In order to construct implementable optimum designs on the unit ball for any m, we replace the uniform distribution on each spherical shell by a distribution on a finite subset of the same shell, with the same moments. It will suffice to consider the shell of radius one. If the measure γ assigns mass α to each of the 2^q points having coordinates $\pm q^{-1/2}$, and mass β to each of the $2q$ points $(\pm 1, 0, \cdots, 0)$, $(0, \pm 1, \cdots, 0)$, $\cdots$, $(0, 0, \cdots, \pm 1)$, then the values $\alpha = q(q+2)^{-1}2^{-q}$, $\beta = 1/q(q+2)$ satisfy

$$\begin{aligned} \alpha + \beta &= 1, \\ \int x_1^2 \gamma(dx) &= 1/q, \qquad (4.6) \\ \int x_1^2 x_2^2 \gamma(dx) &= 1/q(q+2), \\ \int x_1^4 \gamma(dx) &= 3/q(q+2). \end{aligned}$$

Clearly all other moments of order >0 and <4 are zero.

The set of 2^q points with all coordinates $\pm q^{-1/2}$ may be replaced by a subset which is an orthogonal array of strength 4 consisting of $0(q^3)$ points, by (3.43). Thus we know how to construct optimum designs on $0(q^3)$ points. As in section 3.4, if the order of (3.42) were attainable, we could achieve $0(q^2)$ here; and perhaps less symmetric points and weights can also help to achieve a lower order than $0(q^3)$.

Using this method of construction with orthogonal arrays, and data provided by E. Seiden and described in section 3.4, we obtain the following table III giving the number of points of support for ξ in optimum designs for $q = 3, \cdots, 17$, when $m = 2$. There being one point at the origin, we obtain a design on

$$2q + 1 + n_q \tag{4.7}$$

points in this case. The values, of course, compare favorably with those of table I, where the same values of k and p apply.

TABLE III

q	(4.7)
3	23
4	25
5	43
6	45
7	79
8	81
9	147
10	149
11	151
12	281
13	283
14	285
15	287
16	289
17	291

That these designs may not be the best possible, even among designs of quite symmetric construction, is illustrated by an example of Box and Behnken [2]. Using their construction for $q = 7$, one obtains a design with 56 points on the unit sphere plus one at the origin, for a total of 57 points.

REFERENCES

[1] R. C. Bose and K. A. Bush, "Orthogonal arrays of strength two and three," *Ann. Math. Statist.*, Vol. 23 (1952), pp. 508–524.

[2] G. E. P. Box and D. W. Behnken, "Simplex-sum designs: a class of second order rotatable designs derivable from those of first order," *Ann. Math. Statist.*, Vol. 31 (1960), pp. 838–864.

[3] P. G. Guest, "The spacing of observations in polynomial regression," *Ann. Math. Statist.*, Vol. 29 (1958), pp. 294–299.

[4] P. G. Hoel, "Efficiency problems in polynomial regression," *Ann. Math. Statist.*, Vol. 29 (1958), pp. 1134–1145.

[5] J. Kiefer, "Optimum experimental designs," *J. Roy. Soc. Ser. B*, Vol. 21 (1959), pp. 273–319.

[6] ———, "Optimum experimental designs V, with applications to systematic and rotatable designs," *Proceedings of the Fourth Berkeley Symposium on Mathematical Statistics and Probability*, 1961, Vol. 1, pp. 381–405.

[7] ———, "Optimum designs in regression problems II," *Ann. Math. Statist.*, Vol. 32 (1961), pp. 298–325.

[8] J. KIEFER and J. WOLFOWITZ, "The equivalence of two extremum problems," *Canad. J. Math*, Vol. 12 (1960), pp. 363–366.

[9] K. KONO, "Optimum design for quadratic regression on the k-cube," *Mem. Fac. Sci. Kyushu Univ. Ser. A.*, Vol. 16 (1962), pp. 114–122.

[10] E. SEIDEN and R. ZEMACH, "On orthogonal arrays," to be published in 1965.

[11] H. URANISI, written communication concerning the special cubic on the q-simplex, 1962.

[12] R. ZEMACH, "On orthogonal arrays of strength four and their application," Michigan State University N. I. H. Report.

ON BASIC CONCEPTS OF STATISTICS

JAROSLAV HÁJEK
MATHEMATICAL INSTITUTE OF THE CZECHOSLOVAK ACADEMY OF SCIENCES,
and CHARLES UNIVERSITY, PRAGUE

1. Summary

This paper is a contribution to current discussions on fundamental concepts, principles, and postulates of statistics. In order to exhibit the basic ideas and attitudes, mathematical niceties are suppressed as much as possible. The heart of the paper lies in definitions, simple theorems, and nontrivial examples. The main issues under analysis are *sufficiency, invariance, similarity, conditionality, likelihood,* and their mutual relations.

Section 2 contains a definition of sufficiency for a subparameter (or sufficiency in the presence of a nuisance parameter), and a criticism of an alternative definition due to A. N. Kolmogorov [11]. In that section, a comparison of the principles of sufficiency in the sense of Blackwell-Girschick [2] and in the sense of A. Birnbaum [1] is added. In theorem 3.5 it is shown that for nuisance parameters introduced by a group of transformations, the sub-σ-field of invariant events is sufficient for the respective subparameter.

Section 4 deals with the notion of similarity in the x-space as well as in the (x, θ)-space, and with related notions such as ancillary and exhaustive statistics. Confidence intervals and fiducial probabilities are shown to involve a postulate of "independence under ignorance."

Sections 5 and 6 are devoted to the principles of conditionality and of likelihood, as formulated by A. Birnbaum [1]. Their equivalence is proved and their strict form is criticized. The two principles deny gains obtainable by mixing strategies, disregarding that, in non-Bayesian conditions, the expected maximum conditional risk is generally larger than the maximum overall risk. Therefore, the notion of "correct" conditioning is introduced, in a general enough way to include the examples given in the literature to support the conditionality principle. It is shown that in correct conditioning the maximum risk equals the expected maximum conditional risk, and that in invariant problems the sub-σ-field of invariant events yields the deepest correct conditioning.

A proper field of application of the likelihood principle is shown to consist of families of experiments, in which the likelihood functions, possibly after a common transformation of the parameter, have approximately normal form with constant variance. Then each observed likelihood function allows computing the risk without reference to the particular experiment.

In section 7, some forms of the Bayesian approach are touched upon, such as those based on diffuse prior densities, or on a family of prior densities.

In section 8, some instructive examples are given with comments.
References are confined to papers quoted only.

2. Sufficiency

The notion of sufficiency does not promote many disputes. Nonetheless, there are two points, namely sufficiency for a subparameter and the principle of sufficiency, which deserve a critical examination.

2.1. *Sufficiency for a subparameter.* Let us consider an experiment (X, θ), where X denotes the observations and θ denotes a parameter. The random element X takes its values in an x-space, and θ takes its values in a θ-space. A function τ of θ will be called a subparameter. If θ were replaced by a σ-field, then τ would be replaced by a sub-σ-field. Under what conditions may we say that a statistic is sufficient for τ? Sufficiency for τ may also be viewed as sufficiency in the presence of a nuisance parameter. The present author is aware of only one attempt in this direction, due to Kolmogorov [11].

DEFINITION 2.1. *A statistic $T = t(X)$ is called sufficient for a subparameter τ, if the posterior distribution of τ, given $X = x$, depends only on $T = t$ and on the prior distribution of θ.*

Unfortunately, the following theorem shows that the Kolmogorov definition is void.

THEOREM 2.1. *If τ is a nonconstant subparameter, and if T is sufficient for τ in the sense of definition 2.1, then T is sufficient for θ as well.*

PROOF. For simplicity, let us consider the discrete case only. Let T be not sufficient for θ, and let us try to show that it cannot be sufficient for τ. Since T is not sufficient for θ, there exist two pairs, (θ_1, θ_2) and (x_1, x_2), such that

$$T(x_1) = T(x_2) \tag{2.1}$$

and

$$\frac{P_{\theta_1}(X = x_1)}{P_{\theta_2}(X = x_1)} \neq \frac{P_{\theta_1}(X = x_2)}{P_{\theta_2}(X = x_2)}. \tag{2.2}$$

If $\tau(\theta_1) \neq \tau(\theta_2)$, let us consider the following prior distribution: $\nu(\theta = \theta_1) = \nu(\theta = \theta_2) = \frac{1}{2}$. Then, for $\tau_1 = \tau(\theta_1)$ and $\tau_2 = \tau(\theta_2)$,

$$P(\tau = \tau_1 | X = x_1) = P_{\theta_1}(X = x_1)/[P_{\theta_1}(X = x_1) + P_{\theta_2}(X = x_1)] \tag{2.3}$$

and

$$P(\tau = \tau_1 | X = x_2) = P_{\theta_1}(X = x_2)/[P_{\theta_1}(X = x_2) + P_{\theta_2}(X = x_2)]. \tag{2.4}$$

Obviously, (2.2) entails $P(\tau = \tau_1 | X = x_1) \neq P(\tau = \tau_1 | X = x_2)$ which implies, in view of (2.1), that T is not sufficient for τ.

If $\tau(\theta_1) = \tau(\theta_2)$ held, we would choose θ_3 such that $\tau(\theta_3) \neq \tau(\theta_1) = \tau(\theta_2)$. Note that the equations

$$\frac{P_{\theta_1}(X = x_1)}{P_{\theta_3}(X = x_1)} = \frac{P_{\theta_1}(X = x_2)}{P_{\theta_3}(X = x_2)} \tag{2.5}$$

and

$$\frac{P_{\theta_2}(X = x_1)}{P_{\theta_3}(X = x_1)} = \frac{P_{\theta_2}(X = x_2)}{P_{\theta_3}(X = x_2)} \tag{2.6}$$

are not compatible with (2.2). Thus either (2.5) or (2.6) does not hold, and the above reasoning may be accomplished either with (θ_1, θ_3) or (θ_2, θ_3), both pair satisfying the condition $\tau(\theta_1) \neq \tau(\theta_3)$, $\tau(\theta_2) \neq \tau(\theta_3)$. Q.E.D.

Thus we have to try to define sufficiency in the presence of nuisance parameters in some less stringent way.

DEFINITION 2.2. *Let $\mathcal{P}_\tau$ be the convex hull of the distributions $\{P_\theta, \tau(\theta) = \tau\}$ for all possible τ-values. We shall say that T is sufficient for τ if*

(i) *the distribution of T will depend on τ only, that is,*

$$P_\theta(dt) = P_\tau(dt), \tag{2.7}$$

and

(ii) *there exist distributions $Q_\tau \in \mathcal{P}_\tau$ such that T is sufficient for the family $\{Q_\tau\}$.*

In the same manner we define a sufficient sub-σ-field for τ.

Now we shall prove an analogue of the well-known Rao-Blackwell theorem. For this purpose, let us consider a decision problem with a *convex* set D of decisions d, and with a loss function $L(\tau, d)$, which is *convex* in d for each τ, and depends on θ only through τ, $L(\theta, d) = L(\tau(\theta), d)$. Applying the minimax principle to eliminate the nuisance parameter, we associate with each decision function $\delta(x)$ the following risk:

$$R(\tau, \delta) = \sup_{\tau(\theta) = \tau} \int L[\tau(\theta), \delta(x)] P_\theta(dx), \tag{2.8}$$

where the supremum is taken over all θ-values such that $\tau(\theta) = \tau$. Now, if T is sufficient for τ in the sense of definition 2.2, we can associate with each decision function $\delta(x)$ another decision function $\delta(t)$ defined as follows:

$$\bar{\delta}(t) = \int \delta(x) Q_\tau(dx | T = t) \tag{2.9}$$

provided that $\delta(x)$ is integrable. Note that the right side of (2.9) does not depend on τ, since T is sufficient for $\{Q_\tau\}$, according to definition 2.2, and that $\bar{\delta}(t) \in D$ for every t in view of convexity of D. Finally put

$$\delta^*(x) = \bar{\delta}(t(x)). \tag{2.10}$$

THEOREM 2.2. *Under the above assumptions,*

$$R(\tau, \delta^*) \leq R(\tau, \delta) \tag{2.11}$$

holds for all τ.

PROOF. Since the distribution of T depends on τ only, we have

$$\begin{aligned} R(\tau, \delta^*) &= \int L[\tau(\theta), \delta^*(x)] P_\theta(dx) \\ &= \int L[\tau(\theta), \delta^*(x)] Q_\tau(dx) \end{aligned} \tag{2.12}$$

for all θ such that $\tau(\theta) = \tau$. Furthermore, since $L(\tau, d)$ is convex in d,

$$\int L[\tau(\theta), \delta^*(x)]Q_\tau(dx) \leq \int L[\tau(\theta), \delta(x)]Q_\tau(dx) \leq \sup_{\tau(\theta)=\tau} \int L[\tau(\theta), \delta(x)]P_\theta(dx) = R(\tau, \delta). \tag{2.13}$$

This concludes the proof.

Thus, adopting the minimax principle for dealing with nuisance parameters, and under the due convexity assumptions, one may restrict himself to decision procedures depending on the sufficient statistic only. This finds its application in points estimation and in hypothesis testing, for example.

REMARK 2.1. Le Cam [13] presented three parallel ways of defining sufficiency. All of them could probably be used in giving equivalent definitions of sufficiency for a subparameter. For example, Kolmogorov's definition 2.1 could be reformulated as follows: T is called sufficient for τ if it is sufficient for each system $\{P_{\theta'}\}$ such that $\{\theta'\} \subset \{\theta\}$ and $\theta_1' \neq \theta_2' \Rightarrow \tau(\theta_1') \neq \tau(\theta_2')$.

REMARK 2.2. We also could extend the notion of sufficiency by adding some "limiting points." For example, we could introduce ϵ-sufficiency, as in Le Cam [13], and then say that T is sufficient if it is ϵ-sufficient for every $\epsilon > 0$, or if it is sufficient for every compact subset of the τ-space.

REMARK 2.3. (Added in proof.) Definition 2.2 does not satisfy the natural requirement that T should be sufficient for τ, if it is sufficient for some finer subparameter τ', $\tau = \tau(\tau')$. To illustrate this point, let us consider a sample from $N(\mu, \sigma^2)$ and put $T = (\bar{x}, s^2)$, $\tau = \mu$, $\tau' = (\mu, \sigma^2)$. Then T is not sufficient for τ in the sense of definition 2.2, since its distribution fails to be dependent on μ only. On the other hand, T is sufficient for τ' as is well known. The definition 2.2 should be corrected as follows.

DEFINITION 2.2*. *A statistic $T = t(X)$ is called sufficient for a subparameter τ, if it is sufficient in the sense of definition 2.2 for some subparameter τ' such that $\tau = \tau(\tau')$.*

REMARK 2.4. (Added in proof.) A more stringent (and, therefore, more consequential) definition of sufficiency for a parameter is provided by Lehmann [14] in problem 31 of chapter III: T is sufficient for τ if, first, $\theta = (\tau, \eta)$, second, $P_\theta(dt) = P_\tau(dt)$, third, $P_\theta(dx|T = t) = P_\eta(dx|T = t)$. If T is sufficient in this sense, it is also sufficient in the sense of definition 2.2 where we may take $Q_\tau = P_{\tau,\eta_1}$ for any particular η_1.

2.2. *The principle of sufficiency.* If comparing the formulations of this principle as given in Blackwell-Girshick [2] and in A. Birnbaum [1], one feels that there is an apparent discrepancy. According to Birnbaum's sufficiency principle, we are not allowed to use randomized tests, for example, while no such implication follows from the Blackwell-Girshick sufficiency principle. The difference is serious but easy to explain: Blackwell and Girshick consider only convex situations (that is, convex D and convex $L(\theta, d)$ for each θ), where the Rao-Blackwell theorem can be proved, while A. Birnbaum has in mind any possible situation. However, what may be supported in convex situations by a theorem is a rather stringent postulate in a general condition. (If, in Blackwell-Girshick, the situation is not convex, they make it convex by allowing randomized decisions.)

In estimating a real parameter with $L(\theta, d) = (\theta - d)^2$, the convexity conditions are satisfied and no randomization is useful. If, however, θ would run through a discrete subset of real numbers, randomization might bring the same gains from the minimax point of view as in testing a simple hypothesis against a simple alternative. And even in convex situations the principle may not exclude any decision procedure as inadmissible. To this end it is necessary for $L(\tau, d)$ to be strictly convex in d, and not, for example, linear, as in randomized extensions of nonconvex problems.

3. Invariance

Most frequently, the nuisance parameter is introduced by a group of transformations of the x-space on itself. Then, if we have a finite invariant measure on the group, we can easily show that the sub-σ-field of invariant events is sufficient in the presence of the corresponding nuisance parameter.

Let $G = \{g\}$ be a group of one-to-one transformations of the x-space on itself. Take a probability distribution P of X and put

$$P_g(X \in A) = P(gX \in A). \tag{3.1}$$

Then obviously, $P_h(X \in g^{-1}A) = P_{gh}(X \in A)$. Let μ be a σ-finite measure and denote by μg the measure such that $\mu g(A) = \mu(gA)$.

THEOREM 3.1. *Let $P \ll \mu$ and $\mu g \ll \mu$ for all $g \in G$. Then $P_g \ll \mu$. Denoting $p(x, g) = dP_g/d\mu$ and $p(x) = dP/d\mu$, then*

$$p(x, g) = p(g^{-1}x) \frac{d\mu g^{-1}}{d\mu}(x) \tag{3.2}$$

holds. More generally,

$$p(x, h^{-1}g) = p(h(x), g) \frac{d\mu h}{d\mu}(x). \tag{3.3}$$

PROOF. According to the definition, one can write

$$\begin{aligned} P_g(X \in A) = P(X \in g^{-1}A) &= \int_{g^{-1}A} p(x)\, d\mu = \int_A p(g^{-1}(y))\, d\mu g^{-1}(y) \\ &= \int_A p(g^{-1}(y)) \frac{d\mu g^{-1}}{d\mu}(y)\, d\mu(y). \end{aligned} \tag{3.4}$$

CONDITION 3.1. *Let $\mathcal{G}$ be a σ-field of subsets of G, and let $\mathcal{A}$ be the σ-field of subsets of the x-space. Assume that*

(i) *$\mu g \ll \mu$ for all $g \in G$,*

(ii) *$p(x, g)$ is $\mathcal{A} \times \mathcal{G}$-measurable,*

(iii) *functions $\phi_h(g) = hg$ and $\psi_h(g) = gh$ are $\mathcal{G}$-measurable,*

(iv) *there is an invariant probability measure ν on $\mathcal{G}$, that is, $\nu(Bg) = \nu(gB) = \nu(B)$ for all $g \in G$ and $B \in \mathcal{G}$.*

THEOREM 3.2. *Under condition 3.1, let us put*

$$\bar{p}(x) = \int p(x, g)\, d\nu(g). \tag{3.5}$$

Then, for each $h \in G$,

$$\bar{p}(h(x)) = \left[\frac{d\mu h}{d\mu}(x)\right]^{-1} \bar{p}(x). \tag{3.6}$$

REMARK 3.1. Note that the first factor on the right-hand side does not depend on p. Obviously $\bar{p}(x, g) = \bar{p}(x)$ for all $g \in G$.

PROOF. In view of (3.3), we have

$$\begin{aligned} \bar{p}(h(x)) &= \int p(h(x), g)\, d\nu(g) = \left[\frac{d\mu h}{d\mu}(x)\right]^{-1} \int p(x, h^{-1}g)\, d\nu \\ &= \left[\frac{d\mu h}{d\mu}(x)\right]^{-1} \int p(x, f)\, d\nu h(f) \\ &= \left[\frac{d\mu h}{d\mu}(x)\right]^{-1} \int p(x, f)\, d\nu(f) \\ &= \left[\frac{d\mu h}{d\mu}(x)\right]^{-1} \bar{p}(x), \end{aligned} \tag{3.7}$$

Q.E.D.

We shall say that an event is G-invariant, if $gA = A$ for all $g \in G$. Obviously, the set of G-invariant events is a sub-σ-field $\mathcal{B}$, and a measurable function f is $\mathcal{B}$-measurable if and only if $f(g(x)) = f(x)$ for all $g \in G$. Consider now two distributions P and P_0 and seek the derivative of P relative to P_0 on $\mathcal{B}$, say $[dP/dP_0]^{\mathcal{B}}$. Assume that $P \ll \mu$ and $P_0 \ll \mu$, denote $p = dP/d\mu$, $p_0 = dP_0/d\mu$, and introduce $\bar{p}(x)$ and $\bar{p}_0(x)$ by (3.5).

THEOREM 3.3. *Under condition* 3.1 *and under the assumption that* $\bar{p}_0(x) = 0$ *entails* $\bar{p}(x) = 0$ *almost* μ-*everywhere, we have* $P \ll P_0$ *on* $\mathcal{B}$ *and*

$$[dP/dP_0]^{\mathcal{B}} = \bar{p}(x)/\bar{p}_0(x). \tag{3.8}$$

PROOF. Put $l(x) = \bar{p}(x)/\bar{p}_0(x)$. Theorem 3.2 entails $l(g(x)) = l(x)$ for all $g \in G$, namely, $l(x)$ is $\mathcal{B}$-measurable. Further, for any $B \in \mathcal{B}$, in view of (3.2) and of $[p = 0] \Rightarrow [p_0 = 0]$,

$$\begin{aligned} \int_B l(x)\, dP_0 &= \int_B l(x) p_0(x)\, d\mu = \int_B l(x) p_0(g^{-1}(x))\, d\mu g^{-1} \\ &= \int_B l(x) p_0(x, g)\, d\mu = \int_B l(x) \bar{p}_0(x)\, d\mu = \int_B \bar{p}(x)\, d\mu \end{aligned} \tag{3.9}$$

holds. On the other hand,

$$P(B) = \int_B p(x)\, d\mu = \int_B p(g^{-1}(x))\, d\mu g^{-1} = \int_B p(x, g)\, d\mu = \int_B \bar{p}(x)\, d\mu. \tag{3.10}$$

Thus $P(B) = \int_B l(x)\, dP_0$, $B \in \mathcal{B}$, which concludes the proof.

THEOREM 3.4. *Let the statistic* $T = t(X)$ *have an expectation under* P_h, $h \in G$. *Let condition* 3.1 *be satisfied. Then the conditional expectation of* T *relative to the sub-*σ*-field* $\mathcal{B}$ *of* G*-invariant events and under* P_h *equals*

$$E_h(T|\mathcal{B}, x) = \int t(g^{-1}(x)) p(x, gh)\, d\nu(g)/\bar{p}(x). \tag{3.11}$$

PROOF. The proof would follow the lines of the proofs of theorems 3.2 and 3.3.

Consider now a dominated family of probability distributions $\{P_\tau\}$ and define $P_{\tau,g}$ by (3.1) for each τ. Putting $\theta = (\tau, g)$, we can say that τ is a subparameter of θ.

THEOREM 3.5. *Under condition* 3.1 *the sub-σ-field $\mathcal{B}$ of G-invariant events is sufficient for τ in the sense of definition* 2.2.

PROOF. First, the G-invariant events have a probability depending on τ only, that is, $P_{\tau,g}(B) = P_\tau(B)$. Second, for

$$Q_\tau(A) = \int P_{\tau,g}(A)\, d\nu(g) \tag{3.12}$$

we have

$$\begin{aligned} Q_\tau(A) &= \int \left[\int_A p_\tau(x, g)\, d\mu\right] d\nu(g) \\ &= \int_A \int p_\tau(x, g)\, d\nu(g)\, d\mu \\ &= \int_A \bar{p}_\tau(x)\, d\mu. \end{aligned} \tag{3.13}$$

Now let P_0 be some probability measure such that $P_\tau \ll P_0 \ll \mu$, and introduce $\bar{p}_0(x)$ by (3.5). Then, according to theorem 3.3, $\bar{p}_\tau(x)/\bar{p}_0(x)$ is $\mathcal{B}$-measurable for all τ. Thus $\bar{p}(x) = [\bar{p}_\tau(x)/\bar{p}_{\tau 0}(x)]\bar{p}_0(x)$, and $\mathcal{B}$ is sufficient for $\{Q_\tau\}$, according to the factorization criterion.

REMARK 3.2. Considering right-invariant and left-invariant *probability* measures on $\mathcal{G}$ does not provide any generalization. Actually, if ν is right-invariant, then $\bar{\nu}(B) = \int_A \nu(gB)\, d\nu(g)$ is invariant, that is, both right-invariant and left-invariant.

REMARK 3.3. Theorem 3.3 sometimes remains valid even if there exists only a right-invariant countably finite measure ν, as in the case of the groups of location and/or scale shifts (see [6], chapter II).

4. Similarity

The concept of similarity plays a very important role in classical statistics, namely in contributions by J. Neyman and R. A. Fisher. It may be applied in the x-space as well as in the (x, θ)-space.

4.1. *Similarity in the x-space.* Consider a family of distributions $\{P_\theta\}$ on measurable subsets of the x-space. We say that an *event* A is *similar*, if its probability is independent of θ:

$$P_\theta(A) = P(A) \qquad \text{for all } \theta. \tag{4.1}$$

Obviously the events "whole x-space" and "the empty set" are always similar.

The class of similar events is closed under complementation and formation of countable *disjoint* unions. Such classes are called λ-fields by Dynkin [3]. A λ-field is a broader concept than a σ-field. The class of similar events is usually not a σ-field, which causes ambiguity in applications of the conditionality principle, as we shall see. Generally, if both A and B are similar and are not disjoint, then $A \cap B$ and $A \cup B$ may not be similar. Consequently, the system of σ-fields contained in a λ-field may not include a largest σ-field.

Dynkin [3] calls a system of subsets a π-field, if it is closed under intersection, and shows that a λ-field containing a π-field contains the smallest σ-field over the π-field. Thus, for example, if for a vector statistic $T = t(X)$ the events $\{x : t(x) < c\}$ are similar for every vector c, then $\{x : T \in A\}$ is similar for every Borel set A.

More generally, a *statistic* $V = v(X)$ is called *similar*, if $E_\theta V = \int v(x) P_\theta(dx)$ exists and is independent of θ. We also may define similarity with respect to a nuisance parameter only. The notion of similarity forms a basis for the definition of several other important notions.

Ancillary statistics. A statistics $U = u(x)$ is called ancillary, if its distribution is independent of θ, that is, if the events generated by U are similar. (We have seen that it suffices that the events $\{V < c\}$ be similar.)

Correspondingly, a sub-σ-field will be called ancillary, if all its events are similar.

Exhaustive statistics. A statistics $T = t(x)$ is called exhaustive if (T, U), with U an ancillary statistic, is a minimal sufficient statistic.

Complete families of distributions. A family $\{P_\theta\}$ is called complete if the only similar statistics are constants.

Our definition of an exhaustive statistic follows Fisher's explanation (ii) in ([5], p. 49), and the examples given by him.

Denote by I_θ^X, I_θ^T, $I_\theta^T(u)$ Fisher's information for the families $\{P_\theta(dx)\}$, $P_\theta(dt)\}$, $\{P_\theta(dt) | U = u)\}$, respectively. Then, since (T, U) is sufficient and U is ancillary,

$$EI_\theta^T(U) = I_\theta^X. \tag{4.2}$$

If $I_\theta^T < I_\theta^X$, Fisher calls (4.2) "recovering the information lost." What is the real content of this phrase?

The first interpretation would be that $T = t$ contains all information supplied by $X = x$, provided that we know $U = u$. But knowing both $T = t$ and $U = u$, we know $(T, U) = (t, u)$, that is, we know the value of the sufficient statistics. Thus this interpretation is void, and, moreover, it holds even if U is not ancillary.

A more appropriate interpretation seems to be as follows. Knowing $\{P_\theta(dt | U = u)\}$, we may dismiss the knowledge of $P(du)$ as well as of $\{P_\theta(dt) | U = u')\}$ for $u' \neq u$. This, however, expresses nothing else as the conditionality principle formulated below.

Fisher makes a two-field use of exhaustive statistics: first, for extending the scope of fiducial distributions to cases where no appropriate sufficient statistic exists, and, second, to a (rather unconvincing) eulogy of maximum likelihood estimates.

The present author is not sure about the appropriateness of restricting ourselves to "minimal" sufficient statistics in the definition of exhaustive statistics. Without this restriction, there would rarely exist a minimal exhaustive statistic, and with this restriction we have to check, in particular cases, whether the employed sufficient statistic is really minimal.

4.2. *Similarity in the* (x, θ)*-space.* Consider a family of distributions $\{P_\theta\}$ on the x-space, the family of all possible prior distributions $\{\nu\}$ on the θ-space, and the family of distributions $\{R_\nu\}$ on the (x, θ)-space given by

$$R_\nu(dx\, d\theta) = \nu(d\theta)P_\theta(dx). \tag{4.3}$$

Then we may introduce the notion of similarity in the (x, θ)-space in the same manner as before in the x-space, with $\{P_\theta\}$ replaced by $\{R_\nu\}$. Consequently, the prior distribution will play the role of an unknown parameter. Thus an event Λ in the (x, θ)-space will be called similar, if

$$R_\nu(\Lambda) = R(\Lambda) \tag{4.4}$$

for all prior distributions ν.

To avoid confusion, we shall call measurable functions of (x, θ) *quantities* and not statistics. A quantity $H = h(X, \Theta)$ will be called similar if its expectation $EH = \int h(x, \theta)\nu(d\theta)P_\theta(dx)$ is independent of ν. Ancillary statistics in the (x, θ)-space are called *pivotal* quantities or *distribution-free* quantities. Since only the first component of (x, θ) is observable, the applications of similarity in the (x, θ)-space are quite different from those in the x-space.

Confidence regions. This method of region estimation dwells on the following idea. Having a similar event Λ, whose probability equals $1 - \alpha$, with α very small, and knowing that $X = x$, we can feel confidence that $(x, \theta) \in \Lambda$, namely, that the unknown θ lies within the region $S(x) = \{\theta\colon (x, \theta) \in \Lambda\}$. Here, our confidence that the event Λ has occurred is based on its high probability, and this confidence is assumed to be unaffected by the knowledge of $X = x$. Thus we assume a sort of independence between Λ and X, though their joint distribution is indeterminate. The fact that this intrinsic assumption may be dubious is most appropriately manifested in cases when $S(x)$ is either empty, so that we know that Λ did not occur, or equals the whole θ-space, so we are sure that Λ has occurred. Such a situation arises in the following.

EXAMPLE 1. For this example, $\theta \in [0, 1]$, x is real, $P_\theta(dx)$ is uniform over $[0, \theta + 2]$, and

$$\Lambda = \{(x, \theta)\colon \theta + \alpha < x < \theta + 2 - \alpha\}. \tag{4.5}$$

Then $S(x) = \varnothing$ for $3 - \alpha < x < 3$ or $0 < x < \alpha$, and $S(x) = [0, 1]$ for $1 + \alpha < x < 2 - \alpha$. Although this example is somewhat artificial, the difficulty involved seems to be real.

Fiducial distribution. Let $F(t, \theta)$ be the distribution function of T under θ, and assume that F is continuous in t for every θ. Then $H = F(T, \Theta)$ is a pivotal quantity with uniform distribution over $[0, 1]$. Thus $R(F(T, \Theta) \le x) = x$. Now, if $F(t, \theta)$ is strictly decreasing in θ for each t, and if $F^{-1}(t, \theta)$ denotes its inverse for fixed t, then $F(T, \Theta) \le x$ is equivalent to $\Theta > F^{-1}(T, x)$. Now, again, if we know that $T = t$, and if we feel that it does not affect the probabilities concerning $F(T, \Theta)$, we may write

$$(4.6) \qquad x = R(F(T, \Theta) \leq x = R(\Theta \geq F^{-1}(T, x)) = R(\Theta \geq F^{-1}(T, x)|T = t) = R(\Theta \geq F^{-1}(t, x)|T = t)$$

namely, for $\theta = F^{-1}(t, x)$,

$$(4.7) \qquad R(\Theta < \theta|T = t) = 1 - F(t, \theta).$$

As we have seen, in confidence regions as well as in fiducial probabilities, a peculiar postulate of independence in involved. The postulate may be generally formulated as follows.

Postulate of independence under ignorance. Having a pair (Z, W) such that the marginal distribution of Z is known, but the joint distribution of (Z, W), as well as the marginal distribution of W, are unknown, and observing $W = w$, we assume that

$$(4.8) \qquad P(Z \in \Lambda|W = w) = P(Z \in \Lambda).$$

J. Neyman gives a different justification of the postulate than R. A. Fisher. Let us make an attempt to formulate the attitudes of both these authors.

Neyman's justification. Assume that we perform a long series of independent replications of the given experiments, and denote the results by $(Z_1, w_1), \cdots, (Z_N, w_N)$, where the w_i's are observed numbers and the Z_i's are not observable. Let us decide to accept the hypothesis $Z_i \in \Lambda$ at each replication. Then our decisions will be correct approximately in $100P\%$ of cases with $P = P(Z \in \Lambda)$. Thus, in a long series, our mistakes in determining $P(Z \in \Lambda|W = w)$ by (4.8) if any, will compensate each other.

Fisher's justification. Suppose that the only statistics $V = v(W)$, such that the probabilities $P(Z \in \Lambda|V = v)$ are well-determined, are constants. Then we are allowed to take $P(Z \in \Lambda|W = w) = P(Z \in \Lambda)$, since our absence of knowledge prevents us from doing anything better.

The above interpretation of Fisher's view is based on the following passage from his book ([5], pp. 54–55):

"The particular pair of values of θ and T appropriate to a particular experimenter certainly belongs to this enlarged set, and within this set the proportion of cases satisfying the inequality

$$(4.9) \qquad \theta > \frac{T}{2n} \chi^2_{2n}(P)$$

is certainly equal to the chosen probability P. It might, however, have been true . . . that in some recognizable subset, to which his case belongs, the proportion of cases in which the inequality was satisfied should have some value other than P. It is the stipulated absence of knowledge *a priori* of the distribution of θ, together with the exhaustive character of the statistic T, that makes the recognition of any such subset impossible, and so guarantees that in his particular case . . . the general probability is applicable."

To apply our general scheme to the case considered by R. A. Fisher, we should put $Z = 1$ if (4.9) is satisfied and $Z = 0$ otherwise, and $W = T$.

REMARK 4.1. We can see that Fisher's argumentation applies to a more specific situation than described in the above postulate. He requires that conditional probabilities are not known for any statistics $V = v(W)$ except for $V =$ const. Thus the notion of fiducial probabilities is a more special notion than the notion of confidence regions, since Neyman's justification does not need any such restrictions.

Fisher's additional requirement is in accord with his requirement that fiducial probabilities should be based on the minimal sufficient statistics, and in such a case does not lead to difficulties.

However, if the fiducial probabilities are allowed to be based on exhaustive statistics, then his requirement is contradictory, since no minimal exhaustive statistics may exist. Nonetheless, we have the following.

THEOREM 4.1. *If a minimal sufficient statistic $S = s(X)$ is complete, then it is also a minimal exhaustive statistic.*

PROOF. If there existed an exhaustive statistic $T = t(S)$ different from S, there would exist a nonconstant ancillary statistic $U = u(S)$, which contradicts the assumed completeness of S. Q.E.D.

If S is not complete, then there may exist ancillary statistics $U = u(S)$ and the family corresponding to exhaustive statistics contains a minimal member if and only if the family of U's contains a maximal member.

REMARK 4.2. Although the two above justifications are unconvincing, they correspond to habits of human thinking. For example, if one knows that an individual comes from a subpopulation of a population where the proportion of individuals with a property A equals P, and if one knows nothing else about the subpopulation, one applies P to the given individual without hesitation. This is true even if we know the "name" of the individual, that is, if the subpopulation consists of a single individual. Fisher is right, if he claims that we would not use P if the given subpopulation would be a part of a larger one, in which the proportion is known, too. He does not say, however, what to do if there is no minimal such larger subpopulation.

In confidence regions the subpopulation consists of pairs (X, Θ) such that $X = x$. It is true that we know the "proportion" of elements of that subpopulation for which the event Λ occurs, but we do not know how much of the probability mass each element carries. Thus we can utilize the knowledge of this proportion only if it equals 0 or 1, which is exemplified by example 1 but occurs rather rarely in practice. The problem becomes still more puzzling if the knowledge of the proportion is based on estimates only, and if these estimates become less reliable as the subpopulation from which they are derived becomes smaller. Is there any reasonable recommendation as to what to do if we know $P(X \in A|U_1 = u_1)$ and $P(X \in A|U_2 = u_2)$, but we do not know $P(X \in A|U_1 = u_1, U_2 = u_2)$?

REMARK 4.3. Fisher attacked violently the Bayes postulate as an adequate form of expressing mathematically our ignorance. He reinforced this attitude in ([5], p. 20), where we may read: "It is evidently easier for the practitioner of natural science to recognize the difference between knowing and not knowing than it seems to be for the more abstract mathematician." On the other hand, as we have seen, he admits that the absence of knowledge of proportions in a subpopulation allows us to act in the same manner as if we knew that the proportion is the same as in the whole population. The present author suspects that this new postulate, if not stronger than the Bayes postulate, is by no means weaker. Actually, if a proportion P in a population is interpreted as probability for an individual taken from this population, we tacitly assume that the individual has been selected according to the uniform distribution.

One cannot escape the feeling that all attempts to avoid expressing in a mathematical form the absence of our prior knowledge about the parameter have been eventually a failure. Of course, the mathematical formalization of ignorance should be understood broadly enough, including not only prior distributions, but also the minimax principle, and so on.

4.3. *Estimation.* Similarity in the (x, θ)-space is widely utilized in estimation. For example, an estimate $\hat{\theta}$ is unbiased if and only if the quantity $H = \hat{\theta} - \theta$ is similar with zero expectation. Further, similarity provides the following class of estimation methods: starting with a similar quantity $H = h(X, \Theta)$ with $EH = c$, and observing $X = x$, we take for θ the solution of equation

$$h(x, \theta) = c. \tag{4.10}$$

This class includes the method of maximum likelihood for

$$h(x, \theta) = (\partial/\partial\theta) \log p_\theta(x). \tag{4.11}$$

Another method is offered by the pivotal quantity $H = F(T, \Theta)$ considered in connection with fiducial probabilities. Since $EH = \frac{1}{2}$, we may take for $\hat{\theta}$, given $T = t$, the solution if any of

$$F(t, \theta) = \tfrac{1}{2}, \tag{4.12}$$

that is, the parameter value for which the observed value is the median.

5. Conditionality

If the prior distribution ν of θ is known, all statisticians agree that the decisions given $X = x$ should be made on the basis of the conditional distribution $R_\nu(d\theta | X = x)$. This exceptional agreement is caused by the fact that then there is only one distribution in the (x, θ)-space, so that the problems are transferred from the ground of statistics to the ground of pure probability theory.

The problems arise in situations where conditioning is applied to a family of distributions. Here two basically different situations must be distinguished according to whether the conditioning statistic is ancillary or not. Conditioning with respect to an ancillary statistic is considered in the conditionality principle as formulated by A. Birnbaum [1]. On the other hand, for example, conditioning with respect to a sufficient statistic for a nuisance parameter, successfully used in constructing most powerful similar tests (see Lehmann [14]), is a quite different problem and will not be considered here.

Our discussion will concentrate on the conditionality principle, which may be formulated as follows.

The principle of conditionality. Given an ancillary statistic U, and knowing $U = u$, statistical inference should be based on conditional probabilities $P_\theta(dx|U = u)$ only; that is, the probabilities $P_\theta(dx|U = u')$ for $u' \neq u$ and $P(du)$ should be disregarded.

This principle, if properly illustrated, looks very appealing. Moreover, all *proper* Bayesian procedures are concordant with it. We say that a Bayesian procedure is proper, if it is based on a prior distribution ν established independently of the experiment. A Bayesian procedure, which is not proper, is exemplified by taking for the prior distribution the measure ν given by

$$\nu(d\theta) = \sqrt{I_\theta}\, d\theta, \tag{5.1}$$

where I_θ denotes the Fisher information associated with the particular experiment. Obviously, such a Bayesian procedure is not compatible with the principle of conditionality. (Cf. [4].)

The term "statistical inference" used in the above definition is very vague. Birnbaum [1] makes use of an equally vague term "evidential meaning." The present author sees two possible interpretations within the framework of the decision theory.

Risk interpretation. A decision procedure δ defined on the original experiment should be associated with the same risk as its restriction associated with the partial experiment given $U = u$. Of course this transfer of risk is possible only after the experiment has been performed and u is known.

Decision rules interpretation. A decision function δ should be interpreted as a function of two arguments, $\delta = \delta(E, x)$, with E denoting an experiment from a class $\mathcal{E}$ of experiments and x denoting one of its outcomes. The principle of conditionality then restricts the class of "reasonable" decision functions to those for which $\delta(E, x) = \delta(F, x)$ as soon as F is a subexperiment of E, and x belongs to the set of outcomes of F (F being a subexperiment of E means that F equals "E given $U = u$" where U is some ancillary statistics and u some of its particular value).

In this section we shall analyze the former interpretation, returning to the latter in the next section.

The principle of conditionality, as it stands, is not acceptable in non-Bayesian conditions, since it denies possible gains obtainable by randomization (mixed strategies). On the other hand, some examples exhibited to support this principle (see [1], p. 280) seem to be very convincing. Before attempting to delimit the area of proper applications of the principle, let us observe that the principle is generally ambiguous. Actually, since generally no maximal ancillary statistic exists, it is not clear which ancillary statistic should be chosen for conditioning.

Let us denote the conditional distribution given $U = u$ by $P_\theta(dx|U = u)$ and the respective conditional risk by

$$R(\theta, U, u) = \int L(\theta, \delta(x))P_\theta(dx|U = u). \tag{5.2}$$

In terms of a sub-σ-field $\mathcal{B}$, the same will be denoted by

$$R(\theta, \mathcal{B}, x) = \int L(\theta, \delta(x))P_\theta(dx|\mathcal{B}, x), \tag{5.3}$$

where $R(\theta, \mathcal{B}, x)$, as a function x, is $\mathcal{B}$-measurable. Since the decision function δ will be fixed in our considerations, we have deleted it in the symbol for the risk.

DEFINITION 5.1. *A conditioning relative to an ancillary statistic U or an ancillary sub-σ-field $\mathcal{B}$ will be called* correct, *if*

$$R(\theta, U, u) = R(\theta)b(u) \tag{5.4}$$

or

$$R(\theta, \mathcal{B}, x) = R(\theta)b(x), \tag{5.5}$$

respectively. In (5.4) *and* (5.5), *$R(\theta)$ denotes the overall risk, and the function $b(x)$ is $\mathcal{B}$-measurable. Obviously, $Eb(U) = Eb(X) = 1$.*

The above definition is somewhat too strict, but it covers all examples exhibited in literature to support the conditionality principle.

THEOREM 5.1. *If the conditioning relative to an ancillary sub-σ-field $\mathcal{B}$ is correct, then*

$$E[\sup_\theta R(\theta, \mathcal{B}, X)] = \sup_\theta R(\theta). \tag{5.6}$$

PROOF. The proof follows immediately from (5.5). Obviously, for a noncorrect conditioning we may obtain $E[\sup_\theta R(\theta, \mathcal{B}, X)] > \sup_\theta R(\theta)$, so that, from the minimax point of view, conditional reasoning generally disregards possible gains obtained by mixing (randomization), and, therefore, is hardly acceptable under non-Bayesian conditions.

THEOREM 5.2. *As in section* 3, *consider the family $\{P_g\}$ generated by a probability distribution P and a group G of transformations g. Further, assume the loss function L to be invariant, namely, such that $L(hg, \delta(h(x)) = L(g, \delta(x))$ holds for all h, g, and x. Then the conditioning relative to the ancillary sub-σ-field $\mathcal{B}$ of G-invariant events is correct. That is, $R(g) = R$ and*

$$R(g, \mathcal{B}, x) = R(\mathcal{B}, x) = \int L[g, \delta(h^{-1}(x))]p(x, hg)\, d\nu(h)/\bar{p}(x). \tag{5.7}$$

PROOF. For $B \in \mathcal{B}$,

$$\begin{aligned} \int_B L(g, \delta(x))p(x, g)\, d\mu(x) &= \int_{h^{-1}B} L[g, \delta(h(y))]p(h(y), g)\, d\mu h(y) \\ &= \int_B L(h^{-1}g, \delta(y))p(y, h^{-1}g)\, d\mu(y). \end{aligned} \tag{5.8}$$

The theorem easily follows from theorem 3.4 and from the above relations.

The following theorem describes an important family of cases of ineffective conditioning.

THEOREM 5.3. *If there exists a complete sufficient statistic $T = t(X)$, then $R(\theta, U, u) = R(\theta)$ holds for every ancillary statistic U and every δ which is a function of T.*

PROOF. The theorem follows from the well-known fact (see Lehmann [14], p. 162) that all ancillary statistics are independent of T, if T is complete and sufficient.

DEFINITION 5.2. *We shall say an ancillary sub-σ-field $\mathcal{B}$ yield the* deepest correct conditioning, *if for every nonnegative convex function ψ and every other ancillary sub-σ-field $\overline{\mathcal{B}}$ yielding a correct conditioning,*

$$E\psi[b(X)] \geq E\psi[\bar{b}(X)] \tag{5.9}$$

holds, with b and $\bar{b}$ corresponding to $\mathcal{B}$ and $\overline{\mathcal{B}}$ by (5.5), respectively.

THEOREM 5.4. *Under condition 3.1 and under the conditions of theorem 5.2, the ancillary sub-σ-field $\mathcal{B}$ of G-invariant events yields the deepest correct conditioning. Further, if any other ancillary sub-σ-field possesses this property, then $R(\overline{\mathcal{B}}, x) = R(\mathcal{B}, x)$ almost μ-everywhere.*

PROOF. Let ν denote the invariant measure on $\mathcal{G}$. Take a $C \in \overline{\mathcal{B}}$ and denote by $\chi_C(x)$ its indicator. Then

$$\phi_C(x) = \int \chi_C(gx)\, d\nu(g) \tag{5.10}$$

is $\mathcal{B}$-measurable and

$$P(C) = \int \phi_C(x) p(x)\, d\mu. \tag{5.11}$$

Further, since the loss function is invariant,

$$\int_C L(g, \delta(x)) p(x, g)\, d\mu = \int \chi_C(g(y)) L(g_1, \delta(y)) p(y)\, d\mu(y) \tag{5.12}$$

where g_1 denotes the identity transformation. Consequently, denoting by $R(g, C)$ the conditional risk, given C, we have from (5.10) and (5.12),

$$P(C) \int R(g, C)\, d\nu(g) = \int \phi_C(x) L(g_1, \delta(x)) p(x)\, d\mu. \tag{5.13}$$

Now, since $R(g) = R$, according to theorem 5.2, and since the correctness of $\overline{\mathcal{B}}$ entails $R(g, C) = R\bar{b}_C$, we have

$$\int R(g, C)\, d\nu(g) = R\bar{b}_C. \tag{5.14}$$

Note that

$$P(C)\bar{b}_C = \int_C \bar{b}(x) p(x)\, d\mu. \tag{5.15}$$

Further, since $\phi_C(x)$ is $\mathcal{B}$-measurable,

$$\begin{aligned} \int \phi_C(x) L(g_1, \delta(x)) p(x)\, d\mu &= \int \phi_C(x) R(\mathcal{B}, x) p(x)\, d\mu \\ &= \int \phi_C(x) R b(x) p(x)\, d\mu. \end{aligned} \tag{5.16}$$

By combining together (5.13) through (5.16), we obtain

$$P(C)\bar{b}_C = \int \phi_C(x)b(x)p(x)\,d\mu. \tag{5.17}$$

Now, let us assume $E\psi(\bar{b}(X)) < \infty$. Then, given a $\epsilon > 0$, we choose a finite partition $\{C_k\}$, $C_k \in \bar{\mathfrak{B}}$, such that

$$E\psi(\bar{b}(X)) < \sum_k P(C_k)\psi(\bar{b}_{C_k}) + \epsilon, \tag{5.18}$$

and we note that, in view of (5.11), (5.17), and of convexity of ψ,

$$P(C_k)\psi(\bar{b}_{C_k}) \leq \int \phi_{C_k}(x)\psi(b(x))p(x)\,d\mu, \tag{5.19}$$

and, in view of (5.10),

$$\sum_k \phi_{C_k}(x) = 1 \tag{5.20}$$

for every x. Consequently, (5.18) through (5.20) entail

$$E\psi(\bar{b}(X)) < E\psi(b(X)) + \epsilon. \tag{5.21}$$

Since $\epsilon > 0$ is arbitrary, (5.9) follows. The case $E\psi(\bar{b}(X)) = \infty$ could be treated similarly.

The second assertion of the theorem follows from the course of proving the first assertion. Actually, we obtain $E\psi(\bar{b}(X)) < E\psi(b(X))$ for some ψ, unless $\bar{b}$ is a function of b a.e. Also conversely, b must be a function of $\bar{b}$, because the two conditionings are equally deep. Q.E.D.

A restricted conditionality principle. If all ancillary statistics (sub-σ-fields) yielding the deepest correct conditioning give the same conditional risk, and if there exists at least one such ancillary statistic (sub-σ-field), then the use of the conditional risk is obligatory.

6. Likelihood

Still more attractive than the principle of conditionality appears the principle of likelihood, which may be formulated in lines of A. Birnbaum [1], as follows.

The principle of likelihood. Statistical inferences should be based on the likelihood functions only, disregarding the other structure of the particular experiments. Here, again, various interpretations are possible. We suggest the following.

Interpretation. For a given θ-space, the particular decision procedures should be regarded as functionals on a space $\mathfrak{L}$ of likelihood functions, where all functions differing in a positive multiplicator are regarded as equivalent. Given an experiment E such that for all outcomes x the likelihood functions $l_x(\theta)$ belong to $\mathfrak{L}$, and a decision procedure δ in the above sense, we should put

$$\delta(x) = \delta[l_x(\cdot)] \tag{6.1}$$

where $l_x(\theta) = p_\theta(x)$.

If $\mathfrak{L}$ contains only unimodal functions, an estimation procedure of the above kind is the maximum likelihood estimation method.

A. Birnbaum [1] proved that the principle of conditionality joined with the principle of sufficiency is equivalent to the principle of likelihood. He also conjectured that the principle of sufficiency may be left out in his theorem, and gave a hint, which was not quite clear, for proving it. But his conjecture is really true if we assume that *statistical inferences should be invariant under one-to-one transformations of the x-space to some other homeomorph space.* In the following theorem we shall give the "decision interpretation" to the principle of conditionality, and the class $\mathcal{E}$ of experiments will be regarded as the class of all experiments such that all their possible likelihood functions belong to some space $\mathcal{L}$.

THEOREM 6.1. *Under the above stipulations, the principle of conditionality is equivalent to the principle of likelihood.*

PROOF. If F is a subexperiment of E, and if x belongs to the space of possible outcomes of F, the likelihood function $l_x(\theta|E)$ and $l_x(\theta|F)$ differ by a positive multiplicator only, namely, they are identical. Thus the likelihood principle entails the conditionality principle.

Further, assume that the likelihood principle is violated for a decision procedure δ, that is there exist in $\mathcal{E}$ two experiments $E_1 = (\mathcal{X}_1, P_{1\theta})$ and $E_2 = (\mathcal{X}_2, P_{2\theta})$ such that for some points x_1 and x_2,

$$\delta(E_1, x_1) \neq \delta(E_2, x_2), \tag{6.2}$$

whereas

$$P_{1\theta}(X_1 = x_1) = cP_{2\theta}(X_2 = x_2) \qquad \text{for all} \quad \theta, \tag{6.3}$$

with $c = c(x_1, x_2)$, but independent of θ. We here assume the spaces $\mathcal{X}_1$ and $\mathcal{X}_2$ finite and disjoint, and the σ-fields to consist of all subsets. Then let us choose a number λ such that

$$0 < \lambda < \frac{1}{c+1} \tag{6.4}$$

and consider the experiment $E = (\mathcal{X}_1 \cup \mathcal{X}_2, P_\theta)$, where

$$\begin{aligned} P_\theta(X = x) &= \lambda P_{1\theta}(x) && \text{if} \quad x \in \mathcal{X}_1, \\ &= \lambda c P_{2\theta}(x) && \text{if} \quad x \in \mathcal{X}_2 - \{x_2\}, \\ &= 1 - \lambda(c + 1 - P_{1\theta}(x_1)) && \text{if} \quad x = x_2. \end{aligned} \tag{6.5}$$

Then E conditioned by $x \in \mathcal{X}_1$ coincides with E_1, and E conditioned by $x \in (\mathcal{X}_2 - \{x_2\}) \cup \{x_1\}$ coincides with $\tilde{E}_2$ which is equivalent to E_2 up to a one-to-one transformation $\phi(x) = x$, if $x \in \mathcal{X}_2 - \{x_2\}$ and $\phi(x_2) = x_1$. Thus we should have $\delta(E_1, x_1) = \delta(E, x_1) = \delta(\tilde{E}_2, x_1) = \delta(E_2, x_2)$, according to the conditionality principle. In view of (6.2) this is not true, so that the conditionality principle is violated for δ, too. Q.E.D.

Given a fixed space $\mathcal{L}$, the likelihood principle says what decision procedures of wide scope (covering many experimental situations) are permissible. Having any two such procedures, and wanting to compare them, we must resort either to some particular experiment and compute the risk, or to assume some a priori distribution $\nu(d\theta)$ and to compute the conditional risk. In both cases we leave the proper ground of the likelihood principle.

However, for some special spaces $\mathcal{L}$, all conceivable experiments with likelihood functions in $\mathcal{L}$, give us the same risk, so that the risk may be regarded as independent of the particular experiment. The most important situation of this kind is treated in the following.

THEOREM 6.2. *Let θ be real and let $\mathcal{L}_\sigma$ consist of the following functions:*

$$l(\theta) = c \exp\left[-\tfrac{1}{2} \frac{(\theta - t)^2}{\sigma^2}\right], \qquad -\infty < t < \infty, \tag{6.6}$$

with σ fixed and positive. Then for each experiment with likelihood functions in $\mathcal{L}_\sigma$ there exists a complete sufficient statistic $T = t(X)$, which is normally distributed with expectation θ and variance σ^2.

PROOF. We have a measurable space $(\mathfrak{X}, t)$ with a σ-finite measure $\mu(dx)$ such that the densities with respect to μ allow the representation

$$p_\theta(x) = c(x) \exp\left[-\tfrac{1}{2} \frac{(\theta - t(x))^2}{\sigma^2}\right]. \tag{6.7}$$

This relation shows that $T = t(X)$ is sufficient, by the factorization criterion. Further,

$$1 = \int p_\theta(x)\mu(dx) = \int e^{-1/2\,\theta^2 + \theta t}\bar{\mu}(dt) \tag{6.8}$$

where $\bar{\mu} = \mu^* t^{-1}$ and $\mu^*(dx) = c(x) \exp\left[-\tfrac{1}{2}t^2(x)\right]\mu(dx)$. Now, assuming that there exist two different measures $\bar{\mu}$ satisfying (6.8), we easily derive a contradiction with the completeness of exponential families of distributions (see Lehmann [14], theorem 1, p. 132). Thus $\bar{\mu}(dt) = e^{-1/2t^2}\,dt$ and the theorem is proved.

The whole work by Fisher suggests that he associated the likelihood principle with the above family of likelihoods, and, asymptotically, with families, which in the limit shrink to $\mathcal{L}_\sigma$ for some $\sigma > 0$ (see example 8.7). If so, the present author cannot see any serious objections against the principle, and particularly, against the method of maximum likelihood. Outside of this area, however, the likelihood principle is misleading, because the information about the kind of experiment does not lose its value even if we know the likelihood.

7. Vaguely known prior distributions

There are several ways of utilizing a vague knowledge of the prior distribution $\nu(d\theta)$. Let us examine three of them.

7.1. *Diffuse prior distributions.* Assume that $\nu(d\theta) = \sigma^{-1}g(\theta/\sigma)\,d\theta$ where $g(x)$ is continuous and bounded. Then, under general conditions the posterior density $p(\theta|X = x, \sigma)$ will tend to a limit for $\sigma \to \infty$, and the limits will, independently of g, correspond to $\nu(d\theta) = d\theta$. This will be true especially if the likelihood function $l(\theta|X = x)$ are strongly unimodal, that is, if $-\log l(\theta|X = x)$ is convex in θ for every x, in which case also all moments of finite order will converge. There are, however, at least two difficulties connected with this approach.

Difficulty 1. However large a fixed σ is, for a nonnegligible portion of

θ-values placed at the tails of the prior distribution, the experiment will lead to such results x that the posterior distribution with $\nu(d\theta) = \sigma^{-1}g(\theta/\sigma)\, d\theta$ will differ significantly from that with $\nu(\theta) = d\theta$. Thus, independently of the rate of diffusion, our results will be biased in a nonnegligible portion of cases.

Difficulty 2. If interested in rapidly increasing functions of θ, say e^θ, and trying to estimate them by the posterior expectation, the expectation will diverge for $\sigma \longrightarrow \infty$ under usual conditions (see example 8.5).

7.2. *A family of possible prior distributions.* Given a family of prior distributions $\{\nu_\alpha(d\theta)\}$, where α denotes some "metaparameter," we may either

(i) estimate α by $\hat{\alpha} = \hat{\alpha}(X)$, or

(ii) draw conclusions which are independent of α.

7.3. *Making use of non-Bayesian risks.* If we know the prior distribution $\nu(d\theta)$ exactly, we could make use of the Bayesian decision function δ_ν, associating with every outcome x the decision $d = d(x)$ that minimizes

$$R(\nu, x, d) = \int L(\tau(\theta), d)\nu(d\theta|X = x). \tag{7.1}$$

Simultaneously, the respective minimum

$$R(\nu, x) = \min_{d' \in D} R(\nu, x, d') \tag{7.2}$$

could characterize the risk, given $X = x$.

If the knowledge of ν is vague, we can still utilize δ_ν as a more or less good solution, the quality of which depends on our luck with the choice of ν. Obviously, in such a situation, the use of $R(\nu, x)$ would be too optimistic. Consequently, we had better replace it by an estimate $\hat{R}(\theta, \delta_\nu) = r(X)$ of the usual risk

$$R(\theta, \delta_\nu) = \int L(\theta, \delta_\nu(x))P_\theta(dx). \tag{7.3}$$

Then our notion of risk will remain realistic even if our assumptions concerning ν are not. Furthermore, comparing $R(\nu, x)$ with $\hat{R}(\theta, \delta_\nu) = r(x)$, we may obtain information about the appropriateness of the chosen prior distribution.

8. Examples

EXAMPLE 8.1. Let us assume that $\theta_i = 0$ or 1, $\theta = (\theta_1, \cdots, \theta_N)$ and $\tau = \theta_1 + \cdots + \theta_N$. Further, put

$$p_\theta(x_1, \cdots, x_N) = \prod_{i=1}^{N} [f(x_i)]^{\theta_i}[g(x_i)]^{1-\theta_i}, \tag{8.1}$$

where f and g are some one-dimensional densities. In other words, $X = (X_1, \cdots, X_N)$ is a random sample of size N, each member X_i of which comes from a distribution with density either f or g. We are interested in estimating the number of the X_i' associated with the density f. Let $T = (T_1, \cdots, T_N)$ be the order statistic, namely $T_1 \leq \cdots \leq T_N$ are the observations $X_1, \cdots, X_N$ rearranged in ascending magnitude.

PROPOSITION 8.1. *The vector T is a sufficient statistic for τ in the sense of definition 2.2, and*

$$p_\tau(t) = p_\tau(t_1, \cdots, t_N) = \tau!(N - \tau)! \sum_{s_\tau \in S_\tau} \prod_{i \in s_\tau} f(t_i) \prod_{j \notin s_\tau} g(t_j), \tag{8.2}$$

where S_τ denotes the system of all subsets s_τ of size τ from $\{1, \cdots, N\}$.

PROOF. A simple application of theorem 3.5 to the permutation group.

PROPOSITION 8.2. *For every t*

$$p_{\tau+1}(t)p_{\tau-1}(t) = p_\tau^2(t) \tag{8.3}$$

holds.

PROOF. See [7]. Relation (8.3) means that $-\log p_\tau(t)$ is convex in τ; that is, the likelihoods are (strongly) unimodal. Thus we could try to estimate τ by the method of maximum likelihood, or still better by the posterior expectation for the uniform prior distribution, if $X_i = 0$ or 1, the T is equivalent to $T' = X_1 + \cdots + X_N$.

This kind of problem occurs in compound decision making. See H. Robbins [17].

EXAMPLE 8.2. Let $0 < \sigma^2 < K$, μ real, $\theta = (\mu, \sigma^2)$, and

$$p_\theta(x_1, \cdots, x_n) = \sigma^{-n}(2\pi)^{-1/2n} \exp\left\{-\tfrac{1}{2} \sum_{i=1}^{n} (x_i - \mu)^2\sigma^{-2}\right\}. \tag{8.4}$$

PROPOSITION 8.3. *The statistic*

$$s^2 = (n-1)^{-1} \sum_{i=1}^{n} (x_i - \bar{x})^2$$

is sufficient for σ^2.

PROOF. For given $\sigma^2 < K$ we take for the mixing distribution of μ the normal distribution with zero expectation and the variance $(K - \sigma^2)/n$. Then we obtain the mixed density.

$$q_\sigma(x_1, \cdots, x_N) = \sigma^{-n+1} (2\pi)^{-n/2} \exp\left\{-\tfrac{1}{2}\frac{n\bar{x}^2}{K} - \tfrac{1}{2} \sum_{i=1}^{n} (x_i - \bar{x})^2\sigma^{-2}\right\} \tag{8.5}$$

and apply definition 2.2.

REMARK 1. If $K \to \infty$, then the mixing distribution tends to the uniform distribution over the real line, which is not finite. For practical purposes the bound K means no restriction, since such a K always exists.

REMARK 2. A collection of examples appropriate for illustrating the notion of sufficiency for a subparameter could be found in J. Neyman and E. Scott [16].

EXAMPLE 8.4. If τ attains only two values, say τ_0 and τ_1, and if it admits a sufficient statistic in the sense of definition 2.2, then the respective distributions Q_{τ_0} and Q_{τ_1} are *least favorable* for testing the composite hypothesis $\tau = \tau_0$ against the composite alternative $\tau = \tau_1$ (see Lehmann [14]). In particular, if T is ancillary under $\tau(\theta) = \tau_0$, and if

$$p_\theta(x) = r_\tau(t(x))p_{\bar{b}}(x), \qquad \theta = (\tau, \bar{b}) \tag{8.6}$$

where the range of $\bar{b}$ is independent of the range of τ, then T is sufficient for τ in the sense of definition 2.2, and the respective family $\{Q_\tau\}$ may be defined so that we choose an arbitrary b, say b_0, and then put $Q_\tau(dx) = r_\tau(t(x))p_{b_0}(x)\mu(dx)$. In this way asymptotic sufficiency of the vector of ranks for a class of testing problems is proved in [6]. (See also remark 2.4.)

EXAMPLE 8.5. The standard theory of probability sampling from finite populations is linear and nonparametric. If we have any information about the *type* of distribution in the population, and if this type could make nonlinear estimates preferable, we may proceed as follows.

Consider the population values $Y_1, \cdots, Y_N$ as a sample from a distribution with two parameters. Particularly, assume that the random variables $X_i = h(Y_i)$, where h is a known strictly increasing function, are normal (μ, σ^2). For example, if $h(y) = \log y$, then the Y_i's are log-normally distributed. Now a simple random sample may be identified with the partial sequence $Y_1, \cdots, Y_n$, $2 < n < N$. Our task is to estimate $Y = Y_1 + \cdots + Y_N$, or, equivalently, as we know $Y_1 + \cdots + Y_n$, to estimate

$$Z = Y_{n+1} + \cdots + Y_N. \tag{8.7}$$

Now the minimum variance unbiased estimate of Z equals

$$\hat{Z} = (N - n)\,[B\,(\tfrac{1}{2}n - 1, \tfrac{1}{2}n - 1)]^{-1} \int_0^1 h^{-1}[\bar{x} + (2v - 1)s(n - 1)n^{-1/2}][v(1 - v)]^{1/2n-2}\,dy \tag{8.8}$$

where

$$\bar{x} = \frac{1}{n}\sum_{i=1}^{n} h(y_i), \qquad s^2 = \frac{1}{n-1}\sum_{i=1}^{n}[h(y_i) - \bar{x}]^2. \tag{8.9}$$

On the other hand, choosing the usual "diffuse" prior distribution $(d\mu\, d\sigma) = \sigma^{-1}\,d\mu\,d\sigma$, we obtain for the conditional expectation of Z, given $Y_i = y_i$, $i = 1, \cdots, n$, the following result:

$$\tilde{Z} = E(Z|Y_i = y_i, 1 \le i \le n,\nu) = (N - n)(n - 1)^{-1/2} [B\,(\tfrac{1}{2}\,n - \tfrac{1}{2}, \tfrac{1}{2})]^{-1} \int_{-\infty}^{\infty} h^{-1}(\bar{x} + vs(1 + 1/n)^{1/2}) \left(1 + \frac{v^2}{n-1}\right)^{-1/2n} dv. \tag{8.10}$$

However, for most important functions h, for example for $h(y) = \log y$, that is $h^{-1}(x) = e^x$, we obtain $\tilde{Z} = \infty$. Thus the Bayesian approach should be based on some other prior distribution. In any case, however, the Bayesian solution would be too sensitive with respect to the choice of $\nu(d\mu\, d\sigma)$.

Exactly the same unpleasant result (8.10) obtains by the method of fiducial prediction recommended by R. A. Fisher ([5], p. 116). For details, see [9].

EXAMPLE 8.6. Consider the following method of sampling from a finite population of size N: each unit is selected by an independent experiment, and the probability of its being included in the sample equals n/N. Then the sample size K is a binomial random variable with expectation n. To avoid empty samples, let us reject samples of size $<k_0$, where k_0 is a positive integer, so that K will have

truncated binomial distribution. Further, let us assume that we are estimating the population total $Y = y_1 + \cdots + y_N$ by the estimator

$$\check{Y} = \frac{N}{K} \sum_{i \in s_K} y_i, \tag{8.11}$$

where s_K denotes a sample of size K. Then

$$E((\check{Y} - Y)^2 | K = k) = \frac{N(N-k)}{k} \sigma^2 \tag{8.12}$$

holds where σ^2 is the population variance $\sigma^2 = (N-1)^{-1} \sum_{i=1}^N (y_i - \overline{Y})^2$. Thus K yields a correct conditioning. Further, the deepest correct conditioning is that relative to K.

On the other hand, simple random sampling of fixed size does not allow effective correct conditioning, unless we restrict somehow the set of possible $(y_1, \cdots, y_N)$-values.

EXAMPLE 8.7. Let $f(x)$ be a continuous one-dimensional density such that $-\log f(x)$ is convex. Assume that

$$I = \int_{-\infty}^{\infty} \left[\frac{f'(x)}{f(x)}\right]^2 f(x)\, dx < \infty. \tag{8.13}$$

Now for every integer N put

$$p_\theta(x_1, \cdots, x_N) = \prod_{i=1}^N f(x_i - \theta) \tag{8.14}$$

and

$$l_x(\theta) = p_\theta(x_1, \cdots, x_N), \qquad -\infty < \theta < \infty. \tag{8.15}$$

Let $t(x)$ be the mode (or the mid-mode) of the likelihood $l_x(\theta)$, that is,

$$l_x(t(x)) \geq l_x(\theta), \qquad -\infty < \theta < \infty. \tag{8.16}$$

Since $f(x)$ is strictly unimodal, $t(x)$ is uniquely defined. Then put

$$l_x^*(\theta) = c_N I^{1/2} (2\pi)^{-1/2} \exp\left[-\tfrac{1}{2}(\theta - t(x))^2 I\right] \tag{8.17}$$

where c_N is chosen so that

$$\int_{-\infty}^{\infty} l_x^*(\theta)\, dx_1 \cdots dx_N = 1. \tag{8.18}$$

Then $c_N \to 1$ and

$$\lim_{N \to \infty} \int |l_x(\theta) - l_x^*(\theta)|\, dx_1 \cdots dx_N = 0, \tag{8.19}$$

the integrals being independent of θ. Thus, for large N, the experiment with likelihoods l may be approximated by an experiment with normal likelihoods l^* (see [8]). Similar results may be found in Le Cam [12] and P. Huber [10].

EXAMPLE 8.7. Let $\theta = (\theta_1, \cdots, \theta_N)$,

$$p_\theta(x_1, \cdots, x_N) = (2\pi)^{-1/2N} \exp\left(-\tfrac{1}{2} \sum_{i=1}^n (x_i - \theta_i)^2\right), \tag{8.20}$$

and let $(\theta_1, \cdots, \theta_N)$ be regarded as a sample from a normal distribution (μ, σ^2). If μ and σ^2 were known, we would obtain the following joint density of (x, θ):

$$(8.21)\quad r_\nu(x_1, \cdots, x_N, \theta_1, \cdots, \theta_N) = \sigma^{-N}(2\pi)^{-N} \exp\left(-\tfrac{1}{2}\sum_{i=1}^{N}(x_i - \theta_i)^2 - \tfrac{1}{2}\sigma^2 \sum_{i=1}^{N}(\theta_i - \mu)^2\right).$$

Consequently, the best estimator of $(\theta_1, \cdots, \theta_N)$ would be $(\hat{\theta}_1, \cdots, \hat{\theta}_N)$ defined by

$$(8.22)\quad \hat{\theta}_i = \frac{\mu + \sigma^2 x_i}{1 + \sigma^2}.$$

Now, in the prior experiment (μ, σ^2) can be estimated by (θ, s_θ^2), where

$$(8.23)\quad \bar{\theta} = \frac{1}{N}\sum_{i=1}^{N}\theta_i, \qquad s_\theta^2 = \frac{1}{N-1}\sum_{i=1}^{N}(\theta_i - \bar{\theta})^2.$$

In the x-experiment, in turn, a sufficient pair of statistics for $(\bar{\theta}, s_\theta^2)$ is $(\bar{x}, s^2)$, where

$$(8.24)\quad \bar{x} = \frac{1}{N}\sum_{i=1}^{N}x_i, \qquad s^2 = \frac{1}{N-1}\sum_{i=1}^{N}(x_i - \bar{x})^2.$$

Now, while $\bar{\theta}$ may be estimated by $\bar{x}$, the estimation of s_θ^2 must be accomplished by some more complicated function of s^2. We know that the distribution of $(N-1)s^2$ is noncentral χ^2 with $(N-1)$ degrees of freedom and the parameter of noncentrality $(N-1)s_\theta^2$. Denoting the distribution function of that distribution by $F_{N-1}(x, \delta)$, where δ denotes the parameter of noncentrality, we could estimate s_θ^2 by $\hat{s}_\theta^2 = h(s^2)$, where $h(s^2)$ denotes the solution of

$$(8.25)\quad F_{N-1}((N-1)s^2, (N-1)s_\theta^2) = \tfrac{1}{2}.$$

(See section 4.3.) On substituting the estimate in (8.21), one obtains modified estimators

$$(8.26)\quad \tilde{\theta}_i = \frac{\bar{x} + \hat{s}_\theta^2 x_i}{1 + \hat{s}_\theta^2}.$$

The estimators (8.26) will be for large N nearly as good as the estimators (8.26) (cf. C. Stein [18]).

The estimators (8.26) could be successfully applied to estimating the averages in individual stratas in sample surveys. They represent a compromise between estimates based on observations from the same stratum only, which are unbiased but have large variance, and the overall estimates which are biased but have small variance.

The same method could be used for $\theta_i = 1$ or 0, and $(\theta_1, \cdots, \theta_N)$ regarded as a sample from an alternative distribution with unknown p. The parameter p could then be estimated in lines of example 8.1 (cf. H. Robbins [17]).

9. Concluding remarks

The genius of classical statistics is based on skillful manipulations with the notions of sufficiency, similarity and conditionality. The importance of similarity increased after introducing the notion of completeness. An adequate imbedding of similarity and conditionality into the framework of general theory of decision making is not straightforward. Classical statistics provided richness of various methods and did not care too much about criteria. The decision theory added a great deal of criteria, but stimulated very few new methods. From the point of view of decision making, risk is important only before we make a decision. After the decision has been irreversibly done, the risk is irrelevant, and for instance, its estimation or speculating about the conditional risk makes no sense. Such an attitude seems to be strange to the spirit of classical statistics. If regarding statistical problems as games against Nature, one must keep in mind that besides randomizations introduced by the statistician, there are randomizations (ancillary statistics) involved intrinsically in the structure of experiment. Such randomizations may make the transfer to conditional risks facultative.

REFERENCES

[1] ALLAN BIRNBAUM, "On the foundations of statistical inference," *J. Amer. Statist. Assoc.*, Vol. 57 (1962), pp. 269–326.

[2] D. BLACKWELL and M. A. GIRSCHICK, *Theory of Games and Statistical Decisions*, New York, Wiley, 1954.

[3] E. B. DYNKIN, *The Foundations of the Theory of Markovian Processes*, Moscow, Fizmatgiz, 1959. (In Russian.)

[4] BRUNO DE FINETTI and LEONARD J. SAVAGE, "Sul modo di scegliere le probabilita iniziali," *Sui Fondamenti della Statistica*, Biblioteca del Metron, Series C, Vol. I (1962), pp. 81–147.

[5] R. A. FISHER, *Statistical Methods and Scientific Inference*, London, Oliver and Boyd, 1956.

[6] J. HÁJEK and Z. SIDAK, *The Theory of Rank Tests*, Publishing House of the Czech Academy of Sciences, to appear.

[7] S. M. SAMUELS, "On the number of successes in independent trials," *Ann. Math. Statist.*, Vol. 36 (1965), pp. 1272–1278.

[8] J. HÁJEK, "Asymptotic normality of maximum likelihood estimates," to be published.

[9] ———, "Parametric theory of simple random sampling from finite populations," to be published.

[10] P. J. HUBER, "Robust estimation of a location parameter," *Ann. Math. Statist.*, Vol. 35 (1964), pp. 73–101.

[11] A. N. KOLMOGOROV, "Sur l'estimation statistique des parameters de la loi de Gauss," *Izv. Akad. Nauk SSSR Ser. Mat.*, Vol. 6 (1942), pp. 3–32.

[12] L. LE CAM, "Les propriétés asymptotiques des solutions de Bayes," *Publ. Inst. Statist. Univ. Paris*, Vol. 7 (1958), pp. 3–4.

[13] ———, "Sufficiency and approximate sufficiency," *Ann. Math. Statist.*, Vol. 35 (1964), pp. 1419–1455.

[14] E. L. LEHMANN, *Testing Statistical Hypotheses*, New York, Wiley, 1959.

[15] J. NEYMAN, "Two breakthroughs in the theory of decision making," *Rev. Inst. Internat. Statist.*, Vol. 30 (1962), pp. 11–27.

[16] J. NEYMAN and E. L. SCOTT, "Consistent estimates based on partially consistent observations," *Econometrica*, Vol. 16 (1948), pp. 1–32.

[17] H. ROBBINS, "Asymptotically subminimax solutions of compound decision problems," *Proceedings of the Second Berkeley Symposium on Mathematical Statistics and Probability*, Berkeley and Los Angeles, University of California Press, 1950, pp. 131–148.

[18] CHARLES STEIN, "Inadmissibility of the usual estimator for the mean of a multivariate normal distribution," *Proceedings of the Third Berkeley Symposium on Mathematical Statistics and Probability*, Berkeley and Los Angeles, University of California Press, 1956, Vol. I, pp. 197–206.

EFFICIENCY IN NORMAL SAMPLES AND TOLERANCE OF EXTREME VALUES FOR SOME ESTIMATES OF LOCATION

J. L. HODGES, JR.
UNIVERSITY OF CALIFORNIA, BERKELEY

1. Summary

This paper presents a number of separate but interrelated results concerned with estimates for the symmetric one-sample location problem. (1) Devices are discussed which, in the normal case, increase the information obtainable by random sampling experiments by a factor of hundreds or thousands. (2) Using these devices, sampling evidence is presented that supports the asymptotic theory for a recently introduced estimate, here called T. (3) A linear estimate, called W, is proposed as a natural analog of T, and is used to check the sampling experiment. (4) The estimate T is recognized as a member of a class of estimates, and the class is explored for other members that are easier to compute. (5) One of the simplest of these, called D, is seen to correspond to the one-sample analog of Galton's test, whose null distribution is given. (6) The same samples used with T are applied to D, with closely similar results. (7) A simple numerical measure of tolerance to extreme values is proposed, and methods of evaluating it are presented in two classes of cases that cover the estimates here discussed. (8) A number of estimates, including $\overline{X}$, T, D, and the trimmed and Winsorized means, are compared with regard to normal efficiency, ease of computation, and extreme value tolerance.

2. Introduction

Consider the problem of estimating the center μ of a symmetric population on the basis of a sample $X_1, \cdots, X_n$. It was pointed out by Hodges and Lehmann [6] that, in a natural way, an estimate for μ could be formed from any of a class of rank tests of the value of μ. Perhaps the most interesting of the estimates there considered is the one which corresponds to the Wilcoxon one-sample test. This estimate, denoted here by T and defined in the next section, was shown to be asymptotically normal as $n \to \infty$, and to have attractive large-

Prepared with the partial support of the Office of Naval Research under Contract No. Nonr-222(43).

sample properties relative to $\overline{X} = (X_1 + \cdots + X_n)/n$. If we denote the asymptotic efficiency of T relative to $\overline{X}$ by $ae(T)$, then $ae(T)$ is identical with the asymptotic efficiency of the Wilcoxon test relative to the t-test. This means, in particular, that $ae(T)$ will exceed one for populations with tails somewhat heavier than the normal. Further, it is shown in [5] that no matter what the shape of the population, $ae(T) \geq .864$. Finally, in the ideal normal case for which $\overline{X}$ is the optimum estimate, $ae(T) = 3/\pi = .955$.

Since these properties are all asymptotic, it becomes important to know to what extent they hold good in samples of moderate size. For example, one would like to know, for moderate samples drawn from a normal population, how close to .955 is the efficiency of T? How near is its distribution to the limiting normal form? Some results relevant to the first of these questions may be quoted from the literature. In [6] it was noted that when $n = 3$, T becomes a linear function of the three order statistics, and has an efficiency (in the sense of variance ratio) of .979. The possibly related question of the power efficiency of the Wilcoxon test against normal shift was investigated numerically by Klotz [8]. He found, for sample sizes $5 \leq n \leq 10$ and significance levels near .05, values near .955 and even usually somewhat above this limit value.

These results are encouraging, but the case $n = 3$ is rather degenerate, and test efficiency need not reflect estimate efficiency. It was thought worthwhile to carry out a sampling experiment for an intermediate value of n, and the value $n = 18$ was chosen for the reasons given in section 4. In general, the precise study of properties of an estimate by sampling would require enormous numbers of samples. Taking advantage of special features of the normal population, it was possible to obtain with the aid of only 100 samples the close estimate of the efficiency of T reported in section 7, and also to examine the approach of the distribution of T to the normal shape (section 8). (Corresponding results are given for a variant U of the estimate T.) These findings, together with the asymptotic theory and the facts for very small samples stated in section 3, suggest that the asymptotic properties of T may be trusted for normal samples of any size. The adequacy of the sampling is supported by its use on an estimate W of known variance; this estimate has independent interest because it may be viewed as a linear analogue to the nonlinear T (section 5).

The main drawback of the estimate T is that it is troublesome to compute except when n is small. For this reason, it seems desirable to try to simplify the estimate while retaining its good distributional properties. To this end, we give in section 9 a class of estimates of which T is a typical member, and consider two methods of finding in the class estimates that are simpler than T, but which on intuitive grounds should have similar distributional properties.

One of these, called D, is examined in section 10, using the same 100 samples. It appears that, at least for normal samples of 18, the estimate D is about as good as the more laborious T. It is shown that, according to the general principle cited above, D corresponds to a rank test that is analogous to the Galton test, and the null distribution of this test is derived.

Behavior of an estimate such as T or D in normal samples is, of course, only part of the story, and one would like to know, for example, how well it stands up in the presence of extreme values. As a contribution to this difficult problem, a simple measure of tolerance of extreme values is introduced in section 11, and this measure is evaluated for two classes of estimates, including those considered in this paper and also the trimmed and Winsorized means. Finally, in section 12 various estimates are compared with regard both to extreme value tolerance and normal efficiency. The estimate D appears to have attractive properties, and to be worth further study. In particular, it would be very desirable to discover its asymptotic distribution.

3. The estimates T and U

Let us denote the sample $X_1, \cdots, X_n$, when arranged in order of increasing size, by $Y_1 \leq \cdots \leq Y_n$. Denote by $\mathcal{A}$ the set of all pairs (i, j) such that $1 \leq i \leq j \leq n$. For each $(i, j) \in \mathcal{A}$, form the mean $M_{ij} = \frac{1}{2}(Y_i + Y_j)$. The statistic T is defined as the median of these means,

$$T = \text{med}\,\{M_{ij}\colon (i, j) \in \mathcal{A}\}. \tag{3.1}$$

If the number $\#(\mathcal{A}) = \frac{1}{2}n(n + 1)$ of these means is even, we shall, as is customary, define the median T as the value midway between the two central values of the set of means.

It seems natural to consider also the slightly different estimate that results when the identity means $M_{ii} = Y_i$ are excluded. Let $\mathcal{B}$ denote the set of all pairs (i, j) with $1 \leq i < j \leq n$, and define

$$U = \text{med}\,\{M_{ij}\colon (i, j) \in \mathcal{B}\}, \tag{3.2}$$

with the same convention if $\#(\mathcal{B}) = \frac{1}{2}n(n - 1)$ is even.

Both T and U can, of course, be defined directly in terms of the unordered observations, but the definitions as given will unify the treatment with that of section 9. Furthermore, when computing the estimates (section 6) it is more convenient to work with the ordered sample.

For certain very small sample sizes, the estimates T and U degenerate to linear functions of the order statistics. In these cases, and for normal samples, their variances can be computed from table I of Sarhan and Greenberg [10].

TABLE I

LINEAR CASES

n	T	$e(T)$	U	$e(U)$
1	Y_1	1	undefined	—
2	$\frac{1}{2}(Y_1 + Y_2)$	1	$\frac{1}{2}(Y_1 + Y_2)$	1
3	$\frac{1}{4}Y_1 + \frac{1}{2}Y_2 + \frac{1}{4}Y_3$	.979	$\frac{1}{2}(Y_1 + Y_3)$	.920
4	nonlinear	?	$\frac{1}{4}(Y_1 + Y_2 + Y_3 + Y_4)$	1

To anchor the lower end of the range of n, these cases are summarized in table I, where $e(T)$ and $e(U)$ are the efficiencies of T and U relative to $\overline{X}$, as defined by variance ratio.

The fact that $e(U)$ is substantially higher at $n = 4$ than at $n = 3$ may be related to a "parity effect" which was influential in choosing the value of n for the sampling experiment. At $n = 3$, $\#(\mathcal{B}) = 3$ is odd, so U is a "pure" median; at $n = 4$, $\#(\mathcal{B}) = 6$ is even, and U is the average of two means M_{ij}. It seems intuitive that such averaging would improve the estimate when sampling from a normal population.

This phenomenon is easier to display in the case of the sample median $\tilde{X}$, efficiency values of which are given in table II. When n is even, $\tilde{X}$ is the average of two order statistics, and its normal efficiency is substantially higher than indicated by the adjacent odd values, by about $.6/n$ for $10 \leq n \leq 20$. For these reasons, it was thought that T and U would have a somewhat higher normal efficiency when $\#(\mathcal{A})$ and $\#(\mathcal{B})$ are even than when they are odd.

TABLE II

NORMAL EFFICIENCY OF $\tilde{X}$ FOR $n \leq 20$

n	$e(\tilde{X})$	n	$e(\tilde{X})$	n	$e(\tilde{X})$	n	$e(\tilde{X})$
1	1.000 000	6	.776 123	11	.662 784	16	.691 561
2	1.000 000	7	.678 828	12	.709 122	17	.653 257
3	.742 935	8	.743 247	13	.658 594	18	.685 630
4	.838 365	9	.668 936	14	.699 130	19	.651 454
5	.697 268	10	.722 928	15	.655 557	20	.680 855

4. The sampling design

Since precise information about normal efficiency for one value of n seemed more valuable than diffuse information for several, it was decided to concentrate on a single value, taking one large enough to escape the degenerate behavior of very small sample sizes and to reflect the sort of moderate sizes often encountered in practice. The linear check W described in section 5 was possible only for $n \leq 20$ because of the range of the Sarhan-Greenberg table. The value chosen, $n = 18$, is the largest in this range for which both T and U are pure medians; fortunately, for this value the estimate D considered in section 10 is also an unaveraged median. As explained in section 3, this choice should lead to conservative results.

The motivation for the somewhat complex sampling design will be given in terms of T, with similar remarks applying to U, W, and D. Since a linear transformation applied to the sample also affects T and $\overline{X}$, there is no loss of generality in using the standard normal population $\mathfrak{N}(0, 1)$, for which extensive tables of random deviates are available.

To obtain precise estimates of the distribution of T would require an enormous number of samples if we proceeded directly, by simply finding the values of T in each such sample. Fortunately, special features of the normal population permit a drastic economy.

Consider a sample of n from $\mathfrak{N}(0, 1)$ and let $\bar{X}$ and T denote estimates computed from this sample. Write $\Delta = T - \bar{X}$, so that $T = \bar{X} + \Delta$. If each observation X_i is increased by c, so are T and $\bar{X}$, and hence Δ is unchanged. Therefore Δ is a function only of the sample differences $X_i - X_1$, $i = 2, \cdots, n$, which implies that Δ is independent of $\bar{X}$. The distribution of T is the convolution of the distribution of Δ with the known distribution $\mathfrak{N}(0, 1/n)$ of $\bar{X}$. We may therefore proceed indirectly as follows: Draw a number of samples of size n from $\mathfrak{N}(0, 1)$. For each sample compute $\bar{X}$ and T, and thus find the value of Δ. Use these observed Δ-values to estimate the distribution of Δ. Finally, convolute the estimated distribution of Δ with the known distribution of $\bar{X}$ to obtain an estimate of the distribution of T.

The advantage of this indirect method resides in the fact that T is a highly efficient estimate, and therefore highly correlated with $\bar{X}$. This means that Δ has a spread much smaller than that of $\bar{X}$. Consequently, an estimate for the distribution of Δ, based on few samples and crude, relative to the spread of Δ, can give us an estimate for the distribution of T which is precise, relative to its much larger spread.

This indirect approach depends on special features of the normal population, but an additional refinement is of more general applicability. When estimating the distribution of Δ, we may classify the samples into strata, and draw separate samples from each stratum. If k_s samples are drawn from stratum s, resulting in values Δ_{sj} of Δ, $j = 1, 2, \cdots, k_s$, these values may be used to estimate the conditional distribution of Δ in stratum s. If P_s is the probability that a random sample of n comes from stratum s, then the weights P_s may be used to combine the estimates of the conditional distributions, to produce an estimate of the (unconditional) distribution of Δ.

Such stratification is feasible only if two conditions are met: it must be possible to calculate the probabilities P_s, and it must be possible to obtain samples which are randomly drawn conditionally from each stratum. By general principles of stratified sampling, stratification is really effective only if a third condition holds: the conditional distributions of Δ in the different strata must be substantially different.

An outstanding feature of T is its insensitivity to extreme values (section 11). On the other hand, $\bar{X}$ is rather sensitive to them, especially if they occur on only one end of the sample. We may therefore expect $|\Delta|$ to be large when the sample has values far out in one tail but not in the other. In contrast, $\bar{X}$ and T will tend to be close, and $|\Delta|$ to be small, if the sample is nearly symmetric, and especially if it lies in a narrow range.

A method of stratification which meets all three conditions consists in dividing the axis into a finite number of intervals, and classifying the samples into strata according to the numbers H_i of observations from the various intervals. The probabilities P_s of such strata can be calculated from the multinomial distribution. It is easy to see at a glance in which stratum a sample falls. Finally, the values of H_i in the extreme intervals will characterize the range and asymmetry of the sample.

In the experiment as performed, the axis was divided into seven intervals by the points ± 2.5, ± 2.0, ± 1.8. Let $H_1, H_2, \cdots$ denote the numbers of the $n = 18$ observations falling in $(-\infty, -2.5)$, $(-2.5, -2.0)$, $\cdots$. In terms of these numbers, twelve strata were defined by the following conditions:

Stratum 1: $H_3 + H_4 + H_5 = 18$,
Strata 2–7: $H_1 + H_7 = 0$, $H_2 + H_6 > 0$;
Stratum 2: $(H_2 = 1, H_3 > 0, H_6 = 0)$ or $(H_6 = 1, H_5 > 0, H_2 = 0)$,
Stratum 3: $(H_2 = 1, H_3 = 0, H_6 = 0)$ or $(H_6 = 1, H_5 = 0, H_2 = 0)$,
Stratum 4: $H_2 = H_6 = 1$,
Stratum 5: $(H_2 = 2, H_6 = 0)$ or $(H_6 = 2, H_2 = 0)$,
Stratum 6: $(H_2 = 3, H_6 = 0)$ or $(H_6 = 3, H_2 = 0)$,
Strata 8-12: $H_1 + H_7 > 0$;
Stratum 8: $(H_1 = 1, H_2 > 0, H_7 = 0)$ or $(H_7 = 1, H_6 > 0, H_1 = 0)$,
Stratum 9: $(H_1 = 1, H_2 = 0, H_7 = 0)$ or $(H_7 = 1, H_6 = 0, H_1 = 0)$,
Stratum 10: $H_1 = H_7 = 0$,
Stratum 11: $(H_1 = 2, H_7 = 0)$ or $(H_7 = 2, H_1 = 0)$.

The probability of any stratum can easily be expressed in terms of the standard normal cumulative Φ by simple formulas. For example,

$$P_4 = P_5 = 306[\Phi(2.5) - \Phi(2.0)]^2[\Phi(2.0) - \Phi(-2.0)]^{16}. \tag{4.1}$$

The probabilities of the twelve strata are shown in table III.

TABLE III

STRATA PROBABILITIES AND SAMPLE NUMBERS

s	P_s	k_s	s	P_s	k_s	s	P_s	k_s
1	.432 479	15	5	.039 740	8	9	.135 645	15
2	.056 798	10	6	.003 673	2	10	.009 661	2
3	.213 000	25	7	.013 128	2	11	.009 661	4
4	.039 740	4	8	.045 116	10	12	.001 359	3

One hundred samples were drawn, allocated among the strata as shown by the numbers k_s in table III. This allocation was governed by the desire to insure adequate representation of the strata in which it was anticipated that $|\Delta|$ would be large and variable, at the expense of strata where $|\Delta|$ might tend to be small and constant. Thus stratum 1, consisting of samples from the interval $(-2, 2)$, constitutes 43% of all samples. But, since $|\Delta|$ should tend to be small here, only $k_1 = 15$ of the 100 samples were taken from stratum 1. On the other hand, the samples in stratum 11 have two extreme observations, $(|X_i| > 2.5)$ at one end not balanced by any at the other end. Here $|\Delta|$ should be large and variable, so $k_{11} = 4$ samples were drawn; a number that by proportional allocation would correspond to an experiment of $k_{11}/P_{11} = 414$ samples instead of 100.

A random sample from one of the strata can be obtained by either of two methods. One may draw unrestricted random samples of size 18 until a sample is obtained which falls in the stratum. Alternatively, one may independently draw observations from the various intervals and combine them. Both methods were used. The 87 samples from strata 1–5, 8, and 9 were obtained from [12] as follows. Beginning on page 142, the first 18 deviates in each column were regarded as a sample from $\mathfrak{N}(0, 1)$. The first $k_1 = 15$ of these samples which fell in $(-2, 2)$ were used as the samples from stratum 1, and so on. In table V, the page and column numbers of each of these 87 samples are given, permitting the reader to check any value. Because samples of the types of strata 6, 7, and 10–12 are rare, these 13 samples were drawn by the alternative method. For strata 6 and 7, the few extreme values needed were drawn from page 1 of [12] taking the first observations with $2 < |X_i| < 2.5$; these were combined with the required numbers of observations $|X_i| < 2$ from the columns indicated in table V. Similarly the values $|X_i| < 2$ for strata 10 and 11 were taken from pages 3 and 4, and the values for stratum 12 from page 1. For the five samples in the catch-all strata 7 and 12, an independent randomization was used to determine the number of extreme values to be taken.

5. A linear estimate

It is always comforting when using random samples to be able to check the quality of the samples by using them to estimate a known quantity, especially one that is closely related to the quantity under investigation. From the tables of variances and covariances of normal order statistics, one can readily compute the variance of any linear combination $W = \sum w_i Y_i$ of the 18 order statistics. If $\sum w_i = 1$, the variance of W can be estimated by the method discussed in section 4, and then be compared with the correct value.

The check is relevant to the extent that weights w_i can be chosen to make W behave like T (and hence like the closely related U). An heuristic argument, given below, suggests that in the normal case a good choice is $w_i \propto \varphi(x_i)$ where $x_i = E(Y_i)$. Using for simplicity weights that closely approximate these, let us define

$$(5.1) \qquad W = \frac{1}{66}(2M_{1,18} + 4M_{2,17} + 6M_{3,16} + 7M_{4,15} + 8M_{5,14} + 9M_{6,13} + 10M_{7,12} + 10M_{8,11} + 10M_{9,10}).$$

The success of this choice of weights is reported in section 7. We now give its heuristic motivation, dealing with the more general problem of a smooth positive density f symmetric about zero. This argument has independent interest in that it suggests the reason for the good properties of the estimate T.

Let us consider the points (i, j) as arranged in rows, associating with each diagonal point $(i, n + 1 - i)$ the points $(i + a, n + 1 - i + a)$ for $a = 0, \pm 1, \pm 2, \cdots$. For a large, but small compared with n, there is a simple approximate relation between the means at $(i, n + 1 - i)$ and at $(i + a, n + 1 - i + a)$. It can be seen that $Y_{i+a} - Y_i$ and $Y_{n+1-i+a} - Y_{n+1-i}$ are both approximately equal to $a/nf(x_i)$. Hence,

$$M_{i+a,n+1-i+a} \doteq M_{i,n+1-i} + a/nf(x_i). \tag{5.2}$$

Now $M_{i+a,n+1-i+a}$ is an increasing function of a, rising from below T to above T. Let $b(i)$ denote that value of a for which the mean is closest to T, so that $M_{i+b(i),n+1-i+b(i)} \doteq T$. Substitution in (5.2) gives

$$b(i) \doteq nf(x_i)[T - M_{i,n+1-i}]. \tag{5.3}$$

By the definition of T, there are as many points with $M_{ij} > T$ as with $M_{ij} < T$. Imposing this condition on (5.3) gives T as, approximately, a weighted average of the Y_i with weights proportional to $f(x_i)$.

While no attempt has been made to rigorize this intuitive discussion, it does help to explain the success of the estimate T, by suggesting that it will tend to behave like a weighted average of the order statistics with weights that are small in the tails of a population with long tails. In practice, of course, f is unknown; the virtue of T is that it accomplishes a reasonable weighting without requiring knowledge of f.

6. The sampling results

For each of the 100 samples of 18, drawn as described in section 4, the estimates $\overline{X}$, U, T, and W were computed. The values of the differences $U - \overline{X}$, $T - \overline{X}$, and $W - \overline{X}$, multiplied by 1000, are shown in table V, as well as the values of $1000(D - \overline{X})$ for the estimate D to be described in section 10. To simplify the table, the signs of all four differences were changed if $U - \overline{X}$ happened to be negative; this is permissible because of the symmetry of all distributions about zero.

The computations of T and U were carried out by the method of Høyland [7], and may be illustrated on the first sample of stratum 1. The first 18 deviates of ([12], p. 142, column 2) (each multiplied by 1000 to avoid decimals) are arranged in decreasing order as shown in table IV, from $Y_{18} = 1.686$ to $Y_1 = -1.632$. Each number in the body of the table is the sum S_{ij} of Y_i below it and Y_j to its left. (The arrangement of the Y_i's in a "broken line" pattern instead of diagonally saves space and also brings the Y's nearer their sums.) Linear combinations of the nine diagonal sums $S_{9,10} = -.004, \cdots, S_{1,18} = .054$ provide the values $\overline{X} = -.0049$ and $W = .0195$, and hence $W - \overline{X} = .024$.

To find U, one must find the median of the sums S_{ij}, $i < j$, and this is done by trial and error, guided by the fact that $S_{ij} \geq S_{k\ell}$ whenever $i \geq k$ and $j \geq \ell$. The line in table IV, passing through the sum $S_{7,12} = .100$, goes as often above as below the row of diagonal sums, and thus divides the 153 sums into 76.5

TABLE IV

Computation of T and U

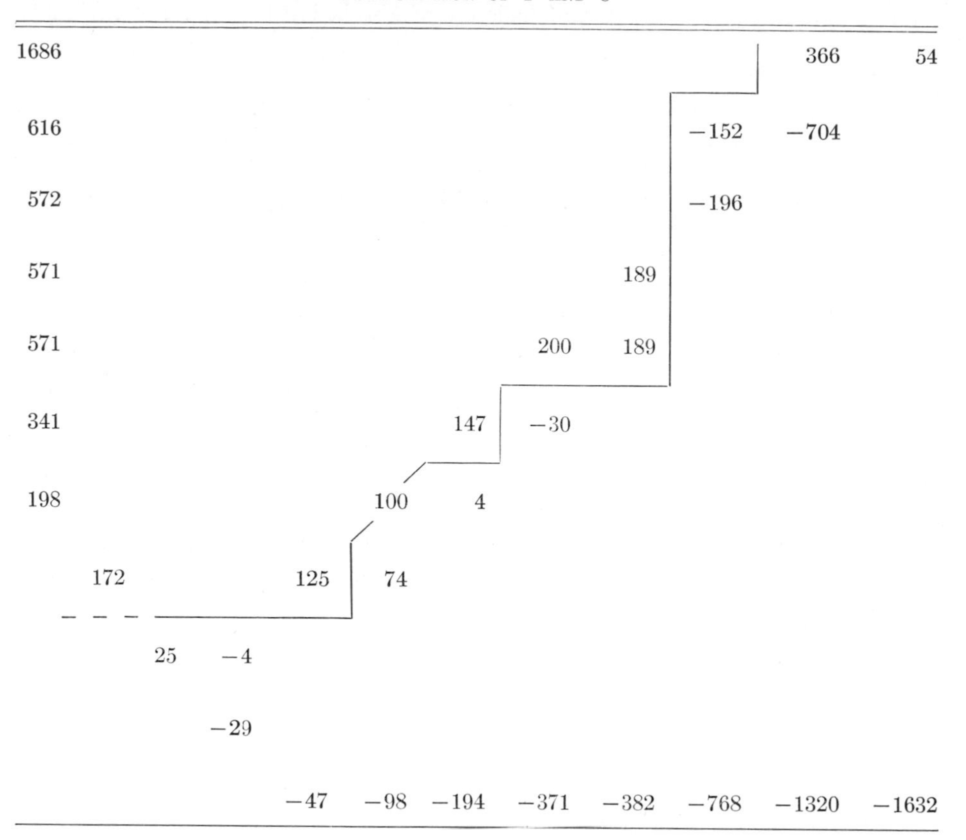

1686										366	54
616									−152	−704	
572									−196		
571								189			
571							200	189			
341						147	−30				
198					100	4					
	172			125	74						
		25	−4								
			−29								
				−47	−98	−194	−371	−382	−768	−1320	−1632

that are above the line, all at least equal to .100, and the 76.5 that are below the line, none greater than .100. Therefore $U = \frac{1}{2}(.100) = .0500$ and $U - \overline{X} = .055$.

To find T, we add in the 18 identity sums $S_{ii} = 2Y_i$. The dashed extension of the line passes between $S_{10,10} = 2(.025)$ and $S_{11,11} = 2(.172)$. As this line has 8 of the newly added sums above it and 10 below it, the median of the augmented set will be the sum next smaller than $S_{7,12}$. By inspection this is seen to be $S_{7,11} = .074$, so that $T = \frac{1}{2}(.074) = .0370$, and $U - \overline{X} = .042$.

As $U - \overline{X}$ is positive, no sign change is required. The entries in the first row of table V give the differences, multiplied by 1000 to simplify the setting. These entries form the bases of the computations reported in sections 7, 8, and 10.

7. Efficiencies

As a first application of the sampling results reported in section 6, we give estimates of the efficiencies of T and U for normal samples of size $n = 18$. Again, the theoretical discussion will be given in terms of the estimate T, with analogous remarks holding for U.

We define, as is customary, the efficiency of T relative to $\overline{X}$ by the variance ratio $e(T) = \text{Var}(\overline{X})/\text{Var}(T) = 1/n \text{Var}(T)$. Because of the independence of $\overline{X}$ and $\Delta = T - X$, $\text{Var}(T) = 1/n + \text{Var}(\Delta)$. Suppose we have an estimate $\mathcal{E}$ for $\text{Var}(\Delta)$, to which is attached an (estimated) standard error, say $\mathcal{E} \pm \sigma$. Then $1/n + \mathcal{E} \pm \sigma$ is an estimate for $\text{Var}(T)$. Since σ is small compared to $1/n + \mathcal{E}$, we may expand linearly and use

$$\frac{1}{1 + n\mathcal{E}} \pm \frac{n\sigma}{(1 + n\mathcal{E})^2} \tag{7.1}$$

as an estimate for $e(T)$.

The desired estimates $\mathcal{E}$ and σ are obtained from our stratified sample as follows. Since Δ is symmetrically distributed about zero, $\text{Var}(\Delta) = E(\Delta^2)$. If Δ_s denotes the random variable Δ conditioned to lie in stratum s, then

$$E(\Delta^2) = \sum_s P_s E(\Delta_s^2). \tag{7.2}$$

The values of Δ_s, say $\Delta_{s1}, \cdots, \Delta_{sk_s}$, observed in the sampling experiment, give in

$$\mathcal{E}_s = \sum_j \Delta_{sj}^2/k_s \tag{7.3}$$

an (unbiased) estimate for $E(\Delta_s^2)$, with variance $\text{Var}(\Delta_s^2)/k_s$. Combining these estimates gives

$$\mathcal{E} = \sum P_s\mathcal{E}_s \tag{7.4}$$

as the desired (unbiased) estimate for $\text{Var}(\Delta)$.

Since the estimates $\mathcal{E}_s$ are independent, the variance of $\mathcal{E}$ is

$$\sum P_s^2 \text{Var}(\mathcal{E}_s) = \sum P_s^2 \text{Var}(\Delta_s^2)/k_s. \tag{7.5}$$

Regarding the Δ_{sj}^2 as observed values of Δ_s^2,

$$\sigma_s^2 = [k_s \sum_j \Delta_{sj}^4 - (\sum_j \Delta_{sj}^2)^2]/k_s(k_s - 1) \tag{7.6}$$

is the usual (unbiased) estimate for $\text{Var}(\Delta_s^2)$. Hence we may use

$$\sigma^2 = \sum P_s^2\sigma_s^2/k_s \tag{7.7}$$

as an (unbiased) estimate for the variance of the estimate $\mathcal{E}$. Using the observed values recorded in table V, these calculations were performed for the estimates T, U, and W, giving the following estimated efficiencies:

$$\begin{aligned} e(T) &= .949 \pm .007, \\ e(U) &= .956 \pm .006, \\ e(W) &= .969 \pm .004. \end{aligned} \tag{7.8}$$

The standard errors are crude, but may serve to indicate the high precision of these estimates.

The check provided by the linear estimate W now comes into play. From

TABLE V

Values of 1000Δ for Four Estimates

p	c	U	T	D	W
		Stratum 1			
142	2	55	42	55	24
	4	48	48	52	42
	5	13	13	−9	8
	8	3	16	43	−18
	9	20	56	−32	12
143	2	17	18	18	−2
	3	35	35	35	12
	5	25	25	20	26
144	4	13	27	−41	33
	8	32	32	35	8
145	2	5	−4	−4	0
	3	73	73	73	54
	4	21	21	−2	−2
	5	31	32	32	15
	6	6	6	27	22
		Stratum 2			
143	4	73	46	−15	19
144	1	62	62	159	89
147	3	7	7	60	−6
148	3	82	88	123	72
152	6	51	51	88	14
153	8	11	83	−19	51
154	1	18	14	18	4
155	10	112	112	84	74
156	4	4	0	15	−15
157	2	71	71	3	88
		Stratum 3			
143	8	30	30	−4	28
	9	22	31	13	5
145	8	1	29	29	−20
147	5	7	16	7	41
	7	20	43	−43	53
	8	59	59	59	21
148	1	8	12	12	10
149	3	26	26	26	−33
	7	9	9	8	9
	9	16	15	−12	41
	10	95	159	76	98
150	3	23	27	−2	36
	5	3	3	3	10
	9	11	47	6	18
151	6	107	146	163	101
	7	70	91	1	86
	9	54	14	−9	26
152	2	69	71	7	40
153	2	25	13	8	22
	3	38	38	38	40
	5	8	8	−11	30
	6	43	34	4	31
	7	19	4	27	−8
154	7	17	11	11	19
	9	21	23	20	38
		Stratum 4			
142	10	50	50	36	21
146	7	12	12	12	−26
147	6	13	27	13	2
150	7	22	19	19	22
		Stratum 5			
142	6	95	103	−52	45
143	10	20	61	61	76
144	5	24	26	26	40
145	1	18	−2	−2	0
147	1	67	67	114	60
148	6	68	68	64	17
149	1	75	84	50	41
162	4	20	24	16	27
		Stratum 6			
100	1	48	48	−117	66
100	2	5	−4	14	−78
		Stratum 7			
101	1	33	33	33	23
101	2	21	21	−6	−58
		Stratum 8			
143	6	87	104	24	93
144	7	66	66	85	50
	9	133	118	116	96
	10	79	97	3	82
148	2	2	2	−6	50
152	7	61	65	65	76
159	1	155	164	70	79
160	9	73	73	76	52
161	4	36	59	−33	25
162	8	62	66	120	88
		Stratum 9			
142	1	45	45	51	70
	3	35	46	26	56
	7	65	66	111	66
143	1	67	67	67	49
	7	64	64	62	50
144	2	70	63	36	59
	3	42	55	42	43
	6	78	75	78	56
145	10	55	55	−86	8
146	1	54	54	59	48
	3	31	31	18	25
	4	17	0	−14	−8
147	2	83	83	60	75
	9	153	153	142	115
148	5	72	72	72	57
		Stratum 10			
120	1	1	1	9	2
	2	152	152	152	100

TABLE V (Continued)

p	c	U	T	D	W	p	c	U	T	D	W
		Stratum 11						Stratum 12			
120	3	171	178	14	109	130	1	64	64	16	44
	4	59	59	29	27		2	164	183	−80	94
	5	207	226	237	126		3	23	−2	−4	39
	6	92	92	72	69						

table I of [10], it is possible to compute the actual variance of W, and hence to find that $e(W) = .9649$. The efficiency estimated from the samples is within one standard error.

It will be recalled that the estimate W was chosen in an attempt to find a linear estimate that would be highly correlated with T and U. If we write $W = \overline{X} + \Gamma$, then the pair (Δ, Γ) is independent of $\overline{X}$, and $E(\Delta) = E(\Gamma) = 0$, so that

$$\text{Cov}\,(T, W) = \frac{1}{n} + E(\Delta\Gamma) = \frac{1}{n} + \sum P_s E(\Delta_s \Gamma_s). \tag{7.9}$$

This formula permits us to use the sampling results to estimate the correlation between T and W, and similarly for U and W. The computed results are

$$\hat{\rho}(T, W) = .994, \qquad \hat{\rho}(U, W) = .995. \tag{7.10}$$

These high correlations indicate that W does indeed behave very similarly to T and U, which lends relevance to the success of the estimate for $e(W)$. (Since the estimate for $e(W)$ is high by .004, one may wish to lower the estimates for $e(T)$ and $e(U)$ by this amount, which does not change the picture appreciably.)

The sampling results suggest that U is somewhat more efficient than T. (Since table V makes it clear that $T - \overline{X}$ and $U - \overline{X}$ are highly correlated, the estimated value of $e(U) - e(T)$ is more precise than the separate standard errors indicate.) This agrees with intuition. The estimate T differs from U by including, in the set of means whose median is taken, the identity means $M_{ii} = Y_i$. This inclusion may be said to move T from U in the direction of the median $\tilde{X}$, whose efficiency is only .686. This should, however, also mean that T has slightly less sensitivity than U to extreme values, as the investigation of section 11 shows to be the case.

The effectiveness of our indirect method of estimating the efficiencies is made clear by noting how many samples would be required to obtain estimates of similar precision by directly observing k values of the estimates themselves. For illustration, suppose V is normally distributed about zero and has efficiency $e(V)$. If $V_1, \cdots, V_k$ are observations on V, this efficiency may be estimated by $k/n \sum V_i^2$. For this estimate to have standard error τ, we would need $k = 2e^2(V)/\tau^2$. For the values $e(U) = .956$ and $\tau = .006$, corresponding to our estimate for $e(U)$, this gives $k = 50{,}000$, compared with the 100 samples actually used.

8. Normality

Asymptotically, as $n \to \infty$, the estimates T and U are known to become normally distributed. It is the approximate normality of the estimates which, to a large extent, justifies our use of variance ratio as a measure of efficiency. If the estimate T based on n observations has (approximately) the distribution $\mathfrak{N}(0, \text{Var}(T))$, and if the efficiency of T relative to the arithmetic mean is defined to be $e = 1/n \text{Var}(T)$, then T will have (approximately) the same distribution as the mean of ne observations. It is accordingly desirable to know, when using $1/n\hat{\sigma}^2(T)$ as an indication of the efficiency of T, the extent to which the actual distribution of T agrees with the normal approximation $\mathfrak{N}(0, \hat{\sigma}^2(T))$. Questions of this kind are usually difficult to answer with precision, but the special features of the normal population again make it possible to use our samples to throw a good deal of light on it for normal samples of size 18.

Because T is symmetrically distributed about zero, it suffices to compare $P(|T| > c)$ with the corresponding probability for the fitted normal $\mathfrak{N}(0, \hat{\sigma}^2(T))$. Let

$$\rho(\delta, c) = P(|\overline{X} + \delta| > c) = \Phi(-\sqrt{n}(c + \delta)) + \Phi(-\sqrt{n}(c - \delta)). \tag{8.1}$$

Then, because of the independence of $\overline{X}$ and $\Delta = T - \overline{X}$, integration with respect to the distribution of Δ gives

$$P(|T| > c) = E\rho(\Delta, c). \tag{8.2}$$

Conditioning on the strata gives

$$E\rho(\Delta, c) = \sum_s P_s E\rho(\Delta_s, c). \tag{8.3}$$

Using the observed values Δ_{sj} of our stratified sample, we therefore have in

$$\sum_s P_s[\sum_j \rho(\Delta_{sj}, c)/k_s] \tag{8.4}$$

an unbiased estimate for $P(|T| > c)$. Clearly,

$$\sum P_s^2\{k_s \sum \rho^2(\Delta_{sj}, c) - [\sum \rho(\Delta_{sj}, c)]^2\}/k_s^2(k_s - 1) \tag{8.5}$$

is an unbiased estimate for the variance of (8.4).

The comparison between the actual distribution of T and the fitted normal is especially worthwhile in the tails of the distribution, since experience indicates that the approach to normality is usually slowest there. The calculations were made for both estimates, for $c = .7$ and $c = .9$, corresponding to about 2.9 and 3.7 standard deviations for T and U. The results are shown below.

TABLE VI

PROBABILITY OUTSIDE $(-c, c)$

c	Estimate	From (8.4)	From Fitted Normal
.7	T	.00386 ± .00013	.00382
	U	.00371 ± .00010	.00368
.9	T	.000205 ± .000012	.000200
	U	.000192 ± .000009	.000188

This excellent agreement, even in the extreme tails, indicates that the asymptotic normality has taken hold at $n = 18$, and also supports the relevance of the estimated efficiencies given in section 7.

The power of the methods we are using is even more striking here than in the estimation of efficiency. To illustrate this point, consider the direct estimation of $P(|U| > .9)$ by observing the frequency with which $|U| > .9$. To produce an estimate as good as that reported above would require something like 2,400,000 such samples, compared with the 100 samples we have used.

9. A family of estimates

While T and U have excellent distributional properties, they are laborious to compute when n is large. This disadvantage may be mitigated by the development of computer programs, but not every potential user will have easy access to a computer. In this section we present a general class of estimates of which T and U are special cases, and indicate other members of the class which are simpler to compute and which also seem to have good distributional properties.

Corresponding to any nonempty subset $\mathcal{S}$ of $\mathcal{A}$, we may define an estimate S by

$$S = \text{med}\,\{M_{ij}\colon (i, j) \in \mathcal{S}\}. \tag{9.1}$$

We shall assume throughout that $\mathcal{S}$ enjoys the symmetry property:

$$(i, j) \in \mathcal{S} \quad \text{implies} \quad (n + 1 - j, n + 1 - i) \in \mathcal{S}; \tag{9.2}$$

in view of the symmetry of the population about its center μ, this assures that S is symmetrically distributed about μ. Note that T and U are special cases of (9.1) corresponding to $\mathcal{S} = \mathcal{A}$ and $\mathcal{S} = \mathcal{B}$.

The labor of computing S depends primarily on the number $\#(\mathcal{S})$ of means whose median is sought, and rises somewhat faster than $\#(\mathcal{S})$. It is the fact that $\#(\mathcal{A})$ and $\#(\mathcal{B})$ both increase at speed n^2 which makes T and U difficult to compute when n is large. Roughly speaking, the work required by these estimates is proportional to n^3. As indicated in section 6, it is not difficult to compute T and U by hand at $n = 18$. The labor would be discouragingly heavy at $n = 50$.

Can we choose $\mathcal{S}$ so that $\#(\mathcal{S})$ is small, permitting S to be computed easily, and still have S share the good distributional properties of T and U? We suggest two approaches to this problem.

(i) *Representative order statistics.* A large sample drawn from a smooth population can be adequately represented by a modest number of its order statistics. The basic reasons for this are as follows. If we are given the values of two order statistics, say $Y_i = u$ and $Y_{i+a+1} = v$, then the intermediate order statistics $Y_{i+1}, \cdots, Y_{i+a}$ are conditionally distributed like an ordered random sample drawn from that portion of the population in the interval (u, v). If the population density is smooth and $v - u$ is small, this conditional distribution is nearly rectangular, so that the intermediate observations contain little information about μ. If a/n is small, $Y_{i+a+1} - Y_i$ is likely to be small, even if a itself is large. If we choose a subset $\mathcal{J}$ of the sequence $1, 2, \cdots, n$, about equally spaced throughout the sequence, we may therefore expect $\{Y_i: i \in \mathcal{I}\}$ to carry nearly all of the information in the sample, even if $\#(\mathcal{I})$ is modest.

Now let $\mathcal{A}(\mathcal{I})$ denote the set of pairs $(i, j) \in \mathcal{A}$ for which $i \in \mathcal{I}$ and $j \in \mathcal{I}$, and let $T(\mathcal{I})$ denote the estimate (9.1) with $\mathcal{S}$ replaced by $\mathcal{A}(\mathcal{I})$. If $\#(\mathcal{I})$ is modest, the hand computation of $T(\mathcal{I})$ will be feasible even if n is quite large. Suppose for example $n = 1000$ and $\#(\mathcal{I}) = 18$. Once the representative order statistics $\{Y_i: i \in \mathcal{I}\}$ have been noted, the work of finding $T(\mathcal{I})$ is the same as that for finding T with $n = 18$, as illustrated in section 6. To the extent to which the 18 selected order statistics represent the sample, we may expect $T(\mathcal{I})$ to have distributional properties similar to those of T computed from the entire set of 1000 observations.

In an entirely analogous way we may define the estimate $U(\mathcal{I})$. For two special cases, this estimate coincides with estimates studied by Mosteller [9]. He proposed as an estimate for μ the arithmetic mean, say $\overline{Y}_k$, of k selected order statistics. From table I we see that when $\#(\mathcal{I}) = 2$, $U(\mathcal{I}) = \overline{Y}_2$, and when $\#(\mathcal{I}) = 4$, $U(\mathcal{I}) = \overline{Y}_4$. Mosteller investigated the asymptotic efficiencies of his estimates $\overline{Y}_k$ for the normal population, for several different methods of spacing the selected order statistics. The asymptotic efficiency of $\overline{Y}_2$ was found to range from .793 to .810, while that of $\overline{Y}_4$ ranged from .896 to .914, depending on the spacing. These figures, and other figures given in his table II, suggest that in normal samples one may suffer an efficiency loss of about one or two percent as a result of replacing a large sample by about 20 equally spaced order statistics.

(ii) *Central means.* We may expect a mean M_{ij} to be near μ if and only if Y_i and Y_j are nearly symmetric in the sample, that is, if $i + j$ is near $n + 1$. This suggests that the estimates T and U would be little affected if we eliminated from $\mathcal{A}$ and $\mathcal{B}$ those pairs (i, j) for which $|i + j - n - 1|$ is large. In general, S may be a reasonable estimate if we use in (9.1) a set of central means M_{ij} where (i, j) is near the diagonal $i + j = n + 1$.

As a preliminary, we note that $\mathcal{A}$ and $\mathcal{B}$ may be reduced substantially without affecting T and U at all. The discussion will be given in terms of $\mathcal{B}$, with analogous remarks holding for $\mathcal{A}$.

It is clear that $M_{ij} = \frac{1}{2}(Y_i + Y_j)$ is an increasing function of both i and j. This means that, if i and j are small enough, we may be sure that M_{ij} falls below the central values in the set $\{M_{ij}: (i, j) \in \mathcal{B}\}$, and hence below U, whatever the values $\{Y_i\}$ may be. The number of means M_{ab} in $\mathcal{B}$ with $a \geq i$, $b \geq j$ is easily seen to be $N_{ij} = (n + 1 - j)[\frac{1}{2}(n + j) - i]$. Let $\mathcal{E}$ denote the set of pairs (i, j) with $N_{ij} \geq \frac{1}{2}[\frac{1}{2}n(n - 1) + 1]$, and let $\mathcal{E}'$ be the set symmetric to $\mathcal{E}$ about the line $i + j = n + 1$. Then it can be seen that any M_{ij} with $(i, j) \in \mathcal{E}$ is below U, while any M_{ij} with $(i, j) \in \mathcal{E}'$ is above U, and hence that

$$U = \text{med}\ \{M_{ij}: (i, j) \in \mathcal{B}^*\} \tag{9.3}$$

where $\mathcal{B}^* = \mathcal{B} - \mathcal{E} - \mathcal{E}'$.

The reduction from $\mathcal{B}$ to $\mathcal{B}^*$ is substantial, and it can be shown that

$$\#(\mathcal{B}^*)/\#(\mathcal{B}) \to \{1 - \sqrt{2} - \log(\sqrt{2} - 1)\} = .4672. \tag{9.4}$$

In a computer program which finds U by comparing each pair of means M_{ij}, the use of $\mathcal{B}^*$ instead of $\mathcal{B}$ would reduce this work by 78%.

Any reduction below $\mathcal{B}^*$ will change the estimate, but in the interest of simplicity this may be worthwhile. The most extreme reduction would consist in using only the diagonal means. The resulting estimate will be considered in the next section.

Each of the estimates (9.1) corresponds to a test according to the method expounded in section 2 of [6]. Without loss of generality we may consider tests for the hypothesis $\mu = 0$. For any $\mathcal{S}$ satisfying (9.2), let Σ denote the number of positive means M_{ij}, where $(i, j) \in \mathcal{S}$. It is obvious that Σ is nondecreasing when all sample values are increased by the same amount, thus meeting condition C of the cited section. By the assumed symmetry of the population, we may under the null hypothesis $\mu = 0$ associate with each sample $X_i, \cdots, X_n$ an equally likely sample $X'_i = -X_i$. In terms of order statistics $Y'_i = -Y_{n+1-i}$ so that $M'_{ij} = -M_{n+1-j,n+1-i}$. In view of (9.2), this means that $\Sigma + \Sigma' = \#(\mathcal{S})$; and since Σ and Σ' are equally likely, this equation implies that Σ is symmetrically distributed about $\frac{1}{2}\#(\mathcal{S})$, as required by condition D. The estimate S corresponds to the test statistic Σ, in the sense of [6].

10. The estimate D

We now consider the estimate (9.1) with $\mathcal{S}$ consisting only of diagonal points. Let

$$D = \text{med } \{M_{ij}: (i, j) \in \mathcal{D}\} \tag{10.1}$$

where $\mathcal{D}$ consists of the diagonal pairs (i, j) with $1 \leq i < j \leq n$ and $i + j = n + 1$. This estimate, the median of the means of symmetric order statistics, is much easier to compute than T or U.

Computations analogous to those of sections 7 and 8 were carried out for D, using the values of $D - \overline{X}$ recorded in table V. The estimated efficiency of D relative to $\overline{X}$, in normal samples of size $n = 18$, is $.954 \pm .007$. The results comparable to those in table VI are the following.

TABLE VII

PROBABILITY OUTSIDE $(-c, c)$

c	From (7.1)	From Fitted Normal
.7	.00375 ± .00013	.00372
.9	.000196 ± .000012	.000192

It must be emphasized that, unlike T and U, the asymptotic properties of D are not known. In particular, it may be questioned whether D is asymptotically normal. However, the sampling experiment suggests strongly that, at least with samples of moderate size drawn from a normal population, the behavior of D

is very close to that of T and U, both with regard to variance and shape of distribution. Because D is substantially easier to compute than are T and U, these sampling results suggest that the estimate D is worth further study. In particular, it would be of interest to find its asymptotic properties.

The test corresponding to D is of some interest. For the case $S = D$, let us denote the statistic Σ of section 9 by G; that is, G is the number of positive symmetric means. To test the hypothesis $\mu = 0$ against the alternative $\mu > 0$, we should reject when G is large. This test may be viewed as a one-sample analogue of the two-sample Galton test discussed in [4], with the upper and lower halves of the single sample playing roles analogous to those of the two equal samples in that test.

To derive the null distribution of G, consider a sample of n drawn from a continuous population symmetric about 0. If $a_1 > a_2 > \cdots > a_n > 0$ are the absolute values in this sample, we may reconstitute a sample distributionally equivalent to that drawn, by assigning independently and at random to these absolute values the signs $+$ and $-$. The sequence of n signs may now be used to determine an n-step path in the manner discussed in chapter III of Feller [3]. The 2^n possible paths are equally likely, and the null distribution of G may be found by relating G to properties of the path.

Let us establish a correspondence between the order statistics of the reconstituted sample and the steps of the path by letting the j-th step of the path correspond to that order statistic which equals $\pm a_j$, and denote by S_i the step path that thus corresponds to Y_i. Then the steps $S_1, S_2, \cdots$ are the successive down-steps of the path, read from left to right, while $S_n, S_{n-1}, \cdots$ are the successive up-steps.

Let us characterize the condition $M_{i,n+1-i} > 0$, for $1 \leq i \leq \frac{1}{2}n$, in terms of the steps. Clearly, $M_{i,n+1-i} > 0$ means that $|Y_i| < Y_{n+1-i}$, which in turn assures that S_{n+1-i} is an up-step; that S_{n+1-i} comes before S_i; and that S_i is preceded by at most $i - 1$ down-steps. From these facts it is clear that both S_i and S_{n+1-i} are positive steps, in the sense that both lie above the horizontal axis drawn through the start of the path. Similarly, one can argue that $M_{i,n+1-i} < 0$ implies that both S_i and S_{n+1-i} are negative steps.

When n is even, say $n = 2k$, we now see that $2G$ is just the number of positive steps in the path. To make this simple relation true also when $n = 2k + 1$, we must modify the definition of G slightly, by adding to its value as previously defined the number $\frac{1}{2}$ in case $Y_{k+1} > 0$, which occurs if and only if S_{k+1} is the $(k + 1)$-st up-step and hence a positive step.

The distribution of the number of positive steps has been investigated. For the case $n = 2k$, it is given in ([3], p. 77) in a simple closed form, which is shown (p. 80) to have the arc sine limit law. The case $n = 2k + 1$ is discussed in [1], and from the remarks made there it is clear that the same limit law holds (as would indeed be the case had we not modified the definition of G). The nonnormality of the limit law of G suggests that the estimate D may also have a nonnormal limit distribution.

While the test G is interesting because of its relations to D, to the Galton test, and to random walks, it suffers from a practical disadvantage: the sample size must be large before the customary small significance levels become available. This fact reflects the remarkable tendency, emphasized by Feller, for random paths to remain always on one side of the axis.

11. Tolerance of extreme values

A principal motive behind the search in recent years for estimates of location alternative to classical $\overline{X}$ has been the realization that $\overline{X}$ is sensitive to extreme values. This motive is explicit in the proposal of trimmed and Winsorized means, and it was also impelling in the development of estimates based on rank tests, such as T. It is the purpose of this section to introduce a simple numerical measure of the degree to which an estimate of location is able to tolerate extreme values, and to use this measure to compare several estimates. No pretense is made that the proposed measure exhausts the complex extreme-value problem; however, it provides in some cases an easily computed solution to one aspect of that problem.

As an example to motivate the definition, consider the trimmed mean

$$R = (Y_3 + Y_4 + Y_5 + Y_6 + Y_7 + Y_8)/6 \tag{11.1}$$

where $n = 9$. It is obvious that $Y_3 \leq R \leq Y_8$, whatever be the sample values. This implies that Y_1, Y_2, and Y_9 may be as extreme as desired without causing R to fall outside the range of the remaining six values. The estimate R is therefore able to withstand two extreme values on the left, or one extreme value on the right, or both, however extreme they may be. It cannot however handle more than this. For example, if there are two extreme values on the right, R will be affected. Indeed, if Y_8 (and hence Y_9) is made to tend to ∞, while the values $Y_1, \cdots, Y_7$ remain fixed at arbitrary values, then $R \to \infty$. Similarly $R \to -\infty$ if $Y_3 \to -\infty$, whatever be the fixed values of $Y_4, \cdots, Y_9$. In these circumstances, it seems natural to say that R can "tolerate" just two extremes on the left and one extreme on the right.

Let us now formalize this idea in a definition. Let V be an estimate based on a sample of size n. Suppose there exist integers $\alpha \geq 0$, $\beta \geq 0$ such that

$$Y_{\alpha+1} \leq V \leq Y_{n-\beta} \qquad \text{for all} \quad Y_1, \cdots, Y_n; \tag{11.2}$$

$$\text{whatever be the fixed values of } Y_{\alpha+2}, \cdots, Y_n, \quad Y_{\alpha+1} \to -\infty \quad \text{implies} \quad V \to -\infty; \tag{11.3}$$

$$\text{whatever be the fixed values of } Y_1, \cdots, Y_{n-\beta-1}, \quad Y_{n-\beta} \to \infty \quad \text{implies} \quad V \to \infty. \tag{11.4}$$

We shall then say that V can tolerate α extreme values on the left and β extreme values on the right. If, as often happens, $\alpha = \beta$, we shall denote their common value by γ. If it is desired to make explicit the dependence on n and V, we may write $\alpha_n(V)$ for α, and so on.

Our quantities α and β do not exist for all estimates. In particular, they are undefined for any estimate which can fall outside the sample range, since in this case no choice of α and β will satisfy (11.2). Examples of such estimates are given in theorem 1 below. Other examples are provided by certain Bayes estimates, which can be arbitrarily far from the sample, if the sample should fall sufficiently far from the region in which the population was believed, a priori, to lie. The measures α and β do however exist for two classes of estimates, as we shall now show.

First, let us consider the class of linear combinations of order statistics, say

$$L = w_1 Y_1 + \cdots + w_n Y_n. \tag{11.5}$$

The behavior of these estimates depends on the signs of the cumulative weights, say

$$\begin{aligned} A_i &= w_1 + w_2 + \cdots + w_i, \\ B_i &= w_{n-i+1} + w_{n-i+2} + \cdots + w_n, \end{aligned} \tag{11.6}$$

where $i = 1, 2, \cdots, n$.

THEOREM 1. *If $A_n \neq 1$, or if any A_i or B_i is negative, α and β do not exist. If $A_n = 1$ and all $A_i \geq 0$ and all $B_i \geq 0$, then $\alpha_n(L)$ is the smallest integer a for which $A_{a+1} > 0$, and $\beta_n(L)$ is the smallest integer b for which $B_{b+1} > 0$.*

PROOF. If $A_n \neq 1$, then $Y_1 = \cdots = Y_n = y \neq 0$ makes $L = A_n y$ fall outside the range. If $A_i < 0$ for some positive integer i, let $Y_1 = \cdots = Y_i = y < 0$ and $Y_{i+1} = \cdots = Y_n = 0$. Then $L = A_i y > 0 = Y_n$ so that L again falls outside the range. A similar argument covers $B_i < 0$, and the first statement of the theorem is proved.

Now suppose $A_n = 1 = B_n$, and $A_i \geq 0$ and $B_i \geq 0$ for all i. Since $A_n > 0$, there must be an integer a, $1 \leq a \leq n$, such that $A_1 = \cdots = A_a = 0$, $A_{a+1} > 0$, and hence $w_1 = \cdots = w_a = 0$ and $w_{a+1} > 0$. Similarly there exists b such that $w_n = \cdots = w_{n-b+1} = 0$ and $w_{n-b} > 0$. We see that $A_{n-b} = 1 = B_{n-a}$. We must verify (11.2)–(11.4) with a, b and L replacing α, β and V.

Since $A_i \geq 0$ and $Y_i \leq Y_{i+1}$, we must have

$$A_i Y_i + w_{i+1} Y_{i+1} \leq A_{i+1} Y_{i+1} \qquad \text{for} \quad i = 1, 2, \cdots, n-1. \tag{11.7}$$

Applying (11.7) inductively on i, we find

$$\begin{aligned} L &\leq A_{a+1} Y_{a+1} + w_{a+2} Y_{a+2} + \cdots + w_n Y_n \\ &\leq A_{n-b} Y_{n-b} + w_{n-b+1} Y_{n-b+1} + \cdots + w_n Y_n. \end{aligned} \tag{11.8}$$

The final term of (11.8) equals $A_{n-b} Y_{n-b} = Y_{n-b}$, hence $L \leq Y_{n-b}$. The analogous argument shows $Y_{a+1} \leq L$, so that (11.2) holds.

To verify (11.3), consider the first inequality of (11.8). If $Y_{a+2}, \cdots, Y_n$ are fixed arbitrarily and $Y_{a+1} \to -\infty$, the middle member of (11.8) will tend to $-\infty$, and therefore so must L. The analogous argument checks (11.4).

Several of the following corollaries are obvious, so proofs are omitted or only sketched.

Tukey has discussed the use of trimmed and Winsorized means for the location problem (see section 14 of [11] and the references there given). The (a, b)-trimmed mean is the arithmetic mean of the $n - a - b$ observations that remain after the a smallest and b largest observations have been removed. For example, the estimate R above is a (2, 1)-trimmed mean.

COROLLARY 1.1. *If R is an (a, b)-trimmed mean, then $\alpha_n(R) = a$ and $\beta_n(R) = b$.*

The (a, b)-Winsorized mean is the arithmetic mean of the n values after each of $Y_1, \cdots, Y_a$ has been replaced by Y_{a+1}, and each of $Y_{n-b+1}, \cdots, Y_n$ has been replaced by Y_{n-b}.

COROLLARY 1.2. *If Z is an (a, b)-Winsorized mean, then $\alpha_n(Z) = a$ and $\beta_n(Z) = b$.*

COROLLARY 1.3. *The mean $\overline{X}$ has zero tolerance for every n.*

PROOF. The mean $\overline{X}$ is the special case of L with each $w_i = 1/n$. As remarked above, $\overline{X}$ cannot tolerate even one extreme value.

COROLLARY 1.4. *The median $\tilde{X}$ has tolerance $\gamma_n(\tilde{X}) = [\frac{1}{2}(n - 1)]$.*

PROOF. We are using the customary notation, where $[u]$ means the greatest integer not greater than u. If n is even, say $n = 2k$, then $\tilde{X} = \frac{1}{2}Y_k + \frac{1}{2}Y_{k+1}$ so that by theorem 1, $\alpha = \beta = k - 1 = [\frac{1}{2}(2k - 1)]$. If n is odd, say $n = 2k + 1$, then $\tilde{X} = Y_{k+1}$ so that $\alpha = \beta = k = [\frac{1}{2}(2k + 1 - 1)]$.

It is clear from (11.2) that, whenever α and β exist, $\alpha + \beta \leq n - 1$. Therefore, γ, whenever it exists, must satisfy $2\gamma \leq n - 1$ or $\gamma \leq [\frac{1}{2}(n - 1)]$. Thus the median $\tilde{X}$ has the maximum possible value of γ, corresponding to the intuitive idea that it tolerates extreme values as well as is possible for any estimate that treats the two extremes symmetrically. (Of course other order statistics may be more tolerant of extremes on one side only.)

COROLLARY 1.5. *The tolerance $\alpha_n(Y_i) = i - 1$ and $\beta_n(Y_i) = n - i$.*

COROLLARY 1.6. *The sample midrange has $\gamma_n = 0$.*

COROLLARY 1.7. *The estimates discussed in section 5, where the weight of Y_i is proportional to $f(EY_i)$, have $\gamma = 0$. This includes the estimate W of section 5.*

As a second class of estimates for which α and β exist, consider the estimates S defined in (9.1), where for the moment we do not impose (9.2). Denote by $\mathcal{K}_a$ the subset of $\mathcal{S}$ consisting of those pairs $(i, j) \in \mathcal{S}$ for which $i \geq a$. Similarly, let $\mathcal{L}_b$ be the subset of $\mathcal{S}$ consisting of those pairs $(i, j) \in \mathcal{S}$ for which $j \leq b$.

THEOREM 2. *The tolerance $\alpha_n(S)$ is the largest integer a such that*

$$\frac{1}{2}\{\#(\mathcal{S}) + 1\} \leq \#(\mathcal{K}_{a+1}). \tag{11.9}$$

We have that $\beta_n(S)$ is the largest integer b such that

$$\frac{1}{2}\{\#(\mathcal{S}) + 1\} \leq \#(\mathcal{L}_{b+1}). \tag{11.10}$$

PROOF. The argument depends slightly on the parity of $\#(\mathcal{S})$; we shall give it for the even case, say $\#(\mathcal{S}) = 2q$. Then, if a is the largest integer for which (11.9) holds, $\#(\mathcal{K}_{a+1}) \geq q + 1$ and $q \geq \#(\mathcal{K}_{a+2})$. It is clear that $Y_{a+1} \leq M_{i,j}$ for every $(i, j) \in \mathcal{K}_{a+1}$. Therefore, Y_{a+1} is not greater than at least $q + 1$ of the $2q$ means, the median of which is S. It follows that $Y_{a+1} \leq S$. Similarly (11.10) implies $S \leq Y_{n-b}$ and (11.2) holds.

Now fix $Y_{a+2}, \cdots, Y_n$ and let $Y_{a+1} \to -\infty$. This implies $M_{ij} \to -\infty$ for every $i \leq a + 1$. At most $\#(\mathcal{K}_{a+2})$ of the $2q + 1$ means whose median is S can avoid tending to $-\infty$, and hence at least $q + 1$ means tend to $-\infty$, implying that $S \to -\infty$. This checks (11.3), and the argument for (11.4) is analogous.

If $\mathcal{S}$ satisfies the symmetry condition (9.2), then clearly $\alpha_n(S) = \beta_n(S) = \gamma_n(S)$. This applies to T, U, and D.

COROLLARY 2.1. *The tolerance* $\gamma_n(D) = [\frac{1}{4}(n - 2)]$.

The argument depends on n mod 4; we give it for $n = 4k$. Then $\#(\mathcal{D}) = 2k$ and $\#(\mathcal{K}_{a+1}) = 2k - a$. The largest integer a for which (11.9) holds is $k - 1 = [\frac{1}{4}(2k - 2)]$.

COROLLARY 2.2. *We have*

$$\gamma_n(U) = \left[n - \tfrac{1}{2} - \tfrac{1}{2}\sqrt{2n^2 - 2n + 5}\right]. \tag{11.11}$$

PROOF. Here $\#(\mathcal{B}) = \frac{1}{2}n(n - 1)$ and $\#(\mathcal{K}_a) = \frac{1}{2}(n - a)(n - a + 1)$. We seek the largest integer a such that

$$\tfrac{1}{4}n(n - 1) + \tfrac{1}{2} \leq \tfrac{1}{2}(n - a - 1)(n - a). \tag{11.12}$$

In the range considered, the right side of (11.12) is a decreasing function of a, treated as continuous. Therefore $\alpha_n(U)$ is $[\tilde{a}]$ where $\tilde{a}$ is the root of the quadratic equation obtained by inserting an equality sign in (11.12).

COROLLARY 2.3. *We have*

$$\gamma_n(T) = \left[n + \tfrac{1}{2} - \tfrac{1}{2}\sqrt{2n^2 + 2n + 5}\right]. \tag{11.13}$$

The proof is similar.

We remark that $\gamma_{n+1}(U) = \gamma_n(T) \geq \gamma_n(U) \geq \gamma_n(D)$.

Table VIII compares the values of $\gamma_n(V)$ for $n = 1(1)20$ and $V = \tilde{X}, T, U$ and D.

TABLE VIII

TOLERANCE OF SEVERAL ESTIMATES

n	$\tilde{X}$	T	U	D	n	$\tilde{X}$	T	U	D
1	0	0	0	0	11	5	3	3	2
2	0	0	0	0	12	5	3	3	2
3	1	0	0	0	13	6	3	3	2
4	1	1	0	0	14	6	4	3	3
5	2	1	1	0	15	7	4	4	3
6	2	1	1	1	16	7	4	4	3
7	3	2	1	1	17	8	5	4	3
8	3	2	2	1	18	8	5	5	4
9	4	2	2	1	19	9	5	5	4
10	4	3	2	2	20	9	5	5	4

If $\lambda(V) = \lim_{n\to\infty} (\gamma_n(V)/n)$ exists, this limit represents the fraction of extreme values which the estimate V can tolerate at each end of a large sample. For large n,

$$\gamma_n(T) = \left(1 - \frac{\sqrt{2}}{2}\right)\left(1 + \frac{1}{2n}\right) + 0\left(\frac{1}{n^2}\right) = \gamma_n(U). \tag{11.14}$$

Thus

$$\lambda(U) = \lambda(T) = 1 - \frac{\sqrt{2}}{2} = .293. \tag{11.15}$$

For comparison, $\lambda(\tilde{X}) = .5$ and $\lambda(D) = .25$.

12. Comparison of several estimates

We conclude by comparing several estimates with regard to three desiderata: (i) efficiency in estimating the center of a normal population; (ii) tolerance of extreme values in the sense of section 11; and (iii) ease of computation. These are of course not the only considerations, but seem important ones. The comparison will be made primarily for sample size $n = 18$, since most is known for that case, but one may suppose that similar results hold for n near 18.

If only (i) is considered, then $\overline{X}$ has a strong claim to be the best estimate, as it is the optimum estimate for the normal location problem according to several criteria. This classical estimate is also easy to compute. However $\overline{X}$ has no tolerance of extreme values, and cannot be considered if (ii) is important.

If only (ii) is considered, then $\tilde{X}$ is the estimate of choice, as it maximizes γ. It is again easy to compute. However, it is poor on criterion (i), with $e(\tilde{X}) = .68563$ at $n = 18$, and $ae(\tilde{X}) = 2/\pi = .63662$. In most situations one would not pay so great a price in efficiency for what may be an unnecessarily great protection against extreme values.

If we require good performance on both (i) and (ii), the estimates T, U and D are all satisfactory. At $n = 18$, all have efficiency near .95 according to the sampling experiments, and $\gamma(T) = \gamma(U) = 5$, $\gamma(D) = 4$. The slightly lower tolerance of D may be balanced against the fact that it is considerably easier to compute than T or U, and if all three desiderata are considered, D is perhaps preferable to $\overline{X}$, $\tilde{X}$, T and U.

Finally, let us compare D with the symmetrically trimmed and Winsorized means. According to corollaries 1.1 and 1.2, these estimates achieve tolerance γ if we trim or Winsorize γ observations at each end. For $n = 18$, table I of [10] permits us to find the variances, and hence the efficiencies, of the (γ, γ)-trimmed and (γ, γ)-Winsorized means for each γ. The results are given in table IX. When $\gamma = 8$, both estimates coincide with $\tilde{X}$; when $\gamma = 0$, both estimates coincide with $\overline{X}$.

It has been pointed out by Dixon [2] that, for $n \leq 20$, the (γ, γ)-Winsorized mean has efficiency that agrees to three figures with the optimum attainable among all weighted averages of the order statistics which assign weight zero to the γ smallest and γ largest. By theorem 1, this implies that, to three figures, the values of efficiency of the Winsorized mean in table IX may be taken to be the efficiency of the optimum linear estimate with tolerance γ.

We see that, at $n = 18$, the efficiency of D is approximately the same as that of the optimum linear estimate with $\gamma = 2$. As $\gamma(D) = 4$, the estimate D provides substantially better tolerance of extreme values, while giving the same normal efficiency. Put another way, if we desire tolerance $\gamma = 4$ the efficiency of the optimum linear estimate is only .889 compared with $e(D) = .955 \pm .007$. This indicates that the restriction to linear estimates entails a substantial cost according to criteria (i) and (ii), and again suggests that the estimate D is worth consideration in the symmetric location problem.

TABLE IX

EFFICIENCIES WITH NORMAL SAMPLES OF 18

γ	Trimmed mean	Winsorized mean
0	1.00000	1.00000
1	.97462	.98116
2	.94084	.95581
3	.90367	.92501
4	.86429	.88896
5	.82314	.84749
6	.78030	.80021
7	.73535	.74649
8	.68563	.68563

REFERENCES

[1] DAVID BLACKWELL and J. L. HODGES, JR., "Geometrical proofs of path counts," unpublished (1965).

[2] W. J. DIXON, "Simplified estimation from censored normal samples," *Ann. Math. Statist.*, Vol. 31 (1960), pp. 385–391.

[3] WILLIAM FELLER, *An Introduction to Probability Theory and Its Applications*, Vol. I, New York, Wiley, 1957 (2d ed.).

[4] J. L. HODGES, JR., "Galton's rank-order test," *Biometrika*, Vol. 42 (1955), pp. 261–262.

[5] J. L. HODGES, JR. and E. L. LEHMANN, "The efficiency of some nonparametric competitors of the *t*-test," *Ann. Math. Statist.*, Vol. 27 (1956), pp. 324–335.

[6] ———, "Estimates of location based on rank tests," *Ann. Math. Statist.*, Vol. 34 (1963), pp. 598–611.

[7] ARNLJOT HØYLAND, "Numerical evaluation of Hodges-Lehmann estimates," *Det Kongelige Norske Videnskabers Selskabs Forhandlinger*, Vol. 37 (1964), pp. 42–47.

[8] JEROME KLOTZ, "Small sample power and efficiency for the one sample Wilcoxon and normal scores tests," *Ann. Math. Statist.*, Vol. 34 (1963), pp. 624–632.

[9] FREDERICK MOSTELLER, "On some useful 'inefficient' statistics," *Ann. Math. Statist.*, Vol. 17 (1946), pp. 377–408.

[10] A. E. SARHAN and B. G. GREENBERG, "Estimation of location and scale parameters by order statistics from singly and doubly censored samples," *Ann. Math. Statist.*, Part I, Vol. 27 (1956), pp. 427–451; Part II, Vol. 29 (1958), pp. 79–105.

[11] JOHN W. TUKEY, "The future of data analysis," *Ann. Math. Statist.*, Vol. 33 (1962), pp. 1–67.

[12] The RAND Corporation, *A Million Random Digits with 100,000 Normal Deviates*, Glencoe, Free Press, 1955.

MOMENTS OF CHI AND POWER OF t

J. L. HODGES, JR.
and
E. L. LEHMANN
UNIVERSITY OF CALIFORNIA, BERKELEY

1. Introduction and summary

This paper is concerned with two related computational problems: the precise calculation of central moments of the chi random variable of ν degrees of freedom, and the use of these moments in computing the power curve of the t-test. Whereas the methods are standard and available in various textbooks, the results have at several points been pushed farther than we have seen them elsewhere. We try to provide the formulas and coefficient tables that would be needed by the computer, but make no attempt to review the extensive literature on chi moments and t power.

Table A gives the coefficients required for obtaining the first twelve moments in terms of the expectation ϵ_ν. In section 3 the general term of an asymptotic series for log ϵ_ν is derived, which provides in table C the early coefficients of the series for ϵ_ν itself. Section 5 presents a formula for the first three terms in the series for moments of arbitrary order, supplemented in table E by additional terms for the first twelve moments. With these coefficients it is relatively easy to obtain precise values of the low moments for large ν.

Section 6 presents a series for the power of the t test in terms of chi moments. In favorable cases this method permits the precise computation of an entire power curve. It also leads to a relatively simple normal approximation for t power, accurate when ν is not too small and the significance level is moderate, and suggests an effective method of interpolation in the noncentral t tables.

2. The moments of chi in terms of its expectation

Let χ denote the chi random variable with ν degrees of freedom, and consider its standardized form $S = \chi/\sqrt{\nu}$. It is well known that

$$ES^p = \Gamma\left(\frac{\nu + p}{2}\right) \bigg/ \left(\frac{\nu}{2}\right)^{p/2} \Gamma\left(\frac{\nu}{2}\right), \qquad p = 1, 2, \cdots, \tag{2.1}$$

and that both the original and central moments of S can be expressed in terms of its expectation,

$$ES = \epsilon_\nu = \Gamma\left(\frac{\nu + 1}{2}\right) \bigg/ \sqrt{\frac{\nu}{2}}\,\Gamma\left(\frac{\nu}{2}\right). \tag{2.2}$$

With the partial support of the Office of Naval Research under Contract No. Nonr-222(43).

The expressions for the moments can be written compactly in terms of the products of successive even or odd integers $\pi_q = (\nu + q - 2)(\nu + q - 4)(\nu + q - 6) \cdots, q = 2, 3, 4, \cdots$, where the last factor is the smallest integer not smaller than ν. That is, $\pi_2 = \nu$, $\pi_3 = \nu + 1$, $\pi_4 = \nu(\nu + 2)$, $\pi_5 = (\nu + 1)(\nu + 3)$, and so on, and for convenience we define $\pi_0 = \pi_1 = 1$. Application of the gamma recursion formula to (2.1) gives, for $a = 0, 1, 2, \cdots$,

$$\nu^a ES^{2a} = \pi_{2a}, \qquad \nu^a ES^{2a+1} = \epsilon_\nu \pi_{2a+1}. \tag{2.3}$$

These formulas give the even moments exactly. For the odd moments, one may use the six decimal table 35 of ϵ_ν in Pearson and Hartley [7] to obtain six-figure values. Greater precision is available by the methods developed in the next two sections.

Consider now the central moments of S, say $M_p = E(S - \epsilon_\nu)^p$. By expanding the binomial and substituting (2.3), it can be seen that, for $a = 1, 2, 3, \cdots$,

$$\begin{aligned} \nu^{a-1} M_{2a} &= \sum_{i=0}^{a} \epsilon_\nu^{2i} \left[-\binom{2a}{2i-1} \pi_{2a-2i+1} + \binom{2a}{2i} \pi_{2a-2i} \right] \nu^{i-1}, \\ \nu^a M_{2a+1}/\epsilon_\nu &= \sum_{i=0}^{a} \epsilon_\nu^{2i} \left[\binom{2a+1}{2i} \pi_{2a-2i+1} - \binom{2a+1}{2i+1} \pi_{2a-2i} \right] \nu^i, \end{aligned} \tag{2.4}$$

where conventionally we put $\binom{2a}{-1} = 0$. The right sides of (2.4) are polynomials in ϵ_ν^2, whose coefficients are themselves polynomials in ν. The coefficients of these latter polynomials are shown in table A for the first twelve central moments. For example, $M_2 = 1 - \epsilon_\nu^2$ and $\nu M_3/\epsilon_\nu = (1 - 2\nu) + 2\nu\epsilon_\nu^2$.

Although table A makes it possible to compute M_p, for $2 \leq p \leq 12$, once ϵ_ν is known, it should be noted that there is a loss of significant figures as p and ν increase. Since the coefficients increase with p, the number of correct decimals in M_p decreases. Further, since M_p tends to 0 as ν increases (see section 5), the number of significant figures may fall off rapidly. For example, M_8 at $\nu = 10$ is not even reliable to one figure if one starts with a six-decimal value of ϵ_{10}.

Thus it may be necessary when using (2.4) to have ϵ_ν correct to many decimal places. In the two following sections we give expansions that provide precise values of ϵ_ν for large ν. For small ν, one may use the following forms of (2.2) which are convenient for use with a table of log factorials:

$$\epsilon_{2a} = \frac{\sqrt{\pi a}(2a)!}{2^{2a}(a!)^2}, \qquad \epsilon_{2a+1} = \frac{2^{2a+1/2}(a!)^2}{\sqrt{\pi}(2a+1)(2a)!}. \tag{2.5}$$

3. An asymptotic series for log ϵ_ν

An asymptotic series of the form

$$\log \epsilon_\nu = \frac{L_1}{\nu} + \frac{L_2}{\nu^2} + \frac{L_3}{\nu^3} + \cdots \tag{3.1}$$

can be developed from the series for log $\Gamma(p)$ in terms of the Bernoulli numbers

TABLE A

COEFFICIENTS OF THE POLYNOMIALS FOR CENTRAL MOMENTS

	1						ϵ_ν^2					
	1	ν	ν^2	ν^3	ν^4	ν^5	1	ν	ν^2	ν^3	ν^4	ν^5
M_2	1						−1					
$\nu M_3/\epsilon_\nu$	1	−2						2				
νM_4	2	1					−4	2				
$\nu^2 M_5/\epsilon_\nu$	3	−6	−4					10	0			
$\nu^2 M_6$	8	6	1				−18	6	9			
$\nu^3 M_7/\epsilon_\nu$	15	−33	−33	−6				63	14	−14		
$\nu^3 M_8$	48	44	12	1			−120	40	96	20		
$\nu^4 M_9/\epsilon_\nu$	105	−256	−310	−92	−8			540	156	−180	−48	
$\nu^4 M_{10}$	384	400	140	20	1		−1050	400	1120	380	35	
$\nu^5 M_{11}/\epsilon_\nu$	945	−2535	−3450	−1310	−195	−10		5775	1760	−2530	−1100	−110
$\nu^5 M_{12}$	3840	4384	1800	340	30	1	−11340	5076	15000	6480	1020	54

	ϵ_ν^4					ϵ_ν^6				ϵ_ν^8			ϵ_ν^{10}		ϵ_ν^{12}
	ν	ν^2	ν^3	ν^4	ν^5	ν^2	ν^3	ν^4	ν^5	ν^3	ν^4	ν^5	ν^4	ν^5	ν^5
νM_4	−3														
$\nu^2 M_5/\epsilon_\nu$		4													
$\nu^2 M_6$	−20	−5				−5									
$\nu^3 M_7/\epsilon_\nu$		35	14				6								
$\nu^3 M_8$	−168	−84	14			−56	−28			−7					
$\nu^4 M_9/\epsilon_\nu$		378	252	0			84	48			8				
$\nu^4 M_{10}$	−1800	−1080	180	90		−756	−588	−42		−120	−75		−9		
$\nu^5 M_{11}/\epsilon_\nu$		4950	3894	198	−132		1386	1188	132		165	110		10	
$\nu^5 M_{12}$	−23100	−14960	2860	2420	275	−11880	−10824	−1584	132	−2376	−2178	−297	−220	−154	−11

(see Cramér [1], p. 129). Substitution of the latter series into (2.2), with p replaced successively by $\nu/2$ and $(\nu + 1)/2$, gives

$$\log \epsilon_\nu = -\tfrac{1}{2} + \frac{\nu}{2} \log\left(1 + \frac{1}{\nu}\right) + \sum_{j=1}^{\infty} \frac{2^{2j-1}B_{2j}}{2j(2j-1)\nu^{2j-1}} \left[\left(1 + \frac{1}{\nu}\right)^{-2j+1} - 1\right]. \tag{3.2}$$

On expanding the logarithm, it is seen that the first two terms on the right side of (3.2) contribute to L_k the amount

$$\frac{(-1)^k}{2(k+1)} = \frac{(-1)^{k+1}}{2k(k+1)} \left[\binom{k+1}{0} B_0 + \binom{k+1}{1} 2B_1\right], \tag{3.3}$$

where we have used $B_0 = 1$, $B_2 = -\frac{1}{2}$. On expanding the binomials, one sees that the third term contributes to L_k the amount

$$\sum_{2j+m-1=k} (-1)^m \frac{2^{2j-1}B_{2j}}{2j(2j-1)} - \binom{2j+m-2}{m}. \tag{3.4}$$

In this sum, m runs over the limits $1 \leq m \leq k - 1$, with $m + k + 1$ even. Thus $(-1)^m = (-1)^{k+1}$. Since $B_3 = B_5 = \cdots = 0$, we may allow m to run over all integers from 1 to $k - 1$. Rearranging the coefficients, and changing variable to $k - m + 1 = q$, (3.4) may be combined with (3.3) to give

$$L_k = \frac{(-1)^{k+1}}{2k(k+1)} \sum_{q=0}^{k} \binom{k+1}{q} 2^q B_q. \tag{3.5}$$

This formula may be simplified with the aid of the identities (see Miller [4], p. 90; Nörlund [6], p. 22)

$$\sum_{q=0}^{k+1} \binom{k+1}{q} 2^q B_q = 2^{k+1} B_{k+1}\left(\tfrac{1}{2}\right) = -2(2^k - 1)B_{k+1}, \tag{3.6}$$

to yield $L_k = (-1)^k(2^{k+1} - 1)B_{k+1}/k(k+1)$. That is, $L_2 = L_4 = \cdots = 0$, whereas for $b = 1, 2, 3, \cdots$,

$$L_{2b-1} = -(4^b - 1)B_{2b}/(2b-1)(2b). \tag{3.7}$$

The Bernoulli numbers have been extensively tabulated (see Peters [8], table 8), making it possible to give explicitly as many terms of (3.1) as could be needed. Table B shows the first twelve nonzero coefficients as fractions in lowest terms.

For higher coefficients there is an effective approximate formula. From the bound on B_{2b} (see [4], p. 101) it is seen that

$$L_{2b-1} \doteq (-1)^{b+1} 2(2b-2)!/\pi^{2b} \tag{3.8}$$

with a relative error not greater than 2^{-2b+1}. At L_{23} this approximation is already correct to 11 figures. Thus, to the terms provided by table B one may add additional terms computed recursively by

$$\frac{L_{2b+1}}{\nu^{2b+1}} \doteq -2b(2b-1)\left(\frac{L_{2b-1}}{\nu^{2b-1}}\right)\left(\frac{1}{(\pi\nu)^2}\right). \tag{3.9}$$

TABLE B

COEFFICIENTS L_{2b-1} OF THE SERIES FOR $\log \epsilon_\nu$

b	Numerator	Denominator
1	−1	4
2	1	24
3	−1	20
4	17	112
5	−31	36
6	691	88
7	−5 461	52
8	929 569	480
9	−3 202 291	68
10	221 930 581	152
11	−4 722 116 521	84
12	968 383 680 827	368

The series (3.1) is asymptotic, not convergent, and terms should not be added beyond the point where they begin to increase. However, it is very effective when ν is not too small. As an illustration, table I shows the computation of $\log \epsilon_{40}$ to

TABLE I

CALCULATION OF $\log \epsilon_{40}$

k	$L_k/(40)^k$
1	−. 006 250
3	651 041 666 6667
5	— 488 281 2500
7	926 4265
9	−3 2849
11	187
13	−2

$\log \epsilon_{40}$ = −. 006 249 349 445 691 4232
ϵ_{40} = . 993 770 137 124 628 880

18 decimal places. The conversion from $\log \epsilon_{40}$ to ϵ_{40} is immediate with the aid of the 18-decimal table of e^{-x} in National Bureau of Standards [5]. The value of ϵ_{40} may be substituted into (2.4) to find central moments of S. For example, M_8 at $\nu = 40$ is

$$(85008 + 1435080\epsilon_{40}^2 + 754880\epsilon_{40}^4 - 1881600\epsilon_{40}^6 - 448000\epsilon_{40}^8)/64000 \\ = .0^5\,258\,683\,675\,02. \tag{3.10}$$

If all the terms in table B are used together with (3.9), $\log \epsilon_{40}$ may be obtained to about 40 decimals. For conversion of $\log \epsilon_\nu$ to ϵ_ν if more than 18 decimals are wanted, the following method may be used. First find a 10-place value of ϵ_ν from the exponential table, and use the method of continued fractions to get a close rational approximation m/n to ϵ_ν, where m and n are four-digit integers. Using table 13 of Peters [8], one can find $\eta = \log \epsilon_\nu - \log m + \log n$, to 48 decimals if desired. Since η is small, only a few terms of $\epsilon_\nu = m(1 + \eta + \frac{1}{2}\eta^2 + \cdots)/n$ will be needed. The same technique gives ϵ_ν^{2i} from $2i \log \epsilon_\nu$ for use in (2.4).

4. An asymptotic series for ϵ_ν

The series (3.1) for $\log \epsilon_\nu$ can be converted into a series

$$\epsilon_\nu = 1 + \frac{E_1}{\nu} + \frac{E_2}{\nu^2} + \cdots \tag{4.1}$$

by formally expanding $\exp(\log \epsilon_\nu)$. The coefficients are

$$E_k = L_k + \frac{1}{2!}\sum\nolimits_2 L_{i_1}L_{i_2} + \frac{1}{3!}\sum\nolimits_3 L_{i_1}L_{i_2}L_{i_3} + \cdots, \tag{4.2}$$

where $\sum_r$ is taken over all $i_1, i_2, \cdots, i_r \geq 1$ having $i_1 + i_2 + \cdots + i_r = k$. We have computed the first twelve coefficients. In order to simplify the denominators, table C gives $4^k E_k$ as a fraction in lowest terms, rather than E_k itself. That is, these are the coefficients for the series in $(4\nu)^{-k}$:

TABLE C

COEFFICIENTS OF THE SERIES FOR ϵ_ν

k	$4^k E_k$		k	E_k
	Numerator	Denominator		
1	−1	1	13	−104. 762 957 37
2	1	2	14	26. 614 715 5
3	5	2	15	1 933. 225 106
4	−21	8	16	−488. 764 02
5	−399	8	17	−47 030. 779 9
6	869	16	18	11 855. 436
7	39 325	16	19	1 458 576. 31
8	−334 477	128	20	−336 973. 7
9	−28 717 403	128	21	−56 169 531
10	59 697 183	256		
11	8 400 372 435	256		
12	−34 429 291 905	1024		

$$\epsilon_\nu = 1 - \frac{1}{(4\nu)} + \frac{1}{2(4\nu)^2} + \frac{5}{2(4\nu)^3} - \frac{21}{8(4\nu)^4} - \cdots. \tag{4.3}$$

The exact coefficients rapidly become cumbersome, and table C gives the coefficients from E_{13} to E_{21} in decimal form.

We do not know a simple closed form for E_k, but can give a fairly simple expression for an approximation for large k. Because of the rapid increase of $|L_{2b-1}|$ (see (3.8)), the sum $\sum_r$ of (4.2) will be dominated by those terms

$L_{i_1}L_{i_2}\cdots L_{i_r}$ where one i_j is a large odd integer. Fix an odd n, $k > n > \frac{1}{2}k$, and collect the terms of (4.2) containing the factor L_n. They are

$$L_n\left\{\frac{1}{2!}2L_{k-n}+\frac{1}{3!}3\sum_{i_1+i_2=k-n}L_{i_1}L_{i_2}+\cdots\right\}=E_{k-n}L_n. \tag{4.4}$$

The form of the next most important terms, whose factor L_n with largest subscript has $n \leq \frac{1}{2}k$, depends on k mod 4. The resulting approximations

$$\begin{aligned} E_{4c} &= E_1L_{4c-1}+E_3L_{4c-3}+\cdots+E_{2c-1}L_{2c+1}+\tfrac{1}{64}L_{2c-1}^2,\\ E_{4c+1} &= L_{4c+1}+E_2L_{4c-1}+\cdots+E_{2c}L_{2c+1}+\tfrac{5}{256}L_{2c-1}^2,\\ E_{4c+2} &= E_1L_{4c+1}+E_3L_{4c-1}+\cdots+E_{2c-1}L_{2c+3}+\tfrac{1}{2}L_{2c+1}^2+\tfrac{1}{32}L_{2c-1}L_{2c+1},\\ E_{4c+3} &= L_{4c+3}+E_2L_{4c+1}+\cdots+E_{2c}L_{2c+3}-\tfrac{1}{8}L_{2c+1}^2, \end{aligned} \tag{4.5}$$

are good to about 8 figures at the limit of table C. Substitution of (3.8) into (4.5) gives the relations

$$\begin{aligned} E_{2b} &\sim -\tfrac{1}{4}L_{2b-1}\left\{1-\frac{5\pi^2}{32(2b-3)(2b-2)}+0(b^{-4})\right\},\\ E_{2b+1} &\sim L_{2b+1}\left\{1-\frac{\pi^2}{32(2b-1)(2b)}+0(b^{-4})\right\}. \end{aligned} \tag{4.6}$$

When computing a precise value of ϵ_ν for large ν by series, one has a choice of (3.1) or (4.1). The latter has the advantage of giving ϵ_ν directly, but the former requires only half as many terms, and with the coefficients here provided is able to give greater precision. For example, (4.1) and table C provides at most 27 decimals in ϵ_{40}, compared with 40 decimals available for log ϵ_{40}. As shown in table II, the direct series requires 14 terms to produce the 18-decimal value of

TABLE II

CALCULATION OF ϵ_{40}

k	$E_k/(40)^k$
0	1.
1	−. 006 250
2	19 531 250
3	610 351 562 5000
4	−4 005 432 1289
5	−475 645 0653
6	3 237 2773
7	915 6065
8	−6 0841
9	−3 2648
10	212
11	187
12	−1
13	−2
ϵ_{40} =	.993 770 137 124 628 880

ϵ_{40} that table I gives with 7 terms. While the L-series is computationally preferable, we shall need the E-series for the development of the following sections.

5. Asymptotic series for M_p

As remarked in section 2, the computation of the central moments M_p by the method of that section becomes awkward when ν and p are large, and we now develop asymptotic series for M_p. Two approaches are used. One may substitute the expansion (4.1) into (2.4), and this method was used to compute the supple-

TABLE D

THREE COEFFICIENTS OF SERIES FOR M_p

a	Even moments			Odd moments		
	P_{2a}	$6C_{2a}^{(1)}$	$540C_{2a}^{(2)}$	P_{2a+1}	$90C_{2a+1}^{(1)}$	$7560C_{2a+1}^{(2)}$
1	1	−3	−270	1	45	−12 285
2	3	−6	135	10	−27	−21 357
3	15	−5	1 575	105	−99	−15 471
4	105	4	2 826	1 260	−151	7 893
5	945	25	1 080	17 325	−163	41 847
6	10 395	62	−7 255	270 270	−115	70 095
7	135 135	119	−24 955	4 729 725	13	69 173
8	2 027 025	200	−51 580	91 891 800	241	12 929
9	34 459 425	309	−80 274	1 964 187 225	589	−120 757
10	654 729 075	450	−93 765	45 831 035 250	1 077	−341 013

TABLE E

SUPPLEMENTARY COEFFICIENTS FOR LOW-ORDER MOMENTS

a	Even moments					
	$7560C_{2a}^{(3)}$	$1680C_{2a}^{(4)}$	$480C_{2a}^{(5)}$	$64C_{2a}^{(6)}$	$128C_{2a}^{(7)}$	$128C_{2a}^{(8)}$
1	4 725	4 830	−1 590	−2 372	5 165	110 123
2	28 350	−3 570	−23 100	1 809	144 646	
3	13 230	−65 870	−11 005	58 669		
4	−142 632	−113 043	224 852			
5	−440 073	130 643				
6	−610 858					

a	Odd moments				
	$1008C_{2a+1}^{(3)}$	$5760C_{2a+1}^{(4)}$	$1280C_{2a+1}^{(5)}$	$1024C_{2a+1}^{(6)}$	$2048C_{2a+1}^{(7)}$
1	−4 725	54 675	87 015	−122 101	−3 371 095
2	819	185 715	−7 921	−737 925	
3	11 349	123 687	−329 165		
4	17 229	−329 729			
5	5 687				

mentary coefficients shown in table E. It is, however, applicable only for small p, and another approach is used to find expressions valid for any p, leading to the coefficients shown in table D. In what follows, we retain only enough terms to make the method clear, although more terms were carried in deriving the values shown in table D.

The chi density $f_\chi(x) \sim x^{\nu-1}e^{-(1/2)x^2}$ may be expanded about $\sqrt{\nu - 1}$ by making the substitution $\chi = \sqrt{\nu - 1} + (Y/\sqrt{2})$. To simplify the notation we write $\sqrt{2(\nu - 1)} = u$, and find

$$\log f_Y(y) = \text{const.} - \frac{y^2}{2} + \frac{y^3}{2\cdot 3u} - \frac{y^4}{2\cdot 4u^2} + \frac{y^5}{2\cdot 5u^3} - \cdots. \tag{5.1}$$

The expansion is, of course, not valid over the entire range $-u < y < \infty$. However, over the interval $|y| < \log \nu$ the remainder after the term in u^{-j} is of smaller order than u^{-j} uniformly in y. Furthermore, the contribution to any moment of Y outside this interval is negligible. This may be seen by comparing the distribution of Y with that of a normal random variable D having $ED = 0$ and var $D = \frac{1}{2}$. It is easy to check that the contribution of $|d| > \log \nu$ to any moment of D is of smaller order than any inverse power of ν. From the fact that

$$\frac{d}{dy} \log \frac{f_Y(y)}{f_D(y)} = \frac{u}{2}\left\{\left(1 + \frac{y}{u}\right)^{-1} - 1\right\} \tag{5.2}$$

decreases monotonely in y, it can be seen that the tails of Y are even less important than those of D. Accordingly, if in our expansions we retain all terms of a given order in ν, the resulting moment series will be valid to that order.

It is notationally convenient to express the series in terms of $\mu = 2\nu$, and we write $\sqrt{\mu} = v$, noting that

$$u^{-j} = v^{-j}\left(1 - \frac{2}{v^2}\right)^{-1/2}. \tag{5.3}$$

Substitution of this expression in (5.1) gives

$$\log f_Y(y) = \text{const.} - \frac{y^2}{2} + \frac{y^3}{6v} - \frac{y^4}{8v^2} + \frac{5y^3 + 3y^5}{30v^3} + \cdots, \tag{5.4}$$

where in general the coefficient of v^{-j} is a polynomial in y of degree $j + 2$ with alternate terms vanishing.

Since we are interested in the central moments of S, we shift the origin by $Z = Y - EY = Y - v\epsilon_\nu + u = Y - (1/2v) + (5/8v^3) + \cdots$, finding

$$\log f_Z(z) = \text{const.} - \frac{z^2}{2} + \frac{-3z + z^3}{6v} + \frac{-1 + 2z^2 - z^4}{8v^2} + \cdots, \tag{5.5}$$

where the coefficient of v^{-j} is a polynomial in z of degree $j + 2$ with alternate terms vanishing. Application of the exponential transformation gives

$$f_Z(z) = C\varphi(z)\left\{1 + \frac{-3z + z^3}{6v} + \frac{-9 + 27z^2 - 15z^4 + z^6}{72v^2} + \cdots\right\} \tag{5.6}$$

where the coefficient of v^{-j} is a polynomial in z of degree $3j$ with alternate terms vanishing. Finally, C may be determined to make the probability equal 1, giving

$$f_Z(z) = \varphi(z)\left\{1 + \frac{-3z + z^3}{6v} + \frac{3 + 27z^2 - 15z^4 + z^6}{72v^2} + \cdots\right\}, \tag{5.7}$$

where the coefficient of v^{-j} has the same structure as before.

From (5.7) it is now easy to express the moments of Z in terms of the even moments of the standard normal distribution, say

$$N_{2a} = \int_{-\infty}^{\infty} z^{2a}\varphi(z)\,dz = 1\cdot 3\cdot 5\cdot \cdots \cdot(2a - 1). \tag{5.8}$$

In the even case,

$$EZ^{2a} = N_{2a} + \frac{1}{72\mu}(3N_{2a} + 27N_{2a+2} - 15N_{2a+4} + N_{2a+6}) + \cdots, \tag{5.9}$$

where in general the coefficient of μ^{-k} is a linear combination of $N_{2a}, N_{2a+2}, \cdots, N_{2a+6k}$. If we factor out N_{2a} and note that $M_{2a} = EZ^{2a}/\mu^a$, it appears that

$$M_{2a} = \frac{P_{2a}}{\mu^a}\left\{1 + \frac{C_{2a}^{(1)}}{\mu} + \frac{C_{2a}^{(2)}}{\mu^2} + \cdots\right\} \tag{5.10}$$

where $P_{2a} = N_{2a}$ and where $C_{2a}^{(k)}$ is a polynomial in a of degree $3k$. The first two of these polynomials are

$$\begin{aligned} 18C_{2a}^{(1)} &= 2a^3 - 6a^2 - 5a, \\ 9720C_{2a}^{(2)} &= 20a^6 - 300a^5 + 512a^4 + 3708a^3 - 4753a^2 - 4047a. \end{aligned} \tag{5.11}$$

The odd case may be handled similarly, giving

$$M_{2a+1} = \frac{P_{2a+1}}{\mu^{a+1}}\left\{1 + \frac{C_{2a+1}^{(1)}}{\mu} + \frac{C_{2a+1}^{(2)}}{\mu^2} + \cdots\right\}, \tag{5.12}$$

where $P_{2a+1} = aN_{2a+2}/3$ and where $C_{2a+1}^{(k)}$ is again a polynomial in a of degree $3k$. We find

$$\begin{aligned} 270C_{2a+1}^{(1)} &= 10a^3 - 60a^2 - 106a + 291, \\ 68040C_{2a+1}^{(2)} &= 28a^6 - 588a^5 + 1372a^4 + 18480a^3 \\ &\quad - 33377a^2 - 114993a + 18513. \end{aligned} \tag{5.13}$$

Table D gives the values of these coefficients for the moments up to order 21, while as mentioned earlier table E provides certain additional coefficients for moments up to order 12. The use of these coefficients is illustrated in table III for M_8 at $\nu = 40$. As always, when using asymptotic series, one must be guided by the rate of decrease of the successive terms in judging the resulting accuracy. The numbers in table III suggest that the value of M_8 should be good to about eight figures. Comparison with the direct calculation (3.10) shows an error of 2 in the ninth figure. In this case, the series computation is much easier than the direct calculation, which requires one to carry about 17 decimals in $\epsilon_{40}^2, \cdots, \epsilon_{40}^8$.

TABLE III

CALCULATION OF M_8 AT $\nu = 40$ BY SERIES

k	$C_8^{(k)}/(80)^k$
0	1.000 000 00
1	8 333 33
2	817 71
3	−36 85
4	−1 64
5	14
	1.009 112 69

$$\times \frac{105}{(80)^4} = .0^5\,258\,683\,673 = M_8$$

6. Power of the *t*-test in terms of chi moments

The power of the t-test with ν degrees of freedom can be expressed as an asymptotic series involving the moments M_p of $\chi/\sqrt{\nu}$. In some cases, this series is an effective way of computing precise values of t power. It also throws some light on approximations and on the problem of interpolation in the noncentral t tables.

Let X have the standard normal distribution, so that $(X + \delta)/S$ has the noncentral t distribution with ν degrees of freedom and noncentrality parameter δ. For both one- and two-sided t-tests, the power can be expressed in terms of the quantity $\beta = P((X + \delta/S) < t) = E\Phi(tS - \delta)$ where Φ is the standard normal cumulative. If we write $w = t\epsilon_\nu - \delta$ and expand $\Phi(tS - \delta) = \Phi(w + t(S - \epsilon_\nu))$ about w, we have the asymptotic series

$$\beta = \Phi(w) + T_2 + T_3 + \cdots, \qquad T_p = \frac{t^p}{p!} M_p \varphi^{(p-1)}(w). \tag{6.1}$$

The term T_p is a product of three factors, each depending on only one of the variables t, ν, w on which β depends. We have just seen how M_p can be computed with considerable precision for low values of p. The factor $t^p/p!$ offers no problems. The normal derivatives are extensively tabled in Harvard Computation Laboratory [2], and this table can, if necessary, be supplemented by the expressions

$$\varphi'(w) = -w\varphi(w), \qquad \varphi''(w) = (w^2 - 1)\varphi(w), \qquad \varphi'''(w) = -(w^3 - 3w)\varphi(w), \cdots \tag{6.2}$$

of normal derivatives in terms of Hermite polynomials. Thus, it is feasible to compute a number of terms of (6.1).

To gain an appreciation of the circumstances in which such a computation will be effective, note that $|\varphi^{(p-1)}(w)|$, and hence $|T_p|$, is bounded in w for any given p, ν, and t. Therefore, for those values of ν and t for which (6.1) works, it should do so for all w and hence for all δ. If ν is not too small, the order of magnitude of M_p will be given by the first term of the series (5.10) or (5.12). With this approximation,

$$(6.3)\qquad \begin{aligned} \max_w |T_{2a}| &\sim \left(\frac{t^2}{\mu}\right)^a R_{2a}, \qquad R_{2a} = \max_w |\varphi^{(2a-1)}(w)|/a!2^a, \\ \max_w |T_{2a+1}| &\sim \frac{1}{t}\left(\frac{t^2}{\mu}\right)^{a+1} R_{2a+1}, \qquad R_{2a+1} = \max_w |\varphi^{(2a)}(w)|/3(a-1)!2^a. \end{aligned}$$

Some values of R_p are shown in table F. For both the even and odd cases, these values change slowly enough so that, roughly speaking, the series of maxima of the terms in (6.1) are like geometric series with ratio t^2/μ. This indicates that (6.1) will work to the extent that t^2/μ is less than one. Furthermore, the entries in table F serve to indicate about how many terms of (6.1) will be required in any given case.

TABLE F

FACTORS FOR MAXIMA OF $|T_p|$

a	R_{2a}	R_{2a+1}
1	.1210	.0665
2	.0688	.0997
3	.0481	.1247
4	.0369	.1454
5	.0300	.1636
6	.0252	.1800
7	.0218	.1950
8	.0192	.2089
9	.0171	.2220
10	.0154	.2343

As an illustration, consider the case $\nu = 40$ and $t = 2$. Here $t^2/\mu = \frac{1}{20}$, so the terms of (6.1) will decrease rapidly. In order to find β as a function of w, we shall compute β for a few equally spaced values of w as a basis for interpolation. The value $w = 0$ is attractive, since here the terms T_{2a} vanish. Since T_{2a} has the same value at w and $-w$, and since T_{2a+1} merely changes sign, it is convenient to use values such as $w = 0$, $w = \pm 1$, $w = \pm 2$. If we carry 12 decimals in the work, terms beyond T_{16} will not be needed. The computations are exhibited in table IV, the necessary values of M_p having been found as indicated in table III. The five values of β are recorded in table V, where the final figure is subject to rounding error.

As a by-product of the series (6.1) and the moment series developed in section 5, one can obtain the Cornish-Fisher development for β. Substitution of (6.2), (5.10) and (5.12) into (6.1) gives

$$(6.4)\qquad \beta = \Phi(w) + \varphi(w)\left[\frac{P_1}{\mu} + \frac{P_2}{\mu^2} + \cdots\right]$$

TABLE IV

ILLUSTRATION OF (5.1) AT $\nu = 40$, $t = 2$, $w = 0, 1, 2$

k	$t^k M_k/k!$	$w = 0$	$w = 1$	$w = 2$
	$\Phi(w)$	.500 000 000 000	.841 344 746 069	.977 249 868 052
2	.024 841 829 119		−6 010 995 390	−2 682 468 728
3	$.0^3$ 209 580 665	−83 610 588		33 946 388
4	$.0^3$ 308 608 225		149 348 312	−33 324 113
5	$.0^5$ 518 651 54	6 207 361	−2 509 970	−1 400 125
6	$.0^5$ 257 823 31		−3 743 142	2 505 623
7	$.0^7$ 641 896	−384 119	248 512	−38 122
8	$.0^7$ 164 243 6		79 484	−76 262
9	$.0^9$ 531 26	22 254	−16 969	7 142
10	$.0^{10}$ 856 35		−580	878
11	$.0^{11}$ 331 7	−1 251	976	−469
12	$.0^{12}$ 382 1		−87	69
13	$.0^{13}$ 167	69	−50	20
14	$.0^{14}$ 150		9	−7
15	$.0^{16}$ 7	−4	2	
16	$.0^{17}$ 5		−1	

where P_k is a polynomial in t and w, of degree $2k - 1$ in w. This, in turn, can be written in the form

$$\Phi^{-1}(\beta) = w + \frac{Q_1}{\mu} + \frac{Q_2}{\mu^2} + \cdots \tag{6.5}$$

where $Q_k = q_{k0} + q_{k1}w + \cdots + q_{k,2k-1}w^{2k-1}$ and the q_{kj} are polynomials in t. In this development, some of the higher terms in t have coefficient 0, and some of the q_{kj} vanish. After some straightforward algebra one finds for the first five Q_k the expressions

$$\begin{aligned}
2q_{11} &= -t^2; \\
6q_{20} &= -t^3, \qquad 8q_{21} = t^2(3t^2 + 2), \qquad 6q_{22} = t^3; \\
12q_{30} &= t^3(3t^2 - 1), \qquad 16q_{31} = -t^2(5t^4 + 6t^2 - 4), \\
&\qquad\qquad 12q_{32} = -t^3(5t^2 - 1); \\
240q_{40} &= -t^3(75t^4 + 18t^2 - 65), \qquad 1152q_{41} \\
&\qquad = t^2(315t^6 + 316t^4 + 108t^2 - 360); \\
240q_{42} &= t^3(175t^4 + 36t^2 - 65), \qquad 72q_{43} = t^4(8t^2 - 9), \\
&\qquad\qquad 40q_{44} = -t^5; \\
480q_{50} &= t^3(175t^6 + 165t^4 - 339t^2 + 375); \\
2304q_{51} &= -t^2(567t^8 - 308t^6 + 2068t^4 - 3240t^2 + 3312); \\
480q_{52} &= -t^3(525t^6 + 427t^4 - 583t^2 + 375); \\
576q_{53} &= -t^4(288t^4 - 316t^2 + 144), \qquad 80q_{54} = 3t^5(3t^2 - 1).
\end{aligned} \tag{6.6}$$

These formulas throw some light on the problem of interpolation with respect to δ in the noncentral t tables of Resnikoff and Lieberman [9] and Locks *et al.* [3].

TABLE V

VALUES OF β FOR $\nu = 40$, $t = 2$

w	β
2	.974 569 020 346
1	.835 477 157 175
0	.499 922 233 722
−1	.164 518 287 827
−2	.025 496 009 322

It is not at first sight obvious how to extract four decimal values from a table in which the consecutive entries are, to take an extreme example, .4123, .0055, and .0000. Because of the vanishing of some of the terms q_{kj}, we see from (6.6) that with an error of order μ^{-4}, $\Phi^{-1}(\beta)$ is a quadratic function of w and hence of δ, whereas it is a quartic function of w with error of order μ^{-6}. Thus, three- or five-point interpolation of $\Phi^{-1}(\beta)$ with respect to δ should give good results in these tables. This is especially convenient in the Locks table since there the entries are equally spaced in δ, permitting the use of Lagrange coefficients.

In the example of table IV, five-point interpolation based on the values at $w = 0, \pm 1, \pm 2$ reproduces the computed values at $w = \pm .5, \pm 1.5, \pm 2.5$ with an error in the eleventh decimal place, while three-point interpolation based on $w = 0, \pm 1$ gives about seven decimals of accuracy. Thus the computation of β by means of (6.1) need be carried out at only a few values of w to provide the entire power curve with high precision, at least when ν is large and t is moderate.

The good results obtained from quadratic interpolation based on the values of β at $w = 0, \pm 1$ suggests a simple approximation for large μ and moderate t. If we fit a quadratic to $\Phi^{-1}(\beta)$ at the points $w = 0, \pm 1$, we obtain from (6.6) the approximation

$$\beta \doteq \Phi(A_0 + A_1 w + A_2 w^2) \tag{6.7}$$

where

$$\begin{aligned} A_0 &= \frac{q_{20}}{\mu^2} + \frac{q_{30}}{\mu^3} + \frac{q_{40}}{\mu^4} + \frac{q_{50}}{\mu^5}, \\ A_1 &= 1 + \frac{q_{11}}{\mu} + \frac{q_{21}}{\mu^2} + \frac{q_{31}}{\mu^3} + \frac{q_{41} + q_{43}}{\mu^4} + \frac{q_{51} + q_{53}}{\mu^5}, \\ A_2 &= \frac{q_{22}}{\mu^2} + \frac{q_{32}}{\mu^3} + \frac{q_{42} + q_{44}}{\mu^4} + \frac{q_{52} + q_{54}}{\mu^5}. \end{aligned} \tag{6.8}$$

In the example of table IV, this approximation appears to have a maximum error of about $.0^7 5$ over the entire range of w. Even if ν is reduced to 10, the approximation is good to the four decimals of the noncentral t tables. Since in those tables it is often necessary to interpolate with respect to all three arguments, the approximation (6.7) is in many cases easier to use than the tables, as well as being more accurate.

REFERENCES

[1] Harald Cramér, *Mathematical Methods of Statistics*, Princeton, Princeton University Press, 1946.

[2] Harvard Computation Laboratory, *Tables of the Error Function and of Its First Twenty Derivatives*, Cambridge, Harvard University Press, 1952.

[3] M. O. Locks, M. J. Alexander, and B. J. Byars, *New Tables of the Noncentral t Distributions*, Aeronaut. Res. Labs., U.S.A.F., 1963.

[4] Kenneth S. Miller, *An Introduction to the Calculus of Finite Differences and Difference Equations*, New York, Holt, 1960.

[5] National Bureau of Standards, *Tables of the Exponential Function e^x*, Applied Mathematics Series 14, Washington, Government Printing Office, 1951.

[6] Niels Erik Nörlund, *Vorlesungen über Differenzenrechnung*, Berlin, Springer, 1924.

[7] E. S. Pearson and H. O. Hartley, *Biometrika Tables for Statisticians*, I, Cambridge, Cambridge University Press, 1954.

[8] J. Peters, *Zehnstellige Logarithmentafel*, I, Berlin, 1922. (Reprinted in New York, Ungar, 1957.)

[9] G. J. Resnikoff and G. J. Lieberman, *Tables of the Non-Central t-Distribution*, Stanford, Stanford University Press, 1957.

ON PROBABILITIES OF LARGE DEVIATIONS

WASSILY HOEFFDING
UNIVERSITY OF NORTH CAROLINA

1. Summary

The paper is concerned with the estimation of the probability that the empirical distribution of n independent, identically distributed random vectors is contained in a given set of distributions. Sections 1–3 are a survey of some of the literature on the subject. In section 4 the special case of multinomial distributions is considered and certain results on the precise order of magnitude of the probabilities in question are obtained.

2. The general problem

Let $X_1, X_2, \cdots$ be a sequence of independent m-dimensional random vectors with common distribution function (d.f.) F. If we want to obtain general results on the behavior of the probability that $X^{(n)} = (X_1, \cdots, X_n)$ is contained in a set A^* when n is large, we must impose some restrictions on the class of sets. One interesting class consists of the sets A^* which are symmetric in the sense that if $X^{(n)}$ is in A^*, then every permutation $(X_{j_1}, \cdots, X_{j_n})$ of the n component vectors of $X^{(n)}$ is in A^*. The restriction to symmetric sets can be motivated by the fact that under our assumption all permutations of $X^{(n)}$ have the same distribution. Let $F_n = F_n(\cdot|X^{(n)})$ denote the empirical d.f. of $X^{(n)}$. The empirical distribution is invariant under permutations of $X^{(n)}$, and for any symmetric set A^* there is at least one set A in the space $\mathcal{G}$ of m-dimensional d.f.'s such that the events $X^{(n)} \in A^*$ and $F_n(\cdot|X^{(n)}) \in A$ are equivalent. The latter event will be denoted by $F_n \in A$ for short. Thus when we restrict ourselves to symmetric sets, we may as well consider the probabilities $P\{F_n \in A\}$, where $A = A_n$ may depend on n. (It is understood that $A \subset \mathcal{G}$ is such that the set $\{x^{(n)}|F_n(\cdot|x^{(n)}) \in A\}$ is measurable.) Since F_n converges to F in a well-known sense (Glivenko-Cantelli theorem), we may say that $P\{F_n \in A_n\}$ is the probability of a large deviation of F_n from F if F is not in A_n and not "close" to A_n, implying that $P\{F_n \in A_n\}$ approaches 0 as $n \to \infty$. For certain classes of sets A_n estimates of $P\{F_n \in A_n\}$

This research was supported in part by the Mathematics Division of the Air Force Office of Scientific Research. Part of the work was done while the author was a visiting professor at the Research and Training School of the Indian Statistical Institute in Calcutta under the United Nations Technical Assistance Program.

(some of which are mentioned below) have been obtained which hold uniformly for "large" and for "small" deviations.

For any two d.f.'s F and G in $\mathcal{G}$ let μ be some sigma-finite measure which dominates the two distributions (for instance, $d\mu = d(F + G)$), and let f and g be the corresponding densities, $dF = f\, d\mu$, $dG = g\, d\mu$. Define

$$I(G, F) = \int (\log (g/f))g\, d\mu, \tag{1}$$

with the usual convention that the integrand is 0 whenever $g = 0$. (The value of $I(G, F)$ does not depend on the choice of μ.) We have $0 \leq I(G, F) \leq \infty$; $I(G, F) = 0$ if and only if $G = F$; $I(G, F) < \infty$ only if F dominates G. Let

$$I(A, F) = \inf_{G \in A} I(G, F), \tag{2}$$

$I(A, F) = +\infty$ if A is empty. Sanov [12] has shown that under certain restrictions on A_n,

$$P\{F_n \in A_n\} = \exp \{-nI(A_n, F) + o(n)\}. \tag{3}$$

If F is discrete and takes only finitely many values, the distribution of F_n may be expressed in terms of a multinomial distribution. In this case the estimate (3), with $o(n)$ replaced by $O(\log n)$, holds under rather mild restrictions on A_n (see [5] and section 4). In [12] (where only sets A independent of n and one-dimensional distributions F are considered) Sanov obtains (3) for a certain class of sets A such that $P\{F_n \in A\}$ can be approximated by multinomial probabilities.

Some necessary conditions for (3) to be true are easily noticed. Let $\mathcal{G}_n(F)$ denote the set of all $G \in \mathcal{G}$ such that nG is integer-valued and $\int_E dF = 0$ implies $\int_E dG = 0$ for every open set $E \subset R^m$. Then $F_n \in \mathcal{G}_n(F)$ with probability one and $P\{F_n \in A\} = P\{F_n \in A \cap \mathcal{G}_n(F)\}$. Let $\mathcal{G}(F)$ denote the set of all G which are dominated by F. Then $I(A, F) < \infty$ only if $A \cap \mathcal{G}(F)$ is not empty. Hence, $\exp \{-nI(A, F)\}$ can be a nontrivial estimate of $P\{F_n \in A\}$ only if both $A \cap \mathcal{G}_n(F)$ and $A \cap \mathcal{G}(F)$ are nonempty. If F is discrete, then $\mathcal{G}_n(F) \subset \mathcal{G}(F)$; if F takes only finitely many values, (3) is always true for $A^{(n)} = A_n \cap G_n(F)$ (see (48), section 4), and (3) holds if $I(A_n^{(n)}, F) - I(A_n, F)$ is not too large. If F is absolutely continuous with respect to Lebesgue measure, then $\mathcal{G}_n(F)$ and $\mathcal{G}(F)$ are disjoint; for (3) to be true and nontrivial, A_n must, as a minimum requirement, contain both values of F_n (which are discrete) and d.f.'s which are dominated by F (hence also by the Lebesgue measure).

In the following two sections the approximation (3) will be related to known results for certain classes of sets, which give more precise estimates of the probability.

3. Half-spaces

Let φ be a real-valued measurable function on R^m, and let

$$H = H(\varphi) = \left\{G \middle| \int \varphi\, dG \geq 0\right\} \tag{4}$$

be the set of all $G \in \mathcal{G}$ such that $\int \varphi\, dG$ is defined and nonnegative. The set H may be called a half-space in $\mathcal{G}$. The asymptotic behavior of $P\{F_n \in H) = P\{\sum_{j=1}^n \varphi(X_j) \geq 0\}$ has been studied extensively. To relate these results to the estimate (3), we first prove the following lemma. We shall write $G[B]$ for $\int_B dG$ and $G[\varphi \in E]$ for $G[\{x|\varphi(x) \in E\}]$.

LEMMA 1. *Let* $H = \{G|\int \varphi\, dG \geq 0\}$, $M(t) = \int \exp(t\varphi)\, dF$.

(A) *We have*

$$I(H, F) = -\log \inf_{t \geq 0} M(t); \tag{5}$$

(B) $0 < I(H, F) < \infty$ *if* $\int \varphi\, dF < 0$, $F[\varphi \geq 0] > 0$, $M(t) < \infty$ *for some* $t > 0$;

(B_1) *if, in addition,* $M'(t^*-) > 0$, *where* $t^* = \sup\{t|M(t) < \infty\}$ *and* $M'(t) = dM(t)/dt$, *then* $\inf_{t \geq 0} M(t) = M(t_\varphi)$, *where* $t_\varphi > 0$ *is the unique root of* $M'(t) = 0$;

(C) $I(H, F) = 0$ *if* $\int \varphi\, dF \geq 0$ *or* $M(t) = \infty$ *for all* $t > 0$;

(D) $I(H, F) = \infty$ *if* $F[\varphi \geq 0] = 0$.

PROOF. If $G \in H$ and $I(G, F) < \infty$, then for $t \geq 0$,

$$\begin{aligned} -I(G, F) &\leq t \int \varphi\, dG - I(G, F) = \int \log(\exp(t\varphi) f g^{-1}) g\, d\mu \\ &\leq \log \int \exp(t\varphi) f\, d\mu = \log M(t) \end{aligned} \tag{6}$$

by Jensen's inequality. Hence, $I(H, F) \geq -\log \inf_{t \geq 0} M(t)$. The equality sign holds in both inequalities in (6) if $\int \varphi\, dG = 0$ and $\exp(t\varphi) f g^{-1} = \text{const. a.e. } (F)$. If $M(t) < \infty$ and $M'(t)$ exists, these conditions are equivalent to $dG = \exp(t\varphi)\, dF/M(t)$ and $M'(t) = 0$. Under the hypothesis of (B), $M'(t)$ exists for $0 < t < t^* = \sup\{t|M(t) < \infty\}$, $M'(0+) < 0$, and $M'(t)$ is increasing. Hence if $M'(t^*-) > 0$, then the root t_φ of $M'(t) = 0$ is unique and positive, and $M(t_\varphi) < 1$. This implies (5), (B), and (B_1) under the condition of (B_1). In particular, if $t^* = \infty$ and the conditions of (B) hold, then that of (B_1) also holds. Next, under the hypothesis of (B), if $F[\varphi > 0] = 0$, then $0 < F[\varphi = 0] < 1$, $\inf_{t \geq 0} M(t) = M(\infty) = F[\varphi = 0]$, and the distribution G with $G[\varphi = 0] = 1$ is in H and $I(G, F) = -\log F[\varphi = 0]$. The remaining case of part (B) is where $t^* < \infty$, $M'(t) < 0$ for $t < t^*$, and $F[\varphi > 0] > 0$. Then $\inf_{t \geq 0} M(t) = M(t^*) < 1$, and we must show that $I(H, F) = -\log M(t^*)$. Let $M_c(t) = \int_{\varphi < c} \exp(t\varphi)\, dF$, which is finite for $t \geq 0$ and $c > \infty$. For c large enough there is a unique number $t(c) > t^*$ such that $M_c'(t(c)) = 0$. It is easy to show that $t(c) \to t^*$ and $M_c(t(c)) \to M(t^*)$ as $c \to \infty$. Let G_c be the d.f. defined by $dG_c = \exp(t(c)\varphi)\, dF/M_c(t(c))$ for $\varphi < c$, $G_c[\varphi \geq c] = 0$. Then $G_c \in H$ and $I(G_c, F) = -\log M_c(t(c)) \to -\log M(t^*)$ as $c \to \infty$, so that $I(H, F) = -\log M(t^*)$. The statements (5), (C), and (D) in the cases $\int \varphi\, dF \geq 0$ and $F[\varphi \geq 0] = 0$ are easily verified, and the part of (C) where $M(t) = \infty$ for all $t > 0$ and $\int \varphi\, dF < 0$ is handled exactly like the last case of part (B), completing the proof.

We have the elementary and well-known inequality

$$P\{F_n \in H\} = P\left\{\sum_{j=1}^n \varphi(X_j) \geq 0\right\} \leq \inf_{t \geq 0} M(t)^n = \exp\{-nI(H, F)\}. \tag{7}$$

Equality is attained only in the trivial cases $F[\varphi \leq 0] = 1$ and $F[\varphi \geq 0] = 1$.

Let $H_c = H(\varphi - c) = \{G | \int \varphi\, dG \geq c\}$, where c is a real number. Then $P\{F_n \in H_c\} = P\{\sum_{j=1}^n \varphi(X_j) \geq nc\}$. If $M(t) < \infty$ for $0 < t < t^*$, and $\int \varphi\, dF < c < L'(t^*-)$, where $L(t) = \log M(t)$, then, by lemma 1,

$$I(H_c, F) = ct(c) - L(t(c)) = I^*(c), \tag{8}$$

say, where $t(c) = t_{\varphi - c}$ is defined by $L'(t(c)) = c$.

A theorem of Cramér [3] as sharpened by Petrov [8] can be stated as follows. Suppose that

$$\int \varphi\, dF = 0, \qquad F[\varphi \neq 0] > 0, \qquad M(t) < \infty \quad \text{if} \quad |t| < t_0 \tag{9}$$

for some $t_0 > 0$. Then for $c = c_n > \alpha n^{-1/2}$, $(\alpha > 0)$, $c = o(1)$ as $n \to \infty$, we have

$$P\left\{\sum_{j=1}^n \varphi(X_j) \geq nc\right\} = b_n(c) \exp\{-nI^*(c)\}(1 + O(c)), \tag{10}$$

where, with $\Phi(x) = (2\pi)^{-1/2} \int_{-\infty}^x \exp(-y^2/2)\, dy$,

$$b_n(c) = (1 - \Phi(x)) \exp(-x^2/2), \qquad x = n^{1/2}c/\sigma, \qquad \sigma^2 = \int \varphi^2\, dF. \tag{11}$$

(Usually the theorem is stated in terms of an expansion of $I^*(\sigma n^{-1/2}x)$ in powers of $n^{-1/2}x$.) Petrov [8] also shows that for any $\epsilon > 0$, equation (10) with $O(c)$ replaced by $r\sigma\epsilon$ holds uniformly for $0 < c < \sigma\epsilon$, where $|r|$ does not exceed an absolute constant. (Compare also the earlier paper of Feller [4].)

Bahadur and Rao [1] have obtained an asymptotic expression for the probability in (10) when c is fixed. It implies that if conditions (9) are satisfied, then for c fixed, $o < c < L'(t^*-)$

$$P\left\{\sum_{j=1}^n \varphi(X_j) \geq nc\right\} \asymp n^{-1/2} \exp\{-nI^*(c)\}. \tag{12}$$

(The notation $a_n \asymp b_n$ means that a_n and b_n are of the same order of magnitude, that is, a_n/b_n is bounded away from zero and infinity from some n on.)

From (11) it is seen that $b_n(c) \asymp x^{-1} \asymp c^{-1}n^{-1/2}$ if $c > \alpha n^{-1/2}(\alpha > 0)$. Hence, the quoted results imply the following uniform estimate of the order of magnitude of the probability under consideration. Let α and β be positive numbers such that $\beta < L'(t^*-)$. If conditions (9) are satisfied, then

$$P\left\{\sum_{j=1}^n \varphi(X_j) \geq nc\right\} \asymp c^{-1}n^{-1/2} \exp\{-nI^*(c)\} \tag{13}$$

uniformly for $\alpha n^{-1/2} < c < \beta$. This also follows from ([1], inequality (57)).

In the case $c \to \infty$, A. V. Nagaev [15] obtained, under certain restrictions on the (assumed) probability density of $\varphi(X_1)$, an asymptotic expression for the probability in (12), which is identical with the leading term of the expansion derived in [1] for c fixed.

Lemma 1 shows that $\exp\{-nI(H, F)\}$ does not approximate $P\{F_n \in H\}$ if $M(t) = \infty$ for all $t > 0$. In this case, S. V. Nagaev [16] showed, under a smoothness condition on $F_\varphi(x) = P\{\varphi(X_1) < x\}$, that $P\{\sum_{j=1}^n \varphi(X_j) \geq nc\} \sim n(1 - F_\varphi(nc))$ if nc increases rapidly enough (see also Linnik [14]), whereas Petrov [9], extending the results of Linnik [7], obtained asymptotic expressions for this probability, of the form (10) but with $I^*(c)$ replaced by a partial sum of its expansion in powers of c, under the assumption that nc does not grow too fast.

For certain sets A the results on half-spaces enable us to obtain upper and/or lower bounds for $P\{F_n \in A\}$ of the form (3). If A is any subset of $\mathcal{G}$, it follows from the definition of $I(A, F)$ that $A \subset B = \{G|I(G, F) \geq I(A, F)\}$. Suppose that $0 < I(A, F) < \infty$ and that there is a $G_0 \in A$ such that $I(G_0, F) = I(A, F)$. If $I(G, F)$ and $I(G, G_0)$ are finite, we have $I(G, F) = \int \log (g_0/f)\, dG + I(G, G_0)$, where $f = dF/d\mu$, $g_0 = dG/d\mu$. Hence, the half-space $H = \{G|\int \log (g_0/f)\, dG \geq I(A, F)\}$ is a subset of B, and $I(H, F) = I(A, F)$. In general, H neither contains nor is contained in A.

Suppose that A contains a half-space H such that $I(H, F) = I(A, F)$. For example, if A is the union of a family of half-spaces $H(\varphi)$, $\varphi \in \Phi$ (so that the complement of A is convex), it is easily seen that $I(A, F) = \inf \{I(H(\varphi), F), \varphi \in \Phi\}$. If the infimum is attained in Φ, the stated assumption is satisfied. Then we have the lower bound $P\{F_n \in A\} \geq P\{F_n \in H\}$, which, under appropriate conditions, can be estimated explicitly, as in (10) or (12), where $I^*(c) = I(A, F)$. If A is contained in a half-space H and $I(H, F) = I(A, F)$, we have analogous upper bounds, including $P\{F_n \in A\} \leq \exp \{-nI(A, F)\}$.

Now suppose that the set A is contained in the union of a finite number $k = k(n)$ of half-spaces H_i, $i = 1, \cdots, k$. Then (using (7))

$$
\begin{aligned}
P\{F_n \in A\} &\leq \sum_{i=1}^{k} P\{F_n \in H_i\} \leq \sum_{i=1}^{k} \exp \{-nI(H_i, F)\} \\
&\leq k \exp \{-n \min_i I(H_i, F)\}.
\end{aligned}
\tag{14}
$$

If $\min I(H_i, F)$ is close to $I(A, F)$ and $k = k(n)$ is not too large, even the crudest of the three bounds in (14) may be considerably better than the upper bound implied by (3). The following example serves as an illustration.

Let

$$
A = \{G| \sup_{x \in R^m} |G(x) - F(x)| \geq c\}, \qquad 0 < c < 1. \tag{15}
$$

The set A is the union of the half-spaces $H_x^+ = \{G|G(x) - F(x) \geq c\}$, $H_x^- = \{G|F(x) - G(x) \geq c\}$, $x \in R^m$. Sethuraman [13] has shown that the estimate (3) holds in the present case with c fixed, and for more general unions of half-spaces.

It follows from lemma 1 by a simple calculation that $I(H_x^+, F) = J(F(x), c)$ and $I(H_x^-, F) = J(1 - F(x), c)$, where

$$
\begin{aligned}
J(p, c) = {} & (p + c) \log ((p + c)/p) \\
& + (1 - p - c) \log ((1 - p - c)/(1 - p))
\end{aligned}
\tag{16}
$$

if $0 < p < 1 - c$, $J(1 - c, c) = -\log (1 - c)$, $J(p, c) = \infty$ if $p = 0$ or $p > 1 - c$.

I shall assume for simplicity that the one-dimensional marginal d.f.'s of F are continuous. Then $F(x)$ takes all values in $(0, 1)$, and we have

$$I(A, F) = \min_p J(p, c) = J(p(c), c) = K(c), \tag{17}$$

say, where $p(c)$ is the unique root in $(0, 1 - c)$ of $\partial J(p, c)/\partial p = 0$. It is easy to show that $((1 - c)/2) < p(c) < \min(\frac{1}{2}, 1 - c)$. For $K'(c) = dK(c)/dc$ we find

$$K'(c) = cp^{-1}(c)[1 - p(c)]^{-1} < 4c/(1 - c^2). \tag{18}$$

For any x with $F(x) = p(c)$ we have

$$P\{F_n \in A\} \geq P\{F_n \in H_x^+\} \geq \binom{n}{r} p(c)^r(1 - p(c))^{n-r}, \tag{19}$$

where $r = r(n, c)$ is the integer defined by $r \geq n(p(c) + c) > r - 1$. An application of Stirling's formula shows that this lower bound is greater than $C_1 n^{-1/2} \exp\{-nJ(p(c), c_n)\}$, where C_1 is a positive constant independent of c and $c_n = (r/n) - p(c) = c + \theta/n$, $0 \leq \theta < 1$. Hence it can be shown that for every $\epsilon > 0$ there is a positive constant C_2 which depends only on ϵ such that for $0 < c < 1 - \epsilon$,

$$P\{\sup_{x \in R^m} |F_n(x) - F(x)| \geq c\} \geq C_2 n^{-1/2} \exp\{-nK(c)\}. \tag{20}$$

Now let k be a positive integer. Since the marginal d.f.'s $F^{(i)}(x^{(i)})$ of $F(x) = F(x^{(1)}, \cdots, x^{(m)})$ are continuous, there are numbers $a_j^{(i)}$,

$$-\infty = a_0^{(i)} < a_1^{(i)} < \cdots < a_{k-1}^{(i)} < a_k^{(i)} = +\infty, \tag{21}$$

$i = 1, \cdots, m$, such that $F^{(i)}(a_j^{(i)}) = j/k$ for all i, j. If

$$a_{j_i-1}^{(i)} \leq x^{(i)} < a_{j_i}^{(i)} \qquad \text{for} \quad i = 1, \cdots, m \tag{22}$$

then

$$G(x) - F(x) \leq G(a) - F(a) + m/k, \tag{23}$$

where $a = (a_{j_1}^{(1)}, \cdots, a_{j_m}^{(m)})$, $1 \leq j_i \leq k$, and we have a similar upper bound for $F(x) - G(x)$. Hence the set A is contained in the union of the $2k^m$ half-spaces $\{G|G(a) - F(a) \geq c - m/k\}$, $\{G|F(a) - G(a) \geq c - m/k\}$, corresponding to the k^m values a. If $c - m/k > 0$, we have for each of these half-spaces H the inequality $I(H, F) \geq K(c - m/k)$. Hence, by (14),

$$P\{F_n \in A\} \leq 2k^m \exp\{-nK(c - m/k)\}. \tag{24}$$

We have $K(c - m/k) = K(c) - (m/k)K'(c - \theta m/k)$, $0 < \theta < 1$. With (18) this implies

$$K(c - m/k) > K(c) - (m/k)4c/(1 - c^2), \tag{25}$$

$$P\{F_n \in A\} < 2(k \exp\{4nc(1 - c^2)^{-1}k^{-1}\})^m \exp\{-nK(c)\}. \tag{26}$$

If we choose k so that $k - 1 \leq 4nc(1 - c^2)^{-1} \leq k$ and take account of the assumption $c > mk^{-1}$, we obtain

$$P\{\sup_{x \in R^m} |F_n(x) - F(x)| \geq c\} < 2e^m\{4cn/(1 - c^2) + 1\}^m \exp\{-nK(c)\} \tag{27}$$

if $4c^2n > m(1 - c^2)$.

For c fixed the bound is of order $n^m \exp\{-nK(c)\}$. The power n^m can be reduced by using the closer bounds in (14). An upper bound of a different form for the probability in (27) has been obtained by Kiefer and Wolfowitz [6].

4. Sums of independent random vectors

Let $\varphi = (\varphi_1, \cdots, \varphi_k)$ be a measurable function from R^m to R^k and consider the set

$$A = \left\{G \middle| \int \varphi \, dG \in D\right\}, \tag{28}$$

where D is a k-dimensional Borel set. Then $P\{F_n \in A\}$ is the probability that the sum $n^{-1} \sum_{j=1}^{n} \varphi(X_j)$ of n independent, identically distributed random vectors is contained in the set D. We have

$$I(A, F) = \inf_{s \in D} I(A(s), F), \qquad A(s) = \left\{G \middle| \int \varphi \, dG = s\right\}. \tag{29}$$

For $t \in R^k$ let $M(t) = \int \exp(t, \varphi) \, dF$, $L(t) = \log M(t)$, where $(t, \varphi) = \sum_{i=1}^{k} t_i \varphi_i$. Let Θ denote the set of points $t \in R^k$ for which $M(t) < \infty$. Suppose that the set Θ_0 of inner points of Θ is not empty. The derivatives $L_i'(t) = \partial L(t)/\partial t_i$ exist in Θ_0. Let Ω_0 denote the set of points $L'(t) = (L_1'(t), \cdots, L_k'(t))$, $t \in \Theta_0$. The following lemma, in conjunction with (29), is a partial extension of lemma 1.

LEMMA 2. *If $s \in \Omega_0$, then*

$$I(A(s), F) = (t(s), s) - L(t(s)) = - \min_{t \in R^k} [L(t) - (t, s)], \tag{30}$$

where $t(s)$ satisfies the equation $L'(t(s)) = s$. Also, $I(A(s), F) = I(G_s, F)$, where G_s is the d.f. in $A(s)$ defined by

$$dG_s = \exp\{(t(s), \varphi)\} \, dF/M(t(s)). \tag{31}$$

PROOF. If $G \in A(s)$, we find as in (6) that $-I(G, F) \leq L(t) - (t, s)$ for all $t \in R^k$, with equality holding only if

$$dG = \exp\{(t, \varphi)\} \, dF/M(t). \tag{32}$$

The d.f. G defined by (32) is in $A(s)$ if and only if $\int \varphi \{\exp(t, \varphi)\} \, dF/M(t) = s$ which for $t \in \Theta_0$ is equivalent to $L'(t) = s$. Since $s \in \Omega_0$, there is at least one point $t(s) \in \Theta_0$ which satisfies this equation. The lemma follows. (If the distribution of the random vector $\varphi(X_1)$ is concentrated on a hyperplane in R^k, the solution $t(s)$ of $L'(t) = s$ is not unique; but the distribution G_s can be shown to be the only $G \in A(s)$ for which $I(G, F) = I(A(s), F)$.)

It is seen from (30) and (31) that if $s \in \Omega_0$, then

$$dF(x) = \exp\{-I(A(s), F) - (t(s), \varphi(x) - s)\} \, dG_s(x), \tag{33}$$

$$\prod_{j=1}^{n} dF(x_j) = \exp\left\{-nI(A(s), F) - n\left(t(s), \int \varphi \, dF_n - s\right)\right\} \prod_{j=1}^{n} dG_s(x_j). \tag{34}$$

(Here the same notation F_n is used for the value $F_n(\cdot|x^{(n)})$ as for the random function $F_n(\cdot|X^{(n)})$.) Hence the distribution of the sum $\int \varphi \, dF_n$ can be symbolically expressed in the form

$$(35)\qquad \int_{\{\int \varphi\, dF_n = s\}} \prod_{j=1}^{n} dF(x_j) = \exp\{-nI(A(s), F)\} \int_{\{\int \varphi\, dF_n = s\}} \prod_{j=1}^{n} dG_s(x_j)$$

for values $s \in \Omega_0$. Here $\{\int \varphi\, dF_n = s\}$ is a shortcut notation for $\{|\int \varphi\, dF_n - s| < \epsilon\}$, $\epsilon \to 0$, and a term which is negligible for $\epsilon \to o$ is suppressed. The integral on the right is the value at s of the distribution of $\int \varphi\, dF_n$ when the $X_1, \cdots, X_n$ have the common distribution G_s, in which case $s = \int \varphi\, dG_s$ is the expected value of $\int \varphi\, dF_n$. The higher moments of this distribution are finite, and the known results on the approximation of the density of a sum of independent random vectors in the center of the distribution can be used to approximate the density (on the left in (35)) at points remote from the center. This, in turn, can be used to approximate $P\{\int \varphi\, dF_n \in D\}$ at least for $D \subset \Omega_0$. This approach has been used by Borovkov and Rogozin [2] to derive an asymptotic expansion of the probability $P\{\int \varphi\, dF_n \in D_n\}$ for an extensive class of sets D_n under the assumption that the distribution of $\int \varphi\, dF_n$ is absolutely continuous with respect to Lebesgue measure in R^k for some n.

Borovkov and Rogozin make the following assumptions concerning D_n. Let $-\psi_n$ denote the essential infimum relative to k-dimensional Lebesgue measure of $I(A(s), F)$ for $s \in D_n$. (Thus $\psi_n = -I(A_n, F)$ where $A_n = \{G | \int \varphi\, dG \in D_n^*\}$ and D_n^* differs from D_n by a set of Lebesgue measure 0.) Let Θ_f be a compact subset of Θ_0 and $\Phi = \{L'(t) | t \in \Theta_f\}$.

ASSUMPTION (A). *For some* $\delta > 0$, $D_n \cap \{s | I(A(s), F) < -\psi_n + \delta\} \in \Phi$.

ASSUMPTION (B). *There is a union* U *of finitely many half-spaces in* R_k *such that*

$$(36)\qquad D_n \cap \{s | I(A(s), F) > -\psi_n + \delta\} \subset U \subset \left\{s | I(A(s), F) > -\psi_n + \frac{\delta}{2}\right\}.$$

Under these assumptions the leading term of the asymptotic expansion obtained in [2] is

$$(37)\qquad P\left\{\int \varphi\, dF_n \in D_n\right\} \sim (2\pi)^{-k/2} n^{k/2} \exp(n\psi_n) \int_0^{\delta} e^{-nu} \varphi_n(u)\, du,$$

where

$$(38)\qquad \varphi_n(u) = \int_{D_n \cap \Gamma(-\psi_n - u)} |\Sigma(s)|^{-1/2}\, ds, \qquad \Gamma(-c) = \{s | I(A(s), F) = c\},$$

$|\Sigma(s)|$ is the determinant of the covariance matrix of $\varphi(X_1)$ when X_1 has the distribution G_s, and the last integral is extended over the indicated surface.

It should be feasible to obtain an analogous expansion for the case of lattice-valued random vectors. An extension of the Euler-Maclaurin sum formula to the case of a function of several variables due to R. Ranga Rao (in a Ph.D. dissertation which is unpublished at this writing; compare [10]), would be useful here. The order of magnitude of the probability $P\{\int \varphi\, dF_n \in D_n\}$ for a fairly extensive class of sets D_n can be determined in a rather simple way, as is shown in section 4 for the multinomial case. Richter [11] derived an estimate of $P\{\int \varphi\, dF_n \in D\}$ for a special class of sets D in the lattice vector case as well as in the absolutely continuous case; it is akin to the Cramér-Petrov estimate for the one-dimensional case but seems to have no simple relation to the Sanov-type estimate (3).

The preceding discussion has an interesting statistical interpretation. Lemma 2 shows that if $s \in \Omega_0$, then the infimum $I(A(s), F)$ is attained in the "exponential" subclass of $\mathcal{G}$ which consists of the distributions G defined by (32). Suppose that $F = F_\theta$ is a member of the class $\{F_\theta, \theta \in \Theta\}$, $dF_\theta = f_\theta \, d\nu$, where

$$f_\theta(x) = \exp \{(\theta, \varphi(x)) - L(\theta)\}, \tag{39}$$

ν is a sigma-finite measure on the m-dimensional Borel sets, φ a function from R^m to R^k, and Θ is the set of points $\theta \in R^k$ for which $\exp L(\theta) = \int \exp (\theta, \varphi) \, d\nu$ is finite. (If the null vector 0 is in Θ, which could be assumed with no loss of generality, then $d\nu = dF_0$.) Let $f_{\theta,n} = f_{\theta,n}(x^{(n)})$ be the density of $X^{(n)}$, so that

$$f_{\theta,n} = \exp n\left\{\left(\theta, \int \varphi \, dF_n\right) - L(\theta)\right\}. \tag{40}$$

Here $\int \varphi \, dF_n$ is a sufficient statistic and it is natural to restrict attention to sets A of the form (28). We have $\int \varphi \, dF_\theta = L'(\theta)$ for $\theta \in \Theta_0$, and

$$I(F_{\theta'}, F_\theta) = (\theta' - \theta, L'(\theta')) - L(\theta') + L(\theta) = I^*(\theta', \theta), \tag{41}$$

say, for $\theta' \in \Theta_0$, $\theta \in \Theta$. From (30) with $F = F_\theta$ we have $I(A(s), F_\theta) = I^*(\theta', \theta)$, where $s = L'(\theta')$.

A maximum likelihood estimator of θ is a function $\hat{\theta}_n$ from R^{mn} into Θ such that $\hat{\theta}_n(x^{(n)})$ maximizes $f_{\theta,n}(x^{(n)})$. If $\int \varphi \, dF_n \in \Omega_0$, then $\hat{\theta}_n$ is a root of $L'(\theta) = \int \varphi \, dF_n$, and we have

$$\max_\theta f_{\theta,n} = f_{\hat{\theta}_n,n} = \exp n\{(\hat{\theta}_n, L'(\hat{\theta}_n)) - L(\hat{\theta}_n)\}. \tag{42}$$

Hence,

$$f_{\theta,n} = f_{\hat{\theta}_n,n} \exp \{-nI^*(\hat{\theta}_n, \theta)\}. \tag{43}$$

Equation (43), which is related to (35), shows that $f_{\theta,n}$ depends on θ only through $I^*(\hat{\theta}_n, \theta)$. Note that the likelihood ratio test for testing the simple hypothesis $\theta = \theta'$ against the alternatives $\theta \neq \theta'$ rejects the hypothesis if $I^*(\hat{\theta}_n, \theta')$ exceeds a constant. For the special case where the distribution of $\int \varphi \, dF_n$ is multinomial the author has shown in [5] that the likelihood ratio test has certain asymptotically optimal properties.

5. Multinomial probabilities

The case where X_1 takes only finitely many values can be reduced to the case where X_1 is a vector of k components and takes the k values $(1, 0, \cdots, 0)$, $(0, 1, 0, \cdots, 0)$, $\cdots$, $(0, \cdots, 0, 1)$ with respective probabilities $p_1, p_2, \cdots, p_k$ whose sum is 1. The sum $nZ^{(n)} = X_1 + \cdots + X_n$ takes the values $nz^{(n)} = (n_1, \cdots, n_k)$, $n_i \geq 0$, $\sum_{i=1}^k n_i = n$, and we have

$$P\{Z^{(n)} = z^{(n)}\} = n! \left(\prod_{i=1}^k n_i!\right)^{-1} \prod_{i=1}^k p_i^{n_i} = P_n(z^{(n)}|p), \tag{44}$$

say. The distribution function F of X_1 and the empirical one, F_n, are respectively determined by the vectors $p = (p_1, \cdots, p_k)$ and $Z^{(n)}$ whose values lie in the simplex

$$\Omega = \left\{(x_1, \cdots, x_k) | x_1 \geq 0, \cdots, x_k \geq 0, \sum_{i=1}^{k} x_i = 1\right\}. \tag{45}$$

It will be convenient to write $I(Z^{(n)}, p)$ for $I(F_n, F)$, where

$$I(x, p) = \sum_{i=1}^{k} x_i \log (x_i/p_i) \tag{46}$$

for x and p in Ω. We have

$$P_n(z^{(n)}|p) = P_n(z^{(n)}|z^{(n)}) \exp \{-nI(z^{(n)}, p)\}, \tag{47}$$

which corresponds to equations (35) and (43).

For any set $A \subset \Omega$, let $I(A, p) = \inf \{I(x, p)|x \in A\}$ and let $A^{(n)}$ denote the set of points $z^{(n)} \in A$. In [5] it is shown that

$$P\{Z^{(n)} \in A\} = \exp \{-nI(A^{(n)}, p) + O(\log n)\} \tag{48}$$

uniformly for $A \subset \Omega$ and $p \in \Omega$. Clearly, $I(A^{(n)}, p) \geq I(A, p)$. Hence, if $\{A_n\}$ is a sequence of sets such that

$$I(A_n^{(n)}, p) \leq I(A_n, p) + O(n^{-1} \log n), \tag{49}$$

then

$$P\{Z^{(n)} \in A_n\} \asymp n^{r_n} \exp \{-nI(A_n, p)\}, \tag{50}$$

where r_n is bounded. Sufficient conditions for (49) to hold are given in the appendix of [5].

Here we shall consider the determination of the order of magnitude of $P\{Z^{(n)} \in A_n\}$, which amounts to the determination of r_n in (50). The point p will be held fixed with $p_i > 0$ for all i. (The results to be derived hold uniformly for $p_i > \epsilon$, $i = 1, \cdots, k$, where ϵ is any fixed positive number.)

LEMMA 3. *For every real m there is a number d (which depends only on m and k) such that uniformly for $A \subset \Omega$,*

$$\begin{aligned} P\{Z^{(n)} \in A\} = {} & P\{Z^{(n)} \in A, I(Z^{(n)}, p) < I(A, p) + dn^{-1} \log n\} \\ & + O(n^{-m} \exp \{-nI(A, p)\}). \end{aligned} \tag{51}$$

The lemma follows from (48) with A replaced by $\{x|I(x, p) \geq I(A, p) + dn^{-1} \log n\}$ and d suitably chosen.

It should be noted that $Z^{(n)} \in A$ implies $I(Z^{(n)}, p) \geq I(A, p)$. Thus if the remainder term in (51) is negligible, the main contribution to $P\{Z^{(n)} \in A\}$ is from the intersection of A with a narrow strip surrounding the (convex) set $\{x|I(x, p) < I(A, p)\}$.

Let Ω_ϵ denote the set of all $x \in \Omega$ such that $x_i > \epsilon$, $i = 1, \cdots, k$. Let for $x \in \Omega$,

$$\Pi_n (x|p) = (2\pi n)^{-(k-1)/2} \left(\prod_{i=1}^{k} x_i\right)^{-1/2} \exp \{-nI(x, p)\}. \tag{52}$$

LEMMA 4. *For $\epsilon > 0$ fixed we have uniformly for $A \subset \Omega_\epsilon$,*

$$P\{Z^{(n)} \in A\} = \sum_{z^{(n)} \in A} \Pi_n (z^{(n)}|p)(1 + O(n^{-1})). \tag{53}$$

This follows from (47) by applying Stirling's formula to

$$P_n(z^{(n)}|z^{(n)}) = (n!/n^n) \prod_{i=1}^{k} (n_i^{n_i}/n_i!). \tag{54}$$

We now approximate the sum in (53) by an integral. To determine the order of magnitude only, a crude approximation will suffice. Let

$$\begin{aligned} R_n(z^{(n)}) &= \{(x_1, \cdots, x_{k-1})|z_i^{(n)} \le x_i < z_i^{(n)} + n^{-1}, i = 1, \cdots, k-1\}, \\ A_n^* &= \bigcup_{z^{(n)} \in A} R_n(z^{(n)}). \end{aligned} \tag{55}$$

LEMMA 5. *For $\epsilon > 0$ we have uniformly for $A \subset \Omega_\epsilon$,*

$$\sum_{z^{(n)} \in A} \Pi_n (z^{(n)}|p) \asymp n^{k-1} \int \underset{A_n^*}{\cdots} \int \Pi_n (x|p)\, dx_1 \cdots dx_{k-1}. \tag{56}$$

PROOF. We have $1 = n^{k-1} \int \cdots \int_{R_n(z^{(n)})} dx_1 \cdots dx_{k-1}$. If $(x_1, \cdots, x_{k-1}) \in R_n(z^{(n)})$, then $|x_i - z_i^{(n)}| < kn^{-1}$ for $i = 1, \cdots, k$, where $x_k = 1 - x_1 \cdots x_{k-1}$. Also, $I(x, p) = I(z, p) + 0(\max_i |x_i - z_i|)$ uniformly for x and $z \in \Omega_\epsilon$. These facts imply the lemma.

Now let $f(x)$ be a function defined on Ω,

$$A(c) = \{x|f(x) \ge c\}, \tag{57}$$

and suppose that for every $\epsilon' > 0$ there is a number $a_1(\epsilon')$ such that

$$|f(z) - f(x)| \le a_1(\epsilon')|z - x| \qquad \text{if} \quad z \in \Omega_{\epsilon'}, \quad x \in \Omega_{\epsilon'}, \tag{58}$$

where $|z - x| = \max_i |z_i - x_i|$. This condition is satisfied if the first partial derivatives of f exist and are continuous in Ω_0 (the set where $x_i > 0$ for all i). Let

$$D(c, \delta) = \{x|f(x) \ge c, \quad I(x, p) \le I(A(c), p) + \delta\}, \tag{59}$$

$$D^*(c, \delta) = \{(x_1, \cdots, x_{k-1})|(x_1, \cdots, x_{k-1}, 1 - x_1 - \cdots - x_{k-1}) \in D(c, \delta)\}, \tag{60}$$

$$V_c(u) = \int \cdots \int_{D^*(c,u)} dx_1 \cdots dx_{k-1}, \tag{61}$$

and, if the derivative $V_c'(u) = dV_c(u)/du$ exists for $0 < u < \delta$,

$$K_n(c, \delta) = \int_0^\delta e^{-nu} V_c'(u)\, du. \tag{62}$$

THEOREM 1. *Let $A(c)$ be defined by (57), where f satisfies (58) for every $\epsilon' > 0$. Let $\{c_n\}$ be a real number sequence and suppose that for every $a' > 0$ there are positive numbers ϵ, δ, and n_0 such that $D(c_n - a'n^{-1}, \delta) \subset \Omega_\epsilon$ for $n > n_0$. Then for every real number m there are positive numbers d and a such that*

$$\begin{aligned} &P\{f(Z^{(n)}) \ge c_n\} \\ &\quad \asymp \exp\{-nI(A(c_n), p)\}\{n^{(k-1)/2}K_n(c_n - \theta a n^{-1}, \delta_n + \theta a n^{-1}) + O(n^{-m})\}, \end{aligned} \tag{63}$$

where $|\theta| \le 1$, $\delta_n = dn^{-1} \log n$, and it is assumed that for each c such that $|c - c_n| \le an^{-1}$ the derivative $V_c'(u)$ exists for $0 < u < \delta$.

PROOF. From lemmas 3, 4, and 5 we obtain

$$P\{f(Z^{(n)}) \geq c_n\} \asymp n^{(k-1)/2} J_{1,n} + 0(n^{-m} \exp\{-nI(A(c_n), p\}), \tag{64}$$

where

$$J_{1,n} = \int \cdots \int_{D_n} \left(\prod_{i=1}^{k} x_i\right)^{-1/2} \exp\{-nI(x, p)\}\, dx_1 \cdots dx_{k-1}, \tag{65}$$

$$D_n^* = \bigcup_{z^{(n)} \in D(c_n, \delta_n)} R_n(z^{(n)}), \tag{66}$$

and $\delta_n = dn^{-1} \log n$. It follows from condition (58), which is also satisfied by $I(\cdot, p)$, that there is a number $a > 0$ such that

$$D^*(c_n + an^{-1}, \delta_n - an^{-1}) \subset D_n^* \subset D^*(c_n - an^{-1}, \delta_n + an^{-1}). \tag{67}$$

Since $(\prod x_i)^{-1/2}$ is bounded in Ω_ϵ, we obtain $J_{1,n} \asymp J_{2,n}(c_n - \theta an^{-1}, \delta_n + \theta an^{-1})$, where

$$J_{2,n}(c, \delta) = \int \cdots \int_{D^*(c,\delta)} \exp\{-nI(x, p)\}\, dx_1 \cdots dx_{k-1} \tag{68}$$

and $|\theta| \leq 1$. If the derivative $V_c'(u)$ exists for $0 < u < \delta$, we can write

$$J_{2,n}(c, \delta) = \exp\{-nI(A(c), p)\} \int_0^\delta e^{-nu} V_c'(u)\, du. \tag{69}$$

The theorem follows.

If we had not suppressed the factor $(\prod x_i)^{-1/2}$, we would have obtained (63) with $V_c(u)$ replaced by

$$V_{1,c}(u) = \int \cdots \int_{D^*(c,u)} \left\{\prod_{i=1}^{k} x_i\right\}^{-1/2} dx_1 \cdots dx_{k-1}. \tag{70}$$

In this form the first term on the right of (63) is analogous to the right side of (37). The integer k in (37) is here replaced by $k - 1$, since the distribution is $(k - 1)$-dimensional.

To apply theorem 1 we need to determine the order of magnitude of $K_n(c_n, \delta_n)$, where $\delta_n \asymp n^{-1} \log n$, so that $\delta_n \to 0$ and $n\delta_n \to \infty$. If, for instance $V_{c_n}'(u) \asymp b(c_n)u^r$ uniformly with respect to n as $u \to 0^+$, then

$$K_n(c_n, \delta_n) \asymp b(c_n) \int_0^{\delta_n} e^{-nu} u^r\, du \asymp b(c_n) n^{-r-1}. \tag{71}$$

Concerning the determination of the order of magnitude of $V_c'(u)$, we observe the following. Note that $I(A(c), p) = 0$ if $f(p) \geq c$. Assume that $f(p) < c$. The continuity condition (58) implies that $I(A(c), p) > 0$. Let Y denote the set of points $y \in A(c)$ such that $I(y, p) = I(A(c), p)$. Then $Y \subset D(c, \delta)$. The assumption $D(c, \delta) \subset \Omega_\epsilon$, condition (58), and the convexity of $I(\cdot, p)$ imply that $f(y) = c$ if $y \in Y$.

Suppose first that the set $A(c)$ is contained in a half-space H such that $I(H, p) = I(A(c), p)$. (This is true if the function $-f(x)$ is convex, so that the set $A(c)$ is convex.) Then the set Y consists of a single point y, and we have

$$\begin{aligned} I(x, p) - I(A(c), p) &= I(x, p) - I(y, p) \\ &= \sum (\log (y_i/p_i))(x_i - y_i) + I(x, y). \end{aligned} \tag{72}$$

Hence, $H = \{x | \sum (\log y_i/p_i)(x_i - y_i) \geq 0\}$, and $x \in A(c)$ implies $I(x, p) - I(A(c), p) \geq I(x, y)$. Therefore, if $x \in D(c, \delta_n)$, then $I(x, y) < \delta_n$. Now $I(x, y) = \frac{1}{2}Q^2(x, y) + O(|x - y|^3)$, where $Q^2(x, y) = \sum (x_i - y_i)^2/y_i$. Hence, if $x \in D(c, \delta_n)$, then $|x - y| = O(\delta_n^{1/2})$, $I(x, y) - \frac{1}{2}Q^2(x, y) = 0(\delta_n^{3/2}) = o(n^{-1})$, and the inequality

$$I(x, p) - I(A(c), p) < \delta_n \tag{73}$$

may be written

$$\sum (\log (y_i/p_i))(x_i - y_i) + \tfrac{1}{2}Q^2(x, y) < \delta_n + o(n^{-1}). \tag{74}$$

An inspection of the proof of theorem 1 shows that in the present case the theorem remains true if in the domain of integration $D(c, u)$ of the integral $V_c(u)$, the left-hand side of (73) is replaced by the left-hand side of (74).

Now suppose further that the partial derivatives $f'_i(x) = \partial f(x)/\partial x_i$, $f''_{ij}(x) = \partial^2 f(x)/\partial x_i \partial x_j$, and the third-order derivatives exist and are continuous in Ω_0. Then

$$f(x) - c = f(x) - f(y) = \sum f'_i(y)(x_i - y_i) + \tfrac{1}{2}F(x - y) + O(|x - y|^3), \tag{75}$$

uniformly for $y \in \Omega_\epsilon$, where $F(x - y) = \sum \sum f''_{ij}(y)(x_i - y_i)(x_j - y_j)$. Hence, if $x \in D(c, \delta_n)$, the inequality $f(x) \geq c$ may be written as

$$\sum f'_i(y)(x_i - y_i) + \tfrac{1}{2}F(x - y) \geq r_n, \qquad r_n = O(\delta_n^{3/2}). \tag{76}$$

Furthermore, the half-space $\{x | \sum f'_i(y)(x_i - y_i) \geq 0\}$ is identical with H. This implies that $y = y(c)$ satisfies the equations

$$\log (y_i/p_i) = t(c)f'_i(y) + s(c), \qquad i = 1, \cdots, k, \tag{77}$$

where $t(c) > 0$ and $s(c)$ are constants, as well as the equations $f(y) = c$ and $\sum y_i = 1$. It follows that under the present assumptions theorem 1 remains true with $K_n(c_n, \delta_n)$ replaced by $K_n(c_n, \delta_n, r_n)$, where

$$K_n(c, \delta, r) = \int_0^\delta e^{-nu} V'_{c,r}(u)\, du \tag{78}$$

and $V'_{c,r}(u)$ is the derivative with respect to u of the volume $V_{c,r}(u)$ of $D^*(c, r, u)$, the set of points $(x_1, \cdots, x_{k-1})$ which satisfy the inequalities

$$\sum f'_i(y)(x_i - y_i) + \tfrac{1}{2}F(x - y) > r, \tag{79}$$

$$t(c) \sum f'_i(y)(x_i - y_i) + \tfrac{1}{2}Q^2(x, y) < u. \tag{80}$$

If we make the substitution $z_i = y_i^{-1/2}(x_i - y_i)$, $i = 1, \cdots, k$, we obtain $Q^2(x, y) = \sum_{i=1}^k z_i^2$, $\sum y_i^{1/2} z_i = 0$ and

$$\sum f'_i(y)(x_i - y_i) = \sum (f'_i(y) - a(c))y_i^{1/2} z_i = \sigma(c) \sum b_i z_i, \tag{81}$$

where $a(c) = \sum y_i f'_i(y)$, $\sigma^2(c) = \sum (f'_i(y) - a(c))^2 y_i$, and $b_i = \sigma^{-1}(c)(f'_i(y) - a(c))$. We have $\sum b_i^2 = 1$, $\sum b_i y_i^{1/2} = 0$. Hence we can perform an orthogonal transformation $(z_1, \cdots, z_k) \to (v_1, \cdots, v_k)$, where $v_1 = \sum b_i z_i = \sigma^{-1}(c) \sum f'_i(y_i)(x_i - y_i)$ and $v_k = \sum y_i^{1/2} z_i = 0$. The inequalities (79), (80) are transformed into

$$\sigma(c)\, v_1 + \tfrac{1}{2}G(v_1, \cdots, v_{k-1}) > r, \tag{82}$$

$$t(c)\sigma(c)v_1 + \tfrac{1}{2}\sum_{i=1}^{k-1} v_i^2 < u, \tag{83}$$

where $G(v_1, \cdots, v_{k-1})$ is a quadratic form in $v_1, \cdots, v_{k-1}$. Thus

$$V_{c,r}(u) = C(c)W_{c,r}(u), \qquad V'_{c,r}(u) = C(c)W'_{c,r}(u), \tag{84}$$

where $C(c)$ is the modulus of the determinant of the linear transformation $(x_1, \cdots, x_{k-1}) \to (v_1, \cdots, v_{k-1})$ and $W_{c,r}(u)$ is the volume of the set defined by (82) and (83). In the estimation of $W'_{c,r}(u)$ we may assume that $u = 0(\delta_n)$ and $r = O(\delta_n^{3/2}) = o(n^{-1})$.

The replacement of $V_c(u)$ by $V_{c,r}(u)$ may be possible under conditions different from those assumed in the two preceding paragraphs. Suppose that $c > f(p)$ is fixed and that the set Y consists of a finite number s of points. Choose $\eta > 0$ so small that the s sets $S_y = \{x| \, |x - y| < \eta\}$, $y \in Y$, are disjoint. Then for δ small enough $D(c, \delta)$ is contained in the union of the sets S_y. If, for each $y \in Y$, the surfaces $f(x) = c$ and $I(x, y) = I(A(c), p)$ are not too close in the neighborhood of y, then $x \in S_y \cap D(c, \delta)$ will imply $|x - y| = 0(\delta^{1/2})$, and we arrive at analogous conclusions as in the preceding case.

If $f(x) = \sum_{i=1}^k a_i x_i$ is a linear function, $nf(z^{(n)})$ is the sum of n independent random variables, each of which takes the values $a_1, \cdots, a_k$ with respective probabilities $p_1, \cdots, p_k$. The following theorem can be deduced from theorem 1 but is a special case of (13).

THEOREM 2. *Let $A(c) = \{x| \sum_{i=1}^k a_i x_i \geq c\}$, where $a_1, \cdots, a_k$ are fixed (not all equal) and $\sum_{i=1}^k a_i p_i = 0$. Then*

$$P\left\{\sum_{i=1}^{k} a_i Z_i^{(n)} \geq c\right\} \asymp c^{-1} n^{-1/2} \exp\{-nI(A(c), p)\} \tag{85}$$

uniformly for $\alpha n^{-1/2} < c < \max a_i - \beta$, where α and β are arbitrary positive constants.

The next theorem gives an analogous uniform estimate for the distribution of $I(Z^{(n)}, p)$.

THEOREM 3. *Let $p_{\min} = \min_i p_i$ and let α and β be arbitrary positive constants. Then*

$$P\{I(Z^{(n)}, p) \geq c\} \asymp (nc)^{(k-3)/2} e^{-nc} \tag{86}$$

uniformly for $\alpha n^{-1} < c < -\log(1 - p_{\min}) - \beta$.

PROOF. In this case $D(c, \delta) = \{x|c \leq I(x, p) < c + \delta\}$. It can be shown that $I(x, p) < -\log(1 - p_{\min})$ implies $x_i > 0$ for all i. The assumption $c < -\log(1 - p_{\min}) - \beta$, $\beta > 0$, implies $D(c, \delta) \subset \Omega_\epsilon$ for some $\epsilon > 0$ if δ is small enough. Let

$$V(u) = \int \cdots \int_{I(x,p) < u} dx_1 \cdots dx_{k-1}. \tag{87}$$

We first prove the following lemma. (Clearly we may assume $p_k = p_{\min}$.)

LEMMA 6. *Let $p_k = p_{\min}$. The derivative $V'(u) = dV(u)/du$ exists, is continuous, and positive for $0 < u < -\log(1 - p_{\min})$ and $V'(u) \asymp u^{(k-3)/2}$ as $u \to 0^+$.*

(Heuristically, as $u \to 0$, $V(u)$ is approximated by the volume of the ellipsoid $Q^2(x, p) < 2u$, which is proportional to $u^{(k-1)/2}$.)

We shall write $I_k(x, p)$ for $I(x, p)$ to indicate the number of components of the arguments x and p, and $V_{k,p}(u)$ for $V(u)$. For $x_k \neq 1$, let $y = (y_1, \cdots, y_{k-1})$, $y_i = x_i/(1 - x_k)$; $z = (z_1, z_2) = (1 - x_k, x_k)$; $q = (q_1, \cdots, q_{k-1})$, $q_i = p_i/(1 - p_k)$; $r = (r_1, r_2) = (1 - p_k, p_k)$. Then we have the identity

$$I_k(x, p) = z_1 I_{k-1}(y, q) + I_2(z, r). \tag{88}$$

Hence, we obtain the recurrence relation

$$V_{k,p}(u) = \int_{I_2(z,r) < u} z_1^{k-2} V_{k-1,q}(z_1^{-1}\{u - I_2(z, r)\})\, dz_1, \qquad k \geq 3. \tag{89}$$

Since $p_k = p_{\min} \leq 1 - p_k$, we have $r_1 \geq r_2$. We also have $V_{2,r}(u) = b(u) - a(u)$ where, for $0 < u < -\log(1 - p_{\min}) = -\log r_2$, $a(u)$ and $b(u)$ are the two roots of the equation $I_2((z_1, 1 - z_1), r) = u$, $0 < a(u) < r_1 < b(u) < 1$. Hence, it is easy to show that the lemma is true for $k = 2$. From (89) we obtain for $k = 3$,

$$V'_{k,p}(u) = \int_{a(u)}^{b(u)} z_1^{k-3} V'_{k-1,q}(z_1^{-1}\{u - I_2(z, r)\})\, dz_1. \tag{90}$$

It now can be shown that the lemma holds for $k = 3$ and, by induction, that equation (90) and the lemma are true for any k.

Under the conditions of theorem 1 we have $V_c(u) = V(c + u) - V(c)$. The lemma implies that $V'_c(u) = V'(c + u) \asymp (c + u)^{(k-3)/2}$ uniformly for $0 < c + u < -\log(1 - p_{\min}) - \beta$. It follows that uniformly for $\alpha n^{-1} < c < -\log(1 - p_{\min}) - \beta$,

$$K_n(c, \delta_n) \asymp \int_0^{\delta_n} e^{-nu}(c + u)^{(k-3)/2}\, du \asymp c^{(k-3)/2} n^{-1}. \tag{91}$$

This establishes the theorem under the restriction $nc > a + \alpha$, where a is the number which appears in (63). That the result holds for $nc > \alpha$ with any $\alpha > 0$ follows from the well-known fact that $2nI(Z^{(n)}, p)$ has a chi-square limit distribution.

Since $A \subset \{x | I(x, p) \geq I(A, p)\}$ for any subset A of Ω, theorem 3 immediately implies the following theorem.

THEOREM 4. *If α and β are any positive numbers, there is a constant $C = C(\alpha, \beta, p)$ such that for any set A which satisfies*

$$\alpha n^{-1} < I(A, p) < -\log(1 - p_{\min}) - \beta, \tag{92}$$

we have

$$P\{Z^{(n)} \in A\} \leq C\{nI(A, p)\}^{(k-3)/2} \exp\{-nI(A, p)\}. \tag{93}$$

REMARK. We have $\max_{x \in \Omega} I(x, p) = -\log p_{\min}$. It seems plausible that the estimate (86) of theorem 3 holds uniformly for $\alpha n^{-1} < c < -\log p_{\min} - \beta$. If so, theorem 4 holds with an analogous modification.

For the functions $f(x) = \sum a_i x_i$ and $f(x) = I(x, p)$ of theorems 2 and 3 the order of magnitude of $P\{f(Z^{(n)}) \geq c\}$ is expressed in the form $c^r n^s$

$\exp\{-nI(A(c), p)\}$ in a wide range of c. That this is not true in general is shown by the following example. Let

$$(94) \qquad A(c) = \{x|Q^2(x, p) \geq c\}, \qquad Q^2(x, p) = \sum (x_i - p_i)^2/p_i.$$

Since $I(x, p) = \frac{1}{2}Q^2(x, p) + 0(|x - p|^3)$, $Q^2(x, p) < c$ implies $I(x, p) = \frac{1}{2}Q^2(x, p) + O(c^{3/2})$ as $c \to 0$. Hence, it follows from theorem 3 that if $c = O(n^{-2/3})$ and $nc > \alpha > 0$, then $P\{Q^2(Z^{(n)}, p) \geq c\}$ is of the same order of magnitude as $P\{I(Z^{(n)}, p) \geq c/2\}$. We have $I(A(c), p) = \frac{1}{2}c + O(c^{3/2})$. By theorem 3 this implies that

$$(95) \qquad P\{Q^2(Z^{(n)}, p) \geq c\} \asymp (nc)^{(k-3)/2} \exp\{-nI(A(c), p)\}$$

uniformly for $\alpha n^{-1} < c < \beta n^{-2/3}$. On the other hand, if c is bounded away from 0 and from $\max_x Q^2(x, p) = p_{\min}^{-1} - 1$, it can be deduced from theorem 1 (see the remarks after the proof of theorem 1 and section 8 of [5]) that

$$(96) \qquad P\{Q^2(Z^{(n)}, p) \geq c\} \asymp n^{-1/2} \exp\{-nI(A(c), p)\}.$$

In this case the probability of the set $A(c)$ is of the same order of magnitude as the probability of any of the half-spaces contained in $B = \{x|I(k, p) \geq I(A(c), p)\}$ and bounded by the supporting hyperplanes of the convex set $B' = \{x|I(x, p) < I(A(c), p)\}$ at the common boundary points y of the sets $A(c)$ and B. This result holds for a wide class of functions f when c is fixed. (However, theorem 4 of [12] is inaccurate in the stated generality, as is seen from theorem 3 above.) An asymptotic expression for $P\{Q^2(Z^{(n)}, p) \geq c\}$ with $c = o(1)$ as $n \to \infty$ has been obtained by Richter [11].

REFERENCES

[1] R. R. BAHADUR and R. RANGA RAO, "On deviations of the sample mean," *Ann. Math. Statist.*, Vol. 31 (1960), pp. 1015–1027.

[2] A. A. BOROVKOV and B. A. ROGOZIN, "On the central limit theorem in the multidimensional case," *Teor. Verojatnost. i Primenen.*, Vol. 10 (1965), pp. 61–69. (In Russian.)

[3] H. CRAMÉR, "Sur un nouveau théorème limite de la théorie des probabilités," *Actualités Sci. Indust.*, No. 736 (1938).

[4] W. FELLER, "Generalization of a probability limit theorem of Cramér," *Trans. Amer. Math. Soc.*, Vol. 54 (1943), pp. 361–372.

[5] WASSILY HOEFFDING, "Asymptotically optimal tests for multinomial distributions," *Ann. Math. Statist.*, Vol. 36 (1965), pp. 369–401.

[6] J. KIEFER and J. WOLFOWITZ, "On the deviations of the empiric distribution function of vector chance variables," *Trans. Amer. Math. Soc.*, Vol. 87 (1958), pp. 173–186.

[7] YU. V. LINNIK, "Limit theorems for sums of independent variables taking into account large deviations, I, II, III," *Teor. Verojatnost. i Primenen.*, Vol. 6 (1961), pp. 145–163 and 377–391; Vol. 7 (1962), pp. 122–134. (In Russian.) (English translation in *Theor. Probability Appl.*, Vol. 6 (1961), pp. 131–148 and 345–360; Vol. 7 (1962), pp. 115–129.)

[8] V. V. PETROV, "Generalization of Cramér's limit theorem," *Uspehi Mat. Nauk* (n.s.), Vol. 9 (1954), pp. 195–202. (In Russian.)

[9] ———, "Limit theorems for large deviations when Cramér's condition is violated, I, II," *Vestnik Leningrad Univ. Ser. Mat. Meh. Astronom.*, Vol. 18 (1963), pp. 49–68; Vol. 19 (1964), pp. 58–75. (In Russian.)

[10] R. Ranga Rao, "On the central limit theorem in R_k," *Bull. Amer. Math. Soc.*, Vol. 67 (1961), pp. 359–361.

[11] W. Richter, "Mehrdimensionale Grenzwertsätze für grosse Abweichungen und ihre Anwendung auf die Verteilung von χ^2," *Teor. Verojatnost. i Primenen.*, Vol. 9 (1964), pp. 31–42.

[12] I. N. Sanov, "On the probability of large deviations of random variables," *Mat. Sbornik* (n.s.), Vol. 42 (1957), pp. 11–44. (In Russian.) (English translation in *Select. Transl. Math. Statist. and Probability*, Vol. 1 (1961), pp. 213–244.)

[13] J. Sethuraman, "On the probability of large deviations of families of sample means," *Ann. Math. Statist.*, Vol. 35 (1964), pp. 1304–1316.

[14] Yu. V. Linnik, "On the probability of large deviations for the sums of independent variables," *Proceedings of the Fourth Berkeley Symposium on Mathematical Statistics and Probability*, Berkeley and Los Angeles, University of California Press, 1961, Vol. 2, pp. 289–306.

[15] A. V. Nagaev, "Large deviations for a class of distributions," *Predel'nye Teoremy Teorii Verojatnostei* (1963), pp. 56–58, Akad. Nauk Uzb. SSR, Tashkent. (In Russian.)

[16] S. V. Nagaev, "An integral limit theorem for large deviations," *Izv. Akad. Nauk UzSSR Ser. Fiz.-Mat. Nauk*, Vol. 6 (1962), pp. 37–43. (In Russian.)

THE BEHAVIOR OF MAXIMUM LIKELIHOOD ESTIMATES UNDER NONSTANDARD CONDITIONS

PETER J. HUBER
SWISS FEDERAL INSTITUTE OF TECHNOLOGY

1. Introduction and summary

This paper proves consistency and asymptotic normality of maximum likelihood (ML) estimators under weaker conditions than usual.

In particular, (i) it is not assumed that the true distribution underlying the observations belongs to the parametric family defining the ML estimator, and (ii) the regularity conditions do not involve the second and higher derivatives of the likelihood function.

The need for theorems on asymptotic normality of ML estimators subject to (i) and (ii) becomes apparent in connection with robust estimation problems; for instance, if one tries to extend the author's results on robust estimation of a location parameter [4] to multivariate and other more general estimation problems.

Wald's classical consistency proof [6] satisfies (ii) and can easily be modified to show that the ML estimator is consistent also in case (i), that is, it converges to the θ_0 characterized by the property $E(\log f(x, \theta) - \log f(x, \theta_0)) < 0$ for $\theta \neq \theta_0$, where the expectation is taken with respect to the true underlying distribution.

Asymptotic normality is more troublesome. Daniels [1] proved asymptotic normality subject to (ii), but unfortunately he overlooked that a crucial step in his proof (the use of the central limit theorem in (4.4)) is incorrect without condition (2.2) of Linnik [5]; this condition seems to be too restrictive for many purposes.

In section 4 we shall prove asymptotic normality, assuming that the ML estimator is consistent. For the sake of completeness, sections 2 and 3 contain, therefore, two different sets of sufficient conditions for consistency. Otherwise, these sections are independent of each other. Section 5 presents two examples.

2. Consistency: case A

Throughout this section, which rephrases Wald's results on consistency of the ML estimator in a slightly more general setup, the parameter set Θ is a locally compact space with a countable base, $(\mathfrak{X}, \mathfrak{A}, P)$ is a probability space, and $\rho(x, \theta)$ is some real-valued function on $\mathfrak{X} \times \Theta$.

Assume that $x_1, x_2, \cdots$ are independent random variables with values in $\mathfrak{X}$ having the common probability distribution P. Let $T_n(x_1, \cdots, x_n)$ be any sequence of functions $T_n: \mathfrak{X}^n \to \Theta$, measurable or not, such that

$$\frac{1}{n}\sum_{i=1}^{n} \rho(x_i, T_n) - \inf_{\theta} \frac{1}{n}\sum_{i=1}^{n} \rho(x_i, \theta) \to 0 \tag{1}$$

almost surely (or in probability—more precisely, outer probability). We want to give sufficient conditions ensuring that every such sequence converges almost surely (or in probability) toward some constant θ_0.

If $dP = f(x, \theta_0)\, d\mu$ and $\rho(x, \theta) = -\log f(x, \theta)$ for some measure μ on $(\mathfrak{X}, \mathfrak{A})$ and some family of probability densities $f(x, \theta)$, then the ML estimator of θ_0 evidently satisfies condition (1).

Convergence of T_n shall be proved under the following set of assumptions.

ASSUMPTIONS.

(A-1). *For each fixed $\theta \in \Theta$, $\rho(x, \theta)$ is $\mathfrak{A}$-measurable, and $\rho(x, \theta)$ is separable in the sense of Doob: there is a P-null set N and a countable subset $\Theta' \subset \Theta$ such that for every open set $U \subset \Theta$ and every closed interval A, the sets*

$$\{x | \rho(x, \theta) \in A, \forall \theta \in U\}, \qquad \{x | \rho(x, \theta) \in A, \forall \theta \in U \cap \Theta'\} \tag{2}$$

differ by at most a subset of N.

This assumption ensures measurability of the infima and limits occurring below. For a fixed P, ρ might always be replaced by a separable version (see Doob [2], p. 56 ff.).

(A-2). *The function ρ is a.s. lower semicontinuous in θ, that is,*

$$\inf_{\theta' \in U} \rho(x, \theta') \to \rho(x, \theta), \qquad \text{a.s.} \tag{3}$$

as the neighborhood U of θ shrinks to $\{\theta\}$.

(A-3). *There is a measurable function $a(x)$ such that*

$$\begin{aligned} &E\{\rho(x, \theta) - a(x)\}^- < \infty \quad \text{for all} \quad \theta \in \Theta, \\ &E\{\rho(x, \theta) - a(x)\}^+ < \infty \quad \text{for some} \quad \theta \in \Theta. \end{aligned} \tag{4}$$

Thus, $\gamma(\theta) = E\{\rho(x, \theta) - a(x)\}$ is well-defined for all θ.

(A-4). *There is a $\theta_0 \in \Theta$ such that $\gamma(\theta) > \gamma(\theta_0)$ for all $\theta \neq \theta_0$.*

If Θ is not compact, let ∞ denote the point at infinity in its one-point compactification.

(A-5). *There is a continuous function $b(\theta) > 0$ such that*

(i) $\displaystyle \inf_{\theta \in \Theta} \frac{\rho(x, \theta) - a(x)}{b(\theta)} \geq h(x)$

for some integrable h;

(ii) $\displaystyle \liminf_{\theta \to \infty} b(\theta) > \gamma(\theta_0)$;

(iii) $\displaystyle E\left\{\liminf_{\theta \to \infty} \frac{\rho(x, \theta) - a(x)}{b(\theta)}\right\} \geq 1.$

If Θ is compact, then (ii) and (iii) are redundant.

EXAMPLE. Let $\Theta = \mathfrak{X}$ be the real axis, and let P be any probability distribution having a unique median θ_0. Then (A-1) to (A-5) are satisfied for $\rho(x, \theta) = |x - \theta|$, $a(x) = |x|$, $b(\theta) = |\theta| + 1$. (This will imply that the sample median is a consistant estimate of the median.)

Taken together, (A-2), (A-3), and (A-5) (i) imply by monotone convergence the following strengthened version of (A-2).

(A-2′). *As the neighborhood U of θ shrinks to $\{\theta\}$,*

$$E\{\inf_{\theta' \in U} \rho(x, \theta') - a(x)\} \to E\{\rho(x, \theta) - a(x)\}. \tag{5}$$

For the sake of simplicity we shall from now on absorb $a(x)$ into $\rho(x, \theta)$. Note that the set $\{\theta \in \Theta | E(|\rho(x, \theta) - a(x)|) < \infty\}$ is independent of the particular choice of $a(x)$; if there is an $a(x)$ satisfying (A-3), then one might choose $a(x) = \rho(x, \theta_0)$.

LEMMA 1. *If* (A-1), (A-3), *and* (A-5) *hold, then there is a compact set $C \subset \Theta$ such that every sequence T_n satisfying* (1) *almost surely ultimately stays in C.*

PROOF. By (A-5) (ii), there is a compact C and a $0 < \epsilon < 1$ such that

$$\inf_{\theta \notin C} b(\theta) \geq \frac{\gamma(\theta_0) + \epsilon}{1 - \epsilon}; \tag{6}$$

by (A-5) (i), (iii) and monotone convergence, C may be chosen so large that

$$E\left\{\inf_{\theta \notin C} \frac{\rho(x, \theta)}{b(\theta)}\right\} \geq 1 - \tfrac{1}{2}\epsilon. \tag{7}$$

By the strong law of large numbers, we have a.s. for sufficiently large n

$$\inf_{\theta \notin C} \left\{\frac{1}{n} \sum_{i=1}^{n} \frac{\rho(x_i, \theta)}{b(\theta)}\right\} \geq \frac{1}{n} \sum_{i=1}^{n} \inf_{\theta \notin C} \frac{\rho(x_i, \theta)}{b(\theta)} \geq 1 - \epsilon; \tag{8}$$

hence,

$$\frac{1}{n} \sum_{i=1}^{n} \rho(x_i, \theta) \geq (1 - \epsilon) b(\theta) \geq \gamma(\theta_0) + \epsilon \tag{9}$$

for $\forall \theta \notin C$, which implies the lemma, since for sufficiently large n

$$\inf_{\theta} \frac{1}{n} \sum_{i=1}^{n} \rho(x_i, \theta) \leq \frac{1}{n} \sum_{i=1}^{n} \rho(x_i, \theta_0) \leq \gamma(\theta_0) + \tfrac{1}{2}\epsilon. \tag{10}$$

By using convergence in probability in (1) and the weak law of large numbers, one shows similarly that $T_n \in C$ with probability tending to 1. (Note that a.s. convergence does not imply convergence in probability, if T_n is not measurable!)

THEOREM 1. *If* (A-1), (A-2′), (A-3), *and* (A-4) *hold, then every sequence T_n satisfying* (1), *and the conclusion of lemma* 1, *converges to θ_0 almost surely. An analogous statement is true for convergence in probability.*

PROOF. We may restrict attention to the compact set C. Let U be an open neighborhood of θ. By (A-2′), γ is lower semicontinuous; hence its infimum on the compact set $C \setminus U$ is attained and is—because of (A-4)—strictly greater than $\gamma(\theta_0)$, say $\geq \gamma(\theta_0) + 4\epsilon$ for some $\epsilon > 0$. Because of (A-2′), each $\theta \in C \setminus U$ admits a neighborhood U_θ such that

$$E\{\inf_{\theta' \in U_\theta} \rho(x, \theta')\} \geq \gamma(\theta_0) + 3\epsilon. \tag{11}$$

Select a finite number of points θ_s such that the $U_s = U_{\theta_s}$, $1 \leq s \leq N$, cover $C \setminus U$. By the strong law of large numbers, we have a.s. for sufficiently large n and all $1 \leq s \leq N$,

$$\inf_{\theta' \in U_s} \frac{1}{n} \sum \rho(x_i, \theta') \geq \frac{1}{n} \sum \inf_{\theta' \in U_s} \rho(x_i, \theta') \geq \gamma(\theta_0) + 2\epsilon \tag{12}$$

and

$$\frac{1}{n} \sum \rho(x_i, \theta_0) \leq \gamma(\theta_0) + \epsilon. \tag{13}$$

It follows that

$$\inf_{\theta \in C \setminus U} \frac{1}{n} \sum \rho(x_i, \theta) \geq \inf_{\theta \in U} \frac{1}{n} \sum \rho(x_i, \theta) + \epsilon, \tag{14}$$

which implies the theorem. Convergence in probability is proved analogously.

REMARKS. (1) If assumption (A-4) is omitted, the above arguments show that T_n a.s. ultimately stays in any neighborhood of the (necessarily compact) set $\{\theta \in \Theta | \gamma(\theta) = \inf_{\theta'} \gamma(\theta')\}$.

(2) Quite often (A-5) is not satisfied—for instance, if one estimates location and scale simultaneously—but the conclusion of lemma 1 can be verified quite easily by ad hoc methods. (This happens also in Wald's classical proof.) I do not know of any fail-safe replacement for (A-5).

3. Consistency: case B

Let Θ be locally compact with a countable base, let $(\mathfrak{X}, \mathfrak{A}, P)$ be a probability space, and let $\psi(x, \theta)$ be some function on $\mathfrak{X} \times \Theta$ with values in m-dimensional Euclidean space R^m.

Assume that $x_1, x_2, \cdots$ are independent random variables with values in $\mathfrak{X}$, having the common probability distribution P. We want to give sufficient conditions that any sequence $T_n: \mathfrak{X}^n \to \Theta$ such that

$$\frac{1}{n} \sum_{i=1}^{n} \psi(x_i, T_n) \to 0 \tag{15}$$

almost surely (or in probability), converges almost surely (or in probability) toward some constant θ_0.

If Θ is an open subset of R^m, and if $\psi(x, \theta) = (\partial/\partial\theta) \log f(x, \theta)$ for some differentiable parametric family of probability densities on $\mathfrak{X}$, then the ML estimate of θ will satisfy (15). However, our ψ need not be a total differential.

Convergence of T_n shall be proved under the following set of assumptions.

ASSUMPTIONS.

(B-1). *For each fixed $\theta \in \Theta$, $\psi(x, \theta)$ is $\mathfrak{A}$-measurable, and $\psi(x, \theta)$ is separable (see* (A-1)).

(B-2). *The function ψ is a.s. continuous in θ:*

$$\lim_{\theta' \to \theta} |\psi(x, \theta') - \psi(x, \theta)| = 0, \qquad \text{a.s.} \tag{16}$$

(B-3). *The expected value* $\lambda(\theta) = E\psi(x, \theta)$ *exists for all* $\theta \in \Theta$, *and has a unique zero at* $\theta = \theta_0$.

(B-4). *There exists a continuous function which is bounded away from zero,* $b(\theta) \geq b_0 > 0$, *such that*

(i) $\sup_\theta \dfrac{|\psi(x, \theta)|}{b(\theta)}$ is integrable,

(ii) $\liminf_{\theta \to \infty} \dfrac{|\lambda(\theta)|}{b(\theta)} \geq 1$,

(iii) $E\left\{\limsup_{\theta \to \infty} \dfrac{|\psi(x, \theta) - \lambda(\theta)|}{b(\theta)}\right\} < 1$.

In view of (B-4) (i), (B-2) can be strengthened to

(B-2′). *As the neighborhood* U *of* θ *shrinks to* $\{\theta\}$

$$E(\sup_{\theta' \in U} |\psi(x, \theta') - \psi(x, \theta)|) \to 0. \tag{17}$$

It follows immediately from (B-2′) that λ is continuous. Moreover, if there is a function b satisfying (B-4), one may obviously choose

$$b(\theta) = \max(|\lambda(\theta)|, b_0). \tag{18}$$

LEMMA 2. *If* (B-1) *and* (B-4) *hold, then there is a compact set* $C \subset \Theta$ *such that any sequence* T_n *satisfying* (15) *a.s. ultimately stays in* C.

PROOF. With the aid of (B-4) (i), (iii), and the dominated convergence theorem, choose C so large that the expectation of

$$v(x) = \sup_{\theta \notin C} \frac{|\psi(x, \theta) - \lambda(\theta)|}{b(\theta)} \tag{19}$$

is smaller than $1 - 3\epsilon$ for some $\epsilon > 0$, and that also (by (B-4) (ii))

$$\inf_{B \notin C} \frac{|\lambda(\theta)|}{b(\theta)} \geq 1 - \epsilon. \tag{20}$$

By the strong law of large numbers, we have a.s. for sufficiently large n,

$$\sup_{\theta \notin C} \frac{|n^{-1} \sum [\psi(x_i, \theta) - \lambda(\theta)]|}{b(\theta)} \leq \frac{1}{n} \sum v(x_i) \leq 1 - 2\epsilon; \tag{21}$$

thus,

$$\left|\frac{1}{n} \sum [\psi(x_i, \theta) - \lambda(\theta)]\right| \leq (1 - 2\epsilon) b(\theta) \leq \frac{1 - 2\epsilon}{1 - \epsilon} |\lambda(\theta)| \leq (1 - \epsilon)|\lambda(\theta)| \tag{22}$$

for $\forall \theta \notin C$, or

$$\left|\frac{1}{n} \sum \psi(x_i, \theta)\right| \geq \epsilon |\lambda(\theta)| \geq \epsilon(1 - \epsilon) b_0 \tag{23}$$

for $\forall \theta \notin C$, which implies the lemma.

THEOREM 2. *If* (B-1), (B-2′), *and* (B-3) *hold, then every sequence* T_n *satisfying* (15) *and the conclusion of lemma* 2 *converges to* θ_0 *almost surely. An analogous statement is true for convergence in probability.*

PROOF. We may restrict attention to the compact set C. For any open neighborhood U of θ_0, the infimum of the continuous function $|\lambda(\theta)|$ on the compact set $C \setminus U$ is strictly positive, say $\geq 5\epsilon > 0$. For every $\theta \in C \setminus U$, let U_θ be a neighborhood of θ such that by (B-2′),

$$E\{\sup_{\theta' \in U_\theta} |\psi(x, \theta') - \psi(x, \theta)|\} \leq \epsilon; \tag{24}$$

hence, $|\lambda(\theta') - \lambda(\theta)| \leq \epsilon$ for $\theta' \in U_\theta$. Select a finite subcover $U_s = U_{\theta_s}$, $1 \leq s \leq N$. Then we have a.s. for sufficiently large n

$$\begin{aligned} \sup_{\theta' \in C \setminus U} \left| \frac{1}{n} \sum [\psi(x, \theta') - \lambda(\theta')] \right| & \\ \leq \sup_{1 \leq s \leq N} \frac{1}{n} \sum \sup_{\theta' \in U_s} |\psi(x, \theta') - \psi(x, \theta_s)| & \\ + \sup_{1 \leq s \leq N} \left| \frac{1}{n} \sum [\psi(x, \theta_s) - \lambda(\theta_s)] \right| + \epsilon & \\ \leq 4\epsilon. & \end{aligned} \tag{25}$$

Since $|\lambda(\theta)| \geq 5\epsilon$ for $\theta \in C \setminus U$, this implies

$$\left| \frac{1}{n} \sum \psi(x_i, \theta) \right| \geq \epsilon \tag{26}$$

for $\theta \in C \setminus U$ and sufficiently large n, which proves the theorem. Convergence in probability is proved analogously.

4. Asymptotic normality

In the following, Θ is an open subset of m-dimensional Euclidean space R^m, $(\mathfrak{X}, \mathfrak{A}, P)$ is a probability space, and $\psi: \mathfrak{X} \times \Theta \to R^m$ is some function.

Assume that $x_1, x_2, \cdots$ are independent identically distributed random variables with values in $\mathfrak{X}$ and common distribution P. We want to give sufficient conditions ensuring that every sequence $T_n = T_n(x_1, \cdots, x_n)$ satisfying

$$(1/\sqrt{n}) \sum_{i=1}^{n} \psi(x_i, T_n) \to 0 \quad \text{in probability} \tag{27}$$

is asymptotically normal.

In particular, this result will imply asymptotic normality of ML estimators: let $f(x, \theta)$, $\theta \in \Theta$ be a family of probability densities with respect to some measure μ on $(\mathfrak{X}, \mathfrak{A})$, $dP = f(x, \theta_0)\, d\mu$ for some θ_0, $\psi(x, \theta) = (\partial/\partial\theta) \log f(x, \theta)$, then the sequence of ML estimators T_n of θ_0 satisfies (27).

Assuming that consistency of T_n has already been proved by some other means, we shall establish asymptotic normality under the following conditions.

ASSUMPTIONS.

(N-1). *For each fixed* $\theta \in \Theta$, $\psi(x, \theta)$ *is* $\mathfrak{A}$*-measurable and* $\psi(x, \theta)$ *is separable* (*see* (A-1)).

Put

$$\lambda(\theta) = E\psi(x, \theta), \qquad (28)$$
$$u(x, \theta, d) = \sup_{|\tau-\theta| \leq d} |\psi(x, \tau) - \psi(x, \theta)|.$$

Expectations are always taken with respect to the true underlying distribution P.

(N-2). *There is a* $\theta_0 \in \Theta$ *such that* $\lambda(\theta_0) = 0$.

(N-3). *There are strictly positive numbers* a, b, c, d_0 *such that*

(i) $|\lambda(\theta)| \geq a \cdot |\theta - \theta_0|$ for $|\theta - \theta_0| \leq d_0$,

(ii) $Eu(x, \theta, d) \leq b \cdot d$ for $|\theta - \theta_0| + d \leq d_0$, $d \geq 0$,

(iii) $E[u(x, \theta, d)^2] \leq c \cdot d$ for $|\theta - \theta_0| + d \leq d_0$, $d \geq 0$.

Here, $|\theta|$ denotes any norm equivalent to Euclidean norm. Condition (iii) is somewhat stronger than needed; the proof can still be pushed through with $E[u(x, \theta, d)^2] \leq o(|\log d|^{-1})$.

(N-4). *The expectation* $E[|\psi(x, \theta_0)|^2]$ *is finite.*

Put

$$Z_n(\tau, \theta) = \frac{\left|\sum_{i=1}^{n} [\psi(x_i, \tau) - \psi(x_i, \theta) - \lambda(\tau) + \lambda(\theta)]\right|}{\sqrt{n} + n|\lambda(\tau)|}. \qquad (29)$$

The following lemma is crucial.

LEMMA 3. *Assumptions* (N-1), (N-2), (N-3) *imply*

$$\sup_{|\tau-\theta_0| \leq d_0} Z_n(\tau, \theta_0) \to 0 \qquad (30)$$

in probability, as $n \to \infty$.

PROOF. For the sake of simplicity, and without loss of generality, take $|\theta|$ to be the sup-norm, $|\theta| = \max(|\theta_1|, \cdots, |\theta_m|)$ for $\theta \in R^m$. Choose the coordinate system such that $\theta_0 = 0$ and $d_0 = 1$.

The idea of the proof is to subdivide the cube $|\tau| \leq 1$ into a slowly increasing number of smaller cubes and to bound $Z_n(\tau, 0)$ in probability on each of those smaller cubes.

Put $q = 1/M$, where $M \geq 2$ is an integer to be chosen later, and consider the concentric cubes

$$C_k = \{\theta \mid |\theta| \leq (1-q)^k\}, \qquad k = 0, 1, \cdots, k_0. \qquad (31)$$

Subdivide the difference $C_{k-1} \setminus C_k$ into smaller cubes (see figure 1) with edges of length

$$2d = (1-q)^{k-1}q, \qquad (32)$$

such that the coordinates of their centers ξ are odd multiples of d, and

$$|\xi| = (1-q)^{k-1}\left(1 - \frac{q}{2}\right). \qquad (33)$$

For each value of k there are less than $(2M)^m$ such small cubes, so there are $N < k_0 \cdot (2M)^m$ cubes contained in $C_0 \setminus C_{k_0}$; number them $C_{(1)}, \cdots, C_{(N)}$.

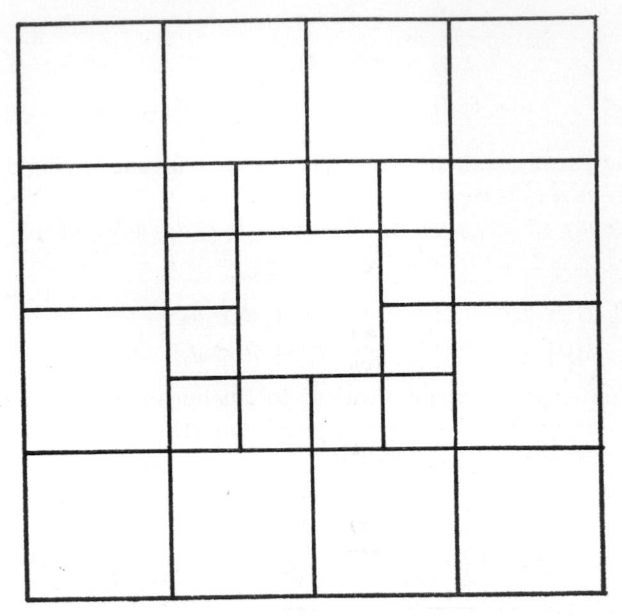

Figure 1

Now let $\epsilon > 0$ be given. We shall show that for a proper choice of M and of $k_0 = k_0(n)$, the right-hand side of

$$P(\sup_{\tau \in C_0} Z_n(\tau, 0) \geq 2\epsilon) \leq P(\sup_{\tau \in C_{k_0}} Z_n(\tau, 0) \geq 2\epsilon) + \sum_{j=1}^{N} P(\sup_{\tau \in C_{(j)}} Z_n(\tau, 0) \geq 2\epsilon) \tag{34}$$

tends to 0 with increasing n, which establishes lemma 1.

Actually, we shall choose

$$M \geq (3b)/(\epsilon a), \tag{35}$$

and $k_0 = k_0(n)$ is defined by

$$(1 - q)^{k_0} \leq n^{-\gamma} < (1 - q)^{k_0 - 1}, \tag{36}$$

where $\frac{1}{2} < \gamma < 1$ is an arbitrary fixed number. Thus

$$k_0(n) - 1 < \frac{\gamma \cdot \log n}{|\log (1 - q)|} \leq k_0(n), \tag{37}$$

hence

$$N = O(\log n). \tag{38}$$

Now take any of the cubes $C_{(j)}$, with center ξ and edges of length $2d$ according to (32) and (33). For $\tau \in C_{(j)}$ we have then by (N-3),

$$|\lambda(\tau)| \geq a|\tau| \geq a \cdot (1 - q)^k, \tag{39}$$

$$|\lambda(\tau) - \lambda(\xi)| \leq Eu(x, \xi, d) \leq bd \leq b(1 - q)^k q. \tag{40}$$

We have

$$Z_n(\tau, 0) \leq Z_n(\tau, \xi) + \frac{\left|\sum_{i=1} [\psi(x_i, \xi) - \psi(x_i, 0) - \lambda(\xi)]\right|}{\sqrt{n} + n|\lambda(\tau)|}, \tag{41}$$

hence

$$\sup_{\tau \in C_{(j)}} Z_n(\tau, 0) \leq U_n + V_n \tag{42}$$

with

$$U_n = \frac{\sum_{i=1}^{n} [u(x_i, \xi, d) + Eu(x, \xi, d)]}{na(1-q)^k}, \tag{43}$$

$$V_n = \frac{\left|\sum_{i=1}^{n} [\psi(x_i, \xi) - \psi(x_i, 0) - \lambda(\xi)]\right|}{na(1-q)^k}. \tag{44}$$

Thus,

$$\begin{aligned} P(U_n \geq \epsilon) \\ = P\left\{\sum_{i=1}^{n} [u(x_i, \xi, d) - Eu(x, \xi, d)] \geq \epsilon na(1-q)^k - 2nEu(x, \xi, d)\right\}. \end{aligned} \tag{45}$$

In view of (40) and (35),

$$\epsilon a(1-q)^k - 2Eu(x, \xi, d) \geq \epsilon a(1-q)^k - 2bq(1-q)^k \geq bq(1-q)^k, \tag{46}$$

hence (N-3) (iii) and Chebyshev's inequality yield

$$P(U_n \geq \epsilon) \leq \frac{c}{b^2 q(1-q)} \cdot \frac{1}{n(1-q)^{k-1}}. \tag{47}$$

In a similar way,

$$P(V_n \geq \epsilon) \leq \frac{c}{9b^2 q^2 (1-q)^2} \cdot \frac{1}{n(1-q)^{k-1}}. \tag{48}$$

Hence, we obtain from (36), (42), (47), (48) that

$$P(\sup_{\tau \in C_{(j)}} Z_n(\tau, 0) \geq 2\epsilon) \leq K \cdot n^{\gamma - 1} \tag{49}$$

with

$$K = \frac{c}{b^2 q(1-q)} + \frac{c}{9b^2 q^2 (1-q)^2}. \tag{50}$$

Furthermore,

$$\sup_{\tau \in C_{k_0}} Z_n(\tau, 0) \leq \frac{\sum_{1}^{n} [u(x_i, 0, d) + Eu(x, 0, d)]}{\sqrt{n}} \tag{51}$$

with $d = (1-q)^{k_0} \leq n^{-\gamma}$. Hence,

$$\begin{aligned} P(\sup_{\tau \in C_{k_0}} Z_n(\tau, 0) \geq 2\epsilon) \\ \leq P(\sum_{i=1}^{n} [u(x_i, 0, d) - Eu(x_i, 0, d)] \geq 2\sqrt{n}\epsilon - 2nEu(x, 0, d)). \end{aligned} \tag{52}$$

Since $Eu(x, 0, d) \leq bd \leq bn^{-\gamma}$, there is an n_0 such that for $n \geq n_0$,

$$2\sqrt{n}\epsilon - 2nEu(x, 0, d) \geq \sqrt{n}\epsilon; \tag{53}$$

thus, by Chebyshev's inequality,

$$P(\sup_{\tau \in C_{k_0}} Z_n(\tau, 0) \geq 2\epsilon) \leq c \cdot \epsilon^{-2} \cdot n^{-\gamma}. \tag{54}$$

Now, putting (34), (38), (49) and (54) together, we obtain

$$P(\sup_{\tau \in C_0} Z_n(\tau, 0) \geq 2\epsilon) \leq O(n^{-\gamma}) + O(n^{\gamma-1} \log n), \tag{55}$$

which proves lemma 3.

THEOREM 3. *Assume that* (N-1) *to* (N-4) *hold and that* T_n *satisfies* (27). *If* $P(|T_n - \theta_0| \leq d_0) \to 1$, *then*

$$(1/\sqrt{n}) \sum_{i=1}^{n} \psi(x_i, \theta_0) + \sqrt{n}\lambda(T_n) \to 0 \tag{56}$$

in probability.

PROOF. Assume again $\theta_0 = 0$, $d_0 = 1$. We have

$$\sum_1^n \psi(x_i, T_n) = \sum_1^n [\psi(x_i, T_n) - \psi(x_i, 0) - \lambda(T_n)] + \sum_1^n [\psi(x_i, 0) + \lambda(T_n)]. \tag{57}$$

Thus, with probability tending to 1

$$\frac{\left|\sum_1^n [\psi(x_i, 0) + \lambda(T_n)]\right|}{\sqrt{n} + n|\lambda(T_n)|} \leq \sup_{|\tau| \leq 1} Z_n(\tau, 0) + (1/\sqrt{n}) \left|\sum_1^n \psi(x_i, T_n)\right|. \tag{58}$$

The terms on the right-hand side tend to 0 in probability (lemma 1 and assumption (27)), so the left-hand side does also.

Now let $\epsilon > 0$ be given. Put $K^2 = 2E(|\psi(x, 0)|^2)/\epsilon$; then, by Chebyshev's inequality and for sufficiently large n, say $n \geq n_0$, the inequalities

$$\left|(1/\sqrt{n}) \sum_1^n \psi(x_i, 0)\right| \leq K \tag{59}$$

and

$$\left|\sum_1^n [\psi(x_i, 0) + \lambda(T_n)]\right| \leq \epsilon \cdot (\sqrt{n} + n|\lambda(T_n)|) \tag{60}$$

are violated with probabilities $\leq \frac{1}{2}\epsilon$; so both hold simultaneously with probability exceeding $1 - \epsilon$. But (60) implies

$$\sqrt{n}|\lambda(T_n)|(1 - \epsilon) \leq \epsilon + (1/\sqrt{n}) \left|\sum_1^n \psi(x_i, 0)\right|, \tag{61}$$

hence, by (59), $\sqrt{n}|\lambda(T_n)| \leq (K + \epsilon)/(1 - \epsilon)$. Thus,

$$\left|(1/\sqrt{n}) \sum_1^n \psi(x_i, 0) + \sqrt{n}\lambda(T_n)\right| \leq (K + 1)\epsilon/(1 - \epsilon) \tag{62}$$

holds with probability exceeding $1 - \epsilon$ for $n \geq n_0$. Since the right-hand side of (62) can be made arbitrarily small by choosing ϵ small enough, the theorem follows.

COROLLARY. *Under the conditions of theorem 3, assume that λ has a nonsingular derivative Λ at θ_0 (that is, $|\lambda(\theta) - \lambda(\theta_0) - \Lambda\cdot(\theta - \theta_0)| = o(|\theta - \theta_0|)$). Then $\sqrt{n}(T_n - \theta_0)$ is asymptotically normal with mean 0 and covariance matrix $\Lambda^{-1}C(\Lambda')^{-1}$, (where C stands for the covariance matrix of $\psi(x, \theta_0)$, and Λ' is the transpose of Λ).*

PROOF. The proof is immediate.

REMARK. We worked under the tacit assumption that the T_n are measurable. Actually, this is irrelevant, except that for nonmeasurable T_n, some careful circumlocutions involving inner and outer probabilities are necessary.

Efficiency. Consider now the ordinary ML estimator, that is, assume that $dP = f(x, \theta_0)\, d\mu$ and that $\psi(x, \theta) = (\partial/\partial\theta) \log f(x, \theta)$ (derivative in measure). Assume that $\psi(x, \theta)$ is jointly measurable, and that (N-1), (N-3), and (N-4) hold locally uniformly in θ_0, and that the ML estimator is consistent.

We want to check whether the ML estimator is efficient. That is, we want to show that $\lambda(\theta_0) = 0$ and $\Lambda = -C = -I(\theta_0)$, where $I(\theta_0)$ is the information matrix. This implies, by the corollary to theorem 3, that the asymptotic variance of the ML estimator is $I(\theta_0)^{-1}$.

Obviously, with $\theta_t = \theta_0 + t\cdot(\theta - \theta_0)$, $0 \leq t \leq 1$,

$$
\begin{aligned}
(63) \qquad 0 &\geq \int [\log f(x, \theta) - \log f(x, \theta_0)] f(x, \theta_0)\, d\mu \\
&= \iint_0^1 \psi(x, \theta_t)\, dt\, f(x, \theta_0)\, d\mu\cdot(\theta - \theta_0) \\
&= \int_0^1 \lambda(\theta_t)\, dt\cdot(\theta - \theta_0).
\end{aligned}
$$

(The interchange of the order of the integrals is legitimate, since $\psi(x, \theta_t)$ is bounded in absolute value by the integrable $|\psi(x, \theta_0)| + u(x, \theta_0, |\theta - \theta_0|)$.)

Since λ is continuous, $\int_0^1 \lambda(\theta_t)\, dt \to \lambda(\theta_0)$ for $\theta \to \theta_0$, hence $\lambda(\theta_0)\cdot\eta \leq 0$ for any vector η, thus $\lambda(\theta_0) = 0$.

Now consider

$$
\begin{aligned}
(64) \qquad \lambda(\theta) - \lambda(\theta_0) &= \int \psi(x, \theta) f(x, \theta_0)\, d\mu \\
&= -\int \psi(x, \theta)[f(x, \theta) - f(x, \theta_0)]\, d\mu \\
&= -\int \psi(x, \theta) \int_0^1 \psi(x, \theta_t) f(x, \theta_t)\, dt\, d\mu\cdot(\theta - \theta_0).
\end{aligned}
$$

But

$$
\begin{aligned}
(65) \qquad \int \psi(x, \theta) \int_0^1 \psi(x, \theta_t) f(x, \theta_t)\, dt\, d\mu &= \iint_0^1 \psi(x, \theta_t)\psi(x, \theta_t) f(x, \theta_t)\, dt\, d\mu + r(\theta) \\
&= \int_0^1 I(\theta_t)\, dt + r(\theta),
\end{aligned}
$$

with

$$|r(\theta)| \leq \int_0^1 \int u(x, \theta_t, |\theta - \theta_0|) |\psi(x, \theta_t)| f(x, \theta_t) \, d\mu \, dt \leq O(|\theta - \theta_0|^{1/2}) \tag{66}$$

by Schwarz's inequality.

If $I(\theta)$ is continuous at θ_0, the assertion $\Lambda = -I(\theta_0)$ follows. If $I(\theta)$ should be discontinuous, a slight refinement of the above argument shows that $I(\theta_0)^{-1}$ is an upper bound for the asymptotic variance. I do not know whether (N-3) actually implies continuity of $I(\theta)$.

5. Examples

EXAMPLE 1. Let $\mathfrak{X} = \Theta = R^m (m \geq 2)$, and let $\rho(x, \theta) = |x - \theta|^p$, $1 \leq p \leq 2$, where $| \ |$ denotes Euclidean norm. Define T_n by the property that it minimizes $\sum_{i=1}^n \rho(x_i, T_n)$; or, if we put

$$\psi(x, \theta) = -\frac{1}{p} \frac{\partial}{\partial \theta} \rho(x, \theta) = |x - \theta|^{p-2}(x - \theta), \tag{67}$$

by the property $\sum_{i=1}^n \psi(x_i, T_n) = 0$. This estimator was considered by Gentleman [3].

A straightforward calculation shows that both u and u^2 satisfy Lipschitz conditions

$$u(x, \theta, d) \leq c_1 \cdot d \cdot |x - \theta|^{p-2} \tag{68}$$

$$u^2(x, \theta, d) \leq c_2 \cdot d \cdot |x - \theta|^{p-2} \tag{69}$$

for $0 \leq d \leq d_0 < \infty$. Thus, conditions (N-3) (ii) and (iii) are satisfied, provided that

$$E|x - \theta|^{p-2} \leq K < \infty \tag{70}$$

in some neighborhood of θ_0, which certainly holds if the true distribution has a bounded density with respect to Lebesgue measure. Furthermore, under the same condition (70),

$$\frac{\partial}{\partial \theta} \lambda(\theta) = E \frac{\partial \psi(x, \theta)}{\partial \theta}. \tag{71}$$

Thus

$$\operatorname{tr} \frac{\partial \lambda}{\partial \theta} = E \operatorname{tr} \frac{\partial \psi}{\partial \theta} = -(m + p - 2) E|x - \theta|^{p-2} < 0, \tag{72}$$

hence also (N-3) (i) is satisfied. Condition (N-1) is immediate, (N-2) and (N-4) hold if $E|x|^{2p-2} < \infty$, and consistency of T_n follows either directly from convexity of ρ, or from verifying (B-1) to (B-4) (with $b(\theta) = \max(1, |\theta|^{p-1})$).

EXAMPLE 2. (Cf. Huber [4], p. 79.)

Let $\mathfrak{X} = \theta = R$, and let $\rho(x, \theta) = \frac{1}{2}(x - \theta)^2$ for $|x - \theta| \leq k$, $\rho(x, \theta) = \frac{1}{2}k^2$ for $|x - \theta| > k$. Condition (A-4) of section 2, namely unicity of θ_0, imposes a restriction on the true underlying distribution; the other conditions are trivially satisfied (with $a(x) \equiv 0$, $b(\theta) \equiv \frac{1}{2}k^2$, $h(x) \equiv 0$). Then, the T_n minimizing $\sum \rho(x_i, T_n)$ is a consistent estimate of θ_0.

Under slightly more stringent regularity conditions, it is also asymptotically normal. Assume for simplicity $\theta_0 = 0$, and assume that the true underlying distribution function F has a density F' in some neighborhoods of the points $\pm k$, and that F' is continuous at these points. Conditions (N-1), (N-2), (N-3) (ii), (iii), and (N-4) are obviously satisfied with $\psi(x, \theta) = (\partial/\partial\theta)\rho(x, \theta)$; if

$$\int_{-1}^{+k} F(dx) - kF'(k) - kF'(-k) > 0, \tag{73}$$

also (N-3) (i) is satisfied. One checks easily that the sequence T_n defined above satisfies (27), hence the corollary to theorem 3 applies.

Note that the consistency proof of section 3 would not work for this example.

REFERENCES

[1] H. E. DANIELS, "The asymptotic efficiency of a maximum likelihood estimator," *Proceedings of the Fourth Berkeley Symposium on Mathematical Statistics and Probability*, Berkeley and Los Angeles, University of California Press, 1961, Vol. 1, pp. 151–163.

[2] J. L. DOOB, *Stochastic Processes*, New York, Wiley, 1953.

[3] W. M. GENTLEMAN, Ph.D. thesis, Princeton University, 1965, unpublished.

[4] P. J. HUBER, "Robust estimation of a location parameter," *Ann. Math. Statist.*, Vol. 35 (1964), pp. 73–101.

[5] YU. V. LINNIK, "On the probability of large deviations for the sums of independent variables," *Proceedings of the Fourth Berkeley Symposium on Mathematical Statistics and Probability*, Berkeley and Los Angeles, University of California Press, 1961, Vol. 2, pp. 289–306.

[6] A. WALD, "Note on the consistency of the maximum likelihood estimate," *Ann. Math. Statist.*, Vol. 20 (1949), pp. 595–601.

THE CLASSICAL PROBLEM OF INFERENCE— GOODNESS OF FIT

OSCAR KEMPTHORNE
IOWA STATE UNIVERSITY

1. General introduction

As a preface to my lecture, I find it necessary to discuss in general terms the status of the statistical art and what we should mean by the term "inference." It seems to me that over the whole history of human thought there have been two basic underlying ideas of inference:

(a) what may best be expressed, perhaps, by the colloquialism "making sense of data";
(b) the choice of an action in a prespecified class of possible actions on the basis of data, costs, risks, and opinions.

Of course, to attempt to characterize the whole of statistics in some such way as the preceding is rather like attempting to characterize mathematics by a few brief common sense statements, and this is obviously foredoomed to failure. But the attempt has been made by others, who with a zeal approaching that of religious fanatics attempt to convince the world that there is one true religion, the one they are preaching. We should feel a considerable debt to van Dantzig [10], [11] for calling attention to the phenomenon of "Statistical Priesthood" with which our profession is now plagued. He gave us just two examples and pointed out the moral. It is curious that even in its activities unrelated to ethics, humanity searches for a religion. At the present time, the religion being "pushed" the hardest is Bayesianism. A few years ago it was decision theory. The actions of the proponents are like those of the religious evangelist. It is characteristic of new religions that they are intolerant of the old ones. It seems obvious that the only religion we should uphold is that there is no true religion. I find myself quite intolerant of the several cults.

My own preference is to say that the bulk of the activities of statisticians is encompassed by one or other of the two basic ideas expressed above, and I like to denote them by (a) Statistical Inference and (b) the Theory of Decision-making. It is remarkable that over the years we have had many papers and even books which take the view that statistical inference is a part of decision-making.

Some aspects of the reported work were supported by NSF Grant G-14237. Journal Paper No. J-5461 of the Iowa Agricultural and Home Economics Station, Ames, Iowa. Project 890.

Such people presumably take the view that the writing down of conclusions can be regarded as an act in a specified class of actions, to which costs and risks can be applied. The problem here is partially one of semantics: what does one mean by an action, for instance? My main criticism of the point of view which takes (a) to be a part of (b) is that it has in my opinion proved to be rather sterile with regard to the general problem of making sense of data, the problem of condensation of data, and so on. Before 1940 we had in the total history of mankind perhaps 100 man years of intellectual effort in the direction of statistics; in the Forties perhaps another 100; in the Fifties perhaps 200, and in the Sixties so far perhaps 400. But the advances in the art of making sense of data have not been at all commensurate. There have, however, been great advances in the theory of decision-making. Our knowledge of possible rules for terminal decisions has expanded very rapidly. It seems, however, to have escaped attention that there is a vast difference between what one is entitled to think on the basis of the data and what action one should take on the basis of the data.

What is the main problem of data interpretation? I believe it to be the development of a condensation of the data which in some imprecise way does not throw away any of the information in the data. It may be said that I am using vague expressions which cannot be given precise meaning and I would agree. But I would then say that the history of Science is full of examples of the dangers of narrowly prescribed specifications and frameworks of thought.

It is a consequence of the above view that one of the basic problems of data interpretation is the problem of model specification. I am highly amused by a statement of Sir Ronald Fisher ([4], p. 314): "As regards problems of specification, these are entirely a matter for the practical statistician." But this curious remark is balanced by a later one on the same page: "The possibility of developing complete and self-contained tests of goodness of fit deserves very careful consideration." It is probably fairly generally agreed that the beginning of what we call statistical inference in the sense (a) was the development of the χ^2 goodness-of-fit test. This was really a most remarkable piece of work and stands out in my opinion as one of the great ideas of human thought. It is true that Pearson made some errors with regard to the concept of degrees of freedom, and it was necessary for Fisher to clear them up. But this should not detract from the magnitude of Pearson's step.

To emphasize the problem of model specification let me give a common example. An experimenter has compared 6 treatments in a randomized block design with regard to their effect on the growth of mice. The statistician says, "Ah, yes! Randomized Blocks, so *the* model is $y_{ij} = \mu + b_i + t_j + e_{ij}$ with the errors normally and independently distributed around zero with common variance σ^2." I have difficulty imagining a more blatant travesty of common sense, let alone scientific method. It is really appalling. But this is taught extensively. Books say, "the appropriate model is. . . ." Fortunately, in recent years there has been some deeply considered attack on the problem, particularly by Anscombe and Tukey.

My interest in the problem of goodness of fit was stimulated by consideration of the problem of transformations, primarily in connection with the analysis of comparative experiments. I was then led to the simplest situation, the case of a single sample. The problem of whether observations should be analyzed on an arithmetic scale or a logarithmic scale is surely one of the most elementary problems, but surprisingly little has been done on it. The application of some goodness-of-fit procedures seemed appropriate, and the χ^2 goodness-of-fit test seemed a reasonable candidate.

I hope that a little of what I have to say is new. But if not, a representation of old material may still be of value. A paper by Slakter [9] has numerical results closely related to some of the results I shall present.

2. A brief historical review

The literature on goodness-of-fit tests is vast. Shapiro and Wilk [8] for instance, give a list of about 70 papers on the subject. I found the review by David [3] very helpful. Even the literature on the χ^2 goodness-of-fit test is very extensive. The procedure is, of course, to divide the distribution into mutually exclusive cells and to form the criterion

$$\sum_{\text{cells}} \frac{(O_i - E_i)^2}{E_i} \tag{2.1}$$

where O_i is the observed frequency, and E_i the expected frequency in the i-th cell. In the case of a discrete data, the discrete classification of the distribution provides a partition into classes for application of the χ^2 procedure. It is customary, however, to invoke some rule such as that the expectations of cells should be greater than 5, or in some cases greater than 2. For example, Cramér [1] says:

"When the χ^2 test is applied in practice, and all the expected frequencies np_i are ≥ 10, the limiting χ^2-distribution tabulated in table III gives as a rule the value χ^2_p corresponding to a given $P = p/100$ with an approximation sufficient for ordinary purposes. If some of the np_i are <10, it is usually advisable to pool the smaller groups, so that every group contains at least 10 expected observations, before the test is applied. When the observations are so few that this cannot be done, the χ^2 tables should not be used, but some information may still be drawn from the values of $E(\chi^2)$ and $D(\chi^2)$ calculated according to (30.1.1)."

In fact, this matter is quite obscure and there is considerable arbitrariness in the application of the procedure to discrete data, as anyone who has looked at data discovers. One can vary the "answer" that is obtained by the choice of grouping the possible classes. Individual judgment is always called into play. In the continuous distribution case, which should always be characterized by some phrase such as the "so-called" continuous case because we can never observe a continuous random variable except with a grouping error, the situation is much worse with regard to applying the χ^2 goodness-of-fit test. "How is one to make up the cells?", "Where are they centered?", "How many cells should one use?" are questions on which the personal arbitrariness of the tester seems to enter. Again, anyone who has used the procedure has met these questions, and answered them in his own somewhat arbitrary way. Even if one has the rule that every cell should have an expectation greater than or equal to 5, there is the matter of placement of the classes. Also, the imposition of such rules has apparently led to the view that the χ^2 goodness-of-fit test is not consistent. One attack on this question was by Mann and Wald [6] who were led to the rule that asymptotic maximum power in a certain sense of the χ^2 test was achieved when the number of cells is proportional to $2(N-1)^{1/5}$, where N is the number of observations.

The purpose of the present paper is to describe and evaluate partially a completely objective simple rule, namely divide the distribution, fitted on the basis of N observations, into N equal parts each with probability $1/N$. This gives N cells each with expectation equal to unity. Count the number x_i in each cell. Then the χ^2 criterion

$$K = \sum_i \frac{(x_i - e_i)^2}{e_i} \tag{2.2}$$

becomes

$$\begin{aligned} K &= \sum_i (x_i - 1)^2 = \sum_i x_i^2 - 2 \sum x_i + N \\ &= \sum_i x_i^2 - N \\ &= \sum_i x_i(x_i - 1). \end{aligned} \tag{2.3}$$

The evaluation of the criterion is then made by reference to the χ^2 distribution with degrees of freedom equal to $(N - 1 - p)$, where p is the number of parameters fitted. In evaluating the criterion, however, it is to be noted that K can take only even integral values, so that one obtains the probability of exceeding $(K - 1)$ for the mathematical χ^2 distribution. It is to be noted that in the case of continuous data this rule is objective, and there is no room for personal choice on number and location of cells. It is assumed that the resulting classes are still wide relative to the grouping interval of observations, though clearly this will not be true with very large samples. In very large samples the grouping error of observations would have to be considered.

The present paper gives some preliminary results on the above procedure. It contains a discussion of the distribution of the Pearsonian criterion with k equally likely classes in the case when no parameters are estimated, this discussion being relevant and appropriate to the case of any continuous distribution. Then some Monte Carlo results are given on the distribution of K for the case of the normal distribution, in which the mean and variance are estimated. Finally a few power comparisons are made, and some discussion on the relevance of power is presented.

3. The distribution of K_k with a completely specified distribution

We consider the partition of a continuous distribution into equiprobable classes each with probability $1/k$, and denote the χ^2 criterion by K_k. In the case of N observations with k equal to N, the criterion is the K given above. If we denote the observed numbers in a total sample of N, which are in the classes, $i = 1, 2, \cdots, k$, by x_i, the probability of $x_1, x_2, \cdots, x_k$ is

$$\frac{N!}{x_i!}\left(\frac{1}{k}\right)^N. \tag{3.1}$$

The χ^2 quantity is equal to

$$K_k = \sum_i \frac{(x_i - (N/k)^2}{(N/k)} \tag{3.2}$$

so that

$$\begin{aligned} \frac{N}{k} K_k &= \sum_i \left(x_i - \frac{N}{k}\right)^2 \\ &= \sum x_i^2 - 2\frac{N}{k}\sum x_i + \frac{N^2}{k} \\ &= \sum x_i^2 - \frac{N^2}{k}, \end{aligned} \tag{3.3}$$

or

$$\frac{N}{k}(K_k + N) = \sum x_i^2 = S \text{ (say)}. \tag{3.4}$$

Hence, to get the moments of K_k we first obtain the moments of S. The first two moments of K_k are well known, but we include them for completeness.

First moment of K_k. One has

$$S = \sum x_i^2 = \sum x_i(x_i - 1) + \sum x_i \tag{3.5}$$

and

$$Ex_i(x_i - 1) = \frac{N_2}{k^2} \tag{3.6}$$

where $N_s = N(N-1)(N-2)\cdots(N-s+1)$. Hence $E(S) = N((N-1)/k) + N$ and

$$\begin{aligned} E(K_k) &= (N-1) + k - N \\ &= (k-1). \end{aligned} \tag{3.7}$$

Variance of K_k. The term S^2 is equal to

$$S^2 = \sum_i x_i^4 + \sum_{i,i'}^{\neq} x_i^2 x_{i'}^{\prime 2}, \tag{3.8}$$

but $x^4 = x_4 + 6x_3 + 7x_2 + x_1$ where, as before,

$$x_s = x(x-1)(x-2)\cdots(x-s+1). \tag{3.9}$$

Such relationships are verified easily by writing, for example,

$$x^6 = x_6 + \alpha_5 x_5 + \alpha_4 x_4 + \alpha_3 x_3 + \alpha_2 x_2 + \alpha_1 x_1 + \alpha_0 \tag{3.10}$$

and successively placing x equal to 0, 1, 2, 3, 4, and 5. This device is helpful throughout. The reason for using this mode of expression is that $E(x_s) = (N_s/k^s)$, and $Ex_{i(s)}x_{i'(s')} = (N_{s+s'}/k^{s+s'})$, and so on. Hence,

$$E(S^2) = k\left\{\frac{N_4}{k^4} + 6\frac{N_3}{k^3} + 7\frac{N_2}{k^2} + \frac{N_1}{k}\right\} + k(k-1)\left\{\frac{N_4}{k^4} + 2\frac{N_3}{k^3} + \frac{N_2}{k^2}\right\} \tag{3.11}$$

so that $V(S) = N(N-1)2((k-1)/k^2)$ and

$$V(K_k) = 2(k-1)(N-1)/N. \tag{3.12}$$

Third moment of K_k. One can write

$$S^3 = \sum_i x_i^6 + 3\sum_{i,i'}^{\neq} x_{i'}^4 x_i^2 + \sum_{i,i',i''}^{\neq} x_i^2 x_{i'}^2 x_{i''}^2, \tag{3.13}$$

but

$$x^6 = x_6 + 15x_5 + 65x_4 + 90x_3 + 31x_2 + x_1, \tag{3.14}$$

$$x^4 = x_4 + 6x_3 + 7x_2 + x_1, \tag{3.15}$$

and $x^2 = x_2 + x_1$, so that

$$\begin{aligned} E(S^3) = {} & k\left\{\frac{N_6}{k^6} + 15\frac{N_5}{k^5} + 65\frac{N_4}{k^4} + 90\frac{N_3}{k^3} + 31\frac{N_2}{k^2} + \frac{N_1}{k}\right\} \\ & + 3k(k-1)\left\{\frac{N_6}{k^6} + 6\frac{N_5}{k^5} + 7\frac{N_4}{k^4} + \frac{N_3}{k^3} + \frac{N_5}{k^5} + 6\frac{N_4}{k^4} + 7\frac{N_3}{k^3} + \frac{N_2}{k^2}\right\} \\ & + k(k-1)(k-2)\left\{\frac{N_6}{k^6} + 3\frac{N_5}{k^5} + 3\frac{N_4}{k^4} + \frac{N_3}{k^3}\right\}, \end{aligned} \tag{3.16}$$

and the third moment of S is $\mu_3(S) = E(S^3) - 3V(S)E(S) - E^3(S)$ so that, after some tedious algebra,

$$\begin{aligned} \mu_3(S) &= 8N_3\frac{(k-1)}{k^3} + 4N_2\frac{(k-1)(k-2)}{k^3} \\ &= \frac{8N(N-1)(N-2)(k-1)}{k^3} + \frac{4N(N-1)(k-1)(k-2)}{k^3}. \end{aligned} \tag{3.17}$$

Hence,

$$\mu_3(K_k) = 8(k-1)\frac{(N-1)(N-2)}{N^2} + \frac{4(k-1)(k-2)(N-1)}{N^2}. \tag{3.18}$$

Fourth moment of K_k. Obtaining the fourth moment was very tedious. Using relationships such as

$$x^8 = x_8 + 28x_7 + 266x_6 + 1050x_5 + 1701x_4 + 966x_3 + 127x_2 + x_1 \tag{3.19}$$

and expressing polynomials in N as linear functions of N_s, I obtained

$$(3.20) \qquad \mu_4(S) = 12(k-1)(k+3)\frac{N_4}{k^4} + \left[144k - 384 + \frac{240}{k}\right]\frac{N_3}{k^3} + \left[8k - 32 + \frac{48}{k} - \frac{24}{k^2}\right]\frac{N_2}{k^2}.$$

Hence,

$$(3.21) \qquad \mu_4(K_k) = 12(k-1)(k+3)\frac{N_4}{N^4} + \left[144k - 384 + \frac{240}{k}\right]k\frac{N_3}{N^4} + \left[8k - 32 + \frac{48}{k} - \frac{24}{k^2}\right]k^2\frac{N_2}{N^4}.$$

If N is large and k fixed, the moments are very close to the moments of the theoretical χ^2 distribution with $(k-1)$ degrees of freedom, which is part of the basis for the use of the theoretical χ^2 table. The inconsistency of the k-group χ^2 test arises because the test detects only deviations from the multinomial obtained by grouping the continuous distribution which is being examined. This inconsistency can, however, be removed by letting k increase with N, and the purpose of the tedious calculation of the moments of K_k was to examine this matter. The obvious candidate mentioned above is to let k equal N. For this case the second moment about the mean is $2((N-1)^2/N)$ which is equal to $2[N - 2 + (1/N)]$, the third moment is $12((N-1)^2(N-2)/N^2)$, which is $12[N - 4 + (5/N) - (2/N^2)]$, and the fourth moment is equal to $12N^2 + 120N + 0(1)$.

The ratios of these moments to the moments of χ^2 with $(N-1)$ degrees of freedom are

Moment	*Ratio*
First	1
Second	$1 - \frac{1}{N}$
Third	$\frac{3}{2}\left(1 - \frac{1}{N}\right)$
Fourth	$\left(1 + \frac{8}{N}\right)$.

Obviously with k equal to N, the first, second, and fourth moments go to those of the χ^2 distribution quite rapidly. The ratio of third moments, however, tends to $\frac{3}{2}$. With both N and k large, the third moment is essentially

$$(3.22) \qquad 8(k-1) + 4\frac{(k-1)(k-2)}{N},$$

so the ratio to the theoretical moment is $1 + \frac{1}{2}((k-2)/N)$, or if k is equal to rN and both large, the ratio is $1 + (r/2)$. It might appear, therefore, that unless r is small, the distribution would not tend to the theoretical one. However, another aspect "saves the day," namely that as the degrees of freedom become large, the χ^2 distribution tends to the normal distribution. The skewness of the distribution of K_k tends to

$$\frac{\sqrt{2}(2 + r)}{\sqrt{k}} \tag{3.23}$$

which with increasing k, and fixed r tends to zero. It therefore appears that if the number of equiprobable classes k used in the goodness-of-fit test is equal to rN, a reasonable approximation for r sizeable relative to unity is to suppose that K_k is normally distributed with a mean of $(k - 1)$ and a variance $2(k - 1)[1 - (1/N)]$.

The results given above seem to tell us that for testing the goodness of fit to continuous distributions, the variety of rules in the literature, stating that cell expectations should be greater than 10, or greater than 5, or greater than 2, seem to be quite irrelevant. In fact, the distribution of K_k is not disturbed appreciably, apparently, if the number of classes is of the order of the number of observations. If k is equal to N, the distribution of K_k is asymptotically normal. It is clear that if k is of greater order, peculiar results obtain. If, for instance, k equals N^2, then the third moment is approximately $4N^3 + 8N^2$, and the skewness would be

$$\frac{(4N^3 + 8N^2)}{2N^2\sqrt{2N}} = \sqrt{2} \tag{3.24}$$

which does not go to zero with increasing N. Similarly, the fourth moment would be approximately $12N^2 + 144N^3 + 8N^4$ with kurtosis of approximately 2 for indefinitely large N.

The whole question of choice of the number of classes for the goodness-of-fit test seems therefore to be still quite an open one. My initial view was that having the number of classes equal to N, the number of observations would be a good choice. Of course this would be modified as soon as the inevitable grouping error of observations from a continuous distribution is met. It seems clear, however, that the larger the number of cells, the greater is the sensitivity of the test to deviations in the tails. It is extremely unlikely that any particular choice can be shown to be best for all circumstances.

The following sections give a few empirical results on the case k equal to N, for a null composite hypothesis of normality with data arising from a normal distribution and from two distributions for which a generating program could be written very quickly. Obviously much more computation needs to be done as well as some theoretical work on the whole matter.

4. Monte Carlo results on the distribution of K

The mathematical results above hold for the case of a completely specified distribution, which we may note has no conditions on its dimensionality. In the case of a multivariate distribution, one merely splits up the distribution into N equiprobable regions, and an intuitively reasonable way of doing this is to base the regions on the equiprobability contours. The grouping error of observations will, however, cause problems.

I have not yet obtained any mathematical results on the effect on the distribution of K when parameters of the distribution are estimated. I imagine that exactly the same type of result as was obtained by Fisher will hold, but the mathematics are not so easy because in the present case the cell expectations remain constant. In the cases considered by Pearson and Fisher, the cells were fixed, and asymptotically the cell frequencies increase and have a multivariate normal distribution, whose exponent is distributed as χ^2.

In envisaging the test described, I felt that it was primarily to be regarded as a test of distribution shape and not of location or scale, though obviously it has power asymptotically with regard to *any* alternatives. I therefore had computations performed for the case when parameters are estimated. Also my initial interest was in tests of normality, so I had computations made on samples from a normal distribution. Obviously, however, the test can be applied to *any* continuous distribution, and there is no reason to surmise that the distribution of K depends on the nature of the true continuous distribution from which the samples originate. Of course, it will be necessary that the method of "estimation" of parameters must be in some sense efficient or else the value of χ^2 will be too large (Fisher [5]).

Monte Carlo computations of the distribution of K were done by drawing sample (y) of size 10, 20, and 50 from normal distributions $N(\mu, \sigma^2)$ estimating the mean and variance of the normal distribution by

$$\begin{aligned} \hat{\mu} &= \text{ave } y, \\ \hat{\sigma}^2 &= \frac{1}{N-1} \sum (y - \text{ave } y)^2, \end{aligned} \tag{4.1}$$

and then comparing the actual sample with the fitted distribution.

Case I. Samples of size 10. As stated above, the possible values of K are even integers, so the obvious continuity correction was made. In view of the projected use of the test as a tail area test, the most appropriate comparison is to compare tail frequencies of the empirical distribution with those of the χ^2 distribution with 7 degrees of freedom. Of course, a test of goodness of fit would use the cell frequencies and not the tail frequencies. The reduction from 9 to 7 degrees of freedom was made because two parameters are estimated. Theoretical frequencies were taken from table 7 of Pearson and Hartley [7]. I have not bothered to make a goodness-of-fit test of the Monte Carlo results because the agreement and lack of agreement is quite obvious. In the case of samples of size 10, the frequency of the class "16 or over" observed was .0258, whereas the theoretical value is .0360; the expected number is therefore 180 and the observed number was 129, which is clearly discrepant. However, from the viewpoint of use, the reporting of a significance level as .0360, when it is close to .0258, cannot be regarded as a serious defect.

It is worth noting that the mean value of K was found to be 7.43, and the variance of K was 12.3. The above theory and the usual rule of subtracting unity for each parameter estimated suggests that the mean would be 7.0 and the variance would be 12, so the agreement is really rather good. Actually the mean is significantly greater than 7, but the tail areas do not seem to have been disturbed seriously by the change in mean and variance from the theoretical values for the χ^2 distribution with 7 degrees of freedom.

TABLE I

MONTE CARLO DISTRIBUTION OF K FOR 5000 SAMPLES OF SIZE 10 FROM A NORMAL POPULATION

	Proportion $\geq K$	
Value of K	Observed	Expected
0	1.000	
2	.995	
4	.941	
6	.732	
8	.499	
10	.284	.253
12	.133	.139
14	.080	.072
16	.0258	.0360
18	.0122	.0174
20	.0078	.0082
22	.0040	.0038
24+	.0004	.0017

Case II. Samples of size 20. One thousand samples of size 20 from a normal distribution were generated, and the distribution of K was estimated. The comparison is given in table II. Actually, the agreement of tail areas for the empirical distribution and the χ^2 distribution with 17 degrees of freedom seems quite remarkable. The mean of the distribution of K was estimated to be 17.4, as opposed to the theoretical value 17, and the variance as 30.90, which is to be compared with an expected value from the theory presented above of 32 and the value 34 for the theoretical χ^2 distribution for 17 degrees of freedom. Apparently, the discrepancy in mean and variance do not affect the tail areas appreciably.

Case III. Samples of size 50. Five hundred samples of size 50 were used, and the comparison with the χ^2 distribution for 47 degrees of freedom is given in table III. Because Pearson and Hartley [7] give tail areas for 46 and for 48 degrees of freedom and for even valued abscissa, simple linear interpolation was used. Actually, better interpolation could be done, but it was not deemed necessary with the sample size considered. The agreement is really quite remarkable. The discrepancy in mean was 47.37 compared to a theoretical value of 47, and in variance 87.6 compared to 94 (or approximately 92, suggested by the theory given above). It is again quite curious that discrepancy in mean

TABLE II

Monte Carlo Distribution of K for 1000 Samples of Size 20 from a Normal Population

	Proportion $\geq K$	
Value of K	Observed	Expected
6	1.000	
8	.996	
10	.972	
12	.905	
14	.788	
16	.627	
18	.460	
20	.331	.329
22	.217	.226
24	.153	.149
26	.092	.095
28	.061	.058
30	.039	.034
32	.025	.020
34	.014	.011
36	.008	.006
38	.005	.003
40	.001	.001

and variance do not affect the upper tail areas appreciably. There may be a small discrepancy with the extreme upper tail, and a larger sample could be examined to check on this point. The moral is, however, quite obvious, namely

TABLE III

Monte Carlo Distribution of K for 500 Samples of Size 50 from a Normal Population

	Proportion $\geq K$			Proportion $\geq K$	
Value of K	Observed	Expected	Value of K	Observed	Expected
26	1.000		54	.232	.26
28	.998		56	.188	.20
30	.994		58	.130	.15
32	.982		60	.102	.11
34	.962		62	.084	.08
36	.936		64	.066	.06
38	.884		66	.048	.04
40	.818		68	.034	.03
42	.742		70	.026	.02
44	.674		72	.018	.014
46	.574	.55	74	.018	.009
48	.452	.45	76	.012	.006
50	.380	.39	78	.010	.004
52	.310	.32	80	.010	.003

that the use of the theoretical χ^2 distribution is quite unlikely to be even slightly misleading.

5. Power of the K test

As stated above, it is obvious that the K test has some power with regard to any alternative. The test is obviously consistent. A detailed examination of power has not yet been made. Obviously there is the possibility of theoretical development. Shapiro and Wilk [8] report their W-test for normality and some comparison of power they made with the χ^2-test (it is not clear how this was defined), $\sqrt{b_1}$, b_2, Kolmogorov-Smirnov, Cramér-Von Mises, and a weighted Cramér-Von Mises test. Their table 10 suggests that the W_n-test is greatly superior to the others. A comparison of the K tests with all these is planned but has not been done because of lack of funds. Two cases are particularly easy to program; when the parent distribution is the triangular distribution and when it is χ^2 with 1 degree of freedom. In most cases in Shapiro and Wilk [8], the b_2 test was fairly good relative to most of the others, except the W_n-test. In order to get evidence cheaply on power of the K test, I have therefore applied the K test to 1000 samples from each of the two distributions named above.

(a) *Triangular distribution.* Five hundred samples were drawn from a triangular distribution and tested for normality. We take size of test from table II and obtain the results in table IV.

TABLE IV

SENSITIVITY OF K TEST FOR NORMALITY WITH SAMPLES FROM A TRIANGULAR DISTRIBUTION

Value of K	Size of test	Estimated power
24	.153	.956
26	.092	.900
28	.061	.814
30	.039	.712
34	.014	.458
38	.005	.290
40	.001	.208

(b) *Chi-square distribution with one d.f.* Results were obtained similarly for testing of normality, when the data originate from the χ^2 distribution with 1 degree of freedom, and are given in table V. The preliminary results show, by using the results given by Shapiro and Wilk [8] that the K test merits further examination and consideration.

6. Concluding remarks

I now return to the matters discussed at the beginning of my lecture.

In a repetitive situation, like acceptance sampling, in which decisions are

TABLE V

SENSITIVITY OF K TEST OF NORMALITY WITH SAMPLES FROM χ^2 (ONE D.F.) DISTRIBUTION

Value of K	Size of test	Estimated power
24	.153	.946
26	.092	.922
28	.061	.866
30	.039	.818
34	.014	.726
38	.005	.616
40	.001	.572

terminal, such as accepting or rejecting the lot, it is obvious that one has to map the sample space onto the decision space and that one has to consider the properties in repetitions of the mapping rule. Clearly, if a most powerful decision rule exists, it should be used. Also the costs of observation and risks of erroneous decisions have to be included in the formulation. Also it seems quite obvious that the decision-maker's personal opinions about the class of repetitive situations he will meet are relevant and should be included in the whole formulation. Even in a so-called repetitive situation goodness of fit of model is relevant. It is not sufficient merely to assume that a particular statistical model fits the situation.

It seems clear to me, however, that the accumulation of knowledge is not a repetitive process but rather an evolutionary one. No new situation is exactly like a previous situation except with regard to some parameter values. An essential part of the application of any model to data is the application of goodness-of-fit evaluations. It might be hoped that there would be one way of evaluating goodness of fit which is superior to all others, but obviously such a hope is foredoomed to failure. The literature on goodness of fit suggests that particular goodness-of-fit procedures are in some sense best with regard to the lack of fit they are designed to detect. It appears, however, that optimality in one direction is always accomplished at the expense of optimality in other directions. We are studying numerically the joint behavior of several goodness-of-fit tests, but I do not have any results yet. Such results seem to be essential for an overall intelligent approach to the problem.

The goodness-of-fit tests so far proposed seem to fall into essentially four main categories:

(a) those based on occupancy of cells determined by the hypothesized or fitted distribution, of which the Pearson chi-square test is the classic case;

(b) those based on the comparison of the cumulative sample frequency and the cumulative population frequence, like the Kolmogorov-Smirnov test, or the tests based on the differences of population cumulative frequencies between successive sample points;

(c) the comparison of the ordered statistic with the expected value of the ordered statistic, as is done in one way by Shapiro and Wilk [8];
(d) comparison of functions of moment statistics with theoretical values.

The comparison of tests is not easy. Asymptotic theory has been developed for some cases, and hopefully it gives a reliable indication of the sensitivity of the tests in various "directions." I have, however, an uneasy feeling that much asymptotic theory is based on the premise that observations of unlimited accuracy are possible. Of course, this is not the case, and with largish samples and a "reasonable" grouping error, ties will occur. This seems to affect all the tests but not with the same force. Many test criteria "blow up" when there are ties. From one point of view this is not unreasonable. If observations have unlimited accuracy, the hypothesis that they arise from a continuous distribution is untenable as soon as ties occur. The occurrence of "ties" appears to be an unavoidable embarrassment to the person who develops theory for so-called continuous observations. I find the discussion in the literature on this quite unsatisfying. The "answer" one obtains by some tests depends critically on how "ties" are broken. To suggest using some extraneous device to break ties, amounts to basing one's opinions with regard to a situation on an independent source of noise and this seems totally repugnant. This is quite unrelated to the use of a coin-toss to make up one's mind when one is unable to do so in any other way.

Of the classes of tests outlined above, many of the tests of classes (b) and (c) encounter the problems arising from grouping of observations in a violent form. The tests of class (d) are relatively not bothered by this, because we have strong intuitions that a moment-like function of a grouped sample is very close in behavior to that of an ungrouped sample. The tests based on occupancy of cells will encounter difficulties from grouping error, but it would appear that these difficulties are mild, and that reasonable smoothing devices will not disturb the distribution of the test criterion very much.

I am inclined to the view that with small samples, when ties will be very infrequent with a reasonable grouping error, tests based on order statistics will prove to be reasonably sensitive in many diverse directions. With intermediate and large samples, I am inclined to think that occupancy type tests give the data analyzer generally satisfactory answers. The question is not "Do the data come from such and such a distribution?", because one can be sure they never do, but "Is such and such a distribution a reasonable model for the description of the data?". It may well prove to be the case that a good overall procedure will combine an occupancy test and some sort of extreme value test. [See David [2] in connection with such a possibility.]

I wish to make one final point on inference. When faced with a sample the statistician makes a goodness-of-fit test for normality, and then constructs some limits on the parameters. The probability stated to be associated with these limits is always stated to be that conditional on normality, say. If, instead, one nominates a test of goodness of fit, and then delimits the parameter values for which there is a fit within a specified significance level, one obtains distributions which are consonant with the whole data. Such limits will be different, and may be wider and might therefore be thought to be not as good as the conditional ones. But they seem to me at least to give an answer to the informational question. This seems to be an example of how the notion of "most powerful," and even the notion of sufficiency have led our profession astray.

ACKNOWLEDGMENT

I am indebted to H. T. David for several informative discussions, to J. D. Atkinson for help in computation, and to L. Jordan for checking the algebra.

REFERENCES

[1] HARALD CRAMÉR, *Mathematical Methods of Statistics*, Princeton, Princeton University Press, 1946.

[2] H. T. DAVID, "Order statistics and statistics of structure (*d*)," *Ann. Math. Statist.*, Vol. 36 (1965), pp. 897–906.

[3] ———, "Goodness of fit," to be published in *International Encyclopedia of the Social Sciences*, 1966.

[4] R. A. FISHER, "On the mathematical foundations of theoretical statistics," *Philos. Trans. Roy. Soc. London Ser. A*, Vol. 222 (1922), pp. 309–368.

[5] ———, "The conditions under which χ^2 measures the discrepancy between observations and hypothesis," *J. Roy. Statist. Soc.*, Vol. 87 (1924), pp. 442–450.

[6] H. J. MANN and A. WALD, "On the choice of the number of class intervals in the application of the chi-square test," *Ann. Math. Statist.*, Vol. 13 (1942), pp. 306–317.

[7] E. S. PEARSON and H. O. HARTLEY, *Biometrika Tables for Statisticians*, Vol. 1, Cambridge, Cambridge University Press, 1945.

[8] S. SHAPIRO and M. B. WILK, "An analysis of variance test for normality (complete samples)," *Biometrika*, Vol. 52 (1965), pp. 591–611.

[9] M. J. SLAKTER, "A comparison of the Pearson chi-square and Kolmogorov goodness-of-fit tests with respect to validity," *J. Amer. Statist. Assoc.*, Vol. 60 (1965), pp. 854–858.

[10] D. VAN DANTZIG, "Statistical Priesthood" (Savage on Personal Probabilities 1), *Statistica Neerlandica*, Vol. 11, Nr. 1 (1957), pp. 1–16.

[11] ———, "Statistical Priesthood II" (Sir Ronald on Scientific Inference), *Statistica Neerlandica*, Vol. 11, Nr. 4 (1957), pp. 185–200.

ON PARTIAL PRIOR INFORMATION AND THE PROPERTY OF PARAMETRIC SUFFICIENCY

HIROKICHI KUDŌ
OSAKA CITY UNIVERSITY and
UNIVERSITY OF CALIFORNIA, BERKELEY

1. Summary

The problem of statistical decisions when there is a partial lack of prior information is considered, and a definition of the optimality of a statistical procedure in such a case is given. This optimality is a generalization of both the minimax property and the Bayes property, in the sense that the former property yields optimality in the case of a complete lack of prior information, whereas the latter coincides with the optimality in the case of complete prior information. A characterization of the sufficiency of a sub-σ-field $\mathcal{B}$ of a σ-field $\mathcal{A}$ of the parameter space is developed from this point of view. The sufficiency of $\mathcal{B}$ is defined as the property that a prior distribution on $\mathcal{B}$ induces the same optimal procedure as a prior distribution on $\mathcal{A}$. In the case of testing hypotheses, there is shown a connection of this concept with that of the parametric sufficiency due to E. W. Barankin [1].

2. Introduction

For some time there have existed characterizations of the sufficiency of a statistic (or a σ-field in a sample space) from the standpoint of decision functions (see [2], [3], [4], and [5]). According to these characterizations, a statistic $t(x)$ is sufficient if and only if in a certain statistical problem the risk by a decision procedure through the observation of the sample x is not increased at all by the restriction to the observation of the statistic $t(x)$. We shall attempt here to give a parallel discussion in the case of parametric sufficiency, a concept introduced by Barankin [1]. A function $u(\theta)$ on a parameter space Θ is called a sufficient parameter if for any measurable set A the probability $P_\theta(A)$ of occurrence of the observed sample x in A when θ is the true parameter is a function of $u(\theta)$. Looking at "the function on the parameter space" more closely, we understand that this idea represents an amount of prior information. Let us consider this problem by example. Suppose a statistician is informed of nothing but the prior

Based on research supported by the National Science Foundation.

probabilities of two parts, ω and ω^c, of Θ before any statistical experiment takes place. The prior information given to the statistician could be considered as a function $u(\theta) = 0$ on ω; $u(\theta) = 1$ on ω^c and a probability distribution on $\{0, 1\}$. Thus a pair of a parametric σ-field $\mathcal{B}$ and a probability distribution on $\mathcal{B}$ is considered to be a kind of representation of prior information.

Suppose that a statistician is supplied with a partial prior information $\{\mathcal{F}, \xi\}$, where $\mathcal{F}$ is a σ-field generated by a finite disjoint partition $\Theta = \bigcup_{i=1}^{k} F_i$ of Θ and ξ is a probability distribution $\xi(F_1), \cdots, \xi(F_k)$. It seems to be reasonable that he will choose, as an optimal procedure in this situation, the procedure $\delta = \delta*$ (if it exists) which minimizes

$$\sum_{i=1}^{k} (\sup_{\theta \in F_i} r(\theta, \delta))\xi(F_i), \tag{2.1}$$

where $r(\theta, \delta)$ is a risk function of a procedure δ when θ is a true value. Such an optimality is a generalization of both the minimax property and the Bayes property.

In section 3 we give a definition of the *mean-max risk* which is a generalization of the formula (2.1), and we also give a useful inequality. In section 4 we define the optimality of procedures with respect to a partial prior information. In section 5 we give a definition of the sufficiency of a sub-σ-field. This section also contains an important theorem on the measurability of the risk function of the optimal procedure. In section 6 we restrict ourselves to the case of testing hypotheses, and give the main theorem that under some conditions a sub-σ-field $\mathcal{B}$ is sufficient if and only if the distribution of the sample x is $\mathcal{B}$-measurable for any fixed event A, that is, the sufficiency in our sense is equivalent to that in Barankin's sense. In the last section, we give some miscellaneous remarks.

3. Mean-max risk of a procedure

Consider a statistical game $(\Theta, \mathcal{D}, r)$, where Θ is the space of the parameter θ, and $\mathcal{D}$ is the space of procedures δ. The number $r(\theta, \delta)$ is a risk imposed on the statistician when he adopts a procedure δ, and θ is the true value. We shall associate with Θ a fixed σ-field $\mathcal{A}$ of subsets of Θ.

ASSUMPTION. *The risk $r(\theta, \delta)$ is a nonnegative function, and for each fixed δ it is $\mathcal{A}$-measurable and bounded in θ.*

Consider a sub-σ-field $\mathcal{B}$ of $\mathcal{A}$ and a prior distribution ξ defined on $\mathcal{A}$. The pair $(\mathcal{B}, \xi)$ is called a *partial prior information.* By this terminology we mean that the statistician will be informed of only the value of ξ on $\mathcal{B}$ before the experimental results are observed, so that he can use this information for the choice of procedures. For example, suppose that the statistician knows the complete symmetry of a die and by using this die he is going to allocate 6 different plants to 6 plots. In this case he knows that the chance of all allocations of the plants to the plots are the same. So he has a partial prior information $(1/6!, \cdots, 1/6!)$ for the 6! permutation of the allocation (or 6! parts of the parameter space).

DEFINITION. *For a sub-σ-field* $\mathcal{B}$ *of* $\mathcal{A}$ *and a prior probability measure* ξ *on* $\mathcal{A}$, *the* mean-max risk *is defined as*

$$r(\mathcal{B}, \xi, \delta) = \inf_{\mathcal{F}\subset\mathcal{B}} \sum_{i=1}^{k} (\sup_{\theta\in F_i} r(\theta, \delta))\cdot\xi(F_i) \tag{3.1}$$

where $\mathcal{F}$ *is a sub-σ-field generated by a finite* $\mathcal{B}$-*measurable disjoint partition* $\{F_1, F_2, \cdots, F_k\}$, $\bigcup_{j=1}^{k} F_j = \Theta$, $F_j \in \mathcal{B}$, $F_i \cap F_j = \varnothing$, $(i \neq j)$, *of* Θ.

According to Saks' definition [6] of the integral, we have

$$r(\mathcal{B}, \xi, \delta) = \int r(\theta, \delta)\xi(d\theta), \tag{3.2}$$

when $r(\theta, \delta)$ is $\mathcal{B}$-measurable on Θ. Hence, it always holds that

$$r(\mathcal{A}, \xi, \delta) = \int r(\theta, \delta)\xi(d\theta). \tag{3.3}$$

It follows directly from the definition of the mean-max risk that if $\mathcal{C}$ is a sub-σ-field of $\mathcal{B}$, then

$$r(\mathcal{C}, \xi, \delta) \geq r(\mathcal{B}, \xi, \delta) \tag{3.4}$$

for every ξ and δ.

We shall denote by $E_\xi[f(\cdot)|\mathcal{B}]$ the conditional expectation given $\mathcal{B}$ of a bounded $\mathcal{A}$-measurable function $f(\theta)$ of θ with respect to a prior distribution ξ.

LEMMA 1. *The following inequality holds:*

$$\tfrac{1}{2}\int |r(\theta, \delta) - E_\xi[r(\cdot, \delta)|\mathcal{B}]|\xi(d\theta) \leq r(\mathcal{B}, \xi, \delta) - \int r(\theta, \delta)\xi(d\theta). \tag{3.5}$$

PROOF. Since $\int_B r(\theta, \delta)\xi(d\theta) = \int_B E_\xi[r(\cdot, \delta)|\mathcal{B}]\xi(d\theta)$ for $B \in \mathcal{B}$, we have

$$\begin{aligned}\tfrac{1}{2}\int_B &|r(\theta, \delta) - E_\xi[r(\cdot, \delta)|\mathcal{B}]|\xi(d\theta)\\ &= \int_{B_+} (r(\theta, \delta) - E_\xi[r(\cdot, \delta)|\mathcal{B}])\xi(d\theta),\end{aligned} \tag{3.6}$$

where $B_+ = \{\theta: r(\theta, \delta) \geq E_\xi[r(\cdot, \delta)|\mathcal{B}]\} \cap B$. The fact that $\sup_{\theta\in B} r(\theta, \delta) \geq E_\xi[r(\cdot, \delta)|\mathcal{B}]$, ξ-almost everywhere on B, implies the following inequality for every $B \in \mathcal{B}$:

$$\begin{aligned}\int_{B_+} &(r(\theta, \delta) - E_\xi[r(\cdot, \delta)|\mathcal{B}])\xi(d\theta)\\ &\leq \int_{B_+} (\sup_{\theta\in B} r(\theta, \delta) - E_\xi[r(\cdot, \delta)|\mathcal{B}])\xi(d\theta)\\ &\leq \int_B (\sup_{\theta\in B} r(\theta, \delta) - E_\xi[r(\cdot, \delta)|\mathcal{B}])\xi(d\theta).\end{aligned} \tag{3.7}$$

Hence, combining (3.6) with (3.7), we have

$$\begin{aligned}\tfrac{1}{2}\int_B &|r(\theta, \delta) - E_\xi[r(\cdot, \delta)|\mathcal{B}]|\xi(d\theta)\\ &\leq (\sup_{\theta\in B} r(\theta, \delta))\xi(B) - \int_B r(\theta, \delta)\xi(d\theta).\end{aligned} \tag{3.8}$$

Let $\mathcal{F}$ be a sub-σ-field of $\mathcal{B}$ generated by a finite $\mathcal{B}$-measurable disjoint partition

$\{B_1, B_2, \cdots, B_k\}$ of Θ. Since (3.8) holds for every B_i, we have, by substituting B_i for B and adding both sides of (3.8),

$$\frac{1}{2}\int_{\Theta}|r(\theta, \delta) - E_{\xi}[r(\cdot, \delta)|\mathcal{B}]|\xi(d\theta) \leq r(\mathcal{F}, \xi, \delta) - \int_{\Theta} r(\theta, \delta)\xi(d\theta). \tag{3.9}$$

This holds for every sub-σ-field $\mathcal{F}$ generated by a finite $\mathcal{B}$-measurable disjoint partition. Taking the infimum of the right side of (3.9), we have the required inequality.

LEMMA 2. *The function* $r(\theta, \delta)$ *is* $\mathcal{B}$*-measurable except for a set of* ξ*-measure zero if* $r(\mathcal{B}, \xi, \delta) = \int r(\theta, \delta)\xi(d\theta)$.

PROOF. The proof is clear from lemma 1.

4. The optimality with respect to partial prior information

DEFINITION. *Write* $R(\mathcal{B}, \xi) = \inf_{\delta \in \mathcal{D}} r(\mathcal{B}, \xi, \delta)$. *A procedure* $\delta^* \in \mathcal{D}$ *is called* optimal *with respect to a prior information* $(\mathcal{B}, \xi)$, *or simply* $(\mathcal{B}, \xi)$-optimal, *if* δ^* *satisfies* $r(\mathcal{B}, \xi, \delta^*) = R(\mathcal{B}, \xi)$.

This concept of optimality is similar to the modified minimax property defined by Wesler [7] from the slicing principle point of view. Let $\mathcal{O}$ be a sub-σ-field of $\mathcal{A}$ which consists only of the whole space Θ and the empty set. Clearly, optimal procedures with respect to $(\mathcal{O}, \xi)$ and $(\mathcal{A}, \xi)$ correspond to minimax and ξ-Bayes procedures, respectively.

It is quite reasonable that if two probability measures ξ and η on $\mathcal{A}$ coincide with each other on $\mathcal{B}$, then $r(\mathcal{B}, \xi, \delta) = r(\mathcal{B}, \eta, \delta)$. This property of the mean-max risk implies that the optimality with respect to $(\mathcal{B}, \xi)$ depends only on the marginal distribution of ξ on $\mathcal{B}$. In other words, the optimality with respect to $(\mathcal{B}, \xi)$ does not depend on the conditional probability measure of ξ, given $\mathcal{B}$. For instance, the minimax procedure does not depend on any prior distribution.

5. Definition of parametric sufficiency

DEFINITION. *A sub-σ-field* $\mathcal{B}$ *of* $\mathcal{A}$ *is said to be* parametric ξ-sufficient *with respect to* $(\Theta, \mathcal{A}, \mathcal{D}, r)$ (*for the sake of brevity we shall simply call* $\mathcal{B}$ *a* ξ-sufficient σ-field *if no confusion occurs*) *if* $R(\mathcal{B}, \xi) = R(\mathcal{A}, \xi)$. *And if* $\mathcal{B}$ *is a* ξ*-sufficient* σ*-field for every prior probability measure* ξ *on* $(\Theta, \mathcal{A})$, $\mathcal{B}$ *is said to be* sufficient *with respect to* $(\Theta, \mathcal{A}, \mathcal{D}, r)$.

It is a direct implication from the definition that if $\mathcal{C}$ is a sub-σ-field of a sub-σ-field $\mathcal{B}$ of $\mathcal{A}$ and $\mathcal{C}$ is a ξ-sufficient sub-σ-field of $\mathcal{A}$, then $\mathcal{B}$ is also a ξ-sufficient sub-σ-field of $\mathcal{A}$.

Concepts analogous to ξ-sufficiency have appeared implicitly in some previous papers. One such concept is that of the least favorable distribution: in a strictly determined statistical game, the ξ-sufficiency of the sub-σ-field $\mathcal{O}$ of $\mathcal{A}$ is equivalent to the fact that ξ is least favorable. Another example of this appeared in Blyth's paper [8] and Hodges-Lehmann's paper [9]. They considered statistical problems with two risk functions. According to them, if a procedure δ_0 minimizes an average risk $\alpha_1 \int r_1(\theta, \delta)\, d\xi_1 + \alpha_2 \int r_2(\theta, \delta)\, d\xi_2$ for some $\alpha_1 > 0$ and $\alpha_2 > 0$ and if

$$\int r_2(\theta, \delta_0)\, d\xi_2 = \sup_{\theta \in \Theta} r_2(\theta, \delta_0), \tag{5.1}$$

then δ_0 is a Bayes solution relative to ξ_1 (with respect to the risk $r_1(\theta, \delta)$) within the class of δ's for which $\sup_{\theta \in \Theta} r_2(\theta, \delta) \leq \sup_{\theta \in \Theta} r_2(\theta, \delta_0)$. To compare this result with our definition of ξ-sufficiency, we introduce a new parameter space $\Theta^* = \Theta \times \{1, 2\}$ and a risk function $r^*(\theta^*, \delta) = r^*((\theta, i), \delta) = r_i(\theta, \delta)$ on Θ^*, $i = 1$ and 2. Then, regarding $\xi = (\alpha_1, \alpha_2, \xi_1, \xi_2)$ as a prior distribution on Θ^*, the condition (5.1) will correspond to the ξ-sufficiency of the sub-σ-field {the empty set, $\Theta \times \{2\}$, (all measurable sets of Θ) $\times \{1\}$, Θ^*}. A similar consideration will be effective for the minimax procedure within a restricted class and for more general cases.

The following lemma is stated for the purpose of later use.

LEMMA. *Let ξ be a prior probability measure on $\mathcal{A}$, and $\mathcal{B}$ a ξ-sufficient sub-σ-field of $\mathcal{A}$ with respect to $(\Theta, \mathcal{A}, \mathcal{D}, r)$. Let ω be a $\mathcal{B}$-measurable subset of Θ, $1 > \xi(\omega) > 0$, and $s(\theta)$ an $\mathcal{A}$-measurable function on ω such that $0 \leq s(\theta) \leq 1$ and*

$$E_\xi[s(\theta) | \mathcal{B}] = \text{constant } c(\neq 0, 1), \qquad \xi\text{-a.e. on } \omega. \tag{5.2}$$

We shall write

$$\begin{aligned} r_1(\delta) &= \frac{1}{1 - \xi(\omega)} \int_{\Theta - \omega} r(\theta, \delta) \xi(d\theta), \\ r_2(\delta) &= \frac{1}{c\xi(\omega)} \int_\omega r(\theta, \delta) s(\theta) \xi(d\theta), \\ r_3(\delta) &= \frac{1}{(1 - c)\xi(\omega)} \int_\omega r(\theta, \delta)(1 - s(\theta)) \xi(d\theta). \end{aligned} \tag{5.3}$$

Let $\Theta^* = \{1, 2, 3\}$, $\mathcal{A}^* =$ the σ-field of all subsets of Θ^*, $\mathcal{D}^* = \mathcal{D}$, $r^*(i, \delta) = r_i(\delta)$, $\mathcal{B}^* =$ {empty set, Θ^*, $\{1\}$, $\{2, 3\}$} and $\xi^*(1) = 1 - \xi(\omega)$, $\xi^*(2) = c\xi(\omega)$, $\xi^*(3) = (1 - c\xi(\omega)$. Then $\mathcal{B}^*$ is ξ^*-sufficient with respect to $(\Theta^*, \mathcal{A}^*, \mathcal{D}^*, r^*)$.

PROOF. For every disjoint finite $\mathcal{B}$-measurable partition $\{F_1, \cdots, F_k\}$ of ω, we have

$$\begin{aligned} &\sum_{i=1}^k (\sup_{\theta \in F_i} r(\theta, \delta)) \xi(F_i) \\ &\quad \geq \sum_{i=1}^k \xi(F_i) \max \left\{ \int_{F_i} r(\theta, \delta) s(\theta) \xi(d\theta) \Big/ \int_{F_i} s(\theta) \xi(d\theta), \right. \\ &\qquad\qquad \left. \int_{F_i} r(\theta, \delta)(1 - s(\theta)) \xi(d\theta) \Big/ \int_{F_i} (1 - s(\theta)) \xi(d\theta) \right\} \end{aligned} \tag{5.4}$$

$$\geq \max\left\{\sum_{i=1}^{k} \xi(F_i) \int_{F_i} r(\theta, \delta)s(\theta)\xi(d\theta)/c\xi(F_i),\right.$$

$$\left.\sum_{i=1}^{k} \xi(F_i) \int_{F_i} r(\theta, \delta)(1 - s(\theta))\xi(d\theta)/(1 - c)\xi(F_i)\right\}$$

$$= \max\left\{\frac{1}{c}\int_{\omega} r(\theta, \delta)s(\theta)\xi(d\theta), \frac{1}{1 - c}\int_{\omega} r(\theta, \delta)(1 - s(\theta))\xi(d\theta)\right\}$$

$$= \xi(\omega) \times \max\{r_2(\delta), r_3(\delta)\}.$$

Since $r^*(\mathcal{B}^*, \xi^*, \delta) = \xi^*(1)r_1(\delta) + \xi(\omega) \max\{r_2(\delta), r_3(\delta)\}$ and

$$r(\mathcal{B}, \xi, \delta) = \inf_{\mathfrak{F}'} \sum_{j=1}^{k'} (\sup_{\theta \in F'_j} r(\theta, \delta))\xi(F'_j) + \inf_{\mathfrak{F}''} \sum_{i=1}^{k''} (\sup_{\theta \in F''_i} r(\theta, \delta))\xi(F''_i) \tag{5.5}$$

for finite partitions $\mathfrak{F}'$ of ω and $\mathfrak{F}''$ of $\Theta - \omega$, we have $r(\mathcal{B}, \xi, \delta) \geq r^*(\mathcal{B}^*, \xi^*, \delta)$. Since $r^*(\mathcal{A}^*, \xi^*, \delta) = \xi^*(1)r_1(\delta) + \xi^*(2)r_2(\delta) + \xi^*(3)r_3(\delta) = \int r(\theta, \delta)\xi(d\theta) = r(\mathcal{A}, \xi, \delta)$, we have $r(\mathcal{B}, \xi, \delta) \geq r^*(\mathcal{B}^*, \xi^*, \delta) \geq r^*(\mathcal{A}^*, \xi^*, \delta) = r(\mathcal{A}, \xi, \delta)$. From this inequality it is clear that the ξ-sufficiency of $\mathcal{B}$ in $\mathcal{A}$ implies the ξ^*-sufficiency of $\mathcal{B}^*$ with respect to $(\Theta^*, \mathcal{A}^*, \mathfrak{D}^*, r^*)$.

The following diagram is instructive for relations among the concepts of sufficiency and optimality:

$$\begin{array}{ccccc} & & (A) & & \\ & r(\mathcal{A}, \xi, \delta^*) & \geq & R(\mathcal{A}, \xi) & \\ (C) & \wedge\!\!| & & \wedge\!\!| & (B). \\ & r(\mathcal{B}, \xi, \delta^*) & \geq & R(\mathcal{B}, \xi) & \\ & & (D) & & \end{array} \tag{5.6}$$

In this diagram the equality symbols show us that:

(i) on (A), δ^* is a ξ-Bayes solution,
(ii) on (B), $\mathcal{B}$ is ξ-sufficient,
(iii) on (C), $r(\theta, \delta^*)$ is $\mathcal{B}$-measurable, ξ-a.e.,
(iv) on (D), δ^* is $(\mathcal{B}, \xi)$-optimal.

From these facts we have theorem 1.

THEOREM 1. *Let* $(\Theta, \mathcal{A}, \mathfrak{D}, r)$ *be a statistical problem. Suppose* δ^* *is a procedure in* $\mathfrak{D}$ *and* $\mathcal{B}$ *a sub-*σ*-field of* $\mathcal{A}$.

(i) *If* $\mathcal{B}$ *is* ξ*-sufficient and* δ^* *is* $(\mathcal{B}, \xi)$*-optimal, then* $r(\theta, \delta^*)$ *is* $\mathcal{B}$*-measurable,* ξ*-a.e., and* δ^* *is a* ξ*-Bayes solution.*

(ii) *If* $r(\theta, \delta^*)$ *is* $\mathcal{B}$*-measurable and* δ^* *is a* ξ*-Bayes solution, then* $\mathcal{B}$ *is* ξ*-sufficient and* δ^* *is* $(\mathcal{B}, \xi)$*-optimal.*

(iii) *If* $\mathcal{B}$ *is a* ξ*-complete sub-*σ*-field (that is, all sets of* ξ*-measure zero in* $\mathcal{A}$ *belong to* $\mathcal{B}$*), then* $\mathcal{B}$ *is* ξ*-sufficient and* δ^* *is* $(\mathcal{B}, \xi)$*-optimal if and only if* $r(\theta, \delta^*)$ *is* $\mathcal{B}$*-measurable and* δ^* *is a* ξ*-Bayes solution.*

As a special case of theorem 1, we shall consider a strictly determined statistical game and put $\mathcal{B} = \mathcal{O}$. Then we obtain the following statement: (i) If ξ is least favorable and δ^* is minimax, then $r(\theta, \delta^*)$ is constant, ξ-a.e., and δ^* is a ξ-Bayes procedure. (ii) If $r(\theta, \delta)$ is constant and δ^* is a ξ-Bayes procedure, then ξ is least favorable and δ^* is minimax (cf. [10], theorems 3.9 and 3.10).

The next theorem is more interesting.

THEOREM 2. *Suppose $\mathcal{B}$ is a sub-σ-field of $\mathcal{A}$ and there exists a ξ-Bayes procedure δ^* in $\mathcal{D}$. If $\mathcal{B}$ is sufficient with respect to $(\Theta, \mathcal{A}, \mathcal{D}, r)$, then $r(\theta, \delta^*)$ is $\mathcal{B}$-measurable, ξ-a.e.,*

PROOF. Let

$$\begin{aligned} \omega_1 &= \{\theta : r(\theta, \delta^*) > E_\xi[r(\cdot, \delta^*)|\mathcal{B}]\}, \\ \omega_2 &= \{\theta : r(\theta, \delta^*) = E_\xi[r(\cdot, \delta^*)|\mathcal{B}]\}, \\ \omega_3 &= \{\theta : r(\theta, \delta^*) < E_\xi[r(\cdot, \delta^*)|\mathcal{B}]\}, \end{aligned} \tag{5.7}$$

and

$$\omega = \{\theta : \xi(\omega_2 \cup \omega_3|\mathcal{B}) = 0\} (\in \mathcal{B}). \tag{5.8}$$

Since

$$\xi(\omega \cap (\omega_2 \cup \omega_3)) = \int_\omega \xi(\omega_2 \cup \omega_3|\mathcal{B})\xi(d\theta) = 0, \tag{5.9}$$

we have

$$\begin{aligned} &\int_{\omega \cap \omega_1} \{r(\theta, \delta^*) - E_\xi[r(\cdot, \delta^*)|\mathcal{B}]\}\xi(d\theta) \\ &= \int_\omega \{r(\theta, \delta^*) - E_\xi[r(\cdot, \delta^*)|\mathcal{B}]\}\xi(d\theta) - \int_{\omega \cap (\omega_2 \cup \omega_3)} \{r - E_\xi[r|\mathcal{B}]\}\xi(d\theta) \\ &= \int_\omega \{r(\theta, \delta^*) - E_\xi[r(\cdot, \delta^*)|\mathcal{B}]\}\xi(d\theta) = 0. \end{aligned} \tag{5.10}$$

Therefore $\xi(\omega \cap \omega_1) = 0$, and so $\xi(\omega) = \xi(\omega \cap \omega_1) + \xi(\omega \cap (\omega_2 \cup \omega_3)) = \xi(\omega \cap \omega_1) = 0$, which means that $\xi(\omega_2 \cup \omega_3|\mathcal{B}) > 0$, ξ-a.e.

Take the indicator function $\chi(\theta)$ of the set $\omega_2 \cup \omega_3$ and consider a probability measure $\eta(\sigma)$ on $\mathcal{A}$:

$$\eta(\sigma) = \int_\sigma \xi(\omega_2 \cup \omega_3|\mathcal{B})^{-1}\chi(\theta)\xi(d\theta), \qquad \sigma \in \mathcal{A}. \tag{5.11}$$

For any $\mathcal{B}$-measurable set τ we have

$$\begin{aligned} \eta(\tau) &= \int_\tau \xi(\omega_2 \cup \omega_3|\mathcal{B})^{-1}\chi(\theta)\xi(d\theta) \\ &= \int_\tau \xi(\omega_2 \cup \omega_3|\mathcal{B})^{-1}E_\xi[\chi|\mathcal{B}]\xi(d\theta) \\ &= \int_\tau \xi(\omega_2 \cup \omega_3|\mathcal{B})^{-1}\xi(\omega_2 \cup \omega_3|\mathcal{B})\xi(d\theta) \\ &= \xi(\tau). \end{aligned} \tag{5.12}$$

Therefore, two measures ξ and η coincide with each other on $\mathcal{B}$, and so we have

$$r(\mathcal{B}, \xi, \delta) = r(\mathcal{B}, \eta, \delta) \qquad \text{for every} \quad \delta \in \mathcal{D}. \tag{5.13}$$

On the other hand,

$$(5.14)\qquad r(\mathcal{A}, \eta, \delta) = \int r(\theta, \delta)\, d\eta$$
$$= \int r(\theta, \delta)\chi(\theta)\xi(\omega_2 \cup \omega_3|\mathcal{B})^{-1}\xi(d\theta)$$
$$= \int_{\omega_2 \cup \omega_3} r(\theta, \delta)\xi(\omega_2 \cup \omega_3|\mathcal{B})^{-1}\xi(d\theta).$$

By the definition of ω_2 and ω_3, we have

$$(5.15)\qquad \int_{\omega_2 \cup \omega_3} r(\theta, \delta^*)\xi(\omega_2 \cup \omega_3|\mathcal{B})^{-1}\xi(d\theta)$$
$$\leq \int_{\omega_2 \cup \omega_3} E_\xi[r(\theta, \delta^*)|\mathcal{B}]\xi(\omega_2 \cup \omega_3|\mathcal{B})^{-1}\xi(d\theta)$$
$$= \int E_\xi[r(\theta, \delta^*)|\mathcal{B}]\xi(\omega_2 \cup \omega_3|\mathcal{B})^{-1}\chi(\theta)\xi(d\theta)$$
$$= \int E_\xi[r(\theta, \delta^*)|\mathcal{B}]\, d\eta$$
$$= \int E_\xi[r(\theta, \delta^*)|\mathcal{B}]\, d\xi = r(\mathcal{A}, \xi, \delta^*),$$

where the equality sign in the second row holds if and only if $\xi(\omega_3) = 0$. Thus we have

$$(5.16)\qquad r(\mathcal{A}, \eta, \delta^*) \leq r(\mathcal{A}, \xi, \delta^*),$$

where the equality sign holds if and only if $\xi(\omega_3) = 0$. Here the reader should notice that $\xi(\omega_3) = 0$ is equivalent to $\xi(\omega_1) = 0$.

Since $\mathcal{B}$ is sufficient by assumption, we have

$$(5.17)\qquad \begin{aligned} R(\mathcal{B}, \xi) &= R(\mathcal{A}, \xi), \\ R(\mathcal{B}, \eta) &= R(\mathcal{A}, \eta), \end{aligned}$$

and from (5.13) we also have $R(\mathcal{B}, \xi) = R(\mathcal{B}, \eta)$. Hence $R(\mathcal{A}, \xi) = R(\mathcal{A}, \eta)$. Since δ^* is ξ-Bayes in $\mathcal{D}$, $r(\mathcal{A}, \xi, \delta^*) = R(\mathcal{A}, \xi)$, and hence $r(\mathcal{A}, \xi, \delta^*) = R(\mathcal{A}, \eta) \leq r(\mathcal{A}, \eta, \delta^*)$. Therefore, it follows from (5.16) and the above inequality that $r(\mathcal{A}, \xi, \delta^*) = r(\mathcal{A}, \eta, \delta^*)$. This shows that $\xi(\omega_1) = \xi(\omega_3) = 0$, that is,

$$(5.18)\qquad r(\theta, \delta^*) = E_\xi[r(\theta, \delta^*)|\mathcal{B}], \qquad \xi\text{-a.e.}$$

COROLLARY. *If $\mathcal{B}$ is sufficient with respect to $(\Theta, \mathcal{A}, \mathcal{D}, r)$, and is ξ-complete in $\mathcal{A}$, then the ξ-Bayes property of a procedure in $\mathcal{D}$ is equivalent to $(\mathcal{B}, \xi)$-optimality.*

PROOF. The implication of $(\mathcal{B}, \xi)$-optimality from ξ-Bayes property is easily seen from theorem 2, whereas the inverse implication follows from theorem 1.

6. The case of testing hypotheses

Let $(X, \mathbf{A})$ be a measurable space, with the sample space X having an associated σ-field $\mathbf{A}$. And let the parameter space Θ, having an associated σ-field $\mathcal{A}$, be a collection of θ's, to each of which corresponds a probability measure P_θ on

$(X, \mathbf{A})$ in such a manner that, for any subset $A \in \mathbf{A}$ of X, $P_\theta(A)$ is an $\mathcal{A}$-measurable function on Θ. Let ω be an $\mathcal{A}$-measurable, nonempty and true subset of Θ' and then consider a problem of testing a hypothesis "$\theta \in \omega$" against the alternative "$\theta \notin \omega$." By Φ we shall denote the set of all test functions φ, namely the set of all $\mathbf{A}$-measurable functions φ on X satisfying $0 \leq \varphi(x) \leq 1$. The problem described above will be denoted by $(X, \mathbf{A}, \Theta, \mathcal{A}, P_\theta, \omega)$. Here the risk function $r(\theta, \varphi)$ of φ is automatically understood as

$$r(\theta, \varphi) = \begin{cases} E_\theta[\varphi] & \text{for} \quad \theta \in \omega, \\ 1 - E_\theta[\varphi] & \text{for} \quad \theta \notin \omega, \end{cases} \tag{6.1}$$

where E_θ stands for the average operator with respect to the probability distribution P_θ on $(X, \mathbf{A})$. As is easily seen, for any prior probability measure ξ on $(\Theta, \mathcal{A})$ there exists at least one ξ-Bayes test $\varphi*$.

Let $\mathcal{B}$ be a sub-σ-field of $\mathcal{A}$. Obviously $P_\theta(A)$ is $\mathcal{B}$-measurable for every $A \in \mathbf{A}$ if and only if $E_\theta[\varphi]$ is $\mathcal{B}$-measurable for every $\varphi \in \Phi$. We shall discuss below the relation between the $\mathcal{B}$-measurability of $P_\theta(A)$ and the sufficiency of $\mathcal{B}$ with respect to $(X, \mathbf{A}, \Theta, \mathcal{A}, P_\theta, \omega)$, provided that ω is $\mathcal{B}$-measurable.

First we shall observe a corollary of theorem 2.

COROLLARY. *If $\mathcal{B}$ is a sufficient sub-σ-field of $\mathcal{A}$ with respect to $(X, A, \Theta, \mathcal{A}, P_\theta, \omega)$ and ω is $\mathcal{B}$-measurable, then, for any ξ-Bayes test $\varphi*$, $E_\theta[\varphi*]$ is $\mathcal{B}$-measurable, ξ-a.e., and φ^* is $(\mathcal{B}, \xi)$-optimal whenever $\mathcal{B}$ is ξ-complete in $\mathcal{A}$.*

As preparation for obtaining the main theorem, we shall give some lemmas without proof, concerning the problem of testing simple hypotheses. In these lemmas we shall use notations Q_0, Q_1, Q_2, and so on, for measures defined on $(X, \mathbf{A})$, and E_i for the average operation with respect to Q_i $(i = 0, 1, \cdots)$. And moreover, by $(Q_i{:}Q_j)$ we mean the problem of testing a simple hypothesis Q_i against a simple alternative Q_j.

LEMMA 1. *For the problem $(Q_1{:}Q_2)$ there is a system $\{\varphi_\alpha\}_{0 \leq \alpha \leq 1}$ of most powerful test functions for the hypothesis Q_1 against Q_2 such that $E_1[\varphi_\alpha] = \alpha$ and $\varphi_\alpha(x) \leq \varphi_{\alpha'}(x)$ on X if $\alpha < \alpha'$. Moreover, for any such system $\{\varphi_\alpha\}$ we can choose a nonnegative function $k(\alpha) \leq \infty$ on $[0, 1]$ such that the inequalities*

$$\begin{aligned} k(\alpha)E_1[(1 - \varphi_\alpha)f] &\geq E_2[(1 - \varphi_\alpha)f], \\ k(\alpha)E_1[\varphi_\alpha g] &\leq E_2[\varphi_\alpha g], \end{aligned} \tag{6.2}$$

hold for all nonnegative $\mathbf{A}$-measurable functions f and g.

LEMMA 2. *Let $\{\varphi_\alpha\}$ and $\{\psi_\alpha\}$ be systems of the most powerful test functions for the problems $\{Q_1{:}Q_2\}$ and $\{Q_1{:}Q_3\}$, respectively, which are the systems defined in lemma 1. If, for any $\beta \in [0, 1]$, there are nonnegative numbers $k(\beta) \leq \infty$ and $\alpha \in (0, 1]$ such that $k(\beta)$ satisfies the same condition for $\{\psi_\beta\}$ as does $k(\alpha)$ for $\{\varphi_\alpha\}$ in lemma 1 and*

$$\begin{aligned} k(\beta)E_1[(1 - \varphi_\alpha)\psi_\beta] &= E_2[(1 - \varphi_\alpha)\psi_\beta], \\ k(\beta)E_1[\varphi_\alpha(1 - \psi_\beta)] &= E_2[\varphi_\alpha(1 - \psi_\beta)], \end{aligned} \tag{6.3}$$

then $\{\varphi_\alpha\}$ is, in turn, a system of the most powerful test functions for the problem $\{Q_1{:}Q_3\}$.

LEMMA 3. *With the same notation as in lemma 2, we suppose that* $Q_2 + Q_3$ *is absolutely continuous with respect to* Q_1. *Then it is a necessary and sufficient condition for* $Q_2 = Q_3$ *that there be a set* $\{\varphi_\alpha\}_{0 \leq \alpha \leq 1}$ *of* **A**-*measurable functions on* X *such that* $\{\varphi_\alpha\}_{0 \leq \alpha \leq 1}$ *is a system of the most powerful test functions for* $\{Q_1:Q_2\}$ *as well as for* $\{Q_1:Q_3\}$, *and* $E_2[\varphi_\alpha] = E_3[\varphi_\alpha]$ *holds for all* $\alpha \in [0, 1]$.

THEOREM 3. *Denote by* T *a problem* $(X, \mathbf{A}, \Theta, \mathcal{A}, P_\theta, \omega)$, *where* $\{P_\theta:\theta \in \Theta\}$ *is mutually absolutely continuous. Let* $\mathcal{B}$ *be a sub-*σ*-field of* $\mathcal{A}$ *and* ω *a* $\mathcal{B}$*-measurable nonempty and true subset of* Θ.

(i) *If* $P_\theta(A)$ *is a* $\mathcal{B}$*-measurable function of* θ *for any* **A**-*measurable subset* A *of* X, *then* $\mathcal{B}$ *is sufficient with respect to* T.

(ii) *If* $\mathcal{B}$ *is sufficient with respect to* T, *then* $P_\theta(A)$ *is* $\mathcal{B}$*-measurable,* ξ*-a.e., as a function of* θ *for any fixed* **A**-*measurable subset* $A \subset X$, *and for any prior distribution* ξ *on* $(\Theta, \mathcal{A})$ *for which* $1 > \xi(\omega) > 0$.

PROOF. Assertion (i) is clear from the definitions of the mean-max risk and sufficiency of $\mathcal{B}$ and the $\mathcal{B}$-measurability of ω.

For (ii), suppose that $\mathcal{B}$ is sufficient in $\mathcal{A}$ and $P_\theta(A_0)$ is not $\mathcal{B}$-measurable, ξ-a.e., for some **A**-measurable subset A_0 of the sample space X, that is,

$$\xi\{\theta:P_\theta(A_0) \neq E_\xi[P.(A_0)|\mathcal{B}]\} > 0. \tag{6.4}$$

Without any loss of generality we may assume that

$$\xi\{\theta \in \omega:P_\theta(A_0) \neq E_\xi[P.(A_0)|\mathcal{B}]\} > 0. \tag{6.5}$$

We shall show here that it is possible to take an $\mathcal{A}$-measurable function $s(\theta)$ on ω such that $0 \leq s(\theta) \leq 1$ and

$$E_\xi[s(\theta)|\mathcal{B}] = \tfrac{1}{2}, \qquad \xi\text{-a.e. on } \omega, \tag{6.6}$$

and

$$\int_\omega s(\theta)P_\theta(A_0)\xi(d\theta) < \tfrac{1}{2}\int_\omega P_\theta(A_0)\xi(d\theta). \tag{6.7}$$

For any $\mathcal{B}$-measurable nonnegative function $k(\theta)$ on ω, let us write

$$S_k = \{\theta \in \omega:P_\theta(A_0) < k(\theta)E_\xi[P.(A_0)|\mathcal{B}]\} \tag{6.8}$$

and

$$T_k = \{\theta \in \omega:P_\theta(A_0) > k(\theta)E_\xi[P.(A_0)|\mathcal{B}]\}. \tag{6.9}$$

Denote by $\mathcal{K}$ the collection of all $k(\theta)$ such that $\xi(S_k|\mathcal{B}) \leq \frac{1}{2}$ holds ξ-a.e. We can easily see that $\mathcal{K}$ is not empty, because $k \equiv 0$ belongs to $\mathcal{K}$. Since for any k_1 and k_2 in $\mathcal{K}$

$$\xi(S_{k_1 \vee k_2}|\mathcal{B}) = \max\{\xi(S_{k_1}|\mathcal{B}), \xi(S_{k_2}|\mathcal{B})\}, \qquad \xi\text{-a.e.}, \tag{6.10}$$

we have $k_1 \vee k_2 = \max\{k_1, k_2\} \in \mathcal{K}$. Therefore we have a max k_α for any chain $k_1 \prec k_2 \prec \cdots \prec k_\alpha \prec \cdots$ of elements of $\mathcal{K}$, where the notation $k_\nu \prec k_\mu$ means that $k_\nu(\theta) \leq k_\mu(\theta)$ ξ-almost everywhere on ω and $\xi(S_{k_\mu} - S_{k_\nu}|\mathcal{B}) > 0$, ξ-a.e. By Zorn's lemma we can find a maximal element k_0 in $\mathcal{K}$ which belongs also to $\mathcal{K}$, that is,

$$\xi(S_{k_0}|\mathcal{B}) \leq \tfrac{1}{2}, \qquad \xi\text{-a.e.}, \tag{6.11}$$

and there exists no $k \in \mathcal{K}$ such that $k_0 \prec k$. Write

$$c(\theta) = \frac{\frac{1}{2} - \xi(S_{k_0}|\mathcal{B})}{\xi(\Theta - T_{k_0} - S_{k_0}|\mathcal{B})} \tag{6.12}$$

if the denominator does not equal 0, and let $c(\theta) = 0$ if the denominator is zero, and define

$$s(\theta) = \begin{cases} 1, & \text{on } S_{k_0}, \\ c(\theta), & \text{on } \Theta - S_{k_0} - T_{k_0}, \\ 0, & \text{on } T_{k_0}, \end{cases} \tag{6.13}$$

which is our desired function.

Write, for every A in $\mathbf{A}$,

$$Q_1(A) = \frac{1}{1 - \xi(\omega)} \int_{\omega^c} P_\theta(A)\xi(d\theta), \tag{6.14}$$

$$Q_2(A) = \frac{2}{\xi(\omega)} \int_{\omega} s(\theta)P_\theta(A)\xi(d\theta), \tag{6.15}$$

$$Q_3(A) = \frac{2}{\xi(\omega)} \int_{\omega} (1 - s(\theta))P_\theta(A)\xi(d\theta), \tag{6.16}$$

and

$$Q_0(A) = \tfrac{1}{2}(Q_2(A) + Q_3(A)) = \frac{1}{\xi(\omega)} \int_{\omega} P_\theta(A)\xi(d\theta). \tag{6.17}$$

These Q_0, Q_1, Q_2, and Q_3 are all probability measures on $(X, \mathbf{A})$.

Consider a problem T^* of testing a simple hypothesis Q_1 against a composite alternative $\{Q_2 \text{ or } Q_3\}$. By the lemma in section 5, the sub-σ-field $\mathcal{B}^* = \{$the empty set, $\{1\}$, $\{2, 3\}$, $\Theta^* = \{1, 2, 3\}\}$ is ξ^*-sufficient with respect to T^*, where $\xi^* = (\xi^*(1), \xi^*(2), \xi^*(3))$, $\xi^*(1) = 1 - \xi(\omega)$, $\xi^*(2) = \xi^*(3) = \frac{1}{2}\xi(\omega)$. However, the assumption (6.5) and the definition of Q_1, Q_2, Q_3 are independent of the value $\xi(\omega)$ as long as we have $0 < \xi(\omega) < 1$. From this fact it follows that the sufficiency of $\mathcal{B}$ with respect to T implies the ξ^*-sufficiency of $\mathcal{B}^*$ with respect to T^* for every ξ^* with $\xi^*(1) > 0$, $\xi^*(2) > 0$ and $\xi^*(3) > 0$. In the case where $\xi(\omega) = 0$ or 1, it is obvious that $\mathcal{B}^*$ is ξ^*-sufficient with respect to T^*. Therefore, $\mathcal{B}^*$ is sufficient with respect to T^*.

From the above argument, our theorem is reduced to the following lemma.

LEMMA 4. *Suppose that Q_1, Q_2, and Q_3 are mutually absolutely continuous. If the σ-field $\mathcal{B}^*$ defined above is sufficient with respect to the problem T^*, then Q_2 coincides with Q_3.*

PROOF. Suppose that $\mathcal{B}^*$ is sufficient and that Q_2 does not coincide with Q_3. Let $\{\varphi_\alpha\}$, $0 \leq \alpha \leq 1$, be a system of the most powerful tests of level α for the problem T_1 of testing a simple hypothesis Q_1 against a simple alternative Q_2 and satisfying the condition that $\alpha < \alpha'$ implies $\varphi_\alpha(x) \leq \varphi_{\alpha'}(x)$. We shall take another system $\{\psi_\alpha\}$, $0 \leq \alpha \leq 1$, of the most powerful tests of level α for the problem T_2 of testing a simple hypothesis Q_1 against a simple alternative Q_0 and satisfying a similar condition: $\alpha < \alpha'$ implies $\psi_\alpha(x) \leq \psi_{\alpha'}(x)$.

We shall show first that there are a $\beta \in (0, 1)$ and a $k \in (0, \infty)$ such that, for any $\alpha \in (0, 1)$, the two following inequalities hold with at least one of them being a strict inequality:

$$\begin{aligned} kE_1[(1 - \varphi_\alpha)\psi_\beta] &\leq E_0[(1 - \varphi_\alpha)\psi_\beta], \\ kE_1[\varphi_\alpha(1 - \psi_\beta)] &\geq E_0[\varphi_\alpha(1 - \psi_\beta)]. \end{aligned} \tag{6.18}$$

The existence of a k for which the above formulas hold is guaranteed by lemma 1 (no trouble for $k = 0$ or ∞ occurs, because of the absolute continuity assumption). Suppose that for every $\beta \in (0, 1)$ there is an $\alpha \in (0, 1)$ such that both of the above formulas hold with the equality signs. Then by lemma 2, we can choose $\{\psi_\alpha\}$ as $\varphi_\alpha(x) = \psi_\alpha(x)$ for all α. On the other hand, the most powerful tests $\psi_\alpha(1 > \alpha > 0)$ for $T_2 = (Q_1 \colon Q_0)$ are η^*-Bayes tests for T^*, where $\eta^* = (\eta_1^*, \eta_2^*, \eta_3^*)$, $\eta_i^* > 0$ $(i = 1, 2, 3)$. Since $\mathfrak{B}^*$ is sufficient with respect to T^*, it follows from theorem 2 that the risks at Q_2 and Q_3 are equal, and hence, $E_2[\psi_\alpha] = E_3[\psi_\alpha] = E_0[\psi_\alpha]$ for $0 < \alpha < 1$. Therefore, from lemma 3 we have $Q_2 = Q_0 = Q_3$, which contradicts our assumption.

Thus there is a $\beta \in (0, 1)$ such that for every $\alpha \in (0, 1)$

$$\begin{aligned} &E_0[\varphi_\alpha(1 - \psi_\beta)] - E_0[(1 - \varphi_\alpha)\psi_\beta] \\ &\qquad < k\{E_1[\varphi_\alpha(1 - \psi_\beta)] - E_1[(1 - \varphi_\alpha)\psi_\beta]\}. \end{aligned} \tag{6.19}$$

Therefore, we have

$$\begin{aligned} E_0[\psi_\beta] &> E_0[\varphi_\beta] - k\{E_1[\varphi_\beta] - E_1[\psi_\beta]\} \\ &= E_0[\varphi_\beta], \end{aligned} \tag{6.20}$$

and obviously,

$$E_1[\psi_\beta] = E_1[\varphi_\beta] = \beta. \tag{6.21}$$

Now we shall consider the closed convex subset

$$C = \{(E_1[\varphi], 1 - E_2[\varphi], 1 - E_3[\varphi]) \colon 0 \leq \varphi(x) \leq 1, \varphi(x) \colon \mathbf{A}\text{-measurable}\} \tag{6.22}$$

of the 3-dimensional Euclidean space, and two points $p = (E_1[\varphi_\beta], 1 - E_2[\varphi_\beta], 1 - E_3[\varphi_\beta])$ and $q = (E_1[\psi_\beta], 1 - E_2[\psi_\beta], 1 - E_3[\psi_\beta])$ in C. By (6.21), p and q have the equal first coordinates. Denote by π the plane which is orthogonal to the first coordinate axis and passes through p and q. Inequality (6.20) makes it possible to determine a pair of positive numbers η_2 and η_3 such that $\eta_2 + \eta_3 < 1$ and

$$\eta_2 E_2[\psi_\beta] + \eta_3 E_3[\psi_\beta] > \eta_2 E_2[\varphi_\beta] + \eta_3 E_3[\varphi_\beta]. \tag{6.23}$$

Let $\varphi*$ be a test function such that the point

$$p^* = (E_1[\varphi^*], 1 - E_2[\varphi^*], 1 - E_3[\varphi^*]) \tag{6.24}$$

in C is located on the plane π and p^* is a supporting point on π in the direction (η_2, η_3), that is,

$$E_1[\varphi^*] = E_1[\varphi_\beta] = \beta \tag{6.25}$$

and

$$(6.26)\qquad \eta_2 E_2[\varphi^*] + \eta_3 E_3[\varphi^*] = \max\,(\eta_2 E_2[\varphi] + \eta_3 E_3[\varphi]\colon 0 \le \varphi(x) \le 1),$$

where φ is **A**-measurable and $E_1[\varphi] = \beta$. Since there is a nonnegative number η_1 such that p^* is also a supporting point of C in the direction (η_1, η_2, η_3) in the 3-dimensional Euclidean space, we have

$$(6.27)\qquad \begin{aligned}&\eta_1 E_1[\varphi^*] + \eta_2(1 - E_2[\varphi^*]) + \eta_3(1 - E_3[\varphi^*])\\ &\quad = \min\,\{\eta_1 E_1[\varphi] + \eta_2(1 - E_2[\varphi]) + \eta_3(1 - E_3[\varphi])\colon\\ &\qquad\qquad 0 \le \varphi(x) \le 1,\ \varphi\colon \mathbf{A}\text{-measurable}\}.\end{aligned}$$

Therefore we have, by (6.21), (6.23), and (6.27),

$$(6.28)\qquad \begin{aligned}&\eta_1 E_1[\varphi*] + \eta_2(1 - E_2[\varphi*]) + \eta_3(1 - E_3[\varphi*])\\ &\quad \le \eta_1 E_1[\psi_\beta] + \eta_2(1 - E_2[\psi_\beta]) + \eta_3(1 - E_3[\psi_\beta])\\ &\quad < \eta_1 E_1[\varphi_\beta] + \eta_2(1 - E_2[\varphi_\beta]) + \eta_3(1 - E_3[\varphi_\beta]).\end{aligned}$$

Since φ_β and φ^* are Bayes tests with respect to T^* and $\mathcal{B}^*$ is sufficient, it follows from theorem 2 that

$$(6.29)\qquad E_2[\varphi^*] = E_3[\varphi^*] \quad\text{and}\quad E_2[\varphi_\beta] = E_3[\varphi_\beta].$$

From (6.28) and (6.29), it follows that

$$(6.30)\qquad (\eta_2 + \eta_3)E_2[\varphi^*] - \eta_1 E_1[\varphi^*] > (\eta_2 + \eta_3)E_2[\varphi_\beta] - \eta_1 E_1[\varphi_\beta].$$

Combining this inequality with (6.25) gives

$$(6.31)\qquad E_2[\varphi^*] > E_2[\varphi_\beta].$$

This inequality shows, with (6.25), that φ_β is not the most powerful test function of level β for the problem T_1 of testing simple hypothesis Q_1 against the alternative Q_2. This is a contradiction.

7. Remarks

(1) A functional $F_\xi[f] = \inf_{\mathfrak{F} \subset \mathcal{B}} \sum_{i=1}^{k} (\sup_{\theta \in F_i} f(\theta))$, $\mathfrak{F} = \{F_1, \cdots, F_k\}$, of an $\mathcal{A}$-measurable function $f(\theta)$ on Θ is also defined as

$$(7.1)\qquad F_\xi[f] = \inf_{\substack{u(\epsilon)\mathcal{B}\\ u \ge f}} \int u(\theta)\xi(d\theta),$$

so that $r(\mathcal{B}, \xi, \delta)$ might be regarded as an upper integral of the risk function $r(\theta, \delta)$ with respect to a sub-σ-field $\mathcal{B}$ of $\mathcal{A}$.

(2) Under certain conditions, the $\mathcal{B}$-measurability of an $\mathcal{A}$-measurable function is equivalent to the $\mathcal{B}$-measurability, ξ-a.e., for any prior distribution ξ on Θ. Therefore, in such cases, the assertion of theorem 3 is simply that $\mathcal{B}$ is sufficient if and only if $P_\theta(A)$ is $\mathcal{B}$-measurable for any **A**-measurable subset A of the sample space. For example, if $\mathcal{B}$ is induced by a statistic in the Bahadur sense (see [11]), and the induced σ-field in the range of the statistic contains every singleton, then every ξ-almost $\mathcal{B}$-measurable set for any ξ is $\mathcal{B}$-measurable.

(3) From theorem 3 we can get the following statement: under the assumption that the space $\{P_\theta\}$ of distributions is mutually absolutely continuous, the sufficiency of $\mathcal{B}$ with respect to every decision problem with a bounded $\mathcal{B} \times \mathbf{S}$ measurable loss function $L(\theta, s) \geq 0$ implies the $\mathcal{B}$-measurability of $P_\theta(A)$, ξ-a.e., for any prior measure ξ and for any set $A \in \mathbf{A}$, where $\mathbf{S}$ is a σ-field of subsets of the action space.

Inversely, if $P_\theta(A)$ is $\mathcal{B}$-measurable, then $\mathcal{B}$ is sufficient with respect to every decision problem with a bounded $\mathcal{B} \times \mathbf{S}$ measurable loss function $L(\theta, s) \geq 0$.

This kind of assertion is parallel to the characterization of the sufficiency of a statistic due to Blackwell [2] and also to Le Cam [12].

(4) It is well known that for a set S of a 2-dimensional Euclidean space there are two probability measures Q_1 and Q_2 on a measurable space $(X, \mathbf{A})$ such that $S = \{(\int \varphi(x)Q_1(dx), \int \varphi(x)Q_2(dx)) : \varphi \in \Phi\}$, if and only if (i) S is closed and convex, (ii) $(0, 0)$ and $(1, 1) \in S$, (iii) $S \subset [0, 1; 0, 1]$, and (iv) S is symmetric with respect to the point $(\frac{1}{2}, \frac{1}{2})$. For the n-dimensional space we do not know a nice necessary and sufficient condition for a convex set S to be the range of some n-dimensional vector measure $(n \geq 3)$. However, our lemma 4 gives a partial solution to this problem. Suppose that $n = 3$ and S, the range set of 3-dimensional vector measure, has only one common point with each coordinate axis, and let π be a plane parallel to the second and third coordinate axes. If every section of S by each of such a plane π is contained in the relative first quadrant, then these sections lie entirely on the plane "the second coordinate = the third coordinate," so that S collapses from three dimensions to two dimensions.

(5) As an example of a sufficient parameter, we can consider the estimable parameters in the linear statistical model

$$\mathbf{X}(n \times 1) = A(n \times k)\boldsymbol{\beta}(k \times 1) + \boldsymbol{\epsilon}(n \times 1), \tag{7.2}$$

where $\mathbf{X}$ and $\boldsymbol{\epsilon}$ are random vectors, A a known matrix, and $\boldsymbol{\beta}$ an unknown vector. Here we assume that the distribution of $\boldsymbol{\epsilon}$ is normal with mean zero-vector and covariance matrix $\sigma^2 I$, I = unit matrix, σ^2 unknown constant. In this problem, $(\boldsymbol{\beta}, \sigma^2)$ is a parameter, and σ^2 together with a system of linearly independent estimable parameters are sufficient. (This example is due to Goro Ishii).

(6) Let $\mathcal{B}$ be a sub-σ-field of $\mathcal{A}$, and $\mathbf{A}(\mathcal{B})$ the family of all $\mathbf{A}$-measurable subsets A of X for which $P_\theta(A)$ is $\mathcal{B}$-measurable. For this family $\mathbf{A}(\mathcal{B})$, analogous assertions to the family of ancillary events in Basu's paper [13] hold. If a sub-σ-field $\mathbf{B}$ of $\mathbf{A}$ is contained in $\mathbf{A}(\mathcal{B})$, then $\mathcal{B}$ is sufficient with respect to every problem of statistical decisions with sample space $(X, \mathbf{B})$. In the case where $\mathcal{B}$ is induced by a function $u(\theta)$ of the parameter θ and $\mathbf{B}$ is induced by a statistic $t(x)$, we could say that $u(\theta)$ is sufficient for the statistic $t(x)$. For example, in the model (7.2) σ^2 is a sufficient parameter for the statistic $t = X'(I - P_A)X$, where P_A is a projection operator of R^n onto the hyperplane spanned by the column vectors of the matrix A. Although t is partially sufficient for σ^2 in Fraser's sense [14] in this case, such an inverse statement is not always true. Our concept of

the parametric sufficiency of $u(\theta)$ for $t(x)$ corresponds to Basu's concept [15] of "φ-free" of $t(x)$ if $\theta = (\varphi, u(\theta))$.

I wish to thank Professor E. W. Barankin for valuable conversations on my problem.

REFERENCES

[1] E. W. BARANKIN, "Sufficient parameters: solution of the minimal dimensionality problem," *Ann. Inst. Statist. Math.*, Vol. 12 (1960), pp. 91–118.
[2] D. BLACKWELL, "Equivalent comparison of experiments," *Ann. Math. Statist.*, Vol. 24 (1953), pp. 265–272.
[3] R. R. BAHADUR, "A characterization of sufficiency," *Ann. Math. Statist.*, Vol. 26 (1955), pp. 286–292.
[4] H. KUDŌ, "Dependent experiments and sufficient statistics," *Natur. Sci. Rep. Ochanomizu Univ.*, Vol. 4 (1953), pp. 151–163.
[5] ———, "On sufficiency and completeness of statistics," *Sûgaku*, Vol. 8 (1957), pp. 129–138. (In Japanese.)
[6] S. SAKS, *Theory of the Integral*, Monografie Matematyczne, Warszawa, 1937.
[7] O. WESLER, "Invariance theory and a modified minimax principle," *Ann. Math. Statist.*, Vol. 30 (1959), pp. 1–20.
[8] COLIN R. BLYTH, "On minimax statistical decision procedures and their admissibility," *Ann. Math. Statist.*, Vol. 22 (1951), pp. 22–42.
[9] J. L. HODGES, JR. and E. L. LEHMANN, "The use of previous experience in reaching statistical decisions," *Ann. Math. Statist.*, Vol. 23 (1952), pp. 396–407.
[10] ABRAHAM WALD, *Statistical Decision Functions*, New York, Wiley, 1950.
[11] R. R. BAHADUR, "Sufficiency and statistical decision functions," *Ann. Math. Statist.*, Vol. 25 (1954), pp. 423–462.
[12] L. LE CAM, "Sufficiency and approximate sufficiency," *Ann. Math. Statist.*, Vol. 35 (1964), pp. 1419–1455.
[13] D. BASU, "The family of ancillary statistics," *Sankhyā*, Vol. 21 (1959), pp. 247–256.
[14] E. L. LEHMANN, *Testing Statistical Hypotheses*, New York, Wiley, 1959.
[15] D. BASU, "Problems relating to the existence of maximal and minimal elements in some families of statistics (subfields)," *Proceedings of the Fifth Berkeley Symposium on Mathematical Statistics and Probability*, Berkeley and Los Angeles, University of California Press, 1966, Vol. I, pp. 41–50.

ON THE ELIMINATION OF NUISANCE PARAMETERS IN STATISTICAL PROBLEMS

YU. V. LINNIK
MATHEMATICAL INSTITUTE OF THE ACADEMY OF SCIENCES, LENINGRAD

1. Introduction

Consider a family of the distributions $\mathcal{P}_\theta$ characterized by the probability densities $\ell(x, \theta)$ with respect to a dominating measure $\mu(x)$ on the σ_n-algebra of a measurable space $(\mathfrak{X}, \mathfrak{A})$ with a parameter $\theta \in \Omega$. Later on $\mathfrak{X} \subset E_n$ will be a parallelepiped of the n-dimensional Euclidean space, $\mathfrak{A}$ the Borel σ-algebra; $\Omega \subset E_s$, in general, a compact in the s-dimensional Euclidean space, $\ell(x, \theta)$ being a function continuous with respect to θ for a fixed x, and for a given θ—almost everywhere continuous with respect to $\mu(x)$, the Lebesgue measure.

We shall consider the problems of hypothesis testing and unbiased estimation. The first class of problems will be formulated as follows.

Let $\Pi_1(\theta), \cdots, \Pi_r(\theta)$ be continuous functions of $\theta \in \Omega \subset E_s$; $r < s$. The hypothesis H_0 to be tested is composite and consists of the equations

$$\Pi_1(\theta) = 0, \cdots, \Pi_r(\theta) = 0 \tag{1.1}$$

which determine the set Ω_0 in the set Ω.

The alternative H_1 to H_0 consists of the inclusion $\theta \in \Omega \backslash \Omega_0$. Sometimes a Bayes distribution $B(\theta)$ on $\mathrm{y} \backslash \mathrm{y}_0$ is given; this converts H_1 into a simple hypothesis. The last set-up is perhaps not quite natural, but it is convenient for the primary investigation of the composite hypothesis H_0.

We study the tests of H_0 against the alternative H_1.

The problem of unbiased estimation will consist in the investigation of the behavior of the statistics $\xi(x)$ possesing the mathematical expectation $E(\xi|\theta) = F(\theta)$ unbiased with respect to this function in the presence of the relations (1.1).

The way the question is presented above does not, of course, cover all the important problems of hypotheses testing and unbiased estimation. For instance, the problems of sequential analysis are not covered in this way. But in the set-up described above we can find a series of problems which are very interesting and deep from the analytical point of view; some of these will be considered below. The proofs of the theorems formulated below are rather long and complicated; therefore it is not possible to exhibit them in this article. (See [12] for the simplest cases.)

2. Verifiable and nonverifiable functions

In general, let $\phi(x)$ be any randomized test of the hypothesis H_0. Form the power function $\varphi(\theta) = E(\phi(x)|\theta)$. The hypothesis H_0 relates only to the values of the functions $\Pi_1(\theta), \cdots, \Pi_r(\theta)$. If the function $\varphi(\theta)$ is not trivial (that is not constant), it is characterized only by the deviations of $\Pi_1, \cdots, \Pi_r$ from the values prescribed by H_0 for all $\theta \in \Omega_0$. In this case we shall say that the hypothesis H_0 is verifiable in an invariant way by means of the test ϕ. If such a test $\varphi(\theta)$ with a nontrivial power function $\varphi(\theta)$ exists for the vector function $(\Pi_1(\theta), \cdots, \Pi_r(\theta))$, we shall say that the function $(\Pi_1(\theta), \cdots, \Pi_r(\theta))$ is verifiable in an invariant way (more concisely, verifiable). The question which now arises, given a family $\mathcal{P}_0 = \{\ell(x, \theta)\}$, is how to describe the verifiable functions $(\Pi_1(\theta), \cdots, \Pi_r(\theta))$.

This set-up for the first time appeared in 1940 in the work of G. Dantzig [1]. He studied Student's problem (for the repeated normal sample $x_1, \cdots, x_n \in N(a, \sigma^2)$ to test the hypothesis H_0: $a = a_0$), and proved that, in the terminology introduced above, $\Pi_1 = a$ is nonverifiable. In 1945 a well-known work of Charles Stein appeared; this showed that the function $\Pi_1 = a$ becomes verifiable in the setup of sequential analysis.

At present we have little information on verifiable and nonverifiable functions. For a repeated normal sample $x_1, \cdots, x_n \in N(a, \sigma^2)$, with the constant sample size n we can prove the following theorem.

THEOREM 2.1. *The functions* $\Pi_1 = (a/\sigma^\rho)$ *are nonverifiable for* $\rho < 1$.

The result of G. Dantzig follows from this theorem for $\rho = 0$. On the other hand, it is easy to prove that for $\rho = 1$ this property fails to hold; the function $\Pi_1 = a/\sigma$ is verifiable.

To test the hypothesis H_0: $(a/\sigma) = \gamma_0$ in an invariant way, it is sufficient to apply the nonrandomized test ϕ with the critical zone depending only upon the ratio

$$\frac{\bar{x}}{\left(\sum_{i=1}^{n} (x_i - \bar{x})^2\right)^{1/2}}. \tag{2.1}$$

Further, for the Behrens-Fisher problem, given two independent repeated normal samples H_0: $a_1 - a_2 = 0$, the function $a_1 - a_2$ proves to be nonverifiable. However, besides these separate results nothing more is known about the verifiable functions. For the case where $\mathcal{P}_\theta$ is an exponential family (cf. [3], pp. 50–59) the problem on the verifiable functions is reducible to certain rather peculiar questions of the theory of the multiple Laplace transforms.

3. Similar tests

The above results, however fragmentary, lead to the conjecture that the verifiable functions are seldom encountered in the usual problems, and if encountered, the corresponding tests $\phi(x)$ form a narrow class among all the tests and do not have desirable properties with respect to their power function. In view of this we can try to find tests for H_0, eliminating the nuisance parameters from the power function $E(\phi(x)|\theta) = \varphi(\theta)$ only for the null hypothesis H_0, so that $\varphi(\theta)$ becomes constant due to the relations (0, 1) implied by the null hypothesis H_0. We can pose the problem of describing a sufficiently complete class of these tests and of singling out the optimal ones in some sense.

This problem is linked in a direct way to the theory of similar regions introduced into statistics in a well-known work of J. Neyman and E. Pearson [4], and the Neyman structures [5], [3]. The last ones can be constructed if, after introducing the relations (1.1), the family $\mathcal{P}_\theta$ is converted into the family $\mathcal{P}_\theta^0$ depending upon the $(s - r)$ dimensional nuisance parameter in the Euclidean space E_{s-r} and admitting for this parameter suitable sufficient statistics. In particular cases there are no relations ($r = 0$), and all the parameters are nuisance parameters. This problem appears in a natural way if it is required to test whether a family of distributions $\{\ell(x_1, \theta)\}$ has a density with the function ℓ of a given type on the evidence of N independent observations $x_1, \cdots, x_N$, where the parameter values θ are different for each observation (the problem of N small samples). For instance, suppose that we are given N samples from an exponential family, characterized by the density

$$p_\theta(x) = \exp - (\theta_1 T_1(x) + \cdots + \theta_s T_s(x))h(x) \tag{3.1}$$

where $x \in E_n$, and $T_1(x), \cdots, T_s(x)$ are sufficient statistics, and $s < n$. If there are no relations between the parameters $\theta_1, \cdots, \theta_s$, we have the complete exponential family, and all the similar regions have Neyman structure (see E. Lehmann, H. Scheffé [7], E. Lehmann [3]). In the problem of N small samples we can take as an alternative the hypothesis H_1 that the family is of the type

$$p_\theta(x) = \exp - [(\theta_1 T_1(x) + \cdots + \theta_s T_s(x)) + \epsilon V(x)]h(x) \tag{3.2}$$

where ϵ is a small number and $V(x)$ a suitable statistic. Here we shall have the same sufficient statistics $T_1(x), \cdots, T_s(x)$.

The case where the parameters are those of the affine transformations of the repeated sample and similar statistics—the affine invariants—was considered in detail by A. A. Petrov [13].

Note the possibility of construction of similar domains for certain mixtures of distributions when we have only trivial sufficient statistics (see [8], [10]). These mixtures are of the type

$$p_\theta(x) = R_1(T_1(x), \theta)r_1(x) + \cdots + R_q(T(x), \theta)r_q(x) \tag{3.3}$$

where $x \in E_m$, $\mathcal{P}(x) \in E_n$, $(n < m)$, and the densities $p_\theta(x)$ are taken with respect to a dominating measure $\mu(x)$; $q \geq 1$ is an integer, and R_j, r are measurable for $j = 1, 2, \cdots, q$, and

$$\int_{E_m} |R_j(T(x), \theta)(r_j(x)| d\mu(x) < \infty, \qquad (j = 1, 2, \cdots, n). \tag{3.4}$$

For $q = 1$, the $p_\theta(x)$ become the well-known families possessing nontrivial sufficient statistics.

We return now to the family $\mathcal{P}_\theta$ subject to the relations (1.1). If we can construct under such requirements nontrivial similar regions, they will lead us to the nonrandomized similar tests $\phi_1(x)$ of a given level $\alpha \in (0, 1)$ for which we have

$$E(\phi_1(x)|H_0) = \alpha. \tag{3.5}$$

For a randomized similar test $\phi(x)$ only this condition is obligatory; in general, its distribution may depend on nuisance parameters. As is well known (see [3]), the randomized similar test can be converted into a nonrandomized one of the same level α, if a new random variable $\mathcal{U}$ is introduced which is independent of the observation x and uniformly distributed on the segment $[0, 1]$. Then the test ϕ^* with the critical region $\mathcal{U} - \phi(x) < 0$ will be nonrandomized in the space $\{\mathcal{U}\} \times \{x\}$ and will be similar with the same level α as the initial test.

Let our family $\mathcal{P}_\theta = \{p_\theta(x)\}$ admit nontrivial sufficient statistics

$$T = (T_1, \cdots, T_k) \tag{3.6}$$

where $k < n$ (we suppose that no pathological situations as described by D. Basu in [9] arise).

Then the transition from the randomized tests $\phi(x)$ to the nonrandomized ones does not require the supplementary random and uniformly distributed variable $\mathcal{U}$; it can be constructed by means of the observations relevant to the problem.

Suppose we are given a randomized test $\phi(x)$ of level α. Form the expression

$$\phi_1(T) = E(\phi(x)|T). \tag{3.7}$$

It will also be a similar level α test measurable with respect to the σ-algebra of sufficient statistics. Moreover, take a measurable scalar function $V(x)$ and form the conditional distribution

$$F(y|T) = P(V < y|T). \tag{3.8}$$

This distribution does not depend on θ. Suppose that for almost all the values of T, the function $F(y|T)$ is strictly monotone with respect to y. Then, as is well known, the transformation $\mathcal{U} = F(V|T)$ gives a random variable $\mathcal{U}$ which for almost all given T, is uniformly distributed on $[0, 1]$.

If we define now the nonrandomized test $\phi^*(x)$ with the critical region $\mathcal{U} - \phi_1(T) < 0$, then it will depend only on the observations $x \in \mathcal{X}$ and will be a similar level α test. In general, its power function will be the same as that of the initial randomized test $\phi(x)$.

Hence, in the construction of the tests which do not depend only on sufficient statistics, there are no essential differences between the randomized and non-randomized tests; one can easily pass (in principle, at least) from the former to the latter without changing the power.

However, if we restrict ourselves to the tests depending only on sufficient statistics, this difference becomes essential, and the construction of nonrandomized similar tests becomes difficult, as we shall see further on in the example of the Behrens-Fisher problem.

If there are nontrivial sufficient statistics $T = (T_1, \cdots, T_n)$, then instead of arbitrary tests $\phi(x)$ we can consider the tests $\phi_1(T)$ equivalent to them with respect to power. These are determined by the formula (3.7) and depend only on sufficient statistics. In what follows, we shall consider only such tests, and this basically for incomplete exponential families.

4. Similar and unbiased tests for incomplete exponential families

For incomplete exponential families, the principal analytical tool for describing the similar and unbiased tests and unbiased estimates can be obtained from the theory of the ideals of holomorphic functions and the theory of analytical sheaves connected with them. Similar tests generate an analytical sheaf of ideals, and their description can be effectuated by means of "theorem B" of H. Cartan on the behavior of the first cohomology group [11].

To single out the cases where the similar tests are described rather simply, we first impose certain requirements on the structure of exponential families and the relations formed by the null hypothesis, and then we shall weaken these requirements.

Conditions upon the exponential family. The exponential family is given by the density, with respect to the Lebesgue measure, of its sufficient statistics $T_1, \cdots, T_s$,

$$P_\theta(T_1, \cdots, T_s) = C(\theta) \exp \{\theta_1 T_1 + \cdots + \theta_s T_s\} h(T_1, \cdots, T_s). \tag{4.1}$$

Denote by $\mathfrak{I} \subset E_s$ the range of values of the sufficient statistics $(T_1, \cdots, T_s)$. The conditions required are the following.

(I) There is a number $s_1 \leq s$ (s_1 might be equal to 0) such that $h(T_1, \cdots, T_s) = 0$ if at least one of the variables $T_j < 0$, $(j = 1, 2, \cdots, s_1)$. This defines the carrier $\mathfrak{I}$ of the function $h(T_1, \cdots, T_s)$.

(II) In all interior points of $\mathfrak{I}$, the function $h(T_1, \cdots, T_s)$ does not vanish and has there continuous partial derivatives. Moreover, in the domain $\mathfrak{I}_\epsilon \subset \mathfrak{I}$ defined by the inequalities $T_1 \leq \epsilon$ and $T_s \geq \epsilon$ for any $\epsilon > 0$, we have the estimate

$$\left|\frac{\partial \ln h}{\partial T_1}\right| + \cdots + \left|\frac{\partial \ln h}{\partial T_s}\right| = 0\left(\frac{1}{\epsilon^a} + 1\right) \tag{4.2}$$

where $a \leq 1$ is a constant.

(III) The integral $\int \underset{\mathfrak{I}}{\cdots} \int P_\theta(T_1, \cdots, T_s)\, dT_1 \cdots dT_s$ is absolutely convergent for $\theta = (\theta_1, \cdots, \theta_s) \in \mathcal{P}$ where $\mathcal{P}$ is the product $\mathcal{P} = R_1 \times \cdots \times R_{S_1} \times S_{S_1+1} \times \cdots \times S_s$ of s_1 right half-planes $\operatorname{Re} \theta_j > 0$ and $(s - s_1)$ strips $0 < \operatorname{Re} \theta_j < A_j$. (Of course, any open vertical strip can be reduced to the type of the strips S_j by shifting the parameter values.)

Conditions upon the null hypothesis H_0. For real points $\theta = (\theta_1, \cdots, \theta_s) \in \mathcal{P}$, the null hypothesis H_0 is determined by $r < s$ relations

$$\Pi_1(\theta_1, \cdots, \theta_s) = 0; \qquad \Pi_r = (\theta_1, \cdots, \theta_s) = 0. \tag{4.3}$$

The functions $\Pi_1, \cdots, \Pi_r$ must be real for real $(\theta_1, \cdots, \theta_s)$. After multiplying by $[(\theta_1 + 1) \cdots (\theta_s + 1)]^{-N}$ where N is a suitable number, the functions $\Pi_1, \cdots, \Pi_r$ must become functions of $(1/\theta_1 + 1), \cdots, (1/\theta_s + 1)$ holomorphic on the closure of $\mathcal{P}$, that is on $\bar{\mathcal{P}}$ (including the points with $\theta_j = \infty$).

(Such conditions are always fulfilled, for instance, for the case of polynomial relations, which often appear in statistics.)

The null hypothesis H_0 consists in the fulfillment of the relations (4.3) on the compact Ω of real numbers, defined by the inequalities $\epsilon_j \leq \theta_j \leq E_j$; $E_j < A_j$, $(j = 1, 2, \cdots, s)$, for ϵ_j sufficiently small. The corresponding set of points will be denoted by Ω_0. The alternatives to H_0 consist in the inclusion $(\theta_1, \cdots, \theta_s) \in \Omega\backslash\Omega_0$. The alternatives are provided with a Bayes probability measure $B(\theta)$ defined on $\Omega\backslash\Omega_0$ which converts them into a simple hypothesis. For the test ϕ similar with respect to H_0 we introduce the Bayes gain

$$W(\phi|B) = \int_{\Omega\backslash\Omega_0} E(\phi|\theta)\, dB(\theta). \tag{4.4}$$

Further conditions on the relations. (I) The equations (4.3) considered in the complex domain must generate there an analytical set of points $V_{\pi_1 \cdots \pi_r}$ which can be decomposed into a finite number of disjoint components $V^q_{\pi_1 \cdots \pi_r}$, each of complex dimension $(s - r)$ and each containing inside a connected set $\mathcal{R}^q_{\pi_1 \cdots \pi_r}$ of real points entering into Ω_0 and having a real dimension $(s - r)$.

(II) rank $\left\| \dfrac{\partial \Pi_i}{\partial \theta_j} \right\| = r$ inside $\mathcal{P}$ $(i = 1, 2, \cdots, r; j = 1, 2, \cdots, s)$.

We can consider the similar tests $\phi(T_1, \cdots, T_s)$ of the null hypothesis H_0 depending upon the sufficient statistics only. As was explained above, any similar test is equivalent to such a test. We can now give a description of an "everywhere dense" family of similar tests.

A simple case of an analogous set-up was considered by Robert A. Wijsman [27].

THEOREM 4.1. *For a given* $\epsilon > 0$ *as small as we please and for a given level* α *similar test* $\phi = \phi(T_1, \cdots, T_s)$, *we can indicate the similar test* $\phi_\epsilon = \phi_\epsilon(T_1, \cdots, T_s)$ *such that*

$$|W(\phi_\epsilon|B) - W(\phi|B)| \leq \epsilon \tag{4.5}$$

for which we have the representation

$$\phi_\epsilon(T_1, \cdots, T_s) = \alpha + \frac{1}{h}(A_1^* H_1 + \cdots + A_r^* H_r). \tag{4.6}$$

Here $A_j = A_j(T)$, $(j = 1, 2, \cdots, r)$ *are pre-images of the functions*

$$\Pi_j(\theta_1, \cdots, \theta_s)^{-(N+1)}$$

*for a one-sided Laplace transform and the asterisk * is the convolution sign. The functions* $H_1, \cdots, H_r$ *have a prescribed number of partial derivatives and have the estimate*

$$(4.7) \qquad H_j(T_1, \cdots, T_s) = 0(\exp \zeta(|T_1| + \cdots + |T_s|); \qquad j = 1, 2, \cdots, r$$

where $\zeta > 0$ *is as small as we please. If* i *is one of the numbers* $1, 2, \cdots, s_1$ *and* $T_i < 0$, *then* $H_j(T_1, \cdots, T_s)$ *vanishes.*

This theorem enables us to solve, at least in principle, the problem of the choice of the "ϵ-optimal" similar test for a given Bayes distribution B. To find an ϵ_n-optimal "cotest" $\psi_\epsilon = \phi_\epsilon - \alpha$, we look for $H_1, \cdots, H_r$ such that

$$(4.8) \qquad \psi_\epsilon = \frac{1}{h}(A_1^*H_1 + \cdots + A_r^*H_r)$$

under the restrictions

$$(4.9) \qquad -\alpha \leq \psi_\epsilon \leq 1 - \alpha$$

gives the largest possible value to

$$(4.10) \qquad W(\psi_\epsilon|B) = \int_{\Omega \setminus \Omega_0} E(\psi_\epsilon|B)\, dB(\theta).$$

We thus obtain a variational problem with restrictions. For its solution one can apply the methods of linear programming (see [12]).

We return now to the numerous requirements imposed upon $h(T_1, \cdots, T_s)$ and the relations (4.3), implied by the null hypothesis H_0. Among these conditions the rather restrictive one is the requirement that $h(T_1, \cdots, T_s) \neq 0$ inside $\mathfrak{J}$.

If we reject this condition, the difference will be only that the functions $H_1, \cdots, H_r$ in the formula should be chosen so that the expression

$$(4.11) \qquad A_1^*H_1 + \cdots + A_r^*H_r$$

vanishes at the points where $h(T_1, \cdots, T_s) = 0$. In this domain we put $\phi = \alpha$.

We pass now to the structure of unbiased tests of H_0 against the alternative H_1. By virtue of the known theorems of test theory (see [3]) in our set-up, the unbiased tests will form a part of the set of similar tests. Under the conditions of theorem 4.1 we can describe an everywhere dense set among all the unbiased tests—the set of all sufficiently smooth unbiased tests $\mathcal{G}$. Namely, for any unbiased test ϕ there exists a $\phi_\epsilon \in \mathcal{G}$ such that the condition (4.5) will be fulfilled. The general form of the tests of the set $\mathcal{G}$ is given by the following theorem.

THEOREM 4.2. *Under the conditions of theorem* 4.1, *any sufficiently smooth unbiased test can be represented in the form*

$$(4.12) \qquad \phi = \alpha + \frac{1}{h}\sum_{i,j=1}^{r} A_i^*A_j^*H_{ij},$$

A_i *being the functions defined above,* H_{ij}, $(i, j = 1, 2, \cdots, r)$ *functions of the same type as the* H_j *introduced above.*

If the condition about the nonvanishing of h in the domain $\mathfrak{Z}$ *is violated, we must require that* $\sum_{i,j=1}^{r} A_i^* A_j^* H_{ij}$ *vanish at the points where* $h = 0$, *and put* $\phi = \alpha$ *at these points.*

5. Unbiased estimates

As is well known from the theorem of C. R. Rao [15] and D. Blackwell, in a sense the unbiased estimates cannot deteriorate if "projected" into the space of sufficient statistics. We shall consider the exponential families (4.1) and sufficiently smooth statistics depending only upon the sufficient statistics $T_1, \cdots, T_s$ and fulfilling the condition

$$\xi(T_1, \cdots, T_s) = O(\exp \zeta(|T_1| + \cdots + |T_s|) \tag{5.1}$$

for any $\zeta > 0$. Each such statistic will be an unbiased estimate of $E(\xi|\theta) = \ell(\theta)$. If there are no relations, so that the family is complete, the unbiased estimate of $\ell(\theta)$ is unique with probability 1. If there are relations (4.3), then all the unbiased estimates of $\ell(\theta)$ differ by unbiased estimates of zero χ, that is, the statistics satisfying the conditions $E(\chi|\theta) = 0$ and called U.E.Z. for short. The set of smooth U.E.Z. with the growth condition (5.1) are described by the following theorem.

THEOREM 5.1. *Under the conditions of theorem* 4.1, *U.E.Z., which are sufficiently smooth and fulfill the growth condition* (5.1) *are described by the formula*

$$\chi = \frac{1}{h} (A_1^* H_1 + \cdots + A_r^* H_r) \tag{5.2}$$

in the notation of section 4.

If h vanishes inside $\mathfrak{Z}$, we must, as indicated earlier, choose H_j so that the numerator of the fraction (5.2) vanishes at the corresponding points. Taking into account the description of all sufficiently smooth unbiased estimates of zero (5.2) for the case of incomplete exponential families considered by us, one can establish certain cases of inadmissibility of unbiased estimates. For instance, A. M. Kagan established that for a repeated sample $x_1, \cdots, x_n$ of a one-parameter family

$$P_\theta(x) = C_0 \exp - (x - \theta)^{2k}, \qquad k \geq 2 \tag{5.3}$$

the sample moments $\overline{x}$, $a_m = (1/n) \sum_{i=1}^{n} x_i^m$ for $2 \leq m \leq 2k - 2$ are unbiased estimates of the corresponding moments of the distribution which are inadmissible on any compact set of values of the parameter θ. (For the estimate of $\overline{x}$ this follows from the well-known theorem of C. R. Rao [15].)

6. Investigations on the Behrens-Fisher problem

In this section are expounded the investigations of Leningrad statisticians in the period of 1963–1965 on the Behrens-Fisher problem—a classical problem on the elimination of nuisance parameters.

First, we shall present the results obtained by applying to the Behrens-Fisher problem the theory of similar tests for incomplete exponential families expounded above. The corresponding normal samples will be denoted $(x_1, \cdots, x_{n_1}) \in N(a_1, \sigma_1^2)$, $(y_1, \cdots, y_{n_2}) \in N(a_2, \sigma_2^2)$, and the sufficient statistics $\bar{x}$, $\bar{y}$, s_1^2, s_2^2. Consider the similar tests ϕ depending only on $|\bar{x} - \bar{y}|$, s_1^2, s_2^2.

Introducing instead of $\bar{x} - \bar{y}$, s_1^2, s_2^2 the proportional variables

$$X = (\bar{x} - \bar{y})\sqrt{n_1 n_2}; \qquad u = n_1 s_1^2; \qquad v = n_2 s_2^2 \tag{6.1}$$

and putting

$$m_1 = \frac{n_1 - 3}{2}, \qquad m_2 = \frac{n_2 - 3}{2}, \qquad n_1, n_2 \geq 3 \tag{6.2}$$

$$F_0 = n_2 u + n_1 v - x,$$

we can represent all sufficiently smooth cotests $\phi - \alpha$ in the form

$$\begin{aligned} \psi(x, u, v) &= x^{1/2} u^{-m_1} v^{-m_2} F_0^* H \\ &= x^{1/2} u^{-m_1} v^{-m_2} \int_0^x \int_0^u \int_0^v F_0(x - \xi, u - \eta, v - \zeta) \cdot H(\xi, \eta, \zeta)\, d\xi\, d\eta\, d\zeta \end{aligned} \tag{6.3}$$

where $H = H(\xi, \eta, \zeta)$ is a sufficiently smooth function of the three variables. The cotest $\psi(x, u, v)$ is to fulfill the restrictions

$$-\alpha \leq \psi(x, u, v) \leq 1 - \alpha. \tag{6.4}$$

For a given Bayes probability measure $B(\theta)$ on the alternatives, the formula (6.3) gives an ϵ-complete family of tests.

We now pass to the properties of similar tests $\phi = \phi(\bar{x}, \bar{y}, s_1^2, s_2^2)$ depending on the sufficient statistics only.

In his well-known article [16], A. Wald considers nonrandomized tests for the Behrens-Fisher problem, subject to four axioms. The first one requires that the tests depend on the sufficient statistics only. The second one requires the invariance of the critical zone with respect to one and the same shift of all the sample elements. The third axiom requires the invariance of the critical zone with respect to the contraction or expansion of all the sample elements by the same scale factor. The fourth axiom of Wald will be formulated later; we consider now the first three. It is easy to deduce from them that the nonrandomized test ϕ must be of the form:

$$\phi = \phi\left(\frac{\bar{x} - \bar{y}}{s_2}, \frac{s_1}{s_2}\right). \tag{6.5}$$

We shall call the general (randomized) test of this form homogeneous; the description of all homogeneous randomized or nonrandomized tests (tests of the form (6.5)) we shall call the homogeneous Behrens-Fisher problem.

The fourth axiom of Wald leads to the conclusion that the critical zone of the nonrandomized test is of the form

$$\frac{|\bar{x} - \bar{y}|}{s_2} > \psi\left(\frac{s_1}{s_2}\right) \tag{6.6}$$

where ψ is a Lebesgue measurable function. The tests of this type were studied by R. A. Fisher [17] and B. Welch [18]. Therefore, we shall call them nonrandomized Fisher-Welch-Wald tests. The problem of the existence of nontrivial similar tests of this type is not yet solved.

In the work [16] cited above, A. Wald considers the tests of the type (6.6) and constructs approximately similar tests of this type. Raising the question of the existence of exact similar tests, he makes an attempt to construct tests with analytical boundary for the critical zone; his calculations are made for samples of the same size.

The investigations expounded in [19] prove that there are no such tests. Denote $\xi = (\bar{x} - \bar{y}/s_2)$, $\eta = (s_1/s_2)$. Then the boundary of the critical zone is of the type

$$|\xi| = \psi(\eta). \tag{6.7}$$

THEOREM 6.1. *For the case of two samples of equal sample size $n \geq 4$ there exists no nonrandomized Fisher-Welch-Wald test with boundary for the critical zone* (6.7) *possessing a finite first derivative in the open interval* (0, 1) *and fulfilling the Lipschitz condition in a sufficiently large segment of the type* $[0, \eta_0]$.

Here one can define the number η_0 in the following way: $\eta_0 > 1$ so that the function $\psi(\eta)$ is continuous in $[0, 1]$. Denote $\sup_{0 \leq \eta \leq 1} \psi(\eta) = M$; then one can take $\eta_0 = 2M + 1$.

In article [19] the condition $n \geq 4$ was omitted by an oversight.

I. L. Romanovskaia [20] transferred this result to the case of the samples of unequal sizes. She proved the nonexistence of the nonrandomized Fisher-Welch-Wald test with critical zone of the type

$$\frac{|\bar{x} - \bar{y}|\sqrt{n_1 n_2}}{\sqrt{n_2 s_1^2 + n_1 s_2^2}} \geq \psi(\eta) \tag{6.8}$$

where $\eta = (n_2 s_1^2/n_1 s_2^2)$ and $n_2 \geq 4$. The test is supposed to be similar with respect to a bounded countable set of values of (σ_1/σ_2). The function $\psi(\eta)$ must be continuous and must satisfy the Lipschitz condition on the segment $[0, \eta_0]$ and have a finite first derivative in the open segment (0, 1). The number $\eta_0 > 0$ is defined as in theorem 6.1. It is not known whether these conditions upon $\psi(\eta)$ can be replaced by continuity or measurability only; the results expounded below cause one to have doubts about it. The method applied in [19] to prove theorem 6.1 to all appearances can also be applied to study this question, but the question still remains unanswered.

At any rate, if the nonrandomized test of the Fisher-Behrens-Welch type exists, it must evidently have a "pathological structure" and bad statistical properties.

However, for the equal sample sizes one can construct a randomized homogeneous test of the type which is similar to the one mentioned above and has good statistical characteristics. This test can be obtained by projecting the well-known Bartlett test (paired sample test) on the space of sufficient statistics and therefore has the properties of the Bartlett test. There exists a whole family of such tests. Denote $\xi = (\bar{x} - \bar{y}/s_2)$, $\eta = (s_1/s_2)$ as was done earlier; let $c \geq 0$ be any constant; form the expression

$$z = \tfrac{1}{2}\left(\eta + \frac{1}{\eta} - \frac{1}{c^2}\frac{\xi^2}{\eta}\right). \tag{6.9}$$

The test $\phi = \phi(\xi, \eta)$ is constructed in the following way: if $|\xi| \leq c|\eta - 1|$, let $\phi = 0$ (the null hypothesis is accepted with probability 1). If $|\xi| > c|\eta - 1|$, then $z < 1$. In this case we introduce the function

$$\phi(\xi, \eta) = \int_{\max(z,-1)}^{1} \ell_n(r)\, dr \tag{6.10}$$

where

$$\ell_n(r) = \frac{\Gamma\left(\frac{n-1}{2}\right)}{\Gamma\left(\frac{n-2}{2}\right)\sqrt{\pi}}(1 - r^2)^{\frac{n-4}{2}}. \tag{6.11}$$

For the test obtained in this way, by changing the constant c, we can get any level $\alpha = E(\phi|H_0)$. Note that the zone where $\phi = 0$ is bounded by segments of straight lines passing through the point (0, 1).

We can consider the families of nonrandomized homogeneous tests with the critical regions of the type

$$t\left(\frac{\bar{x} - \bar{y}}{s_2}, \frac{s_1}{s_2}\right) \geq c \tag{6.12}$$

where the constant c is arbitrary. For the case where t is a continuous function of both variables, the best results were obtained by O. V. Shalaevsky [21].

THEOREM 6.2. *There exist no continuous functions t of two variables which generate for each c a similar test of the type* (6.12), *except for the trivial case $t =$* const.

It would seem that the continuity condition of the function t could be replaced by a measurability condition. But in 1964, several Leningrad statisticians proved simultaneously [22], [23] that this is not so and that for any level $\alpha \in (0, 1)$ there exist measurable nonrandomized homogeneous similar tests for the Behrens-Fisher problem. More exactly, the following theorem was proved.

THEOREM 6.3. *For any level $\alpha \in (0, 1)$ and pairs of samples of sizes n_1 and n_2, one even and another uneven, there exists a nonrandomized similar homogeneous test for the Behrens-Fisher problem with the critical zone depending only on* $(\bar{x} - \bar{y}/s_2)$, (s_1/s_2).

Theorem 6.3 can be improved a little.

THEOREM 6.4. *Suppose that we are given a finite number K of pairs of samples of sizes n_{1i}, n_{2i}, $(i = 1, 2, \cdots, K)$, one even and another uneven. Then there exists a measurable nonrandomized homogeneous test*

$$\phi = \phi\left(\frac{|\bar{x} - \bar{y}|}{s_2}, \frac{s_1}{s_2}\right) \tag{6.13}$$

which is similar for all these sample pairs simultaneously and has a prescribed level $\alpha \in (0, 1)$.

Note that the sufficient statistics $\bar{x}$, $\bar{y}$, s_1, s_2 are chosen for the given sample pair so that the test of theorem 6.4 will depend on the number i of the sample pair through this.

The articles [22] and [23] constructing different variants of tests of theorems 6.3 and 6.4 are based on a lemma proved by I. V. Romanovsky and V. N. Sudakov.

LEMMA OF I. V. ROMANOVSKY AND V. N. SUDAKOV. *Suppose that on a rectangle* θ: $a \leq \alpha \leq b; c \leq y \leq d$ *is given a finite number of measurable probability densities* $p_m(x, y)$, $(m = 1, 2, \cdots, M)$. *Then for any given* $\alpha \in (0, 1)$ *there exists a measurable function* $I(x, y)$ *taking only the values* 0 *and* 1 *such that for almost all values of* x *(correspondingly, almost all values of* y*)*

$$E^{(m)}(I(x, y)|x) = \alpha; \quad E^{(m)}(I(x, y)|y) = \alpha \qquad \text{for} \quad m = 1, 2, \cdots, M. \tag{6.14}$$

Here $E^{(m)}(\cdot|\cdot)$ *is the symbol for the conditional expectation for the densities* $p_m(x, y)$.

The homogeneous similar tests constructed have apparently bad statistical properties.

For the sample sizes of equal parity the question on the existence of the tests of the described type remains unanswered.

We remark that the requirement of similarity of the test (and this is all the more true for the stronger requirement of unbiasedness) cannot always be conciliated with some other conditions of statistical expediency. In particular, it is natural to require that the homogeneous test ϕ accepts the null hypothesis H_0 with probability 1 if $|\bar{x} - \bar{y}|$ is small in comparison with $\sqrt{s_1^2 + s_2^2}$. The following theorem (see [24]) shows that such a condition cannot be conciliated with the similarity property of the test.

THEOREM 6.5. *There exists no randomized homogeneous nontrivial similar test* ϕ, *accepting the null hypothesis* H_0 *with probability* 1 *if*

$$\frac{|\bar{x} - \bar{y}|}{\sqrt{s_1^2 + s_2^2}} < \epsilon_0 \tag{6.15}$$

where $\epsilon > 0$ *is a given arbitrarily small constant.*

In [24] this theorem was proved in a somewhat stronger form. We require from the test ϕ to accept H_0 if in addition to (6.15) we also have $(|\bar{x} - \bar{y}|/s_2) \leq 1 + \eta_0$ where $\eta_0 > 0$ is a given arbitrarily small constant.

7. Characterization of the tests of Bartlett-Scheffé type

The Bartlett-Scheffé tests for the Behrens-Fisher problem (see [25]) are based on the introduction of a new random object which for the null hypothesis H_0: $a_1 = a_2$, admits the description of all similar tests as Neyman's structures.

Namely, linear forms depending on the observations $x_1, \cdots, x_{n_1}; y_1, \cdots, y_n$ are introduced which are independent and whose variances differ only by constant factors for all values of the variances σ_1^2 and σ_2^2. For the simplest case of the samples of equal size ($n_1 = n_2$), we can take, for instance,

$$\chi = \bar{x} - \bar{y}; \qquad \ell_i = (x_i - \bar{x}) - (y_i - \bar{y}), \qquad (i = 1, 2, \cdots, n) \tag{7.1}$$

and form a test with the critical zone

$$|\chi| \left(\sum_{i=1}^{n} \ell_i^2 \right)^{-1/2} \geq C_0 \tag{7.2}$$

where C_0 is chosen corresponding to a prescribed level α (the Bartlett test). We see that $\mathfrak{D}(\bar{x} - \bar{y}) = (1/n)(\sigma_1^2 + \sigma_2^2)$; $\mathfrak{D}(\ell_i) = (1 - (1/n))(\sigma_1^2 + \sigma_2^2)$. Hence, the random vector $(X, \ell_1, \cdots, \ell_n)$ generates an exponential family of distributions with one parameter $(\sigma_1^2 + \sigma_2^2)$ and one sufficient statistic which is a quadratic form and the test (6.2) is a Neyman structure test for this family. The more general Scheffé tests possess the same properties. This proves that such tests can be characterized by the requirements which are very simple and natural from the statistical point of view (see [26]).

Let a randomized test $\phi = \phi(x_1, \cdots, x_{n_1}; y_1, \cdots, y_{n_2})$ be defined in the space of the linear forms $\chi, \ell_1, \ell_2, \cdots, \ell_\mu$ where $\mu \leq n_1 + n_2 - 1$; moreover, these forms are statistically independent and

$$E(\chi | H_0) = E(\ell_i | H_0) = 0, \qquad (i = 1, 2, \cdots, \mu). \tag{7.3}$$

THEOREM 7.1. *Let there exist a small* $\epsilon_0 > 0$ *such that the test* ϕ *accepts the null hypothesis* H_0 *with probability* 1 *if for at least one of the numbers* $i = 1, 2, \cdots, \mu$, *we have*

$$\frac{|\chi|}{|\ell_i|} \leq \epsilon_0 \tag{7.4}$$

(but it can also accept H_0 *in other cases).*

Then the fractions $(\mathfrak{D}(\chi)/\mathfrak{D}(\ell_i)) = a_i$, $(i = 1, 2, \cdots, \mu)$ *do not depend on* σ_1 *and* σ_2; $\theta = \chi^2 + \sum_{i=1}^{\mu} (\ell_i^2/a_i)$ *is the sufficient statistic in the space* $(\chi, \ell_1, \cdots, \ell_\mu)$, *and the test* ϕ *is a Neyman structure for this space.*

We see that the conditions of theorem 7.1 are fulfilled for the Bartlett-Scheffé tests. Hence, their characterization as Neyman structures follows from a simple property connected with (7.3).

REFERENCES

[1] G. DANTZIG, "On the non-existence of tests of 'Student's' hypothesis having power function independent of σ," *Ann. Math. Statist.*, Vol. 11 (1940), pp. 186–191.

[2] C. STEIN, "A two-sample test for a linear hypothesis whose power is independent of the variance," *Ann. Math. Statist.*, Vol. 16 (1945), pp. 243–258.

[3] E. LEHMANN, *Testing Statistical Hypotheses*, New York, Wiley, 1959.

[4] J. NEYMAN and E. PEARSON, "On the problem of the most efficient tests of statistical hypotheses," *Philos. Trans. Roy. Soc. London Ser. A*, Vol. 231 (1933), pp. 289–337.

[5] J. NEYMAN, "Sur la vérification des hypotheses statistiques composées," *Bull. Soc. Math. France*, Vol. 63 (1935), pp. 346–366.

[6] ———, "Current problems of mathematical statistics," *Proc. Internat. Congress of Mathematicians*, Amsterdam, 1954, Vol. 1 (1957), pp. 1–22.

[7] E. LEHMANN and H. SCHEFFÉ, "Completeness, similar regions and unbiased estimates," *Sankhyā*, Vol. 10 (1950), pp. 305–340.

[8] A. M. KAGAN and YU. V. LINNIK, "A class of families admitting similar zones," *Vestnik Leningrad. Univ.*, Vol. 13 (1914), pp. 25–27. (In Russian.)

[9] D. BASU, "On statistics independent of sufficient statistics," *Sankhyā*, Vol. 20 (1958), pp. 3–4.

[10] YU. V. LINNIK, "On the theory of statistically similar zones," *Dokl. Akad. Nauk SSSR*, Vol. 146 (1962), pp. 300–302. (In Russian.)

[11] H. CARTAN, "Variétés analytiques complexes et cohomologie," Colloque de Liege, 1959.

[12] YU. V. LINNIK, "On the construction of optimal similar solutions for the Behrens-Fisher problem," *Trudy Mat. Inst. Steklov*, Vol. 79 (1965), pp. 26–32.

[13] A. A. PETROV, "Testing statistical hypotheses on the distribution type by use of small samples," *Teor. Verojatnost. i Primenen.*, Vol. I (1956), pp. 248–270.

[14] C. R. RAO, "Information and the accuracy attainable in the estimation of statistical parameters," *Bull. Calcutta Math. Soc.*, Vol. 37 (1945), pp. 81–91.

[15] ———, "Some theorems on the minimum variance estimates," *Sankhyā*, Vol. 12 (1952), pp. 1–2, 27–42.

[16] A. WALD, "Testing the difference between the means of two normal populations with unknown standard deviations," *Selected Papers in Probability and Statistics*, New York, McGraw-Hill (1955), pp. 669–695.

[17] R. A. FISHER, "The comparison of samples with possibly unequal variances," *Ann. of Eugenics*, Vol. 9 (1939), pp. 174–180.

[18] B. L. WELCH, "The generalization of 'Student's' problem when several different populations are involved," *Biometrika*, Vol. 34 (1947), pp. 28–35.

[19] YU. V. LINNIK, "On A. Wald's test for comparison of two normal samples," *Teor. Verojatnost. i Primenen.*, Vol. 9 (1964), pp. 16–30. (In Russian.)

[20] J. L. ROMANOVSKAIA, "On the Fisher-Welch-Wald test," *Sibirsk. Mat. Ž.*, Vol. 5 (1964), pp. 1344–1350. (In Russian.)

[21] O. V. SHALAEVSKY, "On the non-existence of the regularly changing tests for the Behrens-Fisher problem," *Dokl. Akad. Nauk SSSR*, Vol. 151 (1950), pp. 509–510. (In Russian.)

[22] A. M. KAGAN and O. V. SHALAEVSKY, "The Behrens-Fisher problem: similar zones in the algebra of sufficient statistics," *Dokl. Akad. Nauk SSSR*, Vol. 155 (1964), pp. 1250–1252. (In Russian.)

[23] YU. V. LINNIK, I. V. ROMANOVSKY, and V. N. SUDAKOV, "Non-randomized homogeneous test in the Behrens-Fisher problem," *Dokl. Akad. Nauk SSSR*, Vol. 155 (1964), pp. 1262–1264. (In Russian.)

[24] ———, "On randomized homogeneous tests for the Behrens-Fisher problem," *Izv. Akad. Nauk SSSR*, Vol. 28 (1964), pp. 1–12. (In Russian.)

[25] H. SCHEFFÉ, "On solutions of the Behrens-Fisher problem based on the *t*-distribution," *Ann. Math. Statist.*, Vol. 14 (1943), pp. 35–44.

[26] YU. V. LINNIK, "On the characterization of the tests of Bartlett-Scheffé type," *Trudy Mat. Inst. Steklov*, Vol. 79 (1965), pp. 32–39. (In Russian.)

[27] R. A. WIJSMAN, "In complete statistics and similar tests," *Ann. Math. Statist.*, Vol. 29 (1958), pp. 1028–1075.

SOME METHODS FOR CLASSIFICATION AND ANALYSIS OF MULTIVARIATE OBSERVATIONS

J. MacQUEEN
UNIVERSITY OF CALIFORNIA, LOS ANGELES

1. Introduction

The main purpose of this paper is to describe a process for partitioning an N-dimensional population into k sets on the basis of a sample. The process, which is called 'k-means,' appears to give partitions which are reasonably efficient in the sense of within-class variance. That is, if p is the probability mass function for the population, $S = \{S_1, S_2, \cdots, S_k\}$ is a partition of E_N, and u_i, $i = 1, 2, \cdots, k$, is the conditional mean of p over the set S_i, then $w^2(S) = \sum_{i=1}^{k} \int_{S_i} |z - u_i|^2 \, dp(z)$ tends to be low for the partitions S generated by the method. We say 'tends to be low,' primarily because of intuitive considerations, corroborated to some extent by mathematical analysis and practical computational experience. Also, the k-means procedure is easily programmed and is computationally economical, so that it is feasible to process very large samples on a digital computer. Possible applications include methods for similarity grouping, nonlinear prediction, approximating multivariate distributions, and nonparametric tests for independence among several variables.

In addition to suggesting practical classification methods, the study of k-means has proved to be theoretically interesting. The k-means concept represents a generalization of the ordinary sample mean, and one is naturally led to study the pertinent asymptotic behavior, the object being to establish some sort of law of large numbers for the k-means. This problem is sufficiently interesting, in fact, for us to devote a good portion of this paper to it. The k-means are defined in section 2.1, and the main results which have been obtained on the asymptotic behavior are given there. The rest of section 2 is devoted to the proofs of these results. Section 3 describes several specific possible applications, and reports some preliminary results from computer experiments conducted to explore the possibilities inherent in the k-means idea. The extension to general metric spaces is indicated briefly in section 4.

The original point of departure for the work described here was a series of problems in optimal classification (MacQueen [9]) which represented special

This work was supported by the Western Management Science Institute under a grant from the Ford Foundation, and by the Office of Naval Research under Contract No. 233(75), Task No. 047-041.

cases of the problem of optimal information structures as formulated by Marschak [11], [12]. (For an interesting treatment of a closely related problem, see Blackwell [1].) In one instance the problem of finding optimal information structures reduces to finding a partition $S = \{S_1, S_2, \cdots, S_k\}$ of E_N which will minimize $w^2(S)$ as defined above. In this special model, individual A observes a random point $z \in E_N$, which has a known distribution p, and communicates to individual B what he has seen by transmitting one of k messages. Individual B interprets the message by acting as if the observed point z is equal to a certain point $\hat{z}$ to be chosen according to the message received. There is a loss proportional to the squared error $|z - \hat{z}|^2$ resulting from this choice. The object is to minimize expected loss. The expected loss becomes $w^2(S)$, where the i-th message is transmitted if $z \in S_i$, since the best way for B to interpret the information is to choose the conditional mean of p on the set associated with the message received. The mean, of course, minimizes the squared error. Thus the problem is to locate a partition minimizing $w^2(S)$. This problem was also studied by Fisher [5], who gives references to earlier related works.

The k-means process was originally devised in an attempt to find a feasible method of computing such an optimal partition. In general, the k-means procedure will not converge to an optimal partition, although there are special cases where it will. Examples of both situations are given in section 2.3. So far as the author knows, there is no feasible, general method which always yields an optimal partition. Cox [2] has solved the problem explicitly for the normal distribution in one dimension, with $k = 2, 3, \cdots, 6$, and a computational method for finite samples in one dimension has been proposed by Fisher [5]. A closely related method for obtaining reasonably efficient 'similarity groups' has been described by Ward [15]. Also, a simple and elegant method which would appear to yield partitions with low within-class variance, was noticed by Edward Forgy [7] and Robert Jennrich, independently of one another, and communicated to the writer sometime in 1963. This procedure does not appear to be known to workers in taxonomy and grouping, and is therefore described in section 3. For a thorough consideration of the biological taxonomy problem and a discussion of a variety of related classification methods, the reader is referred to the interesting book by Sokal and Sneath [14]. (See *Note added in proof* of this paper.)

Sebestyen [13] has described a procedure called "adaptive sample set construction," which involves the use of what amounts to the k-means process. This is the earliest explicit use of the process with which the author is familiar. Although arrived at in ignorance of Sebestyen's work, the suggestions we make in sections 3.1, 3.2, and 3.3, are anticipated in Sebestyen's monograph.

2. K-means; asymptotic behavior

2.1. *Preliminaries.* Let $z_1, z_2, \cdots$ be a random sequence of points (vectors) in E_N, each point being selected independently of the preceding ones using a fixed probability measure p. Thus $P[z_1 \in A] = p(A)$ and $P[z_{n+1} \in A | z_1, z_2, \cdots, z_n] =$

$p(A)$, $n = 1, 2, \cdots$, for A any measurable set in E_N. Relative to a given k-tuple $x = (x_1, x_2, \cdots, x_k)$, $x_i \in E_N$, $i = 1, 2, \cdots, k$, we define a *minimum distance partition* $S(x) = \{S_1(x), S_2(x), \cdots, S_k(x)\}$ of E_N, by

$$\begin{aligned} S_1(x) &= T_1(x), S_2(x) = T_2(x)S_1'(x), \cdots, \\ S_k(x) &= T_k(x)S_1'(x)S_2'(x) \cdots S_{k-1}'(x), \end{aligned} \tag{2.1}$$

where

$$T_i(x) = \{\xi \colon \xi \in E_N, |\xi - x_i| \leq |\xi - x_j|, j = 1, 2, \cdots, k\}. \tag{2.2}$$

The set $S_i(x)$ contains the points in E_N nearest to x_i, with tied points being assigned arbitrarily to the set of lower index. Note that with this convention concerning tied points, if $x_i = x_j$ and $i < j$ then $S_j(x) = \emptyset$. Sample k-means $x^n = (x_1^n, x_2^n, \cdots, x_k^n)$, $x_i^n \in E_N$, $i = 1, \cdots, k$, with associated integer weights $(w_1^n, w_2^n, \cdots, w_k^n)$, are now defined as follows: $x_i^1 = z_i$, $w_i^1 = 1$, $i = 1, 2, \cdots, k$, and for $n = 1, 2, \cdots$, if $z_{k+n} \in S_i^n$, $x_i^{n+1} = (x_i^n w_i^n + z_{n+k})/(w_i^n + 1)$, $w_i^{n+1} = w_i^n + 1$, and $x_j^{n+1} = x_j^n$, $w_j^{n+1} = w_j^n$ for $j \neq i$, where $S^n = \{S_1^n, S_2^n, \cdots, S_k^n\}$ is the minimum distance partition relative to x^n.

Stated informally, the k-means procedure consists of simply starting with k groups each of which consists of a single random point, and thereafter adding each new point to the group whose mean the new point is nearest. After a point is added to a group, the mean of that group is adjusted in order to take account of the new point. Thus at each stage the k-means are, in fact, the means of the groups they represent (hence the term k-means).

In studying the asymptotic behavior of the k-means, we make the convenient assumptions, (i) p is absolutely continuous with respect to Lebesgue measure on E_N, and (ii) $p(R) = 1$ for a closed and bounded convex set $R \subset E_N$, and $p(A) > 0$ for every open set $A \subset R$. For a given k-tuple $x = (x_1, x_2, \cdots, x_k)$—such an entity being referred to hereafter as a k-point—let

$$\begin{aligned} W(x) &= \sum_{i=1}^{k} \int_{S_i} |z - x_i|^2 \, dp(z), \\ V(x) &= \sum_{i=1}^{k} \int_{S_i} |z - u_i(x)|^2 \, dp(z), \end{aligned} \tag{2.3}$$

where $S = \{S_1, S_2, \cdots, S_k\}$ is the minimum distance partition relative to x, and $u_i(x) = \int_{S_i} z \, dp(z)/p(S_i)$ or $u_i(x) = x_i$, according to whether $p(S_i) > 0$ or $p(S_i) = 0$. If $x_i = u_i(x)$, $i = 1, 2, \cdots, k$ we say the k-point x is *unbiased*.

The principal result is as follows.

THEOREM 1. *The sequence of random variables $W(x^1)$, $W(x^2)$, $\cdots$ converges a.s. and $W_\infty = \lim_{n\to\infty} W(x^n)$ is a.s. equal to $V(x)$ for some x in the class of k-points $x = (x_1, x_2, \cdots, x_k)$ which are unbiased, and have the property that $x_i \neq x_j$ if $i \neq j$.*

In lieu of a satisfactory strong law of large numbers for k-means, we obtain the following theorem.

THEOREM 2. *Let $u_i^n = u_i(x^n)$ and $p_i^n = p(S_i(x^n))$; then*

$$\sum_{n=1}^{m} \left(\sum_{i=1}^{k} p_i^n |x_i^n - u_i^n| \right) \Big/ m \underset{\text{a.s.}}{\longrightarrow} 0 \quad \text{as} \quad m \to \infty. \tag{2.4}$$

2.2. *Proofs.* The system of k-points forms a complete metric space if the distance $\rho(x, y)$ between the k-points $x = (x_1, x_2, \cdots, x_k)$ and $y = (y_1, y_2, \cdots, y_k)$, is defined by $\rho(x, y) = \sum_{i=1}^{k} |x_i - y_i|$. We designate this space by M and interpret continuity, limits, convergence, neighborhoods, and so on, in the usual way with respect to the metric topology of M. Of course, every bounded sequence of k-points contains a convergent subsequence.

Certain difficulties encountered in the proof of theorem 1 are caused by the possibility of the limit of a convergent sequence of k-points having some of its constituent points equal to each other. With the end in view of circumventing these difficulties, suppose that for a given k-point $x = (x_1, x_2, \cdots, x_k)$, $x_i \in R$, $i = 1, 2, \cdots, k$, we have $x_i = x_j$ for a certain pair i, j, $i < j$, and $x_i = x_j \neq x_m$ for $m \neq i, j$. The points x_i and x_j being distinct in this way, and considering assumption (ii), we necessarily have $p(S_i(x)) > 0$, for $S_i(x)$ certainly contains an open subset of R. The convention concerning tied points means $p(S_j(x)) = 0$. Now if $\{y^n\} = \{(y_1^n, y_2^n, \cdots, y_k^n)\}$ is a sequence of k-points satisfying $y_i^n \in R$, and $y_i^n \neq y_j^n$ if $i \neq j$, $n = 1, 2, \cdots$, and the sequence y^n approached x, then y_i^n and y_j^n approach $x_i = x_j$, and hence each other; they also approach the boundaries of $S_i(y^n)$ and $S_j(y^n)$ in the vicinity of x_i. The conditional means $u_i(y^n)$ and $u_j(y^n)$, however, must remain in the interior of the sets $S_i(y^n)$ and $S_j(y^n)$ respectively, and thus tend to become separated from the corresponding points y_i^n and y_j^n. In fact, for each sufficiently large n, the distance of $u_i(y^n)$ from the boundary of $S_i(y^n)$ *or* the distance of $u_j(y^n)$ from the boundary of $S_j(y^n)$, will exceed a certain positive number. For as n tends to infinity, $p(S_i(y^n)) + p(S_j(y^n))$ will approach $p(S_i(x)) > 0$—a simple continuity argument based on the absolute continuity of p will establish this—and for each sufficiently large n, at least one of the probabilities $p(S_i(y^n))$ or $p(S_j(y^n))$ will be positive by a definite amount, say δ. But in view of the boundedness of R, a convex set of p measure at least $\delta > 0$ cannot have its conditional mean arbitrarily near its boundary. This line of reasoning, which extends immediately to the case where some three or more members of $(x_1, x_2, \cdots, x_k)$ are equal, gives us the following lemma.

LEMMA 1. *Let $x = (x_1, x_2, \cdots, x_k)$ be the limit of a convergent sequence of k-points $\{y^n\} = \{(y_1^n, y_2^n, \cdots, y_k^n)\}$ satisfying $y_i^n \in R$, $y_i^n \neq y_j^n$ if $i \neq j$, $n = 1, 2, \cdots$. If $x_i = x_j$ for some $i \neq j$, then $\liminf_n \sum_{i=1}^{k} p(S_i(y^n))|y_i^n - u_i(y^n)| > 0$. Hence, if $\lim_{n\to\infty} \sum_{i=1}^{k} p(S_i(y^n))|y_i^n - u_i(y^n)| = 0$, each member of the k-tuple $(x_1, x_2, \cdots, x_k)$ is distinct from the others.*

We remark that if each member of the k-tuple $x = (x_1, x_2, \cdots, x_k)$ is distinct from the others, then $\pi(y) = (p(S_1(y)), p(S_2(y)), \cdots, p(S_k(y)))$, regarded as a mapping of M onto E_k, is continuous at x—this follows directly from the absolute continuity of p. Similarly, $u(y) = (u_1(y), u_2(y), \cdots, u_k(y))$ regarded as a mapping from M onto M is continuous at x—because of the absolute continuity of p and the boundness of R (finiteness of $\int z\, dp(z)$ would do). Putting this remark together with lemma 1, we get lemma 2.

LEMMA 2. *Let $x = (x_1, x_2, \cdots, x_k)$ be the limit of a convergent sequence of k-points $\{y^n\} = \{(y_1^n, y_2^n, \cdots, y_k^n)\}$ satisfying $y_i^n \in R$, $y_i^n \neq y_j^n$ if $i \neq j$, $n = 1, 2,$*

$\cdots$. *If* $\lim_{n\to\infty} \sum_{i=1}^k p(S_i(y^n))|y_i^n - u_i(y^n)| = 0$, *then* $\sum_{i=1}^k p(S_i(x))|x_i - u_i(x^n)| = 0$ *and each point* x_i *in the k-tuple* $(x_1, x_2, \cdots, x_k)$ *is distinct from the others.*

Lemmas 1 and 2 above are primarily technical in nature. The heart of the proofs of theorems 1 and 2 is the following application of martingale theory.

LEMMA 3. *Let* $t_1, t_2, \cdots$, *and* $\xi_1, \xi_2, \cdots$, *be given sequences of random variables, and for each* $n = 1, 2, \cdots$, *let* t_n *and* ξ_n *be measurable with respect to* β_n *where* $\beta_1 \subset \beta_2 \subset \cdots$ *is a monotone increasing sequence of* σ-*fields (belonging to the underlying probability space). Suppose each of the following conditions holds a.s.:* (i) $|t_n| \lesssim K < \infty$, (ii) $\xi_n \geq 0$, $\sum \xi_n < \infty$, (iii) $E(t_{n+1}|\beta_n) \leq t_n + \xi_n$. *Then the sequences of random variables* $t_1, t_2, \cdots$ *and* $s_0, s_1, s_2, \cdots$, *where* $s_0 = 0$ *and* $s_n = \sum_{i=1}^n (t_i - E(t_{i+1}|\beta_i))$, $n = 1, 2, \cdots$, *both converge a.s.*

PROOF. Let $y_n = t_n + s_{n-1}$ so that the y_n form a martingale sequence. Let c be a positive number and consider the sequence $\{\tilde{y}_n\}$ obtained by stopping y_n (see Doob [3], p. 300) at the first n for which $y_n \leq -c$. From (iii) we see that $y_n \geq -\sum_{i=1}^{n-1} \xi_i - K$, and since $y_n - y_{n-1} \geq 2K$, we have $\tilde{y}_n \geq \max(-\sum_{i=1}^{n-1} \xi_i - K, -(c + 2K))$. The sequence $\{\tilde{y}\}$ is a martingale, so that $E\tilde{y}_n = E\tilde{y}_1$, $n = 1, 2, \cdots$, and being bounded from below with $E|\tilde{y}_1| \leq K$, certainly $\sup_n E|\tilde{y}_n| < \infty$. The martingale theorem ([3], p. 319) shows $\tilde{y}_n$ converges a.s. But $y_n = \tilde{y}_n$ on the set A_c where $-\sum_{i=1}^{\infty} \xi_i > -c - K$, $i = 1, 2, \cdots$, and (ii) implies $P[A_c] \to 1$ as $c \to \infty$. Thus $\{y_n\}$ converge a.s. This means $s_n = y_{n+1} - t_{n+1}$ is a.s. bounded. Using (iii) we can write $-s_n = \sum_{i=1}^n \xi_i - \sum_{i=1}^n \Delta_i$ where $\Delta_i \geq 0$. But since s_n and $\sum_1^n \xi_i$ are a.s. bounded, $\sum \Delta_i$ converges a.s., s_n converges a.s., and finally, so does t_n. This completes the proof.

Turning now to the proof of theorem 1, let ω_n stand for the sequence $z_1, z_2, \cdots, z_{n-1+k}$, and let A_i^n be the event $[z_{n+k} \in S_i^n]$. Since S^{n+1} is the minimum distance partition relative to x^{n+1}, we have

$$
\begin{aligned}
(2.5)\qquad E[W(x^{n+1})|\omega_n] &= E\left[\sum_{i=1}^k \int_{S_i^{n+1}} |z - x_i^{n+1}|^2\, dp(z)|\omega_n\right] \\
&\leq E\left[\sum_{i=1}^k \int_{S_i^n} |z - x_i^{n+1}|^2\, dp(z)|\omega_n\right] \\
&= \sum_{j=1}^k E\left[\sum_{i=1}^k \int_{S_i^n} |z - x_i^{n+1}|^2\, dp(z)|A_j^n, \omega_n\right] p_j^n.
\end{aligned}
$$

If $z_{n+k} \in S_j^n$, $x_i^{n+1} = x_i^n$ for $i \neq j$. Thus we obtain

$$
\begin{aligned}
(2.6)\qquad E[W(x^{n+1})|\omega_n] \leq W(x^n) &- \sum_{j=1}^k \left(\int_{S_j^n} |z - x_j^n|^2\, dp(z)\right) p_j^n \\
&+ \sum_{j=1}^k E\left[\int_{S_j^n} |z - x_j^{n+1}|^2\, dp(z)|A_j^n, \omega_n\right] p_j^n.
\end{aligned}
$$

Several applications of the relation $\int_A |z - x|^2\, dp(z) = \int_A |z - u|^2\, dp(z) + p(A)|x - u|^2$, where $\int_A (u - z)\, dp(z) = 0$, enables us to write the last term in (2.6) as

$$(2.7)\qquad \sum_{j=1}^{k} \Big[\int_{S_j^n} |z - x_j^n|^2 \, dp(z) \, p_j^n - (p_j^n)^2 |x_j^n - u_j^n|^2 + (p_j^n)^2 |x_j^u - u_j^n|^2 (w_j^n/(w_j^n + 1))^2 + \int_{S_j^n} |z - u_j^n|^2 \, dp(z) \, p_j^n/(w_j^n + 1)_2\Big].$$

Combining this with (2.6), we get

$$(2.8)\qquad E(W(x^{n+1})|\omega_n] \leq W(x^n) - \sum_{j=1}^{k} |x_j^n - u_j^n|^2 (p_j^n)^2 (2w_j^n + 1)/(w_j^n + 1)^2 + \sum_{j=1}^{k} \sigma_{n,j}^2 (p_j^n)^2/(w_j^n + 1)^2,$$

where $\sigma_{n,j}^2 = \int_{S_j^n} |z - u_j^n|^2 \, dp(z)/p_j^n$.

Since we are assuming $p(R) = 1$, certainly $W(x^n)$ is a.s. bounded, as is $\sigma_{n,j}^2$. We now show that

$$(2.9)\qquad \sum_{n} (p_j^n)^2/(w_j^n + 1)^2$$

converges a.s. for each $j = 1, 2, \cdots, k$, thereby showing that

$$(2.10)\qquad \sum_{n} \Big(\sum_{j=1}^{k} [\sigma_{n,j}^2 (p_j^n)^2/(w_j^n + 1)^2]\Big)$$

converges a.s. Then lemma 3 can be applied with $t_n = W(x^n)$ and $\xi_n = \sum_{j=1}^{k} \sigma_{n,j}^2 (p_j^n)^2/(w_j^n + 1)^2$.

It suffices to consider the convergence of

$$(2.11)\qquad \sum_{n \geq 2} (p_j^n)^2/[(\beta + 1 + w_j^n)(\beta + 1 + w_j^{n+1})]$$

with $\beta > 0$, since this implies convergence of (2.9). Also, this is convenient, for $E(I_j^n|\omega_n) = p_j^n$ where I_j^n is the characteristic function of the event $[z_{n+k} \in S_j^n]$, and on noting that $w_j^{n+1} = 1 + \sum_{i=1}^{n} I_j^i$, an application of theorem 1 in [4], p. 274, says that for any positive numbers α and β,

$$(2.12)\qquad P\Big[\beta + 1 + w_j^{n+1} \geq 1 + \sum_{i=1}^{n} p_j^i - \alpha \sum_{i=1}^{n} v_j^i \textit{ for all } n = 1, 2, \cdots\Big] > 1 - (1 + \alpha\beta)^{-1},$$

where $v_j^i = p_j^i - (p_j^i)^2$ is the conditional variance of I_j^i given ω_i. We take $\alpha = 1$, and thus with probability at least $1 - (1 + \beta)^{-1}$ the series (2.11) is dominated by

$$(2.13)\qquad \sum_{n \geq 2} (p_j^n)^2 \Big/ \Big[\Big(1 + \sum_{i=1}^{n-1} (p_j^i)^2\Big)\Big(1 + \sum_{i=1}^{n} (p_j^i)^2\Big)\Big] = \sum_{n \geq 2} \Big[1 \Big/ \Big(1 + \sum_{i=1}^{n-1} (p_j^i)^2\Big) - 1 \Big/ \Big(1 + \sum_{i=1}^{n} (p_j^i)^2\Big)\Big],$$

which clearly converges.

The choice of β being arbitrary, we have shown that (2.9) converges a.s. Application of lemma 3 as indicated above proves $W(x^n)$ converges a.s.

To identify the limit W_∞, note that with t_n and ξ_n taken as above, lemma 3 entails a.s. convergence of $\sum_n [W(x^n) - E[W(x^{n+1})|\omega_n]]$, and hence (2.8) implies a.s. convergence of

$$\sum_n \left(\sum_{j=1}^{k} |x^n - u_j^n|^2 (p_j^n)^2 (2w_j^n + 1)/(w_j^n + 1)^2 \right). \tag{2.14}$$

Since (2.14) dominates $\sum_n (\sum_{j=1}^{k} p_j^n |x_j^n - u_j^n)/kn$, the latter converges a.s., and a little consideration makes it clear that

$$\sum_{j=1}^{k} p_j^n |x_j^n - u_j^n| = \sum_{j=1}^{k} p(S_j(x^n))|x_j^n - u_j(x^n)| \tag{2.15}$$

converges to zero on a subsequence $\{x^{n_s}\}$ and that this subsequence has itself a convergent subsequence, say $\{x^{n_t}\}$. Let $x = (x_1, x_2, \cdots, x_k) = \lim_{t\to\infty} x^{n_t}$. Since $W(x) = V(x) + \sum_{j=1}^{k} p(S_j(x))|x_j - u(x)|^2$ and in particular,

$$W(x^n) = V(x^n) + \sum_{j=1}^{k} p(S_j(x^n))|x_j^n - u(x_j^n)|^2, \tag{2.16}$$

we have only to show

(a) $\lim_{t\to\infty} W(x^{n_t}) = W_\infty = W(x)$, and
(b) $\lim_{t\to\infty} \sum_{j=1}^{k} p(S_j(x^{n_t}))|x_j^{n_t} - u(x_j^{n_t})|^2 = 0 = \sum_{j=1}^{k} p(S_j(x))|x_j - u_j(x)|^2$.

Then $W(x) = V(x)$ and x is a.s. unbiased. (Obviously, $\sum_{i=1}^{k} p_i|a_i| = 0$ if and only if $\sum_{i=1}^{k} p_i|a_i|^2 = 0$, where $p_i \geq 0$.)

We show that (a) is true by establishing the continuity of $W(x)$. We have

$$\begin{aligned} W(x) &\leq \sum_{j=1}^{k} \int_{S_j(y)} |z - x_j|^2 \, dp(z) \\ &\leq \sum_{j=1}^{k} \int_{S_j(y)} |z - y_j|^2 + \sum_{j=1}^{k} [p(S_j(y))|x_j - y_j|^2 \\ &\quad + 2|x_j - y_j| \int_{S_j(y)} |z - x_j| \, dp(z)], \end{aligned} \tag{2.17}$$

with the last inequality following easily from the triangle inequality. Thus $W(x) \leq W(y) + o(\rho(x, y))$, and similarly, $W(y) \leq W(x) + o(\rho(x, y))$.

To establish (b), lemma 2 can be applied with $\{y^n\}$ and $\{x^{n_t}\}$ identified, for a.s. $x_i^n \neq x_j^n$ for $i \neq j$, $n = 1, 2, \cdots$. It remains to remark that lemma 2 also implies a.s. $x_i \neq x_j$ for $i \neq j$. The proof of theorem 1 is complete.

Theorem 2 follows from the a.s. convergence of $\sum_n(\sum_{i=1}^{k} p_i^n|x_i^n - u_i^n|)/nk$ upon applying an elementary result (c.f. Halmos [8], theorem C, p. 203), which says that if $\sum a_n/n$ converges, $\sum_{i=1}^{n} a_i/n \to 0$.

2.3. *Remarks.* In a number of cases covered by theorem 1, all the unbiased k-points have the same value of W. In this situation, theorem 1 implies $\sum_{i=1}^{k} p_i^n|x_i^n - u_i^n|$ converges a.s. to zero. An example is provided by the uniform distribution over a disk in E_2. If $k = 2$, the unbiased k-point (x_1, x_2) with $x_1 \neq x_2$ consist of the family of points x_1 and x_2 opposite one another on a diameter, and at a certain fixed distance from the center of the disk. (There is one unbiased k-point with $x_1 = x_2$, both x_1 and x_2 being at the center of the disk in this case.)

The k-means thus converge to some such relative position, but theorem 1 does not quite permit us to eliminate the interesting possibility that the two means oscillate slowly but indefinitely around the center.

Theorem 1 provides for a.s. convergence of $\sum_{i=1}^{k} p_i^n|x_i^n - u_i^n|$ to zero in a slightly broader class of situations. This is where the unbiased k-points $x = (x_1, x_2, \cdots, x_k)$ with $x_i \neq x_j$ for $i \neq j$, are all *stable* in the sense that for each such x, $W(y) \geq W(x)$ (and hence $V(y) \geq V(x)$) for all y in a neighborhood of x. In this case, each such x falls in one of finitely many equivalence classes such that W is constant on each class. This is illustrated by the above example, where there is only a single equivalence class. If each of the equivalence classes contains only a single point, theorem 1 implies a.s. convergence of x^n to one of those points.

There are unbiased k-points which are not stable. Take a distribution on E_2 which has sharp peaks of probability at each corner of a square, and is symmetric about both diagonals. With $k = 2$, the two constituent points can by symmetrically located on a diagonal so that the boundary of the associated minimum distance partition coincides with the other diagonal. With some adjustment, such a k-point can be made to be unbiased, and if the probability is sufficiently concentrated at the corners of the square, any small movement of the two points off the diagonal in opposite directions, results in a decrease in $W(x)$. It seems likely that the k-means *cannot* converge to such a configuration.

For an example where the k-means converge with positive probability to a point x for which $V(x)$ is not a minimum, take equal probabilities at the corner points of a rectangle which is just slightly longer on one side than the other. Number with 1 the corner points, and 2 at the end points of one of the short edges, and 3 and 4, at the end points of the other short edge, with 1 opposite 3 on the long edge. Take $k = 2$. If the first four points fall at the corner points 1, 2, 3, 4 in that order, the two means at this stage are directly opposite one another at the middle of the long edges. New points falling at 1 and 3 will always be nearer the first mean, and points falling at 2 and 4 will always be nearer the second mean, unless one of the means has an excursion too near one of the corner points. By the strong law of large numbers there is positive probability this will *not* happen, and hence with positive probability the two means will converge to the midpoints of the long edges. The corresponding partition clearly does not have minimum within-class variance.

3. Applications

3.1. *Similarity grouping: coarsening and refining.* Perhaps the most obvious application of the k-means process is to the problem of "similarity grouping" or "clustering." The point of view taken in this application is *not* to find some unique, definitive grouping, but rather to simply aid the investigator in obtaining qualitative and quantitative understanding of large amounts of N-dimensional data by providing him with reasonably good similarity groups. The method should be used in close interaction with theory and intuition. Consequently, the computer program actually prepared for this purpose involved several modifications of the k-means process, modifications which appear to be helpful in this sense.

First, the program involves two parameters: C for 'coarsening,' and R for 'refinement.' The program starts with a user specified value of k, and takes the first k points in the sample as initial means. The k-means process is started, each subsequent sample point being assigned to the nearest mean, the new mean computed, and so on, except that after each new point is added, and for the initial means as well, the program determines the pair of means which are nearest to each other among all pairs. If the distance between the members of this pair is less than C, they are averaged together, using their respective weights, to form a single mean. The nearest pair is again determined, their separation compared with C, and so on, until all the means are separated by an amount of C or more. Thus k is reduced and the partition defined by the means is coarsened. In addition, as each new point is processed and its distance from the nearest of the current means determined, this distance is compared with R. If the new point is found to be further than R from the nearest mean, it is left by itself as the seed point for a new mean. Thus k is increased and the partition is refined. Ordinarily we take $C \leq R$. After the entire sample is processed in this way, the program goes back and reclassifies all the points on the basis of nearness to the final means. The points thus associated with each mean constitutes the final grouping. The program prints out the points in each group along with as many as 18 characters of identifying information which may be supplied with each point. The distance of each point from its nearest mean, the distances between the means, the average for each group, of the squared distance of the points in each group from their respective defining means, and the grand average of these quantities over groups, are all printed out. The latter quantity, which is not quite the within-group variance, is called the within-class variation for purposes of the discussion below. If requested, the program determines frequencies of occurrence within each group of the values of discrete variables associated with each point. Up to twelve variables, with ten values for each variable, can be supplied. This makes it convenient to determine whether or not the groups finally obtained are related to other attributes of interest. (Copies of this experimental program are available from the author on request.)

The program has been applied with some success to several samples of real data, including a sample of five dimensional observations on the students' environment in 70 U.S. colleges, a sample of twenty semantic differential measurements on each of 360 common words, a sample of fifteen dimensional observations on 760 documents, and a sample of fifteen physiological observations on each of 560 human subjects. While analysis of this data is still continuing, and will be reported in detail elsewhere, the meaningfulness of the groups obtained is suggested by their obvious pertinence to other identifiable properties of the objects classified. This was apparent on inspection. For example, one group of colleges contained Reed, Swarthmore, Antioch, Oberlin, and Bryn

Mawr. Another group contained the Universities of Michigan, Minnesota, Arkansas, and Illinois, Cornell, Georgia Tech, and Purdue. Selecting at random a half-dozen words from several groups obtained from the semantic differential data, we find in one group the words calm, dusky, lake, peace, sleep, and white; in another group the words beggar, deformed, frigid, lagging, low; and in another group the words statue, sunlight, time, trees, truthful, wise.

When the sample points are rearranged in a new random order, there is some variation in the grouping which is obtained. However, this has not appeared to be a serious concern. In fact, when there are well separated clusters, as determined by inspection of the between-mean distances in relation to the within-class variation, repeated runs give virtually identical groupings. Minor shifts are due to the unavoidable difficulty that some points are located between clusters.

A degree of stability with respect to the random order in which the points are processed is also indicated by a tendency for the within-class variation to be similar in repeated runs. Thus when a sample of 250 points in five dimensions with $k = 18$, was run three times, each time with the points in a different random order, the within-class variation (see above) changed over the three runs by at most 7%. A certain amount of stability is to be expected simply because the within-class variation is the mean of k dependent random variables having the property that when one goes up the others generally go down. We can reasonably expect the within-class stability to generally increase with k and the sample size. Actually, it will usually be desirable to make several runs, with different values of C and R, and possibly adding, deleting, or rescaling variables, and so on, in an effort to understand the basic structure of the data. Thus any instabilities due to random ordering of the sample will be quickly noted. Being able to make numerous classifications cheaply and thereby look at the data from a variety of different perspectives is an important advantage.

Another general feature of the k-means procedure which is to be expected on intuitive grounds, and has been noted in practice, is a tendency for the means and the associated partition to avoid having the extreme of only one or two points in a set. In fact, there is an appreciable tendency for the frequency to be evenly split over groups. If there are a few relatively large groups, these tend to have relatively low within-class variation, as would be expected from a tendency for the procedure to approximate minimum variance partitions.

Running times of the above program on the IBM 7094 vary with C, R, the number of dimensions, and the number of points. A conservative estimate for 20-dimensional data, with C and R set so that k stays in the vicinity of 20, is one minute for two hundred sample points. Most of this computation time results from the coarsening and refining procedure and the auxiliary features. A limited amount of experience indicates the undecorated k-means procedure with $k = 20$ will process five hundred points in 20 dimensions in something like 10 seconds.

3.2. *Relevant classifications.* Suppose it is desired to develop a classification scheme on the basis of a sample, so that knowing the classification of a new point, it will be possible to predict a given dependent variable. The values of the dependent variable are known for the sample. One way to do this, closely related to a procedure proposed by Fix and Hodges [6], is illustrated by the following computer experiment. A sample of 250 four-dimensional random vectors was prepared, with the values on each dimension being independently and uniformly distributed on the integers 1 through 10. Two of the dimensions were then arbitrarily selected, and if with respect to these two dimensions a point was either 'high' (above 5) on both *or* 'low' (5 or less) on both, it was called an A; otherwise, it was called a B. This gave 121 A's and 129 B's which were related to the selected dimensions in a strongly interactive fashion. The k-means with $k = 8$ were then obtained for the A's and B's separately. Finally, using the resulting 16 (four-dimensional) means, a prediction, A or B, was made for each of a *new* sample of 250 points on the basis of whether or not each point was nearest to an A mean or a B mean. These predictions turned out to be 87% correct.

As this example shows, the method is potentially capable of taking advantage of a highly nonlinear relationship. Also, the method has something to recommend it from the point of view of simplicity, and can easily be applied in many dimensions and to more than two-valued dependent variables.

3.3. *Approximating a general distribution.* Suppose it is desired to approximate a distribution on the basis of a sample of points. First the sample points are processed using the k-means concept or some other method which gives a minimum distance partition of the sample points. The approximation, involving a familiar technique, consists of simply fitting a joint normal distribution to the points in each group, and taking as the approximation the probability combination of these distributions, with the probabilities proportional to the number of points in each group.

Having fitted a mixture of normals in this way, it is computationally easy (on a computer) to do two types of analysis. One is predicting unknown coordinates of a new point given the remaining coordinates. This may be done by using the regression function determined on the assumption that the fitted mixture is the true distribution. Another possible application is a kind of nonlinear discriminant analysis. A mixture of k normals is fitted in the above fashion to two samples representing two given different populations; one can then easily compute the appropriate likelihood ratios for deciding to which population a new point belongs. This method avoids certain difficulties encountered in ordinary discriminant analysis, such as when the two populations are each composed of several distinct subgroups, but with some of the subgroups from one population actually between the subgroups of the other. Typically in this situation, one or several of the k-means will be centered in each of the subgroups—provided k is large enough—and the fitted normals then provide a reasonable approximation to the mixture.

To illustrate the application of the regression technique, consider the artificial sample of four-dimensional A's and B's described in the preceding section. On a fifth dimension, the A's were arbitrarily given a value of 10, and the B's a value of 0. The k-means procedure with $k = 16$ was used to partition the combined sample of 250 five-dimensional points. Then the mixture of 16 normal distributions was determined as described above for this sample. The second sample of 250 points was prepared similarly, and predictions were made for the fifth dimension on the basis of the original four. The standard error of estimate on the new sample was 2.8. If, in terms of the original A-B classification, we had called a point on A if the predicted value exceeded 5, and a B otherwise, 96% of the designations would have been correct on the new sample. The mean of the predictions for the A's was 10.3, and for B's, 1.3.

Considering the rather complex and highly nonlinear relationship involved in the above sample, it is doubtful that any conventional technique would do as well. In the few instances which were tested, the method performed nearly as well as linear regression on normally distributed samples, provided k was not too large. This is not surprising inasmuch as with $k = 1$ the method *is* linear regression. In determining the choice of k, one procedure is to increase k as long as the error of estimate drops. Since this will probably result in "over fitting" the sample, a cross validation group is essential.

3.4. *A scrambled dimension test for independence among several variables.* As a general test for relationship among variables in a sample of N-dimensional observations, we propose proceeding as follows. First, the sample points are grouped into a minimum distance partition using k-means, and the within-class variance is determined. Then the relation among the variables is destroyed by randomly associating the values in each dimension; that is, a sample is prepared in which the variables are unrelated, but which has exactly the same marginal distributions as the original sample. A minimum distance partition and the associated within-class variance is now determined for this sample. Intuition and inspection of a few obvious examples suggest that on the average this "scrambling" will tend to *increase* the within-class variance, more or less regardless of whatever type of relation might have existed among the variables, and thus comparison of the two variances would reveal whether or not any such relation existed.

To illustrate this method, a sample of 150 points was prepared in which points were distributed uniformly outside a square 60 units on a side, but inside a surrounding square 100 units on a side. This gave a sample which involves essentially a zero correlation coefficient, and yet a substantial degree of relationship which could not be detected by any conventional quantitative technique known to the author (although it could be detected immediately by visual inspection). The above procedure was carried out using k-means with $k = 12$. As was expected, the variance after scrambling was increased by a factor of 1.6. The within-class variances were not only larger in the scrambled data, but were apparently more variable. This procedure was also applied to the five-dimensional sample described in the preceding section. Using $k = 6$, 12, and 18, the within-class variance increased after scrambling by the factors 1.40, 1.55, and 1.39, respectively.

A statistical test for nonindependence can be constructed by simply repeating the scrambling and partitioning a number of times, thus obtaining empirically a sample from the conditional distribution of the within-class variance under the hypothesis that the variables are unrelated *and* given the marginal values of the sample. Under the hypothesis of independence, the unscrambled variance should have the same (conditional) distribution as the scrambled variance. In fact, the rank of the unscrambled variance in this empirical distribution should be equally likely to take on any of the possible values $1, 2, \cdots, n + 1$, where n is the number of scrambled samples taken, regardless of the marginal distributions in the underlying population. Thus the rank can be used in a nonparametric test of the hypothesis of independence. For example, if the unscrambled variance is the lowest in 19 values of the scrambled variance, we can reject the hypothesis of independence with a Type I error of .05.

A computer program was not available to do the scrambling, and its being inconvenient to set up large numbers of scrambled samples using punched cards, further testing of this method was not undertaken. It is estimated, however, that an efficient computer program would easily permit this test to be applied at, say, the .01 level, on large samples in many dimensions.

The power of this procedure remains to be seen. On the encouraging side is the related conjecture, that for fixed marginal distributions, the within-class variance for the optimal partition as defined in section 1 is maximal when the joint distribution is actually the product of the marginals. If this is true (and it seems likely that it is, at least for a large class of reasonable distributions), then we reason that since the k-means process tends to give a good partition, this difference will be preserved in the scrambled and unscrambled variances, particularly for large samples. Variation in the within-class variance due to the random order in which the points are processed, can be reduced by taking several random orders, and averaging their result. If this is done for the scrambled runs as well, the Type I error is preserved, while the power is increased somewhat.

3.5. *Distance-based classification trees.* The k-means concept provides a number of simple procedures for developing lexigraphic classification systems (filing systems, index systems, and so on) for a large sample of points. To illustrate, we describe briefly a procedure which results in the within-group variance of each of the groups at the most refined level of classification being no more than a specified number, say R. The sample k-means are first determined with a selected value of k, for example, $k = 2$. If the variance of any of the groups of points nearest to these means is less than R, these groups are not subclassified further. The remaining groups are each processed in the same way, that is, k-means are determined for each of them, and then for the points nearest each of these, and so on. This is continued until only groups with within-group variance less than R remain. Thus for each mean at the first level, there is associated several means at the second level, and so on. Once the means at each level are determined from the sample in this fashion, the classification of a new point is defined by the rule: first, see which one of the first level k-means the point is nearest; then see which one of the second-level k-means associated with that mean the point is nearest, and so on; finally the point is assigned to a group which in the determining sample has variance no more than R.

This procedure has some promising features. First, the amount of computation required to determine the index is approximately linear in the sample size and the number of levels. The procedure can be implemented easily on the computer. At each stage during the construction of the classification tree, we are employing a powerful heuristic, which consists simply of putting points which are near to each other in the same group. Each of the means at each level is a fair representation of its group, and can be used for certain other purposes, for instance, to compare other properties of the points as a function of their classification.

3.6. *A two-step improvement procedure.* The method of obtaining partitions with low within-class variance which was suggested by Forgy and Jennrich (see section 1.1) works as follows. Starting with an arbitrary partition into k sets, the means of the points in each set are first computed. Then a new partition of the points is formed by the rule of putting the points into groups of the basis of nearness to the first set of means. The average squared distance of the points in the new partition from the first set of means (that is, from their nearest means) is obviously less than the within-class variance of the first partition. But the average within-class *variance* of the new partition is even lower, for the variance of the squared distance of the points in each group from their respective means, and the mean, of course, is that point which minimizes the average squared distance from itself. Thus the new partition has lower variance. Computationally, the two steps of the method are (1) compute the means of the points in each set in the initial partition and (2) reclassify the points on the basis of nearness to these means, thus forming a new partition. This can be iterated and the series of the partitions thus produced have decreasing within-class variances and will converge in a finite number of steps.

For a given sample, one cycle of this method requires about as much computation as the k-means. The final partition obtained will depend on the initial partition, much as the partition produced by k-means will depend on random variation in the order in which the points are processed. Nevertheless, the procedure has much to recommend it. By making repeated runs with different initial starting points, it would seem likely that one would actually obtain the sample partition with minimum within-class variance.

4. General metric spaces

It may be something more than a mere mathematical exercise to attempt to extend the idea of k-means to general metric spaces. Metric spaces other than Euclidian ones do occur in practice. One prominent example is the space of binary sequences of fixed length under Hamming distance.

An immediate difficulty in making such an extension is the notion of mean itself. The arithmetic operations defining the mean in Euclidian space may not be available. However, with the communication problem of section 1 in mind, one thinks of the problem of representing a population by a point, the goal being to have low average error in some sense. Thus we are led to proceed rather naturally as follows.

Let M be a compact metric space with distance ρ, let $\mathfrak{F}$ be the σ-algebra of subsets of M, and let p be a probability measure on $\mathfrak{F}$. For the measure p, a centroid of order $r \geq 0$ is any point in the set $\mathcal{C}^r$ of points x^* such that $\int \rho^r(x^*, z)\, dp(z) = \inf_x \int \rho^r(x, z)\, dp(z)$. The quantity $\int \rho^r(x^*, z)\, dp(z)$ is the r-th moment of p. The compactness and the continuity of ρ guarantee that $\mathcal{C}^r$ is nonempty. For finite samples, sample centroids are defined analogously, each point in the sample being treated as having measure $1/n$ where n is the sample size; namely, for a sample of size n, the sample centroid is defined up to an equivalence class $\mathcal{C}_n^r$ which consists of all those points $\hat{x}_n$ such that $\sum_{i=1}^n \rho^r(\hat{x}_n, z_i) = \inf_x \sum_{i=1}^n \rho^r(x, z_i)$, where $z_1, z_2, \cdots, z_n$ is the sample.

Note that with M the real line, and ρ ordinary distance, $r = 2$ yields the ordinary mean, and $r = 1$ yields the family of medians. As r tends to ∞, the elements of $\mathcal{C}_r$ will tend to have (in a manner which can easily be made precise) the property that they are centers for a spherical covering of the space with minimal radius. In particular, on the line, the centroid will tend to the mid-range. As r tends to zero, one obtains what may with some justification be called a mode, for on a compact set, $\rho^r(x, y)$ is approximately 1 for small r, except where x and y are very near, so that minimizing $\int \rho^r(x, y)\, dp(y)$ with respect to x, involves attempting to locate x so that there is a large amount of probability in its immediate vicinity. (This relationship can also be made precise.)

We note that the optimum communication problem mentioned in section 1.1 now takes the following general form. Find a partition $S = \{S_1, S_2, \cdots, S_k\}$ which minimizes $w = \sum_{i=1}^k \int_{S_i} \rho^r(x_i^*, y)\, dp(y)$, where x_i^* is the centroid of order r with respect to the (conditional) distribution on S_i. If there is any mass in a set S_i nearer to x_j than to x_i, $j \neq i$, then w can be reduced by modifying S_i and S_j so as to reassign this mass to S_j. It follows that in minimizing w we can restrict attention to partitions which are minimum distance partitions, analogous to those defined in section 2, that is, partitions of the form $S(x) = \{S_1(x), S_2(x), \cdots, S_k(x)\}$ where $x = (x_1, x_2, \cdots, x_k)$ is a k-tuple of points in M, and $S_i(x)$ is a set of points at least as near x_i (in terms of ρ) as to x_j if $j \neq i$. In keeping with the terminology of section 2, we may say that a k-tuple, or "k-point," $x = (x_1, x_2, \cdots, x_k)$ is unbiased if x_i, $i = 1, 2, \cdots, k$, belongs to the class of points which are centroids within $S_i(x)$.

It is now clear how to extend the concept of k-means to metric spaces; the notion of centroid replaces the more special concept of mean. The first 'k-centroid' $(x_1^1, x_2^1, \cdots, x_k^1)$ consists of the first k points in the sample, and thereafter as each new point is considered, the nearest of the centroids is determined. The new point is assigned to the corresponding group and the centroid of that group modified accordingly, and so on.

It would seem reasonable to suppose that the obvious extension of theorem 1 would hold. That is, under independent sampling, $\sum_{i=1}^k \int_{S_i(x^n)} \rho^r(z, x_i^n)\, dp(z)$ will converge a.s., and the convergent subsequences of the sequence of sample k-centroids will have their limits in the class of unbiased k-points. This is true, at any rate, for $k = 1$ and $r = 1$, for if $z_1, z_2, \cdots, z_n$ are independent, $\sum_{i=1}^{n} \rho(z_i, y)/n$ is the mean of independent, identically distributed random variables, which because M is compact, are uniformly bounded in y. It follows (cf. Parzen [13]) that $\sum_{i=1}^{n} \rho(z_i, y)/n$ converges a.s. to $\int \rho(z, y)\, dp(z)$ uniformly in y. By definition of the sample centroid, we have $\sum_{i=1}^{n} \rho(z_i, x^*)/n \geq \sum_{i=1}^{n} \rho(z_i, \hat{x}_n)/n$; hence, $\int \rho(z, x^*)\, dp(z) \geq \limsup \sum_{i=1}^{n} \rho(z_i, \hat{x}_n)/n$ with probability 1. On the other hand, from the triangle inequality, $\sum_{i=1}^{n} \rho(z_i, y)/n \leq \sum_{i=1}^{n} \rho(z_i, \hat{x}_n)/n + \rho(\hat{x}_n, y)$. Using this inequality on a convergent subsequence $\hat{x}_{n_1}, \hat{x}_{n_2}, \cdots$, chosen so that

$$\lim_{t\to\infty} \sum_{i=1}^{n_t} \rho(z_i, \hat{x}_{n_t})/n_t = \liminf \sum_{i=1}^{n} \rho(z_i, \hat{x}_n)/n, \tag{4.1}$$

we see that with probability 1,

$$\int \rho(z, x^*)\, dp(z) \leq \int \rho(z, y)\, dp(z) \leq \liminf \sum_{i=1}^{n} \rho(z_i, \hat{x}_n)/n, \tag{4.2}$$

where $y = \lim_{t\to\infty} \hat{x}_{n_t}$.

Provided the necessary computations can be accomplished, the methods suggested in sections 3.1, 3.2, 3.4, 3.5, and 3.6 can all be extended to general metric spaces in a quite straightforward fashion.

ACKNOWLEDGMENTS

The author is especially indebted to Tom Ferguson, Edward Forgy, and Robert Jennrich, for many valuable discussions of the problems to which the above results pertain. Richard Tenney and Sonya Baumstein provided the essential programming support, for which the author is very grateful. Computing facilities were provided by the Western Data Processing Center.

Note added in proof. The author recently learned that C. S. Wallace of the University of Sidney and G. H. Ball of the Stanford Research Institute have independently used this method as a part of a more complex procedure. Ball has described his method, and reviewed earlier literature, in the interesting paper "Data analysis in the social sciences: What about the details?", *Proceedings of the Fall Joint Computer Conference*, Washington, D.C., Spartan Books, 1965.

REFERENCES

[1] David Blackwell, "Comparison of experiments," *Proceedings of the Second Berkeley Symposium on Mathematical Statistics and Probability*, Berkeley and Los Angeles, University of California Press, 1951, pp. 93–102.

[2] D. R. Cox, "Note on grouping," *J. Amer. Statist. Assoc.*, Vol. 52 (1957), pp. 543–547.
[3] J. L. Doob, *Stochastic Processes*, New York, Wiley, 1953.
[4] L. E. Dubins and L. J. Savage, "A Tchebycheff-like inequality for stochastic processes," *Proc. Nat. Acad. Sci. U.S.A.*, Vol. 53 (1965), pp. 274–275.
[5] W. D. Fisher, "On grouping for maximum homogeneity," *J. Amer. Statist. Assoc.*, Vol. 53 (1958), pp. 789–798.
[6] Evelyn Fix and J. L. Hodges, Jr., "Discriminatory Analysis," USAF Project Report, School of Aviation Medicine, Project Number 21-49-004, No. 4 (1951).
[7] Edward Forgy, "Cluster analysis of multivariate data: efficiency vs. interpretability of classifications," abstract, *Biometrics*, Vol. 21 (1965), p. 768.
[8] Paul R. Halmos, *Measure Theory*, New York, Van Nostrand, 1950.
[9] J. MacQueen, "The classification problem," Western Management Science Institute Working Paper No. 5, 1962.
[10] ———, "On convergence of k-means and partitions with minimum average variance," abstract, *Ann. Math. Statist.*, Vol. 36 (1965), p. 1084.
[11] Jacob Marschak, "Towards an economic theory of organization and information," *Decision Processes*, edited by R. M. Thrall, C. H. Coombs, and R. C. Davis, New York, Wiley, 1954.
[12] ———, "Remarks on the economics of information," Proceedings of the scientific program following the dedication of the Western Data Processing Center, University of California, Los Angeles, January 29–30, 1959.
[13] Emanuel Parzen, "On uniform convergence of families of sequences of random variables," *Univ. California Publ. Statist.*, Vol. 2, No. 2 (1954), pp. 23–54.
[14] George S. Sebestyen, *Decision Making Process in Pattern Recognition*, New York, Macmillan, 1962.
[15] Robert R. Sokal and Peter H. Sneath, *Principles of Numerical Taxonomy*, San Francisco, Freeman, 1963.
[16] Joe Ward, "Hierarchical grouping to optimize an objective function," *J. Amer. Statist. Assoc.*, Vol. 58 (1963), pp. 236–244.

CLASSIFICATION BASED ON DISTANCE IN MULTIVARIATE GAUSSIAN CASES

KAMEO MATUSITA
THE INSTITUTE OF STATISTICAL MATHEMATICS, TOKYO

1. Introduction

The author previously treated the problem of classification in discrete cases, employing the notion of distance [1]. The purpose of this paper is to treat that problem for multivariate Gaussian cases from the same point of view.

Now, the classification problem is formulated as follows. Let $\{\omega_\nu\}$ be a class of sets of distributions, and let X be a random variable under consideration. Then the problem is to decide which ω_ν is considered to contain the distribution of X. We, of course, assume here that ω_ν and ω_μ have no common distributions when $\nu \neq \mu$. Further, for efficient decision making we assume that for a suitable distance $d(\cdot, \cdot)$ in the space of distributions concerned, we have $d(\omega_\nu, \omega_\mu) > \alpha$ (> 0), $(\nu \neq \mu)$. In some cases, when $d(\omega_\nu, \omega_\mu) = 0$, we can represent each of those ω_ν by a single distribution F_ν so that $d(F_\nu, F_\mu) > 0$. For such F_ν, we can consider the averaged distribution of ω_ν by an adequate distribution over ω_ν.

When the distributions concerned are all known, the decision rule for the above problem runs as follows. Let S_n be an 'empirical' distribution based on n observations on X. We compare the magnitudes of $d(S_n, \omega_\nu)$, and take the set which minimizes $d(S_n, \omega_\nu)$ as the set which contains the distribution. Then the problem is to evaluate the success rate or error rate of this procedure. In this paper, however, we shall treat the case where the distributions concerned are unknown. When the distributions concerned are unknown, we have to estimate them from observations. For that, the number of distributions concerned is required to be finite. Therefore, we assume that each ω_ν consists of a single distribution F_ν and the number of F_ν is finite.

In the present paper, we do not explicitly take into account a priori probabilities and costs of misclassification. However, our procedure will also apply with a slight modification to the case where they need to be considered.

2. Decision rule based on distance

Let X be the random variable under consideration, and S_n an 'empirical' distribution based on n observations on X. Suppose that X has one of $F_1, \cdots, F_\ell$ as its distribution. Let S_{ν,n_ν} denote the 'empirical' distribution based on a sample of n_ν from F_ν which has the same form as S_n. Then we consider $d(S_n, S_{\nu,n_\nu})$ and take F_{ν_0} when $S_{\nu_0,n_{\nu_0}}$ minimizes $d(S_n, S_{\nu,n_\nu})$.

Since the case of a finite number of distributions can reduce to the case where the number of distributions concerned is 2, we shall confine our consideration to this case.

Let F_1, F_2 be the distributions concerned, and let S'_r, S''_s be the 'empirical' distributions determined by observations on F_1 and F_2, respectively. Then the decision rule for this case is the following:

(i) when $d(S_n, S'_r) < d(S_n, S''_s)$, we decide on F_1;

and

(ii) when $d(S_n, S'_r) > d(S_n, S''_s)$, we decide on F_2.

(iii) For the case $d(S_n, S'_r) = d(S_n, S''_s)$, we determine in advance to take either of F_1, F_2, say F_1.

The success rate is given by

$$P(d(S_n, S'_r) \leq d(S_n, S''_s)|F_1), \tag{1}$$

where "$|F$" in the parentheses means "under the condition that X has F," and

$$P(d(S_n, S'_r) > d(S_n, S''_s)|F_2). \tag{2}$$

Now, when $d(\cdot, \cdot)$ satisfies the triangle axiom, we obtain

$$d(S_n, S'_r) \leq d(S_n, F_1) + d(S'_r, F_1), \tag{3}$$

$$d(S_n, S''_s) \geq d(F_1, F_2) - d(F_1, S_n) - d(F_2, S''_s), \tag{4}$$

and

$$d(S_n, S''_s) - d(S_n, S'_r) \geq d(F_1, F_2) - 2d(S_n, F_1) - d(F_1, S'_r) - d(F_2, S''_s). \tag{5}$$

Therefore, when $d(F_1, F_2) \geq \delta\ (> 0)$, we have

$$d(S_n, S''_s) - d(S_n, S'_r) \geq \delta - 2d(S_n, F_1) - d(F_1, S'_r) - d(F_2, S''_s), \tag{6}$$

and further, when $2d(S_n, F_1) + d(F_1, S'_r) + d(F_2, S''_s) \leq \delta$, we have $d(S_n, S'_r) \leq d(S_n, S''_s)$. As a result we obtain

$$\begin{aligned} P(d(S_n, S'_r) &\leq d(S_n, S''_s)|F_1) \\ &\geq P(2d(S_n, F_1) + d(F_1, S'_r) + d(F_2, S''_s) < \delta|F_1) \\ &\geq P\left(d(S_n, F_1) < \frac{\delta}{4}, d(F_1, S'_r) < \frac{\delta}{4}, d(F_2, S''_s) < \frac{\delta}{4} \middle| F_1\right) \\ &= P\left(d(S_n, F_1) < \frac{\delta}{4} \middle| F_1\right) \cdot P\left(d(F_1, S'_r) < \frac{\delta}{4}\right) \cdot P\left(d(F_2, S''_s) < \frac{\delta}{4}\right). \end{aligned} \tag{7}$$

Thus, for evaluation of the success rate it is sufficient to know about

$$P\left(d(S_n, F) < \frac{\delta}{4} \middle| F\right). \tag{8}$$

3. Distance and 'test' statistics

Let F_1, F_2 be distributions defined in space R, and let $p_1(x)$, $p_2(x)$ be their density functions with respect to a measure m in R. Then the distance between distributions which we employ here is

(9) $$d(F_1, F_2) = \left[\int_R (\sqrt{p_1(x)} - \sqrt{p_2(x)})^2\, dm\right]^{1/2}.$$

This distance satisfies the metric space axioms. When we define

(10) $$\rho(F_1, F_2) = \int_R \sqrt{p_1(x)}\, \sqrt{p_2(x)}\, dm,$$

we have

(11) $$d^2(F_1, F_2) = 2(1 - \rho(F_1, F_2)).$$

The quantity $\rho(\cdot, \cdot)$ expresses the closeness between distributions, and we can use $\rho(\cdot, \cdot)$ in place of $d(\cdot, \cdot)$.

Now, let us turn to the multivariate Gaussian case.

Let R be a k-dimensional space, and let F_1, F_2 be k-dimensional Gaussian distributions with density functions

(12) $$p_1(x) = \frac{|A|^{1/2}}{(2\pi)^{k/2}} \exp\left[-\tfrac{1}{2}(A(x - a), (x - a))\right],$$

(13) $$p_2(x) = \frac{|B|^{1/2}}{(2\pi)^{k/2}} \exp\left[-\tfrac{1}{2}(B(x - b), (x - b))\right],$$

where A, B are positive-definite matrices of degree k, and x, a, b are k-dimensional (column) vectors. Then we obtain

(14)
$$\rho(F_1, F_2) = \frac{|AB|^{1/4}}{|\frac{1}{2}(A + B)|^{1/2}} \exp\left[-\tfrac{1}{4}\{-((A + B)^{-1}(Aa + Bb), (Aa + Bb)) + (Aa, a) + (Bb, b)\}\right]$$

(see [2]). When $A = B$,

(15) $$\rho(F_1, F_2) = \exp\left[-\tfrac{1}{8}(A(a - b), (a - b))\right].$$

When $a = b$,

(16) $$\rho(F_1, F_2) = \frac{|AB|^{1/4}}{|\frac{1}{2}(A + B)|^{1/2}}.$$

Let $X_1, X_2, \cdots, X_n$ be n (≥ 2) observations on a random variable X with a k-dimensional Gaussian distribution. Define

(17) $$\overline{X} = \frac{1}{n} \sum_{i=1}^{n} X_i,$$

(18) $$V = \frac{1}{n - 1} \sum_{i=1}^{n} (X_i - \overline{X})(X_i - \overline{X})',$$

and let S_n be the k-dimensional Gaussian distribution with mean $\overline{X}$ and covariance matrix V, that is, $S_n = N(\overline{X}, V)$. Set $U = V^{-1}$. Similarly, concerning F_1 and F_2, let

(19) $$S_r' = N(\overline{X}_{(1)}, V_{(1)}), \qquad U_{(1)} = V_{(1)}^{-1},$$

(20) $$S_s'' = N(\overline{X}_{(2)}, V_{(2)}), \qquad U_{(2)} = V_{(2)}^{-1}.$$

Then we have

$$\rho(S_n, S_r') = \frac{|UU_{(1)}|^{1/4}}{|\frac{1}{2}(U + U_{(1)})|^{1/2}} \exp\left[-\tfrac{1}{4}\{-((U + U_{(1)})^{-1}(U\overline{X} + U_{(1)}\overline{X}_{(1)}), (U\overline{X} + U_{(1)}\overline{X}_{(1)})) + (U\overline{X}, \overline{X}) + (U_{(1)}\overline{X}_{(1)}, \overline{X}_{(1)})\}\right], \tag{21}$$

$$\rho(S_n, S_s'') = \frac{|UU_{(1)}|^{1/4}}{|\frac{1}{2}(U + U_{(2)})|^{1/2}} \exp\left[-\tfrac{1}{4}\{-((U + U_{(2)})^{-1}(U\overline{X} + U_{(2)}\overline{X}_{(2)}), (U\overline{X} + U_{(2)}\overline{X}_{(2)})) + (U\overline{X}, \overline{X}) + (U_{(2)}X_{(2)}, \overline{X}_{(2)})\}\right]. \tag{22}$$

Using these statistics we can make a decision; that is, when $\rho(S_n, S_r') \geq \rho(S_n, S_s'')$, we decide that X has F_1, and when $\rho(S_n, S_r') < \rho(S_n, S_s'')$, we decide that X has F_2. When it is known in advance that $A = B$, we consider

$$\rho_1(S_n, S_r') = \exp\left[-\tfrac{1}{4}(U(\overline{X} - \overline{X}_{(1)}), (\overline{X} - \overline{X}_{(1)}))\right], \tag{23}$$

$$\rho_1(S_n, S_s'') = \exp\left[-\tfrac{1}{4}(U(\overline{X} - \overline{X}_{(2)}), (\overline{X} - \overline{X}_{(2)}))\right] \tag{24}$$

for the case where A $(= B)$ is unknown, and

$$\rho_2(S_n, S_r') = \exp\left[-\tfrac{1}{4}(A(\overline{X} - \overline{X}_{(1)}), (\overline{X} - \overline{X}_{(1)}))\right], \tag{25}$$

$$\rho_2(S_n, S_s'') = \exp\left[-\tfrac{1}{4}(A(\overline{X} - \overline{X}_{(2)}), (\overline{X} - \overline{X}_{(2)}))\right] \tag{26}$$

for the case where A $(= B)$ is known.

When the problem is concerned only with the covariance matrix, we consider

$$\rho_3(S_n, S_r') = \frac{|UU_{(1)}|^{1/4}}{|\frac{1}{2}(U + U_{(1)})|^{1/2}}, \tag{27}$$

$$\rho_3(S_n, S_s'') = \frac{|UU_{(2)}|^{1/4}}{|\frac{1}{2}(U + U_{(2)})|^{1/2}}. \tag{28}$$

For instance, when

$$\frac{|UU_{(1)}|^{1/4}}{|\frac{1}{2}(U + U_{(1)})|^{1/2}} \geq \frac{|UU_{(2)}|^{1/4}}{|\frac{1}{2}(U + U_{(2)})|^{1/2}}, \tag{29}$$

we decide that X has F_1, and when

$$\frac{|UU_{(1)}|^{1/4}}{|\frac{1}{2}(U + U_{(1)})|^{1/2}} < \frac{|UU_{(2)}|^{1/4}}{|\frac{1}{2}(U + U_{(2)})|^{1/2}}, \tag{30}$$

we decide that X has F_2. (For the case where these two statistics are equal, we can, of course, determine in advance to take F_2.)

As to the success rate, we obtain

$$P(\rho(S_n, S_r') > \rho(S_n, S_s'')|F_1) \geq P\left(\rho(S_n, F_1) > \frac{1-\delta}{16}\middle|F_1\right) \times P\left(\rho(S_r', F_1) > \frac{1-\delta}{16}\right) \cdot P\left(\rho(S_s'', F_2) > \frac{1-\delta}{16}\right), \tag{31}$$

and from this relation we can get an evaluation of the success rate, when we have the value of $P(\rho(F, S_n) > \delta | F)$. Thus the next problem is to evaluate $P(\rho(F, S_n) > \delta | F)$ $(\delta < 1)$.

Assume that X is distributed according to $N(a, \Sigma)$. First, concerning $\rho_1(F, S_n)$, $\rho_2(F, S_n)$, we have

$$-8n \log \rho_1(F, S_n) = n(V^{-1}(\overline{X} - a), (\overline{X} - a)), \tag{32}$$

$$-8n \log \rho_2(F, S_n) = n(\Sigma^{-1}(\overline{X} - a), (\overline{X} - a)), \tag{33}$$

and, as is well known, the right-hand sides have a noncentral F and a chi-square distribution, and we have no problem here.

Concerning $\rho_3(F, S_n)$, we have

$$\rho_3(F, S_n) = \frac{|\Sigma^{-1}U|^{1/4}}{|\frac{1}{2}(\Sigma^{-1} + U)|^{1/2}}, \tag{34}$$

and

$$P(\rho_3(F, S_n) > \delta) \geq \left[P\left(\frac{4Z}{(1 + Z)^2} > \delta\right)\right]^k, \tag{35}$$

where Z is a random variable such that nZ has the chi-square distribution with n degrees of freedom (see [2]). Therefore, for given positive δ and ϵ (<1), there exists an integer n_0 such that $P(\rho_3(F, S_n) > \delta) \geq 1 - \epsilon$ uniformly in F for $n \geq n_0$.

Now we will present the general case. Let δ_1, δ_2 be positive numbers such that $\delta = \delta_1 \exp[-(1/4)\delta_2]$, $\delta_1 < 1$. Then we get

$$\begin{aligned} &P(\rho(F, S_n) > \delta) \\ &\geq P\left(\frac{|\Sigma^{-1}W^{-1}|^{1/4}}{|\frac{1}{2}(\Sigma^{-1} + W^{-1})|^{1/2}} > \delta_1\right) P(\beta(\Sigma^{-1}(X - a), (X - a)) < 2\delta_2) \end{aligned} \tag{36}$$

where

$$W = \begin{pmatrix} \frac{1}{n}\sum_{i=1}^{n} X_{i1}^2 & & 0 \\ & \ddots & \\ 0 & & \frac{1}{n}\sum_{i=1}^{n} X_{ik}^2 \end{pmatrix}, \qquad X_i = \begin{pmatrix} X_{i1} \\ \vdots \\ X_{ik} \end{pmatrix}, \tag{37}$$

$$\beta = \frac{2}{\delta_1^4} - 1 + \frac{2}{\delta_1^4}\sqrt{1 - \delta_1^4} \tag{38}$$

(see [2]). By taking δ_1 (accordingly δ_2) so that the right-hand side becomes maximum, we can get an evaluation (from below) of $P(\rho(F, S_n) > \delta | F)$. When we want to have $P(\rho(F, S_n) > \delta) > 1 - \epsilon$, let $1 - \epsilon = \alpha_1\alpha_2$, $\alpha_1, \alpha_2 > 0$ and take n large so that

$$P\left(\frac{|\Sigma^{-1}W^{-1}|^{1/4}}{|\frac{1}{2}(\Sigma^{-1} + W^{-1})|^{1/2}} > \delta_1\right) \geq \alpha_1, \tag{39}$$

$$P((\Sigma^{-1}(\overline{X} - a), (\overline{X} - a)) < 2\delta_2) \geq \alpha_2. \tag{40}$$

4. Classification by a linear function of vector components

In this section we consider the classification problem by a linear function of components of a random vector.

Let $N(a^{(1)}, \Sigma_1)$, $N(a^{(2)}, \Sigma_2)$ be k-dimensional Gaussian distributions, and let $X = (X_1, \cdots, X_k)$ be a k-dimensional random vector. The problem is to decide which one of $N(a^{(1)}, \Sigma_1)$, $N(a^{(2)}, \Sigma_2)$ is the distribution of X. For this problem we consider a linear function of the components of X of the form $(c, X) = c_1X_1 + \cdots + c_kX_k$, where c is a constant vector ($\neq 0$). The decision procedure is as follows. Let $X^{(1)}$, $X^{(2)}$ be samples from $N(a^{(1)}, \Sigma_1)$, $N(a^{(2)}, \Sigma_2)$, and let F_{1c}, F_{2c} be the distributions of $(c, X^{(1)})$ and $(c, X^{(2)})$. Further, let E_1, E_2 be optimal regions (on the real line) for classifying an observation from F_{1c}, or F_{2c}. (For instance, E_1, E_2 can be defined by the probability ratio rule.) Then, when (c, X) lies in E_1, we decide that X has $N(a^{(1)}, \Sigma_1)$, and when (c, X) lies in E_2, we decide that X has $N(a^{(2)}, \Sigma_2)$. Therefore, for reducing the probability of misclassification, it is necessary to find an adequate c.

Now, we have

$$
\begin{aligned}
E(c, X^{(1)}) &= (c, a^{(1)}),\\
V(c, X^{(1)}) &= (c, \Sigma_1 c),\\
E(c, X^{(2)}) &= (c, a^{(2)}),\\
V(c, X^{(2)}) &= (c, \Sigma_2 c),
\end{aligned} \tag{41}
$$

and

$$
\rho(F_{1c}, F_{2c}) = \left[\frac{2(c, \Sigma_1 c)^{1/2}(c, \Sigma_2 c)^{1/2}}{(c, \Sigma_1 c) + (c, \Sigma_2 c)}\right]^{1/2} \cdot \exp\left[-\frac{1}{4}\frac{(c, a^{(1)} - a^{(2)})^2}{(c, (\Sigma_1 + \Sigma_2)c)}\right]. \tag{42}
$$

Therefore, from our standpoint, we should choose a c that minimizes $\rho(F_{1c}, F_{2c})$ when $a^{(1)}$, $a^{(2)}$, Σ_1, Σ_2 are known. When $a^{(1)}$, $a^{(2)}$, Σ_1, Σ_2 are unknown, we use in place of them their estimates obtained from samples.

For example, when it is known beforehand that $\Sigma_1 = \Sigma_2$, we consider

$$
\frac{(c, a^{(1)} - a^{(2)})^2}{(c, \Sigma_1 c)} \tag{43}
$$

and determine c so as to maximize this value. (This is a familiar procedure in multivariate analysis.)

REFERENCES

[1] K. MATUSITA, "Decision rule, based on the distance, for the classification problem," *Ann. Inst. Statist. Math.*, Vol. 8 (1956), pp. 67–77.

[2] ———, "A distance and related statistics in multivariate analysis," *Proceedings of the International Symposium on Multivariate Analysis*, New York, Academic Press, 1966.

ON EMPIRICAL MULTIPLE TIME SERIES ANALYSIS

EMANUEL PARZEN
STANFORD UNIVERSITY

1. Introduction

Like probability theory, modern time series analysis has the feature that many of its most elementary theorems are based on rather deep mathematics, while many of its most advanced theorems are known and understood by research workers who do not have the mathematical background to understand the proofs. It is natural to think of the theory of time series analysis as composed of two parts, *foundations* (a probabilistic part involving deep mathematics and based on the unrealistic assumption that one knows the probability law of the time series) and *empirical* (in which one considers statistical and computational procedures). While the probabilistic theory of time series can be pursued for the sake of its great beauty, it would be a mistake if the statistical theory were to be developed only for its elegance. The ultimate aim of the statistical theory of time series analysis must be to provide *data-handling procedures* for achieving the aim of time series analysis, *synthesis of stochastic models* which can be used to describe and perhaps to control the mechanisms generating each time series and relating various time series. For this reason, one may define a field which may be called "empirical time series analysis" with aims such as the following:

(1) to develop the statistical theory in such a way that it provides a philosophy for judging and interpreting the statistical data reduction which can be provided by computers;

(2) to develop efficient computer programs for the statistical analysis of empirical time series;

(3) to obtain experience in the small sample applicability and robustness of statistical procedures derived from asymptotic theory;

(4) to focus attention on theoretical questions requiring further investigation.

One of my concerns in recent years has been to develop a computer program for empirical time series analysis. There were several reasons motivating this concern:

(1) I discovered that when a researcher came to me for advice on time series analysis, I could do him the most good by (in addition to telling him which formulas to use) making available to him a computer program for carrying out the analysis.

Prepared with the partial support of the Office of Naval Research and the National Science Foundation. Reproduction is permitted for any purpose of the United States Government.

(2) I was curious to see if there were any truth to the proposition that a statistician interested in data analysis need not be interested in theorems, since experience with computer output would provide all the insights he needs; my experience leads me to conclude that a knowledge of relevant theorems is indispensable if one desires to be able to interpret as many features of the computer output as possible.

(3) I desired to develop an approach to empirical time series analysis. Before describing this approach, let me quote some recent remarks of John Tukey (who in my view is "The Father of Empirical Time Series Analysis"). Tukey notes ([30], p. 1284) that, "It is a commonplace of science that where one can, one learns faster by deliberately reaching in and changing something, by seeing what happens when something is varied in a controlled way." Unfortunately, time series often arise in the field sciences rather than in the laboratory sciences; and it is nature rather than the observer who determines the conditions under which the data will be observed. Nevertheless, quoting Tukey again, "How can at least some of the advantages of reaching in be had when one can only sit and look?" One important answer is given by Tukey, "The answer is simple and well known: look in two [or more] places and try to assess the relationship of the things observed." He sums up this point of view in the maxim, "Look here, look there, compare, and interrelate." I believe there is another way in which to compare and interrelate; this is by varying the way in which one analyzes the data. One should consider a variety of models for the observed time series. For each model one should estimate the parameters which represent the incompletely specified characteristics of the probability law. Comparing the analyses often provides rough tests of hypotheses concerning which model provides a better fit to the data.

As in statistics, so in time series analysis, one may distinguish three main problems:

(1) *estimation* of the parameters of a given model for the observed time series (in particular, in any field where the properties of the phenomenon being studied can be characterized in terms of its behavior in the frequency domain, one needs to estimate spectral density functions and other spectral characteristics associated with stationary multiple time series);

(2) *hypothesis testing* and hypothesis suggesting (testing the fit of various models and suggesting possible models to fit); and

(3) *description* (to provide measurements about a phenomenon, which together with other kinds of measurements, represent the observational regularities which it is a purpose of any theory of the phenomenon to explain).

Some techniques for fitting models to single time series have been discussed in previous papers (see Parzen [22], [24]). Fitting models to multiple time series seems a much harder problem. While cross-spectral analysis is clearly one of the main tools for fitting models to multiple time series, it is not yet clear what are the sample cross-spectral functions which should be routinely computed. To an observed sample of a multiple time series one can associate a bewildering array of cross-spectral density quantities involving such adjectives as "co-spectral," "quadrature-spectral," "partial cross-spectral," and such nouns as "amplitude," "phase," "coherence," and "gain."

The aim of this paper is to sketch a unified exposition of cross-spectral analysis. The exposition is not entirely rigorous, but rather attempts to indicate the theoretical questions which require further investigation. Among the results believed to be new are those in the section on the asymptotic sampling theory of partial cross-spectra. Rigorous proofs of all results stated are given in the Stanford Ph.D. thesis of Grace Wahba, June 1966.

Applied statisticians, actually computing sample spectra, often complain that papers written on spectral analysis are highly mathematical and offer no guide on how to proceed in practice. I am willing to grant some merit to this complaint in general. While this paper is not by itself a guide to how to proceed in practice, I hope that it will be of value as a discussion of some of the main mathematical considerations which need to be borne in mind in order to interpret sample cross-spectra. Excellent introductions to general considerations in empirical time series analysis are given by Jenkins [12] and Tukey [29].

2. Sample cross-spectra

Observed time series come in a variety of shapes. Economic and social time series often have the typical shapes shown in figure 1. In analyzing observed time series, I have found it valuable to distinguish two consecutive stages: (i) time series transformation and detrending, and (ii) correlation and spectral computations. In forming sample correlations and covariances, one should not automatically subtract out sample means (or fitted straight lines, and so on); any such subtractions should be done in the time series transformation and detrending stage. Consequently, given finite samples of r real time series

$$\{X_1(t), t = 1, 2, \cdots, T\}, \cdots, \{X_r(t), t = 1, 2, \cdots, T\}, \tag{2.1}$$

we make the following definitions.

The sample cross-covariance $R_{hj;T}(v)$ of lag v between $X_h(\cdot)$ and $X_j(\cdot)$ is defined to be

$$\begin{aligned} R_{hj;T}(v) &= \frac{1}{T}\sum_{t=1}^{T-v} X_h(t)X_j(t+v) && \text{for } v = 0, 1, \cdots, T-1, \\ R_{hj;T}(v) &= \frac{1}{T}\sum_{t=-v+1}^{T} X_h(t)X_j(t+v) && \text{for } v = -1, -2, \cdots, -(T-1), \\ R_{hj;T}(v) &= 0 && \text{otherwise.} \end{aligned} \tag{2.2}$$

Since

$$R_{jh;T}(v) = \frac{1}{T}\sum_{t=1}^{T-v} X_j(t)X_h(t+v) = \frac{1}{T}\sum_{s=v+1}^{T} X_h(t)X_j(s-v) = R_{hj;T}(-v), \tag{2.3}$$

it suffices to compute all the cross-covariances for positive lags v to know them for all v.

The sample cross-spectral density function between $X_h(\cdot)$ and $X_j(\cdot)$ is defined by (writing i for $\sqrt{-1}$)

$$f_{hj;T}(\omega) = \frac{1}{2\pi T} \sum_{s=1}^{T} e^{i\omega s} X_h(s) \sum_{t=1}^{T} e^{-i\omega t} X_j(t). \tag{2.4}$$

It may be verified that these quantities form a pair of Fourier transforms:

$$R_{hj;T}(v) = \int_{-\pi}^{\pi} e^{i\omega v} f_{hj;T}(\omega)\, d\omega, \tag{2.5}$$

$$f_{hj;T}(\omega) = \frac{1}{2\pi} \sum_{|v|<T} e^{-iv\omega} R_{hj;T}(v). \tag{2.6}$$

The sample cross-correlation function $\rho_{hj;T}(v)$ is defined by

$$\rho_{hj;T}(v) = R_{hj;T}(v) \div \{R_{hh;T}(0) R_{jj;T}(0)\}^{1/2}. \tag{2.7}$$

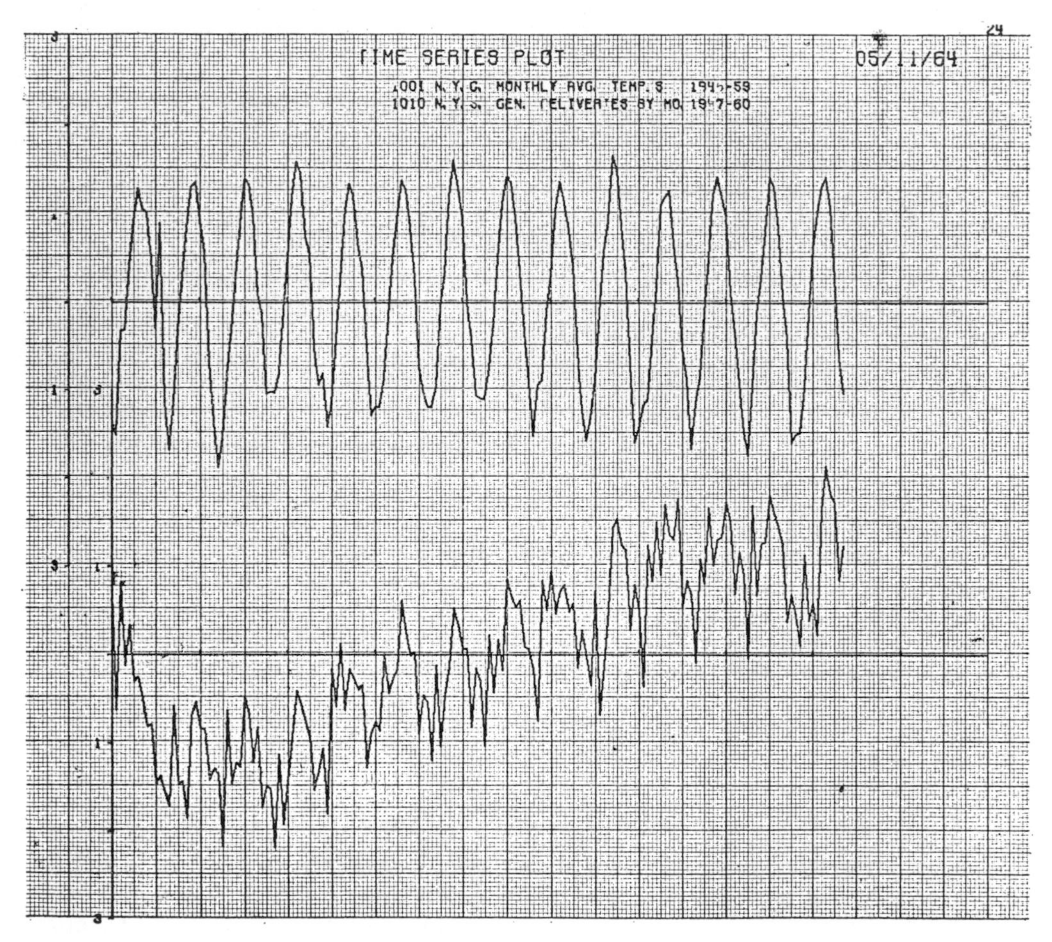

FIGURE 1

The normalized sample spectral density function $\bar{f}_{hj;T}(\omega)$ is defined by

$$\bar{f}_{hj;T}(\omega) = f_{hj;T}(\omega) \div \{R_{hh;T}(0)R_{jj;T}(0)\}^{1/2}. \tag{2.8}$$

The sample cross-spectral density function is generally complex-valued. The following notation and terminology, due to Tukey, is used to describe the real and negative imaginary parts of $f_{hj;T}(\omega)$:

$$\begin{aligned} c_{hj;T}(\omega) &= \operatorname{Re} f_{hj;T}(\omega), \quad \textit{sample co-spectral density,} \\ q_{hj;T}(\omega) &= -\operatorname{Im} f_{hj;T}(\omega), \quad \textit{sample quadrature spectral density.} \end{aligned} \tag{2.9}$$

In the foregoing definitions, we are guided by the idea that when the observed time series are *zero mean* covariance stationary time series, the sample cross-spectral quantities should provide sample versions of corresponding population cross-spectra. However, these quantities can be defined for *any* sample, and their statistical characteristics can be investigated for *any* model that one may want to consider for an observed set of time series. Consequently, one can interpret sample cross-spectra without necessarily making the assumption of zero mean covariance stationarity.

One important class of models for time series for which one desires to understand the properties of sample cross-spectra is the following: for $j = 1, 2, \cdots, r$

$$X_j(t) = m_j(t) + Z_j(t), \qquad t = 1, 2, \cdots, \tag{2.10}$$

where $m_j(\cdot)$ is the mean value function of $X_j(\cdot)$,

$$m_j(t) = E[X_j(t)], \tag{2.11}$$

and $\mathbf{Z}(\cdot) = (Z_1(\cdot), \cdots, Z_r(\cdot))$ has zero means, is jointly normal, and is covariance stationary with covariance functions

$$R_{hj}(v) = E[Z_h(t)Z_j(t+v)] \tag{2.12}$$

(for $h, j = 1, \cdots, r; t = 1, 2, \cdots$; and $v = 0, \pm 1, \pm 2, \cdots$), and *spectral density matrix*

$$\mathbf{f}(\omega) = \begin{bmatrix} f_{11}(\omega) & \cdots & f_{1r}(\omega) \\ \cdot & & \cdot \\ \cdot & & \cdot \\ \cdot & & \cdot \\ f_{r1}(\omega) & \cdots & f_{rr}(\omega) \end{bmatrix} \tag{2.13}$$

satisfying

$$R_{hj}(v) = \int_{-\pi}^{\pi} e^{iv\omega} f_{hj}(\omega)\, d\omega. \tag{2.14}$$

The diagonal element $f_{jj}(\omega)$ is called the spectral density function of the series $X_j(\cdot)$; the (h, j)-th element $f_{hj}(\omega)$ of the spectral density matrix is called the *cross-spectral density* of the series $X_h(\cdot)$ and $X_j(\cdot)$. Following the terminology introduced by Tukey, the real and negative imaginary parts of $f_{hj}(\omega)$ are called, respectively, the *co-spectral* density, denoted $c_{hj}(\omega)$, and *quadrature spectral* density, denoted $q_{hj}(\omega)$.

We do not demand that time series submitted for spectral analysis have vanishing mean value functions. Therefore, in studying the behavior of sample spectra, we must distinguish two general cases: the observed time series are jointly covariance stationary with absolutely continuous spectrum and have:

(1) zero means,
(2) possibly nonzero means.

We call the second case the *mixed spectrum* case.

The problem of mixed spectra has been extensively discussed for univariate time series (see Hext [10] for a history of the problem). The theory of mixed cross-spectral analysis of multiple time series is not discussed in this paper (which is already too long) but will be discussed in a separate paper. (In her Ph.D. thesis, Grace Wahba gives a rigorous derivation of the small sample distribution theory of sample cross-spectral estimates for jointly stationary normal multiple time series with bounded nonzero mean value functions.)

In order to study the properties of the sample cross-spectral density function in the mixed spectrum case, one would introduce the sample cross-spectral density function of $m_h(\cdot)$ and $m_j(\cdot)$, defined by

$$f_{hj;m,T}(\omega) = \frac{1}{2\pi T} \sum_{t=1}^{T} e^{i\omega t} m_h(t) \sum_{t=1}^{T} e^{-i\omega t} m_j(t). \tag{2.15}$$

The sample cross-spectral density function of the time series $Z_h(\cdot)$ is defined similarly:

$$f_{hj;Z,T}(\omega) = \frac{1}{2\pi T} \sum_{t=1}^{T} e^{i\omega t} Z_h(t) \sum_{t=1}^{T} e^{-i\omega t} Z_j(t). \tag{2.16}$$

One use of these expressions is in writing the mean of a sample cross-spectral density:

$$E[f_{hj;X,T}(\lambda)] = E[f_{hj;Z,T}(\lambda)] + f_{hj;m,T}(\lambda). \tag{2.17}$$

It is important to note that in order to study the properties of sample spectra it is *not* necessary to assume that the sample cross-spectral density function of the mean value functions, defined by (2.15), possesses a limit as T tends to ∞.

3. Windowed sample cross-spectra

As is well known, if one is seeking to estimate the spectral density functions of covariance stationary time series, one cannot use the sample spectral density functions but must use *windowed* sample spectra.

Given a kernel $k(v)$ and truncation point M, the *windowed cross-spectral density function*, denoted $f_{hj;T,M}(\omega)$, is defined by

$$f_{hj;T,M}(\omega) = \frac{1}{2\pi} \sum_{|v|<M} e^{-iv\omega} k\left(\frac{v}{M}\right) R_{hj;T}(v). \tag{3.1}$$

Its real and negative imaginary parts, denoted $c_{hj;T,M}(\omega)$ and $q_{hj;T,M}(\omega)$, are called respectively the windowed sample co-spectral density function and the windowed sample quadrature spectral density function. The *windowed normalized cross-*

spectral density function, denoted $\bar{f}_{hj;T,M}(\omega)$, is defined similarly in terms of the sample cross-correlation function:

$$\bar{f}_{hj;T,M}(\omega) = \frac{1}{2\pi} \sum_{|v|<M} e^{-iv\omega} k\left(\frac{v}{M}\right) \rho_{hj;T}(v). \tag{3.2}$$

For ease of comparing sample spectra arising from different time series, I believe it is wisest to *compute and plot normalized versions of these functions.* Indeed, I believe that normalization is vital for interpretation and that it facilitates the exchange of ideas among research workers concerned with time series arising in quite different fields. It should be noted that the theory of normalized spectra is more difficult than the unnormalized theory.

There is an extensive literature (in particular, see *Technometrics* [28] and Jenkins [12]) concerning the choice of the function $k(\cdot)$, called the lag window, and the integer $M(<T)$, called the *truncation* point (since it represents the number of sample correlations actually used in computing the spectrum). It should be noted that most methods of computing sample spectra can be essentially represented in the form (3.1) even if a formula of this kind is not explicitly employed. An extensive comparison of the effects of different choices of $k(\cdot)$ and M is beyond the scope of this paper (although an empirical comparison of a few windows is given in the next section).

At this point, let us merely note the choices of $k(\cdot)$ and M we normally make. In our work we use mainly the following lag window:

$$\begin{aligned} k(u) &= 1 - 6u^2 + 6|u|^3, && |u| \leq 0.5, \\ &= 2(1 - |u|)^3 && 0.5 \leq |u| \leq 1.0, \\ &= 0, && |u| \geq 1. \end{aligned} \tag{3.3}$$

A kernel widely used in existing spectral analysis programs is one suggested by Tukey (see Blackman and Tukey [5], p. 14):

$$\begin{aligned} k(u) &= \tfrac{1}{2}(1 + \cos \pi u), && |u| < 1, \\ &= 0, && \text{otherwise.} \end{aligned} \tag{3.4}$$

This lag window is not used in our work because the corresponding windowed spectrum is not necessarily nonnegative (and the corresponding estimates of coherence are not necessarily between 0 and 1).

Two other kernels which might be considered are one generally known as the Bartlett kernel,

$$\begin{aligned} k(u) &= 1 - |u|, && |u| \leq 1, \\ &= 0, && \text{otherwise,} \end{aligned} \tag{3.5}$$

and one which we call the Bohman kernel (after Bohman [6] who introduced it in connection with the numerical inversion of characteristic functions to compute distribution functions),

$$
\begin{aligned}
(3.6)\qquad k(u) &= (1-u)\cos \pi u + \frac{1}{\pi}\sin \pi u, && 0 < u < 1,\\
&= 0, && u > 1,\\
&= k(-u), && u < 0.
\end{aligned}
$$

The *spectral window* of a windowed sample spectrum of the form of (3.2) is defined to be the function

$$
(3.7)\qquad K_M(\omega) = \frac{1}{2\pi} \sum_{|v| \leq M} e^{-iv\omega} k\left(\frac{v}{M}\right);
$$

the *spectral window* generator is defined to be the Fourier transform

$$
(3.8)\qquad K(\omega) = \frac{1}{2\pi} \int_{-\infty}^{\infty} e^{-iu\omega} k(u)\, du.
$$

For the lag window (3.4), it may be shown that

$$
(3.9)\qquad K_M(\omega) = \frac{3}{8\pi M^3} \left\{ \frac{\sin (M\omega/4)}{\frac{1}{2}\sin \frac{\omega}{2}} \right\}^4 \left\{ 1 - \tfrac{2}{3}\left(\sin \frac{\omega}{2}\right)^2 \right\},
$$

$$
(3.10)\qquad K(\omega) = \frac{3}{8\pi} \left\{ \frac{\sin (\omega/4)}{\omega/4} \right\}^4.
$$

It may be shown that a windowed sample spectral density function $f_{hj;T,M}(\omega)$ is the convolution of the sample cross-spectral density function $f_{hj;T,}(\omega)$, and the spectral window $K_M(\omega)$,

$$
(3.11)\qquad f_{hj;T,M}(\omega) = \int_{-\pi}^{\pi} K_M(\omega - \lambda) f_{hj;T}(\lambda)\, d\lambda.
$$

Therefore, its mean is also a convolution,

$$
(3.12)\qquad E[f_{hj;T,M}(\omega)] = \int_{-\pi}^{\pi} K_M(\omega - \lambda) E[f_{hj;T}(\lambda)]\, d\lambda.
$$

While it is more difficult to justify universal advice on the choice of the truncation point, my experience leads me to believe that it is necessary and sufficient to use *three* truncation points, M_1, M_2, M_3, satisfying a condition of the following kind:

$$
(3.13)\qquad 5\% \leq \frac{M_1}{T} \leq 10\%, \qquad 10\% \leq \frac{M_2}{T} \leq 25\%, \qquad 25\% \leq \frac{M_3}{T} \leq 75\%.
$$

I have several justifications for this advice: (1) in general, if one is in doubt as to which of two ways to perform an analysis, one should do it both ways and decide by a comparison of results which way was right; (2) the three truncation points given in (3.13) span the range of possible truncation points, and not too much additional information can be obtained by using additional truncation points; and (3) the presence of peaks and the smoothness of spectra can be determined by comparing spectra corresponding to different truncation points.

Given finite samples of r time series, for brevity we often denote by $\hat{f}_{hj}(\omega)$ the windowed normalized cross-spectral density function defined by (3.2). We let

$$\hat{\mathbf{f}}(\omega) = \begin{bmatrix} \hat{f}_{11}(\omega) & \cdots & \hat{f}_{1r}(\omega) \\ \cdot & & \cdot \\ \cdot & & \cdot \\ \cdot & & \cdot \\ \hat{f}_{r1}(\omega) & \cdots & \hat{f}_{rr}(\omega) \end{bmatrix} \tag{3.14}$$

denote the windowed sample cross-spectral density matrix. A variety of derived spectral quantities may be computed which hopefully will provide insight into the relations among the observed time series. For two series $X_h(\cdot)$ and $X_j(\cdot)$, one can form the following derived sample spectral quantities: the *sample regression transfer function*

$$\hat{B}_{h;j}(\omega) = \frac{\hat{f}_{hj}(\omega)}{\hat{f}_{jj}(\omega)} = \frac{\hat{c}_{hj}(\omega)}{\hat{f}_{jj}(\omega)} - i\,\frac{\hat{q}_{hj}(\omega)}{\hat{f}_{jj}(\omega)} = \hat{\alpha}_{h;j}(\omega) + i\hat{\beta}_{h;j}(\omega), \tag{3.15}$$

the *sample residual spectral density function*

$$\hat{f}_{h;j}(\omega) = \hat{f}_{hh}(\omega)\{1 - \hat{W}_{hj}(\omega)\} \tag{3.16}$$

where $\hat{W}_{hj}(\omega)$, called the *sample coherence* between series h and series j, is defined by

$$\hat{W}_{hj}(\omega) = \frac{|\hat{f}_{hj}(\omega)|^2}{\hat{f}_{hh}(\omega)\hat{f}_{jj}(\omega)} = \frac{\hat{c}_{hj}^2(\omega) + \hat{q}_{hj}^2(\omega)}{\hat{f}_{hh}(\omega)\hat{f}_{jj}(\omega)}. \tag{3.17}$$

The regression transfer function may be written as

$$\hat{B}_{h;j}(\omega) = \hat{G}_{h;j}(\omega)e^{-i\hat{\varphi}_{h;j}(\omega)} \tag{3.18}$$

where

$$\hat{G}_{h;j}(\omega) = \frac{|\hat{f}_{hj}(\omega)|}{\hat{f}_{jj}(\omega)}, \tag{3.19}$$

called the *sample gain* at frequency ω of the predictor of $X_h(\cdot)$ given $X_j(\cdot)$, and

$$\begin{aligned} \hat{\varphi}_{h;j}(\omega) &= \tan^{-1}\left(\frac{\hat{q}_{hj}(\omega)}{\hat{c}_{hj}(\omega)}\right) & &\text{if } \hat{c}_{hj}(\omega) \geq 0, \\ \hat{\varphi}_{h;j}(\omega) &= \left\{\tan^{-1}\left(\frac{\hat{q}_{hj}(\omega)}{\hat{c}_{hj}(\omega)}\right) + \pi \operatorname{sign}\left[\hat{q}_{hj}(\omega)\right]\right\} & &\text{if } \hat{c}_{hj}(\omega) < 0, \end{aligned} \tag{3.20}$$

called the *sample phase* difference between the two series at frequency ω. Interpretations and generalizations of these quantities are given in section 7; their sampling theory is discussed in section 8.

4. A comparison of spectral windows

In interpreting windowed sample cross-spectra of observed time series, it is valuable to compare them with similarly computed windowed sample cross-spectra of artificially generated time series. We present here examples of sample cross-spectra for a few simple artificial time series. Our aim is first to gain some idea of what sample cross-spectra look like, and second, to see some of the ways in which the choice of lag window affects the results.

A time series identically equal to 1,

$$X(t) = 1 \qquad \text{for all} \quad t, \tag{4.1}$$

has as its sample covariance function

$$R_T(v) = \frac{1}{T} \sum_{t=1}^{T} X(t)X(t+v) = \left(1 - \frac{v}{T}\right). \tag{4.2}$$

The corresponding windowed sample spectral density is

$$K_{M,T}(\omega) = \frac{1}{2\pi} \sum_{|v|<T} e^{-iv\omega} \left(1 - \frac{|v|}{T}\right) k\left(\frac{v}{M}\right). \tag{4.3}$$

For many purposes it can be verified that approximately

$$K_{M,T}(\omega) \doteq K_M(\omega). \tag{4.4}$$

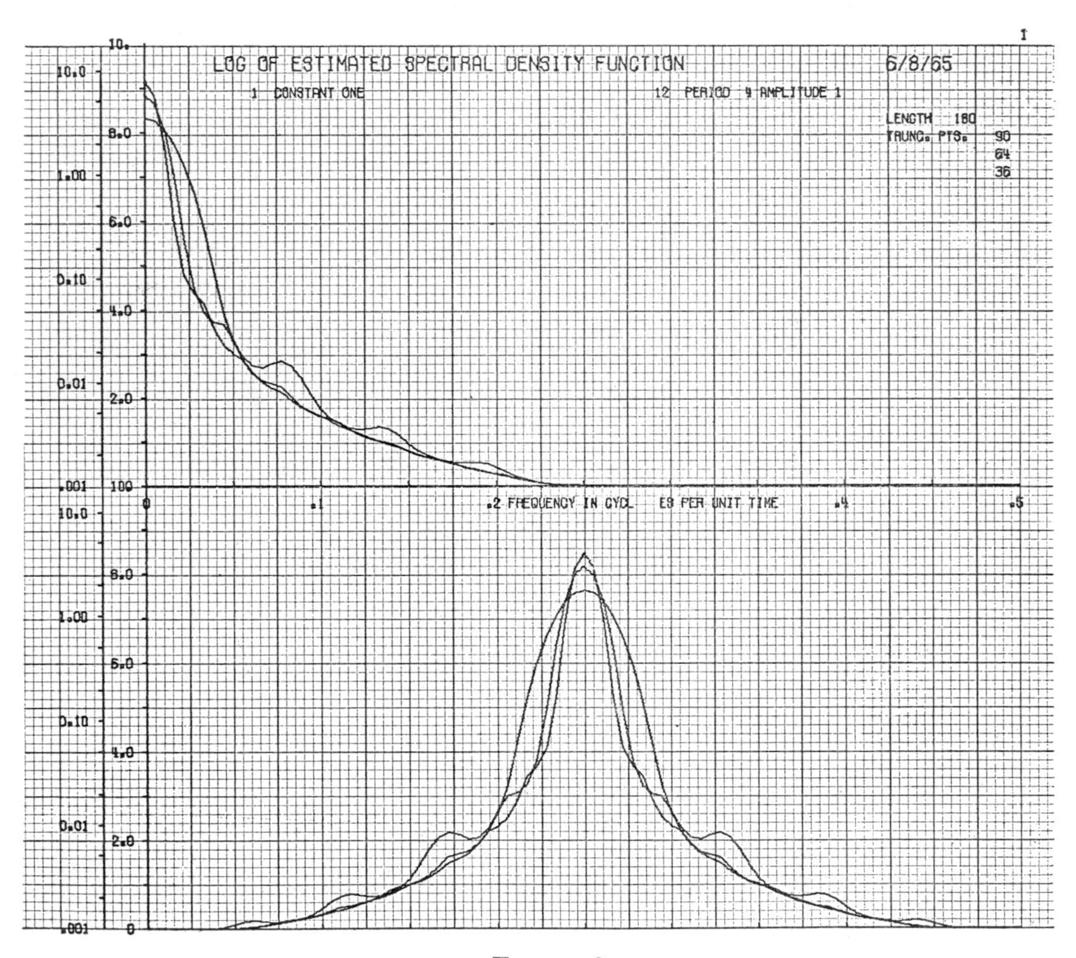

FIGURE 2

Figures 2–5 plot (in the top half) the function $K_{M,T}(\omega)$ for $T = 180$, $M = 90$, 64, 36, and for the four lag windows we have mentioned. The bottom half of each of these figures plots the windowed sample spectral density function of the time series

$$X(t) = \cos \frac{2\pi}{4} t. \tag{4.5}$$

The horizontal axis of these figures measures frequency on an axis from 0 to 0.5, representing ν (cycles per unit time) rather than ω (radians per unit time). The function plotted is not the windowed sample spectral density $\hat{f}(\omega)$, but rather

$$1000 \log_e \hat{f}\left(\frac{\omega}{2\pi}\right). \tag{4.6}$$

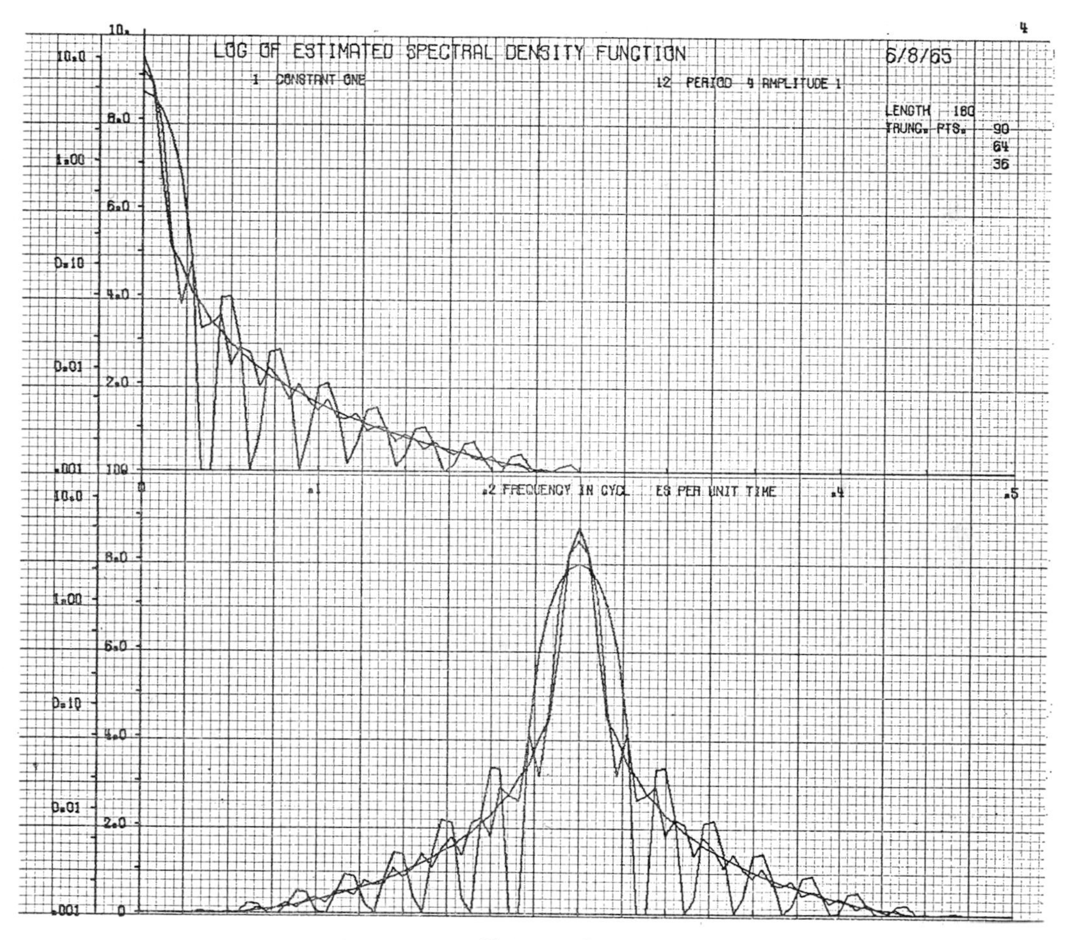

FIGURE 3

By white noise we mean any time series of uncorrelated random variables with zero means; it is covariance stationary with covariance function

$$
\begin{aligned}
R(v) &= 1 \qquad \text{if} \quad v = 0, \\
R(v) &= 0 \qquad \text{if} \quad v \neq 0,
\end{aligned} \tag{4.7}
$$

when normalized to have unit variance. The corresponding spectral density function is

$$
f(\omega) = \frac{1}{2\pi}, \qquad -\pi \leq \omega \leq \pi. \tag{4.8}
$$

Now

$$
1000 \log_e \frac{1}{2\pi} = 5.07. \tag{4.9}
$$

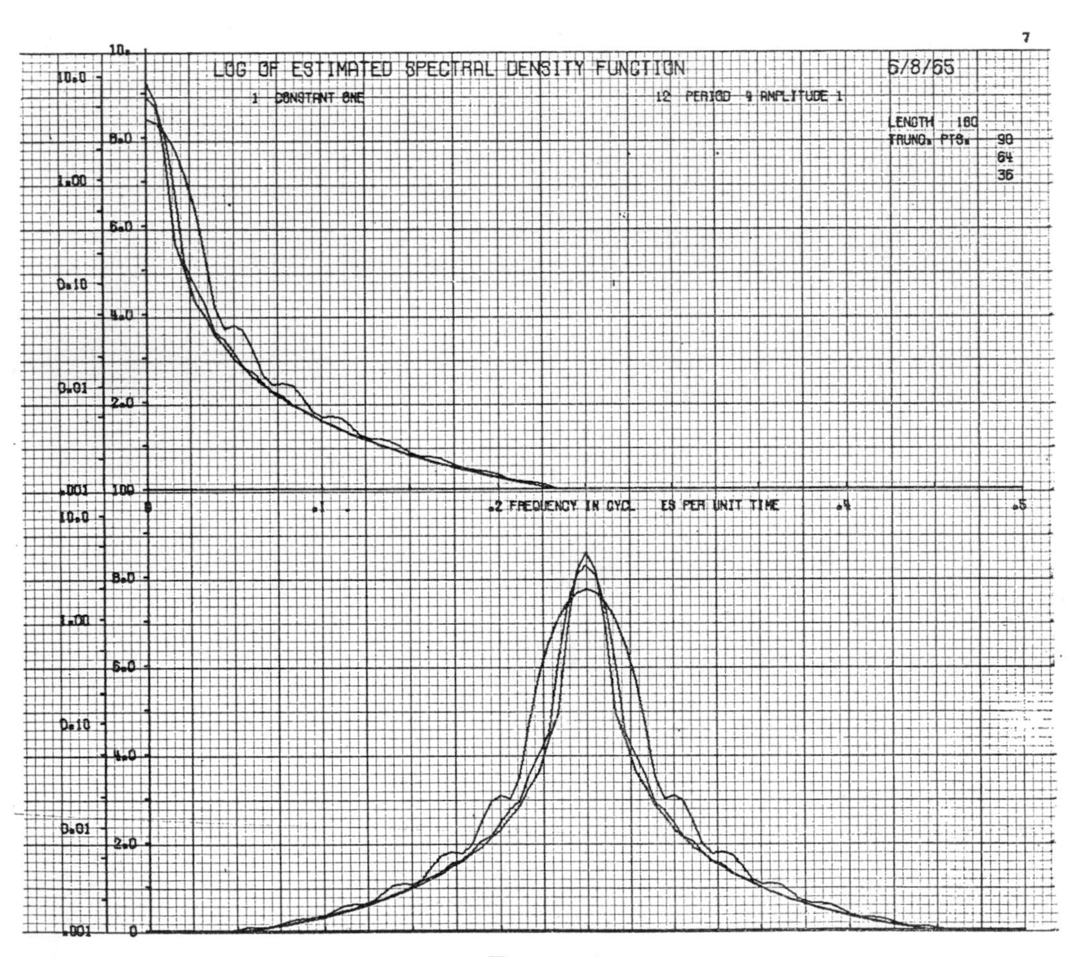

FIGURE 4

Therefore, in figures 2–5 the normalized windowed sample spectral density functions are plotted on a scale that goes from 1 to 10. A sample of white noise would be expected to oscillate about the middle of the graph.

Various numerical measures for comparing properties of various spectral windows have been introduced in previous papers [see Parzen [21], [23]). Here, taking an empirical attitude, we study the computer output one obtains from an empirical time series analysis of various series. Comparing in figures 2–5 the window sample spectra of a pure sine wave, one sees that figure 3 (the Tukey kernel) is more oscillatory than figure 4 (the Bohman kernel), which in turn is slightly more oscillatory than figure 2 (the Parzen kernel). Figure 5 (the Bartlett kernel) shows strong oscillations as well as a very unsatisfactory failure to damp down. These differences in behavior hold more for small truncation points than for large truncation points.

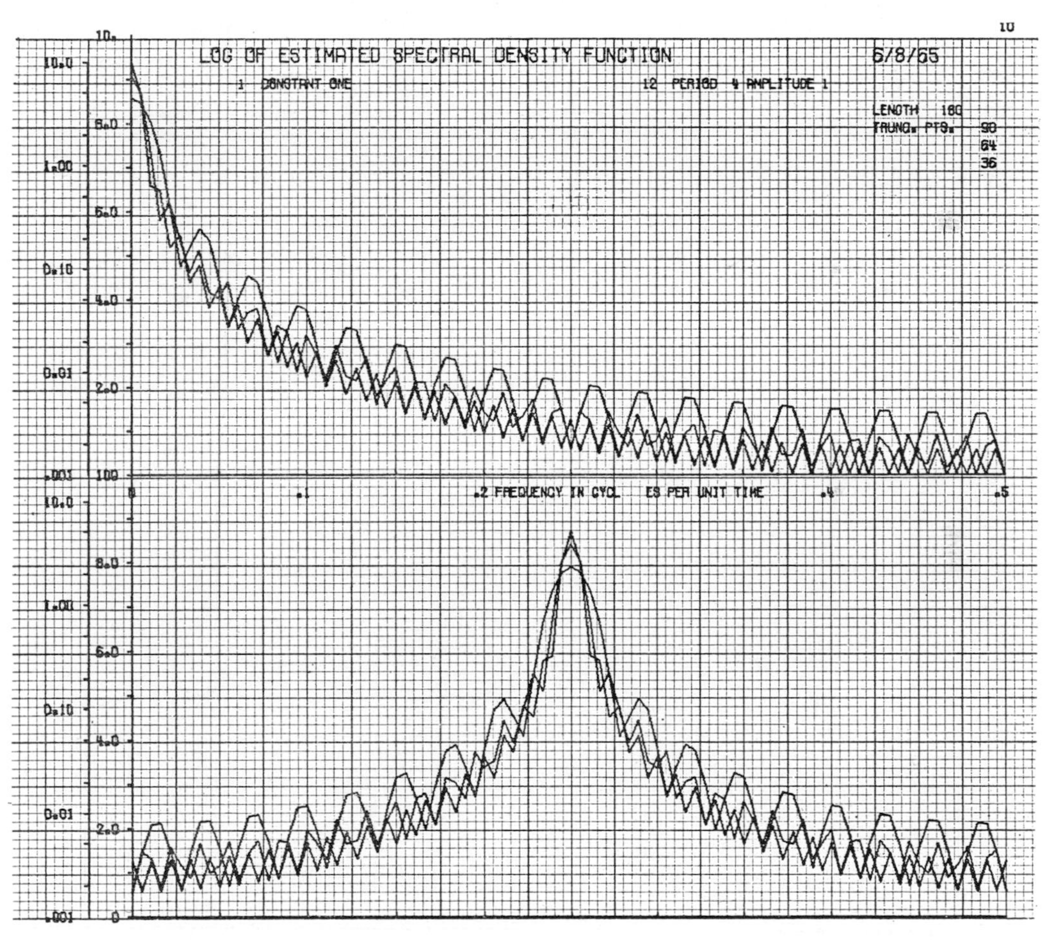

FIGURE 5

Figures 6–9 present windowed sample spectra of time series $N(t) + \cos(2\pi/4)t$ and $N(t)$, where $N(t)$ is a "sample of white noise" internally generated by the computer. For these time series the differences between the various windows is much less pronounced than in figures 2–5.

However, there seems to be much difference between the graphs of the sample coherence, plotted in figures 10–13 for the various windows. Figure 10 (the Parzen window) seems to have the smoothest behavior.

Although spectral distribution functions are not discussed in this paper, plots of windowed sample distribution functions are given in figure 14.

5. Sampling theory of sample cross-spectra

In this section we outline the properties of the windowed sample cross-spectra when the observed time series are jointly covariance stationary with absolutely continuous spectrum and have zero means.

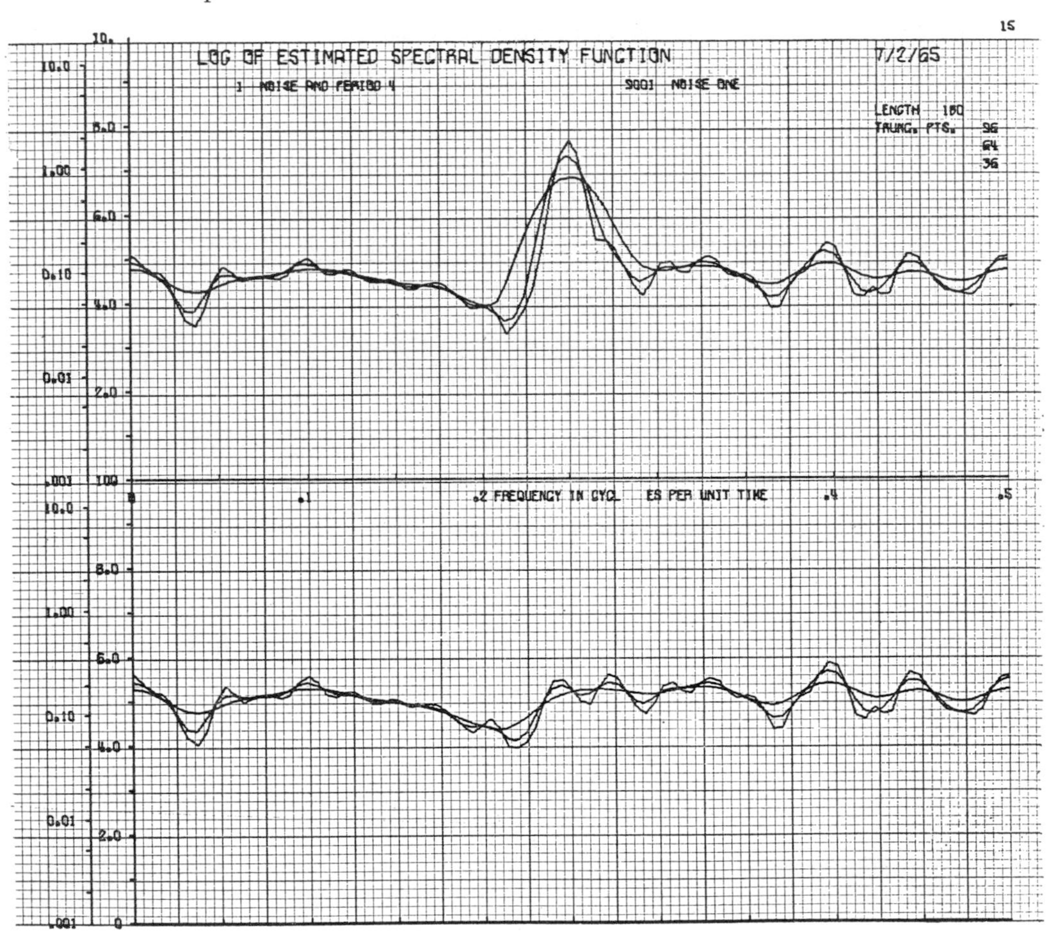

FIGURE 6

We first consider the mean of a windowed sample cross-spectral density

$$E[\hat{f}_{hj}(\omega)] = \frac{1}{2\pi} \sum_{|v|<M} e^{-iv\omega} k\left(\frac{v}{M}\right)\left(1 - \frac{|v|}{T}\right) R_{hj}(v) \tag{5.1}$$
$$= \int_{-\pi}^{\pi} K_{M,T}(\omega - \lambda) f_{hj}(\lambda)\, d\lambda,$$

where $K_{M,T}(\omega)$ is defined by (4.3). Assuming that (4.4) holds, we obtain the following approximation for the mean of a sample cross-spectral density

$$E[\hat{f}_{hj}(\omega)] \doteq \int_{-\pi}^{\pi} K_M(\omega - \lambda) f_{hj}(\lambda)\, d\lambda. \tag{5.2}$$

To evaluate this integral it is often assumed that, in the neighborhood of ω, the real and imaginary parts $f_{hj}(\lambda)$ are both varying slowly compared to $K_M(\omega - \lambda)$; then approximately

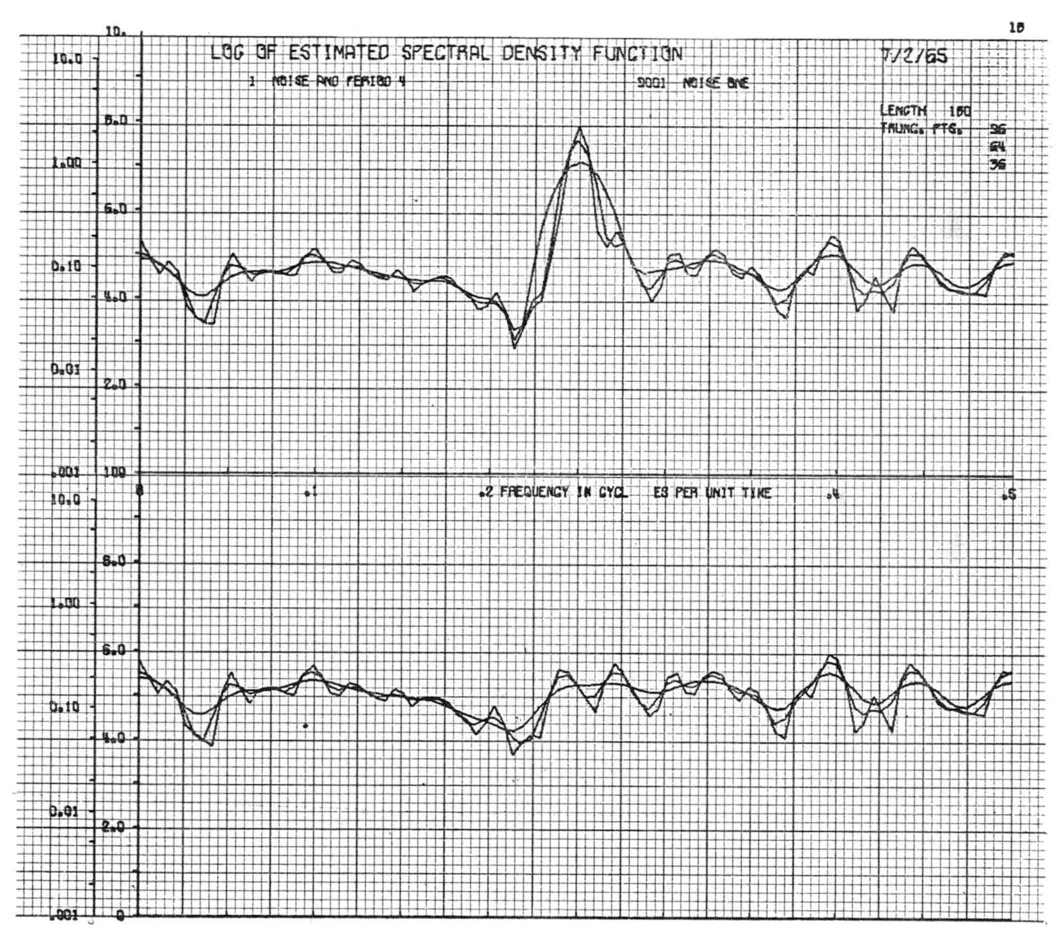

FIGURE 7

$$\int_{-\pi}^{\pi} K_M(\omega - \lambda) f_{hj}(\lambda)\, d\lambda \doteq f_{hj}(\omega) \int_{-\pi}^{\pi} K_M(\omega - \lambda)\, d\lambda = f_{hj}(\omega). \tag{5.3}$$

It is thus implied that to a first-order approximation (as $M \to \infty$), the windowed sample cross-spectral density is an *unbiased* estimate of the true cross-spectral density (when the observed time series is zero mean covariance stationary). While this is a correct statement from the asymptotic point of view, for finite samples there is a bias in cross-spectral estimates not present in auto-spectral estimates; this bias is discussed in section 9 since in order to discuss it, we need to first introduce the notions of gain and phase. A comprehensive and rigorous discussion of bias in cross-spectral estimates is given by Nigel Nettheim in his 1966 Stanford Ph.D. thesis.

Much of the mathematical literature on cross-spectral analysis has been concerned with variability rather than bias. One can investigate the sampling theory

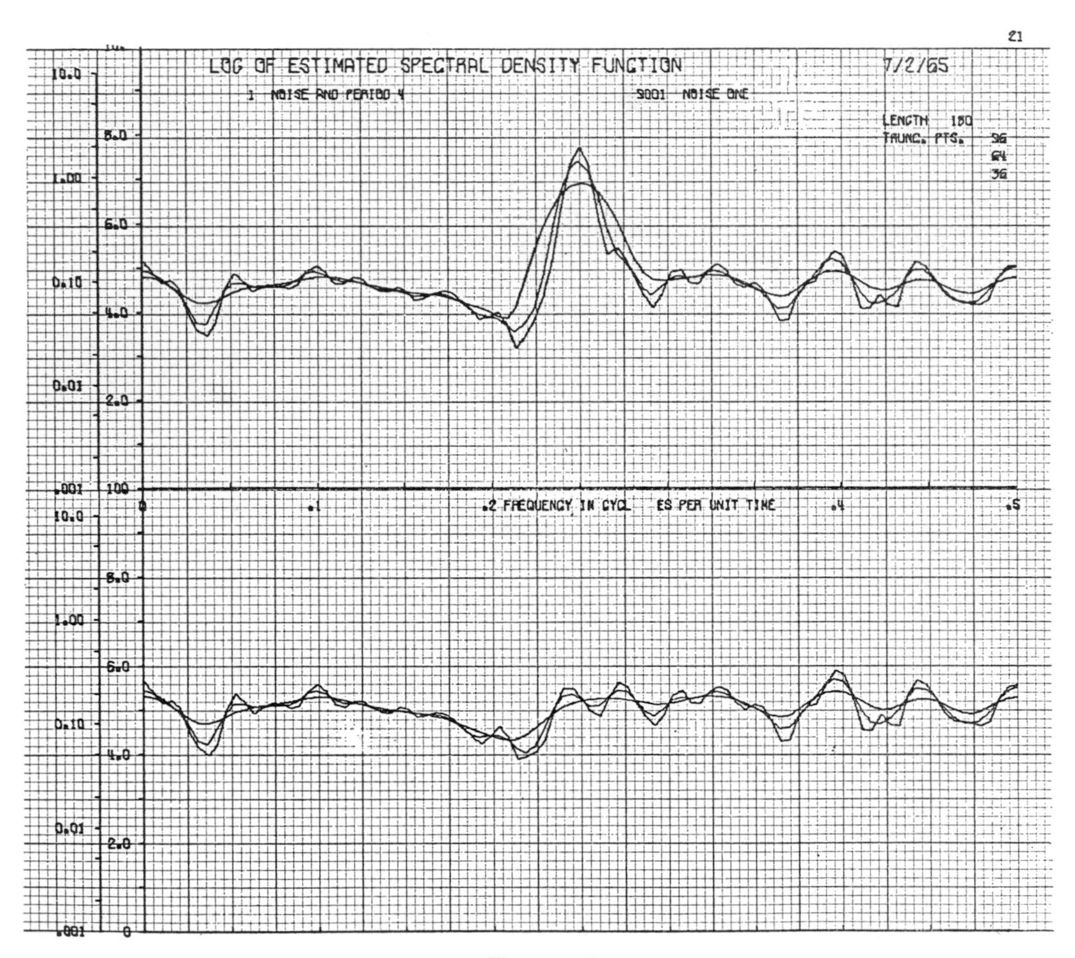

FIGURE 8

of sample cross-spectra from an asymptotic point of view or from a small sample point of view in the case that the observed time series are assumed to be zero mean, normal, jointly covariance stationary, and possessing spectral density functions.

The basic formula of the asymptotic point of view is (under suitable conditions on the kernel $k(u)$ and assuming that the same kernel and truncation are used in every estimate)

$$\operatorname{cov}\,[\hat{f}_{hj}(\omega), \hat{f}_{kn}(\omega)] \doteq C f_{hk}(\omega) \bar{f}_{jn}(\omega), \qquad 0 < \omega < \pi, \tag{5.4}$$

writing $\bar{z}$ to denote the complex conjugate of a complex number z and defining

$$C = \frac{M}{T} \int_{-\infty}^{\infty} k^2(u)\, du. \tag{5.5}$$

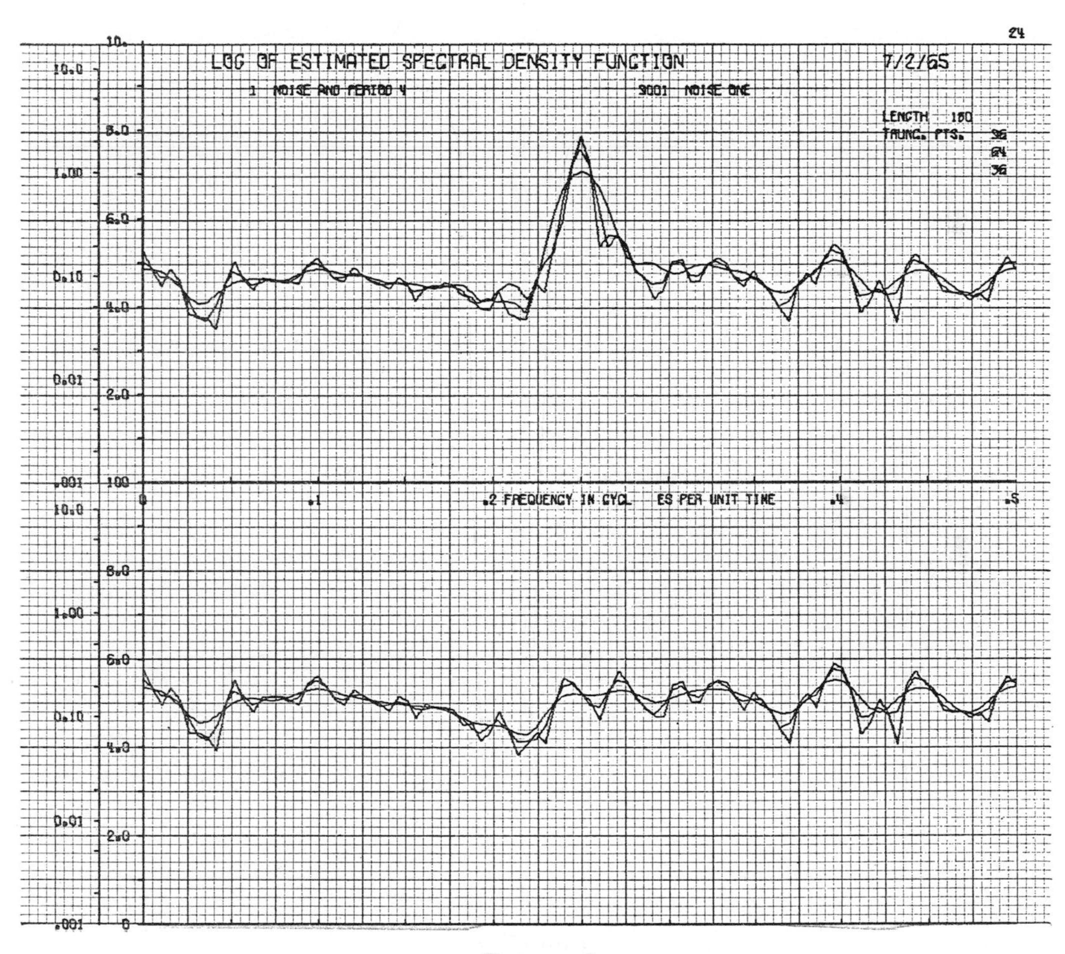

FIGURE 9

To prove (5.4) let us first note, without proof, that (compare Rosenblatt [27])

$$\text{cov}\,[\hat{f}_{11}(\omega), \hat{f}_{22}(\omega)] \doteq C|f_{12}(\omega)|^2, \qquad 0 < \omega < \pi. \tag{5.6}$$

From (5.6) one derives (5.4) as follows: consider arbitrary linear combinations of the observed time series $Z_j(t)$,

$$Y_1(t) = \sum_j a_j Z_j(t), \qquad Y_2(t) = \sum_k b_k Z_k(t). \tag{5.7}$$

Their sample spectra can be written (using $\hat{f}_{11}(\omega)$ with two meanings, as the windowed sample spectral density of both $Y_1(t)$ and $Z_1(t)$; similarly for $\hat{f}_{22}$ and f_{12})

$$\hat{f}_{11} = \sum_{h,j} a_h \hat{f}_{hj} \bar{a}_j, \qquad \hat{f}_{22} = \sum_{k,n} b_k \hat{f}_{kn} \bar{b}_n; \tag{5.8}$$

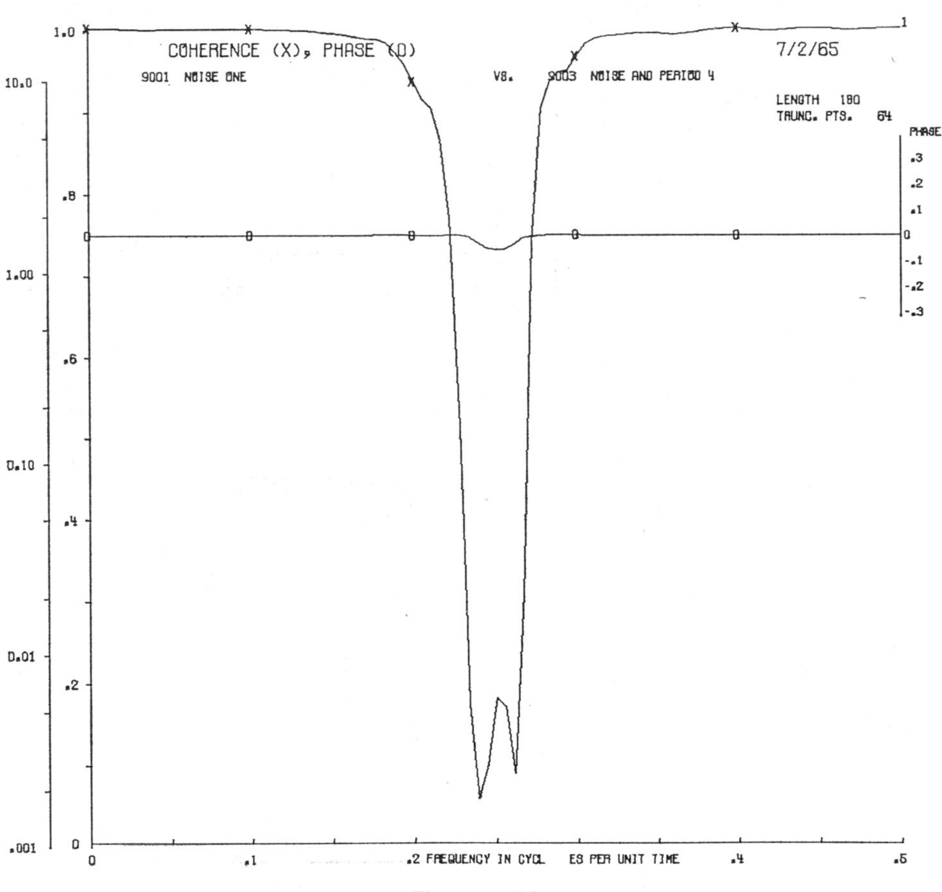

FIGURE 10

consequently,

$$\begin{aligned}(5.9)\qquad \operatorname{cov}\,[\hat{f}_{11}(\omega), \hat{f}_{22}(\omega)] &= \sum_{h,j,k,n} a_h\bar{a}_j \operatorname{cov}\,[\hat{f}_{hj}(\omega), \hat{f}_{kn}(\omega)]\bar{b}_k b_n, \\ &\doteq C|f_{12}(\omega)|^2, \\ &= C|\sum_{h.k} a_h f_{hk}(\omega)\bar{b}_k|^2, \\ &= C \sum_{h,j,k,n} a_h\bar{a}_j f_{hk}(\omega)\bar{f}_{jn}(\omega)\bar{b}_k b_n.\end{aligned}$$

One may now infer (5.4).

The meaning of (5.4) is best understood by writing out the variance-covariance matrix of the estimates $\hat{f}_{11}(\omega)$, $\hat{f}_{22}(\omega)$, $\hat{f}_{12}(\omega)$, $\hat{f}_{21}(\omega)$.

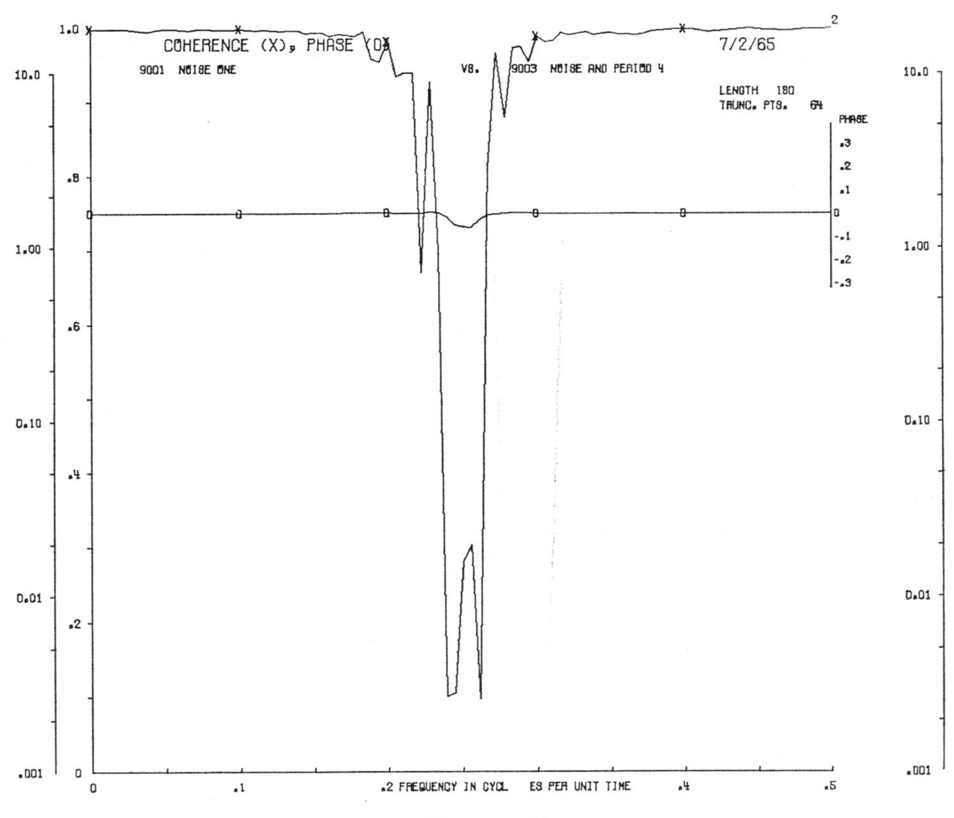

FIGURE 11

(5.10)

	$\hat{f}_{11}(\omega)$	$\hat{f}_{22}(\omega)$	$\hat{f}_{12}(\omega)$	$\hat{f}_{21}(\omega)$
$\hat{f}_{11}(\omega)$	$C\|f_{11}(\omega)\|^2$	$C\|f_{12}(\omega)\|^2$	$Cf_{11}(\omega)\bar{f}_{12}(\omega)$	$Cf_{11}(\omega)\bar{f}_{21}(\omega)$
$\hat{f}_{22}(\omega)$		$C\|f_{22}(\omega)\|^2$	$Cf_{22}(\omega)\bar{f}_{12}(\omega)$	$Cf_{22}(\omega)\bar{f}_{21}(\omega)$
$\hat{f}_{12}(\omega)$			$Cf_{11}(\omega)f_{22}(\omega)$	$C\{f_{12}(\omega)\}^2$
$\hat{f}_{21}(\omega)$				$Cf_{11}(\omega)f_{22}(\omega)$

From (5.10) one obtains the covariances of $\hat{f}_{11}(\omega)$, $\hat{f}_{22}(\omega)$, $\hat{c}_{12}(\omega) = \operatorname{Re}\hat{f}_{12}(\omega)$, and $\hat{q}_{12}(\omega) = -\operatorname{Im}\hat{f}_{12}(\omega)$. In writing the following table we have *omitted* from every entry the factor C defined by (5.5).

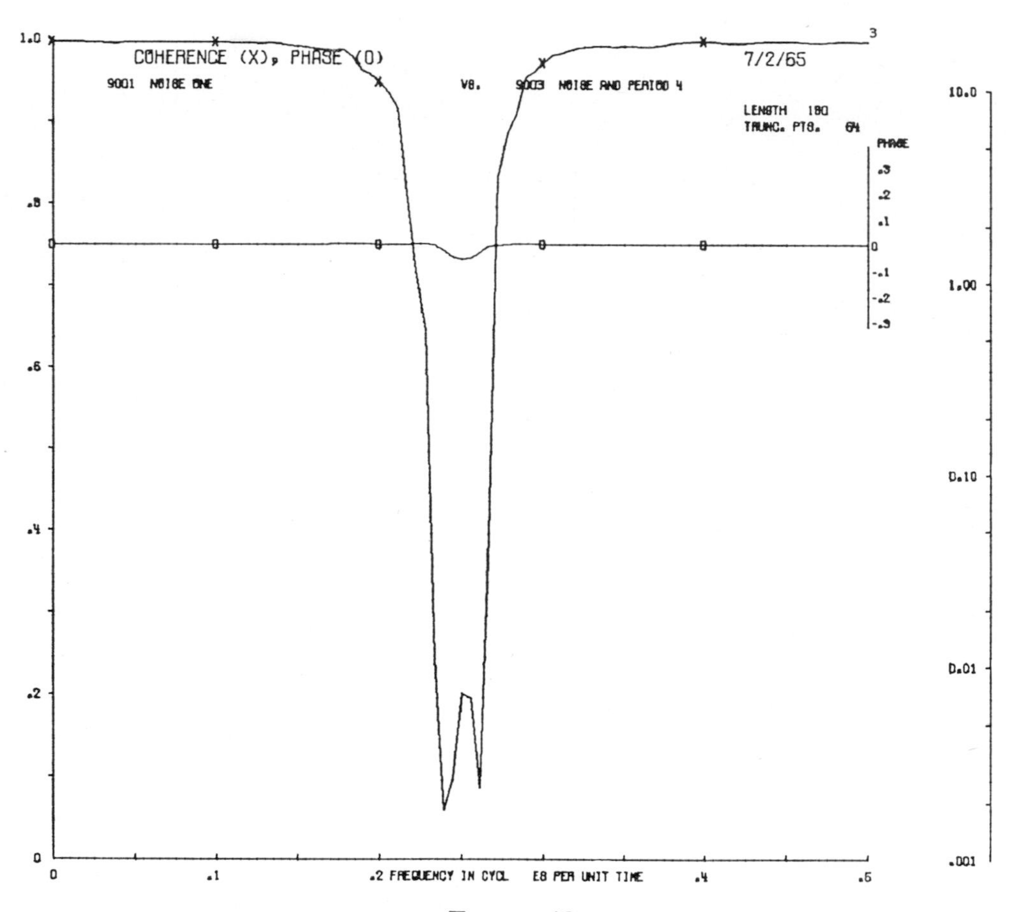

FIGURE 12

(5.11)

	$\hat{f}_{11}(\omega)$	$\hat{f}_{22}(\omega)$	$\hat{c}_{12}(\omega)$	$\hat{q}_{12}(\omega)$
$\hat{f}_{11}(\omega)$	$\lvert f_{11}(\omega)\rvert^2$	$c_{12}^2(\omega) + q_{12}^2(\omega)$	$f_{11}(\omega)c_{12}(\omega)$	$-f_{11}(\omega)q_{12}(\omega)$
$\hat{f}_{22}(\omega)$		$\lvert f_{22}(\omega)\rvert^2$	$f_{22}(\omega)c_{12}(\omega)$	$-f_{22}(\omega)q_{12}(\omega)$
$\hat{c}_{12}(\omega)$			$\frac{1}{2}\{f_{11}(\omega)f_{22}(\omega) + c_{12}^2(\omega) - q_{12}^2(\omega)\}$	$c_{12}(\omega)q_{12}(\omega)$
$\hat{q}_{12}(\omega)$				$\frac{1}{2}\{f_{11}(\omega)f_{22}(\omega) + q_{12}^2(\omega) - c_{12}^2(\omega)\}$

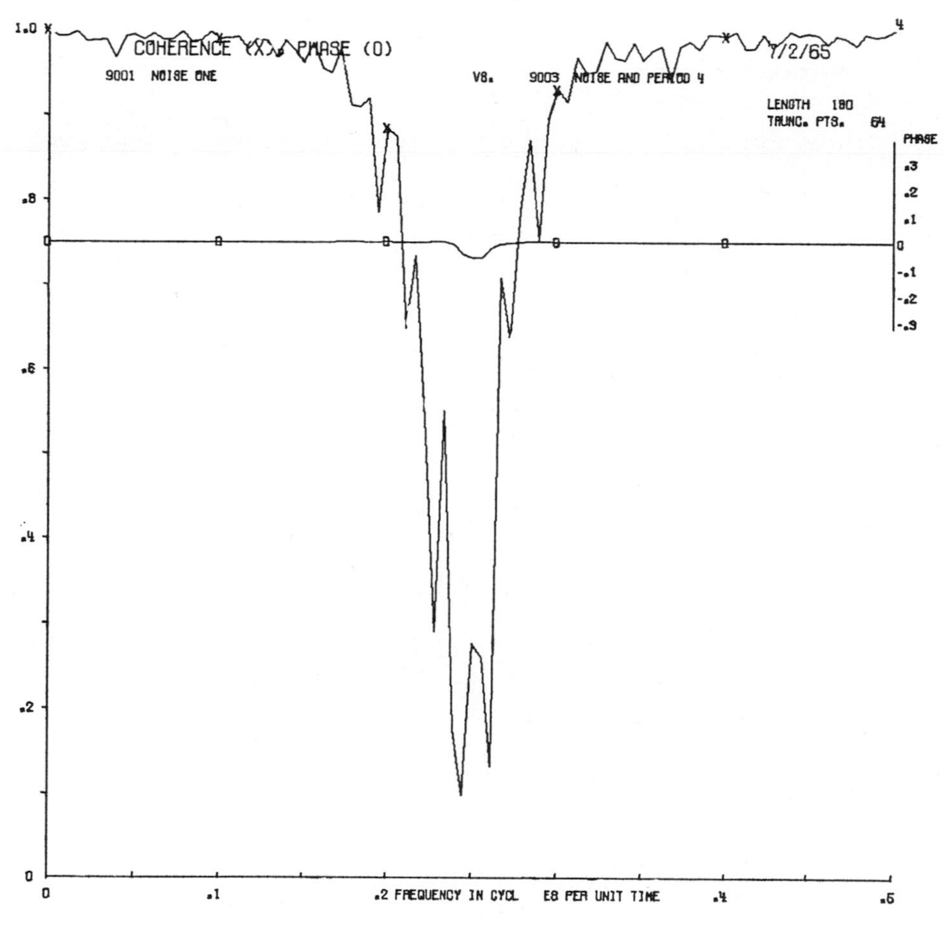

FIGURE 13

One approach to the "small sample" distribution theory of sample cross-spectra is due to Goodman [9] who uses an analogy between windowed sample cross-spectra and a sample covariance matrix

$$\hat{K}_{jk} = \frac{1}{n} \sum_{t=1}^{n} X_j(t)\overline{X}_k(t), \qquad j, k = 1, \cdots, r, \tag{5.12}$$

where for $t = 1, 2, \cdots, n$ $\{X_j(t), j = 1, \cdots, r\}$ are independent *complex* random vectors identically distributed as the vector $\{X_j, j = 1, \cdots, r\}$ which is assumed to (i) be normally distributed, (ii) have zero means, and (iii) satisfy the conditions for any indices j and k,

$$E[X_j X_k] = 0, \qquad E[X_j \overline{X}_k] = K_{jk}. \tag{5.13}$$

One important case in which the first equation in (5.13) holds is when

$$X_j = U_j + iV_j, \qquad X_k = U_k + iV_k, \tag{5.14}$$

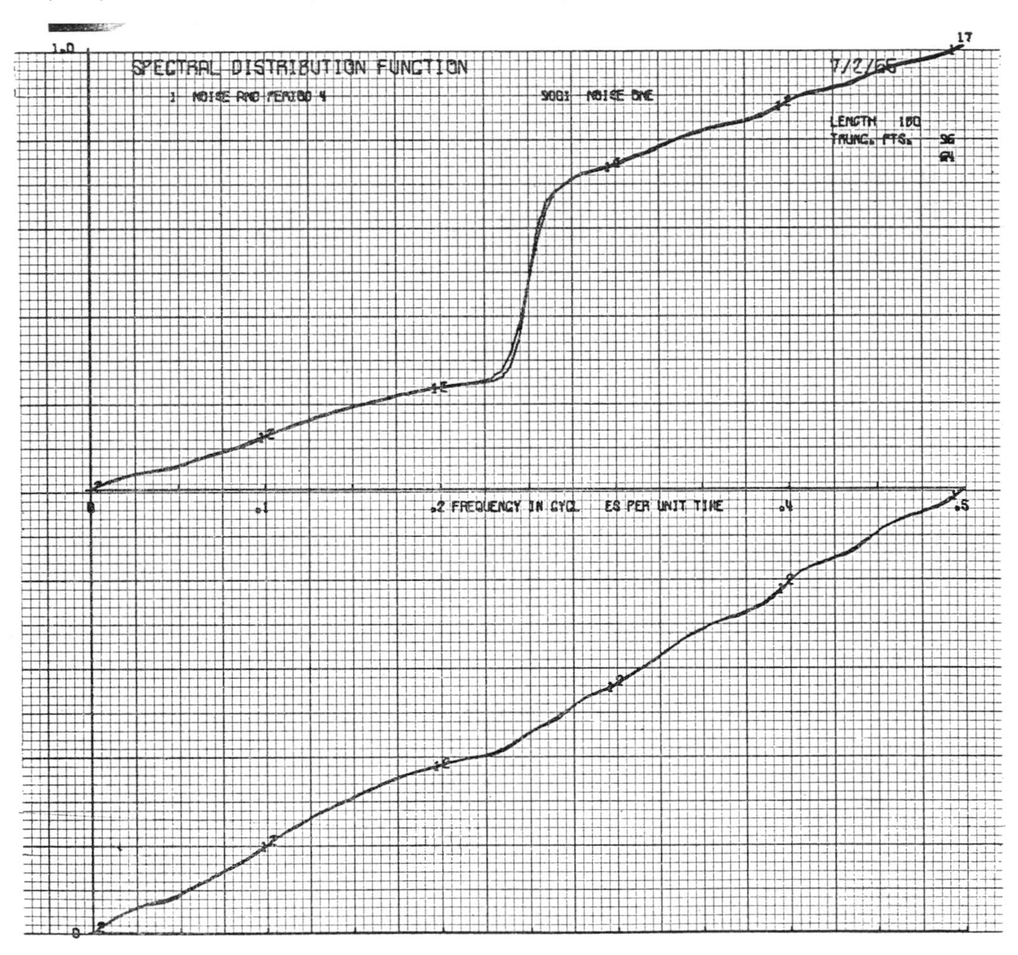

FIGURE 14

where U_j, V_j, U_k, V_k are jointly normal random variables with zero means and covariances

$$(5.15)\qquad \begin{array}{c} \\ U_j \\ V_j \\ U_k \\ V_k \end{array} \begin{array}{c} \begin{array}{cccc} U_j & V_j & U_k & V_k \end{array} \\ \begin{bmatrix} \sigma_j^2 & 0 & \alpha_{jk} & \beta_{jk} \\ 0 & \sigma_j^2 & -\beta_{jk} & \alpha_{jk} \\ \alpha_{jk} & -\beta_{jk} & \sigma_k^2 & 0 \\ \beta_{jk} & \alpha_{jk} & 0 & \sigma_k^2 \end{bmatrix} \end{array}.$$

From (5.15) it follows that

$$(5.16)\qquad \begin{aligned} \operatorname{cov}[X_h\overline{X}_j, X_k\overline{X}_m] &= E[X_h\overline{X}_j\overline{X}_kX_m] - E[X_h\overline{X}_j]E[\overline{X}_kX_m], \\ &= E[X_h\overline{X}_k]E[\overline{X}_jX_m] + E[X_hX_m]E[\overline{X}_j\overline{X}_k], \\ &= K_{hk}K_{jm}. \end{aligned}$$

We thus obtain a basic formula for the covariances of the sample covariance function (true for every n, and not only asymptotically):

$$(5.17)\qquad \operatorname{cov}[\hat{K}_{hj}, \hat{K}_{km}] = \frac{1}{n} K_{hk}\overline{K}_{jm}.$$

This formula is reminiscent of (5.4), identifying $1/n$ with C.

To illustrate the application of this result, let us consider two time series of length $T = 180$ whose windowed sample cross-spectra are computed for a truncation point $M = 64$. Then

$$(5.18)\qquad \begin{aligned} n = 1 \div C &= T \div M \int_{-\infty}^{\infty} k^2(u)\,du \\ &= \begin{cases} 5.21 & \text{for Parzen window,} \\ 3.75 & \text{for Tukey window,} \end{cases} \end{aligned}$$

since

$$(5.19)\qquad \int_{-\infty}^{\infty} k^2(u)\,du = \begin{cases} 0.54 & \text{for Parzen window,} \\ 0.75 & \text{for Tukey window.} \end{cases}$$

From Goodman's small sample approximation to the sampling theory of windowed sample cross-spectra, one can derive significance levels for the sample coherence. For T/M approximately 3, the 95% significance levels to test the hypothesis that true coherence at a given frequency is *zero* are, respectively, 0.464 (for Parzen window) and 0.632 (for the Tukey window); these values are obtained from Amos and Koopmans [4]).

6. Prediction filters and partial cross-spectra

Increasingly, techniques of regression analysis, correlation analysis, and multivariate analysis are being applied by research workers in various disciplines as a means of studying the relations between various variables. A lucid discussion of the basic methodology is given by Kendall and Stuart [13]. The aim of this section is to indicate how these ideas generalize to time series. Rigorous developments of some of these ideas have been given by Koopmans [14], [15].

The notions introduced in multivariate analysis to describe the relations between a family of random variables may be generalized in two ways to time series, depending on whether one uses two-sided or one-sided prediction filters. The generalization using two-sided prediction filters (which for technical reasons is the one most often considered) is discussed in detail. The generalization using one-sided prediction filters is briefly mentioned.

Let $X_1(\cdot), \cdots, X_r(\cdot)$ be r time series. Let P be a subset of the set $D = \{1, 2, \cdots, r\}$ of indices, and let j be any index in D. We define a new time series, denoted

$$X_{j;P}(t), \qquad t = 0, \pm 1, \cdots, \tag{6.1}$$

and called the minimum mean square error linear predictor of $X_j(t)$, given $\{X_k(s), s = 0, \pm 1, \cdots, k \in \mathrm{P}\}$, as follows: $X_{j,P}(t)$ is a linear combination of the predictor random variables, which we write

$$X_{j;P}(t) = \sum_{k \in P} \sum_{s=-\infty}^{\infty} b_{jk;P}(t-s) X_k(s) \tag{6.2}$$

whose mean square prediction error as an estimate of $X_j(t)$ is a minimum (that is, does not exceed the mean square prediction error of any other predictor of $X_j(t)$ which is a linear combination of $\{X_k(t), t = 0, \pm 1, \cdots, k \in P\}$). It should be noted that in general $X_{j;P}(t)$ cannot be written as an infinite series; this assumption is made only for ease of exposition. As shown by Koopmans [15], the conclusions given may be shown to hold under somewhat more general conditions using the Hilbert space theory of time series.

The coefficient $b_{jk;P}(t-s)$ is a function only of the time difference $t-s$ because of the joint stationarity; similarly, the mean square prediction error is independent of t. We call $b_{jk;P}(t-s)$ the *partial regression coefficient* of $X_j(t)$ on $X_k(s)$ given $\{X_k(\cdot), k \in P\}$. These coefficients are determined by the conditions

$$E[X_{j;P}(t) X_h(u)] = E[X_j(t) X_h(u)], \qquad \text{for} \quad u = 0, \pm 1, \cdots, h \in P, \tag{6.3}$$

which lead to the normal equations

$$\sum_{k \in P} \sum_{s=-\infty}^{\infty} b_{jk;P}(t-s) E[X_k(s) X_h(u)] = E[X_j(t) X_h(u)], \tag{6.4}$$

which may be written in terms of covariance functions

$$\sum_{k \in P} \sum_{v=-\infty}^{\infty} b_{jk;P}(v) R_{kh}(u+v-t) = R_{jh}(u-t), \tag{6.5}$$

and in terms of spectral density functions

$$\sum_{k \in P} \sum_{s=-\infty}^{\infty} b_{jk;P}(v) \int_{-\pi}^{\pi} e^{i\omega(u+v-t)} f_{kh}(\omega)\, d\omega = \int_{-\pi}^{\pi} e^{i\omega(u-t)} f_{jh}(\omega)\, d\omega. \tag{6.6}$$

The Fourier transform of the partial regression coefficients is called the *partial regression transfer function* and denoted

$$B_{jk;P}(\omega) = \sum_{v=-\infty}^{\infty} b_{jk;P}(v)e^{-iv\omega}; \tag{6.7}$$

its interpretation is discussed in section 7. Writing (6.6) in the form

$$\int_{-\pi}^{\pi} d\omega\, e^{i\omega(u-t)} \left\{ \sum_{k \in P} \overline{B}_{jk;P}(\omega) f_{kh}(\omega) - f_{jh}(\omega) \right\} = 0, \tag{6.8}$$

we obtain a system of *normal equations* for the regression transfer functions. For each ω in $-\pi \leq \omega \leq \pi$ and h in P,

$$\sum_{k \in P} \overline{B}_{jk;P}(\omega) f_{kh}(\omega) = f_{jh}(\omega). \tag{6.9}$$

The *partial covariance function* between two time series $X_h(\cdot)$ and $X_j(\cdot)$, given predictors $\{X_k(t),\ t = 0, \pm 1, \cdots, k \in P\}$, is denoted by $K_{hj;P}(v)$ and is defined by

$$K_{hj;P}(t_2 - t_1) = E[\epsilon_{h;P}(t_1)\epsilon_{j;P}(t_2)] \tag{6.10}$$

where

$$\epsilon_{h;P}(t) = X_h(t) - X_{h;P}(t) \tag{6.11}$$

is the *residual* series of $X_h(\cdot)$ given the predictors. We next show that the partial covariance function depends on t_1 and t_2 only through the time difference $t_2 - t_1$ by a method which also obtains a *spectral representation* for the partial covariance function $E[\epsilon_{h;P}(t_1)\epsilon_{j;P}(t_2)] = E[\epsilon_{h;P}(t_1)X_j(t_2)]$. This is equal to

$$\begin{aligned} (6.12) \quad & E[X_h(t_1)X_j(t_2)] - E\left[\left\{\sum_{k \in P} \sum_{s=-\infty}^{\infty} b_{hk;P}(s)X_k(t_1 - s)\right\} X_j(t_2)\right] \\ &= R_{hj}(t_2 - t_1) - \sum_{k \in P} \sum_{s=-\infty}^{\infty} b_{hk;P}(s)R_{kj}(t_2 - t_1 + s) \\ &= \int_{-\pi}^{\pi} d\omega\, e^{i\omega(t_2 - t_1)} \left\{ f_{hj}(\omega) - \sum_{k \in P} \sum_{s=-\infty}^{\infty} b_{hk;P}(s) f_{kj}(\omega) e^{i\omega s} \right\} \\ &= \int_{-\pi}^{\pi} d\omega\, e^{i\omega(t_2 - t_1)} \left\{ f_{hj}(\omega) - \sum_{k \in P} \overline{B}_{hk;P}(\omega) f_{kj}(\omega) \right\}. \end{aligned}$$

The *partial spectral density function* of two series $X_h(\cdot)$ and $X_j(\cdot)$ given predictors $\{X_k(t),\ t = 0, \pm 1, \cdots, k \in P\}$, denoted $f_{hj;P}(\omega)$, may be defined by the condition that it provides a spectral representation for the partial covariance function

$$K_{hj;P}(t_2 - t_1) = \int_{-\pi}^{\pi} e^{i\omega(t_2 - t_1)} f_{hj;P}(\omega)\, d\omega. \tag{6.13}$$

From (6.12) we obtain the *basic* formula

$$f_{hj;P}(\omega) = f_{hj}(\omega) - \sum_{k \in P} \overline{B}_{hk;P}(\omega) f_{kj}(\omega). \tag{6.14}$$

To interpret the partial spectral density function, let us first consider the properties of the residual series $\epsilon_{j;P}(t)$. Its spectral density function, called the *residual spectral density* function of the series $X_j(\cdot)$ given the predictors $\{X_k(\cdot), k \in P\}$, is given by

$$f_{jj;P}(\omega) = f_{jj}(\omega) - \sum_{k \in P} \overline{B}_{jk;P}(\omega) f_{kj}(\omega), \tag{6.15}$$
$$= f_{jj}(\omega)\{1 - W_{j;P}(\omega)\},$$

defining

$$W_{j;P}(\omega) = 1 - \frac{f_{jj;P}(\omega)}{f_{jj}(\omega)}, \tag{6.16}$$
$$= \sum_{k \in P} \overline{B}_{jk;P}(\omega) f_{kj}(\omega) \div f_{jj}(\omega).$$

One calls $W_{j;P}(\omega)$ the *multiple coherence* function of the series $X_j(\cdot)$ given the predictors $\{X_k(t), t = 0, \pm 1, \cdots, k \in P\}$. It is analogous to the squared multiple correlation coefficient and is a measure of the predictability of the component of $X_j(\cdot)$ at frequency ω from the components of $\{X_k(\cdot), k \in P\}$ at frequency ω. The analogue of the square of the partial correlation coefficient is called the *partial coherence* between series $X_h(\cdot)$ and $X_j(\cdot)$, given $\{X_k(\cdot), k \in P\}$; it is denoted, and given, by

$$W_{hj;P}(\omega) = |f_{hj;P}(\omega)|^2 \div f_{hh;P}(\omega) f_{jj;P}(\omega). \tag{6.17}$$

It is instructive to consider the case where only a single series $\{X_k(t), t = 0, \pm 1, \cdots\}$ is used as the predictor. The regression transfer function $B_{jk;\{k\}}(\omega)$ will be denoted $B_{j;k}(\omega)$, the partial spectral density function $f_{hj;\{k\}}(\omega)$ will be denoted $f_{hj;k}(\omega)$, and the multiple spectral density function $W_{j;\{k\}}(\omega)$ will be denoted $W_{j;k}(\omega)$. We obtain the following formulas: $\overline{B}_{j;k}(\omega) f_{kk}(\omega) = f_{jk}(\omega)$, so that the regression transfer function is given by (assuming $f_{kk}(\omega)$ never vanishes)

$$\overline{B}_{j;k}(\omega) = \frac{f_{jk}(\omega)}{f_{kk}(\omega)}. \tag{6.18}$$

The partial cross-spectral density function is given by

$$f_{hj;k}(\omega) = f_{hj}(\omega) - \overline{B}_{h;k}(\omega) f_{kj}(\omega), \tag{6.19}$$
$$= f_{hj}(\omega) - \frac{f_{hk}(\omega) f_{kj}(\omega)}{f_{kk}(\omega)}.$$

In particular, the residual spectral density function is given by

$$f_{jj;k}(\omega) = f_{jj}(\omega) - \frac{|f_{jk}(\omega)|^2}{f_{kk}(\omega)} = f_{jj}(\omega)\{1 - W_{j;k}(\omega)\} \tag{6.20}$$

where

$$W_{j;k}(\omega) = \frac{|f_{jk}(\omega)|^2}{f_{jj}(\omega) f_{kk}(\omega)}. \tag{6.21}$$

If one examines the formula for the multiple coherence function of $X_j(\cdot)$ given $X_k(\cdot)$, one sees that the indices j and k play a symmetrical role. We therefore define the symbol

$$W_{jk}(\omega) = \frac{|f_{jk}(\omega)|^2}{f_{jj}(\omega) f_{kk}(\omega)}, \tag{6.22}$$

which is called the *coherence* between the series $X_j(\cdot)$ and $X_k(\cdot)$. It is denoted by the letter W in commemoration of Norbert Wiener who first introduced the notion of coherence. The coherence is related to a frequency decomposition of the residual series when one uses either of the series $X_j(\cdot)$ and $X_k(\cdot)$ to predict the other. It remains an open question whether it is more informative to plot the coherence $W_{jk}(\omega)$ or the residual spectral density functions $f_{jj;k}(\omega)$ and $f_{kk;j}(\omega)$.

Inductive formulas. Partial spectral densities and regression transfer functions are best computed by adding a variable at a time. Let P be an index set. By $P + m$ we mean the index set $\{j, j \in P \text{ or } j = m\}$; it is understood in this case that m does not belong to P. By $P - m$ we mean the index set $\{j, j \in P \text{ and } j \neq m\}$; it is understood in this case that m belongs to P.

From innovation theory one obtains the basic formula

$$X_{j;P+m}(t) = X_{j;P}(t) + \sum_{s=-\infty}^{\infty} b_{jm;P+m}(t-s)\{X_m(s) - X_{m;P}(s)\}. \tag{6.23}$$

The regression coefficients $b_{jm;P+m}(t-s)$ are determined by the conditions (for $u = 0, \pm 1, \cdots$)

$$E[X_j(t)\{X_m(u) - X_{m;P}(u)\}] = E[X_{j;P+m}(t)\{X_m(u) - X_{m;P}(u)\}] \tag{6.24}$$

which lead to the formulas

$$\begin{aligned} K_{jm;P}(u-t) &= \sum_{s=-\infty}^{\infty} b_{jm;P+m}(t-s)K_{mm;P}(u-s), \\ f_{jm;P}(\omega) &= \overline{B}_{jm;P+m}(\omega) f_{mm;P}(\omega). \end{aligned} \tag{6.25}$$

Thus

$$\overline{B}_{jm;P+m}(\omega) = \frac{f_{jm;P}(\omega)}{f_{mm;P}(\omega)}. \tag{6.26}$$

Similarly, one derives other inductive formulas from (6.23):

$$\begin{aligned} B_{jk;P+m}(\omega) &= B_{jk;P}(\omega) - B_{jm;P+m}(\omega)B_{mk;P}(\omega), \\ f_{hj;P+m}(\omega) &= f_{hj;P}(\omega) - \frac{f_{hm;P}(\omega)f_{jm;P}(-\omega)}{f_{mm;P}(\omega)}. \end{aligned} \tag{6.27}$$

More generally, one can conveniently compute a matrix A_P defined as follows. Fix a subset P of indices, and let Q denote the set of indices in D but not in P. Define A_P by

(6.28)

$j \backslash k$	P	Q
P	inverse matrix of $\{f_{jk}(\omega), j, k \in P\}$, denoted $\{g_{jk;P}(\omega)\}$	$\overline{B}_{kj;P}(\omega)$, conjugate of regression transfer function at ω
Q	$-B_{jk;P}(\omega)$, negative of regression transfer function	$f_{jk;P}(\omega)$, partial spectral density at ω

.

Given A_P, and an index m not in P, it can be shown that one forms $A_{P+m} = \{a_{jk;P+m}\}$ by the formulas

$$(6.29) \qquad \begin{aligned} a_{mm;P+m} &= 1 \div a_{mm;P}, \\ a_{mj;P+m} &= a_{mj;P} \div a_{mm;P} && \text{for } j \neq m, \\ a_{jm;P+m} &= -a_{jm;P} \div a_{mm;P} && \text{for } j \neq m, \\ a_{jk;P+m} &= a_{jk;P} - a_{jm;P+m}a_{mk;P} && \text{for } j \neq m \text{ and } k \neq m. \end{aligned}$$

Similarly, given A_P and an index m in P, the same formulas yield a matrix whose entries contain the regression transfer functions and partial cross-spectra for the set of predictors $\{X_k, k \in P, \text{ but } k \neq m\}$.

To prove (6.29), one needs the following formulas for the inverse matrix $\{g_{hj;P+m}(\omega)\} = \{f_{hj}(\omega), h, j \in P + m\}^{-1}$ when one adds an index m to a predictor set P (the argument ω is omitted for ease of writing):

$$(6.30) \qquad \begin{aligned} g_{mm;P+m} &= 1 \div f_{mm;P}, \\ g_{mh;P+m} &= \bar{g}_{hm;P+m} = -B_{mh;P} \div f_{mm;P} && \text{for } h \in P, \\ g_{hj;P+m} &= g_{hj;P} + \frac{\overline{B}_{mh;P}B_{mj;P}}{f_{mm;P}}, \\ &= g_{hj;P} - g_{hm;P+m}B_{mj;P} && \text{for } h, j \in P. \end{aligned}$$

One-sided prediction filters. The predictors considered in the foregoing are two-sided. One often desires to examine one-sided prediction filters.

Let $X_1(\cdot), \cdots, X_q(\cdot)$ be q time series. Let P be a subset of the set $D = \{1, 2, \cdots, q\}$ of indices, and let j be any index in D. We define a new time series, denoted $X_{j;P}^{(r)}(t)$, $t = 0, \pm 1, \cdots$ and called the minimum mean square error linear predictor of $X_j(t)$ given $\{X_k(s), s = t - r, t - r - 1, \cdots, k \in P\}$ as follows: $X_{j;P}^{(r)}(t)$ is a linear combination of the predictor random variables up to time $t - r$, which we write

$$(6.31) \qquad X_{j;P}^{(r)}(t) = \sum_{k \in P} \sum_{s=-\infty}^{t-r} b_{jk;P}^{(r)}(t - s)X_k(s) = \sum_{k \in P} \sum_{v=r}^{\infty} b_{jk;P}^{(r)}(v)X_k(t - v)$$

whose mean square prediction error as an estimate of $X_j(t)$, denoted

$$(6.32) \qquad K_{jj;P}^{(r)} = E[|X_j(t) - X_{j;P}^{(r)}(t)|^2],$$

is minimized. The regression coefficients $b_{jk;P}^{(r)}(t - s)$ are now determined by the conditions

$$(6.33) \qquad E[X_{j;P}^{(r)}(t)X_h(u)] = E[X_j(t)X_h(u)] \qquad \text{for } u \leq t - r \text{ and } h \text{ in } P,$$

which lead to the normal equations

$$(6.34) \qquad \begin{aligned} &\sum_{k \in P} \sum_{v=r}^{\infty} b_{jk;P}^{(r)}(v)E[X_k(t - v)X_h(u)] = E[X_j(t)X_h(u)] \\ & && \text{for } u \leq t - r, \\ &\sum_{k \in P} \sum_{v=r}^{\infty} b_{jk;P}^{(r)}(v)R_{kh}(u - t - v) = R_{jh}(u - t) && \text{for } u - t \leq -r, \\ &\sum_{k \in P} \sum_{v=r}^{\infty} b_{jk;P}^{(r)}(v)R_{hk}(s + v) = R_{hj}(s) && \text{for } s \geq r. \end{aligned}$$

Solving this system of equations for $b_{jk;P}^{(r)}(v)$ is the well-known Wiener-Hopf problem. The regression transfer functions

$$B_{jk;P}^{(r)}(\omega) = \sum_{v=r}^{\infty} b_{jk;P}^{(r)}(v)e^{-iv\omega} \tag{6.35}$$

can be obtained by a method involving factorization of the spectral density functions which is difficult to carry out (see Whittle [31]). If one is content with a numerical solution on a computer, rather than an analytical solution, one can find the regression coefficients $b_{jk}^{(r)}(v)$ directly and then compute the regression transfer function.

7. Gain and phase

Let $X_1, \cdots, X_r$ be jointly normal random variables. For any subset P of $D = \{1, \cdots, r\}$ and index j not in P, one can form (i) the regression coefficients $\{b_{jk;P}, k \in P\}$, (ii) the partial covariances $K_{hj;P}$ and partial correlation coefficients, and (iii) multiple correlation coefficient.

Similarly, for jointly covariance stationary time series $X_1(\cdot), \cdots, X_r(\cdot)$ one can form (i) the regression transfer functions $\{B_{jk;P}(\omega), k \in P\}$, (ii) the partial spectral density functions $f_{hj;P}(\omega)$ and partial coherence functions $W_{hj;P}(\omega)$, and (iii) the multiple coherence function $W_{j;P}(\omega)$.

The regression transfer function $B_{jk;P}(\omega)$ is best interpreted by regarding it as the frequency transfer function of a filter and introducing its gain and phase.

A discrete time invariant filter is described by its pulse response sequence $\{b_s, s = 0, \pm 1, \cdots\}$ or its frequency transfer function

$$B(\omega) = \sum_{s=-\infty}^{\infty} b_s e^{-is\omega}. \tag{7.1}$$

In terms of pulse response function, the output $\hat{X}(t)$ of the filter corresponding to an input $X(t)$ is given by

$$\hat{X}(t) = \sum_{s=-\infty}^{\infty} b_s X(t-s), \qquad s = 0, \pm 1, \cdots. \tag{7.2}$$

For a sinusoidal input $X(t) = e^{i\omega t}$, the output is $\hat{X}(t) = B(\omega)e^{i\omega t}$. Therefore, for an input which is a superposition of harmonics,

$$X(t) = \int_{-\pi}^{\pi} e^{it\omega}\, dZ(\omega), \tag{7.3}$$

the output is

$$\hat{X}(t) = \int_{-\pi}^{\pi} e^{it\omega}B(\omega)\, dZ(\omega). \tag{7.4}$$

The frequency response function $B(\omega)$ of a filter is a complex number which we can write

$$B(\omega) = \alpha(\omega) + i\beta(\omega) = G(\omega)e^{-i\varphi(\omega)} \tag{7.5}$$

where

$$\alpha(\omega) = \mathrm{Re}B(\omega), \qquad \beta(\omega) = \mathrm{Im}B(\omega). \tag{7.6}$$

The *gain* $G(\omega)$ and *phase* $\varphi(\omega)$ of a filter are defined by

$$G(\omega) = \sqrt{\alpha^2(\omega) + \beta^2(\omega)}, \tag{7.7}$$

$$\begin{aligned} \varphi(\omega) &= \arctan\{-\beta(\omega)/\alpha(\omega)\} & \text{if } \alpha(\omega) \geq 0, \\ &= \arctan\{-\beta(\omega)/\alpha(\omega)\} + \pi \operatorname{sign}\{-\beta(\omega)\} & \text{if } \alpha(\omega) < 0. \end{aligned} \tag{7.8}$$

To interpret the gain and phase of a filter, consider an input signal

$$X(t) = f(t)e^{i\omega_0 t} \tag{7.9}$$

whose frequency spectrum is nonvanishing only in a neighborhood of the frequency ω_0. Further, assume that in this region the gain of the filter is essentially constant and the phase is essentially a linear function of ω. Then the output signal will be a delayed but undistorted replica of the original (see Mason and Zimmerman [16], p. 367):

$$\hat{X}(t) = G(\omega_0)f(t - t_1)e^{i\omega_0(t-t_0)} \tag{7.10}$$

where

$$t_0 = \left.\frac{\varphi(\omega)}{\omega}\right|_{\omega=\omega_0}, \quad \text{carrier delay or phase delay;} \tag{7.11}$$

$$t_1 = \varphi'(\omega)|_{\omega=\omega_0}, \quad \text{envelope delay or group delay.} \tag{7.12}$$

The terminology "carrier delay" and "envelope delay" is used in the communication theory literature (for example, Mason and Zimmerman [16]). The terminology "phase delay" and "group delay" is used by Robinson ([26], p. 31), who extensively discusses these concepts.

In summary, one way to describe the relations between time series is by describing the characteristics of various regression transfer functions. There are a number of characteristics which need to be looked at: gain, logarithm of gain (or attenuation), phase, phase delay, and group delay. One of the problems of empirical multiple time series analysis is to determine which of these characteristics is most wisely used in routine statistical data reduction of multiple time series.

8. Sampling theory of sample partial and derived cross-spectra

Given a windowed sample spectral density matrix, one can form estimated partial regression transfer functions

$$\begin{aligned} \hat{B}_{jk;P}(\omega) &= \hat{\alpha}_{jk;P}(\omega) + i\hat{\beta}_{jk;P}(\omega), \\ &= \hat{G}_{jk;P}(\omega) \exp\left[-i\hat{\varphi}_{jk;P}(\omega)\right]. \end{aligned} \tag{8.1}$$

This estimate is computed by the methods of section 6; by analogy with (6.25) the estimates of $B_{jk;P}(\omega)$ can be explicitly written as

$$\hat{B}_{jk;P}(\omega) = \bar{\hat{f}}_{jk;P-k}(\omega) \div \hat{f}_{kk;P-k}(\omega). \tag{8.2}$$

By analogy with results of the usual theory of partial correlation (see Kendall and Stuart ([13], p. 333)), one might conjecture that for normal stationary time series with zero means

$$(8.3) \qquad (\hat{B}_{jk;P}(\omega) - B_{jk;P}(\omega)) \left\{ \frac{f_{kk;P-k}(\omega)}{Cf_{jj;P}(\omega)} \right\}^{1/2}$$

is asymptotically complex normal with mean 0 and variance 1.

To establish the plausibility of (8.3), let us relate it to certain established results for ordinary cross-spectral analysis (compare Jenkins [11]). In the case that the prediction set P contains only the predictor k, we write

$$(8.4) \qquad \hat{B}_{j;k}(\omega) = \hat{\alpha}_{j;k}(\omega) + i\hat{\beta}_{j;k}(\omega)$$

for the sample regression transfer function. The estimates are formed by

$$(8.5) \qquad \hat{\alpha}_{j;k}(\omega) = \frac{\hat{c}_{jk}(\omega)}{\hat{f}_{kk}(\omega)}, \qquad \hat{\beta}_{j;k}(\omega) = \frac{\hat{q}_{jk}(\omega)}{\hat{f}_{kk}(\omega)}.$$

The variance of $\hat{\alpha}$ can be derived by the well-known delta method (compare Kendall and Stuart ([13], vol. I, p. 231)); writing $\hat{c}$ and $\hat{f}$, respectively, for the numerator and denominator of $\hat{\alpha}$,

$$(8.6) \qquad \text{var}\,[\hat{\alpha}] = \frac{\text{var}\,[\hat{c}]}{E^2[\hat{f}]} - \frac{2\,\text{cov}\,[\hat{c}, \hat{f}]E[\hat{c}]}{E^3[\hat{f}]} + \frac{\text{var}\,[\hat{f}]E^2[\hat{c}]}{E^4[\hat{f}]},$$

one obtains the asymptotic covariances

$$(8.7) \qquad \text{var}\,[\hat{\alpha}_{j;k}(\omega)] = \text{var}\,[\hat{\beta}_{j;k}(\omega)] = \tfrac{1}{2}C\frac{f_{jj}(\omega)}{f_{kk}(\omega)}\{1 - W_{jk}(\omega)\},$$

$$(8.8) \qquad \text{cov}\,[\hat{\alpha}_{j;k}(\omega), \hat{\beta}_{j;k}(\omega)] = 0.$$

Therefore,

$$(8.9) \qquad \begin{aligned} \text{var}\,[\hat{B}_{j;k}(\omega)] &= C\frac{f_{jj}(\omega)}{f_{kk}(\omega)}\{1 - W_{jk}(\omega)\}, \\ &= C\frac{f_{jj;P}(\omega)}{f_{kk;P-k}(\omega)}, \end{aligned}$$

which agrees with (8.3).

Under the assumptions $\text{var}\,[\hat{\alpha}] = \text{var}\,[\hat{\beta}]$ and $\text{cov}\,[\hat{\alpha}, \hat{\beta}] = 0$, the gain $\hat{G}$ and phase $\hat{\varphi}$ defined by

$$(8.10) \qquad \hat{G}e^{-i\varphi} = \hat{\alpha} + i\hat{\beta}$$

have asymptotic variances (by the delta method)

$$(8.11) \qquad \text{var}\,[\hat{G}] = \text{var}\,[\hat{\alpha}], \qquad \text{var}\,[\hat{\varphi}] = \frac{1}{G^2}\,\text{var}\,[\hat{\alpha}]$$

where $Ge^{-i\varphi} = \alpha + i\beta$. In view of (8.3) and (8.7), we conjecture that

$$
\begin{aligned}
\operatorname{var}[\hat{\alpha}_{jk;P}(\omega)] &= \operatorname{var}[\hat{\beta}_{jk;P}(\omega)], \\
&= \tfrac{1}{2}C\,\frac{f_{jj;P}(\omega)}{f_{kk;P-k}(\omega)}, \\
&= \tfrac{1}{2}C\left|\frac{f_{jk;P-k}(\omega)}{f_{kk;P-k}(\omega)}\right|^2 \frac{f_{jj;P}(\omega)f_{kk;P-k}(\omega)}{|f_{jk;P-k}(\omega)|^2}, \qquad (8.12)\\
&= \tfrac{1}{2}C|G_{jk;P}(\omega)|^2\,\frac{f_{jj;P}(\omega)\div f_{jj;P-k}(\omega)}{W_{jk;P-k}(\omega)}, \\
&= \tfrac{1}{2}C|G_{jk;P}(\omega)|^2\left\{\frac{1}{W_{jk;P-k}(\omega)} - 1\right\}.
\end{aligned}
$$

From (8.11) and (8.12) one obtains expressions for the asymptotic variances of the partial gain and phase. In particular,

$$
\begin{aligned}
\operatorname{var}[\hat{\varphi}_{jk;P}(\omega)] &= \operatorname{var}[\log_e \hat{G}_{jk;P}(\omega)], \qquad (8.13)\\
&= \tfrac{1}{2}C\left\{\frac{1}{W_{jk;P-k}(\omega)} - 1\right\}.
\end{aligned}
$$

One may interpret (8.13) in words as follows: the variability of the estimated partial attenuation (log gain) and phase is determined by the partial coherency $W_{jk;P-k}(\omega)$; in particular, the variance tends to 0 as the partial coherence tends to 1. These results provide one interpretation of partial coherency.

To actually compute partial regression functions and their sampling error, one should use the algorithm (6.21), since using (6.30) one can rewrite (8.3): for any index k in P, asymptotic variance of $\hat{B}_{jk;P}(\omega)$ is $Cf_{jj;P}(\omega)\ g_{kk;P}(\omega)$. Stopping rules for selecting a significant set P of indices remain to be investigated.

9. Mean and bias of cross-spectral estimates

The behavior and interpretation of windowed sample cross-spectral density functions cannot be understood on the basis of their variability theory alone. Their means must be investigated.

To study the means of windowed sample spectra, one needs to consider two possible assumptions for the observed time series: (i) they are jointly covariance stationary with zero means, (ii) they are the sum of mean value functions and jointly covariance stationary zero mean fluctuations. Only case (i) is discussed in this paper.

We consider separately auto-spectra and cross-spectra. Asymptotic expressions for the means of windowed sample auto-spectral density functions have been studied by many writers, especially Parzen [18] and Hext [10]. We consider only the case that the spectral window satisfies the assumptions $K_{M,1}(\omega) = 0$ and $K_{M,2}(\omega) > 0$, defining

$$
K_{M,\nu}(\omega) = \int_{-\pi}^{\pi} (\lambda - \omega)^\nu K_M(\lambda - \omega)\, d\lambda. \qquad (9.1)
$$

Then the mean of a windowed sample auto-spectral density function may be approximated

$$(9.2)\qquad \begin{aligned} E[\hat{f}_{jj}(\omega)] &\doteq \int_{-\pi}^{\pi} K_M(\lambda - \omega) f_{jj}(\lambda)\, d\lambda, \\ &\doteq f_{jj}(\omega) - \frac{1}{2M^2} k''(0) f''(\omega), \end{aligned}$$

where

$$(9.3)\qquad k''(0) = -\int_{-\infty}^{\infty} \omega^2 K(\omega)\, d\omega$$

is the value at 0 of the second derivative of the covariance kernel $k(u) = \int_{-\infty}^{\infty} e^{iu\omega} K(\omega)\, d\omega$. We digress for a moment to note that some authors (Daniels [7], Akaike [1]) have suggested that the spectral window K_M be chosen so that $K_{M,\nu}(\omega) = 0$ for as many values of ν as possible. While this reduces the bias, it necessarily leads to possibly negative estimates which may lead to difficulties of interpretation of spectral estimates.

In evaluating the mean of a windowed sample cross-spectral density function,

$$(9.4)\qquad E[\hat{f}_{jk}(\omega)] \doteq \int_{-\pi}^{\pi} K_M(\lambda - \omega) f_{jk}(\lambda)\, d\lambda,$$

it is most convenient to express $f_{jk}(\lambda)$ in terms of the true regression transfer function

$$(9.5)\qquad \overline{B}_{j;k}(\lambda) = \frac{f_{jk}(\lambda)}{f_{kk}(\lambda)} = G_{j;k}(\lambda) \exp\,[i\varphi_{j;k}(\lambda)]$$

by

$$(9.6)\qquad f_{jk}(\lambda) = f_{kk}(\lambda) G_{j;k}(\lambda) \exp\,[i\varphi_{j;k}(\lambda)].$$

To understand the special sources of bias in cross-spectral estimation, let us first find the leading term of the mean $E[\hat{f}_{jk}(\omega)]$ by assuming that in the region $|\omega - \lambda| \leq B$ where $K_M(\omega - \lambda)$ is appreciably nonzero, both the auto-spectral density $f_{kk}(\cdot)$ and the gain are practically constant while the phase is linear; then approximately

$$(9.7)\qquad \begin{aligned} f_{jk}(\lambda) &= f_{kk}(\omega) G_{j;k}(\omega) \exp\,[i\{\varphi_{j;k}(\omega) + (\lambda - \omega)\varphi'_{j;k}(\omega)\}], \\ &= f_{jk}(\omega) \exp\,[i(\lambda - \omega)\varphi'_{j;k}(\omega)]; \end{aligned}$$

recall that the phase derivative $\varphi'_{j;k}(\omega)$ may be interpreted as a group delay or carrier delay. From (9.7) it follows that

$$(9.8)\qquad \begin{aligned} E[\hat{f}_{jk}(\omega)] &= f_{jk}(\omega) \int_{-\pi}^{\pi} K_M(\lambda - \omega) \exp\,[i(\lambda - \omega)\varphi'_{j;k}(\omega)]\, d\lambda, \\ &\doteq f_{jk}(\omega) \int_{-\infty}^{\infty} MK(M\mu) \exp\,[i\mu\varphi'_{j;k}(\omega)]\, d\mu, \\ &= f_{jk}(\omega) k(\varphi'_{j;k}(\omega)/M). \end{aligned}$$

In words, if the truncation point M is not chosen large compared to the group delay, there will be an appreciable bias in estimating the cross-spectral density function.

A possible method of avoiding this source of bias in cross-spectral density estimation is to use *shifted* cross-spectral estimates, which we now define (this method is due to Akaike [3]).

Let L be an integer (positive or negative). Define the shifted windowed sample cross-spectral density function with shift L by

$$f_{jk;T,M,L}(\omega) = \frac{1}{2\pi} \sum_{|v|<M} e^{-iv\omega} k\left(\frac{v}{M}\right) R_{jk;T}(v+L). \tag{9.9}$$

One may verify that its mean is approximately given by

$$\begin{aligned} E[f_{jk;T,L,M}(\omega)] &\doteq \int_{-\pi}^{\pi} d\lambda\, f_{jk}(\lambda) e^{i\lambda L} K_M(\omega-\lambda), \\ &\doteq \int_{-\pi}^{\pi} d\lambda\, K_M(\omega-\lambda) e^{i\lambda L} f_{kk}(\lambda) G_{j;k}(\lambda) e^{i\varphi_{j;k}(\lambda)\, d\lambda}. \end{aligned} \tag{9.10}$$

Using the same approximations as before, one may show that the mean is approximately equal to

$$f_{jk}(\omega) e^{i\omega L} \int_{-\pi}^{\pi} d\lambda\, K_M(\lambda-\omega) e^{i(\lambda-\omega)L} e^{i(\lambda-\omega)\varphi'_{j;k}(\omega)}. \tag{9.11}$$

Finally, one obtains the following approximation:

$$E[f_{jk;T,L,M}(\omega)] = f_{jk}(\omega) e^{i\omega L} k\left(\frac{L+\varphi'(\omega)}{M}\right) + \text{terms in } \frac{1}{M^2}. \tag{9.12}$$

If L is so chosen that

$$L + \varphi'_{j;k}(\omega) \ll M, \tag{9.13}$$

then an approximately unbiased estimate of $f_{jk}(\omega)$ is given by

$$e^{-i\omega L} f_{jk;T,M,L}(\omega). \tag{9.14}$$

The question of how to choose L remains; it may vary with ω and may have to be estimated from the sample phase. As a first guess, it could be taken to be the lag at which the sample cross-covariance function $R_{jk;T}(v)$ achieves its maximum absolute value.

We do not discuss here the terms in the bias of cross-spectral estimates which are of the order of $1/M^2$; they are analogous to the bias of auto-spectral estimates. It should be noted that the foregoing derivations are very heuristic; a complete and rigorous discussion is given by Nigel Nettheim in his Stanford Ph.D. thesis.

If one investigates (using the delta method) how the bias in cross-spectral density estimates propagates into the estimates of derived cross-spectral quantities, one finds that the bias is present in the estimated coherence but is absent in the estimated phase. It would seem that corrections for bias could be introduced using the estimated phase derivative. It remains to be investigated whether it would not be wise to directly estimate the phase derivative (group delay)

$$\varphi'_{j;k}(\omega) = \frac{d}{d\omega} \arctan \frac{q_{jk}(\omega)}{c_{jk}(\omega)}, \tag{9.15}$$
$$= \frac{c_{jk}(\omega)q'_{jk}(\omega) - q_{jk}(\omega)c'_{jk}(\omega)}{c^2_{jk}(\omega) + q^2_{jk}(\omega)},$$

by directly estimating the derivatives of the co-spectral and quadrature-spectral density functions.

The group delay (or phase derivative) should be routinely estimated in cross-spectral analysis since it seems easier to interpret than the phase. Further, the phase may be estimated without ambiguities modulo 2π by integrating (by Simpson's rule) the phase derivative.

REFERENCES

[1] H. Akaike, "On the design of lag window for the estimation of spectra," *Ann. Inst. Statist. Math.*, Vol. 14 (1962), pp. 1–21.

[2] H. Akaike and Y. Yamanouchi, "On the statistical estimation of frequency response function," *Ann. Inst. Statist. Math* , Vol. 14 (1962), pp. 23–56.

[3] H. Akaike, "Statistical measurement of frequency response function," *Ann. Inst. Statist. Math.*, Supplement III, Vol. 15 (1964), pp. 5–17.

[4] D. E. Amos and L. H. Koopmans, "Tables of the distribution of the coefficient of coherence for stationary bivariate Gaussian processes," (1963), Albuquerque, New Mexico, Sandia Corporation (available from the Office of Technical Services, Department of Commerce, Washington, D. C.).

[5] R. B. Blackman and J. Tukey, *The Measurement of Power Spectra from the Point of View of Communication Engineering*, New York, Dover, 1959.

[6] H. Bohman, "Approximate Fourier analysis of distribution functions," *Ark. Mat.*, Vol. 4 (1960), pp. 99–157.

[7] H. E. Daniels, "The estimation of spectral densities," *J. Roy. Statist. Soc. Ser. B*, Vol. 24 (1962), pp. 185–198.

[8] W. F. Freiberger, "Approximate distributions of cross-spectral estimates for Gaussian processes," *Time Series Analysis Symposium Proceedings*, edited by M. Rosenblatt, New York, Wiley, 1963, pp. 244–259.

[9] N. R. Goodman, "Statistical analysis based on a certain multivariate complex Gaussian distribution," *Ann. Math. Statist.*, Vol. 34 (1963), pp. 152–177.

[10] G. Hext, "A new approach to time series with mixed spectra," Ph.D. thesis, Statistics Department, Stanford University, 1966.

[11] G. M. Jenkins, "Cross-spectral analysis and the estimation of linear open loop transfer functions," *Time Series Analysis Symposium Proceedings*, edited by M. Rosenblatt, New York, Wiley, 1963, pp. 267–278.

[12] ———, "Some examples of and comments on spectral analysis," *Proceedings IBM Scientific Computing Symposium on Statistics*, White Plains, IBM, 1965, pp. 205–246.

[13] M. G. Kendall and A. Stuart, *Advanced Theory of Statistics*, Vols. 1 and 2, London, Griffin, 1958 and 1961.

[14] L. H. Koopmans, "On the coefficient of coherence for weakly stationary stochastic processes," *Ann. Math. Statist.*, Vol. 35 (1964), pp. 532–549.

[15] ———, "On the multivariate analysis of weakly stationary stochastic processes," *Ann. Math. Statist.*, Vol. 35 (1964), pp. 1765–1780.

[16] S. J. Mason and H. J. Zimmerman, *Electronic Circuits, Signals, and Systems*, New York, Wiley, 1960.

[17] V. K. Murthy, "Estimation of the cross spectrum," *Ann. Math. Statist.*, Vol. 34 (1963), pp. 1012–1021.
[18] E. Parzen, "On consistent estimates of the spectrum of a stationary time series," *Ann. Math. Statist.*, Vol. 28 (1957), pp. 329–348.
[19] ———, "On choosing an estimate of the spectral density function of a stationary time series," *Ann. Math. Statist.*, Vol. 28 (1957), pp. 921–932.
[20] ———, "On asymptotically efficient consistent estimates of the spectral density function of a stationary time series," *J. Roy. Statist. Soc. Ser. B*, Vol. 20 (1958), pp. 303–322.
[21] ———, "Mathematical considerations in the estimation of spectra," *Technometrics*, Vol. 3 (1961), pp. 167–190.
[22] ———, "An approach to empirical time series analysis," *Radio Science*, Vol. 68D (1964), pp. 937–957.
[23] ———, "On statistical spectral analysis," *Proceedings of Symposia in Applied Mathematics*, Vol. XVI, Providence, American Mathematical Society, 1964, pp. 221–246.
[24] ———, "The role of spectral analysis in time series analysis," International Statistical Institute (Belgrade meeting, September, 1965).
[25] ———, *Empirical Time Series Analysis*, San Francisco, Holden-Day, 1967.
[26] E. A. Robinson, *Random Wavelets and Cybernetic Systems*, London, Griffin, 1962.
[27] M. Rosenblatt, "Statistical analysis of stochastic processes with stationary residuals," *Probability and Statistics* (Cramér volume), edited by U. Grenander, New York, Wiley, 1959, pp. 300–330.
[28] *Technometrics*, Papers on Spectral Analysis of Time Series, Vol. 3 (1961), pp. 133–268.
[29] J. W. Tukey, "An introduction to the measurement of spectra," *Probability and Statistics* (Cramér volume), edited by U. Grenander, New York, Wiley, 1959, pp. 1283–1289.
[30] ———, "Data analysis and the frontiers of geophysics," *Science*, Vol. 148 (1965), pp. 1283–1289.
[31] P. Whittle, "On the fitting of multivariate autoregressions, and the approximate canonical factorization of a spectral density matrix," *Biometrika*, Vol. 50 (1963), p. 129.

SOME CHARACTERIZATION PROBLEMS IN STATISTICS

YU. V. PROHOROV
V. A. STEKLOV INSTITUTE, MOSCOW

1. Introduction

In this paper we shall discuss problems connected with tests of the hypothesis that a theoretical distribution belongs to a given class, for instance, the class of normal distributions, or uniform distribution or Poisson distribution. The statistical data consist of a large number of small samples (see [1]).

2. Reduction to simple hypotheses

Let $(\mathcal{X}, \mathcal{A})$ be a measurable space ($\mathcal{X}$ is a set and $\mathcal{A}$ is a σ-algebra of subsets of $\mathcal{X}$). Let $\mathcal{P}$ be a set of probability distributions defined on $\mathcal{A}$, let $(\mathcal{Y}, \mathcal{B})$ be another measurable space, and let $Y = f(X)$, $X \in \mathcal{X}$, be a measurable mapping of $(\mathcal{X}, \mathcal{A})$ into $(\mathcal{Y}, \mathcal{B})$. With this mapping every distribution P induces on $\mathcal{B}$ a corresponding distribution which we shall denote by Q_P^Y. We will be interested in the mappings (statistics) Y which possess the following two properties:

(1) Q_P^Y is the same for all $P \in \mathcal{P}$; in this case we will simply write $Q_{\mathcal{P}}^Y$.

(2) If for some P' on $\mathcal{A}$ one has $Q_{P'}^Y = Q_{\mathcal{P}}^Y$, then $P' \in \mathcal{P}$.

Sometimes it is expedient to formulate requirement (2) in the weakened form:

(2a) If $P' \in \mathcal{P}' \supset \mathcal{P}$ and $Q_{P'}^Y = Q_{\mathcal{P}}^Y$, then $P' \in \mathcal{P}$. In other words, we can assert in this case only that the equation $Q_{P'}^Y = Q_{\mathcal{P}}^Y$ implies $P' \in \mathcal{P}$ for some a priori restrictions ($P' \in \mathcal{P}'$) on P'.

If Y is a statistic satisfying (1) and (2), then it is clear that the hypothesis that the distribution of X belongs to class $\mathcal{P}$ is equivalent to the hypothesis that the distribution of Y is equal to $Q_{\mathcal{P}}^Y$.

Let us consider some examples. In these examples $(\mathcal{X}, \mathcal{A})$ is an n-dimensional Euclidean space of points $X = (x_1, \cdots, x_n)$ with the σ-algebra of Borel sets. The distributions belonging to $\mathcal{P}$ have a probability density of the form

$$p(x_1, \theta)p(x_2, \theta) \cdots p(x_n, \theta) \tag{2.1}$$

where p is a one-dimensional density and θ a parameter taking values in a parameter space.

EXAMPLE 1 (I. N. Kovalenko [2]). *Translation parameter.* Let $p(x; \theta) = p(x - \theta)$, with $-\infty < \theta < \infty$ (*additive type*). Here obviously it is necessary to take the $(n - 1)$-dimensional statistic $Y = (x_1 - x_n, \cdots, x_{n-1} - x_n)$. Of course, we can take any uniquely invertible function, for example $Y' = (x_1 - \bar{x}, \cdots, x_n - \bar{x})$ where $\bar{x} = (1/n) \sum_1^n x_k$.

In [2] it is shown that for $n \geq 3$, the distribution of Y determines the characteristic function $f(t) = \int_{-\infty}^{\infty} e^{itx} p(x)\, dx$ to within a factor of the form $e^{i\gamma t}$, on every interval where $f(t) \neq 0$. In particular, if $f(t) \neq 0$ for every t, then for $n \geq 3$, the statistic Y satisfies conditions (1) and (2) of section 2. This is also true if $f(t)$ is uniquely determined by its values in some neighborhood of zero (for example, if $f(t)$ is analytic in some neighborhood of zero).

In this paper, for every n there is given a pair of distributions, not belonging to the same additive type, for which the distribution of the statistic Y is the same for samples of size n. In section 4 these results are extended to a sample from a multidimensional population, and in section 5 to the case of a scale parameter.

REMARK. Let us assume that a distribution with density $p(x)$ has four finite moments: $m_1 = 0$, m_2, m_3, m_4, and that $p(x) \leq A$. Let us denote by $F(x)$ the corresponding distribution function and let $G(x)$ be another distribution function such that the distribution of Y is the same for F and G. Then it can be shown that

$$\inf_{\theta} \sup_{x} |G(x) - F(x - \theta)| \leq C(A, m_2, m_3, m_4) \frac{1}{\sqrt{n}}. \tag{2.2}$$

That is, if the sample size n is large, all the additive types corresponding to a given distribution of the statistic Y must be close to each other.

EXAMPLE 2 (A. A. Zinger, Yu. V. Linnik [3], [4]). Let $\theta = (a, \sigma)$, $-\infty < a < \infty$, $\sigma > 0$, and let

$$p(x, \theta) = \frac{1}{\sigma} \varphi\left(\frac{x - a}{\sigma}\right) \tag{2.3}$$

where φ is a normal (0, 1) density. Here it is natural to take the $(n - 2)$-dimensional statistic $Y = (y_1, \cdots, y_n)$, $y_k = (x_k - \bar{x})/s$, where $s^2 = \sum_{k=1}^{n} (x_k - \bar{x})^2$, $s > 0$. The sum of the components y_k of the vector Y is equal to zero, and the sum of their squares is unity. Thus the distribution of Y is concentrated on an $(n - 2)$-dimensional sphere $\sum y_k = 0$, $\sum y_k^2 = 1$.

It is known [1], [3] that for $p(x, \theta)$ defined by formula (2.3) the distribution of Y is uniform on this sphere. In [3] it is shown that for $n \geq 6$, the statistic Y possesses properties (1) and (2) of section 2; that is, from uniformity of the distribution of Y on the corresponding sphere it follows that the x's are normally distributed. This result is extended to distributions different from the normal in section 6.

It is clear for both examples cited that the choice of the statistic Y is based on considerations of invariance. Namely, there exists a group $\mathcal{G}$ of one-to-one (or almost one-to-one) mappings of the sample space onto itself ($X = (x_1, \cdots, x_n) \rightarrow (x_1 - a, \cdots, x_n - a)$ in the first example and $X \rightarrow ((x_1 - a)/\sigma), \cdots, (x_n - a)/\sigma))$ in the second) having the property that distributions of "random elements" X and gX, $g \in \mathcal{G}$, simultaneously belong to or do not belong to $\mathcal{P}$. In addition, for any two distributions P_1 and P_2 there exists $g \in \mathcal{G}$ such that for every $\mathcal{A}$, $P_2(\mathcal{A}) = P_1(g\mathcal{A})$.

In this case it is natural to take for Y a maximal invariant of the group $\mathcal{G}$.

Obviously, for $P \in \mathcal{P}$, Y possesses property (1) of section 2. The question of when Y possesses property (2) is related to a number of very difficult questions of analytic statistics. (See below for the problem of characterization of multidimensional distributions; see also [5]).

EXAMPLE 3 (L. N. Bolshev [6]). In the case when $x_1, x_2, \cdots, x_n$, take only values $0, 1, 2, \cdots$ with probabilities $p(x, \theta) = (\theta^x/x!)e^{-\theta}$ with $\theta > 0$, the considerations of invariance appear useless. Another approach based on utilization of sufficient statistics appears suitable. This approach will be discussed in detail in another paper.

3. Multidimensional location parameter

Now we shall consider the case of a family $\mathcal{P}$, given by formula (2.1) under the assumption that the x_j are ℓ-dimensional vectors: $x_j = (x_j^{(1)}, \cdots, x_j^{(\ell)})$ and

$$p(x, \theta) = p(x - \theta) \tag{3.1}$$

where it is known beforehand that θ lies in a k-dimensional subspace π_k of the space R^ℓ. Without loss of generality we shall suppose that π_k is defined by the relations

$$\theta_{k+1} = \theta_{k+2} = \cdots = \theta_\ell = 0. \tag{3.2}$$

As usual, we say that the density p satisfies the condition of Cramér if the integral $\int_{R^\ell} e^{(h,x)}p(x)\, dx$ is finite for all h lying in some neighborhood of zero of the space R^ℓ.

THEOREM 1. *Let $X = (x_1, \cdots, x_n)$ be a sample from the distribution* (3.1) *with conditions* (3.2). *We let $x_j' = (x_j^{(1)}, \cdots, x_j^{(k)}, 0, \cdots, 0)$. Then the statistic $Y = (Y_1, Y_2)$ where $Y_1 = x_1 - x_3'$, $Y_2 = x_2 - x_3'$, satisfies conditions* (1) *and* (2) *of section* 2.

PROOF. Let $t = (t^{(1)}, \cdots, t^{(\ell)})$, $\tau = (\tau^{(1)}, \cdots, \tau^{(\ell)})$, and let t' and τ' be defined in terms of t and τ in the same way that x_j' is defined in terms of x_j. Let $f(t) = Ee^{i(t,x_j)}$.

We note first of all that $Y_1 - Y_2 = x_1 - x_2$, and therefore the characteristic function of $x_1 - x_2$, that is $|f(t)|^2$, is uniquely defined by the distribution of the statistic Y. Now let f_1 and f_2 be two different characteristic functions of the x's, constituting a solution of the problem. Then the characteristic function of Y is equal to

$$E \exp [i(t, x_2 - x_3') + i(\tau, x_2 - x_3')] = f_u(t)f_u(\tau)\overline{f_u(t' + \tau')}, \quad u = 1, 2. \tag{3.3}$$

We take $\delta > 0$ so small that in a δ-neighborhood of zero, the functions $f_u(t)$, $u = 1, 2$, do not vanish. In what follows we will assume that t, τ and $t + \tau$ lie in this neighborhood. For these t and τ the principal value $A_u(t)$ of the argument of the function $f(t)$ satisfies the equation

$$A_u(t) + A_u(\tau) - A_u(t' + \tau') = \text{given function}. \tag{3.4}$$

Let us consider the corresponding homogeneous equation

$$a(t) + a(\tau) - a(t' + \tau') = 0. \tag{3.5}$$

We are interested in its real continuous solutions with $a(0) = 0$ (actually, from the assumption of the theorem it follows that $A_u(t)$ is infinitely differentiable in the neighborhood of zero which we are considering, and therefore it can be assumed that $a(t)$ is infinitely differentiable). We have $a(t') + a(\tau') = a(t' + \tau')$. Therefore,

$$a(t') = \sum_{j=1}^{k} \gamma_j t^{(j)}. \tag{3.6}$$

Further, from $a(t) + a(\tau') = a(t' + \tau')$, it follows that $a(t) = a(t')$. In such a way, in the neighborhood of zero which we are considering

$$A_1(t) - A_2(t) = \sum_{j=1}^{k} \gamma_j t^{(j)} \tag{3.7}$$

and

$$f_1(t) = f_2(t) \exp\left\{i \sum_{j=1}^{k} \gamma_j t^{(j)}\right\}. \tag{3.8}$$

Because of the analyticity of f_1 and f_2, this equation holds for all values of t.

REMARK. If $f(t) \neq 0$ for every t, then equation (3.8) is obtained without the condition stated in theorem 1.

4. Scale parameter in a multidimensional population

Now we shall consider the case of a family $\mathcal{P}$, given by formula (2.1), under the assumption that x_j is an ℓ-dimensional vector and

$$p(x, \theta) = p\left(\frac{x}{\theta}\right)\frac{1}{\theta^{\ell}}. \tag{4.1}$$

Let $X = (x_1, x_2, \cdots, x_n)$ be a sample from the distribution (4.1) $x_j = (x_j^{(1)}, \cdots, x_j^{(\ell)}$. The distribution of the 2 ℓ-dimensional vectors

$$V_j = (\ln |x_j^{(1)}|, \cdots, \ln |x_j^{(\ell)}|, \qquad \text{sign } x_j^{(1)}, \cdots, \text{sign } x_j^{(\ell)}) \tag{4.2}$$

belongs to the 2 ℓ-dimensional additive type with density

$$q(v, \hat{\theta}) = q(v^{(1)} - \hat{\theta}, \cdots, v^{(\ell)} - \hat{\theta}, \qquad v^{(\ell+1)}, \cdots, v^{(2\ell)}) \tag{4.3}$$

where $\hat{\theta} = \ln \theta$. The following theorem is easily derived from the result of the preceding section.

THEOREM 2. *Assume that $p(x)$ is bounded and satisfies Cramér's condition. Then the statistic $Y = (V_1 - V_3', V_2 - V_3')$, where V_3' is defined in terms of V_3 according to the rule of theorem* 1 (*with replacement of ℓ by 2ℓ and k by ℓ*), *possesses properties* (1) *and* (2).

The proof consists of verifying that the distribution of V_j satisfies the conditions of theorem 1.

5. One-dimensional linear type

We return to the one-dimensional case analogous to that considered in example 2, section 2. Let $\theta = (a, b)$, $-\infty < a < \infty$, $b > 0$ and

$$p(x, \theta) = \frac{1}{b} p\left(\frac{x-a}{b}\right). \tag{5.1}$$

We will call a type symmetric if it is possible to choose the function p to be even. Let $\bar{x}$, s, and y_k keep the same meaning as in example 2, section 2. Let us denote by $\mathcal{P}'$ the family of distributions (2.1) which corresponds to symmetric types.

THEOREM 3. *If p is symmetric and bounded and satisfies Cramér's condition, then for $n \geq 6$, the statistic*

$$Y^* = \left[\left(\frac{y_4 - y_3}{y_2 - y_1}\right)^2, \left(\frac{y_6 - y_5}{y_2 - y_1}\right)^2\right] \tag{5.2}$$

possesses properties (1) *and* (2a) *of section* 2.

PROOF. We have

$$Y_1^* = \left(\frac{y_4 - y_3}{y_2 - y_1}\right)^2 = \left(\frac{x_4 - x_3}{x_2 - x_1}\right)^2 \tag{5.3}$$

and an analogous equality for the second component Y_2^* of the vector Y^*. Let p' be a symmetric density, different from p and such that $Q_{p'}^{Y^*} = Q_p^{Y^*}$. From the fact that p satisfies Cramér's condition and is bounded, it follows easily that $\ln Y_1^*$ and at the same time $\ln (x_n - x_1)^2$ satisfy Cramér's condition (both for p and for p'). To the sample of size 3 made up of the variables $\ln (x_2 - x_1)^2$, $\ln (x_4 - x_3)^2$, $\ln (x_6 - x_5)^2$, one can apply what was said in the remark on example 1, section 2. Consequently, the distribution of Y^* determines the distribution of $\ln (x_2 - x_1)^2$ to within a translation parameter, and the distribution of $(x_2 - x_1)^2$ to within a scale parameter. Since the variable $x_2 - x_1$ is symmetrically distributed, its distribution also is determined to within a scale parameter. We note that thus far we have not made use anywhere of the symmetry of p'. If, for example, p is normal, then the distribution of $x_2 - x_1$ under p' is normal, and by Cramér's theorem x_1 is normal. In the general case, for a symmetric density p', the distribution of x_1 is uniquely determined except for a translation parameter by the distribution of $x_2 - x_1$. The theorem is proved.

Without the assumptions of symmetry, the formulation must be changed.

THEOREM 4. *Assume that p is bounded and satisfies Cramér's condition. Then for $n \geq 9$, the statistic $Y^{**} = (Y_1^{**}, Y_2^{**})$ where*

$$\begin{aligned} Y_1^{**} &= \left(\ln \left|\frac{y_3 - y_1}{y_9 - y_7}\right|, \ln \left|\frac{y_2 - y_1}{y_8 - y_7}\right|, \operatorname{sign}(y_3 - y_1), \operatorname{sign}(y_2 - y_1)\right), \\ Y_2^{**} &= \left(\ln \left|\frac{y_6 - y_4}{y_9 - y_7}\right|, \ln \left|\frac{y_5 - y_4}{y_8 - y_7}\right|, \operatorname{sign}(y_6 - y_4), \operatorname{sign}(y_5 - y_4)\right) \end{aligned} \tag{5.4}$$

possesses properties (1) *and* (2), *section* 2.

PROOF. The distribution of the vector $(x_3 - x_1, x_2 - x_1)$ belongs to the multiplicative type. Using a sample of size 3, namely $(x_3 - x_1, x_2 - x_1)$, $(x_6 - x_4, x_5 - x_4)$, $(x_9 - x_7, x_8 - x_7)$, this multiplicative type is determined uniquely by the distribution of the statistic mentioned in the formulation of the theorem. Knowing the distribution of $(x_3 - x_1, x_2 - x_1)$, we determine the additive type of the distribution of x_1. The theorem is proved.

6. Property of stability

We shall consider now the question of continuity of the correspondence $\mathcal{P} \Leftrightarrow Q_{\mathcal{P}}^{Y}$ assuming that the statistic Y satisfies conditions (1) and (2). In order to avoid unwieldy formulas, we will consider at first the case of the one-dimensional additive type

$$p(x, \theta) = p(x - \theta), \qquad -\infty < \theta < \infty, \tag{6.1}$$

and the one-dimensional linear type

$$p(x, \theta) = \frac{1}{b} p\left(\frac{x - a}{b}\right). \tag{6.2}$$

Let us recall the concept of convergence of types. One says (see [7]) that a sequence of types $T^{(N)}$ converges weakly to type T ($T^{(N)} \Rightarrow T$) if there exist $\mathfrak{F}^{(N)} \in T^{(N)}$ converging weakly to $\mathfrak{F} \in T$.

In the case of linear types, one usually considers convergence *to proper types only.*

Let $p^{(N)}$ and p be probability densities. Assume that $T(p^{(N)}) \Rightarrow T(p)$. Then from the property of weak convergence, it follows immediately that $Q_{p}^{Y}(N) \Rightarrow Q_{p}^{Y}$ where $Y = (x_1 - x_n, \cdots, x_{n-1} - x_n)$ for the case (6.1), and

$$Y = \left(\frac{x_1 - \bar{x}}{s}, \cdots, \frac{x_n - \bar{x}}{s}\right) \tag{6.3}$$

for the case (6.2). The reciprocal assertion gives the following theorem.

THEOREM 5. *For the situation described by* (6.1) *or* (6.2), *suppose that*

$$Q_{p}^{Y}(N) \Rightarrow Q_{p}^{Y}, \tag{6.4}$$

where the type $T(p)$ is uniquely determined by p. Then the sequence of types generated by $p^{(N)}$ converges to the type generated by p: $T(p^{(N)}) \Rightarrow T(p)$. For the case of linear types we assume in addition that we have a sample size $n \geq 4$.

PROOF. A. *Additive type.* From the convergence (6.4) follows, as is easily seen, that $|f^{(N)}(t)|^2 \to |f(t)|^2$, from which follows (see [8]) "shift compactness" of $p^{(N)}$. (This means that for appropriately chosen θ_N the sequence of distributions with densities $p^{(N)}(x - \theta_N)$ is weakly compact.) Now if the distributions with densities $p^{(N_k)}(x - \theta_{N_k})$ converge weakly to the limit distribution with density p', then

$$Q_{p'}^{Y} \Leftarrow Q_{p}^{Y}(N_k) \Rightarrow Q_{p}^{Y}, \tag{6.5}$$

from which we obtain $p'(x) = p(x - \theta)$.

B. *Linear type.* Let $x_j^{(N)}$, $y_j^{(N)}$, $\cdots$ be values of x, y, and so on, with distributions generated by $p^{(N)}$. From the convergence (6.4) follows convergence of the distributions of

$$Y^{(N)*} = \ln\left(\frac{x_4^{(N)} - x_3^{(N)}}{x_2^{(N)} - x_1^{(N)}}\right)^2 \tag{6.6}$$

to the distribution of

$$Y^* = \ln\left(\frac{x_4 - x_3}{x_2 - x_1}\right)^2. \tag{6.7}$$

From this follows, as is easily seen, the "shift-compactness" of the distributions of $\ln (x_2^{(N)} - x_1^{(N)})^2$. We shall take now an arbitrary sequence $N_k \uparrow$ of natural numbers and choose from it a subsequence M_k for which the distributions of

$$\ln \left(\frac{x_2^{(M_k)} - x_1^{(M_k)}}{b_k} \right)^2, \qquad b_k > 0, \tag{6.8}$$

converge weakly to a limit distribution. Then the distributions of

$$\frac{x_2^{(M_k)}}{b_k} - \frac{x_1^{(M_k)}}{b_k} \tag{6.9}$$

also form a weakly convergent sequence, and the sequence of distributions of $(x_1^{(M_k)})/b_k$ is "shift-compact." From this it is obvious that the convergence (6.4) implies relative compactness of the sequence of types $T(p^{(N)})$. The proof can now be completed in the same way as in part A.

7. Characterization of multidimensional linear types

We shall say that the distributions of random vectors x and y belong to the same type if there exists a nonsingular matrix A and vector b such that

$$gx = Ax + b \tag{7.1}$$

has the same distribution as y.

Let $p(x)$ be any ℓ-dimensional density and $\theta = (A, b)$. We denote by $\mathcal{P} = T(p) = \{p_\theta\}$ the linear type generated in the obvious manner by the density p.

All presently known results on characterization of multidimensional distributions have been obtained under the assumption that the distributions considered belong to the class $\mathcal{P}'$, defined in the following manner. The distribution of an ℓ-dimensional random vector x belongs to class $\mathcal{P}'$ if in some coordinate system its components are independent. The group $\mathcal{G}$ of all transformations (7.1) induces a group $\hat{\mathcal{G}}$ of transformations $\hat{g}X = (gx_1, \cdots, gx_n)$ in the $n\ell$-dimensional space of vectors $X = (x_1, \cdots, x_n)$. A maximal invariant Y of the group $\hat{\mathcal{G}}$ can be expressed in terms of the determinants

(7.2)

$$\Delta_{i_1}, \cdots, {}_{i_\ell} = [x_{i_1} - \bar{x}, \cdots, x_{i_\ell} - \bar{x}], \qquad \text{where} \quad \bar{x} = (1/n)(x_1 + \cdots + x_n)$$

and where $[z_1, \cdots, z_\ell]$ denotes the volume of the oriented parallelepiped constructed on the vectors $z_1, \cdots, z_\ell$.

Let us assume that the sample size $n \geq 6\ell$. We shall take vectors $z_j = x_{2j} - x_{2j-1}$ with components $z_j^{(k)}$, $k = 1, 2, \cdots, \ell$. Let

$$\delta_k = [z_{k\ell+1}, \cdots, z_{k\ell+\ell}]^2, \qquad k = 0, 1, 2 \tag{7.3}$$

$$\hat{Y}_1 = \ln \frac{\delta_1}{\delta_0}, \qquad \hat{Y}_2 = \ln \frac{\delta_2}{\delta_0}, \tag{7.4}$$

$$\hat{Y} = (\hat{Y}_1, \hat{Y}_2). \tag{7.5}$$

It is clear that $\hat{Y}$ is a function of a maximal invariant Y of the group $\hat{\mathcal{G}}$. The following theorem (see [5]) holds.

THEOREM 6. *If a density p satisfies Cramér's condition and if it is bounded, then the statistic Y possesses properties* (1) *and* (2a) *with respect to the class $\mathcal{P}''$ of distributions of random vectors x which can be transformed into vectors with independent, identically distributed, symmetrical components by a transformation of the form* (7.1).

The proof of this theorem is based on a lemma which has independent interest.

LEMMA. *Let $V_j^{(i)}$, $(i, j = 1, \cdots, \ell)$ be independent random variables with the same distribution function $V(x)$, and let $W_j^{(i)}$, $(i, j = 1, \cdots, \ell)$ also be independent and have a distribution function $W(x)$. If all moments of $V(x)$ exist and the distribution of the determinant $\Delta = \det \|V_j^{(i)}\|$ coincides with the distribution of the determinant $\delta = \det \|W_j^{(i)}\|$, then $V = W$.*

8. Application to testing hypothesis

The classical method of testing the hypothesis that the distribution of a sample belongs to a given parametric family (2.1) consists in the construction, based on the results of observations, of an estimate θ^* for θ and in the subsequent test of the significance of the deviation of the empirical distribution from the theoretical with $\theta = \theta^*$. Another statement of the problem will interest us.

A large number s of small samples $X_1, \cdots, X_s$ of sizes $n_1, \cdots, n_s$ respectively is given. The null hypothesis H_0 is that for every j the distribution of X_j is in the family (2.1). If there exist statistics $Y_1, \cdots, Y_s$; $Y_j = f_j(X_j)$, satisfying properties (1) and (2), then the composite hypothesis H_0 is replaced by the simple hypothesis H_0': for every j the distribution of Y_j is equal to Q^{Y_j}. Let the dimensionality of the statistic Y_j be equal to m_j. With the proper transformation one can translate Y_j into z_j, $Y_j = \psi_j(z_j)$, where z_j has a uniform distribution on the unit cube in m_j-dimensional Euclidean space. This transforms the hypothesis H_0' into the equivalent hypothesis H_0'': the components of the $(m_1 + \cdots + m_s)$-dimensional vector $Z = (z_1, \cdots, z_s)$ are independent and uniformly distributed on the interval $[0, 1]$. In this way one can give a *standard form* to the hypothesis H_0. Of course, the first question which arises in connection with such transformations concerns the form taken by the alternative hypotheses. From this point of view the transformations mentioned must be "sufficiently smooth" so that they transform the "alternatives close" to H_0 into the "alternatives close" to H_0''.

For now we shall postpone the corresponding analysis.

REFERENCES

[1] A. A. PETROV, "Tests, based on small samples, of statistical hypotheses concerning the type of a distribution," *Teor. Verojatnost. i Primenen.*, Vol. I (1956), pp. 248–271.

[2] I. N. KOVALENKO, "On the recovery of the additive type of a distribution on the basis of a sequence of series of independent observations," *Proceedings of the All Union Congress on the Theory of Probability and Mathematical Statistics* (Erevan, 1958), Erevan, Press of the Armenian Academy of Sciences, 1960.

[3] A. A. Zinger, "On a problem of A. N. Kolmogorov," *Bulletin of Leningrad University*, Vol. 11, Part I (1956), pp. 53–56.

[4] A. A. Zinger and Yu. V. Linnik, "On a characterization of the normal distribution," *Teor. Verojatnost. i Primenen.*, Vol. IX (1964), pp. 692–695.

[5] C. G. Hahubiya, "Testing the hypothesis of normality of a multidimensional distribution," to appear in *Teor. Verojatnost. i Primenen.*

[6] L. N. Bolshev, "On a characterization of the Poisson distribution, with some statistical applications," to appear in *Teor. Verojatnost. i Primenen.*

[7] B. V. Gnedenko and A. N. Kolmogorov, *Limit Distributions for Sums of Independent Random Variables*, Moscow-Leningrad, State Technical Press, 1949.

[8] M. Loève, "A l'intérieur du problème central," *Publ. Inst. Statist. Univ. Paris*, Vol. 6 (1957), pp. 313–325.

A NOTE ON MAXIMAL POINTS OF CONVEX SETS IN ℓ_∞

ROY RADNER
UNIVERSITY OF CALIFORNIA, BERKELEY

1. Introduction

The problem of characterizing maximal points of convex sets often arises in the study of admissible statistical decision procedures, of efficient allocation of economic resources (cf. Koopmans, [4], chapter 1, and references given there), and of mathematical programming (cf. Arrow, Hurwicz, and Uzawa, [2]).

Let C be a convex set in a finite dimensional vector space, partially ordered coordinate-wise (that is, for $x = (x_i)$ and $z = (z_i)$, $x \geq z$ means that $x_i \geq z_i$ for every coordinate i). Let D be the set of all strictly positive vectors (namely vectors all of whose coordinates are strictly positive); further, let B be the set of vectors in C that maximize $\sum_i y_i x_i$ for some vector $y = (y_i)$ in D. It is obvious that every vector in B is maximal in C with respect to the partial ordering $\leq$. One can also show that every vector that is maximal in C also maximizes $\sum_i y_i x_i$ on C for some nonnegative vector y. Arrow, Barankin, and Blackwell [1] showed further that every vector maximal in C is in the (topological) closure of B. They also gave an example (in 3 dimensions) in which a vector in the closure of B (and in C) is not maximal in C.

The purpose of this note is to generalize the Arrow-Barankin-Blackwell result to the case of ℓ_∞, the space of bounded sequences topologized by the sup norm. In this generalization, however, the set C is assumed to be compact.

2. The theorem

Let X denote ℓ_∞, that is, the Banach space of all bounded sequences of real numbers, with the sup norm topology, where the norm of $x = (x_i)$ in X is

$$\|x\| \equiv \sup_i |x_i|. \tag{2.1}$$

For x in X, I shall say that $x \geq 0$ if $x_i \geq 0$ for every i, and that $x > 0$ if $x \geq 0$ but $x \neq 0$. Also, for $x^1 = (x_i^1)$ and $x^2 = (x_i^2)$ in X, I shall say that $x^1 \geq x^2$ if $x^1 - x^2 \geq 0$ (and so on for $x^1 > x^2$).

A point $\hat{x}$ in a subset C of X will be called *maximal in* C if there is no x in C for which $x > \hat{x}$.

This research was supported in part by the Office of Naval Research under Contract ONR 222(77) with the University of California, and by a grant to the University from the National Science Foundation.

Let Y denote the set of all continuous linear functions on X. For any y in Y, I shall say that $y \geq 0$ if $y(x) \geq 0$ for all $x \geq 0$ in X, and that $y \gg 0$ if $y(x) > 0$ for all $x > 0$. Define

$$S \equiv \{y\colon\ y \in Y,\ \|y\| = 1,\ y \geq 0\},$$
$$S^+ \equiv \{y\colon\ y \in S,\ y \gg 0\}. \tag{2.2}$$

(Recall that for y in Y, $\|y\| \equiv \sup\ \{|y(x)|\colon x \in X.\ \|x\| = 1\}$). It shall be understood that Y has the weak* topology, and that the Cartesian product $X \times Y$ has the corresponding product topology.

If $\hat{y} \gg 0$, and $\hat{x}$ maximizes $\hat{y}(x)$ in a subset C of X, then $\hat{x}$ is clearly maximal in C. On the other hand, if $\hat{x}$ is maximal in a *convex* subset C of X, then there is a $\hat{y} \geq 0$ in Y such that $\hat{x}$ maximizes $\hat{y}(x)$ in C. (To see this, consider the nonnegative orthant of X; this is a convex set with a nonempty interior, and its interior is disjoint from the convex set of all points $(x - \hat{x})$ for which x is in C. The hyperplane that separates these two convex sets corresponds to the required $\hat{y}$.) It is easy to see that there can be maximal points in a convex set C that do not maximize any strictly positive continuous linear function on C. The following theorem gives information about such points in the case in which C is compact.

THEOREM. *If $\hat{x}$ is maximal in a compact convex subset C of X, then there is a $\hat{y}$ in S such that*

(1) *$\hat{x}$ maximizes $\hat{y}(x)$ on C, and*

(2) *$(\hat{x}, \hat{y})$ is the limit of a generalized sequence (x^m, y^m) of points in $C \times S^+$ such that for each m, x^m is maximal in C and maximizes $y^m(x)$ on C.*

LEMMA 1. *Define $f(x, y) \equiv y(x)$; then f is continuous on $X \times S$.*

PROOF. For any x, $\bar{x}$ in X and y, $\bar{y}$ in S,

$$\begin{aligned} |f(x, y) - f(\bar{x}, \bar{y})| &= |y(x - \bar{x}) + y(\bar{x}) - \bar{y}(\bar{x})| \\ &\leq 1 \cdot \|x - \bar{x}\| + |y(\bar{x}) - \bar{y}(\bar{x})|. \end{aligned} \tag{2.3}$$

Hence $\|x - \bar{x}\| < \epsilon/2$ and $|y(\bar{x}) - \bar{y}(\bar{x})| < \epsilon/2$ imply $|f(x, y) - f(\bar{x}, \bar{y})| < \epsilon$, which completes the proof of the lemma.

LEMMA 2. *For any $p \gg 0$ in Y, define*

$$S_p \equiv \{y\colon y \in S,\ y \geq p\}; \tag{2.4}$$

then for every $p \gg 0$ in Y, S_p is convex and compact.

PROOF. The set S_p is immediately seen to be convex, as the intersection of two convex sets, S and $\{y\colon y \in Y.\ y \geq p\}$. Note that the latter set is also closed. The set S can also be characterized as $\{y\colon y \in Y, y \geq 0,\ y(e) = 1\}$, where $e \equiv (1, 1, \cdots, \text{etc.} \cdots)$, and is therefore clearly closed. Thus S is a closed subset of the unit sphere in Y, which, by Alaoglu's theorem, is compact in the weak* topology; hence, S is compact, and therefore also S_p.

LEMMA 3. *If $y(\bar{x}) \geq 0$ for every y in S^+, then $\bar{x} \geq 0$.*

PROOF. Suppose that $\bar{x} = (\bar{x}_i)$ and that for some k, $\bar{x}_k < 0$ Let

$$q_k \equiv \frac{\|\bar{x}\| - (\frac{1}{2})\,\bar{x}_k}{\|\bar{x}\| - \bar{x}_k}, \tag{2.5}$$

let q_j $(j \neq k)$ be any sequence of positive numbers such that

$$\sum_{j \neq k} q_j = 1 - q_k, \tag{2.6}$$

and define $q(x) \equiv \sum_i q_i x_i$. It is easy to verify that $q \gg 0$, $\|q\| = 1$, and $q(\bar{x}) < 0$, which completes the proof of the lemma.

Proof of the theorem. The point $\hat{x}$ is maximal in the compact convex set C if and only if 0 is maximal in the compact convex set $C - \{\hat{x}\}$; hence, without loss of generality we may take $\hat{x} = 0$.

By lemmas 1 and 2, for every $p \gg 0$ in Y, the hypotheses of a minimax theorem of Ky Fan (cf. [3], p. 121) are satisfied for the function f defined on $C \times S_p$. Hence, there exist x^p in C and y^p in S_p such that, for all x in C and y in S_p,

$$y(x^p) \geq y^p(x^p) \geq y^p(x). \tag{2.7}$$

In particular, since 0 is in C,

$$y^p(x^p) \geq 0. \tag{2.8}$$

Let D be the set of all $p \gg 0$ in Y. The family $\mathfrak{N} \equiv \{(x^p, y^p) : p \in D\}$ is a net if D is directed by $\leq$. It was noted in the proof of lemma 2 that S is compact; hence, $\mathfrak{N}$ has a cluster point, say $(\bar{x}, \hat{y})$, in $C \times S$, and a subnet, say $\mathfrak{M}$, of $\mathfrak{N}$ converges to $(\bar{x}, \hat{y})$. Note that for every (x^p, y^p) in $\mathfrak{M}$, inequality (2.7) implies that x^p maximizes $y^p(x)$ on C, and therefore (since $y^p \gg 0$), x^p is maximal in C.

I now show that $\bar{x} = 0$. For every y in S^+ and p in Y such that $0 \ll p \leq y$, we have y in S_p, and hence, by (2.7) and (2.8), $y(x^p) \geq 0$; hence, by continuity, $y(\bar{x}) \geq 0$. In other words, for every y in S^+, $y(\bar{x}) \geq 0$. It follows by lemma 3 that $\bar{x} \geq 0$. Since 0 is maximal in C, $\bar{x} = 0$.

To complete the proof, it suffices to show that the maximum of $\hat{y}(x)$ on C is 0. From (2.7), for every $p \gg 0$ in Y and every x in C,

$$f[(x - x^p), y^p] \leq 0. \tag{2.9}$$

Hence, by the continuity of f (lemma 1), $f(x, \quad) \leq 0$.

Every continuous linear function y on X can be represented as an integral with respect to a finitely additive, finite, measure on the integers. In particular, it can be represented in the form

$$y(x) = \sum_{i < \infty} y_i x_i + y_\infty(x), \tag{2.10}$$

where $\sum_{i<\infty} |y_i| < \infty$, and y_∞ is a continuous linear function such that $y_\infty(x) = 0$ for every x with only a finite number of nonzero coordinates. From this representation, it is clear that $y \gg 0$ if and only if, in (2.10), $y_i > 0$ for every $i < \infty$.

It is an open question whether the theorem can be sharpened by replacing the set S^+ by the set of continuous linear functions of the form (2.10) with $y \gg 0$, $y_\infty = 0$, and $\sum_{i<\infty} y_i = 1$. It is also not known whether the condition that C be compact can be dispensed with.

REFERENCES

[1] K. J. Arrow, E. W. Barankin, and D. Blackwell, "Admissible points of convex sets," *Contributions to the Theory of Games*, Vol. II, edited by Kuhn and Tucker, Princeton, Princeton University Press (1953), pp. 87–92.
[2] K. J. Arrow, L. Hurwicz, and H. Uzawa, *Studies in Linear and Non-Linear Programming*, Stanford, Stanford University Press, 1958.
[3] Ky Fan, *Convex Sets and Their Applications*, Argonne National Laboratory, Applied Mathematics Division, Summer Lectures (mimeographed), 1959.
[4] T. C. Koopmans, *Three Essays on the State of Economic Science*, New York, McGraw-Hill, 1957.

LEAST SQUARES THEORY USING AN ESTIMATED DISPERSION MATRIX AND ITS APPLICATION TO MEASUREMENT OF SIGNALS

C. RADHAKRISHNA RAO
INDIAN STATISTICAL INSTITUTE

1. Introduction

In this paper are considered some problems in the estimation and inference on unknown parameters in a linear model under various assumptions on the error term. We write the linear model in the form

$$\mathbf{Y} = \mathbf{X}\boldsymbol{\tau} + \mathbf{e} \tag{1}$$

where $\mathbf{Y}$ is a $p \times 1$ vector of observable random variables, $\mathbf{X}$ is $p \times m$ matrix of known coefficients, $\boldsymbol{\tau}$ is a $p \times 1$ vector of unknown (nonstochastic) parameters, and $\mathbf{e}$ is a $p \times 1$ vector of errors. If $\boldsymbol{\Sigma}$, the dispersion matrix of $\mathbf{e}$, is known, then there is no problem, as the method of least squares (for the correlated case) can be applied to estimate and draw inferences on linear parametric functions of $\boldsymbol{\tau}$. We shall consider the case where $\boldsymbol{\Sigma}$ is unknown but an estimate $\hat{\boldsymbol{\Sigma}}$ of $\boldsymbol{\Sigma}$ is available, which may be computed from previous data or from the present data without making any assumption on $\boldsymbol{\tau}$, and discuss how this information can be used. In other words, we will discuss the theory of *least squares using an estimated dispersion matrix*. It is shown that the estimator of $\boldsymbol{\tau}$, obtained by merely substituting $\hat{\boldsymbol{\Sigma}}$ for $\boldsymbol{\Sigma}$ in the least squares estimator of $\boldsymbol{\tau}$ when $\boldsymbol{\Sigma}$ is known, is not necessarily the best. Certain improvements can be made depending on the *known or inferred structure* of $\boldsymbol{\Sigma}$.

Let us denote by E, D, and C the operators for expectation, dispersion, and covariance respectively. We consider the following specific structures for $\boldsymbol{\Sigma}$.

Case 1. The matrix $D(\mathbf{Y}) = \boldsymbol{\Sigma}$ is an unknown arbitrary positive definite matrix.

Case 2. The matrix $\boldsymbol{\Sigma} = \mathbf{X}\boldsymbol{\Gamma}\mathbf{X}' + \mathbf{Z}\boldsymbol{\Theta}\mathbf{Z}' + \sigma^2\mathbf{I}$, where $\boldsymbol{\Gamma}$, $\boldsymbol{\Theta}$, and σ^2 are unknown, and $\mathbf{Z}$ is a matrix such that $\mathbf{X}'\mathbf{Z} = \mathbf{0}$. Such a situation arises when we consider the mixed model

$$\mathbf{Y} = \mathbf{X}\boldsymbol{\tau} + \mathbf{X}\boldsymbol{\gamma} + \mathbf{Z}\boldsymbol{\xi} + \mathbf{e} \tag{2}$$

where $\boldsymbol{\gamma}$, $\boldsymbol{\xi}$, and $\mathbf{e}$ are all uncorrelated random vectors such that $E(\boldsymbol{\gamma}) = \mathbf{0}$, $D(\boldsymbol{\gamma}) = \boldsymbol{\Gamma}$, $E(\boldsymbol{\xi}) = \mathbf{0}$, $D(\boldsymbol{\xi}) = \boldsymbol{\Theta}$, and $D(\mathbf{e}) = \sigma^2\mathbf{I}$.

Case 3. The matrix $\boldsymbol{\Sigma} = \mathbf{C}\boldsymbol{\Gamma}\mathbf{C}' + \sigma^2\mathbf{I}$, where $\boldsymbol{\Gamma}$ and σ^2 are unknown and $\mathbf{C}$ is a known matrix. Such a situation arises when we consider the mixed model

$$\mathbf{Y} = \mathbf{X}\boldsymbol{\tau} + \mathbf{C}\boldsymbol{\gamma} + \mathbf{e} \tag{3}$$

where $\boldsymbol{\gamma}$ is a random vector such that $E(\boldsymbol{\gamma}) = \mathbf{0}$, $D(\boldsymbol{\gamma}) = \boldsymbol{\Gamma}$, $D(\mathbf{e}) = \sigma^2\mathbf{I}$ and $C(\boldsymbol{\gamma}, \mathbf{e}) = \mathbf{0}$. A model of the type (3) has been considered by Duncan [3], Henderson *et al.* [5], and others under a different set of assumptions on the variables $\boldsymbol{\gamma}$ and $\mathbf{e}$.

Case 4. The matrix $\boldsymbol{\Sigma} = \mathbf{C}\boldsymbol{\Gamma}\mathbf{C}' + \sigma^2\mathbf{I}$, where $\mathbf{C}$, $\boldsymbol{\Gamma}$, and σ^2 are unknown, but the rank of $\mathbf{C}$ is known or can be inferred from an estimate of $\boldsymbol{\Sigma}$. Or, in other words, the error vector $\mathbf{e}$ in (1) has a factor analytic structure with a common specific variance for thc components.

Case 5. Let $\mathbf{Y}' = (y_1, \cdots, y_p)$ and $\mathbf{e}' = (e_1, \cdots, e_p)$. The component y_t has the representation

$$y_t = P_k(t) + e_t \tag{4}$$

where the nonrandom part is a polynomial of the k-th degree in time, $P_k(t) = \beta_0 + \beta_1 t + \cdots + \beta_k t^k$, and the error terms e_t have an autoregressive scheme

$$e_t = \rho_1 e_{t-1} + \cdots + \rho_m e_{t-m} + \eta_t \tag{5}$$

where η_t are uncorrelated errors with a common variance σ^2. The parameters β_i representing the coefficients of the polynomial $P_k(t)$, the autoregressive parameters ρ_j, and σ^2 are all unknown. The problem is to estimate the parameters β_i from a single series of observations on y_t and in the absence of an independent estimate of $\boldsymbol{\Sigma}$.

In practice, we have the additional problem of checking the accuracy of an assumed model before estimating the unknown parameters. Appropriate tests for this purpose have been suggested in each case. Such tests are possible if an independent estimate of $\boldsymbol{\Sigma}$ is available.

An independent estimate of $\boldsymbol{\Sigma}$ may be available from past data or from multiple observations on vector $\mathbf{Y}$ of model (1). In the latter case, the observations are replaced by the average vector for which model (1) is true and the sample variance covariance (dispersion) matrix provides an estimate of $\boldsymbol{\Sigma}$. Note that if $\overline{\mathbf{Y}}$ and $(n-1)^{-1}\mathbf{S}$ represent the sample average and dispersion matrix, then the model (1) applied to $\overline{\mathbf{Y}}$ is written $\overline{\mathbf{Y}} = \mathbf{X}\boldsymbol{\tau} + \mathbf{e}$, and an estimate of $D(\overline{\mathbf{Y}})$ is $[n(n-1)]^{-1}\mathbf{S}$ where n is the sample size. Thus, the problem is reduced to the standard form with a linear model for a single vector random variable for which an estimate of the dispersion matrix is available.

In the general case we shall represent the dispersion matrix of $\mathbf{Y}$ by $\boldsymbol{\Sigma}$ and its estimator by $f^{-1}\mathbf{S}$. For purposes of tests of significance and computing confidence intervals for unknown parameters we shall assume the following distributions for $\mathbf{Y}$ and $\mathbf{S}$:

$$\mathbf{Y} \sim N_p(\boldsymbol{\tau}, \boldsymbol{\Sigma}), \tag{6}$$

$$\mathbf{S} \sim W_p(f, \boldsymbol{\Sigma}), \tag{7}$$

where $N_p(\boldsymbol{\tau}, \boldsymbol{\Sigma})$ denotes a p-variate normal distribution with mean and dispersion matrix as indicated in the brackets, and $W_p(f, \boldsymbol{\Sigma})$ denotes Wishart's distribution on degrees of freedom and expected matrix as indicated in the brackets. The symbol $\sim$ is used for "distributed as."

We let p denote the dimension (or the number of components) of $\mathbf{Y}$, and $\mathbf{k}$ that of $\boldsymbol{\tau}$. Without loss of generality, we shall assume that rank $\mathbf{X}$ is also k.

The methods discussed in this paper have wide applicability, although the specific problem of signal measurement is considered in the last section. Other areas in which these methods may be applied are in the estimation of polynomial trends of growth curves and time series, and prediction in time series.

Sometimes it may be possible to make a preliminary transformation of model (1) by multiplying both sides by $\boldsymbol{\Sigma}_0^{-1/2}$, where $\boldsymbol{\Sigma}_0$ is a guessed, or an a priori dispersion matrix of $\mathbf{e}$. The new model is $\mathbf{Y}^* = \mathbf{X}^*\boldsymbol{\tau} + \mathbf{e}^*$ where $\mathbf{Y}^* = \boldsymbol{\Sigma}_0^{-1/2}\mathbf{Y}$, $\mathbf{e}^* = \boldsymbol{\Sigma}_0^{-1/2}\mathbf{e}$ and $\mathbf{X}^* = \boldsymbol{\Sigma}_0^{-1/2}\mathbf{X}$. The estimated dispersion matrix of $\mathbf{Y}^*$ is $f^{-1}\mathbf{S}^*$ where $\mathbf{S}^* = \boldsymbol{\Sigma}_0^{-1/2}\mathbf{S}\boldsymbol{\Sigma}_0^{-1/2}$. We can then apply the methods of this paper assuming similar models for $D(\mathbf{Y}^*)$.

2. Some algebraic lemmas

Now we will prove some algebraic lemmas which are used in later sections of the paper.

LEMMA 2a. *Let* $\mathbf{A}$ *be a positive definite matrix partitioned as*

$$\begin{pmatrix} \mathbf{A}_{11} & \mathbf{A}_{12} \\ \mathbf{A}_{21} & \mathbf{A}_{22} \end{pmatrix} \tag{8}$$

with its inverse as

$$\begin{pmatrix} \mathbf{A}^{11} & \mathbf{A}^{12} \\ \mathbf{A}^{21} & \mathbf{A}^{22} \end{pmatrix}. \tag{9}$$

Then $\mathbf{A}^{11} - (\mathbf{A}_{11})^{-1}$ *is nonnegative definite.*

Multiplying (8) by (9) we have

$$\mathbf{A}_{11}\mathbf{A}^{11} + \mathbf{A}_{12}\mathbf{A}^{21} = \mathbf{I}, \qquad \mathbf{A}_{11}\mathbf{A}^{12} + \mathbf{A}_{12}\mathbf{A}^{22} = \mathbf{0}. \tag{10}$$

Multiplying both the equations in (10) by $\mathbf{A}_{11}^{-1}$,

$$\mathbf{A}^{11} + \mathbf{A}_{11}^{-1}\mathbf{A}_{12}\mathbf{A}^{21} = \mathbf{A}_{11}^{-1}, \qquad \mathbf{A}^{12} = -\mathbf{A}_{11}^{-1}\mathbf{A}_{12}\mathbf{A}^{22}. \tag{11}$$

Rearranging the terms in the first and substituting for $\mathbf{A}^{21} = (\mathbf{A}^{12})'$ from the second equation of (11), we have

$$\mathbf{A}^{11} - \mathbf{A}_{11}^{-1} = -\mathbf{A}_{11}^{-1}\mathbf{A}_{12}\mathbf{A}^{21} = \mathbf{A}_{11}^{-1}\mathbf{A}_{12}\mathbf{A}^{22}\mathbf{A}_{21}\mathbf{A}_{11}^{-1}. \tag{12}$$

The last matrix in (12) is nonnegative definite, which proves the required result.

As a corollary we have the following result. Let the partitioned matrix

$$\begin{pmatrix} \mathbf{A}_{11} & \mathbf{A}_{12} & \mathbf{A}_{13} & \cdots \\ \mathbf{A}_{21} & \mathbf{A}_{22} & \mathbf{A}_{23} & \cdots \\ \mathbf{A}_{31} & \mathbf{A}_{32} & \mathbf{A}_{33} & \cdots \\ \cdot & \cdot & \cdot & \cdots \end{pmatrix} \tag{13}$$

be positive definite. Denote by $\mathbf{A}_i^{11}$ the partition in the leading position in the reciprocal of the submatrix of (13) obtained by considering the first i row and column partitions. Then the matrix $\mathbf{A}_i^{11} - \mathbf{A}_j^{11}$ is nonnegative definite for any i and j such that $i \geq j$.

LEMMA 2b. *Let* $\mathbf{X}$ *be a* $p \times k$ *matrix of rank* k, *and let* $\mathbf{Z}$ *be a* $p \times (p - k)$ *matrix of rank* $(p - k)$ *such that* $\mathbf{X}'\mathbf{Z} = \mathbf{0}$. *Then*

$$(\mathbf{X}'\boldsymbol{\Sigma}^{-1}\mathbf{X})^{-1}\mathbf{X}'\boldsymbol{\Sigma}^{-1} = (\mathbf{X}'\mathbf{X})^{-1}\mathbf{X}' - (\mathbf{X}'\mathbf{X})^{-1}\mathbf{X}'\boldsymbol{\Sigma}\mathbf{Z}(\mathbf{Z}'\boldsymbol{\Sigma}\mathbf{Z})^{-1}\mathbf{Z}' \tag{14}$$

where $\boldsymbol{\Sigma}$ *is any* $p \times p$ *positive definite matrix.*

Multiplying both sides of (14) by $\mathbf{X}$, it is easily seen that the equality holds. If multiplication by $\mathbf{Z}$ also results in equality, then (14) is true. Multiplying by $\mathbf{Z}$ from the right and by $(\mathbf{X}'\boldsymbol{\Sigma}^{-1}\mathbf{X})$ from the left we have

$$\mathbf{X}'\boldsymbol{\Sigma}^{-1}\mathbf{Z} = -\mathbf{X}'\boldsymbol{\Sigma}^{-1}\mathbf{X}(\mathbf{X}'\mathbf{X})^{-1}\mathbf{X}'\boldsymbol{\Sigma}\mathbf{Z}[(\mathbf{Z}'\mathbf{Z})^{-1}\mathbf{Z}'\boldsymbol{\Sigma}\mathbf{Z}]^{-1}, \tag{15}$$

$$\mathbf{X}'\boldsymbol{\Sigma}^{-1}\mathbf{Z}[(\mathbf{Z}'\mathbf{Z})^{-1}\mathbf{Z}'\boldsymbol{\Sigma}\mathbf{Z}] = -\mathbf{X}'\boldsymbol{\Sigma}^{-1}\mathbf{X}(\mathbf{X}'\mathbf{X})^{-1}\mathbf{X}'\boldsymbol{\Sigma}\mathbf{Z}, \tag{16}$$

$$\mathbf{X}'\boldsymbol{\Sigma}^{-1}[\mathbf{Z}(\mathbf{Z}'\mathbf{Z})^{-1}\mathbf{Z}' + \mathbf{X}(\mathbf{X}'\mathbf{X})^{-1}\mathbf{X}']\boldsymbol{\Sigma}\mathbf{Z} = 0, \tag{17}$$

which is true, since the expression within the square brackets of (17) is $\mathbf{I}$ and $\mathbf{X}'\mathbf{Z} = 0$

LEMMA 2c. *With* $\mathbf{X}$ *and* $\mathbf{Z}$ *as in lemma 2b, the matrix*

$$(\mathbf{X}'\mathbf{X})^{-1}\mathbf{X}'\boldsymbol{\Sigma}\mathbf{X}(\mathbf{X}'\mathbf{X})^{-1} - (\mathbf{X}'\boldsymbol{\Sigma}^{-1}\mathbf{X})^{-1} \tag{18}$$

is nonnegative definite.

Consider the matrix

$$\begin{pmatrix} \mathbf{X}'\boldsymbol{\Sigma}^{-1}\mathbf{X} & \mathbf{X}'\boldsymbol{\Sigma}^{-1}\mathbf{Z} \\ \mathbf{Z}'\boldsymbol{\Sigma}^{-1}\mathbf{X} & \mathbf{Z}'\boldsymbol{\Sigma}^{-1}\mathbf{Z} \end{pmatrix} = \begin{pmatrix} \mathbf{X}' \\ \mathbf{Z}' \end{pmatrix} \boldsymbol{\Sigma}^{-1}(\mathbf{X} \vdots \mathbf{Z}). \tag{19}$$

The reciprocal of the right-hand side is

$$(\mathbf{X} \vdots \mathbf{Z})^{-1}\boldsymbol{\Sigma}\begin{pmatrix} \mathbf{X}' \\ \mathbf{Z}' \end{pmatrix}^{-1}. \tag{20}$$

But

$$(\mathbf{X} \vdots \mathbf{Z})^{-1} = \begin{pmatrix} (\mathbf{X}'\mathbf{X})^{-1}\mathbf{X}' \\ (\mathbf{Z}'\mathbf{Z})^{-1}\mathbf{Z}' \end{pmatrix}. \tag{21}$$

Substituting the result (21) in (20), we have the leading partition in the reciprocal of (19) as

$$(\mathbf{X}'\mathbf{X})^{-1}\mathbf{X}'\boldsymbol{\Sigma}\mathbf{X}(\mathbf{X}'\mathbf{X})^{-1}. \tag{22}$$

Hence, (18) follows by applying the result of lemma 1a.

Finally we need some results on restricted eigenvectors of a symmetric matrix as developed by the author elsewhere (Rao [12]).

LEMMA 2d. *Let* $\boldsymbol{\Sigma}$ *and* $\mathbf{X}$ *be as in lemma 2b and consider the nonzero eigenvalues and right eigenvectors of the matrix* $(\mathbf{I} - \mathbf{X}(\mathbf{X}'\mathbf{X})^{-1}\mathbf{X}')\boldsymbol{\Sigma}$. *Let* $\lambda_1 \geq \lambda_2, \cdots \geq \lambda_r \geq 0$ *be the eigenvalues and* $\mathbf{L}_1, \cdots, \mathbf{L}_r$ *be the corresponding eigenvectors which can be chosen to be mutually orthogonal. Then we have the following results:*

$$\begin{aligned} \mathbf{L}_i'\mathbf{X} &= 0, \qquad i = 1, \cdots, r, \\ \mathbf{L}_i'\boldsymbol{\Sigma}\mathbf{L}_j &= 0, \qquad i \neq j. \end{aligned} \tag{23}$$

The results are easy to prove. The $\mathbf{L}_i$ are said to be restricted eigenvectors of $\boldsymbol{\Sigma}$ with the restriction $\mathbf{L}_i'\mathbf{X} = 0$.

LEMMA 2e. *Let* $\boldsymbol{\Sigma} = \mathbf{C}\boldsymbol{\Gamma}\mathbf{C}' + \sigma^2\mathbf{I}$ *and let* $\mathbf{L}$ *be a restricted eigenvector of* $\boldsymbol{\Sigma}$ *corresponding to the restricted eigenvalue* σ^2 *so that* $\mathbf{L}'\mathbf{X} = 0$. *Then*

$$\mathbf{L}'\boldsymbol{\Sigma}\mathbf{X} = 0. \tag{24}$$

By definition,

$$(\mathbf{I} - \mathbf{X}(\mathbf{X}'\mathbf{X})^{-1}\mathbf{X}')(\mathbf{C}\boldsymbol{\Gamma}\mathbf{C}' + \sigma^2\mathbf{I})\mathbf{L} = \sigma^2\mathbf{L}, \tag{25}$$

which on simplification gives

$$(\mathbf{I} - \mathbf{X}(\mathbf{X}'\mathbf{X})^{-1}\mathbf{X}')\mathbf{C}\boldsymbol{\Gamma}\mathbf{C}'\mathbf{L} = 0. \tag{26}$$

Multiplying by $\mathbf{L}'$ and putting $\mathbf{L}'\mathbf{X} = 0$, we have

$$\mathbf{L}'\mathbf{C}\boldsymbol{\Gamma}\mathbf{C}'\mathbf{L} = 0 \Rightarrow \mathbf{C}\boldsymbol{\Gamma}\mathbf{C}'\mathbf{L} = 0 \Rightarrow \mathbf{X}'\boldsymbol{\Sigma}\mathbf{L} = 0. \tag{27}$$

3. Covariance adjustment

In this paper, we frequently refer to covariance adjustment in an estimator using another statistic (with zero expectation) as a concomitant variable. The procedure is described as follows.

Let $\mathbf{T}_1$ and $\mathbf{T}_2$ be two vector statistics of orders k and r such that $E(\mathbf{T}_1) = \boldsymbol{\tau}$ and $E(\mathbf{T}_2) = \mathbf{0}$, where $\boldsymbol{\tau}$ is a vector of k unknown parameters. The vector $\mathbf{T}_1$ is an estimator of $\boldsymbol{\tau}$, but if $C(\mathbf{T}_1, \mathbf{T}_2) \neq \mathbf{0}$, then a better estimator of $\boldsymbol{\tau}$ can be found when the dispersion matrix

$$\boldsymbol{\Lambda} = \begin{pmatrix} \boldsymbol{\Lambda}_{11} & \boldsymbol{\Lambda}_{12} \\ \boldsymbol{\Lambda}_{21} & \boldsymbol{\Lambda}_{22} \end{pmatrix} \tag{28}$$

of $(\mathbf{T}_1, \mathbf{T}_2)$ is known. Thus, if we consider the estimator

$$\boldsymbol{\tau}^* = \mathbf{T}_1 - \boldsymbol{\Lambda}_{12}\boldsymbol{\Lambda}_{22}^{-1}\mathbf{T}_2, \tag{29}$$

then

$$D(\boldsymbol{\tau}^*) = \boldsymbol{\Lambda}_{11} - \boldsymbol{\Lambda}_{12}\boldsymbol{\Lambda}_{22}^{-1}\boldsymbol{\Lambda}_{21} = D(\mathbf{T}_1) - \boldsymbol{\Lambda}_{12}\boldsymbol{\Lambda}_{22}^{-1}\boldsymbol{\Lambda}_{21}; \tag{30}$$

that is, $D(\mathbf{T}_1) - D(\boldsymbol{\tau}^*)$ is always nonnegative definite. Hence $\boldsymbol{\tau}^*$ is more efficient than $\mathbf{T}_1$.

If only an estimate of $\boldsymbol{\Lambda}$,

$$\mathbf{U} = \begin{pmatrix} \mathbf{U}_{11} & \mathbf{U}_{12} \\ \mathbf{U}_{21} & \mathbf{U}_{22} \end{pmatrix}, \tag{31}$$

is available, we may substitute $\mathbf{U}_{ij}$ for $\boldsymbol{\Lambda}_{ij}$ in the formula (29) and obtain the (covariance) adjusted estimator as

$$\hat{\boldsymbol{\tau}} = \mathbf{T}_1 - \mathbf{U}_{12}\mathbf{U}_{22}^{-1}\mathbf{T}_2. \tag{32}$$

It is seen that when $\mathbf{U}_{ij}$ are distributed independently of $\mathbf{T}_i$,

$$E(\hat{\tau}) = E(\mathbf{T}_1 - \mathbf{U}_{12}\mathbf{U}_{22}^{-1}\mathbf{T}_2) = \tau \tag{33}$$

so that the adjusted estimator $\hat{\tau}$ is also unbiased for τ. Now

$$\begin{aligned} D(\hat{\tau}) &= D(\mathbf{T}_1) + E[\mathbf{U}_{12}\mathbf{U}_{22}^{-1}D(\mathbf{T}_2)\mathbf{U}_{22}^{-1}\mathbf{U}_{21}] - E[\mathbf{U}_{12}\mathbf{U}_{22}^{-1}C(\mathbf{T}_2, \mathbf{T}_1)] \\ &\qquad - E[C(\mathbf{T}_1, \mathbf{T}_2)\mathbf{U}_{22}^{-1}\mathbf{U}_{21}] \\ &= \Lambda_{11} + E[\mathbf{U}_{12}\mathbf{U}_{22}^{-1}\Lambda_{22}\mathbf{U}_{22}^{-1}\mathbf{U}_{21}] - E[\mathbf{U}_{12}\mathbf{U}_{22}^{-1}\Lambda_{21}] - E[\Lambda_{12}\mathbf{U}_{22}^{-1}\mathbf{U}_{21}] \end{aligned} \tag{34}$$

where the expectations are taken over the variations of $\mathbf{U}_{ij}$. We are no longer in a position to claim that $D(\mathbf{T}_1) - D(\hat{\tau})$ is always nonnegative definite as in the case of $D(\mathbf{T}_1) - D(\tau^*)$. As a matter of fact, it is seen from (34) that if $\Lambda_{21} = \mathbf{0}$, or very nearly $\mathbf{0}$, then $D(\mathbf{T}_1) - D(\hat{\tau})$ is negative definite; that is, $\mathbf{T}_1$ is more efficient than $\hat{\tau}$.

Thus, covariance adjustment can result in a decrease in efficiency when an estimated dispersion matrix is used in the place of the unknown matrix. However, if Λ_{12} is not close to zero, we should expect $D(\mathbf{T}_1) - D(\hat{\tau})$ to be nonnegative definite.

There is, however, an important problem. It is possible that the use of $\mathbf{T}_2$ as a whole for covariance adjustment is not optimum, and a suitable choice of functions of $\mathbf{T}_2$ for this purpose may provide maximum efficiency. In the absence of an exact knowledge about Λ the optimum solution cannot be found. However, an estimate of Λ may provide some guidance in the choice of suitable functions of $\mathbf{T}_2$. We consider such problems in the rest of the sections.

We were able to draw some conclusions on the basis of the formula (34) for $D(\hat{\tau})$ without making an explicit evaluation of the expectations involved. We shall now complete the discussion by making the following assumptions on the distributions of $(\mathbf{T}_1, \mathbf{T}_2)$ and $\mathbf{U}$:

$$\begin{pmatrix}\mathbf{T}_1\\ \mathbf{T}_2\end{pmatrix} \sim N_{k+r}\left[\begin{pmatrix}\tau\\ 0\end{pmatrix}, \Lambda\right], \tag{35}$$

$$f\mathbf{U} \sim W_{k+r}(f, \Lambda). \tag{36}$$

In the rest of the present section we lay down procedures for drawing inferences on τ on the basis of the estimator $\hat{\tau}$ obtained by adjusting a *given* estimator T_1 with respect to a *given* concomitant variable $\mathbf{T}_2$ under the assumptions (35) and (36). An important result in this direction is contained in lemma 3a.

LEMMA 3a. *The conditional distributions of* $\hat{\tau}$ *and* $\mathbf{G} = f(\mathbf{U}_{11} - \mathbf{U}_{12}\mathbf{U}_{22}^{-1}\mathbf{U}_{21})$, *given* $\mathbf{T}_2$ *and* $\mathbf{U}_{22}$, *are independent, and the conditional distributions are*

$$\hat{\tau} \sim N_k(\tau, (1 + T_r^2)\Gamma), \tag{37}$$

$$\mathbf{G} \sim W_k(f - r, \Gamma), \tag{38}$$

where $fT_r^2 = \mathbf{T}_2'\mathbf{U}_{22}^{-1}\mathbf{T}_2$ *and* $\Gamma = \Lambda_{11} - \Lambda_{12}\Lambda_{22}^{-1}\Lambda_{21}$.

As a consequence of lemma 3a, we have the results of lemma 3b.

LEMMA 3b. *Let*

$$U_k = \frac{(\hat{\tau} - \tau)'\mathbf{G}^{-1}(\hat{\tau} - \tau)}{1 + T_r^2}, \tag{39}$$

$$V_k = (\hat{\tau} - \tau)'\mathbf{G}^{-1}(\hat{\tau} - \tau). \tag{40}$$

Then

$$\frac{f - r - k + 1}{k} U_k \sim F(k, f - r - k + 1); \tag{41}$$

that is, a variance ratio distribution on k and $f - r - k + 1$ degrees of freedom, and V_k has the distribution

$$\text{const. } V_k^{k/2}(1 + V_k)^{-(f-k-r+3)/2}\,{}_2F_1\left(\frac{r}{2}, \frac{f - r + 1}{2}; \frac{f + k + 1}{2}, \frac{V_k}{1 + V_k}\right) dV_k. \tag{42}$$

The distributions of lemmas 3a and 3b follow from the basic results derived in earlier papers by the author (Rao [7], [8], [9], [11]).

The $(1 - \alpha)$ probability concentration ellipsoids for the unknown parameter τ based on U_k and V_k are

$$(\hat{\tau} - \tau)'\mathbf{G}^{-1}(\hat{\tau} - \tau) \le \frac{k}{f - k - r + 1} F_\alpha(1 + T_r^2), \tag{43}$$

$$(\hat{\tau} - \tau)'\mathbf{G}^{-1}(\hat{\tau} - \tau) \le V_{k\alpha}, \tag{44}$$

where F_α and $V_{k\alpha}$ are the upper α-probability points of the F and V_k distributions given in (41) and (42), respectively. It has been shown in the earlier paper (Rao [8]) that the inferences based on the F and V_k distributions are not very different, although there is slight advantage in using the V_k distribution. However, the percentage points of the V_k distribution are not yet available.

Observing that

$$(\hat{\tau} - \tau)'\mathbf{G}^{-1}(\hat{\tau} - \tau) = \max_{\mathbf{P}} \frac{(\mathbf{P}'\hat{\tau} - \mathbf{P}'\tau)^2}{\mathbf{P}'\mathbf{G}\mathbf{P}}, \tag{45}$$

we find that the simultaneous confidence intervals for linear functions $\mathbf{P}'\tau$ of τ are provided by

$$\mathbf{P}'\hat{\tau} \pm [\mathbf{P}'\mathbf{G}\mathbf{P}kF_\alpha(1 + T_r^2)/(f - k - r + 1)]^{1/2} \tag{46}$$

$$\mathbf{P}'\hat{\tau} \pm [\mathbf{P}'\mathbf{G}\mathbf{P}V_{k\alpha}]^{1/2} \tag{47}$$

using (43) or (44).

If the confidence interval for a particular linear function $\mathbf{P}'\tau$ is needed, we replace k by unity in the expressions (46) and (47). The number F_α is then the upper α-probability value of F on 1 and $f - r$ degrees of freedom, and $V_{1\alpha}$ is the upper α-probability value of the V_1 distribution.

The knowledge of the actual distributions of $(\mathbf{T}_1, \mathbf{T}_2)$ and $\mathbf{U}$ enable us to find the exact expression for $D(\hat{\tau})$ which is left in a symbolic form in (34). Using the result (37),

$$D(\hat{\tau}|\mathbf{T}_2, \mathbf{U}_{22}) = (1 + T_r^2)\boldsymbol{\Gamma}. \tag{48}$$

Now, observing that

$$f - \frac{r}{r} - 1\, T_r^2 \sim F(r, f - r + 1), \tag{49}$$

we find

$$D(\hat{\boldsymbol{\tau}}) = E[(1 + T_r^2)]\boldsymbol{\Gamma} = \frac{f-1}{f-r-1}\boldsymbol{\Gamma}. \tag{50}$$

When $\boldsymbol{\Lambda}$ is known, the best estimator $\boldsymbol{\tau}^*$ has the dispersion matrix $\boldsymbol{\Gamma}$ so that the loss of efficiency in using an estimate of $\boldsymbol{\Lambda}$ is $r/(f-r-1)$, which is zero when $r = 0$, and which tends to zero as f tends to infinity.

Further, if $\boldsymbol{\Lambda}_{12} = \mathbf{0}$, $\boldsymbol{\Gamma} = \boldsymbol{\Lambda}_{11}$ and therefore $\mathbf{T}_1$ is more efficient than $\hat{\boldsymbol{\tau}}$ for any fixed values of r and f. The situation will be the same if $\boldsymbol{\Lambda}_{12}$ is close to $\mathbf{0}$.

In the cases considered in this paper we shall first reduce the problem to that of making covariance adjustment in an estimator $\mathbf{T}_1$ using a concomitant variable $\mathbf{T}_2$ and an estimated dispersion matrix of $\mathbf{T}_1$, $\mathbf{T}_2$. Then the inference follows on the lines discussed in this section.

4. Case 1: An arbitrary matrix $\boldsymbol{\Sigma}$

4.1. *Test for specification of the model.* Let us recall that the linear model is $\mathbf{Y} = \mathbf{X}\boldsymbol{\tau} + \mathbf{e}$ where $D(\mathbf{e}) = \boldsymbol{\Sigma}$, an arbitrary positive definite matrix. An independent estimate $f^{-1}\mathbf{S}$ of $\boldsymbol{\Sigma}$ is available. Assume that $\mathbf{Y}$ and $\mathbf{S}$ have the distributions (7) and (8) respectively. The theory and appropriate statistical methods in such a case have been worked out in an earlier paper (Rao [11]). However, we shall make some important comments and also discuss an alternative way of expressing the precision of the estimators.

LEMMA 4a. *Let k be the rank of* $\mathbf{X}$, $r = p - k$, *and*

$$T_r^2 = \min_{\boldsymbol{\tau}} (\mathbf{Y} - \mathbf{X}\boldsymbol{\tau})'\mathbf{S}^{-1}(\mathbf{Y} - \mathbf{X}\boldsymbol{\tau}). \tag{51}$$

Then under the assumptions (7) *and* (8) *on the distributions of* $\mathbf{Y}$ *and* $\mathbf{S}$,

$$\frac{f-r+1}{r} T_r^2 \sim F(r, f-r+1). \tag{52}$$

The result of lemma 4a is proved in [11]. The test criterion T_r^2 examines the adequacy of the model, $\mathbf{Y} = \mathbf{X}\boldsymbol{\tau} + \mathbf{e}$, with respect to the nonrandom part.

4.2. *Estimation of parameters.* If $\boldsymbol{\Sigma}$ were known, the least squares estimator of the unknown vector $\boldsymbol{\tau}$ is obtained by minimizing

$$(\mathbf{Y} - \mathbf{X}\boldsymbol{\tau})'\boldsymbol{\Sigma}^{-1}(\mathbf{Y} - \mathbf{X}\boldsymbol{\tau}) \tag{53}$$

with respect to $\boldsymbol{\tau}$. A natural method of estimation when only an estimate of $\boldsymbol{\Sigma}$ is available is to apply the method of least squares, substituting $f^{-1}\mathbf{S}$ for $\boldsymbol{\Sigma}$. Thus, we are led to minimize the expression

$$(\mathbf{Y} - \mathbf{X}\boldsymbol{\tau})'\mathbf{S}^{-1}(\mathbf{Y} - \mathbf{X}\boldsymbol{\tau}), \tag{54}$$

and to obtain the normal equations

$$(\mathbf{X}'\mathbf{S}^{-1}\mathbf{X})\boldsymbol{\tau} = \mathbf{X}'\mathbf{S}^{-1}\mathbf{Y}. \tag{55}$$

Then we have the estimator

$$\hat{\boldsymbol{\tau}} = (\mathbf{X}'\mathbf{S}^{-1}\mathbf{X})^{-1}\mathbf{X}'\mathbf{S}^{-1}\mathbf{Y} \tag{56}$$

under the assumption that rank $\mathbf{X}$ is k, the dimension of $\boldsymbol{\tau}$ (without loss of generality).

To study the properties of $\hat{\boldsymbol{\tau}}$ and to draw inferences on $\boldsymbol{\tau}$, we make the following transformation of the model:

$$\mathbf{T}_1 = (\mathbf{X}'\mathbf{X})^{-1}\mathbf{X}'\mathbf{Y}, \qquad E(\mathbf{T}_1) = \boldsymbol{\tau}, \tag{57}$$

$$\mathbf{T}_2 = \mathbf{Z}'\mathbf{Y}, \qquad E(\mathbf{T}_2) = \mathbf{0}, \tag{58}$$

where $\mathbf{Z}$ is $p \times r$ matrix of rank r such that $\mathbf{Z}'\mathbf{X} = \mathbf{0}$. The estimated dispersion matrix of $(\mathbf{T}_1, \mathbf{T}_2)$ is $\mathbf{U}$ where

$$f\mathbf{U} = \begin{pmatrix} (\mathbf{X}'\mathbf{X})^{-1}\mathbf{X}'\mathbf{S}\mathbf{X}(\mathbf{X}'\mathbf{X})^{-1} & (\mathbf{X}'\mathbf{X})^{-1}\mathbf{X}'\mathbf{S}\mathbf{Z} \\ \mathbf{Z}'\mathbf{S}\mathbf{X}(\mathbf{X}'\mathbf{X})^{-1} & \mathbf{Z}'\mathbf{S}\mathbf{Z} \end{pmatrix}. \tag{59}$$

Then the theory and methods developed in section 3 apply. The adjusted estimator, according to the formula (32), is

$$\begin{aligned} \mathbf{T}_1 - \mathbf{U}_{12}\mathbf{U}_{22}^{-1}\mathbf{T}_2 &= (\mathbf{X}'\mathbf{X})^{-1}\mathbf{X}'\mathbf{Y} - (\mathbf{X}'\mathbf{X})^{-1}\mathbf{X}'\mathbf{S}\mathbf{Z}(\mathbf{Z}'\mathbf{S}\mathbf{Z})^{-1}\mathbf{Z}'\mathbf{Y} \\ &= (\mathbf{X}'\mathbf{S}^{-1}\mathbf{X})^{-1}\mathbf{X}'\mathbf{S}^{-1}\mathbf{Y}, \end{aligned} \tag{60}$$

using the identity of lemma 2b. Thus, the estimators (56) and (60) are the same. The formula (56) is useful in that it provides the estimator directly in terms of given quantities $\mathbf{X}$, $\mathbf{S}$, and $\mathbf{Y}$ (that is, not involving $\mathbf{Z}$).

We shall now apply the formulae of section 3 to obtain explicit expressions in terms of $\mathbf{X}$, $\mathbf{S}$, and $\mathbf{Y}$ for drawing inferences on $\boldsymbol{\tau}$. The quantities that appear in the formulae (43), (44), (46), and (47) providing confidence intervals are

$$T_r^2 = f^{-1}\mathbf{T}_2'\mathbf{U}_{22}^{-1}\mathbf{T}_2 = \mathbf{Y}'\mathbf{S}^{-1}\mathbf{Y} - \hat{\boldsymbol{\tau}}'\mathbf{X}'\mathbf{S}^{-1}\mathbf{Y}, \tag{61}$$

$$\mathbf{G} = f(\mathbf{U}_{11} - \mathbf{U}_{12}\mathbf{U}_{22}^{-1}\mathbf{U}_{21}) = (\mathbf{X}'\mathbf{S}^{-1}\mathbf{X})^{-1}. \tag{62}$$

The identity in (62) is derived from the identity (14) of lemma 2b. We now have $\hat{\boldsymbol{\tau}}$, T_r^2, and $\mathbf{G}$ all expressed in terms of $\mathbf{X}$, $\mathbf{S}$, and $\mathbf{Y}$, and the methods of section 3 can be applied using the computed values of $\hat{\boldsymbol{\tau}}$, T_r^2, and $\mathbf{G}$.

5. A lemma on least squares estimators

In section 4 we have exhibited the estimator (56) as derived from $\mathbf{T}_1 = (\mathbf{X}'\mathbf{X})^{-1}\mathbf{X}'\mathbf{Y}$ after making covariance adjustment with respect to $\mathbf{T}_2 = \mathbf{Z}'\mathbf{Y}$. As a matter of fact, the choice of $\mathbf{T}_1$ can be arbitrary subject to the condition that $E(\mathbf{T}_1) = \boldsymbol{\tau}$, and the choice of $\mathbf{Z}$ defining $\mathbf{T}_2$ can be arbitrary subject to the condition $\mathbf{Z}'\mathbf{X} = \mathbf{0}$; the adjusted estimator using the appropriate dispersion matrix in each case is the same. We have seen that there are situations where covariance adjustment may not lead to better estimators, but such questions cannot be examined unless there are preassigned choices of $\mathbf{T}_1$ and $\mathbf{T}_2$. We can then raise the specific question as to what components or functions of $\mathbf{T}_2$ would be useful for covariance adjustment. We shall consider some special structures for $\boldsymbol{\Sigma}$ which enable us to make a choice of $\mathbf{T}_1$ and $\mathbf{T}_2$, and then look for relevant concomitant variables.

Let us observe that the choice $\mathbf{T}_1 = (\mathbf{X}'\mathbf{X})^{-1}\mathbf{X}'\mathbf{Y}$ is the best linear estimator of $\boldsymbol{\tau}$ when $\boldsymbol{\Sigma}$, the dispersion matrix of $\mathbf{Y}$, has the special form $\sigma^2\mathbf{I}$ (that is, when the components of the error vector $\mathbf{e}$ in the model (1) are uncorrelated). We shall now determine the class of $\boldsymbol{\Sigma}$ matrices for which $(\mathbf{X}'\mathbf{X})^{-1}\mathbf{X}'\mathbf{Y}$ is the best linear estimator of $\boldsymbol{\tau}$. The object is to characterize the class of $\boldsymbol{\Sigma}$ matrices for which covariance adjustment in $\mathbf{T}_1$ results in a loss of efficiency. For given $\boldsymbol{\Sigma}$, the best linear estimator of $\boldsymbol{\tau}$ is $(\mathbf{X}'\boldsymbol{\Sigma}^{-1}\mathbf{X})^{-1}\mathbf{X}'\boldsymbol{\Sigma}^{-1}\mathbf{Y}$. Then the question raised is equivalent to the problem of determining $\boldsymbol{\Sigma}$ so that it satisfies the equation

$$(\mathbf{X}'\boldsymbol{\Sigma}^{-1}\mathbf{X})^{-1}\mathbf{X}'\boldsymbol{\Sigma}^{-1} = (\mathbf{X}'\mathbf{X})^{-1}\mathbf{X}'. \tag{63}$$

Lemma 5a provides the set of solutions to (63).

LEMMA 5a. *Let* $\mathbf{Z}$ *be a* $p \times r$ *matrix of rank* $r = (p - \text{rank } \mathbf{X})$ *such that* $\mathbf{Z}'\mathbf{X} = 0$, *and* R *be the set of* $\boldsymbol{\Sigma}$ *matrices of the form*

$$\boldsymbol{\Sigma} = \mathbf{X}\boldsymbol{\Gamma}\mathbf{X}' + \mathbf{Z}\boldsymbol{\Theta}\mathbf{Z}' + \sigma^2\mathbf{I} \tag{64}$$

where $\boldsymbol{\Gamma}$, $\boldsymbol{\Theta}$, *and* σ^2 *are arbitrary. Then the necessary and sufficient condition that the least squares estimator of* $\boldsymbol{\tau}$, *in the model* $\mathbf{Y} = \mathbf{X}\boldsymbol{\tau} + \mathbf{e}$ *with* $D(\mathbf{e}) = \boldsymbol{\Sigma}$, *is the same as that for the special choice* $D(\mathbf{e}) = \sigma^2\mathbf{I}$ *is that* $\boldsymbol{\Sigma} \in R$.

The result of lemma 5a shows that for any $\boldsymbol{\Sigma}$ of the form (64) the least squares estimator of $\boldsymbol{\tau}$ is $(\mathbf{X}'\mathbf{X})^{-1}\mathbf{X}'\mathbf{Y}$ (which is well known for the special case of uncorrelated errors) and vice versa.

We note that the complete class of linear functions of $\mathbf{Y}$ with zero expectation is provided by $\mathbf{Z}'\mathbf{Y}$. Hence, if $(\mathbf{X}'\mathbf{X})^{-1}\mathbf{X}'\mathbf{Y}$ is the least squares estimate of $\boldsymbol{\tau}$, then

$$\text{cov}\,[(\mathbf{X}'\mathbf{X})^{-1}\mathbf{X}'\mathbf{Y}, \mathbf{Z}'\mathbf{Y}] = \mathbf{0}; \tag{65}$$

that is,

$$(\mathbf{X}'\mathbf{X})^{-1}\mathbf{X}'\boldsymbol{\Sigma}\mathbf{Z} = \mathbf{0} \Leftrightarrow \mathbf{X}'\boldsymbol{\Sigma}\mathbf{Z} = \mathbf{0}. \tag{66}$$

Then it is easy to verify that

$$\mathbf{X}'\boldsymbol{\Sigma}\mathbf{Z} = \mathbf{0} \Leftrightarrow \boldsymbol{\Sigma} = \mathbf{X}\boldsymbol{\Gamma}\mathbf{X}' + \mathbf{Z}\boldsymbol{\Theta}\mathbf{Z}' + \sigma^2\mathbf{I}. \tag{67}$$

In the proof $\mathbf{X}$ is taken to be of full rank. But this is unnecessary as all the steps are valid with a general inverse of $\mathbf{X}'\mathbf{X}$ (Rao [14]). Further, $\boldsymbol{\Sigma}$ may be singular.

As a corollary, we find that for any $\boldsymbol{\Sigma}$ of the form

$$\boldsymbol{\Sigma} = \mathbf{X}\boldsymbol{\Gamma}\mathbf{X}' + \boldsymbol{\Sigma}_0\mathbf{Z}\boldsymbol{\Theta}\mathbf{Z}'\boldsymbol{\Sigma}_0 + \boldsymbol{\Sigma}_0, \tag{68}$$

the least squares estimator is the same as that for Σ_0.

Let us compare the estimators $\mathbf{T}_1 = (\mathbf{X}'\mathbf{X})^{-1}\mathbf{X}'\mathbf{Y}$ and $\hat{\boldsymbol{\tau}} = (\mathbf{X}'\mathbf{S}^{-1}\mathbf{X})^{-1}\mathbf{X}'\mathbf{S}^{-1}\mathbf{Y}$ of $\boldsymbol{\tau}$ under the assumption that $\boldsymbol{\Sigma} \in R$. It is seen that

$$D[(\mathbf{X}'\mathbf{X})^{-1}\mathbf{X}'\mathbf{Y}] = (\mathbf{X}'\boldsymbol{\Sigma}^{-1}\mathbf{X})^{-1}, \tag{69}$$

$$D[(\mathbf{X}'\mathbf{S}^{-1}\mathbf{X})^{-1}\mathbf{X}'\mathbf{S}^{-1}\mathbf{Y}] = \frac{f-1}{f-r-1}(\mathbf{X}'\boldsymbol{\Sigma}^{-1}\mathbf{X})^{-1}, \tag{70}$$

so that the effect of using an estimate of $\boldsymbol{\Sigma}$, when in fact $\boldsymbol{\Sigma} \in R$, or to a slightly extended class, is to decrease efficiency. For $\boldsymbol{\Sigma} \notin R$ equation (70) remains the same, while (69) changes to

(71) $$D[(\mathbf{X}'\mathbf{X})^{-1}\mathbf{X}'\mathbf{Y}] = (\mathbf{X}'\mathbf{X})^{-1}\mathbf{X}'\boldsymbol{\Sigma}\mathbf{X}(\mathbf{X}'\mathbf{X})^{-1}.$$

From the result (18) we find that the estimator using an estimate of $\boldsymbol{\Sigma}$ has a smaller dispersion matrix only if

(72) $$(\mathbf{X}'\mathbf{X})^{-1}\mathbf{X}'\boldsymbol{\Sigma}\mathbf{X}(\mathbf{X}'\mathbf{X})^{-1}$$

is somewhat larger than

(73) $$(\mathbf{X}'\boldsymbol{\Sigma}^{-1}\mathbf{X})^{-1}$$

to compensate for the multiplying factor in (70).

6. Case 2: $D(Y) - \mathbf{X}\boldsymbol{\Gamma}\mathbf{X}' + \mathbf{Z}\,\boldsymbol{\Theta}\,\mathbf{Z}' + \sigma^2\mathbf{I}$

As mentioned in the introduction, the dispersion matrix of case 2 arises from the mixed model

(74) $$\mathbf{Y} = \mathbf{X}\boldsymbol{\tau} + \mathbf{X}\boldsymbol{\gamma} + \mathbf{Z}\boldsymbol{\xi} + \mathbf{e}$$

where $\boldsymbol{\gamma}$, $\boldsymbol{\xi}$, $\mathbf{e}$ are all uncorrelated random vectors with zero expectations and dispersion matrices $\boldsymbol{\Gamma}$, $\boldsymbol{\Theta}$, and $\sigma^2\mathbf{I}$, respectively. These $\boldsymbol{\Gamma}$, $\boldsymbol{\Theta}$, and σ^2 are all unknown, but an independent estimate $f^{-1}\mathbf{S}$ of $\boldsymbol{\Sigma}$ is available. By choosing $\boldsymbol{\Gamma}$ or $\boldsymbol{\Theta}$ or both to be zero we obtain special cases.

It was seen in earlier sections that the problem considered is essentially one of making covariance adjustments in the estimator $(\mathbf{X}'\mathbf{X})^{-1}\mathbf{X}'\mathbf{Y}$ using the concomitant variables $\mathbf{Z}'\mathbf{Y}$. The decrease in efficiency arises when the association between the estimator $(\mathbf{X}'\mathbf{X})^{-1}\mathbf{X}'\mathbf{Y}$ and the concomitant variables $\mathbf{Z}'\mathbf{Y}$ is weak. Our aim is then to make a selection of suitable concomitant variables or their functions for covariance adjustment on the basis of a given structure for $\boldsymbol{\Sigma}$. It is clear that any selection made on the basis of observed association (using $f^{-1}\mathbf{S}$) does not improve the situation, for the effect of such a selection has to be considered in the estimation of precision of the adjusted estimator.

In the case of model (74), we find that

(75) $$C[(\mathbf{X}'\mathbf{X})^{-1}\mathbf{X}'\mathbf{Y}, \mathbf{Z}'\mathbf{Y}] = (\mathbf{X}'\mathbf{X})^{-1}\mathbf{X}'\boldsymbol{\Sigma}\mathbf{Z} = \mathbf{0},$$

and therefore, there is definite loss in efficiency by covariance adjustment. Indeed $\mathbf{T}_1 = (\mathbf{X}'\mathbf{X})^{-1}\mathbf{X}'\mathbf{Y}$ is the least squares estimate of $\boldsymbol{\tau}$ since $\boldsymbol{\Sigma}$ has the structure of lemma 5a. Now

(76) $$\mathbf{T}_1 \sim N_k(\boldsymbol{\tau}, \mathbf{H})$$

where $\mathbf{H} = (\mathbf{X}'\mathbf{X})^{-1}\mathbf{X}'\boldsymbol{\Sigma}\mathbf{X}(\mathbf{X}'\mathbf{X})^{-1}$. For drawing inferences on $\boldsymbol{\tau}$, we need an estimate of $\mathbf{H}$. This is supplied by

(77) $$\hat{\mathbf{H}} = f^{-1}(\mathbf{X}'\mathbf{X})^{-1}\mathbf{X}'\mathbf{S}\mathbf{X}(\mathbf{X}'\mathbf{X})^{-1}$$

with the distribution

(78) $$f\hat{\mathbf{H}} \sim W_k(f, \mathbf{H}).$$

Hence the inference on $\boldsymbol{\tau}$ follows on standard lines using the distributions (76) and (78), as indicated in section 3.

7. Case 3: $D(\mathbf{Y}) = \mathbf{C}\boldsymbol{\Gamma}\mathbf{C}' + \sigma^2\mathbf{I}$ (C known)

Such a situation arises when we consider a mixed model

$$\mathbf{Y} = \mathbf{X}\boldsymbol{\tau} + \mathbf{C}\boldsymbol{\gamma} + \mathbf{e} \tag{79}$$

where $\boldsymbol{\gamma}$ and $\mathbf{e}$ are uncorrelated, $E(\boldsymbol{\gamma}) = \mathbf{0}$, $D(\boldsymbol{\gamma}) = \boldsymbol{\Gamma}$ and $E(\mathbf{e}) = \mathbf{0}$, $D(\mathbf{e}) = \sigma^2\mathbf{I}$. The matrices $\mathbf{X}$ and $\mathbf{C}$ are known. Such a model was studied by a number of authors (see Duncan [2]) under a different set of assumptions on $\boldsymbol{\Gamma}$. In the present problem, lemma 7a provides a first reduction of the problem.

LEMMA 7a. *Let* $\mathbf{B}$ *be a matrix such that* $\mathbf{X}'(\mathbf{C} - \mathbf{XB}) = \mathbf{0}$; *that is,*

$$\mathbf{B} = (\mathbf{X}'\mathbf{X})^{-1}\mathbf{X}'\mathbf{C},$$

and let the rank of $\mathbf{C} - \mathbf{XB}$ *be* m. *Consider the linear functions*

$$\begin{aligned} \mathbf{T}_1 &= (\mathbf{X}'\mathbf{X})^{-1}\mathbf{X}'\mathbf{Y}, \\ \mathbf{T}_2 &= (\mathbf{C} - \mathbf{XB})'\mathbf{Y}, \\ \mathbf{T}_3 &= \mathbf{G}'\mathbf{Y}, \quad \text{where} \quad \mathbf{G}'\mathbf{X} = \mathbf{0}, \quad \mathbf{G}'(\mathbf{C} - \mathbf{XB}) = \mathbf{0}, \end{aligned} \tag{80}$$

which provide a linear transformation of $\mathbf{Y}$. *Then* $E(\mathbf{T}_3) = \mathbf{0}$, $C(\mathbf{T}_1, \mathbf{T}_3) = \mathbf{0}$, *and* $C(\mathbf{T}_2, \mathbf{T}_3) = \mathbf{0}$. *Further,* $\mathbf{T}_1$ *and* $\mathbf{T}_2$ *are correlated, unless* $\mathbf{C}'\mathbf{X} = \mathbf{0}$ *or* $\mathbf{C} - \mathbf{XB} = \mathbf{0}$.

The results are easy to verify. Lemma 7a shows that $\mathbf{T}_3$ does not throw any information on $\boldsymbol{\tau}$ and should be discarded. Since $E(\mathbf{T}_1) = \boldsymbol{\tau}$, $E(\mathbf{T}_2) = \mathbf{0}$, and $\mathbf{T}_1$ and $\mathbf{T}_2$ are possibly correlated, covariance adjustment in $\mathbf{T}_1$ using $\mathbf{T}_2$ as concomitant variable might provide good estimators. The estimated dispersion matrix of $\mathbf{T}_1$, $\mathbf{T}_2$ is

$$\mathbf{U} = f^{-1}\begin{pmatrix} (\mathbf{X}'\mathbf{X})^{-1}\mathbf{X}'\mathbf{S}\mathbf{X}(\mathbf{X}'\mathbf{X})^{-1} & (\mathbf{X}'\mathbf{X})^{-1}\mathbf{X}'\mathbf{S}(\mathbf{C} - \mathbf{XB}) \\ (\mathbf{C} - \mathbf{XB})'\mathbf{S}\mathbf{X}(\mathbf{X}'\mathbf{X})^{-1} & (\mathbf{C} - \mathbf{XB})'\mathbf{S}(\mathbf{C} - \mathbf{XB}) \end{pmatrix} \tag{81}$$

with the distribution

$$f\mathbf{U} \sim W_{k+m}(f, \boldsymbol{\Lambda}) \tag{82}$$

where $\boldsymbol{\Lambda}$ is the true dispersion matrix of $(\mathbf{T}_1, \mathbf{T}_2)$. The inference on $\boldsymbol{\tau}$, then proceeds on standard lines making covariance adjustment in $(\mathbf{X}'\mathbf{X})^{-1}\mathbf{X}'\mathbf{Y}$ using the concomitant variable $(\mathbf{C} - \mathbf{XB})'\mathbf{Y}$ and the estimated dispersion matrix (81).

Note 1. The difference between case 1 and case 3 is that $(\mathbf{C} - \mathbf{XB})'\mathbf{Y}$ constitutes a subset of all linear functions of $\mathbf{Z}'\mathbf{Y}$. In case 1, no selection out of $\mathbf{Z}'\mathbf{Y}$ was possible, as nothing was known about the structure of $\boldsymbol{\Sigma}$. The structure for $\boldsymbol{\Sigma}$, as in case 3, enabled us to choose suitable functions of $\mathbf{Z}'\mathbf{Y}$ for covariance adjustment.

Note 2. The validity of the structure for $\boldsymbol{\Sigma}$ as in case 3 can be examined on the basis of $\mathbf{S}$ by testing the equality of the last $(p - b)$ eigenvectors of $\mathbf{S}$, where b is the rank of $\mathbf{C}$. Appropriate test criteria for this purpose have been given by Bartlett [1] and Rao [10].

8. Case 4: $D(\mathbf{Y}) = \mathbf{C}\boldsymbol{\Gamma}\mathbf{C}' + \sigma^2\mathbf{I}$ (C unknown)

Let us first consider the case of $\boldsymbol{\Sigma}$ known. Then we can determine the restricted eigenvalues and eigenvectors of $\boldsymbol{\Sigma}$ subject to the condition that the eigenvectors are orthogonal to the columns of $\mathbf{X}$, as considered in lemmas 2d and 2e of section 2. The restricted eigenvalues and eigenvectors are the nonzero eigenvalues and their corresponding eigenvectors of the matrix $(\mathbf{I} - \mathbf{X}(\mathbf{X}'\mathbf{X})^{-1}\mathbf{X}')\boldsymbol{\Sigma}$.

For the special choice $\boldsymbol{\Sigma} = \mathbf{C}\boldsymbol{\Gamma}\mathbf{C}' + \sigma^2\mathbf{I}$, we have the following results. The smallest nonzero eigenvalue of $(\mathbf{I} - \mathbf{X}(\mathbf{X}'\mathbf{X})^{-1}\mathbf{X}')\boldsymbol{\Sigma}$ is σ^2. Let the multiplicity of this root be m and represent the corresponding eigenvectors by $\mathbf{G}_1, \cdots, \mathbf{G}_m$. Let $\mathbf{G}$ be the matrix with $\mathbf{G}_i$ as its columns.

There are $(p - k)$ nonzero eigenvalues on the total. Let the eigenvectors corresponding to the eigenvalues different from σ^2 be $\mathbf{B}_1, \cdots, \mathbf{B}_{p-k-m}$. Let $\mathbf{B}$ be the matrix with $\mathbf{B}_i$ as its columns.

With $\mathbf{B}$ and $\mathbf{G}$ as defined above, we have from lemma 2e

$$\mathbf{X}'\mathbf{G} = \mathbf{X}'\boldsymbol{\Sigma}\mathbf{G} = 0, \qquad \mathbf{B}'\mathbf{G} = \mathbf{B}'\boldsymbol{\Sigma}\mathbf{G} = 0. \tag{83}$$

The conditions (83) imply that the linear functions

$$\mathbf{T}_3 = \mathbf{G}'\mathbf{Y} \tag{84}$$

have zero expectation and are uncorrelated with the linear functions

$$\mathbf{T}_1 = (\mathbf{X}'\mathbf{X})^{-1}\mathbf{X}'\mathbf{Y}, \qquad \mathbf{T}_2 = \mathbf{B}'\mathbf{Y}. \tag{85}$$

Further, $E(\mathbf{T}_2) = \mathbf{0}$, but $\mathbf{T}_2$ is possibly correlated with $\mathbf{T}_1$. Hence, the best estimate of $\boldsymbol{\tau}$ can be obtained by making covariance adjustment in $\mathbf{T}_1$ using $\mathbf{T}_2$ only as concomitant variables (that is, discarding $\mathbf{T}_3$).

If $\boldsymbol{\Sigma}$ is unknown, we proceed as follows. First determine the eigenvalues and eigenvectors of the matrix $(\mathbf{I} - \mathbf{X}(\mathbf{X}'\mathbf{X})^{-1}\mathbf{X}')\mathbf{S}$ and choose all the (say q) eigenvectors corresponding to dominant roots. Let $\mathbf{B}_1, \cdots, \mathbf{B}_q$ be the eigenvectors chosen and $\mathbf{B}$ the matrix with $\mathbf{B}_i$ as columns. Now consider the linear functions:

$$\mathbf{T}_1 = (\mathbf{X}'\mathbf{X})^{-1}\mathbf{X}'\mathbf{Y}, \qquad \mathbf{T}_2 = \mathbf{B}'\mathbf{Y}. \tag{86}$$

Note that since $\mathbf{B}$ *is an estimated matrix, no exact theory exists for making covariance adjustment using* $\mathbf{T}_2$. For covariance adjustment we consider the estimated dispersion matrix of $\mathbf{T}_1$, $\mathbf{T}_2$, treating $\mathbf{B}$ as fixed, which is

$$f^{-1}\begin{pmatrix} (\mathbf{X}'\mathbf{X})^{-1}\mathbf{X}'\mathbf{S}\mathbf{X}(\mathbf{X}'\mathbf{X})^{-1} & (\mathbf{X}'\mathbf{X})^{-1}\mathbf{X}'\mathbf{S}\mathbf{B} \\ \mathbf{B}'\mathbf{S}\mathbf{X}(\mathbf{X}'\mathbf{X})^{-1} & \mathbf{B}'\mathbf{S}\mathbf{B} \end{pmatrix}. \tag{87}$$

The basic theory of covariance adjustment (in $\mathbf{T}_1$ using $\mathbf{T}_2$ as a concomitant variable) and expressions for the precision of the estimates as discussed in section 3 are applicable, in an approximate way. Note that $\mathbf{B}'\mathbf{Y}$ is selected not on the basis of the observed association with $(\mathbf{X}'\mathbf{X})^{-1}\mathbf{X}'\mathbf{Y}$, but in a manner which does not overestimate precision due to covariance adjustment. The value of q is taken as the number of eigenroots of $(\mathbf{I} - \mathbf{X}(\mathbf{X}'\mathbf{X})^{-1}\mathbf{X})\mathbf{S}$ judged to be dominant by an appropriate test, if necessary.

9. Case 5: Autoregressive errors

The model assumed is

$$y_t = P_k(t) + e_t, \qquad t = 1, \cdots, p \tag{88}$$

where $P_k(t)$ is a k-th degree polynomial in time

$$P_k(t) = \beta_0 + \beta_1 t + \cdots + \beta_k t^k, \tag{89}$$

and e_t have the autoregressive scheme

$$e_t = \rho_1 e_{t-1} + \cdots + \rho_m e_{t-m} + \eta_t. \tag{90}$$

The proposed method of estimation of the coefficients of the polynomial trend and the autoregressive parameters $\rho_1, \cdots, \rho_m$ is an extension of the least squares method considered by Mann and Wald [6]. The estimating equations are obtained by minimizing $\Sigma \eta_t^2$ where

(91)

$$\begin{aligned} \eta_t &= e_t - \rho_1 e_{t-1} - \cdots - \rho_m e_{t-m} \\ &= y_t - P_k(t) - \rho_1[y_{t-1} - P_k(t-1)] - \cdots - \rho_m[y_{t-m} - P_k(t-m)] \\ &= y_t - \rho_1 y_{t-1} - \cdots - \beta_0[1 - \rho_1 - \cdots] - \beta_1[t - \rho_1(t-1) - \cdots] - \cdots. \end{aligned}$$

When ρ_i and β_j are all unknown, η_t is nonlinear in the parameters, and therefore, the estimating equations obtained by minimizing $\Sigma \eta_t^2$ become nonlinear. Fortunately the problem can be reduced to yield a definitive solution. First we transform the parameters in such a way that η_t is linear in the new parameters. Thus we can write

$$\eta_t = y_t - \rho_1 y_{t-1} - \cdots - \rho_m y_{t-m} - \gamma_0 - \gamma_1 t - \cdots - \gamma_k t^k, \tag{92}$$

which is linear in ρ_i and γ_i, where γ_j are defined as follows. Let

$$\begin{aligned} \delta_i &= \rho_1 + 2^i \rho_2 + \cdots + m^i \rho_m \qquad \text{for } i \geq 1 \\ \delta_0 &= 1 - \rho_1 - \cdots - \rho_n \end{aligned} \tag{93}$$

Then

$$\begin{aligned} \gamma_k &= \beta_k \delta_0 \\ \gamma_{k-1} &= \binom{k}{1} \beta_k \delta_1 + \beta_{k-1} \delta_0 \\ \gamma_{k-2} &= -\binom{k}{2} \beta_k \delta_2 + \binom{k-1}{1} \beta_{k-1} \delta_1 + \beta_{k-2} \delta_0 \\ &\cdots\cdots \\ \gamma_{k-i} &= (-1)^{i+1} \binom{k}{i} \beta_k \delta_i + (-1)^i \binom{k-1}{i-1} \beta_{k-1} \delta_{i-1} + \cdots \\ &\cdots\cdots \\ \gamma_0 &= (-1)^{k+1} \beta_k \delta_k + (-1)^k \beta_{k-1} \delta_{k-1} + \cdots. \end{aligned} \tag{94}$$

First we estimate ρ_i and γ_j by minimizing

$$\sum_{m+1}^{p} \eta_t^2 = \sum_{m+1}^{p} (y_t - \rho_1 y_{t-1} - \cdots - \gamma_0 - \gamma_1 t - \cdots)^2 \tag{95}$$

with respect to ρ_i, $i = 1, \cdots, m$, and γ_j, $j = 1, \cdots, k$. Let $\hat{\rho}_i$, $\hat{\gamma}_j$ be the solutions of the normal equations. Then we obtain $\hat{\delta}_i$ from the equation (93), and then successively $\hat{\beta}_k, \hat{\beta}_{k-1}, \cdots, \hat{\beta}_0$ from the equations (94).

The normal equations obtained by minimizing (95) provide consistent estimators of ρ_i, γ_j, and the least sum of squares a consistent estimator of σ^2. The inverse of the matrix of normal equations multiplied by the estimator of σ^2 provides the asymptotic dispersion matrix of the estimators $\hat{\rho}_i$, $\hat{\gamma}_j$. We observe that $\hat{\beta}_r$ are simple functions of $\hat{\rho}_i$, $\hat{\gamma}_j$ and, therefore, the asymptotic dispersion matrix of $\hat{\beta}_r$ can be computed by the well known formula (see [13], p. 322).

We shall give the explicit formula for the estimated dispersion matrix of $\hat{\beta}_k, \cdots, \hat{\beta}_0$ in terms of the estimated dispersion matrix of $\hat{\gamma}_k, \cdots, \hat{\gamma}_0; \hat{\rho}_1, \cdots, \hat{\rho}_m$,

$$\hat{\sigma}^2 \begin{pmatrix} \mathbf{A}_{11} & \mathbf{A}_{12} \\ \mathbf{A}_{21} & \mathbf{A}_{22} \end{pmatrix}. \tag{96}$$

Note that to obtain the matrix in (96), it is convenient to write down the normal equations by deriving (95) with respect to $\gamma_k, \cdots, \gamma_0, \rho_1, \cdots, \rho_m$ in the order indicated. Let $\mathbf{B}$ be the matrix of coefficients of $\beta_k, \cdots, \beta_0$ on the right-hand side of (94), and $\mathbf{C}$ be the matrix of coefficients of $\delta_0, \cdots, \delta_k$ on the right-hand side of (94). Finally, let $\mathbf{F}$ be the matrix

$$\begin{pmatrix} -1 & -1 & \cdots & -1 \\ 1 & 2 & \cdots & m \\ \cdot & \cdot & \cdots & \cdot \\ 1 & 2^k & \cdots & m^k \end{pmatrix}. \tag{97}$$

Then we have the relation connecting the differentials $d\hat{\boldsymbol{\gamma}}$, $d\hat{\boldsymbol{\rho}}$ and $d\hat{\boldsymbol{\beta}}$ as

$$d\hat{\boldsymbol{\gamma}} - \mathbf{CF}\, d\hat{\boldsymbol{\rho}} = \mathbf{B}\, d\hat{\boldsymbol{\beta}} \tag{98}$$

where the vectors $\hat{\boldsymbol{\gamma}}$, $\hat{\boldsymbol{\rho}}$, and $\hat{\boldsymbol{\beta}}$ are

$$\hat{\boldsymbol{\gamma}} = \begin{pmatrix} \hat{\gamma}_k \\ \cdot \\ \cdot \\ \cdot \\ \hat{\gamma}_0 \end{pmatrix}, \quad \hat{\boldsymbol{\rho}} = \begin{pmatrix} \hat{\rho}_1 \\ \cdot \\ \cdot \\ \cdot \\ \hat{\rho}_m \end{pmatrix}, \quad \hat{\boldsymbol{\beta}} = \begin{pmatrix} \hat{\beta}_k \\ \cdot \\ \cdot \\ \cdot \\ \hat{\beta}_0 \end{pmatrix}. \tag{99}$$

From (98),

$$d\hat{\boldsymbol{\beta}} = \mathbf{B}^{-1}\, d\hat{\boldsymbol{\gamma}} - \mathbf{B}^{-1}\mathbf{CF}\, d\hat{\boldsymbol{\rho}} = \mathbf{H}\, d\hat{\boldsymbol{\gamma}} - \mathbf{G}\, d\hat{\boldsymbol{\rho}} \tag{100}$$

writing $\mathbf{H} = \mathbf{B}^{-1}$ and $\mathbf{G} = \mathbf{B}^{-1}\mathbf{CF}$. Then the asymptotic dispersion matrix of $\hat{\boldsymbol{\beta}}$ is

$$D(\hat{\boldsymbol{\beta}}) = \hat{\sigma}^2(\mathbf{HA}_{11}\mathbf{H}' + \mathbf{GA}_{22}\mathbf{G}' - \mathbf{HA}_{12}\mathbf{G}' - \mathbf{GA}_{21}\mathbf{H}'). \tag{101}$$

Let us note that, in practice, the order of the autoregressive scheme for the errors may not be preassigned and may have to be inferred from data. If the chosen order is higher than the true one, then an application of lemma 2a shows that the estimators of γ_i and ρ_j lose in efficiency. For determining the appropriate order, the residual sum of squares providing the estimate of σ^2 has to be examined. If there is no significant reduction in the residual sum of squares by increasing the number of autoregressive parameters, then a lower order is indicated.

We observe that the linearization of the parameters in η_t is not possible if the nonstochastic part in y_t is not a polynomial in t, but is simply linear in the unknown parameters β_j with given coefficients. An appropriate method in such a case is given by Durbin [3].

10. Estimation of phase and amplitude

In the estimation of signal parameters, models of the type (1) are used. Thus, if β and θ represent amplitude and phase of a signal, the t-th observation is written

$$y_t = x_{t1}\beta \sin \theta + x_{t2}\beta \cos \theta + e_t, \qquad t = 1, \cdots, p \tag{102}$$

where x_{ti} are known coefficients. Writing $\tau_1 = \beta \sin \theta$, $\tau_2 = \beta \cos \theta$, and $\boldsymbol{\tau}' = (\tau_1, \tau_2)$, $\mathbf{X} = (x_{ti})$, $\mathbf{Y}' = (y_1, \cdots, y_p)$, and $\mathbf{e}' = (e_1, \cdots, e_p)$, we have the linear model

$$\mathbf{Y} = \mathbf{X}\boldsymbol{\tau} + \mathbf{e}. \tag{103}$$

Let us suppose that under suitable assumption on $\boldsymbol{\Sigma}$, estimates $\hat{\tau}_1$, $\hat{\tau}_2$ of τ_1, τ_2 have been obtained as discussed in the present paper. Further, let

$$\begin{pmatrix} a_{11} & a_{12} \\ a_{21} & a_{22} \end{pmatrix} \tag{104}$$

be the estimated dispersion matrix of the estimates $\hat{\tau}_1$, $\hat{\tau}_2$.

We are interested in estimating β and θ which are nonlinear functions of τ_1 and τ_2. We may estimate β and θ by

$$\begin{aligned} \hat{\beta} &= (\hat{\tau}_1^2 + \hat{\tau}_2^2)^{1/2}, \\ \hat{\theta} &= \tan^{-1} (\hat{\tau}_1/\hat{\tau}_2), \end{aligned} \tag{105}$$

and compute their large sample standard errors (or asymptotic variance). Using the formula (see [13], p. 322) for asymptotic variance

$$\begin{aligned} V(\hat{\beta}) &= (a_{11}\hat{\tau}_1^2 + 2a_{12}\hat{\tau}_1\hat{\tau}_2 + a_{22}\hat{\tau}_2^2)/(\hat{\tau}_1^2 + \hat{\tau}_2^2), \\ V(\hat{\theta}) &= (a_{11}\hat{\tau}_2^2 - 2a_{12}\hat{\tau}_1\hat{\tau}_2 + a_{22}\hat{\tau}_1^2)/(\hat{\tau}_1^2 + \hat{\tau}_2^2)^2. \end{aligned} \tag{106}$$

Exact confidence intervals of a given probability level $(1 - \alpha)$ can be obtained for $\lambda = \tan \theta$ (and hence for θ), provided that under the assumptions made on the errors there exists an exact test for the linear hypothesis $\tau_1 - \lambda\tau_2 = 0$. In all cases, except that of autoregressive errors, such an exact test is possible using a t distribution on appropriate degrees of freedom. Then the confidence limits for λ are obtained by solving the equation in λ

$$\frac{(\hat{\tau}_1 - \lambda\hat{\tau}_2)^2}{a_{11} - 2\lambda a_{12} + \lambda^2 a_{22}} = t_{\alpha/2} \tag{107}$$

where $t_{\alpha/2}$ is the upper $(\alpha/2)$ probability value of t on degrees of freedom applicable for a t test of a linear hypothesis on $\boldsymbol{\tau}$. When an exact test is not available, we may use the upper $(\alpha/2)$ probability value of the standard normal distribution in the place of $t_{\alpha/2}$ in (107). Thus we obtain asymptotic confidence limits for λ from which those of θ can be computed.

There is, however, no exact method for determining confidence intervals for β, even when exact tests of linear hypotheses on $\boldsymbol{\tau}$ are possible. One may have to use the estimator for β as in (105), and the asymptotic variance $V(\hat{\beta})$ as in (106), to obtain an asymptotic confidence interval.

Simultaneous confidence intervals for β and θ may be deduced from a confidence ellipsoid (exact or asymptotic) of τ_1 and τ_2. Let

$$a^{11}(\hat{\tau}_1 - \tau_1)^2 + 2a^{12}(\hat{\tau}_1 - \tau_1)(\hat{\tau}_2 - \tau_2) + a^{22}(\hat{\tau}_2 - \tau_2)^2 \leq c \tag{108}$$

be the confidence ellipsoid of τ_1, τ_2 with a given probability $(1 - \alpha)$. The confidence interval for β with probability $\geq 1 - \alpha$ is

$$(\sqrt{\min (\tau_1^2 + \tau_2^2)}, \sqrt{\max (\tau_1^2 + \tau_2^2)}) \tag{109}$$

where the minimum and maximum are obtained subject to the restriction (108) on τ_1 and τ_2 for given $\hat{\tau}_1$ and $\hat{\tau}_2$. A satisfactory solution to such a problem of finding the extrema of a quadratic form on an ellipsoid in many dimensions has been recently given by Forsyth and Golub [4]. The confidence interval for θ is given by the angles which the pair of tangents from the origin to the ellipsoid (108) make with the τ_2 axis.

REFERENCES

[1] M. S. Bartlett, "Tests of significance in factor analysis," *British J. Psych.* (Statist. Sec.), Vol. 3 (1950), pp. 77–85.

[2] D. B. Duncan, "On some theory of mixed linear models and its application in the estimation of missile trajectory," paper No. 332 from the Department of Biostatistics, The Johns Hopkins University.

[3] J. Durbin, "Estimation of parameters in time-series regression models," *J. Roy. Statist. Soc.*, Vol. 22 (1960), pp. 139–153.

[4] G. E. Forsyth and G. Golub, "On the stationary values of a second-degree polynomial on the unit sphere," *J. SIAM*, to appear.

[5] C. R. Henderson, Oscar Kempthorne, S. R. Searle, and C. M. Krosigk, "The estimation of environmental and genetic trends from records subject to culling," *Biometrics*, Vol. 15 (1959), pp. 192–218.

[6] H. B. Mann and A. Wald, "On the statistical treatment of linear stochastic difference equations," *Econometrics*, Vol. 11 (1943), pp. 173–220.

[7] C. R. Rao, "Tests with discriminant functions in multivariate analysis," *Sankhyā*, Vol. 7 (1946), pp. 407–414.

[8] ———, "On some problems arising out of discrimination with multiple characters," *Sankhyā*, Vol. 9 (1949), pp. 343–364.

[9] ———, "A note on the distribution of $D_{p+q}^2 - D_p^2$ and some computational aspects of D^2 statistic and discriminant function," *Sankhyā*, Vol. 10 (1950), pp. 257–268.

[10] ———, "Estimation and tests of significance in factor analysis," *Psychometrika*, Vol. 20 (1955), pp. 93–111.

[11] ———, "Some problems involving linear hypotheses in multivariate analysis," *Biometrika*, Vol. 46 (1959), pp. 49–58.
[12] ———, "The use and interpretation of principal component analysis in applied research," *Sankhyā*, Vol. 26 (1964), pp. 329–358.
[13] ———, *Linear Statistical Inference and Its Applications*, New York, Wiley, 1965.
[14] ———, "Generalized inverses for matrices and its applications in mathematical statistics," *Research Papers in Statistics* (Festschrift volume for J. Neyman), pp. 263–280, New York, Wiley, 1966.

ON TESTING FOR NORMALITY

KÁROLY SARKADI
MATHEMATICAL INSTITUTE OF THE HUNGARIAN ACADEMY OF SCIENCES
and
UNIVERSITY OF CALIFORNIA, BERKELEY

1. Introduction

The majority of the goodness-of-fit problems arising in practice involves nuisance parameters. On the other hand, the majority of the results which have appeared within this field deal with simple hypothesis testing only. Among the relatively few results concerning the first-mentioned, more general problem, the most known and applied one is the modification given by R. A. Fisher to Karl Pearson's χ^2-test (see for example, [4], pp. 424–434). As it is well known, this modification consists of replacing the unknown parameters by their estimates; the distribution of the modified test statistic (at least its approximate distribution) was determined. The same way was followed by Kac, Kiefer, and Wolfowitz [7] and Darling [5] concerning the Cramér-von Mises test and the Kolmogorov test. Computational difficulties, however, prevented them from providing tables having sufficient range and accuracy for practical purposes.

But even if these difficulties could be overcome in the future, neither they nor Fisher's method work in some nonsimple sample cases.

Considering this fact as well as the disadvantages of the χ^2-test (see [7], pp. 191–192), other solutions of problems of this type seem to be of particular interest. The straightforward way is to find an equivalent simple hypothesis.

The basic theory and the most important results of this approach are dealt with in another paper appearing in this volume [12].

Solutions of this type form the subject of the present paper, but it is confined in a rather special direction. This specialization may be characterized by the aim of avoiding theoretical and computational difficulties and of utilizing the known results of the theory of goodness-of-fit tests as much as possible.

Therefore, we are interested in such equivalent (substitute) hypotheses which are of the form of goodness-of-fit problems. In other words, we want to provide a set of random variables which are distribution-free and independently and identically distributed in the case of the null hypothesis.

As a further specialization, we require that each of these transformed values should represent, in the same way, one of the original sample elements. The purpose of this restriction is that the test statistic made with the transformed variables should approximate the correct test one could form with the knowledge of the unknown constants. Therefore, the properties of the combined test, consisting of the transformation and the testing of the simple hypothesis, will be near that of the test concerning a simple original hypothesis, and this indirect information is the more important since exact investigations of properties of tests relating to the original hypotheses seem very difficult.

Since the test statistics which we can apply are of different types, the problem of what to consider a good representation is the matter of an arbitrary decision. We shall apply, in the cases to be considered (sections 2–5), the correlation coefficient of the original and transformed values (in the case of null hypothesis) as the measure of goodness of representation. Similarly, the minimum value of these measures will be considered the measure of the goodness of the representation if we replace the original set or its subset by a set of transformed variables. In these cases—and in any case when the nuisance parameters are location and scale parameters—this measure does not depend on the nuisance parameters. Furthermore, in these cases the maximization of the correlation coefficient is equivalent to the minimization of the square of difference between the transformed values and the standardized original values.

In section 8 the goodness of representation in a special case is numerically investigated.

The question whether the substitute simple hypothesis is strictly equivalent to the original null hypothesis in the sense that the latter is not only sufficient but also necessary to the former, is important from the point of view of the biasedness of the test. The conditions of this property, in the case of normal distribution with unknown variance, were stated by Prohorov [12].

Solutions of the problem using the method outlined above have been given until now by Durbin [6], Störmer [15], and the present author [13] independently. The purpose of this paper is the comparison, investigation, and improvement of these results.

The main difference between the method of Durbin and that of the other two authors is that Durbin uses random numbers. Consequently, the number of the transformed variables is the same as that of the original variables in Durbin's method, whereas the other method of transformation decreases the number of variables by the number of unknown constants.

Here we deal with the latter method mainly. But first, in section 2, we make a comparison between the two methods in the case where the underlying distribution is normal, with both parameters unknown, and where the transformation is linear for any sample realization. This is the case considered in [6] in detail.

In sections 3–5 some "optimal" transformations are derived for the case of normal distributions. In section 6 the possibilities of generalization are discussed, and section 7 deals with practical applications.

2. Comparison with Durbin's method

Durbin's method is the following [6]: let $x_1, \cdots, x_n$ be independent, normally distributed random variables with unknown expectation and variance, μ and

σ^2, respectively. Let $\bar{x}$ and s^2 be the sample mean and empirical variance, respectively.

Durbin considers the transformation where the transformed values $y_1, \cdots, y_n$ are independent $N(0, 1)$ variables and fulfill the relation

$$\frac{y_i - \bar{y}}{s'} = \frac{x_i - \bar{x}}{s}, \quad (i = 1, \cdots, n). \tag{2.1}$$

The variables $\bar{y}$ and s'^2 are independent and are distributed as the sample mean and empirical variance of a sample of size n from a standardized normal parent population, that is, $\bar{y}$ has the distribution $N(0, 1/n)$, and $(n - 1)s'^2$ has χ^2-distribution with $n - 1$ degrees of freedom.

Durbin generates the values of $\bar{y}$ and s' by the help of random numbers. Similarly, they may be generated as functions of the sample elements (of those defined above and of additional ones) according to the alternative method.

Given $\bar{y}$ and s', $y_1, \cdots, y_n$ can be determined by (2.1).

Durbin proved that the $y_1, \cdots, y_n$ so defined are, in fact, independent $N(0, 1)$ variables. Below we give an alternative, simple proof for this fact. We suppose that $\bar{y}$ and s' are independent, have the above mentioned distribution, and $[y, s']$ is independent of $[(x_1 - \bar{x})/s, \cdots, (x_n - \bar{x})/s]$.

Let

$$\begin{aligned} \mathbf{x} &= [x_1, \cdots, x_n], \\ \mathbf{y} &= [y_1, \cdots, y_n], \\ \mathbf{v} &= \left[\frac{x_1 - \bar{x}}{s}, \cdots, \frac{x_n - \bar{x}}{s}\right], \end{aligned} \tag{2.2}$$

and let us denote by $\mathbf{1}_n$ an n-way vector with unit components. As it is known, $\bar{x}$, s, and $\mathbf{v}$ are completely independent. Let us consider the two following identities:

$$\mathbf{y} = \mathbf{1}_n\bar{y} + \mathbf{v}s', \tag{2.3}$$

$$\frac{(\mathbf{x} - \mu\mathbf{1}_n)}{\sigma} = \frac{\mathbf{1}_n(\bar{x} - \mu)}{\sigma} + \frac{\mathbf{v}s}{\sigma}. \tag{2.4}$$

If $\mathbf{v}$ and $[\bar{y}, s']$ are independent, the right-hand sides are identical in distribution; therefore, $\mathbf{y}$ has the same distribution as $(\mathbf{x} - \mu\mathbf{1}_n)/\sigma$, that is, $y_1, \cdots, y_n$ are independent $N(0, 1)$ variables.

Clearly, condition (2.1) can be expressed equivalently as follows: for any given sample realization, there is a linear relation between the sequences $x_1, \cdots, x_n$ and $y_1, \cdots, y_n$. This seems a very reasonable property of the transformation which preserves the shape of the empirical distribution function.

Denoting $v_i = (x_i - \bar{x})/s$, we obtain for the correlation coefficient of x_i and y_i,

$$r(x_i, y_i) = \frac{\text{cov}(\bar{x}, \bar{y})}{\sigma} + D^2(v_i)\left[E^2(s') + \frac{\text{cov}(s, s')}{\sigma}\right]. \tag{2.5}$$

In the case of Durbin's transformation, $\text{cov}(\bar{x}, \bar{y}) = \text{cov}(s, s') = 0$. Both quantities are positive by appropriate choice of the functions defining $\bar{y}$ and s'

according to the alternative method. This means that the use of the additional sample elements is not merely an alternative way to generate random numbers, the information involved in them is utilized to improve the goodness of the transformation.

3. The case of unknown expectation

In this section we formulate our problem somewhat more generally than in the other cases. Namely, we include cases where the difference of the number of the original and transformed values is larger than 1, the number of unknown parameters.

Let us suppose we have $n + k$ random variables $x_1, \cdots, x_{n+k}$ which are independently normally distributed with common unknown expectation μ and known variance σ^2, and we want to provide the transformed variables $y_1^{(1)}, \cdots, y_n^{(1)}$ which have the expectation and variance σ^2.

Similarly to (2.1), a possible way of defining the variables $y_i^{(1)}$ is to define them by the equation

$$y_i^{(1)} - \bar{y}^{(1)} = x_i - \bar{x} \tag{3.1}$$

where $\bar{y}^{(1)}$ has the distribution $N(0, (\sigma^2/n))$ and $\bar{x} = (x_1 + \cdots + x_n)/n$ as before. The variable $\bar{y}^{(1)}$ may be suitably defined by the formula

$$\bar{y}^{(1)} = (\bar{x} - \bar{x}_k) \sqrt{\frac{k}{n + k}}, \tag{3.2}$$

whereas

$$\bar{x}_k = \frac{x_{n+1} + \cdots + x_{n+k}}{k}. \tag{3.3}$$

It is shown below that the transformation (3.1) is optimal in the adopted sense.

THEOREM 3.1. *Let $x_1, \cdots, x_{n+k}$ be independent random variables with common distribution, having the expectation μ and variance σ^2, and let $y_1^{(1)}, \cdots, y_n^{(1)}$ be defined by* (3.1) *and* (3.2). *Let $z_i = z_i(x_1, \cdots, x_{n+k})$, $(i = 1, \cdots, n)$ functions of $x_1, \cdots, x_{n+k}$ such that if $x_1, \cdots, x_{n+k}$ are normally distributed, then $z_1, \cdots, z_n$ are independent with the common distribution $N(0, \sigma^2)$.*

Then if $x_1, \cdots, x_{n+k}$ are normally distributed or $z_1, \cdots, z_n$ are linear functions of $x_1, \cdots, x_{n+k}$, then

$$\min_{1 \leq i \leq n} r(z_i, x_i) \leq \min_{1 \leq i \leq n} r(y_i^{(1)}, x_i). \tag{3.4}$$

If in (3.4) *equality holds, then it is valid not only for the minima, but for any i.*

PROOF. Suppose $x_1, \cdots, x_{n+k}$ are normally distributed. It follows from a theorem of Lukács [8] that z_i has constant regression on the mean

$$\overline{X} = \frac{x_1 + \cdots + x_{n+k}}{n + k}, \tag{3.5}$$

and therefore, $E(z_i\overline{X}) = E(\overline{X}E(z_i)) = 0$.

Since $z_1, \cdots, z_n$ are independent random variables, we have for any random variable u,

$$\sum_{i=1}^{n} [r(z_i, u)]^2 \leq 1. \tag{3.6}$$

It follows that for some i, $(1 \leq i \leq n)$,

$$r(z_i, u) \leq n^{-1/2} \tag{3.7}$$

where strict inequality holds unless equality holds for any i. We choose

$$u = \sum_{j=1}^{n} x_j - n\overline{X}. \tag{3.8}$$

We have

$$r(x_i - \overline{X}, u) = \sqrt{\frac{k}{n(n+k-1)}}, \tag{3.9}$$

$$r(z_i, x_i - \overline{X}) = [E(x_i, z_i)/\sigma] \sqrt{\frac{n+k}{n+k-1}}. \tag{3.10}$$

Using the triangle inequality which is valid to any triplet of random variables (see [13], p. 271)

$$\text{arc cos } r(x_i - \overline{X}, z_i) \geq |\text{arc cos } r(u, z_i) - \text{arc cos } r(u, x_i - \overline{X})|, \tag{3.11}$$

we obtain

$$r(x_i, z_i) = E(x_i z_i)\sigma \leq \frac{\sqrt{k} + (n-1)\sqrt{n+k}}{n\sqrt{n+k}} \tag{3.12}$$

$$= 1 - \frac{1}{n + k + \sqrt{k(n+k)}},$$

which proves (3.4), provided $x_1, \cdots, x_{n+k}$ are normally distributed. If $z_1, \cdots, z_n$ are linear functions of $x_1, \cdots, x_n$, the left-hand side of (3.4) does not depend on the distribution of $x_1, \cdots, x_n$. Thus theorem 3.1 is proved.

In the case of transformation (3.1) the substitute simple hypothesis is equivalent to the original one; in other words, $y_1^{(1)}, \cdots, y_n^{(1)}$ are normally distributed if and *only if* $x_1, \cdots, x_{n+k}$ are. This follows easily from the theorem of Cramér.

We remark that the normality of $x_1, \cdots, x_{n+k}$ is necessary for the independence of $y_1^{(1)}, \cdots, y_n^{(1)}$ as well. This follows from a theorem of Skitovich [14].

4. The case of unknown variance

Let us consider now the case when (according to the null hypothesis) $x_1, \cdots, x_{n+1}$ are independently normally distributed with common expectation 0 and unknown variance $\sigma^2(> 0)$. We want to obtain independent $N(0, 1)$ normally distributed variables $y_1^{(2)}, \cdots, y_n^{(2)}$. The way corresponding to (2.1) is to define them by the equation

$$\frac{y_i^{(2)}}{s_0'} = \frac{x_i}{s_0}, \qquad (i = 1, \cdots, n) \tag{4.1}$$

where

$$s_0 = \sqrt{\frac{x_1^2 + \cdots + x_n^2}{n}}, \tag{4.2}$$

and s_0' is a function of s_0 and x_{n+1}, and $ns_0'^2$ has χ^2-distribution with n degrees of freedom.

Heuristic arguments lead to the choice

$$s_0' = \psi_n\left(\frac{|x_{n+1}|}{s_0}\right) \tag{4.3}$$

where $\psi_n(t)$ is a monotone decreasing function. This condition, and the distribution of $|x_{n+1}|/s_0$ and s_0', determine the function $\psi_n(t)$ completely (see formula (7.4)).

The transformation gives an indefinite result in the case when $x_1 = \cdots = x_n = 0$; the probability of this event is 0 under the null hypothesis. For this case let us define $y_1^{(2)} = \cdots = y_n^{(2)} = 0$.

The result we give below concerning the optimality of (4.1) is of weaker character than that of the preceding section. Now the class of admitted alternatives is more restricted; we suppose both linearity for any sample realization and scale invariance. The latter supposition means, equivalently, that the transformed values are of structure d'_{n+1} (see [2]).

THEOREM 4.1. *Let $x_1, \cdots, x_{n+1}$ be independent $N(0, \sigma^2)$ random variables, and let $y_1^{(2)}, \cdots, y_n^{(2)}$ be defined by* (4.1), (4.2) *and* (4.3), *and let $z_1, \cdots, z_n$ be defined by*

$$\frac{z_i}{s_0''} = \frac{x_i}{s_0}, \qquad (i = 1, \cdots, n) \tag{4.4}$$

where $s_0'' = s_0''(x_1, \cdots, x_{n+1})$ is a scale invariant function of $x_1, \cdots, x_{n+1}$; that is,

$$s_0''(x_1, \cdots, x_{n+1}) \equiv s_0''(cx_1, \cdots, cx_{n+1}) \qquad \text{for any} \quad c > 0, \tag{4.5}$$

and the function is such as to assure that s_0'' is independent of the vector $[x_1/s_0, \cdots, x_n/s_0]$ and its distribution is a χ^2-distribution with n degrees of freedom. Then

$$r(z_i, x_i) \leq r(y_i^{(2)}, x_i), \qquad (i = 1, \cdots, n). \tag{4.6}$$

PROOF. Let us denote

$$\frac{x_i}{s_0} = \xi_i, \qquad (i = 1, \cdots, n + 1) \tag{4.7}$$

and

$$\zeta = ns_0^2 + x_{n+1}^2. \tag{4.8}$$

It follows from (4.5) that

$$s_0''(x_1, \cdots, x_{n+1}) = s_0''(\xi_1, \cdots, \xi_{n+1}). \tag{4.9}$$

It follows further that the following pairs of random variables (vectors) are pairwise independent:

$$\begin{aligned} &\zeta \text{ and } [\xi_1, \cdots, \xi_{n+1}], \\ (4.10)\qquad &\xi_i \text{ and } [\xi_{n+1}, \zeta], &(i = 1, \cdots, n), \\ &\xi_i \text{ and } s_0''(\xi_1, \cdots, \xi_{n+1}), &(i = 1, \cdots, n). \end{aligned}$$

(However, ξ_i and the vector $[\xi_{n+1}, s_0'']$ do not have to be independent.)

One has

$$\begin{aligned} x_i &= \xi_i \sqrt{\frac{\zeta}{n + \xi_{n+1}^2}}, \\ (4.11)\qquad y_i^{(2)} &= \xi_i \psi_n(|\xi_{n+1}|), \\ z_i &= \xi_i s_0''(\xi_1, \cdots, \xi_{n+1}). \end{aligned}$$

Therefore,

$$(4.12)\qquad E(x_i y_i^{(2)}) = E(\xi_i^2) E(\zeta^{1/2}) E\left(\frac{\psi_n(|\xi_{n+1}|}{\sqrt{n + \xi_{n+1}^2}}\right),$$

$$(4.13)\qquad E(x_i z_i) = E(\xi_i^2) E(\zeta^{1/2}) E\left(E\left(\frac{s_0''(\xi_1, \cdots, \xi_{n+1})}{\sqrt{n + \xi_{n+1}^2}} |\xi_i\right)\right).$$

The distribution of $\psi_n(|\xi_{n+1}|) = s_0'$ agrees with the distribution of

$$(4.14)\qquad s_0''(\xi_1, \cdots, \xi_{n+1}).$$

Since ξ_i is independent of $s_0''(\xi_1, \cdots, \xi_{n+1})$, the conditional distribution of (4.14) given ξ_i is the same. Therefore, since $E(\xi_i^2) > 0$, $E(\zeta^{1/2}) > 0$, it suffices to prove for the conditional correlation coefficient of (4.14) and $(n + \xi_{n+1}^2)^{-1/2}$ given ξ_i that

$$(4.15)\qquad r(s_0''(\xi_1, \cdots, \xi_{n+1}), (n + \xi_{n+1}^2)^{-1/2} | \xi_i) \le r(\psi_n(|\xi_{n+1}|), (n + \xi_{n+1}^2)^{-1/2}).$$

This follows, however, from a theorem of Fréchet and Bass ([1], p. 640) according to which in case of a two-way variable having given marginal distributions, the correlation coefficient is maximal if one of the marginal variables is a monotone increasing function of the other. This is, however, the case between $\psi_n(|\xi_{n+1}|)$ and $(n + \xi_{n+1}^2)^{-1/2}$, and thus, (4.15) is true, and theorem 4.1 is proven.

5. The case of two unknown parameters

Let n be given and let $k = 2$. Let us join to the transformation (3.1), where $\bar{y}^{(1)}$ is defined by (3.2), the equality

$$(5.1)\qquad y_{n+1}^{(1)} = 2^{-1/2}(x_{n+2} - x_{n+1}),$$

and let us denote this transformation by T_1,

$$(5.2)\qquad [y_1^{(1)}, \cdots, y_{n+1}^{(1)}] = T_1[x_1, \cdots, x_{n+2}].$$

Let us denote by T_2 the transformation (4.1) where s_0' is defined by (4.3), that is

$$(5.3)\qquad [y_1^{(2)}, \cdots, y_n^{(2)}] = T_2[x_1, \cdots, x_{n+1}].$$

Let $y_1^{(3)}, \cdots, y_n^{(3)}$ be defined by the formula

$$(5.4)\qquad [y_1^{(3)}, \cdots, y_n^{(3)}] = T_2 T_1[x_1, \cdots, x_{n+2}].$$

Alternatively, (5.4) can be written in the form of (2.1):

$$\frac{y_i^{(3)} - \overline{y}^{(3)}}{s^{(3)}} = \frac{x_i - \overline{x}}{s}, \qquad (i = 1, \cdots, n) \tag{5.5}$$

where

$$\begin{aligned} \overline{y}^{(3)} &= (2\overline{x} - x_{n+1} - x_{n+2})q(2[n+2])^{-1/2}, \\ s^{(3)} &= sq, \end{aligned} \tag{5.6}$$

while

$$n\overline{x} = \sum_{i=1}^{n} x_i, \qquad (n-1)s^2 = \sum_{i=1}^{n} (x_i - \overline{x})^2, \tag{5.7}$$

$$\begin{aligned} q = \psi_n\left(|x_{n+2} - x_{n+1}| \sqrt{\frac{n(n+2)}{2(n-1)(n+2)s^2 + n(2\overline{x} - x_{n+1} - x_{n+2})^2}}\right) \\ \times \sqrt{\frac{2n(n+2)}{2(n-1)(n+2)s^2 + n(2\overline{x} - x_{n+1} - x_{n+2})^2}}. \end{aligned} \tag{5.8}$$

Now we have the following result characterizing the property of the above transformation.

THEOREM 5.1. *Let $x_1, \cdots, x_{n+2}$ be independent $N(\mu, \sigma^2)$ random variables, and let $y_1^{(3)}, \cdots, y_n^{(3)}$ be defined by (5.4) and let $z_1, \cdots, z_n$ be defined by the relation*

$$z_i = \frac{u_i}{(u_1^2 + \cdots + u_n^2)^{1/2}} \varphi(u_1, \cdots, u_{n+1}), \qquad (i = 1, \cdots, n) \tag{5.9}$$

where $u_1, \cdots, u_{n+1}$ are linear functions of $x_1, \cdots, x_{n+2}$ such that they are independently distributed with the common distribution $N(0, \sigma^2)$, and $\varphi(u_1, \cdots, u_{n+1})$ is a scale invariant function (for any $c > 0$, $\varphi(u_1, \cdots, u_{n+1}) = \varphi(cu_1, \cdots, cu_{n+1})$), and such that $z_1, \cdots, z_n$ is a sequence of independent $N(0, 1)$ random variables. Then

$$\min_{1 \le i \le n} r(z_i, x_i) \le \min_{1 \le i \le n} r(y_i^{(3)}, x_i). \tag{5.10}$$

If in this formula equality holds, then it is valid not only for the minima but for any i.

PROOF. Let us introduce the following notation:

$$\xi_i = \frac{y_i^{(1)}}{s_0}, \qquad (i = 1, \cdots, n+1) \tag{5.11}$$

where $y_1^{(1)}, \cdots, y_n^{(1)}$ are defined by formula (4.1), $y_{n+1}^{(1)}$ by formula (5.1),

$$\begin{aligned} \xi_i' &= \frac{u_i n^{1/2}}{(u_1^2 + \cdots + u_n^2)^{1/2}}, \qquad (i = 1, \cdots, n+1), \\ \overline{X} &= \frac{x_1 + \cdots + x_{n+2}}{n+2}, \\ \zeta &= \sum_{i=1}^{n+2} x_i^2 - n\overline{X}^2. \end{aligned} \tag{5.12}$$

It follows (as in the proof of theorem 3.1) that $\overline{X}$ is independent of $u_1, \cdots, u_{n+1}$ which implies that x_i may be written in the form

$$x_i = \sum_{j=1}^{n+1} a_{ij}u_j + a_{in}\overline{X}, \qquad (i = 1, \cdots, n+2) \tag{5.13}$$

and that $u_1^2 + \cdots + u_{n+1}^2 = \zeta$.

Moreover, we have (see the proof of the preceding theorem) that

$$\varphi(u_1, \cdots, u_{n+1}) = \varphi(\xi_1', \cdots, \xi_{n+1}'). \tag{5.14}$$

And further, the following pairs of random variables (vectors) are independent for $(i = 1, \cdots, n)$:

$$\begin{aligned} &[\overline{X}, \zeta] \text{ and } [\xi_1, \cdots, \xi_{n+1}], \\ &[\overline{X}, \zeta] \text{ and } [\xi_1', \cdots, \xi_{n+1}'], \\ &\xi_i \text{ and } [\xi_{n+1}, \zeta], \\ &\xi_i' \text{ and } \xi_{n+1}', \\ &\xi_i' \text{ and } \varphi(\xi_1', \cdots, \xi_{n+1}'). \end{aligned} \tag{5.15}$$

Evidently $E(\xi_i\xi_j) = E(\xi_i'\xi_j') = 0$ for $i \neq j;\ i, j = 1, \cdots, n+1$.

We have

$$\begin{aligned} x_i &= \left(\xi_i - \frac{\sum_{j=1}^{n} \xi_j}{n + 2 + [2(n+2)]^{1/2}}\right)\left(\frac{\zeta}{n + \xi_{n+1}^2}\right)^{1/2} + \overline{X} \\ &= \sum_{j=1}^{n+1} a_{ij}\,\xi_j' \left(\frac{\zeta}{n + \xi_{n+1}'^2}\right)^{1/2} + \overline{X}, \end{aligned} \tag{5.16}$$

$$y_i^{(3)} = \xi_i\psi_n(|\xi_{n+1}|) \tag{5.17}$$

$$z_i = \xi_i'\varphi(\xi_1', \cdots, \xi_{n+1}')/n^{1/2}, \qquad (i = 1, \cdots, n). \tag{5.18}$$

Therefore, for $(i = 1, \cdots, n)$,

$$\begin{aligned} &E(x_iy_i^{(3)}) \\ &\quad = [1 - (n + 2 + [2(n+2)]^{1/2})^{-1}]E(\xi_i^2)E(\zeta^{1/2})E\left(\frac{\psi_n(|\xi_{n+1}|)}{(n + \xi_{n+1}^2)^{1/2}}\right), \end{aligned} \tag{5.19}$$

$$E(x_iz_i) = a_{ii}E(\xi_i'^2)E(\zeta^{1/2})E\left(E\left(\frac{\varphi(\xi_1', \cdots, \xi_{n+1}')}{[n(n + \xi_{n+1}'^2)]^{1/2}}\,|\xi_i'\right)\right). \tag{5.20}$$

Now $a_{ii} = r(u_i, x_i)$ and $E(\xi_i^2) = E(\xi_i'^2)$. Using (3.4) and (4.15) we obtain (5.10). The last assertion of the theorem follows from the similar assertion of theorem 3.1.

6. The general case

This section deals briefly with the possibilities of extending the method to more general cases.

In general, let $x_1, \cdots, x_{n+k}$ be independent random variables, having under the null hypothesis the distribution $F(x, \boldsymbol{\mu}_i)$, $(i = 1, \cdots, n+k)$. Suppose the shape of the function $F(x, \cdot)$ is known, and the parameter vectors $\boldsymbol{\mu}_i$ are to be known functions of the unknown constant vector $\boldsymbol{\theta}$,

$$\boldsymbol{\mu}_i = \boldsymbol{\mu}_i(\boldsymbol{\theta}) \tag{6.1}$$

where $\boldsymbol{\theta}$ is a point of the parameter space Θ, $(\boldsymbol{\theta} \in \Theta)$. Let us suppose there exists a $\theta \in \Theta$ for which $\boldsymbol{\mu}_i(\boldsymbol{\theta}_0)$ does not depend on i, that is

$$\boldsymbol{\mu}_i(\boldsymbol{\theta}_0) = \boldsymbol{\mu}_0, \qquad (i = 1, \cdots, n + k). \tag{6.2}$$

In other words, if $\boldsymbol{\theta} = \boldsymbol{\theta}_0$, the variables $x_1, \cdots, x_{n+k}$ are independent, identically distributed random variables.

Let us suppose, further, the existence of a distribution-free statistic on the first n sample elements. Let us denote it by $\mathbf{T} = \mathbf{T}(\mathbf{X})$, where $\mathbf{X} = [x_1, \cdots, x_n]$.

Let $\boldsymbol{\Omega}$ be a random vector whose distribution is known. Let us define the function $\mathbf{Y}(\mathbf{T}, \boldsymbol{\Omega})$ in such a way that (cf. [6])

$$\begin{aligned} &\text{"cond. distr. of } \mathbf{Y}(\mathbf{T}, \boldsymbol{\Omega}) \text{ given } T\text{"} \\ &= \text{"cond. distr. of } \mathbf{X} \text{ given } \mathbf{T}, \boldsymbol{\theta}_0\text{."} \end{aligned} \tag{6.3}$$

This is in general possible, in many ways. Computational difficulties, however, may arise in determining the conditional distribution of $\mathbf{X}$.

Since, according to (6.3), the conditional distributions of $\mathbf{X}$ and $\mathbf{Y}$ agree for all given values of $\mathbf{T}$, the unconditional distributions have to agree too; that is, its components $y_1, \cdots, y_n$ are independent, identically distributed random variables having the distribution $F(x, \boldsymbol{\mu}_0)$.

The random vector $\boldsymbol{\Omega}$ may be generated by random numbers, or it may be a function of the variables $x_1, \cdots, x_{n+k}$, such that its distribution is independent of $\mathbf{T}$ and $\boldsymbol{\theta}$.

Thus the original null hypothesis may be replaced by the simple hypothesis that the variables $y_1, \cdots, y_n$ are independently distributed with the distribution $F(x, \boldsymbol{\theta}_0)$.

The adequateness of representation of the original hypothesis by this substitute may be investigated in the particular cases. Clearly, it depends on the forms of the function $F(x, \boldsymbol{\mu})$ and the distribution-free statistic $\mathbf{T}$.

EXAMPLES. (a) *Gamma-distribution.* Let $x_1, \cdots, x_{n+1}$ be independent random variables with the common density function

$$F(x, \mu) = \begin{cases} 0 & \text{for } x \le 0, \\ \dfrac{\mu^\lambda}{\Gamma(\lambda)} \displaystyle\int_0^x t^{\lambda-1} e^{-\mu t}\, dt & \text{for } x > 0. \end{cases} \tag{6.4}$$

Here λ is known and μ is the nuisance parameter.

The appropriate transformation is

$$y_i^{(4)} = \frac{x_i}{x_1 + \cdots + x_n} \psi_{\lambda,n}\left(\frac{x_{n+1}}{x_1 + \cdots + x_{n+1}}\right), \qquad (i = 1, 2, \cdots, n) \tag{6.5}$$

where the function $\psi_{\lambda,n}(t)$ is a monotone decreasing function assuring that

$$\psi_{\lambda,n}\left(\frac{x_{n+1}}{x_1 + \cdots + x_{n+1}}\right) \tag{6.6}$$

has the same distribution as $x_1 + \cdots + x_n$ if $\mu = 1$. This condition determines the function $\psi_{\lambda,n}(t)$ completely (see formula (7.13)).

Equality (6.5) yields independent gamma variables with the original shape parameter and unit scale parameter, that is, having the distribution function

$$F(x) = \begin{cases} 0 & \text{if } x \leq 0, \\ \dfrac{1}{\Gamma(\lambda)} \displaystyle\int_0^x t^{\lambda-1} e^{-t}\, dt & \text{if } x > 0. \end{cases} \tag{6.7}$$

A theorem similar to theorem 4.1 can be proved concerning the optimality of this transformation.

(b) *Small subsamples.* In the case of normal null hypothesis, if we have several small subsamples with different nuisance parameters, the transformations given in sections 3–5 can be separately applied for each subsample.

(c) *Analysis of variance model.* In [13] a transformation was given for the case of a two-way classification, one observation per cell, when the distribution of the error term is normal, according to the null hypothesis [13], (2). If the variance is unknown as well, we may apply transformation (4.1) as a second step.

Also, this transformation enables, as mentioned there, to test the homogeneity of variances as well in the mentioned cases. In the cases of alternative hypothesis, the variances of the transformed variables will differ from that of the corresponding original ones, but the transformation preserves the magnitude order of the variances so that the hypothesis concerning the homogeneity of variances in the original and transformed date are equivalent.

An alternative solution for this problem has been given by N. L. Johnson [3].

7. Practical applications

It is convenient to make some slight modifications on the formulae given in sections 3–7 for the purpose of practical application. In the modified formulae we denote by n the original sample size.

(a) If we have a random simple sample $x_1, \cdots, x_n$, which comes, according to the null hypothesis, from a normal distribution (expectation and variance being unspecified), apply the following transformation:

$$y_i = \frac{x_i - \bar{x}''}{S} \psi_{n-2}\left(\frac{|x_{n-1} - x_n|\sqrt{n-2}}{S\sqrt{2}}\right), \qquad (i = 1, \cdots, n-2) \tag{7.1}$$

where

$$\bar{x}'' = \frac{\sum_{i=1}^{n} x_i + \sqrt{\frac{n}{2}}\,(x_{n-1} + x_n)}{n + \sqrt{2n}}, \tag{7.2}$$

$$S = \sqrt{\sum_{i=1}^{n} x_i^2 - \frac{1}{n}\left(\sum_{i=1}^{n} x_i\right)^2 - \frac{1}{2}\,(x_{n-1} - x_n)^2}, \tag{7.3}$$

and the function $\psi_\nu(t)$ is defined by the relation

$$Q([\psi_\nu(t)]^2|\nu) = 2P(t|\nu) - 1, \tag{7.4}$$

while $\psi_\nu(t) \geq 0$,

$$P(t|\nu) = \frac{\Gamma\left(\frac{\nu+1}{2}\right)}{\Gamma\left(\frac{\nu}{2}\right)\sqrt{\nu\pi}} \int_{-\infty}^{t} \left(1 + \frac{u^2}{\nu}\right)^{-\frac{\nu+1}{2}} du, \tag{7.5}$$

$$Q(t|\nu) = \frac{1}{2^{1/2}\Gamma\left(\frac{\nu}{2}\right)} \int_{t}^{\infty} e^{-\frac{u}{2}} u^{\frac{\nu-1}{2}} du. \tag{7.6}$$

The functions $P(t|\nu)$ and $1 - Q(t|\nu)$ are the distributions of the Student distribution and the χ^2-distribution, respectively, each with ν degrees of freedom. The notations ($P(t|\nu)$ and $Q(t|\nu)$ agree with that of the *Biometrika Tables* [9], tables 9 and 7, respectively, where they are tabulated. These tables may be used in applications of (7.1). (The formula (7.1) is an equivalent form of formula (5.4)).

(b) If, according to the null hypothesis, our random simple sample $x_1, \cdots, x_n$ comes from a normal distribution with specified variance (expectation being unspecified), then we have to use the transformation

$$y_i = x_i - \bar{x}' \tag{7.7}$$

where

$$\bar{x}' = \frac{\sum_{i=1}^{n} x_i + x_n\sqrt{n}}{n + \sqrt{n}}. \tag{7.8}$$

This transformation was given in [13] as formula (1) and is the special case of formula (3.1) of the present paper.

(c) If, according to the null hypothesis, our random simple sample $x_1, \cdots, x_n$ comes from a normal distribution with expectation 0 (variance being unspecified), then the transformation to be used is

$$y_i = \frac{x_i}{x_i^2 + \cdots + x_{n-1}^2} \psi_{n-1}\left(\frac{|x_n|\sqrt{n-1}}{\sqrt{x_1^2 + \cdots + x_{n-1}^2}}\right), \tag{7.9}$$

the definition of $\psi_\nu(t)$ being given in (7.4).

(d) Let $x_{i1}, \cdots, x_{in_i}$, $(i = 1, 2, \cdots, m)$ be independent random variables. For each given i the n_i random variables have the common distribution function $F_i(x)$. Some of the variances of the distributions are supposed to be equal but unknown.

The hypothesis to be tested is that the distributions $F_i(x)$ are normal. The unknown variance is not specified in the null hypothesis.

In this case we apply the transformation (7.7) for all but one subsample. It is suitable to choose the notation so that $n_m \geq n_i$, $(i = 1, \cdots, m-1)$ and to apply transformation (7.7) for each of the series $x_{i1}, \cdots, x_{in_i}$, $(i = 1, 2, \cdots, m-1)$ and the modified form of (3.1) with $k = 2$, that is the transformation

$$\text{(7.10)} \qquad \begin{aligned} y_i &= x_i - \bar{x}'' \\ y_{n-1} &= \frac{x_{n-1} - x_n}{\sqrt{2}} \end{aligned}$$

for the series $x_{m1}, \cdots, x_{mn_m}$. Here $\bar{x}''$ is the same as in (7.2). As a second step we apply (7.9) for the sequence of the resultant $\sum_{i=1}^{m} n_i - m$ variables

$$\text{(7.11)} \qquad y_{11}, \cdots, y_{1n_1-1}, y_{21}, \cdots, y_{mn_m-2}, y_{mn_m-1}.$$

(e) *Gamma distribution.* The suitable form of the transformation (6.5) is the following:

$$\text{(7.12)} \qquad y_i = \frac{x_i}{x_1 + \cdots + x_{n-1}} \psi_{\lambda,(n-1)} \left(\frac{x_n}{x_1 + \cdots + x_n} \right)$$

where the function $\psi_{\lambda,\nu}(t)$ is defined by the relation

$$\text{(7.13)} \qquad 1 - I\left(\frac{\psi_{\lambda,\nu}(t)}{\sqrt{\nu\lambda - 1}}, \nu\lambda - 2 \right) = I_t(\lambda, \nu\lambda),$$

whereas

$$\text{(7.14)} \qquad I(u, p) = \frac{\int_0^{u\sqrt{p+1}} e^{-t} t^{p+1}\, dt}{\int_0^{\infty} e^{-t} t^{p+1}\, dt}$$

is a gamma distribution function, tabulated in [11], and

$$\text{(7.15)} \qquad I_t(p, q) = \frac{\int_0^t x^{p-1}(1 - x)^{q-1}\, dx}{\int_0^1 x^{p-1}(1 - x)^{q-1}\, dx}$$

is a beta distribution function, tabulated in [10].

8. A comparative example

The aim of the following example is to illustrate in a special case that the transformation fulfilling our optimality criterion gives, in fact, better representation than the previously known ones.

Let x_{i1}, x_{i2}, $(i = 1, \cdots, m)$ be independent random variables with distributions

$$\text{(8.1)} \qquad \begin{aligned} P(x_{ij} &= a_i) = \tfrac{2}{3}, \\ P(x_{ij} &= -2a_i) = \tfrac{1}{3}, \qquad (a_i > 0), \quad (i = 1, \cdots, m; j = 1, 2). \end{aligned}$$

The first three columns of table I show the distributions of the transformed values yielded, by (4.1) and (3) of [15], respectively, $(n = 2)$:

TABLE I

Probability	Possible values of the transformed values using transformation	
	(4.1)	(3) of [15]
2/9	−1.047	−1.447
1/9	− .675	− .319
2/9	.378	.457
4/9	.675	1.150

(The second factor in the right-hand side of formula (4) of [15] should be read $\varphi_m(y_m/(y_1^2 + \cdots + y_m^2)^{1/2})$.)

If, according to [6], we use random numbers, the distribution of the transformed values will be a continuous distribution, the mixture of four χ-distributions. The density of this distribution is

$$\frac{5x}{18} e^{-5x^2/8} + \frac{2x}{9} e^{-x^2/4} \qquad \text{for} \quad x \geq 0, \tag{8.2}$$

$$-\frac{5x}{36} e^{-5x^2/16} - \frac{x}{18} e^{-x^2/4} \qquad \text{for} \quad x < 0. \tag{8.3}$$

Apparently the first of these three distributions gives the best representation for the original ones.

REFERENCES

[1] JEAN BASS, "Sur la compatibilité des fonctions de répartition," *C. R. Acad. Sci. Paris*, Vol. 240 (1955), pp. 839–841.

[2] C. B. BELL, "Some basic theorems of distribution-free statistics," *Ann. Math. Statist.*, Vol. 35 (1964), pp. 150–156.

[3] T. CALINSKY, "On a certain statistical method of investigating interaction in serial experiment with plant varieties," *Bull. Acad. Polon. Sci. Sér. Sci. Biol.*, Vol. 8 (1960), pp. 565–568.

[4] H. CRAMÉR, *Mathematical Methods of Statistics*, Princeton, Princeton University Press, 1946.

[5] D. A. DARLING, "The Cramér-Smirnov test in the parametric case," *Ann. Math. Statist.*, Vol. 26 (1955), pp. 1–20.

[6] S. DURBIN, "Some methods of constructing exact tests," *Biometrika*, Vol. 48 (1961), pp. 41–55.

[7] M. KAC, J. KIEFER, and J. WOLFOWITZ, "On tests of normality and other tests on goodness of fit based on distance methods," *Ann. Math. Statist.*, Vol. 26 (1955), pp. 189–211.

[8] E. LUKÁCS, "On distribution-free partition statistics for the normal family," *Bull. Inst. Internat. Statist.*, Vol. 36 (1957), pp. 37–42.

[9] E. S. PEARSON and H. O. HARTLEY, *Biometrika Tables for Statisticians*, Vol. I, Cambridge, Cambridge University Press, 1956.

[10] KARL PEARSON, *Tables of the Incomplete Beta-Function*, Cambridge, Cambridge University Press, 1934.

[11] ———, *Tables of the Incomplete* Γ*-Function*, Cambridge, Cambridge University Press, 1934.

[12] YU. PROHOROV, "Some characterization theorems in statistics," *Proceedings of the Fifth Berkeley Symposium on Mathematical Statistics and Probability*, Berkeley and Los Angeles, University of California Press, 1966, Vol. I, pp. 341–349.

[13] K. SARKADI, "On testing for normality," *Publ. Math. Inst. Hungar. Acad. Sci.*, Vol. 5 (1960), pp. 269–275.

[14] V. P. SKITOVICH, "Linear forms of independent random variables and the normal distribution," *Izv. Acad. Nauk USSR*, Vol. 18 (1954), pp. 185–200. (In Russian.)

[15] HORAND STÖRMER, "Ein Test zum Erkennen von Normalverteilungen," *Z. Wahrscheinlichkeitstheorie und Verw. Gebiete*, Vol. 2 (1964), pp. 420–428.

CROSS-SECTIONS OF ORBITS AND THEIR APPLICATION TO DENSITIES OF MAXIMAL INVARIANTS

R. A. WIJSMAN
UNIVERSITY OF ILLINOIS

1. Introduction and summary

Let G be a group of one-to-one transformations of the sample space X onto itself. A maximal invariant is a function constant on orbits and distinguishing orbits. If G leaves a certain statistical problem invariant and an invariant procedure is to be selected, it is necessary first to solve the problem of how to obtain the distribution of a maximal invariant, given any distribution on X. One of the possible methods consists of writing this distribution as an integral over the group G. This method has been promoted notably by Stein [10], [11], Karlin [6], and James [5], but does not seem to have been used very much in the literature (among the exceptions, see [3], [7]) in spite of the fact that the method has several advantages. Unfortunately, although some specific problems have thus been treated, there does not seem to exist much in the form of a general theory. This paper is intended as a step in that direction. Some new theorems will be presented and several examples given.

The principal tool used in this paper that makes things work is the so-called *cross-section* of orbits, local or global (precise definitions of various terms will be given in section 2). A global cross-section is a subset Z of X such that every orbit intersects Z at exactly one point, in addition to a few other properties to be defined in section 2. A local cross-section at x is a global cross-section for an open, invariant neighborhood of the orbit passing through x. If a global cross-section Z exists, it is possible to convert an integral $\int_X p \, d\mu$ (μ is Lebesgue measure) into an iterated integral of the form $\int_Z \nu_Z(dz) \int_G p(gz)\nu_G(dg)$, where ν_Z and ν_G are certain measures on Z, G, respectively. For any global cross-section Z there is a natural maximal invariant, namely the function that associates to every orbit its intersection with Z. For any distribution P on X, with density p with respect to Lebesgue measure, the distribution of the maximal invariant is then a distribution on Z given by $\nu_Z(dz) \int p(gz)\nu_G(dg)$. The exact nature of the measures ν_Z and ν_G will be given in sections 4 and 5.

In many statistical problems the primary interest is in the probability ratio of a maximal invariant, given any two densities p_1 and p_2. It is then not necessary

Research supported, in part, by the National Science Foundation under Grant GP-3814.

to obtain a global cross-section; one can get by with a local cross-section at every x, and the probability ratio at x is then given by $\int p_2(gx)\nu_G(dg)/\int p_1(gx)\nu_G(dg)$.

There is another function served by a cross-section, global or local. If the principle of invariance is invoked and statistical procedures restricted to depend only on a maximal invariant, this amounts to demand that the procedures be measurable with respect to the sigma-field $\mathfrak{A}^I$ of invariant measurable sets. There is a priori no guarantee that this leaves the statistician with enough procedures to choose from. An obvious situation of this kind arises if G is transitive on X, for then X is one orbit and $\mathfrak{A}^I$ is trivial. The same thing could happen without G being transitive on X, if every orbit is dense in X. Such "misbehavior" of orbits is excluded if a cross-section exists, and we have then a guarantee that $\mathfrak{A}^I$ is "rich" enough. This will be shown in more detail in theorem 5.

Among practitioners of invariance it is customary to choose the range space of any maximal invariant to be Euclidean. On the other hand, if a global cross-section Z exists, it is in general not Euclidean (it is an analytic manifold under the conditions to be imposed presently). This seems contradictory, but, in fact, the Euclidean choice is often possible only after removing from X a set of Lebesgue measure 0. For example, let X be Euclidean n-space with the origin deleted, that is, X consists of the points $x = (x_1, \cdots, x_n) \neq 0$. Let G consist of the transformations $x \to cx$, where c runs through the positive reals. The orbits are then the rays emanating from the origin. It is customary to choose as a maximal invariant the function $(x_1/x_n, \cdots, x_{n-1}/x_n, \operatorname{sgn} x_n)$, but this is possible only when the collection of rays with $x_n = 0$ is removed from X. There are, of course, many other choices of removal of a null set from X to make the maximal invariant Euclidean. On the other hand, any $(n-1)$-sphere concentric with the origin is a global cross-section and provides a natural maximal invariant that does not suffer from the defect of the Euclidean maximal invariants mentioned above.

Invariance considerations in statistics have been useful in parametric and in nonparametric problems. In this paper, only the parametric case will be considered. In the bulk of applications to parametric problems X is Euclidean and G is a Lie group consisting of translations and/or linear transformations. The translations are trivial to deal with, so they will not be considered here. We shall therefore make the following restrictions throughout this paper: X is a nonempty open subset of Euclidean n-space E^n, and G is a Lie subgroup of the general linear group $GL(n, R)$ of $n \times n$ real nonsingular matrices. Thus, $x \in X$ is an n-vector (taken to be a column vector), g an $n \times n$ nonsingular matrix, and the transformation of X by g given by $x \to gx$. The subset X is called a *linear G-space.*

It will be understood throughout that all Lie groups of $n \times n$ matrices that arise (including G) are endowed with the usual topology inherited from E^{n^2}.

2. Definitions and notation

The action gx of $g \in G$ on $x \in X$ was already defined in section 1. Then $g \in G$ acts in a natural way on sets, families of sets, measures, and functions. That is, if $A \subset X$, define $gA = \{gx: x \in A\}$; if $\mathfrak{A}$ is a family of sets, define $g\mathfrak{A} = \{gA: A \in \mathfrak{A}\}$; if P is a measure on $\mathfrak{A}$, define gP on $g\mathfrak{A}$ by $gP(gA) = P(A)$; if f is a function on X, define gf by $gf(gx) = f(x)$. We say that $A \subset X$ is invariant if $gA = A$ for every $g \in G$, and that f on X is invariant if $gf = f$ for every $g \in G$. The *orbit* of x (more precisely the G-orbit of x) is $Gx = \{gx: g \in G\}$. Thus, a function f on X is invariant if and only if it is constant on each orbit. The space of orbits, considered as an abstract space, is written X/G. If for any $x \in X$, $Gx = X$, we say that G is transitive on X. For any $A \subset X$, $GA = \{gA: g \in G\}$; GA can also be considered as the union of the orbits that intersect A.

If G_0 is a subgroup of G, then for any $g \in G$, gG_0 is called the left coset of g (relative to G_0), sometimes written $[g]$. If g_1 and g_2 are in the same coset, we shall sometimes write $g_1 \sim g_2$. The space of left cosets relative to G_0 is written G/G_0. The natural map $\varphi: G \to G/G_0$ is defined by $\varphi(g) = [g]$. The group G acts in a natural way on G/G_0 by $g_1[g_2] = [g_1g_2]$. The *isotropy group* G_x of $x \in X$ is defined as $G_x = \{g \in G: gx = x\}$. It is easily verified that there is a one-to-one correspondent between Gx and G/G_x.

We shall denote by $\mathfrak{A}$ the sigma-field of Borel subsets of X. Since for every $g \in G$, $x \to gx$ is a homeomorphism, $g\mathfrak{A} = \mathfrak{A}$. Define $\mathfrak{A}^I = \{A: A \in \mathfrak{A}$ and A is invariant$\}$, then $\mathfrak{A}^I$ is a sub-sigma-field of $\mathfrak{A}$. For any P on $\mathfrak{A}$, let P^I be the restriction of P to $\mathfrak{A}^I$. A *maximal invariant* is a triple $(Z, \mathcal{B}, t)$, Z a space, $\mathcal{B}$ a sigma-field of subsets of Z, and t a function: $X \to$ onto Z such that if $z = t(x)$, $x = X$, then $t^{-1}(z) = Gx$, and such that $t^{-1}\mathcal{B} = \mathfrak{A}^I$. Thus, t is invariant and takes different values on different orbits. Consequently, Z is in one-to-one correspondence with X/G. Furthermore, the sets of $\mathcal{B}$ are in one-to-one correspondence with those of $\mathfrak{A}^I$. Given any distribution P on $\mathfrak{A}$, t induces on $\mathcal{B}$ the distribution $Pt^{-1} = P^It^{-1}$, which we shall denote by P^Z.

If G acts on two spaces, X and Y, then $f: X \to Y$ is called *equivariant* if $f(gx) = gf(x)$ for all $x \in X$, $g \in G$ (that is, f commutes with g). A *map* is a continuous function. Before defining cross-sections, it is convenient to define first the somewhat more general object of a *slice*. (A good reference is Palais [9], who also gives references to earlier work. For references to cross-sections see [4] and [8].) Definitions 1, 4, and 5 are taken from Palais [9].

DEFINITION 1. *A slice at x is a set $Z \subset X$ such that* (i) $x \in Z$; (ii) GZ *is open in* X; (iii) *there exists an equivariant map* $f: GZ \to G/G_x$ *such that* $f^{-1}(G_x) = Z$.

A slice Z has the property that $G_xZ = Z$, and if $g \notin G_x$, then $gZ \cap Z = \varnothing$. However, an orbit may intersect a slice in more than one point. This is not allowed in a local cross-section.

DEFINITION 2. *A local cross-section at x is a slice Z at x such that if $z \in Z$ and $gz \in Z$ for some $g \in G$, then $gz = z$.*

DEFINITION 3. *A global cross-section is a local cross-section Z such that $GZ = X$.*

DEFINITION 4. *A neighborhood V of x is called* thin *if the closure of $\{g \in G: gV \cap V \neq \varnothing\}$ is compact.*

DEFINITION 5. *The space X is called a* Cartan *G-space if every $x \in X$ has a thin neighborhood.*

We shall say that a k-dimensional slice or cross-section is *flat* if it is contained in the translate of a k-dimensional linear space. We shall denote n-dimensional Lebesgue measure by μ_n, left Haar measure on G by μ_G. Let G_0 be a compact subgroup of G, $Y = G/G_0$, and φ the natural map $G \to Y$. Then φ induces on Y the left Haar measure $\mu_Y = \mu_G\varphi^{-1}$. Finally, $|g|$ stands for the absolute value of the determinant of g, and the $n \times n$ identity matrix will be denoted I_n or e.

3. Existence of local cross-sections

In section 4 it will be shown how to use local cross-sections in integration. Here we shall deal with their existence. One cannot expect to be able to put a local cross-section at every x; for example, any point that lies on an orbit of less than maximum dimension has no local cross-section. For the purpose of integration with respect to Lebesgue measure μ_n, it would be sufficient to show that the set of exceptional points is of μ_n measure 0. Unfortunately, this is not so in general [12]. The extra condition that makes things work is the assumption that X be a Cartan G-space (definition 5). The following theorem is proved in [12].

THEOREM 1. *If $X \subset E^n$ is a linear Cartan G-space, there is an open linear Cartan G-space $X^0 \subset X$ such that $\mu_n(X - X^0) = 0$ and there is a flat local cross-section at every $x \in X^0$.*

In order to apply theorem 1, it is necessary to verify the Cartan condition. In many applications this can be done directly without difficulty, but in others it could conceivably be troublesome. There are no known easy general sufficient conditions (a necessary condition is, of course, that G_x be compact for every $x \in X$ (see definitions 4 and 5)). Fortunately, there is an important class of applications of invariance to problems in multivariate normal analysis where the Cartan property can be proved once and for all. This follows from theorem 2.

THEOREM 2. *Let $X = X_1 \times X_2$, where $X_1 \subset E^{n_1}$, and X_2 is a space of $k \times k$ positive definite matrices, so that $X_2 \subset E^{n_2}$ with $n_2 = k(k+1)/2$. For any $x \in X$ put $x = (r, s)$, $r \in X_1$, $s \in X_2$, so that x is an n-vector, where $n = n_1 + n_2$. Let G^* be a closed subgroup of $GL(k, R)$, F^* a continuous homomorphism of G^*, and let a group G of linear transformations on X be defined by $r \to Br$, $s \to CsC'$, $C \in G^*$, $B = B(C) \in F^*$. Then X is a linear Cartan G-space.*

PROOF. Let $V = V_1 \times V_2$ be a neighborhood of (r, s). A simple argument, using the continuity of $B(C)$, shows that if V_2 is a thin neighborhood of s for the transformations $s \to CsC'$, $C \in G^*$, then V is a thin neighborhood of (r, s). Therefore, in the proof we may assume $X = X_2$. Furthermore, it follows from definition 4 that it is sufficient to give the proof for X being the space of all $k \times k$ positive definite matrices. In order to show that every $s \in X$ has a thin neighborhood, it is sufficient to show this for I_k, which we shall abbreviate I. We have to show that there is a neighborhood V of I in X such that if $M = \{C \in G^*: CVC' \cap V \neq \emptyset\}$, then the closure of M in G^* is compact:

equivalently (since G^* is closed in $GL(k, R)$), that the closure of M in $GL(k, R)$ is compact. Take any $0 < a < 1$ and define $\mathfrak{I}_a$ = set of all $k \times k$ lower triangular matrices whose elements are in absolute value less than a. Let $V = \{(I + T)(I + T)' : T \in \mathfrak{I}_a\}$; then V is a neighborhood of I. If $C \in M$, then there exist $T_1, T_2 \in \mathfrak{I}_a$ such that $(I + T_2)(I + T_2)' = C(I + T_1)(I + T_1)'C'$ so that there exists a $k \times k$ orthogonal matrix Ω such that $C(I + T_1) = (I + T_2)\Omega$; that is $C = (I + T_2)\Omega(I + T_1)^{-1}$. It can easily be verified that the elements of $(I + T)^{-1}$ are uniformly bounded for $T \in \mathfrak{I}_a$ so that M is bounded. Moreover, for $T \in \mathfrak{I}_a$, $(1 - a)^k < |I + T| < (1 + a)^k$ so that $|C| > (1 - a)^{2k} > 0$ if $C \in M$. It follows that the closure of M in $GL(k, R)$ coincides with the closure of M in E^{k^2}, which is compact as a closed, bounded subset of E^{k^2}.

EXAMPLE 1. In the derivation of Hotelling's T^2, r is the sample mean and s the sample covariance matrix in a sample from a multivariate normal distribution. The group $G^* = F^* = GL(k, R)$ so that $gx = (Cr, CsC')$. Since theorem 2 applies, and therefore theorem 1, there is a local cross-section at almost every x.

4. Application of local cross-sections to the probability ratio of a maximal invariant

It is proved in ([12], lemma 3) that if Z is a local cross-section at x, then $G_z = G_x$ for every $z \in Z$. Putting $Y = G/G_x$, every orbit intersecting Z is now a copy of Y. Thus, there is a one-to-one correspondence between GZ and $Y \times Z$.

THEOREM 3. *Let Z be a flat k-dimensional* $(0 < k < n)$ *local cross-section at a point $x_0 \in X$ such that G_{x_0} is compact, and let p be a real-valued function on GZ, integrable with respect to μ_n. Then there exists an analytic, real-valued function ψ on Z such that*

$$\int_{GZ} p(x)\mu_n(dx) = \int_Z \psi(z)\mu_k(dz) \int_G p(gz)|g|\mu_G(dg). \tag{1}$$

PROOF. Denote $Y = G/G_{x_0}$, φ the natural map $G \to Y$, and $\mu_Y = \mu_G\varphi^{-1}$. Being compact, G_{x_0} is conjugate to an orthogonal group [1]. It follows that $|g_0| = 1$ for every $g_0 \in G_{x_0}$ so that $g_1 \sim g_2$ implies $|g_1| = |g_2|$. The common value of $|g|$ for all $g \in \varphi^{-1}y$ will be denoted $|y|$. Now the function $(y, z) \to gz$, where g is any member of $\varphi^{-1}y$, is an analytic homeomorphism of $Y \times Z$ onto GZ ([9], proposition 2.1.2, [12], section 2). This permits writing the integral of p over GZ as an integral over the product space $Y \times Z$, and the latter as an iterated integral, using Fubini's theorem. The volume element $\mu_n(dx)$ is expressible as $\mu_n(dx) = \psi(y, z)\mu_Y(dy)\mu_k(dz)$, $\psi > 0$ analytic. Making the transformation $x \to gx$ which transforms $dx \to |g|\, dx$ and leaves μ_Y and z invariant, we readily deduce $\psi(y, z) = |y|\psi(z)$, $\psi > 0$ analytic on Z. Thus we have $\int_{GZ} p(x)\mu_n(dx) = \int_Z \psi(z)\mu_k(dz) \int_Y p(gz)|y|\mu_Y(dy)$, in which g is any member of $\varphi^{-1}y$. The integral over Y equals $\int_G p(gz)|g|\mu_G(dg)$, and the theorem is proved.

If p is the density with respect to μ_n of a probability distribution P on X, then it follows from (1) that

$$P^Z(dz)/\mu_k(dz) = \psi(z) \int_G p(gz)|g|\mu_G(dg) \tag{2}$$

is the density with respect to μ_k of the maximal invariant defined locally by the local cross-section Z.

THEOREM 4. *Let $X \subset E^n$ be a linear Cartan G-space, and let $p_i \geq 0$, $\int_X p_i(x)\mu_n(dx) = 1$, $i = 1, 2$, be two given probability densities. Then for any maximal invariant $(Z, \mathcal{B}, t)$ its probability ratio is for almost all (μ_n) x given by*

$$\frac{dP_2^Z}{dP_1^Z}(t(x)) = \frac{dP_2^I}{dP_1^I}(x) = \frac{\int p_2(gx)|g|\mu_G(dg)}{\int p_1(gx)|g|\mu_G(dg)}. \tag{3}$$

PROOF. The first inequality in (3) is clear, and it implies that the probability ratio does not depend on the choice of maximal invariant. If at x a local cross-section Z exists we may take a maximal invariant defined locally by Z (with $\mathcal{B}$ the Borel subsets of Z; the measurability question will be settled in theorem 5). Writing (2) for P_2 and P_1 and taking the ratio gives (3). According to theorem 1, we may exclude from X a set of μ_n-measure 0 such that in the remaining X^0 there is a local cross-section at every point x, concluding the proof.

Note that in (3) the extreme left member is constant along each orbit, so that this ought to be true also for the ratio of integrals on the extreme right. That this is indeed so can be verified directly by replacing x with $g_1^{-1}x$, for any fixed $g_1 \in G$; then numerator and denominator are both multiplied by the same constant $|g_1|\ \Delta(g_1)$, where Δ is the modular function.

One of the great advantages of the expression (3) is that it is not necessary to find an explicit expression for a maximal invariant which in some cases may be quite a hard problem. Expression (3) is especially useful in cases where G is not specified completely so that it is out of the question to give an explicit expression for a maximal invariant. Yet, even in such cases, (3) may give sufficient information. For instance, when the p_i also depend on an integer m, we may be able to study the asymptotic behavior of (3) as $m \to \infty$ for arbitrary G (within the restrictions imposed on G). An application of this kind to the question of termination with probability one of a certain class of sequential probability ratio tests of composite hypotheses will be made in a future paper.

5. Global cross-sections

A global cross-section gives more but is also harder to come by than a local one, and the theorems guaranteeing the existence of a global cross-section (theorems 6 and 7) are much more restricted in their generality than theorem 1 on the existence of local cross-sections. First we shall deal with the measurability question.

THEOREM 5. *Let Z be a global cross-section; then Z is closed in X and is therefore a Borel set. Let $\mathcal{B}$ be the sigma-field of Borel subsets of Z. Define $t: X \to Z$ by $t(x) = Gx \cap Z$; then $t^{-1}\mathcal{B} = \mathfrak{A}^I$ so that $(Z, \mathcal{B}, t)$ is a natural maximal invariant.*

PROOF. Let x_0 be an arbitrary point of Z. Then Z is a slice at x_0. With x, in definition 1, replaced by x_0, let f be the equivariant map of definition 1. Suppose $x_m \in Z$, $x_m \to x$; then $G_{x_0} = f(x_m) \to f(x)$ so that $f(x) = G_{x_0}$, proving $x \in Z$. Putting $y = f(x)$, $z = t(x)$, it follows from a result of Palais ([9], proposition 2.1.2; see also [12], section 2) that the one-to-one correspondence $x \leftrightarrow (y, z)$ is a homeomorphism. Under this homeomorphism there is a one-to-one correspondence between the Borel sets of X and those of $Y \times Z$, and to the invariant Borel sets of X correspond sets of the form $Y \times B$, $B \in \mathfrak{B}$, in $Y \times Z$, proving $t^{-1}\mathfrak{B} = \mathfrak{A}^I$.

The basic method for finding a global cross-section is to find another group H that also acts on X and such that the combined action of G and H is transitive on X. Then, under certain additional conditions to be specified in theorems 6 and 7, any H-orbit is a global cross-section.

THEOREM 6. *Let G and H be two commuting Lie groups of linear transformations on X, and x_0 a point of X, such that the following conditions are fulfilled:* (i) *G_{x_0} and H_{x_0} are compact;* (ii) *if $g \in G$, $h \in H$, then $gx_0 = hx_0$ only if $gx_0 = hx_0 = x_0$;* (iii) *the dimensions of the orbits Gx_0 and Hx_0 are positive;* (iv) *GH is transitive on X. Put $Y = G/G_{x_0}$, $Z = H/H_{x_0}$, and identify Z with Hx_0. Then Z is a global cross-section, and if the real-valued function p is μ_n-integrable on X, we have*

$$\int_X p(x)\mu_n(dx) = c \int_Z |h|\mu_Z(dz) \int_G p(ghx_0)|g|\mu_G(dg), \tag{4}$$

in which h is any member of H such that $[h] = z$, and the constant c is given by the Radon-Nikodym derivative

$$c = \mu_n(dx)/\mu_Y(dy)\mu_Z(dz) \qquad \text{evaluated at } x_0. \tag{5}$$

PROOF. Let φ_G be the natural map $G \to Y$, and similarly, $\varphi_H\colon H \to Z$. Suppose x has two representations: $x = ghx_0 = g_1h_1x_0$; then, using the commutativity of G and H, we have $g_1^{-1}gx_0 = h^{-1}h_1x_0$. Since $g_1^{-1}g \in G$, $h^{-1}h_1 \in H$, it follows from (ii) that $g_1^{-1}g \in G_{x_0}$, $h^{-1}h_1 \in H_{x_0}$, that is, $g \sim g_1$ and $h \sim h_1$. Consequently, there is a one-to-one correspondence between $Y \times Z$ and X given by $(y, z) \leftrightarrow ghx_0$, where g is any member of $\varphi_G^{-1}y$, h any member of $\varphi_H^{-1}z$.

We shall show now that it is an analytic homeomorphism. It is sufficient to do this in a neighborhood of x_0. Let $\{K_\alpha\}$, α in a finite set of integers, be a basis for the Lie algebra of G such that $K_1x_0, \cdots, K_kx_0$ are linearly independent and $K_\alpha x_0 = 0$ for the remaining α's. Similarly, let $\{L_\beta\}$ be a basis for the Lie algebra of H such that $L_1x_0, \cdots, L_\ell x_0$ are linearly independent and $L_\beta x_0 = 0$ for the remaining β's. By (iii), $k > 0$, $\ell > 0$. If W is any submanifold of X (such as Y, Z, or X itself), we shall denote by W_x the tangent space to W at x. With a slight abuse of notation, any tangent vector at x_0 to a submanifold of X is of the form $\sum v^i\, \partial/\partial x^i$, where the v^i and x^i are the components of vectors v, x, and the differentiations are to be performed at x_0. For convenience of notation, however, we shall identify such a tangent vector with v. With this convention, the tangent space Y_{x_0} at x_0 to $Y = Gx_0$ is spanned by the vectors $K_1x_0, \cdots, K_kx_0$, Z_{x_0} by $L_1x_0, \cdots, L_\ell x_0$, $(GHx_0)_{x_0}$ by $K_1x_0, \cdots, K_kx_0, L_1x_0, \cdots, L_\ell x_0$, while X_{x_0} is all of E^n.

Although by (iv) GHx_0 and X are the same set of points, we have not shown that they are the same analytic manifold, that is, carry the same analytic structure. Actually, we shall only need to know that GHx_0 as an analytic manifold has dimension n. To show this, suppose $\dim GHx_0 = m < n$ (for the following argument I am indebted to R. L. Bishop and N. T. Hamilton). Each element of GH has a neighborhood V small enough that Vx_0 is homeomorphic to an m-cell, so that Vx_0 is a Borel set of μ_n-measure 0 (since $m < n$). The group GH can be covered by a countable family of such neighborhoods V since the topology of GH is the relativized topology of E^{n^2} (see section 1) and has therefore a countable base. Then GHx_0 is covered by a countable family of sets of the form Vx_0. It would follow that $\mu_n(GHx_0) = 0$ which is impossible since $GHx_0 = X$ as a point set, and $\mu_n(X) > 0$.

It was shown above that $\dim GHx_0 = n$. Since $\dim GHx_0 = \dim (GHx_0)_{x_0}$, the vectors $K_1x_0, \cdots, L_\ell x_0$ must span n-space (so that $k + \ell \geq n$). We shall show now that the vectors are actually linearly independent (implying $k + \ell = n$). Suppose the contrary; then there exist $K = \sum_1^k a_iK_i$ and $L = \sum_1^\ell b_jL_j$ with $Kx_0 \neq 0$, $Lx_0 \neq 0$, and $(K - L)x_0 = 0$. For any real t we have then $e^{t(K-L)}x_0 = x_0$, or $e^{-tL}e^{tK}x_0 = x_0$ (making use of the commutativity of G and H), or $e^{tK}x_0 = e^{tL}x_0$. Now $e^{tK} \in G$ and $e^{tL} \in H$, and then it follows from (ii) that $e^{tK}x_0 = e^{tL}x_0 = x_0$ for every t. Using the latter of these two equalities, it follows that $Lx_0 = 0$, which is a contradiction.

We have shown now that $K_1x_0, \cdots, K_kx_0, L_1x_0, \cdots, L_\ell x_0$ is a basis for E^n. Remembering that $K_1x_0, \cdots, K_kx_0$ is a basis for Y_{x_0}, $L_1x_0, \cdots, L_\ell x_0$ for Z_{x_0}, and keeping in mind that $(Y \times Z)_{x_0} = Y_{x_0} \times Z_{x_0}$ and $X_{x_0} = E^n$, we have established that $(Y \times Z)_{x_0}$ and X_{x_0} are linearly isomorphic. It follows then from ([1], proposition 3, p. 80), that $Y \times Z$ and X are locally analytically homeomorphic at x_0, as was to be proved.

With the correspondence $x \leftrightarrow (y, z)$ define f by $f(x) = y$; then f is continuous by the above result, f is equivariant (for G), and $f^{-1}(G_{x_0}) = Z$. Therefore, f can be taken as the function f in definition 1 (with x in definition 1 replaced by x_0).

We conclude that Z is a slice at x_0. But Z also satisfies definition 2, because the orbit of x intersects Z in the unique point hx_0, where h is any member of $\varphi_H^{-1}z$. Therefore, Z is a local cross-section at x_0, and since $GHx_0 = X$, Z is a global cross-section. The proof of (4) and (5) rests on the fact that $\mu_n(dx) = c|g|\mu_Y(dy)|h|\mu_Z(dz)$ (g any member of $\varphi_G^{-1}y$, h any member of $\varphi_H^{-1}z$) and is essentially the same as the proof of theorem 3.

REMARKS. 1. If the conditions of theorem 6 hold for some x_0, they hold for every $x_0 \in X$ so that every H-orbit is a global cross-section for (X, G). Furthermore, the statement of the theorem is symmetric in G and H, so that every G-orbit is a global cross-section for (X, H).

2. If p in theorem 6 is the density with respect to μ_n of a distribution P on $\mathfrak{X}$, then (4) gives the density of P^Z with respect to $\mu_Z(dz)$ as $c|h| \int p(ghx_0)|g|\mu_G(dg)$.

3. In many applications $H_{x_0} = \{e\}$ in which case $Z = H$.

4. To determine c by (5) amounts essentially to differentiation. Alternatively, c can be determined by integrating the right-hand side of (4), with p any manageable function, and setting the result equal to the left-hand side (which equals 1 if p is a probability density).

Some of the examples that follow have also been treated by Karlin [6], using integration over invariant measures on groups and arriving at the same results along a slightly different path.

EXAMPLE 2 (ratio of two variables, noncentral t). Let $n = 2$, $x = (x_1, x_2)'$, $X = \{x: x_2 > 0\}$. Let G consist of the matrices $g = aI_2$, $a > 0$, with $\mu_G(dg) = da/a$ and $|g| = a^2$. Let H be the group of 2×2 triangular matrices with 1's on the diagonal and b above the diagonal, $-\infty < b < \infty$. Then $\mu_H(dh) = db$ and $|h| = 1$. Clearly, G and H commute, and it is easy to check that GH is transitive on X. Choose $x_0 = (0, 1)'$ so that $x = ghx_0 = (ab, a)'$; G_{x_0} and H_{x_0} are trivial, so $Z = H$, and $gx_0 \neq hx_0$, unless $g = h = e$. Therefore, all conditions of theorem 6 are met. We compute c from $x_1 = ab$, $x_2 = a$, so that at $a = 1$, $b = 0$ we have $dx_1 = db$, $dx_2 = da$ so $dx_1dx_2 = \mu_G(dg)\mu_H(dh)$; hence $c = 1$. Substitution into (4) gives $\int p(x)\mu_2(dx) = \int_{-\infty}^{\infty} db \int_0^{\infty} p(ab, a)a\, da$. We observe that a maximal invariant under G is $x_1/x_2 = b$. If x_1 and x_2 are considered random variables, p their joint density with respect to μ_2, then we read off the density of x_1/x_2 at b with respect to μ_1 as $\int_0^{\infty} p(ab, a)a\, da$. In particular, if x_1 is the sample mean, x_2 the sample standard deviation in a sample from a normal distribution, we get an integral for the noncentral t-density.

EXAMPLE 3 (noncentral Wishart). Consider all $k \times n$ matrices $\tilde{x}$ that are of rank k, $k \leq n$, and let x be the kn-vector obtained from $\tilde{x}$ by writing the elements of $\tilde{x}$ in some arbitrary but fixed order (note that the n in our general theory is replaced by kn). Let X be the totality of all such x. Let G correspond to all transformations $\tilde{x} \to \tilde{x}\Omega$, with Ω an $n \times n$ orthogonal transformation. Haar measure on G can be chosen normalized so that $\mu_G(G) = 1$. Furthermore, $|g| = 1$. Choose H to be the group corresponding to all transformations $\tilde{x} \to T\tilde{x}$, with T a $k \times k$ lower triangular matrix with positive diagonal elements, then $|h| = |T|^n$. For left Haar measure on H we may take $\mu_H(dh) = d(TT')/|TT'|^{(k+1)/2}$ (here $d(TT')$ is short for the product of differentials of the elements on, and on one side of, the diagonal of the symmetric matrix TT'). Choose x_0 to correspond to $\tilde{x}_0 = (I, 0)$, where $I = I_k$ and 0 denotes a $k \times (n - k)$ matrix of 0's. We have then $\tilde{x} = T\tilde{x}_0\Omega$. All conditions of theorem 6 can be verified to hold. In particular, H_{x_0} is trivial so that $Z = H$. Since there is a one-to-one correspondence between $h = T$ and TT', we may take as maximal invariant $TT' = \tilde{x}\tilde{x}'$. This is the Wishart matrix if the columns of $\tilde{x}$ form a sample from a multivariate normal distribution with 0 mean vector, and (4) provides an easy way to evaluate the Wishart density. If the columns of $\tilde{x}$ are independently multivariate normal with common covariance matrix but arbitrary means, then (4) yields an integral for the noncentral Wishart density.

In order to compute c using (5) note that Y corresponds to all $k \times n$ matrices which have orthonormal rows, that is, Y is the Stiefel manifold of k-frames in n-space [5]. Writing an element of this manifold as a $k \times n$ matrix A, with rows a_i', an invariant differential form on Y is given by James ([5], (4.39)), as

$$\prod_{j=1}^{n-k} \prod_{i=1}^{k} b_j' \, da_i \prod_{i<j\leq k} a_j' \, da_i, \tag{6}$$

where the b_j form with the a_i an orthonormal set. At $\tilde{x}_0$ this form reduces to $\prod da_{ij}$ where the product is over all (i, j) with $i < j \leq k$ and all (i, j) with $j > k$. It should also be noted that in order that $A + dA$ has orthonormal rows, we must have $dA = (d\Sigma, dC)$ with $d\Sigma$ skew-symmetric and dC arbitrary. Taking the normalization factor into account, taken from ([5], (5.10)), we find at x_0

$$\mu_Y(dy) = 2^{-k}\left[\prod_{\nu=n-k+1}^{n} \pi^{-\nu/2}\Gamma(\nu/2)\right]\prod da_{ij}. \tag{7}$$

In order to find $\mu_Z(dz) = \mu_H(dh)$, let dT be a lower triangular matrix with elements dt_{ij}, $i \geq j$. Then at x_0 (where $T = I$),

$$\mu_H(dh) = \prod_{i\geq j} (dT + dT')_{ij} = 2^k \prod_{i\geq j} dt_{ij}. \tag{8}$$

Finally, we obtain $\mu_n(dx)$ in terms of the dt_{ij} and da_{ij} as follows: $\tilde{x}_0 + d\tilde{x} = (I + dT)(I + d\Sigma, dC)$, so that $d\tilde{x} = (dT + d\Sigma, dC)$, omitting higher order differentials which will not contribute to the exterior differential form. We get

$$\mu_n(dx) = \prod dx_i = \prod d\tilde{x}_{ij} = \prod dt_{ii} \prod_{i>j} (dt_{ij} + d\sigma_{ij}) \prod_{i<j} d\sigma_{ij} \prod dc_{ij}. \tag{9}$$

Now

$$\prod_{i>j} (dt_{ij} + d\sigma_{ij}) \prod_{i<j} d\sigma_{ij} = \prod_{i>j} dt_{ij} \prod_{i<j} d\sigma_{ij} \tag{10}$$

since $d\sigma_{ij} = -d\sigma_{ji}$ and any exterior differential form with a repeated differential is 0. Therefore,

$$\mu_n(dx) = \prod_{i\geq j} dt_{ij} \prod_{i<j} d\sigma_{ij} \prod c_{ij} = \prod_{i\geq j} dt_{ij} \prod da_{ij}. \tag{11}$$

Substituting all this into (5) yields $c = \prod_{\nu=n-k+1}^{n} \pi^{\nu/2}[\Gamma(\nu/2)]^{-1}$.

The value of c in example 3 can of course be obtained much more simply by relating it to the known multiplicative constant of the central Wishart density. The point of the above computation is to illustrate how (5) can be used directly to compute c, even in a fairly complicated case. In such cases the use of exterior differential forms may be of some practical advantage over the type of computation that uses Jacobians.

A special case of example 3 arises when $k = n$. In that case (4) provides a decomposition of an integral over all $k \times k$ nonsingular matrices into an integration over the orthogonal group and an integration over the identity component of the lower triangular group. This decomposition was also derived by Stein in [10] by a different method.

There is no guarantee that in every problem one is successful in finding a group H such that the hypotheses of theorem 6 are satisfied. The following theorem gives a result similar to (4) (but not as easy to apply) under weaker conditions. Specifically, it is no longer required that G and H commute.

THEOREM 7. *Assume the same conditions as in theorem 6, except that the commutativity of G and H is replaced by the following conditions: G is closed in GH, for each $h \in H$, $hGh^{-1} = G$ (that is, G is a closed normal subgroup of GH), and $hG_{x_0}h^{-1} = G_{x_0}$. Then Z is a global cross-section, and for any μ_n-integrable p we have*

$$\int p(x)\mu_n(dx) = c \int \psi(h)|h|\mu_Z(dz) \int p(ghx_0)|g|\mu_G(dg) \tag{12}$$

in which c is given by (5) *and $\psi(h)$ is the Radon-Nikodym derivative*

$$\psi(h) = \mu_Y(dy)/\mu_Y(hdyh^{-1}) \qquad \text{evaluated at} \quad y = G_{x_0}. \tag{13}$$

PROOF. It can be checked algebraically as in the proof of theorem 6 that there is a one-to-one correspondence $x \leftrightarrow (y, z)$. It is still true that a G-orbit transforms into a G-orbit under the transformation $x \rightarrow hx$, but it is no longer true that y remains constant under this transformation. The hypothesis $hGh^{-1} = G$ implies that each $h \in H$ acts on G by $g \rightarrow hgh^{-1}$, and the hypothesis $hG_{x_0}h^{-1} = G_{x_0}$ implies that each h even acts on G/G_{x_0}, by $h[g]h^{-1} = [hgh^{-1}]$. It is immediately verified that to $x \rightarrow hx$ corresponds $(y, z) \rightarrow (hyh^{-1}, hz)$. The proof of theorem 7 is the same as the proof of theorem 6, except for the factor $\psi(h)$ and the proof of the linear independence of the vectors $K_1x_0, \cdots, L_tx_0$.

To establish (13), let dy and dz be "small" neighborhoods of $[e]$ in Y, Z, respectively. Then the volume element is $\mu_n(dx) = c\mu_Y(dy)\mu_Z(dz)$. Under the transformation $x \rightarrow hx$, $\mu_n(hdx) = |h|\mu_n(dx) = c|h|\mu_Y(dy)\mu_Z(dz)$. On the other hand, under this transformation $dy \rightarrow hdyh^{-1}$, $dz \rightarrow hdz$, so $\mu_n(hdx) = c|h|\psi(h)\mu_Y(hdyh^{-1})\mu_Z(dz)$, where we have made use of the left invariance of μ_Z. Equating the two expressions for $\mu_n(hdx)$ yields (13).

In the part of the proof of theorem 6 where the linear independence of $K_1x_0, \cdots, L_tx_0$ was established we used the fact that $e^{t(K-L)} = e^{-tL}e^{tK}$. This is no longer true in general if G and H do not commute. However, the proof goes through in exactly the same way after we have shown that $e^{t(K-L)} = e^{-tL}g$ for some $g \in G$ (g may depend on t). In order to establish this fact, denote by $\Lambda(G)$ the Lie algebra of G; $\Lambda(H)$ and $\Lambda(GH)$ are similarly defined. Then $\Lambda(GH)$ consists of all $K + L$, $K \in \Lambda(G)$, $L \in \Lambda(H)$. Let φ be the natural map: $GH \rightarrow GH/G$, that is, $\varphi(hg) = hG$. The differential $d\varphi$ maps $\Lambda(GH)$ onto $\Lambda(GH)/\Lambda(G)$ ([1], p. 115, proposition 1; [2], p. 132, theorem 6.6.4). More specifically, if $M \in \Lambda(GH)$, then $d\varphi(M)$ depends on M only through its residue class mod $\Lambda(G)$ ([1], pp. 114–115). That is, if $K \in \Lambda(G)$, then

$$d\varphi(M + K) = d\varphi(M). \tag{14}$$

Furthermore, if $M \in \Lambda(GH)$, we have (see [1], p. 118, proposition 1; [2], p. 129, (26))

$$\varphi(e^M) = e^{d\varphi(M)}. \tag{15}$$

Taking in (14) $M = L \in \Lambda(H)$ and in (15) first $M = K + L$ and then $M = L$, we obtain

$$\varphi(e^{K+L}) = \varphi(e^L). \tag{16}$$

Now $e^L \in H$, so $\varphi(e^L) = e^L G$. Substituting this into (16) we get $\varphi(e^{K+L}) = e^L G$, so that $e^{K+L} = e^L g$ for some g. If t is any real number, replacing K by tK, L by $-tL$, we have $e^{t(K-L)} = e^{-tL} g$ for some $g \in G$, as was to be shown. This concludes the proof of theorem 7.

EXAMPLE 4. Consider all $(p + q) \times (p + q)$ positive definite matrices $\tilde{x}$, partitioned as

$$\tilde{x} = \begin{bmatrix} S_{11} & S_{12} \\ S_{21} & S_{22} \end{bmatrix} \tag{17}$$

in which S_{11} is $p \times p$ positive definite, S_{22} is $q \times q$ positive definite, $S_{12} = S'_{21}$ is $p \times q$, and let x be the corresponding n-vector, where $n = p(p + 1)/2 + pq + q(q + 1)/2$. Let G correspond to the transformations $\tilde{x} \to \tilde{C}\tilde{x}\tilde{C}'$, where

$$\tilde{C} = \begin{bmatrix} I_p & 0 \\ C & I_q \end{bmatrix} \tag{18}$$

and C runs through all $q \times p$ matrices. We can take $\mu_G(dg) = \prod dC_{ij}$ where the C_{ij} are the elements of C. If we take x_0 corresponding to $\tilde{x}_0 = I_{p+q}$, then G_{x_0} is trivial, so $Y = G$. Define H by the transformations $\tilde{x} \to \tilde{D}\tilde{x}\tilde{D}'$, where $\tilde{D} = \operatorname{diag}(A, B)$, and A runs through $GL(p, R)$, B through $GL(q, R)$. All conditions of theorem 7 can be verified to hold. We shall pursue this example only to the extent of computing $\psi(h)$. For notational economy, denote by $(C)^*$ a $(p + q) \times (p + q)$ matrix that has C as its last q rows and first p columns, and zeros otherwise. Then if $dy = dg$ corresponds to $(dC)^*$, and h to $\operatorname{diag}(A, B)$, $hdgh^{-1}$ corresponds to $\operatorname{diag}(A, B)(dC)^* \operatorname{diag}(A^{-1}, B^{-1}) = (BdCA^{-1})^*$. We have then $\mu_G(dg) = \prod dC_{ij}$ and $\mu_G(hdgh^{-1}) = \prod (BdCA^{-1})_{ij} = |B|^p|A|^{-q} \prod dC_{ij}$. Thus, $\psi(h) = |A|^q|B|^{-p}$.

REFERENCES

[1] C. CHEVALLEY, *Theory of Lie Groups* I, Princeton, Princeton University Press, 1946.
[2] P. M. COHN, *Lie Groups*, Cambridge, Cambridge University Press, 1957.
[3] N. GIRI, "On the likelihood ratio test of a normal multivariate testing problem," *Amer. Math. Statist.*, Vol. 35 (1964), pp. 181–189.
[4] A. GLEASON, "Spaces with a compact Lie group of transformations," *Proc. Amer. Math. Soc.*, Vol. 1 (1950), pp. 35–43.
[5] A. T. JAMES, "Normal multivariate analysis and the orthogonal group," *Ann. Math. Statist.*, Vol. 25 (1954), pp. 40–75, and many later papers, mostly integration over the orthogonal group.
[6] S. KARLIN, "Notes on multivariate analysis," Stanford University, unpublished.
[7] S. KARLIN and D. TRUAX, "Slippage problems," *Ann. Math. Statist.*, Vol. 31 (1960), pp. 296–324.
[8] D. MONTGOMERY and L. ZIPPIN, *Topological Transformation Groups*, New York, Interscience, 1955.
[9] R. S. PALAIS, "On the existence of slices for actions of noncompact Lie groups," *Ann. of Math.*, Vol. 73 (1961), pp. 295–323.
[10] C. M. STEIN, "Notes on multivariate analysis," Stanford University, unpublished.
[11] ———, "Some problems in multivariate analysis," Part I, Technical Report No. 6, Department of Statistics, Stanford University, 1956.
[12] R. A. WIJSMAN, "Existence of local cross-sections in linear Cartan G-spaces under the action of non-compact groups," *Proc. Amer. Math. Soc.*, Vol. 17 (1966), pp. 295–301.

ASYMPTOTICALLY POINTWISE OPTIMAL PROCEDURES IN SEQUENTIAL ANALYSIS

PETER J. BICKEL
and
JOSEPH A. YAHAV
UNIVERSITY OF CALIFORNIA, BERKELEY

1. Introduction

After sequential analysis was developed by Wald in the forties [5], Arrow, Blackwell, and Girshick [1] considered the Bayes problem and proved the existence of Bayes solutions. The difficulties involved in computing explicitly the Bayes solutions led Wald [6] to introduce asymptotic sequential analysis in estimation. Asymptotic in his sense, as for all subsequent authors, refers to the limiting behavior of the optimal solution as the cost of observation tends to zero. Chernoff [2] investigated the asymptotic properties of sequential testing. The testing theory was developed further by Schwarz [4] and generalized by Kiefer and Sacks [3]. This paper approaches the asymptotic theory from a slightly different point of view. We introduce the concept of "asymptotic pointwise optimality," and we construct procedures that are "asymptotically pointwise optimal" (**A.P.O.**) for certain rates of convergence [as $n \to \infty$] of the a posteriori risk. The rates of convergence that we consider apply under some regularity conditions to statistical testing and estimation with quadratic loss.

2. Pointwise optimality

Let $\{Y_n, n \geq 1\}$ be a sequence of random variables defined on a probability space (Ω, F, P) where Y_n is F_n measurable and $F_n \subset F_{n+1} \cdots \subset F$ for $n \geq 1$. We assume the following two conditions:

$$P(Y_n > 0) = 1, \tag{2.1}$$

$$Y_n \to 0, \quad \text{a.s.} \tag{2.2}$$

Define

$$X_n(c) = Y_n + nc \qquad \text{for} \quad c > 0. \tag{2.3}$$

Prepared with the partial support of the Ford Foundation Grant given to the School of Business Administration and administered by the Center for Research in Management Science, University of California, Berkeley.

Let T be the class of all stopping times defined on the σ-fields F_n. We say that $s \in T$ is "pointwise optimal" if

$$P\left[\frac{X_s(c)}{X_t(c)} \leq 1\right] = 1 \qquad \text{for all} \quad t \in T. \tag{2.4}$$

Unfortunately, such s's usually do not exist except in essentially deterministic cases. Let us consider two examples of such situations:

$$Y_n = \frac{V}{n}, \qquad \infty > V > 0, \tag{2.5}$$

$$\frac{\log Y_n}{n} = U, \qquad -\infty < U < 0. \tag{2.6}$$

In these examples one easily sees that the pointwise optimal rule is given by the following:

Example 1: stop as soon as $\frac{V}{n(n-1)} \leq c$;

Example 2: stop as soon as $e^{nU} \leq \frac{c}{(1 - e^U)}$.

These examples will play a role in theorem 2.1. In nondeterministic cases one might hope that, under some conditions, we can get **A.P.O.** procedures. Let us define these more formally.

Abusing our notation, in a fashion long used in large sample theory, use the words "stopping rule" to also denote a function from $(0, \infty)$ to T, say $t(\cdot)$, $c \in (0, \infty)$, $t(c) \in T$. Now in analogy to our previous definition we say $s(\cdot)$ is **A.P.O.** if for any other $t(\cdot)$,

$$\limsup_{c \to 0} \frac{X_{s(c)}(c)}{X_{t(c)}(c)} \leq 1, \qquad \text{a.s.} \tag{2.7}$$

Consideration of the deterministic case naturally leads us to hope for asymptotically pointwise optimal solutions in situations where the rate of convergence of Y_n stabilizes. This hope is fulfilled in the following theorem.

THEOREM 2.1. (i) *If condition* (2.1) *holds and* $nY_n \to V$, *a.s. where* V *is a random variable such that* $P(V > 0) = 1$, *the stopping rule, which is determined by "stop the first time that* $(Y_n/n) \leq c$," *is* **A.P.O.**

(ii) *If condition* (2.1) *holds and* $(\log Y_n/n) \to U$, *a.s. where* U *is a random variable and* $P(U < 0) = 1$, *then the rules* (ii)a, *"stop the first time* $Y_n \leq c$" *and* (ii)b, *"stop the first time* $Y_n(1 - Y_n^{1/n}) \leq c$" *are* **A.P.O.**

PROOF. Let $s_1(\cdot)$ be the stopping time defined by rule (i). Let $t(\cdot)$ be any other rule. Then

$$\frac{X_{s_1}}{X_t} = \frac{Y_{s_1} + cs_1}{Y_t + ct} = \frac{\frac{Y_{s_1}}{s_1 c} + 1}{\frac{y_t}{s_1 c} + \frac{t}{s_1}} \leq \frac{2}{\frac{Y_t}{s_1 c} + \frac{t}{s_1}}. \tag{2.8}$$

It suffices to show that $\liminf_{c \to 0} (Y_t/s_1c) + t/s_1 \geq 2$, a.s., but this follows upon remarking that $(x + 1/x) \geq 2$ for $x \geq 0$ and applying the following lemma.

LEMMA 2.1. *If* $c_n \to 0$, $t(c_n)/s_1(c_n) \to x \geq 0$, $t(c_n)$ *converges (possibly to* $+\infty$) *with probability* 1, *then* $\liminf_{c_n \to 0} (Y_{t(c_n)}/c_n s_1(c_n)) \to 1/x$, *a.s.*

PROOF OF LEMMA. It follows from the assumptions of the theorem and the definition of $s_1(c)$ that

$$P[\lim_{c \to 0} s_1(c) = \infty] = 1. \tag{2.9}$$

Suppose first that $P[\lim_{c_n} t(c_n) < \infty] > 0$. On this set $\liminf_{c_n} Y_{t(c_n)} > 0$, and our lemma will follow in this case if we show that $cs_1(c) \to 0$, a.s. as $c \to 0$. We in fact will show the stronger

$$cs_1^2(c) \to V, \qquad \text{a.s.} \tag{2.10}$$

This follows immediately from the inequalities

$$\frac{Y_{s_1}}{s_1} \leq c < \frac{Y_{(s_1-1)}}{(s_1 - 1)}, \tag{2.11}$$

$$s_1 Y_{s_1} \leq s_1^2 c \leq \frac{s_1^2}{(s_1 - 1)^2}(s_1 - 1) Y_{(s_1-1)}, \tag{2.12}$$

and (2.9).

The general case of the lemma, on the set $[t(c) \to \infty]$ is a consequence of the identity

$$\frac{Y_t}{s_1 c} = \frac{s_1}{t} \frac{t Y_t}{c s_1^2} \tag{2.13}$$

and our assumptions.

We prove case (ii)a; case (ii)b follows similarly. Let $s_2(\cdot)$ be the rule defined by (ii)a, $t(\cdot)$ be any other stopping rule. Again we have,

$$\frac{X_{s_2}}{X_t} = \frac{\dfrac{Y_{s_2}}{s_2 c} + 1}{\dfrac{Y_t}{s_2 c} + \dfrac{t}{s_2}}, \quad \text{and} \quad s_2 \to \infty, \qquad \text{a.s.} \tag{2.14}$$

But then, $Y_{s_2}/s_2 c \leq 1/s_2 \to 0$, a.s. In an analogous fashion to lemma 2.1, we use lemma 2.2.

LEMMA 2.2. *If* $c_n \to 0$, $t(c_n)$ *converges a.s. (possibly to* $+\infty$), $t(c_n)/s_2(c_n) \to x < 1$, *then*, $Y_{t(c_n)}/cs_2(c_n) \to \infty$.

PROOF. We prove first that

$$\frac{s_2(c)}{|\log c|} \to \frac{1}{|U|}, \qquad \text{a.s.} \tag{2.15}$$

This is a consequence of inequalities,

$$Y_{s_2} \leq c < Y_{(s_2-1)}, \tag{2.16}$$

$$\frac{\log Y_{s_2}}{s_2} \leq \frac{\log c}{s_2} < \frac{\log Y_{(s_2-1)}}{(s_2 - 1)} \frac{(s_2 - 1)}{s}. \tag{2.17}$$

Now suppose $t(c_n) \to \infty$, a.s. Then,

$$\frac{\log y_t}{cs_2} = t\left(\frac{\log Y_t}{t} - \frac{s_2}{t}\left(\frac{\log s_2}{s_2} + \frac{\log c}{s_2}\right)\right). \tag{2.18}$$

Now,

$$\frac{\log Y_t}{t} \to U, \qquad \text{a.s.,} \tag{2.19}$$

$$\frac{\log c}{s_2} \to U, \qquad \text{a.s.,} \tag{2.20}$$

and

$$\frac{s(c_n)}{t(c_n)} \to \frac{1}{x} > 1 \tag{2.21}$$

by hypothesis and (2.15). Since $U < 0$, the result follows.

The theorem is proved.

COROLLARY 2.1. *Let $N(c)$ be defined as any solution of $X_{N(c)}(c) = \inf_n X_n(c)$. Then, in both cases of theorem* 2.1,

$$\lim_{c \to 0} \frac{X_{s(c)}(c)}{X_{N(c)}(c)} = 1. \tag{2.22}$$

PROOF. Note that in the proof of the theorem, no use was made of the fact that the $t(c)$ is a stopping time.

REMARK. In both cases it may readily be seen that $s_i(c)$ is strictly better than $t(c)$ if $t(c)/s(c) \not\to 1$, a.s. However, although in case (i) the converse holds, that is, $t(c)$ is also asymptotically pointwise optimal if $s(c)/t(c) \to 1$, a.s., this is not true necessarily in case (ii). However, as the existence of rule (ii)b indicates, here too there are many **A.P.O.** rules. We shall see more in the conclusion.

3. Sequential estimation with quadratic loss

The main theorem of this section states that for the one parameter exponential family (Koopman-Darmois, K-D), Bayes estimation with quadratic loss satisfies condition (i) of theorem 2.1, and therefore the rule given in theorem 2.1 (i) is **A.P.O.** This result can in fact be generalized to an arbitrary family of distributions under some regularity conditions. A theorem of this type will be stated at the end of this section. We give the proof only for the K-D family both for ease of exposition and because we hope to weaken the regularity conditions of our general theorem. Let $\{Z_i, i \geq 1\}$ be a sequence of independent identically distributed random variables having density function $f_\theta(z) = e^{q(\theta)T(z) - b(\theta)}$ with respect to some σ-finite nondegenerate measure μ on the real line endowed with the Borel σ-field where $q(\theta)$ and $T(z)$ are real-valued.

We let Θ, the parameter space, be the natural range of θ; that is,

$$\Theta = \left\{\theta\colon \int_{-\infty}^{\infty} e^{q(\theta)T(z)} \mu(dz) < \infty\right\}$$

and endow it also with the Borel σ-field and the usual topology.

It follows that Θ is an interval, finite or infinite, and we assume that (i) $q(\theta)$ possesses at least two continuous derivatives in the interior of Θ and (ii), $q'(\theta) \neq 0$.

It is known that under these conditions the following propositions hold:

(A) $E_\theta[T(Z_1)] = \dfrac{b'(\theta)}{q'(\theta)}$.

(B) If $q(\theta) = \theta$, then $b''(\theta) = \text{var}_\theta\,[T(Z_1)]$.

(C) Let $\Phi(\theta, z) = \log f_\theta(z)$ and $A(\theta) = E_\theta\left[\left(\dfrac{\delta\Phi(\theta, Z_1)}{\delta\theta}\right)\right]$; then $0 < A(\theta) = [q'(\theta)]^2 \text{var}_\theta\,[T(Z_1)] < \infty$;

(D) For θ_0 in the interior of Θ, the equation $\sum_{i=1}^{n} \dfrac{\delta\Phi(\theta, Z_i)}{\delta\theta} = 0$

has eventually a unique solution $\hat{\theta}_n$, the maximum likelihood estimate, and $\hat{\theta}_n \to \theta_0$, a.s. P_{θ_0} where P_{θ_0} is the measure induced on the space of all real sequences $\{z_1, z_2, \cdots\}$ by the density $f_{\theta_0}(z)$.

Let ν be a probability measure on Θ which has a continuous bounded density Ψ with respect to Lebesgue measure such that $\int \theta^2\Psi(\theta)\, d\theta < \infty$. Consider the problem of estimating θ sequentially, where the loss on taking n observations and deciding $\theta = d$ is given by $nc + (d - \theta_0)^2$ when θ_0 is the true value of the parameter. The overall risk, $R(\tilde{\theta}, t)$, for a sequential procedure consisting of a stopping rule t and estimator $\tilde{\theta}(Z_1, \cdots, Z_t)$ is then given by,

$$R(\tilde{\theta}, t) = cE(t) + E[(\tilde{\theta}(Z_1, \cdots, Z_t) - \theta)^2]. \tag{3.1}$$

It follows from the results of Arrow, Blackwell, and Girshick that whatever be the choice of t, the optimal estimate given t is the conditional expectation of θ given the past, $E[\theta|Z_1, \cdots, Z_t] = \tilde{\theta}_B$. Hence, finding optimal procedures for the sequential problem is equivalent to constructing optimal stopping rules for the sequence $\{X_n\}$ where $X_n = Y_n + nc$ and

$$Y_n = E[(\theta - \tilde{\theta}_B)^2|Z_1, \cdots, Z_n] = \text{var}\,(\theta|Z_1, \cdots, Z_n). \tag{3.2}$$

In order to find an **A.P.O.** rule by the method of theorem 2.1 (i), we have to show that $P(Y_n > 0) = 1$ and $nY_n \to V$, a.s. where $P(V > 0) = 1$.

THEOREM 3.1. *For the K-D family obeying assumptions* (i) *and* (ii), *we have* $P(Y_n > 0) = 1$ *and*

$$nY_n \to 1/A(\theta). \tag{3.3}$$

PROOF. Since the a posteriori density exists with probability one, the conditional variance of θ is positive with probability one.

To show (3.3) we will establish

$$P_{\theta_0}\{nE[(\theta - \tilde{\theta}_n)^2|Z_1, \cdots, Z_n] \to 1/A(\theta_0)\} = 1 \tag{3.4}$$

and

$$n^{1/2}(E[\theta|Z_1, \cdots, Z_n] - \hat{\theta}_n) \to 0 \tag{3.5}$$

with probability one, where $\hat{\theta}_n$ is the maximum likelihood esitmate of θ. The theorem readily follows from (3.4), (3.5), and the identity,

$$\begin{aligned}(3.6)\qquad \operatorname{var}(\theta|Z_1, \cdots, Z_n) &= E[(\theta - \hat{\theta}_n)^2|Z_1, \cdots, Z_n] \\ &\quad - [E(\theta|Z_1, \cdots, Z_n) - \hat{\theta}_n]^2.\end{aligned}$$

Let us define, $\Psi^*(t|Z_1, \cdots, Z_n)$ to be the a posteriori density of $n^{1/2}(\theta - \hat{\theta}_n)$. Thus,

$$\begin{aligned}(3.7)\qquad \Psi^*(t|Z_1, \cdots, Z_n) &= \exp\left\{\sum_{i=1}^n \Phi\left(\hat{\theta}_n + \frac{t}{\sqrt{n}}, Z_i\right)\right\} \\ &\quad \cdot \Psi\left(\frac{t}{\sqrt{n}} + \hat{\theta}_n\right)\left[n^{1/2}\int_{-\infty}^{\infty} \exp\left\{\sum_{i=1}^n \Phi(s, Z_i)\right\}\Psi(s)\, ds\right]^{-1}.\end{aligned}$$

Equations (3.4) and (3.5) follow easily from

$$(3.8)\qquad P_{\theta_0}\left[\int_{-\infty}^{\infty} |t|^i|\Psi^*(t|Z_1, \cdots, Z_n) - \sqrt{A(\theta_0)}\phi(t\sqrt{A(\theta_0)})|\, dt \to 0\right] = 1$$

for $i = 1, 2$, where $\phi(x)$ is the standard normal density.

Define the random quantity

$$(3.9)\qquad \nu_n(t) = \exp\left\{\sum_{i=1}^n \Phi\left[\left(\hat{\theta}_n + \frac{t}{\sqrt{n}} Z_i\right) - \phi(\hat{\theta}_n, Z_i)\right]\right\}.$$

To prove (3.8), it suffices to show

$$(3.10)\qquad \int_{-\infty}^{\infty} |t|^i|\nu_n(t) - \sqrt{2\pi}\,\phi(\sqrt{A(\theta_0)}\,t)|\Psi\left(\hat{\theta}_n + \frac{t}{\sqrt{n}}\right) dt \to 0, \qquad \text{a.s. } P_\theta$$

for $i = 0, 1, 2$. To see this, note that by the case $i = 0$ we would have

$$\begin{aligned}(3.11)\qquad &\int_{-\infty}^{\infty} \nu_n(t)\Psi\left(\hat{\theta}_n + \frac{t}{\sqrt{n}}\right) dt \\ &\quad - \int_{-\infty}^{\infty} \sqrt{2\pi}\,\phi(\sqrt{A(\theta_0)}\,t)\Psi\left(\hat{\theta}_n + \frac{t}{\sqrt{n}}\right) dt \to 0.\end{aligned}$$

Now by the dominated convergence theorem, the boundedness and continuity of Ψ, and the consistency of $\hat{\theta}_n$, we have

$$(3.12)\qquad \int_{-\infty}^{\infty} \sqrt{2\pi}\,\phi(\sqrt{A(\theta_0)}\,t)\Psi\left(\hat{\theta}_n + \frac{t}{\sqrt{n}}\right) dt \to \frac{\Psi(\theta_0)\sqrt{2\pi}}{\sqrt{A(\theta_0)}}.$$

Let

$$(3.13)\qquad c_n = \int_{-\infty}^{\infty} \exp\left\{\sum_{i=1}^n \left(\Phi\left(\frac{s}{\sqrt{n}} + \hat{\theta}_n, Z_i\right) - \Phi(\hat{\theta}_n, Z_i)\right)\right\}\Psi\left(\frac{s}{\sqrt{n}} + \hat{\theta}_n\right) ds.$$

Then,

$$(3.14)\qquad \Psi^*(t|Z_1, \cdots, Z_n) = \frac{\nu_n(t)\Psi(\hat{\theta}_n + t/\sqrt{n})}{c_n},$$

and since Ψ^* is a probability density, we have

$$c_n = \int_{-\infty}^{\infty} \nu_n(t)\Psi\left(\hat{\theta}_n + \frac{t}{\sqrt{n}}\right) dt. \tag{3.15}$$

Therefore,

$$c_n \to \frac{\Psi(\theta_0)\sqrt{2\pi}}{\sqrt{A(\theta_0)}}, \tag{3.16}$$

and the sufficiency of (3.10) for our result is clear.

Write (3.10) as,

$$\int_{|t|<\delta^*\sqrt{n}} \Psi\left(\hat{\theta}_n + \frac{t}{\sqrt{n}}\right) |t|^i |\nu_n(t) - \sqrt{2\pi}\,\phi(\sqrt{A(\theta_0)}\,t)|\,dt \tag{3.17}$$
$$+ \int_{|t|\geq\delta^*\sqrt{n}} \Psi\left(\hat{\theta}_n + \frac{t}{\sqrt{n}}\right) |t|^i |\nu_n(t) - \sqrt{2\pi}\,\phi(\sqrt{A(\theta_0)}\,t)|\,dt.$$

We first establish the following lemma.

LEMMA 3.1. *Under the above conditions,*

$$A_n = \int_{|t|\geq\delta^*\sqrt{n}} \Psi\left(\hat{\theta}_n + \frac{t}{\sqrt{n}}\right) |t|^i \nu_n(t)\,dt \to 0, \qquad \text{a.s. } P_{\theta_0}, \tag{3.18}$$

for $i = 0, 1, 2,$ *and all* $\delta^* > 0$.

PROOF. We change variables to $y = \frac{t}{\sqrt{n}}$. Then,

$$A_n = n^{\frac{i+1}{2}} \int_{|y|\geq\delta^*} |y|^i \exp \sum_{i=1}^{n} \{\Phi(y + \hat{\theta}_n, Z_i) - \Phi(\hat{\theta}_n, Z_i)\}\Psi(\hat{\theta}_n + y)\,dy. \tag{3.19}$$

Define,

$$H_n(y) = [q(\hat{\theta}_n + y) - q(\hat{\theta}_n)]\frac{1}{n}\sum_{i=1}^{n} T(Z_i) - [b(\hat{\theta}_n + y) - b(\hat{\theta}_n)]. \tag{3.20}$$

Then, in our case, (3.14) reduces to

$$A_n = n^{\frac{i+1}{2}} \int_{|y|\geq\delta^*} |y|^i \exp\{nH_n(y)\}\Psi(y + \hat{\theta}_n)\,dy. \tag{3.21}$$

By (D), for n sufficiently large the equation $H_n'(y) = 0$ has a unique solution given by $y = 0$, and moreover, 0 is then the unique local maximum of H_n.

Therefore we may conclude that

$$\sup_{|y|\geq\delta^*} H_n(y) = \max\,(H_n(\delta^*), \qquad H_n(-\delta^*)) \leq -\text{M} < 0, \tag{3.22}$$

eventually. Therefore,

$$A_n \leq n^{\frac{i+1}{2}} \exp(-Mn) \int_{-\infty}^{\infty} |y|^i \Psi(y + \hat{\theta}_n)\,dy \to 0, \qquad \text{a.s. } P_\theta, \tag{3.23}$$

since $\hat{\theta}_n$ are bounded a.s., which proves the lemma.

From lemma 3.1, the boundedness of Ψ, and the well-known properties of the normal distribution, it follows that

$$\int_{|t| \geq \delta^* \sqrt{n}} \Psi\left(\hat{\theta}_n + \frac{t}{\sqrt{n}}\right) |t|^i |\nu_n(t) - \sqrt{2\pi}\, \phi(\sqrt{A(\theta_0)}\, t)|\, dt \to 0, \qquad \text{a.s. } P_{\theta_0}. \tag{3.24}$$

We finish the theorem with lemma 3.2.

LEMMA 3.2. *Under the above conditions, there exists a* $\delta^* > 0$ *such that*

$$B_n = \int_{|t| < \delta^* \sqrt{n}} |\nu_n(t) - \sqrt{2\pi}\, \phi(\sqrt{A(\theta_0)}\, t)|\, dt \to 0, \qquad \text{a.s. } P_{\theta_0}. \tag{3.25}$$

PROOF. We expand $\log \nu_n(t)$ formally to get

$$\log \nu_n(t) = \sum_{i=1}^{n} \left\{\frac{\delta \Phi}{\delta \theta} (\hat{\theta}_n, Z_i) \frac{t}{\sqrt{n}} + \frac{\delta^2 \Phi}{\delta \theta^2} (\theta_n^*(t, Z_i), Z_i)) \frac{t^2}{2n}\right\} \tag{3.26}$$

where $\theta_n^*(t, Z_i)$ lies between $\hat{\theta}_n$ and $\hat{\theta}_n + t/\sqrt{n}$.

Of course, $\sum_{i=1}^{n} \delta\Phi/\delta\theta(\hat{\theta}_n, Z_i) = 0$ whenever this expression is valid. In our case, (3.26) is valid and simplifies to

$$\log \nu_n(t) = \frac{t^2}{2}\left\{q''(\theta_n^*(t)) \frac{1}{n} \sum_{i=1}^{n} T(Z_i) - b''(\theta_n^*(t))\right\}. \tag{3.27}$$

Choose $\epsilon > 0$ so that $3\epsilon < A(\theta_0)$. Then, by the continuity of q'', there exists a $\delta^*(\epsilon)$ so that

$$|q''(s) - q''(\theta_0)| < \frac{\epsilon |q'(\theta_0)|}{2|b'(\theta_0)|} \tag{3.28}$$

and $|b''(s) - b''(\theta_0)| < \epsilon$, for $|s - \theta_0| \leq \delta^*(\epsilon)$.

On the other hand, with probability one for n sufficiently large,

$$\left|\frac{1}{n} \sum_{i=1}^{n} T(Z_i) - \frac{b'(\theta_0)}{q'(\theta_0)}\right| < \frac{\epsilon |q'(\theta_0)|}{2|b'(\theta_0)|}, \tag{3.29}$$

and therefore, for such n, $|t/\sqrt{n}| \leq \delta^*(\epsilon)$, we have

$$\left|\log \nu_n(t) - \frac{t^2}{2}\left\{q''(\theta_0) \frac{b'(\theta_0)}{q'(\theta_0)} - b''(\theta_0)\right\}\right| < 3\epsilon. \tag{3.30}$$

But,

$$q''(\theta_0) \frac{b'(\theta_0)}{q'(\theta_0)} - b'(\theta_0) = -A(\theta_0). \tag{3.31}$$

Equality (3.31) follows by double differentiation of the identity

$$\int_{-\infty}^{\infty} e^{q(\theta)T(z) - b(\theta)} \mu(dz) = 1 \tag{3.32}$$

and (C).

Therefore, $\nu_n(t) \leq \exp\{(3\epsilon - A(\theta_0))(t^2/2)\}$ for n sufficiently large, independent of t. But $\nu_n(t) - \sqrt{2\pi}\, \phi(\sqrt{A(\theta_0)}\, t) \to 0$ for each fixed t by (3.27) and (3.31). Applying the dominated convergence theorem, the lemma follows.

The theorem is now an immediate consequence since Ψ is bounded. For reference we now consider the general model and state theorem 3.2. Let Θ be an open subset of the line. Let $\{Z_i, i \geq 1\}$ be distributed according to $f_\theta(x)$, a density with respect to a σ-finite measure μ for $\theta \in \Theta$. Let Ψ be a probability density on Θ with respect to Lebesgue measure satisfying the conditions of this section. We define $\Phi(\theta, x) = \log f_\theta(x)$ as before. We then have the following theorem.

THEOREM 3.2. *If*

(1) $\frac{\delta^2\Phi}{\delta\theta^2}(\theta, x)$ *exists and is continuous for almost all* x;

(2) $E_\theta(\sup_{|s-\theta|\geq\epsilon} [\Phi(s, Z_1) - \Phi(\theta, Z_1)]) < 0$ *for almost all* θ *and all* $\epsilon > 0$;

(3) $E_\theta\left(\sup_{|s-\theta|<\epsilon} \left[\frac{\delta^2\Phi}{\delta\theta^2}(s, Z_1)\right]^+\right) < \infty$ *for some* $\epsilon > 0$, *for almost all* θ;

(4) $E_\theta\left(\frac{\delta^2\Phi(\theta, Z_1)}{\delta\theta^2}\right) = -E_\theta\left[\frac{\delta\Phi(\theta, Z_1)}{\delta\theta}\right]^2$;

(5) *maximum likelihood estimates* $\{\hat{\theta}_n\}$ *of* θ *exist and are consistent;*

(6) Ψ *satisfies the condition of this section, is continuous, bounded, and* $\int \theta^2\Psi(\theta)\, d\theta < \infty$;

then $nY_n \to \frac{1}{A(\theta)}$, *a.s., where* $A(\theta) = E_\theta\left(\left[\frac{\delta\Phi}{\delta\theta}(\theta, Z_1)\right]^2\right)$*—the Fisher information number, and* $Y_n = \text{var}(\theta|Z_1, \cdots, Z_n)$.

4. Sequential testing

The main theorem of this section states that for the one parameter K-D family sequential Bayesian testing satisfies condition (ii) of theorem 2.1, and therefore the rules given in theorem 2.1 (ii)a, (ii)b are **A.P.O.**

Again we shall state a more general theorem at the end of the section whose proof will appear elsewhere.

Without loss of generality, we assume $\{Z_i, i \geq 1\}$ to be distributed according to the density $f_\theta(x) = e^{\theta T(x) - b(\theta)}$ with respect to some nondegenerate σ-finite measure μ. Let μ be as before and let ν be a probability measure on Θ such that ν assigns positive probability to any nonempty open subset of Θ.

As is customary in the testing problem, we have a decomposition of Θ into two disjoint Borel sets H and $\bar{H}$ (H complement), H being the hypothesis. We have a choice of two decisions (accepting or rejecting H); we pay no penalty for the right decision and incur a measurable loss $\ell(\theta) \geq 0$ when θ is the true parameter and we make the wrong decision. We assume that $\int \ell(\theta)\nu(d\theta) < \infty$. In addition, as usual, we pay $c > 0$ for each observation. The overall risk $R(\phi, t)$ for a sequential procedure consisting of a stopping rule t and randomized test $\phi(Z_1, \cdots, Z_t)$ is then given by

$$R(\phi, t) = cE(t) + E[\phi(Z_1, \cdots, Z_t)\ell(\theta)I_H(\theta)] + E[(1 - \phi(Z_1, \cdots, Z_t))\ell(\theta)I_{\bar{H}}(\theta)] \tag{4.1}$$

where $I_A(\theta)$ is 1 if $\theta \in A$, and 0 otherwise. Again by [1], we can separate the final decision problem from the stopping problem; that is, there is an obvious optimal choice of ϕ given t. We may now write $X_n = Y_n + nc$ where

$$Y_n = \phi_B(Z_1, \cdots, Z_n)E[\ell(\theta)[I_H(\theta) - I_{\bar{H}}(\theta)]|Z_1, \cdots, Z_n] + E[\ell(\theta)I_{\bar{H}}(\theta)|Z_1, \cdots, Z_n], \tag{4.2}$$

and ϕ_B is the Bayes test given n observations. Now again the problem is to find optimal stopping rules for the process X_n. We will establish under some regularity conditions on ν that $\log Y_n/n \to U$ and $P(U < 0) = 1$. Let ν^* be the measure defined by $\nu^*(A) = \int_A \ell(\theta)\nu(d\theta)$. We have the following theorem.

THEOREM 4.1. *Assume, in addition to the conditions given beforehand in this section:*

(1) $0 < \nu^*(H) < \nu^*(\Theta)$,

(2) $\nu(\hat{H}) = 0$ *where* $\hat{H}$ *is the boundary of* H, *and* $\nu(U) > 0$ *for all open* U, *and*

(3) $\ell(\theta)$ *is strictly bounded away from zero outside of some compact* K. *Then,*

$$\frac{\log Y_n}{n} \to (\nu^*) \operatorname*{ess\,sup}_{\theta \in \bar{H}} J(\theta, \theta_0)I_H(\theta_0) + (\nu^*) \operatorname*{ess\,sup}_{\theta \in H} J(\theta, \theta_0)I_{\bar{H}}(\theta_0) = B(\theta_0) \tag{4.3}$$

where

$$J(\theta, \theta_0) = E_{\theta_0}(\Phi(\theta, Z_1) - \Phi(\theta_0, Z_1)), \tag{4.4}$$

and $I_A(\theta)$ *is the indicator function of* A.

PROOF. It is well known that $J(\theta, \theta_0) < 0$ if $P_\theta \neq P_{\theta_0}$. This observation and the following lemma will establish that $P[B(\theta_0) < 0] = 1$ in our case.

LEMMA 4.1. *In a* K-D *family as above,* $J(\theta, \theta_0)$ *is concave in* θ *with a unique maximum of* 0 *at* $\theta = \theta_0$.

PROOF. According to condition (A), $J(\theta, \theta_0) = (\theta - \theta_0)b'(\theta_0) + b(\theta_0) - b(\theta)$. Further, $J'(\theta_0, \theta_0) = 0$ and $J''(\theta, \theta_0) = -b''(\theta) < 0$ by (B). The lemma follows.

To prove convergence of $\log Y_n/n$, it evidently suffices to consider $Y_n^{1/n}$ which is given by

$$Y_n^{1/n} = 1\Big/\Big[\int_\Theta \exp nQ_n(\theta)\nu(d\theta)\Big]^{1/n} \min\Big\{\Big(\int_H \ell(\theta) \exp nQ_n(\theta)\nu(d\theta)\Big)^{1/n}, \Big(\int_{\bar{H}} \ell(\theta) \exp nQ_n(\theta)\nu(d\theta)\Big)^{1/n}\Big\} \tag{4.5}$$

where $Q_n(\theta) = [1/n \sum_{i=1}^n T(Z_i)]\theta - b(\theta)$. Let $Q(\theta, \theta_0) = b'(\theta_0) - b(\theta)$.

LEMMA 4.2. *Let* $\{W_n\}$ *be a sequence of essentially bounded random variables such that* ess sup $|W_n - W| \to 0$. *Then* $E^{1/n}|W_n|^n \to$ ess sup W.

PROOF. By Minkowski's inequality,

$$E^{1/n}|W|^n - E^{1/n}|W_n - W|^n \leq E^{1/n}|W_n|^n \leq E^{1/n}|W_n|^n + E^{1/n}|W_n - W|^n. \tag{4.6}$$

Since $E^{1/n}|W_n - W|^n \leq$ ess sup $|W_n - W|$, the lemma follows from the convergence of the L_n norm to the L_∞ norm. Q.E.D.

To establish the theorem, it suffices to show that if $\nu^*(B) > 0$,

$$\Big\{\int_B [\exp Q_n(\theta)]^n \ell(\theta)\nu(d\theta)\Big\}^{1/n} \to (\nu^*) \operatorname*{ess\,sup}_{\theta \in B} [\exp Q(\theta, \theta_0)], \tag{4.7}$$

and in particular,

$$\left\{\int_\Theta [\exp nQ_n(\theta)]\nu(d\theta)\right\}^{1/n} \to (\nu) \operatorname{ess\,sup}_\theta [\exp Q(\theta, \theta_0)] \tag{4.8}$$

since then, a.s. P_{θ_0}, $Y_n^{1/n}$ converges to

$$\frac{\min \{(\nu^*) \operatorname{ess\,sup}_{\theta\in H} \exp Q(\theta, \theta_0), (\nu^*) \operatorname{ess\,sup}_{\theta\in \overline{H}} \exp Q(\theta, \theta_0)\}}{(\nu^*) \operatorname{ess\,sup}_{\theta\in\Theta} \exp Q(\theta, \theta_0)}. \tag{4.9}$$

Now (ν) ess sup $Q(\theta, \theta_0) = Q(\theta_0, \theta_0)$ by lemma 4.1 and condition (2), and $J(\theta, \theta_0) = Q(\theta, \theta_0) - Q(\theta_0, \theta_0)$. We prove (4.7). By lemma (4.1) and condition (3), there exists a compact K such that

$$(\nu^*) \operatorname{ess\,sup}_{\theta\in K\cap B} Q(\theta, \theta_0) = \operatorname{ess\,sup}_{\theta\in B} Q(\theta, \theta_0) \tag{4.10}$$

and

$$(\nu^*) \operatorname{ess\,sup}_{\theta\in \overline{K}\cap B} Q(\theta, \theta_0) < \operatorname{ess\,sup}_{\theta\in B} Q(\theta, \theta_0). \tag{4.11}$$

Now clearly $\nu^*(K \cap B) > 0$. Remark first that

$$\begin{aligned}(\nu^*) \operatorname{ess\,sup}_{\theta\in B} \exp Q_n(\theta) &\geq \left[\int_B \exp nQ_n(\theta)\nu^*(d\theta)\right]^{1/n} \\ &\geq \left[\int_{K\cap B} \exp nQ_n(\theta)\nu^*(d\theta)\right]^{1/n}.\end{aligned} \tag{4.12}$$

But by lemma 4.2,

$$\begin{aligned}&\lim_n \left[\int_{K\cap B} \exp nQ_n(\theta)\nu^*(d\theta)\right]^{1/n} \\ &\qquad = \lim_n \left[1/\nu^*(K\cap B)\cdot \int_{K\cap B} \exp nQ_n(\theta)\nu^*(d\theta)\right]^{1/n} \\ &\qquad = (\nu^*) \operatorname{ess\,sup}_{\theta\in K\cap B} \exp Q(\theta, \theta_0),\end{aligned} \tag{4.13}$$

since $Q_n(\theta) \to Q(\theta, \theta_0)$, a.s. P_{θ_0} uniformly on K by the S.L.L.N. On the other hand, by lemma 4.1, (4.11), and the S.L.L.N.,

$$(\nu^*) \operatorname{ess\,sup}_{\theta\in B} \exp Q_n(\theta) \to (\nu^*) \operatorname{ess\,sup}_{\theta\in B} \exp Q(\theta, \theta_0); \tag{4.14}$$

which completes the proof of the theorem.

As in section 3, we again state a general theorem without proof. Let $\{Z_i, i \geq 1\}$ be distributed according to a density $f_\theta(x)$ with respect to a σ-finite measure μ for $\theta \in \Theta \subset R^p$ for some p, Θ-Borel measurable. Let H be a measurable subset of Θ. Then let ν, $\ell(\theta)$, ν^*, Y_n, $J(\theta, \theta_0)$, $B(\theta)$, $\Phi(\theta, x)$, H be defined as before. We have the following theorem.

THEOREM 4.2. *Suppose that*

(1) $\nu(U) > 0$ *for any open set* U, $0 < \nu^*(H) < \nu^*(\Theta)$;
(2) $\nu^*(\hat{H}) = 0$;
(3) $\ell(\theta) \geq 0$ *and is strictly positive outside a compact;*

(4) $\Phi(\theta, x)$ *is continuous in* θ *for almost all* x;

(5) $E_{\theta_0}\{\sup_{\|s-t\| \leq \Delta(\theta_0)} |\Phi(s, Z_1) - \Phi(t, Z_1)|\} < \infty$ *for some* $\Delta(\theta_0) > 0$, *and for all* θ_0;

(6) $E_{\theta_0}[\Phi(s, Z_1)] > -\infty$ *for all* s;

(7) $E_{\theta_0}\{\sup_{\|\theta-\theta_0\| \geq K(\theta_0)} [\Phi(\theta, Z_1) - \Phi(\theta_0, Z_1)]\} \leq B(\theta_0)$ *for some* $K(\theta_0) < \infty$.

Then, $\log Y_n/n \to B(\theta_0)$, a.s. P_{θ_0}.

This theorem of course covers the multivariate as well as univariate K-D families and also many other examples. The reader will also note the by no means accidental resemblance of our conditions to those of Kiefer and Sacks in [3]. Of course, the conditions required to prove pointwise optimality are less stringent.

6. Conclusion

Some of the procedures suggested in this paper and similar **A.P.O.** procedures for estimation and testing have already appeared in the literature. Thus, Wald [6] proved that under some regularity conditions, similar to those of theorem 3.2, the following procedure is asymptotically minimax: "Stop the first time $(1/(n+1)A(\hat{\theta}_n)) \leq c$" where $\hat{\theta}_n$ is the maximum likelihood estimate and $A(\theta)$ is as before. It is not difficult to verify that this procedure is asymptotically equivalent to the rule given in theorem 2.1 (i) since, under the conditions of theorem 3.2,

$$\left|\frac{(n+1)Y_n}{A(\hat{\theta}_n)} - 1\right| \to 0, \qquad \text{a.s.} \tag{6.1}$$

Schwarz [4] showed the procedure of theorem 2.1 (ii)a to have asymptotically the same shape as the optimal Bayes region for the exponential family under essentially the conditions of theorem 4.1. Kiefer and Sacks [3] extended his results to more general families and strengthened them. They proved, under some regularity conditions, that, in the presence of an indifference region between hypothesis and alternative, the procedure "Stop when Y_n is first $\leq c$" is asymptotically Bayes. It may be shown from their results that the Bayes optimal rule is **A.P.O.** as might be expected.

The procedure given in theorem 2.1 (ii)b seems to be "better" if an indifference region is not assumed. We are, at present, investigating the connection between **A.P.O.** rules and asymptotic Bayes solutions in this and other instances. The results of [3], [4], [6] give the reader some idea of what can be expected. It may be noted that the rules of Wald and one of the asymptotically Bayes rules proposed by Kiefer and Sacks which is **A.P.O.** are essentially independent of the choice of prior distribution. In general (because of dependence on large samples), the concept of asymptotic pointwise optimality seems to be "prior distribution free" a property which augurs well for its application to non-Bayesian and even nonparametric statistics. We hope to explore these questions also in subsequent papers.

Finally, it seems that the results of this paper are answers to interesting examples of a more general question. Suppose we are given a stochastic sequence of processes, $\{X_c(t)\}$ consisting of a deterministic component $\{D_c(t)\}$ and a noise component $\{N_c(t)\}$. Let $d(c)$ denote the time at which $\{D_c(t)\}$ reaches its minimum, and $o(c)$ denote the time at which $\{X_c(t)\}$ reaches its minimum and suppose that $o(c) \to \infty$ as $c \to 0$. Assume further that we can estimate $D_c(t)$ consistently from $X_c(t)$ by $\hat{D}_c(t)$, where consistency refers to the behavior of $\hat{D}$ as $c \to 0$. Let $\hat{d}(c)$ denote the approximation to $d(c)$ based on $\hat{D}_c(t)$. When is it true that $D_c(d(c)) \sim D_c(\hat{d}(c)) \sim X_c(o(c))$? Obviously, $N_c(t) \to 0$ as $c \to 0$, but further investigation is required. We intend to deal with this question in a forthcoming paper.

We would like to thank A. Dvoretzky for a remark which led to corollary 2.1.

REFERENCES

[1] K. J. Arrow, D. Blackwell, and M. A. Girshick, "Bayes and minimax solutions of sequential decision problems," *Econometrica*, Vol. 17 (1949), pp. 213–244.

[2] H. Chernoff, "Sequential design of experiments," *Ann. Math. Statist.*, Vol. 30 (1959), pp. 755–770.

[3] J. Kiefer and J. Sacks, "Asymptotically optimum sequential inference and design," *Ann. Math. Statist.*, Vol. 34 (1963), pp. 705–750.

[4] G. Schwarz, "Asymptotic shapes of Bayes sequential testing regions," *Ann. Math. Statist.*, Vol. 33 (1962), pp. 224–236.

[5] A. Wald, *Sequential Analysis*, New York, Wiley, 1947.

[6] ———, "Asymptotic minimax solutions of sequential point estimation problems," *Proceedings of the Second Berkeley Symposium on Mathematical Statistics and Probability*, Berkeley and Los Angeles, University of California Press, 1950, pp. 1–11.

POSITIVE DYNAMIC PROGRAMMING

DAVID BLACKWELL
UNIVERSITY OF CALIFORNIA, BERKELEY

1. Introduction

A dynamic programming problem is specified by four objects: S, A, q, r, where S is a nonempty Borel set, the set of *states* of some system, A is a nonempty Borel set, the set of *acts* available to you, q is the *law of motion* of the system; it associates (Borel measurably) with each pair (s, a) a probability distribution $q(\cdot \mid s, a)$ on S: when the system is in state s and you choose act a, the system moves to a new state selected according to $q(\cdot|s, a)$, and r is a bounded Borel measurable function on $S \times A \times S$, the *immediate return:* when the system is in state s, and you choose act a, and the system moves to s', you receive an income $r(s, a, s')$. A *plan* π is a sequence $\pi_1, \pi_2, \cdots$, where π_n tells you how to select an act on the n-th day, as a function of the previous history $h = (s_1, a_1, \cdots, a_{n-1}, s_n)$ of the system, by associating with each h (Borel measurably) a probability distribution $\pi_n(\cdot|h)$ on (the Borel subsets of) A.

Any sequence of Borel measurable functions $f_1, f_2, \cdots$, each mapping S into A, defines a plan. When in state s on the n-th day, choose act $f_n(s)$. Plans $\pi = \{f_n\}$ of this type may be called *Markov* plans. A single f defines a still more special kind of plan: whenever in state s, choose act $f(s)$. This plan is denoted by $f^{(\infty)}$, and plans $f^{(\infty)}$ are called *stationary.*

A plan π associates with each initial state s a corresponding *expected n-th period return* $r_n(\pi)(s)$ and an *expected discounted total return*

$$I_\beta(\pi)(s) = \sum_1^\infty \beta^{n-1} r_n(\pi)(s), \tag{1}$$

where β is a fixed discount factor, $0 \leq \beta < 1$.

The problem of finding a π which maximizes I_β was studied in [1]. Three of the principal results obtained were the following.

RESULT (i). *For any probability distribution p on S and any $\epsilon > 0$, there is a stationary plan $f^{(\infty)}$ which is (p, ϵ)-optimal; that is,*

$$p\{I_\beta(f^{(\infty)}) > I_\beta(\pi) - \epsilon\} = 1 \qquad \text{for all } \pi. \tag{2}$$

RESULT (ii). *Any bounded u which satisfies*

$$u(s) \geq \int [r(s, a, \cdot) + \beta u(\cdot)]\, dq(\cdot|s, a) \qquad \text{for all } s, a \tag{3}$$

is an upper bound on incomes;

This research was supported by the Information Systems Branch of the Office of Naval Research under Contract Nonr-222(53).

$$I_\beta(\pi) \le u \qquad \text{for all} \quad \pi. \tag{4}$$

RESULT (iii). *If A is countable, the optimal return u_β^* is the unique bounded fixed point of the operator U_β, mapping the set of bounded functions u on S into itself, defined by*

$$U_\beta u(s) = \sup_a \int [r(s, a, \cdot) + \beta u(\cdot)]\, dq(\cdot|s, a), \tag{5}$$

that is, $u_\beta^* = \sup_\pi I_\beta(\pi)$. Also $U_\beta^n u \to u_\beta^*$ as $n \to \infty$, for every bounded u.

In this paper we consider the positive (undiscounted) case $r \ge 0$, $\beta = 1$, and we are interested in maximizing $I(\pi) = \sum r_n(\pi)$. A weakened form of (i) is proved. Modified forms of (ii) and (iii) are obtained.

THEOREM 1. *For any probability distribution p on S for which*

$$v = \sup_\pi \int I(\pi)\, dp \tag{6}$$

is finite, and any $\epsilon > 0$, there is a stationary plan $f^{(\infty)}$ which is weakly (p, ϵ)-optimal, that is

$$\int I(f^{(\infty)})\, dp > v - \epsilon. \tag{7}$$

THEOREM 2 (Compare [2], theorem 2.12.1). *Any nonnegative u which satisfies*

$$u(s) \ge \int [r(s, a, \cdot) + u(\cdot)]\, dq(\cdot|s, a) \qquad \text{for all} \quad s, a \tag{8}$$

is an upper bound on incomes;

$$I(\pi) \le u \qquad \text{for all} \quad \pi. \tag{9}$$

THEOREM 3. *If A is countable, the optimal return u^* is the smallest nonnegative fixed point of the operator U, taking the set of nonnegative (possibly $+\infty$-valued) functions on S into itself, defined by*

$$Uu(s) = \sup_a \int [r(s, a, \cdot) + u(\cdot)]\, dq(\cdot|s, a). \tag{10}$$

Also $U^n 0 \to u^*$ as $n \to \infty$.

2. Proofs

Theorem 1 is an easy consequence of (i): first choose π so that $\int I(\pi)\, dp > v - \epsilon$; next, choose $\beta < 1$ so that $\int I_\beta(\pi)\, dp > v - \epsilon$. Now invoke (i) to choose $f^{(\infty)}$ in such a way that

$$p\{I_\beta(f^{(\infty)}) > I_\beta(\pi) - \delta\} = 1, \tag{11}$$

where $\delta = \int I_\beta(\pi)\, dp + \epsilon - v$, so that

$$\int I(f^{(\infty)})\, dp \ge \int I_\beta(f^{(\infty)})\, dp > \int I_\beta(\pi)\, dp - \delta = v - \epsilon. \tag{12}$$

Similarly, theorem 2 is an easy consequence of (ii): fix $\beta < 1$, and define $w = \min(u, R/(1 - \beta))$, where $R = \sup_{s,a,s'} r(s, a, s')$. We show that w satisfies the hypothesis of (ii), that is,

(13) $$w(s) \geq \int [r(s, a, \cdot) + \beta w(\cdot)]\, dq(\cdot|s, a) \qquad \text{for all} \quad s, a.$$

First,

(14) $$u(s) \geq \int [r(s, a, \cdot) + u(\cdot)]\, dq(\cdot|s, a)$$
$$\geq \int [r(s, a, \cdot) + w(\cdot)]\, dq(\cdot|s, a).$$

Second, putting $R/(1 - \beta) = c$,

(15) $$c = R + \beta c \geq \int [r(s, a, \cdot) + \beta w(\cdot)]\, dq(\cdot|s, a).$$

Thus,

(16) $$w = \min (u, c) \geq \int [r(s, a, \cdot) + \beta w(\cdot)]\, dq(\cdot|s, a).$$

So, (ii) implies that $I_\beta(\pi) \leq u$ for all π, $\beta < 1$. Letting $\beta \to 1$ yield $I(\pi) \leq u$.

For theorem 3, note that U is monotone: if $u \geq v$, then $Uu \geq Uv$. Hence, $U^n 0 = u_n$ increases with n, say to w. We show that w is a fixed point of U. Define the operator T_a by

(17) $$T_a u(s) = \int [r(s, a, \cdot) + u(\cdot)]\, dq(\cdot|s, a),$$

so that $U = \sup_a T_a$. We have $T_a u_n \leq U u_n = u_{n+1} \leq Uw$, so that $(n \to \infty)$ $T_a w \leq w \leq Uw$ and $(\sup_a)$ $Uw \leq w \leq Uw$. The function w is the smallest non-negative fixed point of U, since $v \geq 0$ and $Uv = v$ imply, applying U n times to $v \geq 0$, $v \geq U^n 0$, so $(n \to \infty)$ $v \geq w$.

To identify the optimal return with w, note that, for any $\beta < 1$, we have $U_\beta^n 0 \leq U^n 0$, so that $(n \to \infty)$ $u_\beta \leq w$. Thus,

(18) $$I_\beta(\pi) \leq w \quad \text{for all } \pi, \quad \text{and} \quad I(\pi) \leq w \quad \text{for all } \pi.$$

On the other hand,

(19) $$\sup_\pi I(\pi) \geq \sup_\pi I_\beta(\pi) = u_\beta^* \geq U_\beta^n(0), \quad \text{so that} \quad (\beta \to 1),$$
$$\sup_\pi I(\pi) \geq U^n 0; \quad \text{and} \quad (n \to \infty), \sup_\pi I(\pi) \geq w.$$

3. Remarks

(1) In the negative case, $r \leq 0$, $\beta = 1$, Dubins and Savage [2] have given an example in which theorem 1 is false. Ralph Strauch has recently studied the negative case extensively in his thesis, finding that it differs substantially in other ways from the positive case.

(2) Here is an example in which no (p, ϵ)-optimal stationary plan exists, showing that (i) cannot be generally extended to the positive case. There is a sequence $p(1)$, $p(2)$, $\cdots$ of primary states, a sequence $s(1)$, $s(2)$, $\cdots$ of secondary states, and a terminal state t. From primary state $p(n)$ we have two choices: (1) move to secondary state $s(2^n - 1)$, or (2) move to the next primary state $p(n + 1)$ with probability $\frac{1}{2}$, and to the terminal state t with probability $\frac{1}{2}$.

The immediate income is 0 no matter what happens. From secondary state $s(n)$, $n \geq 2$, you move to secondary state $s(n-1)$ and receive \$1. In secondary state $s(1)$ you move to t and receive \$1. Once state t is reached, you stay there and receive nothing.

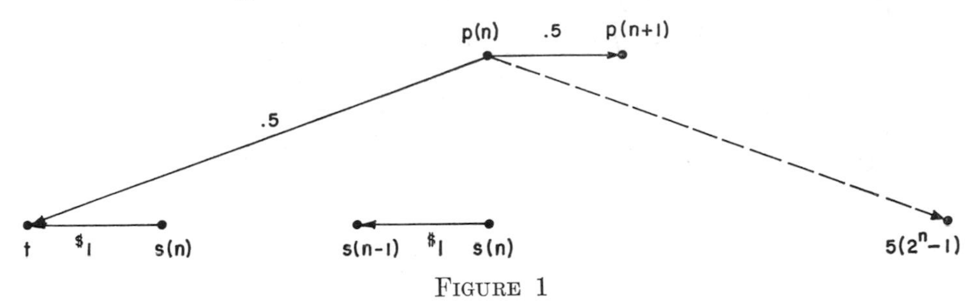

FIGURE 1

The income from $s(n)$ is n, and from t is 0. From $p(n)$, by aiming for $s(2^{n+k}-1)$ via $p(n+k)$, your expected income is $(2^{n+k}-1)/2^k = 2^n - 2^{-k}$, so you can get nearly 2^n from $p(n)$. The function u: $u(p(n)) = 2^n$, $u(s(n)) = n$, $u(t) = 0$ satisfies the hypothesis of theorem 2 (with equality), so is the optimal return: $u = \sup_\pi I(\pi)$. But for any stationary plan $f^{(\infty)}$, either f elects to gamble at every primary state so that $I(f^{(\infty)}) = 0$ for all $p(n)$, or there is a primary state $p(n_0)$ from which f moves to $s(2^{n_0}-1)$ so that $I(f^{(\infty)}) = 2^{n_0} - 1$ at $p(n_0)$, one dollar short of the optimal return at $p(n_0)$. So for any p which assigns positive probability to every primary state and any $\epsilon < 1$, there is no (p, ϵ)-optimal stationary plan.

(3) In the above example, the optimal return is unbounded. Don Ornstein (unpublished) has shown that for a certain class of (positive) problems with bounded optimal return and countable state space, there is for every $\epsilon > 0$ an ϵ-optimal plan $f^{(\infty)}$ which is stationary:

$$I(f^{(\infty)}) > I(\pi) - \epsilon \qquad \text{for all } \pi, s. \tag{20}$$

His method appears to apply to any (positive) problem with bounded optimal return and countable state space. Whether there is a (p, ϵ)-optimal stationary plan in every positive problem with bounded optimal return remains open.

(4) In the discounted case, if there is an optimal plan, there is one which is stationary. Whether this is true in the positive case remains open, even for bounded optimal return.

REFERENCES

[1] DAVID BLACKWELL, "Discounted dynamic programming," *Ann. Math. Statist.*, Vol. 36 (1965), pp. 226–235.

[2] L. E. DUBINS and L. J. SAVAGE, *How to Gamble If You Must*, New York, McGraw-Hill, 1965.

A CLASS OF OPTIMAL STOPPING PROBLEMS

Y. S. CHOW and H. ROBBINS
PURDUE UNIVERSITY and COLUMBIA UNIVERSITY

1. Introduction and summary

Let $x_1, x_2, \cdots$, be independent random variables uniformly distributed on the interval $[0, 1]$. We observe them sequentially, and must stop with some x_i, $1 \leq i < \infty$; the decision whether to stop with any x_i must be a function of the values $x_1, \cdots, x_i$ only. (For a general discussion of optimal stopping problems we refer to [1], [3].) If we stop with x_i we lose the amount $i^\alpha x_i$, where $\alpha \geq 0$ is a given constant. What is the minimal expected loss we can achieve by the proper choice of a stopping rule?

Let C denote the class of all possible stopping rules t; then we wish to evaluate the function

$$v(\alpha) = \inf_{t \in C} E(t^\alpha x_t). \tag{1}$$

If there exists a t in C such that $E(t^\alpha x_t) = v(\alpha)$, we say that t is optimal for that value of α. Let C^N for $N \geq 1$ denote the class of all t in C such that $P[t \leq N] = 1$; then $C^1 \subset C^2 \subset \cdots \subset C$, and hence, defining

$$v^N(\alpha) = \inf_{t \in C^N} E(t^\alpha x_t), \tag{2}$$

we have

$$\tfrac{1}{2} = v^1(\alpha) \geq v^2(\alpha) \geq \cdots \geq v(\alpha) \geq 0. \tag{3}$$

We shall show that as $N \to \infty$,

$$v^N(\alpha) \sim \begin{cases} 2(1-\alpha)/N^{1-\alpha} & \text{for} \quad 0 \leq \alpha < 1, \\ 2/\log N & \text{for} \quad \alpha = 1, \end{cases} \tag{4}$$

from which it follows that

$$v(\alpha) = 0, \qquad \text{for} \quad 0 \leq \alpha \leq 1. \tag{5}$$

(For $\alpha = 0$, J. P. Gilbert and F. Mosteller [4] give the expression $v^N(0) \approx 2/(N + \log(N+1) + 1.767)$; this case is closely related to a problem of optimal selection considered in [2]. It can be shown that $Nv^N(0) \uparrow 2$ as $N \to \infty$.)

Research supported by the Office of Naval Research under Contract Numbers Nonr-1100(26) and Nonr-266(59). Reproduction in whole or part is permitted for any purpose of the United States Government.

We shall show, moreover, that

$$\begin{cases} 0 < v(\alpha) < \tfrac{1}{2}, & \text{for } 1 < \alpha \leq 1.4, \\ \phantom{0 <} v(\alpha) = \tfrac{1}{2}, & \text{for } \phantom{1 <} \alpha \geq 1.5, \end{cases} \tag{6}$$

and that the relation

$$\lim_{N\to\infty} v^N(\alpha) = v(\alpha) \tag{7}$$

holds for all $\alpha \geq 0$. No optimal rule exists for $0 \leq \alpha \leq 1$ by (5), since $E(t^\alpha x_t) > 0$ for every t in C. We shall show that an optimal rule does exist for every $\alpha > 1$; when $v(\alpha) = \frac{1}{2}$ the optimal rule is $t = 1$, but for any α such that $0 < v(\alpha) < \frac{1}{2}$ the optimal rule t is such that $Et = \infty$. The function $v(\alpha)$ is continuous for all $\alpha \geq 0$.

2. Proof of (4)

For any fixed $\alpha \geq 0$ and $N \geq 1$, set $v_{N+1}^N = \infty$ and define

$$v_i^N = E\{\min (i^\alpha x_i, v_{i+1}^N)\} = \int_0^1 \min (i^\alpha x, v_{i+1}^N)\, dx \qquad (i = N, \cdots, 1). \tag{8}$$

The constants v_i^N can be computed recursively from (8), and by a familiar argument it follows that

$$v^N(\alpha) = v_1^N = E(t^\alpha x_t), \tag{9}$$

where

$$t = \text{first} \quad i \geq 1 \quad \text{such that} \quad i^\alpha x_i \leq v_{i+1}^N. \tag{10}$$

For the remainder of this section we shall regard N as a fixed positive integer and α as a fixed constant such that $0 \leq \alpha \leq 1$; for brevity we shall write v_i for v_i^N. Then from (8),

$$v_i \leq E(i^\alpha x_i) = i^\alpha/2, \qquad (i = 1, \cdots, N), \tag{11}$$

so that

$$v_{i+1} i^{-\alpha} \leq \tfrac{1}{2}\left(\frac{i+1}{i}\right)^\alpha \leq \tfrac{1}{2} \cdot 2^\alpha \leq 1, \qquad (i = 1, \cdots, N-1). \tag{12}$$

Hence from (8),

$$\begin{aligned} v_i &= \int_0^{v_{i+1}i^{-\alpha}} i^\alpha x\, dx + (1 - v_{i+1}i^{-\alpha})v_{i+1} \\ &= v_{i+1}\left(1 - \frac{v_{i+1}}{2i^\alpha}\right), \qquad (i = 1, \cdots, N-1). \end{aligned} \tag{13}$$

Noting that $v_i > 0$ for $i = 1, \cdots, N$, we can rewrite (13) as

$$\frac{1}{v_i} = \frac{1}{v_{i+1}} + \frac{1}{2i^\alpha - v_{i+1}} = \frac{1}{v_{i+1}} + \frac{1}{2i^\alpha} + \frac{v_{i+1}}{2i^\alpha(2i^\alpha - v_{i+1})}, \qquad (i = 1, \cdots, N-1). \tag{14}$$

Summing (14) for $i = 1, \cdots, N-1$ and noting that from (8)

$$v_N = \frac{N^\alpha}{2}, \tag{15}$$

we obtain the formula

$$\frac{1}{v_1} = \frac{2}{N^\alpha} + \frac{1}{2}\sum_1^{N-1}\frac{1}{i^\alpha} + \frac{1}{2}\sum_1^{N-1}\frac{v_{i+1}}{i^\alpha(2i^\alpha - v_{i+1})}. \tag{16}$$

We shall show at the end of this section that, setting

$$I_N = \frac{1}{2}\sum_1^{N-1}\frac{1}{i^\alpha}, \qquad J_N = \frac{1}{2}\sum_1^{N-1}\frac{v_{i+1}}{i^\alpha(2i^\alpha - v_{i+1})}, \tag{17}$$

we have as $N \to \infty$

$$J_N = o(I_N), \qquad I_N \sim \begin{cases} N^{1-\alpha}/2(1-\alpha), & (0 \le \alpha < 1), \\ \log N/2, & (\alpha = 1). \end{cases} \tag{18}$$

Relations (4) follow from (9), (16), and (18).

Proof of (18). The second part of (18) follows from the relation

$$I_N \sim \frac{1}{2}\int_1^N \frac{dt}{t^\alpha}. \tag{19}$$

The first part of (18) follows from two lemmas.

Lemma 1. *The following inequality holds:*

$$v_i \le \frac{2N^\alpha}{N - i + 1}, \qquad (i = 1, \cdots, N). \tag{20}$$

Proof. Equation (20) holds for $i = N$ by (15). Suppose it holds for some $i + 1 = 2, \cdots, N$; we shall show that it holds for i also.

(a). If $2N^\alpha/(N - i) > i^\alpha$, then by (11),

$$v_i \le \frac{i^\alpha}{2} \le \frac{N^\alpha}{N - i} \le \frac{2N^\alpha}{N - i + 1}. \tag{21}$$

(b). If $2N^\alpha/(N - i) \le i^\alpha$, then setting

$$f(x) = x\left(1 - \frac{x}{2i^\alpha}\right), \quad f'(x) = 1 - \frac{x}{i^\alpha} \ge 0, \qquad \text{for} \quad x \le i^\alpha, \tag{22}$$

so by (13)

$$v_i = f(v_{i+1}) \le f\left(\frac{2N^\alpha}{N - i}\right) = \frac{2N^\alpha}{N - i}\left(1 - \frac{N^\alpha}{i^\alpha(N - i)}\right) \le \frac{2N^\alpha}{N - i + 1}, \tag{23}$$

which completes the proof.

From (12) and (20) we have

$$J_N = \frac{1}{2}\sum_1^{N-1}\frac{v_{i+1}}{i^\alpha(2i^\alpha - v_{i+1})} \le N^\alpha \sum_1^{N-1}\frac{1}{(N - i)i^{2\alpha}}. \tag{24}$$

To prove the first part of (18), in view of the second part, it will suffice to show the following.

Lemma 2. *As $N \to \infty$,*

$$N^\alpha \sum_1^{N-1}\frac{1}{(N - i)i^{2\alpha}} = \begin{cases} o(N^{1-\alpha}), & (0 \le \alpha \le 1), \\ 0(1), & (\alpha = 1). \end{cases} \tag{25}$$

PROOF. (a). Assume $0 \leq \alpha < 1$. For any $0 < \delta < 1$, the left side of (25) can be written as

$$N^{\alpha}\left(\sum_{1}^{[\delta N]} + \sum_{[\delta N]+1}^{N-1}\right)\frac{1}{(N-i)i^{2\alpha}} \tag{26}$$
$$\leq N^{\alpha}\left(\frac{1}{N(1-\delta)}\sum_{1}^{N-1}\frac{1}{i^{\alpha}} + N(1-\delta)(\delta N)^{-2\alpha}\right)$$
$$\sim N^{\alpha}\left(\frac{1}{N(1-\delta)}\frac{N^{1-\alpha}}{1-\alpha} + N(1-\delta)(\delta N)^{-2\alpha}\right) \sim \frac{(1-\delta)N^{1-\alpha}}{\delta^{2\alpha}}.$$

Hence,

$$\overline{\lim_{N\to\infty}}\frac{J_N}{N^{1-\alpha}} \leq \frac{1-\delta}{\delta^{2\alpha}}. \tag{27}$$

Since δ can be arbitrarily near 1, the left-hand side of (27) must be 0.

(b). Assume $\alpha = 1$. We have for the left-hand side of (25), setting $M = [N/2]$,

$$N\sum_{1}^{N-1}\frac{1}{(N-i)i^{2}} = N\left(\sum_{1}^{M} + \sum_{M+1}^{N-1}\right)\frac{1}{(N-i)i^{2}} \leq 2\sum_{1}^{M} i^{-2} + N\sum_{M+1}^{N-1} i^{-2} \tag{28}$$
$$\leq 2\int_{1/2}^{\infty}\frac{dt}{t^{2}} + N\left(\frac{N}{2}\right)\left(\frac{2}{N}\right)^{2} = 0(1).$$

3. An optimal rule exists for $\alpha > 1$ and $v(\alpha) > 0$

Define $z_n = \inf_{i\geq n}(i^{\alpha}x_i)$. Then for any constant $0 \leq A \leq n^{\alpha}$, we have

$$P[z_n \geq A] = P[i^{\alpha}x_i \geq A; i \geq n] = \prod_{n}^{\infty}\left(1 - \frac{A}{i^{\alpha}}\right). \tag{29}$$

Hence,

$$P\left[z_1 \geq \frac{1}{2}\right] = \prod_{1}^{\infty}\left(1 - \frac{1}{2i^{\alpha}}\right) > 0, \tag{30}$$

and therefore,

$$v(\alpha) \geq Ez_1 > 0. \tag{31}$$

Next, for any $A > 0$,

$$\sum_{1}^{\infty} P[n^{\alpha}x_n \leq A] \leq \sum_{1}^{\infty}\frac{A}{n^{\alpha}} < \infty. \tag{32}$$

Hence, by the Borel-Cantelli lemma,

$$P[\lim_{n\to\infty} n^{\alpha}x_n = \infty] = 1. \tag{33}$$

The existence of an optimal t for $\alpha > 1$ now follows from lemma 4 of [1].

4. For $\alpha \geq \frac{3}{2}$, $v(\alpha) = \frac{1}{2}$

We define for $i = 1, 2, \cdots$, and any fixed $\alpha \geq 0$,

$$v_i = \inf_{t\in C_i} E(t^{\alpha}x_t), \tag{34}$$

where C_i denotes the class of all $t \in C$ such that $P[t \geq i] = 1$. Then $v(\alpha) = v_1 \leq v_2 \leq \cdots$. It can be shown [3], although it is not trivial to prove, that in analogy with (8),

$$v_i = E\{\min(i^\alpha x_i, v_{i+1})\} = \int_0^1 \min(i^\alpha x, v_{i+1})\, dx, \qquad (i \geq 1). \tag{35}$$

It follows that

$$v_i \leq \frac{i^\alpha}{2}, \qquad (i \geq 1). \tag{36}$$

From now on in this section we shall assume that $1 < \alpha \leq \frac{3}{2}$. Then

$$v_{i+1} i^{-\alpha} \leq \frac{1}{2}\left(\frac{i+1}{i}\right)^\alpha \leq \frac{1}{2}\left(\frac{3}{2}\right)^\alpha \leq 1, \qquad (i \geq 2). \tag{37}$$

Hence, as in (13),

$$v_i = v_{i+1}\left(1 - \frac{v_{i+1}}{2i^\alpha}\right), \qquad (i \geq 2), \tag{38}$$

and since $v_1 = v(\alpha) > 0$ for $\alpha > 1$ by (31), we have as in (14),

$$\frac{1}{v_i} = \frac{1}{v_{i+1}} + \frac{1}{2i^\alpha - v_{i+1}}, \qquad (i \geq 2). \tag{39}$$

Summing (39) for $i = n, \cdots, m-1$, we obtain

$$\frac{1}{v_n} = \frac{1}{v_m} + \sum_n^{m-1} \frac{1}{2i^\alpha - v_{i+1}}, \qquad (2 \leq n \leq m). \tag{40}$$

From (29), for any $A > 0$, we have as $m \to \infty$,

$$P[z_m \geq A] = \prod_m^\infty \left(1 - \frac{A}{i^\alpha}\right) \to 1, \tag{41}$$

thus $Ez_m \to \infty$, and since $v_m \geq Ez_m$, it follows that $v_m \to \infty$. Hence from (40),

$$\frac{1}{v_n} = \sum_n^\infty \frac{1}{2i^\alpha - v_{i+1}}, \qquad (n \geq 2). \tag{42}$$

From (42) and (37) we have for $n \geq 1$,

$$\begin{aligned} \frac{1}{(\alpha - 1)n^{\alpha-1}} &\geq \sum_{n+1}^\infty \frac{1}{i^\alpha} \geq \frac{1}{v_{n+1}} \geq \frac{1}{2}\sum_{n+1}^\infty \frac{1}{i^\alpha} \\ &\geq \frac{1}{2}\int_{n+1}^\infty \frac{dt}{t^\alpha} = \frac{1}{2(\alpha-1)(n+1)^{\alpha-1}}, \end{aligned} \tag{43}$$

and hence,

$$\frac{\alpha - 1}{n} \leq \frac{v_{n+1}}{n^\alpha} \leq \frac{2(\alpha-1)}{n+1}\left(\frac{n+1}{n}\right)^\alpha, \qquad (n \geq 1). \tag{44}$$

We shall now show that $v_2 > 1$ for $\alpha = \frac{3}{2}$. It will follow from (35) that $v_1 = \frac{1}{2}$ and that $t = 1$ is optimal for $\frac{3}{2}$; the same is true a fortiori for any $\alpha \geq \frac{3}{2}$.
From (38) we obtain

$$v_{i+1} = i^\alpha - \sqrt{i^{2\alpha} - 2i^\alpha v_i}, \qquad (i \geq 2); \tag{45}$$

the + sign being excluded because of (37). Suppose now that $v_2 \leq 1$ for $\alpha = \frac{3}{2}$. Then by (45),

$$\begin{aligned} v_3 &\leq 2^{3/2} - \sqrt{8 - 2.2^{3/2}} = 1.3, \\ v_4 &\leq 3^{3/2} - \sqrt{27 - 2\sqrt{27}(1.3)} = 1.52, \\ v_5 &\leq 4^{3/2} - \sqrt{64 - 16(1.52)} = 1.7. \end{aligned} \tag{46}$$

On the other hand, by (44) we have for $\alpha = \frac{3}{2}$,

$$\frac{v_{n+1}}{n^{3/2}} \leq \frac{1}{n+1}\left(\frac{n+1}{n}\right)^{3/2} \leq \frac{1}{6}\left(\frac{6}{5}\right)^{3/2} \leq \frac{11}{50}, \qquad (n \geq 5). \tag{47}$$

Hence, from (42) for $\alpha = \frac{3}{2}$,

$$\frac{1}{v_5} = \sum_5^\infty \frac{1}{2i^\alpha - v_{i+1}} = \sum_5^\infty \frac{1}{2i^\alpha\left(1 - \frac{v_{i+1}}{2i^\alpha}\right)} \leq \sum_5^\infty \frac{1}{2i^\alpha\left(1 - \frac{11}{100}\right)} \tag{48}$$

$$\leq \frac{50}{89}\int_{9/2}^\infty \frac{dt}{t^\alpha} = \frac{50}{89}\frac{1}{\alpha - 1}\sqrt{\frac{2}{9}} = \frac{100}{89}\cdot\frac{\sqrt{2}}{3} < \frac{1}{1.7},$$

contradicting (46). Hence $v_2 > 1$ for $\alpha = \frac{3}{2}$.

5. If $1 < \alpha \leq 1.4$, then $v(\alpha) < \frac{1}{2}$

By (44) we have for $\alpha = \frac{7}{5}$,

$$v_3 \leq \tfrac{4}{5}\cdot 3^{2/5} < \tfrac{5}{4}, \tag{49}$$

and hence by (38), $v_2 < \frac{5}{4}(1 - (5/4.2.2^{7/5})) < 1$. Hence by (35), $v_1 = v(\frac{7}{5}) < \frac{1}{2}$.

For $\alpha > 1$, an optimal t exists by section 3, and from ([3], theorem 2), a minimal optimal t is defined by

$$t = \text{first} \quad n \geq 1 \quad \text{such that} \quad x_n \leq \frac{v_{n+1}}{n^\alpha}. \tag{50}$$

Let α be any constant >1 such that $v(\alpha) < \frac{1}{2}$. Then $P[t > 1] > 0$ by (50), and for $\alpha < \frac{3}{2}$ we have from (44) that

$$\frac{v_{n+1}}{n^\alpha} \leq \frac{1}{n+1}\left(\frac{n+1}{n}\right)^2 < \frac{n+1}{n^2} < 1, \qquad \text{for} \quad n \geq 2. \tag{51}$$

Hence, $P[t > N] > 0$ for every $N \geq 1$, so t is not bounded. In fact, if $1 < \alpha = (3 - \epsilon)/2$ for some $\epsilon > 0$, then from (44)

$$\frac{v_{n+1}}{n^\alpha} \leq (1 - \epsilon)\left(\frac{n+1}{n^2}\right) \leq \frac{1}{n}, \qquad \text{for} \quad n \geq \frac{1-\epsilon}{\epsilon}. \tag{52}$$

Hence, if $v(\alpha) < \frac{1}{2}$, so that $P[t > N] > 0$ for every $N \geq 1$, it follows that for $n > N \geq \frac{1-\epsilon}{\epsilon}$ and some $K > 0$,

$$P[t > n] \geq K\left(1 - \frac{1}{N}\right)\left(1 - \frac{1}{N+1}\right)\cdots\left(1 - \frac{1}{n}\right) = K \cdot \frac{N-1}{n}, \tag{53}$$

so that $Et = \sum_0^\infty P[t > n] = \infty$.

We thus have for $\alpha > 1$: either $0 < v(\alpha) < \frac{1}{2}$ and $Et = \infty$, or $v(\alpha) = \frac{1}{2}$ and $t = 1$, where t is optimal for that α. The least value α^* such that $v(\alpha^*) = \frac{1}{2}$ is not known to us, but by the results of this and the previous section, it lies between 1.4 and 1.5.

6. The identification of optimal rules for $1 < \alpha$

For $N = 1, 2, \cdots$, define t_N by (10). Then $t_N \leq t_{N+1} \leq \cdots$. Let $b_i = \lim_{N\to\infty} v_i^N$. Then from (8),

$$b_i = \int_0^1 \min(i^\alpha x, b_{i+1})\, dx, \qquad (i = 1, 2, \cdots). \tag{54}$$

Define

$$s = \text{first } i \geq 1 \text{ such that } i^\alpha x_i \leq b_{i+1} \text{ if such an } i \text{ exists, } = \infty \text{ otherwise.} \tag{55}$$

Then [1] $s = \lim_{N\to\infty} t_N$. Since $v_i^N \geq v_i$ for each N, $b_i \geq v_i$. Therefore $s \leq t$, where t is an optimal rule defined by (50). We shall now show that $s = t$ by showing that $b_i = v_i$ for all $i \geq 1$.

From (54) we have

$$b_i \leq i^\alpha/2, \qquad (i \geq 1), \tag{56}$$

and hence as in (37) and (39), for some $i_0 = i_0(d)$,

$$b_{i+1} i^{-\alpha} \leq 1, \qquad (i \geq i_0),$$

$$\frac{1}{b_i} = \frac{1}{b_{i+1}} + \frac{1}{2i^\alpha - b_{i+1}}, \qquad (i \geq i_0). \tag{57}$$

Since $b_i \geq v_i \to \infty$ as $i \to \infty$, we have, as in (42),

$$\frac{1}{b_n} = \sum_n^\infty \frac{1}{2i^\alpha - b_{i+1}}, \qquad (n \geq i_0). \tag{58}$$

Assume that for some $j \geq 1$, $b_j > v_j$. Then by (35) and (54) this inequality must hold for some $i_1 \geq i_0$ (since if $j < i_0$ and $b_{i_0} \leq v_{i_0}$, then $b_j \leq v_j$), and hence for every $i \geq i_1$. Hence by (42) and (54),

$$\frac{1}{v_{i_1}} = \sum_{i=i_1}^\infty \frac{1}{2i^\alpha - v_{i+1}} < \sum_{i=i_1}^\infty \frac{1}{2i^\alpha - b_{i+1}} = \frac{1}{b_{i_1}}, \tag{59}$$

a contradiction. Hence $b_j = v_j$ for all $j \geq 1$.

It follows from the above that for $1 < \alpha$,

$$v(\alpha) = v_1 = b_1 = \lim_{N\to\infty} v_1^N = \lim_{N\to\infty} v^N(\alpha). \tag{60}$$

That this relation holds also for $0 \leq \alpha \leq 1$ has been shown already.

7. Continuity of $v(\alpha)$

From (60), which holds for any $\alpha \geq 0$, given $\epsilon > 0$ we can find $N = N(\alpha, \epsilon)$ so large that

$$v(\alpha) + \frac{\epsilon}{2} \geq v^N(\alpha) = E(t^\alpha x_t) \tag{61}$$

for some t in C^N. Hence for $\alpha' > \alpha$,

$$v(\alpha) \leq v(\alpha') \leq E(t^{\alpha'} x_t) \leq N^{\alpha'-\alpha} E(t^\alpha x_t) \leq N^{\alpha'-\alpha}\left(v(\alpha) + \frac{\epsilon}{2}\right) \leq v(\alpha) + \epsilon, \tag{62}$$

provided that $\alpha' - \alpha$ is sufficiently small. Hence $v(\alpha)$ is continuous on the right for each $\alpha \geq 0$.

Since $v(\alpha)$ is nondecreasing in α for each fixed $i \geq 1$, we have by the bounded or monotone convergence theorem for integrals from (35)

$$\begin{aligned} v_i(\alpha - 0) &= \lim_{\epsilon \to 0} v_i(\alpha - \epsilon) = \lim_{\epsilon \to 0} \int_0^1 \min\,(i^{\alpha-\epsilon}, v_{i+1}(\alpha - \epsilon))\, dx \\ &= \int_0^1 \min\,(i^\alpha, v_{i+1}(\alpha - 0))\, dx \qquad (i \geq 1), \end{aligned} \tag{63}$$

and by the remark preceding (42), $\lim_{n\to\infty} v_n(\alpha - 0) = \infty$ for $\alpha > 1$. Hence, as in the preceding section, (58) holds with b_n replaced by $v_n(\alpha - 0)$, and the argument shows that $v_n(\alpha - 0) = v_n(\alpha)$. In particular, $v_n(\alpha - 0) = v(\alpha)$, which shows that $v(\alpha)$ is continuous on the left for $\alpha > 1$. Since $v(\alpha) = 0$ for $0 \leq \alpha \leq 1$, it follows that $v(\alpha)$ is continuous on the left for each $\alpha \geq 0$.

REFERENCES

[1] Y. S. Chow and H. Robbins, "On optimal stopping rules," *Z. Wahrscheinlichkeitstheorie und Verw. Gebiete*, Vol. 2 (1963), pp. 33–49.

[2] Y. S. Chow, S. Moriguti, H. Robbins, and S. M. Samuels, "Optimal selection based on relative rank," *Israel J. Math.*, Vol. 2 (1964), pp. 81–90.

[3] Y. S. Chow and H. Robbins, "On values associated with a stochastic sequence," *Proceedings of the Fifth Berkeley Symposium on Mathematical Statistics and Probability*, Berkeley and Los Angeles, University of California Press, 1966, Vol. I, pp. 427–440.

[4] J. P. Gilbert and F. Mosteller, "Recognizing the maximum of a sequence," *J. Amer. Statist. Soc.*, Vol. 61 (1966), pp. 35–73.

ON VALUES ASSOCIATED WITH A STOCHASTIC SEQUENCE

Y. S. CHOW[1]
PURDUE UNIVERSITY
and
H. ROBBINS[2]
COLUMBIA UNIVERSITY

1. Introduction

Let $\{z_n\}_1^\infty$ be a sequence of random variables with a known joint distribution. We are allowed to observe the z_n sequentially, stopping anywhere we please; the decision to stop with z_n must be a function of $z_1, \cdots, z_n$ only (and not of $z_{n+1}, \cdots$). If we decide to stop with z_n, we are to receive a reward $x_n = f_n(z_1, \cdots, z_n)$ where f_n is a known function for each n. Let t denote any rule which tells us when to stop and for which $E(x_t)$ exists, and let v denote the supremum of $E(x_t)$ over all such t. How can we find the value of v, and what stopping rule will achieve v or come close to it?

2. Definition of the γ_n sequence

We proceed to give a more precise definition of v and associated concepts. *We assume given always*

(a) a probability space $(\Omega, \mathfrak{F}, P)$ with points ω;
(b) a nondecreasing sequence $\{\mathfrak{F}_n\}_1^\infty$ of sub-Borel fields of $\mathfrak{F}$;
(c) a sequence $\{x_n\}_1^\infty$ of random variables $x_n = x_n(\omega)$ such that for each $n \geq 1$, x_n is measurable $(\mathfrak{F}_n)$ and $E(x_n^-) < \infty$.

(In terms of the intuitive background of the first paragraph, $\mathfrak{F}_n$ is the Borel field $\mathfrak{B}(z_1, \cdots, z_n)$ generated by $z_1, \cdots, z_n$. Having served the purpose of defining the $\mathfrak{F}_n$ and x_n, the z_n disappear in the general theory which follows.) Any random variable (r.v.) t with values $1, 2, \cdots$ (not including ∞) such that the event $[t = n]$ (that is, the set of all ω such that $t(\omega) = n$) belongs to $\mathfrak{F}_n$ for each $n \geq 1$, is called a *stopping variable* (s.v.); $x_t = x_{t(\omega)}(\omega)$ is then a r.v. Let C denote the class of all t for which $E(x_t^-) < \infty$. We define the *value* of the stochastic sequence $\{x_n, \mathfrak{F}_n\}_1^\infty$ to be

[1] Research supported in part by National Science Foundation Grant NSF-GP-3694 at Columbia University, Department of Mathematical Statistics.
[2] Research supported by the Office of Naval Research under Contract No. Nonr-266(59), Project No. 042-205. Reproduction in whole or part is permitted for any purpose of the United States Government.

$$v = \sup_{t \in C} E(x_t). \tag{1}$$

Similarly, for each $n \geq 1$ we denote by C_n the class of all t in C such that $P[t \geq n] = 1$, and set

$$v_n = \sup_{t \in C_n} E(x_t). \tag{2}$$

Then

$$C = C_1 \supset C_2 \supset \cdots \quad \text{and} \quad v = v_1 \geq v_2 \geq \cdots ; \tag{3}$$

since $t = n \in C_n$, we have $v_n \geq E(x_n) > -\infty$.

For any family $(y_t, t \in T)$ of r.v.'s we define $y = \text{ess sup}_{t \in T}\, y_t$ if (a) y is a r.v. such that $P[y \geq y_t] = 1$ for each t in T, and (b) if z is any r.v. such that $P[z \geq y_t] = 1$ for each t in T, then $P[z \geq y] = 1$. It is known that there always exists a *sequence* $\{t_k\}_1^\infty$ in T such that

$$\sup_k y_{t_k} = \text{ess sup}_{t \in T}\, y_t. \tag{4}$$

We may therefore define for each $n \geq 1$ a r.v. γ_n measurable $(\mathfrak{F}_n)$ by

$$\gamma_n = \text{ess sup}_{t \in C_n} E(x_t|\mathfrak{F}_n); \tag{5}$$

then $\gamma_n \geq x_n$ (equalities and inequalities are understood to hold up to sets of P-measure 0) and $E(\gamma_n^-) \leq E(x_n^-) < \infty$.

It might seem more natural to consider, instead of C_n, the larger class $\tilde{C}_n$ of all s.v.'s t such that $P[t \geq n] = 1$ and $E(x_t)$ exists, that is $E(x_t^-)$ and $E(x_t^+)$ not both infinite. However, this would yield the same v_n and γ_n. For if $t \in \tilde{C}_n$, define

$$t' = \begin{cases} t \text{ if } E(x_t|\mathfrak{F}_n) \geq x_n, \\ n \text{ otherwise.} \end{cases} \tag{6}$$

Then setting $A = [E(x_t|\mathfrak{F}_n) \geq x_n]$, we have

$$E(x_{t'}^-) \leq E(x_n^-) + \int_A x_t^-. \tag{7}$$

But $-\infty < \int_A x_n \leq \int_A x_t$, so $\int_A x_t^- < \infty$. Hence, $E(x_{t'}^-) < \infty$ and $t' \in C_n$. Now $E(x_{t'}|\mathfrak{F}_n) = \max(x_n, E(x_t|\mathfrak{F}_n)) \geq E(x_t|\mathfrak{F}_n)$, and hence $E(x_{t'}) \geq E(x_t)$. *It follows that v_n and γ_n are unchanged if we replace C_n by $\tilde{C}_n$ in their definitions.*

3. Some lemmas

LEMMA 1. *For each $n \geq 1$ there exists a sequence $\{t_k\}_1^\infty$ in C_n such that*

$$x_n \leq E(x_{t_k}|\mathfrak{F}_n) \uparrow \gamma_n \qquad \text{as } k \to \infty. \tag{8}$$

PROOF. Choose $\{t_k\}_1^\infty$ in C_n with $t_1 = n$ such that $\gamma_n = \sup_k E(x_{t_k}|\mathfrak{F}_n)$. By lemmas 2 and 3 below, we can assume that (8) holds.

LEMMA 2. *For any $t \in C_n$, define t' = first $k \geq n$ such that $E(x_t|\mathfrak{F}_k) \leq x_k$. Then*

(a) $t' \leq t$, $t' \in C_n$,

(b) $E(x_{t'}|\mathfrak{F}_n) \geq E(x_t|\mathfrak{F}_n)$,

(c) $t' > j \geq n \Rightarrow E(x_{t'}|\mathfrak{F}_j) > x_j$.

PROOF. If $t = j \geq n$, then $E(x_t|\mathfrak{F}_j) = x_j$, so $t' \leq j$; hence, $t' \leq t$. Now

$$(9) \qquad E(x_{t'}^-) = \sum_{k=n}^{\infty} \int_{[t'=k]} x_k^- \leq \sum_{k=n}^{\infty} \int_{[t'=k]} E^-(x_t|\mathfrak{F}_k) \leq \sum_{k=n}^{\infty} \int_{[t'=k]} E(x_t^-|\mathfrak{F}_k)$$
$$= E(x_t^-) < \infty,$$

so that $t' \in C_n$. Hence (a) holds. For any $A \in \mathfrak{F}_j$ with $j \geq n$,

$$(10) \qquad \int_{A[t'\geq j]} x_{t'} = \sum_{k=j}^{\infty} \int_{A[t'=k]} x_k \geq \sum_{k=j}^{\infty} \int_{A[t'=k]} E(x_t|\mathfrak{F}_k) = \int_{A[t'\geq j]} x_t.$$

Putting $j = n$ gives (b). For $t' > j$ we obtain $E(x_{t'}|\mathfrak{F}_j) \geq E(x_t|\mathfrak{F}_j) > x_j$, which gives (c).

Any $t' \in C_n$ satisfying (c) of lemma 2 will be called *n-regular.*

LEMMA 3. *Let* $\{t_i\}_1^\infty \in C_n$ *be n-regular for some fixed* $n \geq 1$, *and define* $\tau_i = \max(t_1, \cdots, t_i)$. *Then* $\tau_i \in C_n$ *is n-regular and*

$$(11) \qquad \max_{1\leq k\leq i} E(x_{t_k}|\mathfrak{F}_n) \leq E(x_{\tau_i}|\mathfrak{F}_n) \leq E(x_{\tau_{i+1}}|\mathfrak{F}_n).$$

PROOF. That $\tau_i \in C_n$ is clear. For $j \geq n$ and $A \in \mathfrak{F}_j$,

$$(12) \qquad \int_{A[\tau_i\geq j]} x_{\tau_i} = \sum_{k=j}^{\infty} \left(\int_{A[\tau_i=k\geq t_{i+1}]} x_{\tau_{i+1}} + \int_{A[\tau_i=k<t_{i+1}]} x_k \right)$$
$$\leq \sum_{k=j}^{\infty} \left(\int_{A[\tau_i=k\geq t_{i+1}]} x_{\tau_{i+1}} + \int_{A[\tau_i=k<t_{i+1}]} x_{t_{i+1}} \right)$$
$$= \int_{A[\tau_i\geq j]} x_{\tau_{i+1}}.$$

For $j = n$, this gives

$$(13) \qquad E(x_{\tau_{i+1}}|\mathfrak{F}_n) \geq E(x_{\tau_i}|\mathfrak{F}_n) \geq \cdots \geq E(x_{\tau_1}|\mathfrak{F}_n) = E(x_{t_1}|\mathfrak{F}_n),$$

and hence, by symmetry,

$$(14) \qquad E(x_{\tau_i}|\mathfrak{F}_n) \geq \max_{1\leq k\leq i} E(x_{t_k}|\mathfrak{F}_n).$$

To prove that τ_i is n-regular, we observe by the above that

$$(15) \qquad \tau_i \geq j \Rightarrow E(x_{\tau_i}|\mathfrak{F}_j) \leq E(x_{\tau_{i+1}}|\mathfrak{F}_j).$$

Since t_1 is n-regular,

$$(16) \qquad t_1 < j \Rightarrow x_j < E(x_{t_1}|\mathfrak{F}_j) = E(x_{\tau_1}|\mathfrak{F}_j) \leq \cdots \leq E(x_{\tau_i}|\mathfrak{F}_j),$$

and by symmetry,

$$(17) \qquad \tau_i > j \Rightarrow x_j < E(x_{\tau_i}|\mathfrak{F}_j).$$

4. The fundamental theorem

THEOREM 1. *The following relations hold:*

$$\begin{aligned}&\text{(a)}\ \gamma_n = \max(x_n, E(\gamma_{n+1}|\mathfrak{F}_n)),\\ &\text{(b)}\ E(\gamma_n) = v_n.\end{aligned} \qquad (n \geq 1)$$

PROOF. (a). Given any $t \in C_n$, let $t' = \max(t, n+1) \in C_{n+1}$ and set $A = [t = n]$, and I_A = indicator function of A. Then

$$\begin{aligned}E(x_t|\mathfrak{F}_n) &= I_A \cdot x_n + I_{\Omega-A} \cdot E(x_{t'}|\mathfrak{F}_n)\\ &= I_A \cdot x_n + I_{\Omega-A} \cdot E(E(x_{t'}|\mathfrak{F}_{n+1})|\mathfrak{F}_n)\\ &\leq I_A \cdot x_n + I_{\Omega-A} \cdot E(\gamma_{n+1}|\mathfrak{F}_n) \leq \max(x_n, E(\gamma_{n+1}|\mathfrak{F}_n)).\end{aligned} \tag{18}$$

To prove the reverse inequality, choose, by lemma 1, $\{t_k\}_1^\infty \in C_{n+1}$ such that

$$x_{n+1} \leq E(x_{t_k}|\mathfrak{F}_{n+1}) \uparrow \gamma_{n+1} \qquad \text{as} \quad k \to \infty; \tag{19}$$

then by the monotone convergence theorem for conditional expectations,

$$E(\gamma_{n+1}|\mathfrak{F}_n) = E(\lim_{k\to\infty} E(x_{t_k}|\mathfrak{F}_{n+1})|\mathfrak{F}_n) = \lim_{k\to\infty} E(x_{t_k}|\mathfrak{F}_n) \leq \gamma_n. \tag{20}$$

And since $t = n$ is in C_n, $x_n = E(x_n|\mathfrak{F}_n) \leq \gamma_n$. This completes the proof of (a).

(b). Since for each t in C_n, $E(x_t|\mathfrak{F}_n) \leq \gamma_n$, $E(x_t) \leq E(\gamma_n)$, so $v_n \leq E(\gamma_n)$. Now choose $\{t_k\}_1^\infty$ in C_n, according to lemma 1; then

$$E(\gamma_n) = \lim_{k\to\infty} E(x_{t_k}) \leq v_n. \tag{21}$$

LEMMA 4. *If $t \in C$, then*

$$t \geq n \Rightarrow E(x_t|\mathfrak{F}_n) \leq \gamma_n \quad \text{and} \quad E(x_t^-|\mathfrak{F}_n) \geq \gamma_n^-. \tag{22}$$

PROOF. Set $t' = \max(t, n) \in C_n$. By definition of γ_n,

$$t \geq n \Rightarrow E(x_t|\mathfrak{F}_n) = E(x_{t'}|\mathfrak{F}_n) \leq \gamma_n, \tag{23}$$

and hence

$$t \geq n \Rightarrow E(x_t^-|\mathfrak{F}_n) \geq E^-(x_t|\mathfrak{F}_n) \geq \gamma_n^-. \tag{24}$$

5. The r.v. σ

We define the r.v.

$$\sigma = \text{first } n \geq 1 \text{ such that } x_n = \gamma_n \quad (= \infty \text{ if no such } n \text{ exists}). \tag{25}$$

In general, $P[\sigma < \infty] < 1$, so that σ is not always a s.v.

LEMMA 5. *If $t \in C$, then $t' = \min(t, \sigma) \in C$ and $E(x_{t'}) \geq E(x_t)$.*

PROOF. From lemma 4 we have

$$\begin{aligned}E(x_t^-) &= \int_{[t'=t]} x_{t'}^- + \sum_{n=1}^{\infty} \int_{[t>n=\sigma]} x_t^- \geq \int_{[t'=t]} x_{t'}^- + \sum_{n=1}^{\infty} \int_{[t>n=\sigma]} \gamma_n^-\\ &= \int_{[t'=t]} x_{t'}^- + \sum_{n=1}^{\infty} \int_{[t>n=\sigma]} x_n^- = E(x_{t'}^-),\end{aligned} \tag{26}$$

so that $t' \in C$. The same argument without the $^-$ and with reversed inequality proves the inequality $E(x_t) \leq E(x_{t'})$.

A s.v. $t \in C$ is *optimal* if $v = E(x_t)$. A s.v. t in C is *regular* if it is 1-regular; that is, if for each $n \geq 1$, $t > n \Rightarrow E(x_t|\mathfrak{F}_n) > x_n$.

THEOREM 2. (a) *If* $\sigma \in C$ *and is regular, then it is optimal.* (b) *If* $v < \infty$ *and an optimal s.v. exists, then* $\sigma \in C$ *and is optimal and regular; moreover,* σ *is the minimal optimal s.v. and*

$$\sigma \geq n \Rightarrow E(x_\sigma|\mathfrak{F}_n) = E(\gamma_\sigma|\mathfrak{F}_n) = \gamma_n \qquad (n \geq 1). \tag{27}$$

PROOF. (a) If $\sigma \in C$ and is regular, then $\sigma > n \Rightarrow E(x_\sigma|\mathfrak{F}_n) > x_n$ for each $n \geq 1$. And for any $t \in C$, $\sigma = n$, $t \geq n \Rightarrow E(x_t|\mathfrak{F}_n) \leq \gamma_n = x_n$ by lemma 4. Hence by lemma 1 of [1], σ is optimal.

(b) Since $v < \infty$, $v_n = E(\gamma_n) < \infty$ for each $n \geq 1$. Let s in C be any optimal s.v., set $A = [s = n < \sigma]$, and suppose $P(A) > 0$. Then

$$\int_A \gamma_n > \int_A x_n + \epsilon \qquad \text{for some} \quad \epsilon > 0. \tag{28}$$

Choose $\{t_k\}_1^\infty$ in C_n by lemma 1; then $\int_A x_{t_k} \uparrow \int_A \gamma_n$, so that we can find k so large that $\int_A x_{t_k} > \int_A \gamma_n - \epsilon$. Set

$$s' = \begin{cases} s & \text{off } A \\ t_k & \text{on } A \end{cases}; \tag{29}$$

then it is easy to see that s' is a s.v. in C. But

$$E(x_{s'}) = \int_{\Omega - A} x_s + \int_A x_{t_k} > \int_{\Omega - A} x_s + \int_A x_n = E(x_s), \tag{30}$$

a contradiction. Hence $P(A) = 0$, and thus $P[\sigma \leq s] = 1$, so σ is a s.v. By lemma 5, $\sigma = \min(s, \sigma)$ is in C and σ is optimal and minimal.

For any $n \geq 1$, let $A = [E(x_\sigma|\mathfrak{F}_n) < \gamma_n, \sigma > n] \in \mathfrak{F}_n$. If $P(A) > 0$, then $\int_A \gamma_n > \int_A x_\sigma$, since $E(\gamma_n) \leq E(\gamma_1) = v < \infty$. By lemma 1, there exists t in C_n such that $\int_A x_t > \int_A x_\sigma$. Define

$$\tau = \begin{cases} t & \text{on } A \\ \sigma & \text{off } A \end{cases}; \tag{31}$$

then it is easy to see that τ is a s.v. in C and $E(x_\tau) > E(x_\sigma) = v$, a contradiction. Hence $P(A) = 0$, and by lemma 4,

$$\sigma > n \Rightarrow E(\gamma_\sigma|\mathfrak{F}_n) = E(x_\sigma|\mathfrak{F}_n) = \gamma_n > x_n, \tag{32}$$

so σ is regular and the last part of (b) holds.

6. Bounded stopping variables

The r.v.'s γ_n and the constants v_n are in general impossible to compute directly. To this end we define for any $N \geq 1$ and $1 \leq n \leq N$ the expressions

$$C_n^N = \text{all } t \in C_n \text{ such that } P[t \leq N] = 1; \; v_n^N = \sup_{t \in C_n^N} E(x_t); \tag{33}$$

$$\gamma_n^N = \operatorname*{ess\,sup}_{t \in C_n^N} E(x_t|\mathfrak{F}_n). \tag{34}$$

Then

$$-\infty < E(x_n) = v_n^n \leq v_n^{n+1} \leq \cdots \leq v_n \text{ and } x_n = \gamma_n^n \leq \gamma_n^{n+1} \leq \cdots \leq \gamma_n, \tag{35}$$

so that we can define

$$v_n' = \lim_{N\to\infty} v_n^N, \qquad \gamma_n' = \lim_{N\to\infty} \gamma_n^N, \tag{36}$$

and we have

$$-\infty < E(x_n) \leq v_n' \leq v_n, \qquad x_n \leq \gamma_n' \leq \gamma_n. \tag{37}$$

By the argument of theorem 1 applied to the *finite* sequence $\{x_n\}_1^N$, we have

$$\begin{aligned} &\gamma_N^N = x_N, \\ &\gamma_n^N = \max(x_n, E(\gamma_{n+1}^N|\mathfrak{F}_n)), \qquad (n = 1, \cdots, N-1), \end{aligned} \tag{38}$$

and $E(\gamma_n^N) = v_n^N$, so that γ_n^N and v_n^N are computable by recursion. By the monotone convergence theorem for expectations and conditional expectations, $E(\gamma_n') = v_n'$, and

$$\gamma_n' = \max(x_n, E(\gamma_{n+1}'|\mathfrak{F}_n)), \qquad (n \geq 1). \tag{39}$$

Hence $\{\gamma_n'\}_1^\infty$ *satisfies the same recursion relation as does* $\{\gamma_n\}_1^\infty$. (In [2], $\gamma_n^N = \beta_n^N$, $\gamma_n' = \beta_n$.)

THEOREM 3. *If the condition* A^-: $E(\sup_n x_n^-) < \infty$ *holds, then*

$$\gamma_n' = \gamma_n \quad \text{and} \quad v_n' = v_n, \qquad (n \geq 1). \tag{40}$$

PROOF. For any $t \in C_n$ and $A \in \mathfrak{F}_n$,

$$\int_{A[t\leq N]} x_t \leq \int_A x_{\min(t,N)} + \int_{A[t>N]} x_N^-. \tag{41}$$

Since $E(x_{\min(t,N)}|\mathfrak{F}_n) \leq \gamma_n^N \leq \gamma_n'$,

$$\int_{A[t\leq N]} x_t \leq \int_A \gamma_n' + \int_{A[t>N]} (\sup_m x_m^-). \tag{42}$$

Letting $N \to \infty$,

$$\int_A x_t \leq \int_A \gamma_n', \qquad E(x_t|\mathfrak{F}_n) \leq \gamma_n', \qquad \gamma_n \leq \gamma_n', \tag{43}$$

so $\gamma_n = \gamma_n'$ and $v_n = v_n'$.

COROLLARY. *If* A^- *holds and* $\{x_n\}_1^\infty$ *is Markovian, and* $\mathfrak{F}_n = \mathcal{B}(x_1, \cdots, x_n)$, *then* $\gamma_n = E(\gamma_n|x_n)$.

PROOF. The Markovian property of $\{x_n\}_1^\infty$ implies (by downward induction on n) $\gamma_n^N = E(\gamma_n^N|x_n)$ which entails $\gamma_n' = E(\gamma_n'|x_n)$, and then $\gamma_n = E(\gamma_n|x_n)$. (The assumption A^- will be dropped in the corollary to theorem 9.)

7. Supermartingales

A sequence $\{y_n\}_1^\infty$ of r.v.'s is a *supermartingale* (or lower semimartingale) if for each $n \geq 1$, y_n is measurable $(\mathfrak{F}_n)$, $E(y_n)$ exists, $-\infty \leq E(y_n) \leq \infty$, and $E(y_{n+1}|\mathfrak{F}_n) \leq y_n$. We shall denote by D the class of all supermartingales $\{y_n\}_1^\infty$ such that $y_n \geq x_n$ for each $n \geq 1$. The sequences $\{\gamma_n\}_1^\infty$ and $\{\gamma_n'\}_1^\infty$ are in D.

THEOREM 4. *The sequence* $\{\gamma_n'\}$ *is the minimal element of* D.

PROOF. For any $\{y_n\}_1^\infty$ in D,

$$y_n \geq x_n = \gamma_n^n,$$

$$y_{n-1} \geq E(y_n|\mathfrak{F}_{n-1}) \geq E(\gamma_n^n|\mathfrak{F}_{n-1}), \tag{44}$$

$$y_{n-1} \geq \max(x_{n-1}, E(\gamma_n^n|\mathfrak{F}_{n-1})) = \gamma_{n-1}^n, \cdots, y_i \geq \gamma_i^n, \cdots$$

so that

$$y_i \geq \lim_{n\to\infty} \gamma_i^n = \gamma_i', \qquad (i \geq 1). \tag{45}$$

We shall define various types of "regularity" for elements of D, according to the class of s.v.'s t for which $E(y_t)$ is assumed to exist and the relation

$$t \geq n \Rightarrow E(y_t|\mathfrak{F}_n) \leq y_n, \qquad (n \geq 1) \tag{46}$$

to hold. An element $\{y_n\}_1^\infty$ of D is said to be

(a) *regular* if for every s.v. t, $E(y_t)$ exists and (46) holds;

(b) *semiregular* if for every s.v. t such that $E(y_t)$ exists, (46) holds;

(c) *C-regular* if for every s.v. $t \in C$ (for which $E(y_t)$ necessarily exists), (46) holds.

Clearly, for elements of D, regular $\Rightarrow$ semiregular $\Rightarrow$ C-regular.

We shall use the notation A^+: $E(\sup_n x_n^+) < \infty$, A^*: $E(x_t)$ exists for every s.v. t. Clearly, $A^+ \Rightarrow A^* \Leftarrow A^-$.

LEMMA 6. *If A^* holds, then for any $\epsilon > 0$ and $n \geq 1$, there exists $s \in C_n$ such that*

$$E(x_s|\mathfrak{F}_n) > \gamma_n - \epsilon \qquad \text{on} \quad [\gamma_n < \infty]. \tag{47}$$

PROOF. Choose $\{t_k\}_1^\infty$ in C_n by lemma 1. On $[\gamma_n < \infty]$ define α = first $k \geq 1$ such that $E(x_{t_k}|\mathfrak{F}_n) > \gamma_n - \epsilon$, and set

$$s = \begin{cases} t_\alpha \text{ on } [\gamma_n < \infty] \\ n \text{ elsewhere.} \end{cases} \tag{48}$$

Then $E(x_s)$ exists, and on $[\gamma_n < \infty]$, $E(x_s|\mathfrak{F}_n) > \gamma_n - \epsilon$. Hence,

$$E(x_s) \geq \int_{[\gamma_n<\infty]} (\gamma_n - \epsilon) + \int_{[\gamma_n=\infty]} x_n > -\infty, \tag{49}$$

so that $s \in C_n$.

LEMMA 7. (a) *Condition A^- implies $E(\gamma_t^-) = E((\gamma_t')^-) < \infty$ for every s.v. t,* and (b) *condition A^+ implies $E((\gamma_t')^+) \leq E(\gamma_t^+) < \infty$ for every s.v. t.*

PROOF. (a) Since by theorem 3 $x_n \leq \gamma_n' = \gamma_n$, $\gamma_t^- = (\gamma_t')^- \leq \sup x_n^-$.

(b) Since

$$\gamma_n^+ = \operatorname*{ess\,sup}_{t\in C_n} E^+(x_t|\mathfrak{F}_n) \leq E(\sup_j x_j^+|\mathfrak{F}_n), \tag{50}$$

then

$$E((\gamma_t')^+) \leq E(\gamma_t^+) = \sum_{n=1}^\infty \int_{[t=n]} \gamma_n^+ \leq \sum_{n=1}^\infty \int_{[t=n]} E(\sup_j x_j^+|\mathfrak{F}_n) \tag{51}$$

$$= E(\sup_j x_j^+).$$

THEOREM 5. (a) *If* $\{y_n\}_1^\infty \in D$ *and is C-regular, then* $y_n \geq \gamma_n$ *for each* $n \geq 1$;
(b) $A^* \Rightarrow \{\gamma_n\}_1^\infty$ *is semiregular;*
(c) A^- *or* $A^+ \Rightarrow \{\gamma_n\}_1^\infty$ *is regular;*
(d) $\{\gamma_n\}_1^\infty$ *is C-regular.*

PROOF. (a) If $\{y_n\}_1^\infty \in D$ and is C-regular, then

$$\gamma_n = \operatorname*{ess\,sup}_{t \in C_n} E(x_t|\mathfrak{F}_n) \leq \operatorname*{ess\,sup}_{t \in C_n} E(y_t|\mathfrak{F}_n) \leq y_n. \tag{52}$$

(b) Let τ be any s.v. such that $P[\tau \geq n] = 1$ and $E(\gamma_\tau)$ exists. For arbitrary $\epsilon > 0$, $k \geq n$, and $m \geq 1$, setting $A_m = [\gamma_n < m]$, we have

$$m \geq \int_{A_m} \gamma_n \geq \int_{A_m} \gamma_{n+1} \geq \cdots \geq \int_{A_m} \gamma_k \geq \cdots, \tag{53}$$

so that $\gamma_k < \infty$ on A_m. Hence, $\gamma_k < \infty$ on $A = [\gamma_n < \infty]$. By lemma 6, we can choose $t_k \in C_k$ such that

$$E(x_{t_k}|\mathfrak{F}_k) > \gamma_k - \epsilon \qquad \text{on} \quad A. \tag{54}$$

Define

$$t = \begin{cases} t_k & \text{on } A[\tau = k], \\ \tau & \text{off } A. \end{cases} \tag{55}$$

Then $E(x_t)$ exists, and on A,

$$\begin{aligned} E(x_t|\mathfrak{F}_n) &= E\left(\sum_{k=n}^\infty I_{[\tau=k]} \cdot E(x_{t_k}|\mathfrak{F}_k)|\mathfrak{F}_n\right) \geq E\left(\sum_{k=n}^\infty I_{[\tau=k]}(\gamma_k - \epsilon)|\mathfrak{F}_n\right) \\ &= E(\gamma_\tau|\mathfrak{F}_n) - \epsilon; \end{aligned} \tag{56}$$

and therefore on A, by the remark preceding lemma 1,

$$\gamma_n = \operatorname*{ess\,sup}_{t \in \tilde{C}_n} E(x_t|\mathfrak{F}_n) \geq E(\gamma_\tau|\mathfrak{F}_n) - \epsilon \tag{57}$$

(recall that $\tilde{C}_n$ = all s.v.'s $t \geq n$ such that $E(x_t)$ exists). Hence,

$$\gamma_n \geq E(\gamma_\tau|\mathfrak{F}_n) \qquad \text{on} \quad \Omega. \tag{58}$$

Now let t be any s.v. such that $E(\gamma_t)$ exists. Set $\tau = \max(t, n)$. Then if $E(\gamma_t^+) = \infty$, $E(\gamma_t^-) < \infty$, and hence

$$E(\gamma_\tau^-) = \int_{[t>n]} \gamma_t^- + \int_{[t \leq n]} \gamma_n^- < \infty, \tag{59}$$

while if $E(\gamma_t^+) < \infty$, then

$$E(\gamma_\tau^+) = \int_{[t>n]} \gamma_t^+ + \int_{[t \leq n]} \gamma_n^+ < \infty, \tag{60}$$

since

$$\infty > \int_{[t \leq n]} \gamma_t = \sum_{k=1}^n \int_{[t=k]} \gamma_k \geq \sum_{k=1}^n \int_{[t=k]} \gamma_n = \int_{[t \leq n]} \gamma_n. \tag{61}$$

Hence $E(\gamma_\tau)$ exists. By the previous result, $\gamma_n \geq E(\gamma_\tau|\mathfrak{F}_n)$, and hence,

$$t \geq n \Rightarrow \gamma_n \geq E(\gamma_\tau|\mathfrak{F}_n) = E(\gamma_t|\mathfrak{F}_n). \tag{62}$$

(c) This statement follows from (b) and lemma 7.

(d) For $0 \leq b < \infty$, let $x_n(b) = \min(x_n, b)$, and let γ_n^b $(\leq \gamma_n)$ denote γ_n for the sequence $\{x_n(b)\}_1^\infty$. As $b \to \infty$, $-x_n^- \leq \gamma_n^b \uparrow \tilde{\gamma}_n$, say, where $\tilde{\gamma}_n \leq \gamma_n$, and for any t in C_n, $x_t(b) \geq -x_t^-$, so that $E(x_t(b)|\mathfrak{F}_n) \uparrow E(x_t|\mathfrak{F}_n)$. Since $\tilde{\gamma}_n \geq \gamma_n^b \geq E(x_t(b)|\mathfrak{F}_n)$, $\tilde{\gamma}_n \geq E(x_t|\mathfrak{F}_n)$, and hence $\tilde{\gamma}_n \geq \gamma_n$, $\tilde{\gamma}_n = \gamma_n$. Now if $t \in C$, then by (c), $t \geq n \Rightarrow E(\gamma_t^b|\mathfrak{F}_n) \leq \gamma_n^b \leq \gamma_n$. As $b \to \infty$, since $\gamma_t^b \geq -x_t^-$ and $E(x_t^-) < \infty$, $t \geq n \Rightarrow E(\gamma_t|\mathfrak{F}_n) \leq \gamma_n$, so $\{\gamma_n\}_1^\infty$ is C-regular.

COROLLARY 1. (a) *The sequence* $\{\gamma_n\}_1^\infty$ *is the minimal C-regular element of D.*

(b) *Condition* A^* *implies that* $\{\gamma_n\}_1^\infty$ *is the minimal semiregular element of D.*

(c) *Either* A^- *or* A^+ *implies that* $\{\gamma_n\}_1^\infty$ *is the minimal regular element of D.*

We remark that under A^-, $E(\sup_n \gamma_n^-) \leq E(\sup_n x_n^-) < \infty$. Hence, by a well-known theorem, $\{\gamma_n\}_1^\infty$ is regular, and similarly for $\{\gamma_n'\}_1^\infty$. By theorems 4 and 5(a), $\{\gamma_n'\}_1^\infty = \{\gamma_n\}_1^\infty$, which gives an alternative proof of theorem 3.

COROLLARY 2. *If* $\gamma_n^b = \text{ess sup}_{t\in C_n} E(\min(x_t, b)|\mathfrak{F}_n)$, *then*

$$\gamma_n = \lim_{b\to\infty} \gamma_n^b. \qquad (n \geq 1). \tag{63}$$

8. Almost optimal stopping variables

LEMMA 8. *If* $v < \infty$, *then for any* $\epsilon > 0$, $P[x_n \geq \gamma_n - \epsilon, \text{i.o.}] = 1$.

PROOF. Since $\infty > v = E(\gamma_1) \geq E(\gamma_2) \geq \cdots$, we have $P[\gamma_n < \infty] = 1$ for each $n \geq 1$. Choose any $\epsilon > 0$ and $r > 0$, and define for $n \geq 1$,

$$B_n = \left[E(x_{t_n}|\mathfrak{F}_n) > \gamma_n - \frac{\epsilon}{r}\right], \tag{64}$$

where $\{t_n\}_1^\infty$ is chosen by lemma 1 for each $n \geq 1$ so that $t_n \in C_n$ and $P(B_n) > 1 - 1/r$ (convergence a.e. $\Rightarrow$ convergence in probability). Define

$$B = [x_n < \gamma_n - \epsilon \quad \text{for all} \quad n \geq m] \tag{65}$$

where m is any fixed positive integer. Then

$$x_n \leq \gamma_n - \epsilon I_B \qquad \text{for} \quad n \geq m, \tag{66}$$

so on B_n for any $n \geq m$,

$$\begin{aligned} \gamma_n - \frac{\epsilon}{r} < E(x_{t_n}|\mathfrak{F}_n) &\leq E(\gamma_{t_n}|\mathfrak{F}_n) - \epsilon P(B|\mathfrak{F}_n) \\ &\leq \gamma_n - \epsilon P(B|\mathfrak{F}_n) \qquad \text{by theorem 5(d).} \end{aligned} \tag{67}$$

Hence on B_n, $P(B|\mathfrak{F}_n) \leq 1/r$, and therefore $P(BB_n) \leq 1/r$. It follows that $P(B) \leq P(BB_n) + P(\Omega - B_n) \leq (1/r) + (1/r) = (2/r)$. Since r can be arbitrarily large, $P(B) = 0$, and therefore,

$$P[x_n \geq \gamma_n - \epsilon \text{ for some } n \geq m] = 1 \tag{68}$$

and

$$P[x_n \geq \gamma_n - \epsilon, \text{i.o.}] = \lim_{m\to\infty} 1 = 1. \tag{69}$$

THEOREM 6. *For any* $\epsilon \geq 0$, *define*

$$s = \text{first } n \geq 1 \text{ such that } x_n \geq \gamma_n - \epsilon \ (s = \infty \text{ if no such } n \text{ exists}). \tag{70}$$

Assume the following: (a) $P[s < \infty] = 1$,
(b) $E(x_s)$ *exists*,
(c) $\liminf_{n\to\infty} \int_{[s>n]} E^+(\gamma_{n+1}|\mathfrak{F}_n) = 0.$

Then $E(x_s) \geq v - \epsilon$.

PROOF. We can assume $E(x_s) < \infty$. Since $\gamma_s \leq x_s + \epsilon$, $E(\gamma_s) < \infty$. Now

$$\begin{aligned}(71)\qquad v = E(\gamma_1) &= \int_{[s=1]} \gamma_s + \int_{[s>1]} E(\gamma_2|\mathfrak{F}_1)\\ &= \int_{[s=1]} \gamma_s + \int_{[s=2]} \gamma_s + \int_{[s>2]} E(\gamma_3|\mathfrak{F}_2) = \cdots\\ &= \int_{[1\leq s\leq n]} \gamma_s + \int_{[s>n]} E(\gamma_{n+1}|\mathfrak{F}_n) \leq \int_{[1\leq s\leq n]} \gamma_s + \int_{[s>n]} E^+(\gamma_{n+1}|\mathfrak{F}_n).\end{aligned}$$

Letting $n \to \infty$, $v \leq E(\gamma_s) \leq E(x_s) + \epsilon$.

COROLLARY. *For any* $\epsilon \geq 0$, *define* s *by* (70). *Then*

(i) *for* $\epsilon > 0$, $A^+ \Rightarrow P[s < \infty] = 1$ *and* $E(x_s) \geq v - \epsilon$;

(ii) *for* $\epsilon = 0$, $\{A^+, P[s < \infty] = 1\} \Rightarrow E(x_s) = v$.

PROOF. Condition A^+ implies $v < \infty$, and by lemma 8, this implies that $P[s < \infty] = 1$. Condition A^+ also implies (b) and (c).

THEOREM 7. *Let* $\{\alpha_n\}_1^\infty$ *be any sequence of r.v.'s such that* α_n *is* $(\mathfrak{F}_n)$ *measurable and* $E(\alpha_n)$ *exists for each* $n \geq 1$, *and such that*

(a) $\alpha_n = \max(x_n, E(\alpha_{n+1}|\mathfrak{F}_n))$,

(b) $P[x_n \geq \alpha_n - \epsilon \text{ i.o.}] = 1$ *for every* $\epsilon > 0$,

(c) $\{E^+(\alpha_{n+1}|\mathfrak{F}_n)\}_1^\infty$ *is uniformly integrable*,

(d) *either* $E(\sup_n \alpha_n^-) < \infty$, *or* A^+ *holds*.

Then for each $n \geq 1$, $\alpha_n \leq \gamma_n$.

PROOF. For $m \geq 1$, $A \in \mathfrak{F}_m$, and $\epsilon > 0$, define $t =$ first $n \geq m$ such that $x_n \geq \alpha_n - \epsilon$. Then $P[m \leq t < \infty] = 1$. If the first part of (d) holds, then $E(\alpha_t^-) < \infty$, and since $x_t \geq \alpha_t - \epsilon$, it follows that $E(x_t^-) < \infty$, and hence, by theorem 5(d),

$$(72)\qquad \int_A \alpha_t \leq \int_A x_t + \epsilon \leq \int_A \gamma_t + \epsilon \leq \int_A \gamma_m + \epsilon.$$

If A^+ holds, then $E(\alpha_t^+) \leq E(x_t^+) + \epsilon < \infty$, and the same result follows from theorem 5(c). Now

$$\begin{aligned}(73)\qquad \int_A \alpha_m &= \int_{A[t=m]} \alpha_t + \int_{A[t>m]} \alpha_{m+1} = \cdots = \int_{A[m\leq t\leq m+k]} \alpha_t\\ &+ \int_{A[t>m+k]} \alpha_{m+k+1} \leq \int_{A[m\leq t\leq m+k]} \alpha_t + \int_{A[t>m+k]} E^+(\alpha_{m+k+1}|\mathfrak{F}_{m+k}).\end{aligned}$$

Letting $k \to \infty$, it follows from (c) that

$$(74)\qquad \int_A \alpha_m \leq \int_A \alpha_t \leq \int_A \gamma_m + \epsilon,$$

so since ϵ was arbitrarily small, $\int_A \alpha_m \leq \int_A \gamma_m$, and therefore, $\alpha_m \leq \gamma_m$.

COROLLARY. *Assume that A^- holds. If $\{\alpha_n\}_1^\infty$ is any sequence such that α_n is measurable* $(\mathfrak{F}_n)$, *$E(\alpha_n)$ exists for each $n \geq 1$, and* (a), (b), *and* (c) *hold, then*

$$\alpha_n = \gamma_n. \tag{75}$$

PROOF. By theorems 7, 3, and 4, since A^- implies (d),

$$\gamma_n' \leq \alpha_n \leq \gamma_n = \gamma_n'. \tag{76}$$

9. A theorem of Dynkin

We next prove a slight generalization of a theorem of Dynkin [3]. Let $\{z_n\}_1^\infty$ be a homogeneous discrete time Markov process with arbitrary state space Z. For any nonnegative measurable function $g(\cdot)$ on Z, define the function $Pg(\cdot)$ by

$$Pg(z) = E(g(z_{n+1})|z_n = z), \tag{77}$$

and set

$$Qg = \max(g, Pg), \qquad Q_g^{k+1} = Q(Q^k g), \qquad (k \geq 0), \quad Q_g^\circ = g. \tag{78}$$

Then $g \leq Qg \leq Q^2 g \leq \cdots$, so

$$h = \lim_{N\to\infty} Q^N g \tag{79}$$

exists. Let $\mathfrak{F}_n = \mathfrak{B}(z_1, \cdots, z_n)$ and consider the sequence $\{x_n\}_1^\infty$ with $x_n = g(z_n)$.

THEOREM 8. *For the process defined above,* $\sup_t E(g(z_t)) = E(h(z_1))$.

PROOF. By theorem 3,

$$\gamma_1 = \gamma_1' = \lim_{N\to\infty} \gamma_1^N, \tag{80}$$

where

$$\begin{aligned}
\gamma_N^N &= g(z_N),\\
\gamma_{N-1}^N &= \max(g(z_{N-1}), E(g(z_N)|z_{N-1})) = Qg(z_{N-1}),\\
\gamma_{N-2}^N &= \max(g(z_{N-2}), E(Qg(z_{N-1})|z_{N-2})) = \max(g(z_{N-2}), PQg(z_{N-2}))\\
&= \max(g(z_{N-2}), Pg(z_{N-2}), PQg(z_{N-2})) = Q^2 g(z_{N-2}),\\
&\cdot\\
&\cdot\\
&\cdot\\
\gamma_1^N &= Q^{N-1} g(z_1) \to h(z_1) \quad \text{as} \quad N \to \infty.
\end{aligned} \tag{81}$$

Hence $\gamma_1 = h(z_1)$ and $v = E(\gamma_1) = E(h(z_1))$.

10. The triple limit theorem

LEMMA 9. *Assume A^+ holds, and define*

$$\begin{aligned}
x_n(a) &= \max(x_n, -a), \qquad (0 \leq a < \infty),\\
\gamma_n^a &= \operatorname*{ess\,sup}_{P[t \geq n]=1} E(x_t(a)|\mathfrak{F}_n).
\end{aligned} \tag{82}$$

Then

$$\gamma_n = \lim_{a\to\infty} \gamma_n^a. \tag{83}$$

PROOF. Since $\gamma_n^a = \max(x_n(a), E(\gamma_{n+1}^a|\mathfrak{F}_n))$ and $\gamma_n(a) \downarrow \gamma_n^*$, say, as $a \to \infty$, where $\gamma_n^* \geq \gamma_n$, it follows from A^+ that $\gamma_n^* = \max(x_n, E(\gamma_{n+1}^*|\mathfrak{F}_n))$. For any $\epsilon > 0$ and $m \geq 1$, define $s =$ first $n \geq m$ such that $x_n \geq \gamma_n^* - \epsilon$ ($= \infty$ if no such n exists). Then $\{\gamma_{\min(s,n)}^*\}_{n=m}^{\infty}$ is a martingale, since

$$(84) \qquad E(\gamma_{\min(s,n+1)}^*) = I_{[s>n]}E(\gamma_{n+1}^*|\mathfrak{F}_n) + I_{[s\leq n]}E(\gamma_s^*|\mathfrak{F}_n)$$
$$= I_{[s>n]}\cdot\gamma_n^* + I_{[s=m]}\cdot\gamma_m^* + \cdots + I_{[s=n]}\cdot\gamma_n^* = \gamma_{\min(s,n)}^*.$$

Since $E((\gamma_{\min(s,n)}^*)^+) \leq E(\sup_n x_n^+) < \infty$, and since $E((\gamma_m^*)^-) < \infty$, we have by a martingale convergence theorem,

$$(85) \qquad \gamma_{\min(s,n)}^* \to \text{a finite limit} \qquad \text{as} \quad n \to \infty,$$

and hence,

$$(86) \qquad \gamma_n^* \to \text{a finite limit on } [s = \infty] \qquad \text{as} \quad n \to \infty.$$

But on $[s = \infty]$, $\gamma_n^* > x_n + \epsilon$ for $n \geq m$, so

$$(87) \qquad \limsup_n x_n \leq \limsup_n \gamma_n^* - \epsilon \qquad \text{on} \quad [s = \infty].$$

Since $\gamma_n^a \leq E(\sup_{j\geq m} x_j(a)|\mathfrak{F}_n)$ for $n \geq m$,

$$(88) \qquad \limsup_n \gamma_n^* \leq \limsup_n \gamma_n^a \leq \sup_{j\geq m} x_j(a),$$

and hence,

$$(89) \qquad \limsup_n \gamma_n^* \leq \limsup_n x_n(a) = \max(\limsup_n x_n, -a),$$

and

$$(90) \qquad \limsup_n \gamma_n^* \leq \limsup_n x_n,$$

but $\gamma_n^* \geq x_n$. Hence,

$$(91) \qquad \limsup_n \gamma_n^* = \limsup_n x_n,$$

contradicting (87) unless $P[s = \infty] = 0$. Hence,

$$(92) \qquad P[x_n \geq \gamma_n^* - \epsilon, \text{i.o.}] = 1,$$

and by theorem 7, $\gamma_n^* \leq \gamma_n$. Therefore, $\gamma_n^* = \gamma_n$.

THEOREM 9. *The random variables γ_n are equal to*

$$(93) \qquad \gamma_n = \lim_{b\to\infty} \lim_{a\to-\infty} \lim_{N\to\infty} \gamma_n^N(a, b),$$

where

$$(94) \qquad \gamma_n^N(a, b) = \operatorname*{ess\,sup}_{P[n\leq t\leq N]=1} E(x_t(a, b)|\mathfrak{F}_n)$$

and

$$(95) \qquad x(a, b) = \begin{cases} a & \text{if} \quad x < a, \\ x & \text{if} \quad a \leq x \leq b, \\ b & \text{if} \quad x > b. \end{cases}$$

PROOF. This follows from lemma 9, theorem 3, and corollary 2 of theorem 5.

COROLLARY 1. *The values v_n are equal to*

$$\lim_{b\to\infty}\lim_{a\to-\infty}\lim_{N\to\infty} v_n^N(a, b). \tag{96}$$

COROLLARY 2. *If* $\{x_n\}_1^\infty$ *is Markovian and* $\mathfrak{F}_n = \mathfrak{B}(x_1, \cdots, x_n)$, *then*

$$\gamma_n = E(\gamma_n|x_n). \tag{97}$$

If the x_n *are independent, then*

$$E(\gamma_{n+1}|\mathfrak{F}_n) = E(\gamma_{n+1}) = v_{n+1}, \tag{98}$$

and the v_n *satisfy the recursion relation*

$$v_n = E\{\max (x_n, v_{n+1})\}, \qquad (n \geq 1). \tag{99}$$

PROOF. By induction $\gamma_n^N(a, b) = E(\gamma_n^N(a, b)|x_n)$ from $n = N$ down, as in the proof of the corollary of theorem 3. Letting N, a, b become infinite yields (97). Under independence,

$$E(\gamma_{n+1}|\mathfrak{F}_n) = E(E(\gamma_{n+1}|x_{n+1})|\mathfrak{F}_n) = E(\gamma_{n+1}) = v_{n+1}. \tag{100}$$

And from $\gamma_n = \max (x_n, E(\gamma_{n+1}|\mathfrak{F}_n)) = \max (x_n, v_{n+1})$, we obtain (99) on taking expectations.

11. Remarks on the independent case

THEOREM 10. *Let the* $\{x_n\}_1^\infty$ *be independent with* $\mathfrak{F}_n = B(x_1, \cdots, x_n)$. *Set* s = *first* $n \geq 1$ *such that* $x_n \geq \gamma_n - \epsilon$ *for* $\epsilon > 0$ ($= \infty$ *if no such* n *exists*). *Then*

$$v < \infty \Rightarrow P[s < \infty] = 1, \tag{101}$$

and if in addition $E(x_s)$ *exists, then*

$$E(x_s) \geq v - \epsilon. \tag{102}$$

PROOF. By lemma 8 and theorem 6, since by (87)

$$\int_{[s>n]} E^+(\gamma_{n+1}|\mathfrak{F}_n) = \int_{[s>n]} v_{n+1}^+ = v_{n+1}^+ P[s > n] \leq v^+ P[s > n] \to 0. \tag{103}$$

We remark that when $\epsilon = 0$ the conditions $v < \infty$, $P[s < \infty] = 1$, $E(x_s)$ exists, imply $E(x_s) = v$.

THEOREM 11. *Let the* $\{x_n\}_1^\infty$ *be independent with* $\mathfrak{F}_n = \mathfrak{B}(x_1, \cdots, x_n)$, *and let* $\{\alpha_n\}_1^\infty$ *be any sequence of r.v.'s such that* α_n *is measurable* $(\mathfrak{F}_n)$ *and* $E(\alpha_n)$ *exists,* $n \geq 1$. *If*

(a) $\alpha_n = \max (x_n, E(\alpha_{n+1}|\mathfrak{F}_n))$, $(n \geq 1)$,

(b) $P(x_n \geq \alpha_n - \epsilon \text{ i.o.}) = 1$ *for every* $\epsilon > 0$,

(c) $E(\alpha_{n+1}|\mathfrak{F}_n) = c_n$ = *constant, with* $E(\alpha_1) = c_1 < \infty$,

(d) A^+ *holds, or* $\liminf_n E(x_n) > -\infty$,

then

$$\alpha_n \leq \gamma_n, \qquad (n \geq 1). \tag{104}$$

PROOF. Define A and t as in theorem 7. Since

$$c_n = E\{\max(x_{n+1}, c_{n+1})|\mathfrak{F}_n\} \geq c_{n+1}, \tag{105}$$

we have

$$\begin{aligned} \int_A \alpha_m &= \int_{A[m \leq t \leq m+k]} \alpha_t + \int_{A[t > m+k]} \alpha_{m+k+1} \\ &= \int_{A[m \leq t \leq m+k]} \alpha_t + \int_{A[t > m+k]} c_{m+k} \\ &\leq \int_{A[m \leq t \leq m+k]} \alpha_t + c_1 P[t > m + k]. \end{aligned} \tag{106}$$

Hence under A^+ (or A^-),

$$\begin{aligned} \int_A \alpha_m \leq \liminf_{k\to\infty} \int_{A[m \leq t \leq m+k]} \alpha_t &\leq \liminf_{k\to\infty} \int_{A[m \leq t \leq m+k]} x_t + \epsilon \\ \leq \liminf_{k\to\infty} \int_{A[m \leq t \leq m+k]} \gamma_t + \epsilon &= \int_A \gamma_t + \epsilon \leq \int_A \gamma_m + \epsilon \end{aligned} \tag{107}$$

by theorem 5(c), so $\alpha_m \leq \gamma_m$. If the second part of (d) holds, then since $c_n \downarrow c$, say, where $c \geq \liminf_n E(x_n) > -\infty$, and $x_t \geq c_t - \epsilon \geq c - \epsilon$, it follows that $E(x_t^-) < \infty$, so theorem 5(d) yields the same conclusion.

REMARKS. 1. Lemmas 2 and 3 are slight extensions of lemmas 1 and 2 of [2].

2. Theorem 1 has been proved independently by G. Haggstrom [4] when $E|x_n| < \infty$ and $E(\sup_n x_n^+) < \infty$, as have theorem 4, corollary 1(c) of theorem 5 under A^+, and the corollary of theorem 6. The latter was also proved by J. L. Snell [5].

3. We are greatly indebted to Mr. D. Siegmund for improvements in the statement and proof of many of our results. In particular, theorem 9 is largely due to him.

REFERENCES

[1] Y. S. CHOW and H. ROBBINS, "A martingale system theorem and applications," *Proceedings of the Fourth Berkeley Symposium on Mathematical Statistics and Probability*, Berkeley and Los Angeles, University of California Press, 1961, Vol. 1, pp. 93–104.

[2] ———, "On optimal stopping rules," *Z. Wahrscheinlichkeitstheorie und Verw. Gebiete*, Vol. 2 (1963), pp. 33–49.

[3] E. B. DYNKIN, "The optimum choice of the instant for stopping a Markov process," *Dokl. Akad. Nauk SSSR*, Vol. 150 (1963), pp. 238–240; *Soviet Math. Dokl.*, Vol. 4 (1963), pp. 627–629.

[4] G. HAGGSTROM, "Optimal stopping and experimental design," *Ann. Math. Statist.*, Vol. 37 (1966), pp. 7–29.

[5] L. J. SNELL, "Application of martingale system theorems," *Trans. Amer. Math. Soc.*, Vol. 73 (1952), pp. 293–312.

EXISTENCE AND PROPERTIES OF CERTAIN OPTIMAL STOPPING RULES

ARYEH DVORETZKY
THE HEBREW UNIVERSITY OF JERUSALEM
and
COLUMBIA UNIVERSITY

1. Introduction

The main purpose of this note is to prove the existence of optimal stopping rules for certain problems involving sums of independent, identically distributed random variables. A special case was treated by Y. S. Chow and H. E. Robbins [2]. Their problem is very easily stated: let s_n be the excess of the number of heads over the number of tails in the first n tosses of an infinite sequence of independent tosses of a fair coin. Does there exist a stopping variable τ for which the expected average gain is maximal? In other words, does there exist a τ for which the expectation of s_τ/τ is at least as great as the expectation of s_t/t for any other stopping variable t? It turns out that this simple problem is not reducible to any of the available standard results on the existence of optimal stopping rules. Chow and Robbins do prove the existence of an optimal τ by an ingenious method which is, at least in part, suited only to the special case which they consider. Here, following in part the method of [2] and substituting general considerations for the specific ones used there, we establish the existence of an optimal stopping variable, maximizing the expected average gain under the sole assumption that the random variables involved have finite variance.

In section 2 we prove the above result. Our method also yields interesting information on the structure of the optimal τ which we present in section 3. For the sake of clarity, we confined the main exposition to the problem of maximizing the expected average gain; however, the methods developed here can deal with more general situations, and one generalization is presented in section 4. The last section contains various remarks.

Throughout we denote by $(\Omega, \mathcal{B}, P)$ the underlying probability space. Also, E denotes expectation, and we write $\{\cdots\}$ to denote $\{\omega\colon \cdots\}$, the set of ω having the indicated properties. All random variables are, of course, defined only almost surely but, in the interest of brevity, this qualification is usually omitted.

Throughout the paper, $x_1, x_2, \cdots, x_n, \cdots$, is a sequence of *independent, iden-*

Research supported in part by National Science Foundation Grant NSF-GP-3694 at Columbia University, Department of Mathematical Statistics.

tically distributed random variables with zero mean and positive finite variance σ^2. We put

$$s_n = x_1 + \cdots + x_n, \qquad (n = 1, 2, \cdots) \tag{1}$$

and denote by $\mathcal{B}_n$ the σ-field generated by $x_1, \cdots, x_n$. By a *stopping variable* t (relative to the sequence $x_1, x_2, \cdots$) we understand a random variable whose range is the set of positive integers and such that

$$\{t = n\} \in \mathcal{B}_n, \qquad (n = 1, 2, \cdots). \tag{2}$$

This definition implies

$$P(t < \infty) = 1. \tag{3}$$

For technical reasons we find it convenient to consider also *generalized stopping variables.* These are defined as random variables t whose range is the set consisting of the positive integers and $+\infty$, and which satisfy (2).

We shall denote by T the set of all generalized stopping variables, that is, those $t \in T_\infty$ which satisfy (3). If $y_1, y_2, \cdots$ is a sequence of random variables and $t \in T_\infty$, then we define

$$Ey_t = \int_{\{t<\infty\}} y_t \, dP, \tag{4}$$

provided the right side is defined. For $t \in T$, this reduces to the usual definition of the expectation of y_t.

2. Existence of an optimal stopping variable

In this section we shall prove the following result.

THEOREM 1. *There exists a stopping variable* $\tau \in T$ *such that*

$$E\frac{s_\tau}{\tau} = \sup_{t \in T_\infty} E\frac{s_t}{t}. \tag{5}$$

Moreover, we have

$$0 < E\frac{s_\tau}{\tau} < \frac{\pi}{\sqrt{6}}\sigma. \tag{6}$$

Since the proof is somewhat long, we shall break it into several auxiliary assertions.

LEMMA 1. *Let* $t \in T_\infty$, *and let* $t(m)$, $(m = 0, 1, 2, \cdots)$ *be defined by*

$$t(m) = \begin{cases} t & \text{if } t \le m, \\ \infty & \text{otherwise.} \end{cases} \tag{7}$$

Then $t(m) \in T_\infty$ *and*

$$Es_{t(m)}^2 \le m\sigma^2. \tag{8}$$

PROOF. That $t(m)$ is a generalized stopping variable follows immediately from the definition. To establish (8) we proceed as follows:

$$(9)\qquad Es_{t(m)}^2 = \sum_{i=1}^m \int_{\{t=i\}} s_i^2\,dP \le \sum_{i=1}^m \int_{\{t=i\}} (s_i^2 + (s_m - s_i)^2)\,dP$$

$$= \sum_{i=1}^m \int_{\{t=i\}} (s_i + (s_m - s_i))^2\,dP = \int_{\{t\le m\}} s_m^2\,dP \le Es_m^2 = m\sigma^2.$$

In the passage from the first to the second line we used the facts that $\{t = i\} \in \mathcal{B}_i$ and $E(x_j|\mathcal{B}_i) = 0$, for $j > i$.

LEMMA 2. *For all $t \in T_\infty$ and all $a > -1$ we have*

$$(10)\qquad E\left(\frac{s_t}{a+t}\right)^2 < \sum_{i=1}^\infty \frac{\sigma^2}{(a+1)^2}.$$

PROOF. Defining $t(m)$ by (8) we have

$$(11)\qquad E\left(\frac{s_t}{a+t}\right)^2 = \sum_{i=1}^\infty \frac{1}{(a+i)^2}\int_{\{t=i\}} s_i^2\,dP$$

$$= \sum_{i=1}^\infty \frac{1}{(a+i)^2}(Es_{t(i)}^2 - Es_{t(i-1)}^2).$$

Putting, for $i = 1, 2, \cdots$,

$$(12)\qquad v_i = Es_{t(i)}^2 - Es_{t(i-1)}^2,$$

the right side of (11) becomes

$$(13)\qquad \sum_{i=1}^\infty \frac{v_i}{(a+i)^2}$$

with v_i satisfying, by (12) and (8),

$$(14)\qquad v_i \ge 0, \qquad \sum_{i=1}^m v_i \le m\sigma^2, \qquad (m = 1, 2, \cdots).$$

Since $(a + i)^2 > 0$ and is strictly increasing with i, (13) is increased if some v_i is increased and a v_j, with $j > i$, is decreased by the same amount. Hence, the maximum of (13), for v_i satisfying (14), is obtained for, and only for, $v_1 = v_2 = \cdots = \sigma^2$.

This proves (10) with $\le$ instead of the sharp inequality. Though this is quite enough for our purposes, we add the short argument which yields (10). By the preceding, equality of the two sides of (10) would imply $v_1 = v_2 = \cdots = \sigma^2$. But $v_1 = \int_{\{t=1\}} x_1^2\,dP = \sigma^2$ implies that $x_1 \ne 0 \Rightarrow t = 1$ a.s., and thus, by independence, $v_2 = \int_{\{t=2\}} (x_1 + x_2)^2\,dP \le \int_{\{t>1\}} x_2^2\,dP = \sigma^2 P(t > 1) < \sigma^2$.

LEMMA 3. *For all $t \in T_\infty$ and $a > -1$ we have*

$$(15)\qquad E\frac{s_t}{a+t} < \sigma\left(\sum_{i=1}^\infty \frac{1}{(a+i)^2}\right)^{1/2}.$$

In particular we have for $a > 0$,

$$(16)\qquad E\frac{s_t}{a+t} < \frac{\sigma}{\sqrt{a}}.$$

PROOF. Inequality (15) is an immediate consequence of (10) and implies (16), since, for $a > 0$,

$$\sum_{i=1}^{\infty} \frac{1}{(a+i)^2} < \sum_{i=1}^{\infty} \int_{i-1}^{i} \frac{du}{(a+u)^2} = \int_0^{\infty} \frac{du}{(a+u)^2} = \frac{1}{a}. \tag{17}$$

The next lemmas study the expected value of $b + s_t/a + t$ for suitable generalized stopping variables t.

LEMMA 4. *If for some $a > 0$, real b and $t \in T_\infty$ we have*

$$E\frac{b + s_t}{a + t} \geq \frac{b}{a}, \tag{18}$$

then there exists $t' \in T_\infty$ satisfying

$$t' < \infty \Rightarrow s_{t'} > -b \tag{19}$$

and

$$E\frac{b + s_{t'}}{a + t'} \geq \frac{b}{a}. \tag{20}$$

PROOF. Put $t' = t$ if $t < \infty$ and $s_t > -b$; and $t' = \infty$ otherwise.

LEMMA 5. *Let $a > 0$, b real, and $t' \in T_\infty$ satisfy* (19) *and* (20). *Then we have*

$$E\frac{b' + s_{t'}}{a' + t'} \geq \frac{b'}{a'} \tag{21}$$

for all $a' \geq a$ and $b' \leq b$. (Moreover, for $b > 0$ the equality sign occurs in (21) *only for $a' = a$, $b' = b$.)*

PROOF. The derivative, with respect to b of the right side of (18), or (20), is $1/a$, while that of the left side is bounded by $1/(a + 1)$. Hence, (18) implies the same relation (even with strict inequality, unless the left side vanishes identically, that is $P(t = \infty) = 1$, and $b \leq 0$) when b is replaced by $b' < b$.

It remains to prove that (20) implies the same relation (even with strict inequality unless $P(t = \infty) = 1$) when a is replaced by $a' > a$. But (20) is equivalent to $E(a(b + s_{t'})/(a + t')) \geq b$ and, by (19), the left side is increasing in a (strictly, unless $t' = \infty$ almost surely).

LEMMA 6. *Let $a' \geq a > 0$, $b' \leq b$, $t \in T_\infty$, and* (18) *hold. Then there exists, for every $m = 0, 1, 2, \cdots$, a generalized stopping variable $t_m \in T_\infty$ satisfying*

$$t_m \leq m \Rightarrow s_{t_m} > b - b' \tag{22}$$

and

$$E\frac{b' + s_{t_m}}{a' + t_m} \geq \frac{b'}{a'}. \tag{23}$$

PROOF. For $m = 0$ the requirement (22) is vacuously satisfied, and by lemmas 4 and 5 there exists $t_0 = t'$ (the same for all $a' \geq a$, $b' \leq b$) satisfying (23). Having proved the existence of t_i for $i < m$, we put

$$t_0^{(m)}(x_1, x_2, \cdots, x_m, x_{m+1}, x_{m+2}, \cdots) = t_0(x_{m+1}, x_{m+2}, \cdots) \tag{24}$$

and define

$$(25)\qquad t_m = \begin{cases} t_{m-1} + t_0^{(m)} & \text{if} \quad t_{m-1} = m \quad \text{and} \quad s_m \leq b - b', \\ \infty & \text{otherwise.} \end{cases}$$

Since t_0 is a fixed generalized stopping variable, it follows that $t_m \in T_\infty$ and, by definition, satisfies (22). Moreover, it follows from (23) with $m = 0$ that

$$(26)\qquad E\,\frac{b' + s_{t_m}}{a' + t_m} \geq E\,\frac{b' + s_{t_{m-1}}}{a' + t_{m-1}}, \qquad (m = 1, 2, \cdots),$$

and thus (23) is valid for all m.

(We used a somewhat abbreviated notation in (24). The precise meaning is the following: if ω and ω' on Ω are such that $x_i(\omega') = x_{m+i}(\omega)$ for all $i = 1, 2, \cdots$, then $t_0^{(m)}(\omega) = t_0(\omega')$. The function $t_0^{(m)}$ itself is not a generalized stopping variable, since the set $\{t = i\}$ need not belong to $\mathcal{B}_i$, but it belongs to $\mathcal{B}_{m+i}$, and therefore, $t_m \in T_\infty$). It may be remarked that lemmas 4 and 5 hold for arbitrary sequences of random variables, but lemma 6 utilizes the stationarity of the sequence $x_1, x_2, \cdots$. In the following lemma, even more properties of this sequence are used.

LEMMA 7. *Let* (18) *hold for some* $a > 0$, $b > 0$, *and* $t \in T_\infty$. *Then there exists* $t^* \in T_\infty$ *satisfying*

$$(27)\qquad E\,\frac{\tfrac{1}{2}b + s_{t^*}}{a + t^*} > \frac{b}{2a}$$

and

$$(28)\qquad E\,\frac{1}{a + t^*} < \frac{1}{2a} + \frac{2\sigma^2}{b^2}.$$

PROOF. Let t^* be the t_m, whose existence is assured by the previous lemma, corresponding to $a' = a$, $b' = b/2$, and $m = [a]$ (namely, the greatest integer $\leq a$). Then (27) holds and, by (22),

$$(29)\qquad t^* \leq a \Rightarrow \max\,(s_1, s_2, \cdots, s_{[a]}) > b/2.$$

Therefore, by Kolmogorov's inequality,

$$(30)\qquad P(t^* \leq a) \leq \frac{\sigma^2[a]}{(b/2)^2} \leq \frac{4\sigma^2 a}{b^2}.$$

But

$$(31)\qquad E\,\frac{1}{a + t^*} \leq \frac{P(t^* \leq a)}{a + 1} + \frac{P(t^* > a)}{a + [a] + 1}$$

$$< \frac{P(t^* \leq a)}{a} + \frac{P(t^* > a)}{2a} = \frac{1}{2a} + \frac{P(t^* \leq a)}{2a}$$

and (28) follows from (30).

LEMMA 8. *Let* $a > 0$ *and*

$$(32)\qquad b \geq 5\sigma\sqrt{a}.$$

Then

$$(33)\qquad E\,\frac{b + s_t}{a + t} < \frac{b}{a}$$

for all $t \in T_\infty$.

PROOF. If (33) were false, we would have, by (27), (28), and lemma 3, for the t^* of the preceding lemma,

$$\frac{b}{2a} < \frac{b}{2} E \frac{1}{a + t^*} + E \frac{s_{t^*}}{a + t^*} < \frac{b}{4a} + \frac{\sigma^2}{b} + \frac{\sigma}{\sqrt{a}}. \tag{34}$$

Hence, $b^2 - 4\sigma\sqrt{a}b - 4\sigma^2 a < 0$, thus

$$b < 2\sigma\sqrt{a} + \sqrt{4\sigma^2 a + 4\sigma^2 a} = 2(1 + \sqrt{2})\sigma\sqrt{a}, \tag{35}$$

contradicting (32).

This lemma asserts that the conditional expectation of s_t/t, given that $t \geq n$ and $s_n \geq 5\sigma\sqrt{n}$, is maximized by putting $t = n$.

LEMMA 9. *We have*

$$E\left(\sup_{n=1,2,\cdots} \frac{s_n^+}{n}\right) < \infty. \tag{36}$$

PROOF. Denoting the sup in (36) by s, we have for every $u > 0$,

$$\begin{aligned} P(s \geq u) &\leq \sum_{i=1}^{\infty} P\left(\max_{2^{i-1} \leq n < 2^i} \frac{s_n}{n} \geq u\right) \\ &\leq \sum_{i=1}^{\infty} P(\max_{1 \leq n < 2^i} s_n \geq 2^{i-1}u). \end{aligned} \tag{37}$$

Therefore, by Kolmogorov's inequality,

$$P(s \geq u) < \sigma^2 \sum_{i=1}^{\infty} \frac{2^i}{2^{2(i-1)}u^2} = \frac{4\sigma^2}{u^2}, \tag{38}$$

whence

$$Es = -\int_0^\infty u\, dP(s \geq u) = \int_0^\infty P(s \geq u)\, du < \infty. \tag{39}$$

To complete the proof of the theorem we need the following result. It may be found in [1] (see lemma 2 there) however, for the sake of self-containedness we reproduce its proof here.

LEMMA 10. *Let $\bar{t} \in T$, let $y_n (n = 1, 2, \cdots)$ be a sequence of random variables satisfying*

$$E(\sup_{n=1,2,\cdots} y_n^+) < \infty, \tag{40}$$

and let $\overline{T}$ be the family of all stopping variables $t \leq \bar{t}$ (that is, those $t \in T$ for which $t(\omega) \leq \bar{t}(\omega)$ almost surely). Then there exists $\tau \in \overline{T}$ satisfying

$$Ey_\tau = \sup_{t \in \overline{T}} Ey_t. \tag{41}$$

PROOF. The stopping time $t \in T$ is called regular if $t > j \Rightarrow E(y_t|\mathcal{B}_j) > y_j$ for all $j = 1, 2, \cdots$. If t and t' are regular and $t \leq t'$, then we have on $\{t = j\}$ the inequality $E(y_{t'}|\mathcal{B}_j) \geq y_j = y_t$, hence $t \leq t' \Rightarrow Ey_{t'} \geq Ey_t$. Now, if the right side of (41) is $-\infty$, there is nothing to prove. We may therefore assume that it is finite, M say. Then there exists for every $n = 1, 2, \cdots$, a stopping variable $t_n \in \overline{T}$ with $Et_n \geq M - 1/n$. Let $t_n' =$ smallest integer $i \geq 1$ for which

$E(y_{t_n}|\mathcal{B}_i) \leq y_i$. Then $t_n' \in \overline{T}$, $Ey_{t'_n} \geq Ey_{t_n}$, and t_n' is regular. Putting $\bar{t}_n = \max(t_1', \cdots, t_n')$, it is clear that $t_n' \leq \bar{t}_n \leq \bar{t}_{n+1}$, and that $\bar{t}_n \in \overline{T}$ is regular. Finally let $\tau = \lim_{n=\infty} \bar{t}_n$. Because of (40) we have, by Fatou's lemma, $E(y_\tau|\mathcal{B}_j) \geq \limsup_{n=\infty} E(y_{\bar{t}_n}|\mathcal{B}_j)$ and, since $\tau > j \Rightarrow \bar{t}_n > j$ for some n, it follows that $\tau > j \Rightarrow E(y_\tau|\mathcal{B}_j) > y_j$. Hence, τ is regular. Since $\tau \geq \bar{t}_n \geq t_n'$ we have $Ey_\tau > M - 1/n$ for all n, that is, $Ey_\tau = M$.

Proof of (5). Let $\bar{t} \in T_\infty$ be defined as follows: $\bar{t} = j$ when j is the smallest positive integer for which $s_j \geq 5\sigma\sqrt{j}$. Then, by the law of the iterated logarithm, $\bar{t} \in T$. If $t \in T_\infty$ is arbitrary and we put $t' = \min(t, \bar{t})$, then, by lemma 8, $E(s_{t'}/t') \geq E(s_t/t)$. Thus the sup in (5) may be taken only over the class $\overline{T} \subset T$ of all stopping variables $\leq \bar{t}$. Since, by lemma 9, the sequence $y_n = s_n/n$ $(n = 1, 2, \cdots)$ satisfies (40), we can apply lemma 10 and deduce the existence of an optimal $\tau \in \overline{T}$ satisfying (5).

Proof of (6). For the generalized stopping variable t defined by $t = 1$ if $x_1 > 0$ and $t = \infty$ otherwise, we have $E(s_t/t) = Ex_1^+ > 0$. This gives the first inequality of (6). The other inequality follows from (15) with $a = 0$.

For reference in the sequel we state the following slight extension of theorem 1, which is proved in exactly the same way.

Theorem 1′. *Let a and b be real, then there exists a stopping variable $\tau \in T$ for which*

$$E\,\frac{b + s_\tau}{a + \tau} = \sup_{t \in T_\infty} \frac{b + s_t}{a + t}. \tag{42}$$

Since we do not need the analogue of (6), we do not state it here.

3. Structure of the optimal rule

By Theorem 1′, there exists for every $\beta > 0$ and $n = 1, 2, \cdots$ a stopping variable $\tau(\beta, n)$ for which

$$V_n(\beta) = E\,\frac{\beta + s_{\tau(\beta,n)}}{n + \tau(\beta, n)} = \sup_{t \in T_\infty} E\,\frac{\beta + s_t}{n + t}. \tag{43}$$

Since we have for all real β, β', and every $t \in T_\infty$,

$$\left| E\,\frac{\beta' + s_t}{n + t} - E\,\frac{\beta + s_t}{n + t} \right| \leq \frac{\beta' - \beta}{n + 1}, \tag{44}$$

it follows that $V_n(\beta)$ is continuous in β. Consider now the equation

$$V_n(\beta) = \frac{\beta}{n}. \tag{45}$$

Since $V_n(\beta) > 0$ for all β (consider the rule: stop when $s_i > -\beta$ for the first time), and since, by lemma 8, $V_n(\beta) < \beta/n$ for large β, it follows from the continuity of $V_n(\beta)$ that equation (45) is solvable for every n. If β_n is a solution of (45), then it follows from the last sentence of lemma 5 that $V_n(\beta) < \beta/n$ for $\beta > \beta_n$.

The equation

$$\frac{\beta_n}{n} = \sup_{t \in T_\infty} \frac{\beta_n + s_t}{n + t} \tag{46}$$

defines β_n *uniquely.*

By lemma 5, the sequence

$$\beta_1, \beta_2, \cdots, \beta_n, \cdots \tag{47}$$

is a strictly increasing sequence of positive numbers, and by the considerations of section 2, an optimal stopping variable τ_0 satisfying (5) may be defined by

$$\tau_0 = n \Rightarrow s_n \geq \beta_n \quad \text{and} \quad s_i < \beta_i \qquad \text{for} \quad i = 1, \cdots, n-1, \tag{48}$$

that is, by the rule: stop whenever $s_n \geq \beta_n$ for the first time.

The analysis of section 2 also shows that if τ is any stopping variable optimal in the sense of (5), then $P(\tau \geq \tau_0) = 1$, and that there exists an optimal τ with $P(\tau \neq \tau_0) > 0$ if and only if $P(s_n = \beta_n, \tau_0 = n) > 0$ for at least one $n = 1, 2, \cdots$.

Thus, if we identify two stopping variables which are equal almost surely, τ_0 given by (48) is the unique minimal optimal stopping variable. It is interesting to study the rate of growth of the sequence (47), and the following result contains information about this rate.

THEOREM 2. *The numbers* β_n $(n = 1, 2, \cdots)$ *defined by* (46) *satisfy*

$$\limsup_{n=\infty} \frac{\beta_n}{\sqrt{n}} \leq c_1 \sigma \tag{49}$$

and

$$\liminf_{n=\infty} \frac{\beta_n}{\sqrt{n}} \geq c_2 \sigma, \tag{50}$$

where $c_1 = 4.06 \cdots$ *is the infimum for* $0 < \nu < 1$ *of the positive root* c *of the equation*

$$c - \nu c^2 + \frac{\nu}{(1-\nu)^2} \log\left(1 + (1-\nu)^2 c^2\right) = 0 \tag{51}$$

and $c_2 = 0.32 \cdots$ *is the supremum for* $\nu > 0$ *of the positive root* c *of the equation*

$$c - \frac{\sqrt{\nu}}{(1+\nu)\sqrt{2\pi}} \left(e^{-\frac{c^2}{2\nu}} + \frac{c}{\sqrt{\nu}} \int_{-c/\sqrt{\nu}}^{\infty} e^{-u^2/2}\, du \right) = 0. \tag{52}$$

The assertion (49) with c_1 replaced by 5 was proved in lemma 8 (or with c_1 replaced by $4.828 \cdots$ in (35)). Since c_1 is not the best possible constant, our main reason for stating (49) is to introduce the following strengthening of lemma 7, which is of certain independent interest.

LEMMA 11. *Let* (18) *hold for some* $a > 0$, $b > 0$ *and* $t \in T_\infty$, *and let* ν *satisfy* $0 < \nu < 1$. *Then there exists* $t^* \in T_\infty$ *satisfying*

$$E \frac{\nu b + s_{t^*}}{a + t^*} > \frac{\nu b}{a} \tag{53}$$

and

$$E \frac{1}{a + t^*} < \frac{\sigma^2}{(1-\nu)^2 b^2} \log\left(1 + \frac{1 + (1-\nu)^2 b^2/\sigma^2}{a}\right). \tag{54}$$

PROOF. Let $m = [(1-\nu)^2b^2/\sigma^2]$ and define t^* as the t_m given by lemma 6 for this m and $a' = a$, $b' = \nu b$. Then (53) holds and we have

$$E\frac{1}{a+t^*} = \sum_{i=1}^{\infty}\frac{P(t^*=i)}{a+i}. \tag{55}$$

But, by (22), $t^* \leq i$ implies, for $i \leq m$, the inequality $\max(s_1, \cdots, s_i) > (1-\nu)b$, and therefore, by Kolmogorov's inequality,

$$P(t^* \leq i) < \frac{\sigma^2 i}{(1-\nu)^2b^2}, \qquad (i = 1, \cdots, m). \tag{56}$$

The right side of (56) is >1 for $i \geq m+1$, and hence, by a reasoning similar to that which leads from (13) to (10), the infinite sum in (55) is smaller than

$$\frac{\sigma^2}{(1-\nu)^2b^2}\sum_{i=1}^{m+1}\frac{1}{a+i}. \tag{57}$$

Estimating the sum in (57) by an integral, we obtain (54).

PROOF OF (49). We first remark that the left side of (51) vanishes for $c = 0$, is a convex function of c, and tends to $-\infty$ as $c \to \infty$; therefore, the equation (51) has indeed a unique positive root.

From (46), (43), lemma 11, and (16), we have

$$\frac{\nu\beta_n}{n} < \frac{\nu\sigma^2}{(1-\nu)^2\beta_n}\log\left(1 + \frac{1+(1-\nu)^2\beta_n^2/\sigma^2}{n}\right) + \frac{\sigma}{\sqrt{n}} \tag{58}$$

or, putting $c = \beta_n/(\sigma\sqrt{n})$,

$$\nu c^2 < \frac{\nu}{(1-\nu)^2}\log\left(1 + \frac{1}{n} + (1-\nu)^2c^2\right) + c. \tag{59}$$

Letting $n \to \infty$ and remarking that (59) holds for all $0 < \nu < 1$, we obtain (49).

PROOF OF (50). We first remark that the left side of (52) is negative for $c = 0$ and that it has, relative to c, a positive derivative bounded away from zero; therefore, the equation (52) has indeed a unique positive root.

Let now $\nu > 0$ and $c > 0$ be given. For $a > 0$, define b and $t \in T_\infty$ by

$$b = c\sigma\sqrt{a}, \qquad t = \begin{cases}[\nu a] & \text{if } s_{[\nu a]} > -b\\ \infty & \text{otherwise.}\end{cases} \tag{60}$$

Then

$$E\frac{b+s_t}{a+t} = \frac{1}{a+[\nu a]}\int_{-b}^{\infty}(b+u)\,dP(s_{[\nu a]} \leq u). \tag{61}$$

Using the fact that $s_n/(\sigma\sqrt{n})$ is asymptotically normal with zero mean and unit variance and that its first absolute moment converges to the same moment of the limiting distribution, we see from (60) and (61) that

$$\lim_{a=\infty}\frac{a}{b}E\frac{b+s_t}{a+t} = \frac{\sqrt{\nu}}{(1+\nu)c\sqrt{2\pi}}\int_{-c/\sqrt{\nu}}^{\infty}\left(\frac{c}{\sqrt{\nu}}+u\right)e^{-u^2/2}\,du. \tag{62}$$

But, by the definition of c_2, the right side of (62) is smaller than 1 for $c < c_2$ and an appropriate ν. Hence, we have for $c < c_2$ and large n,

$$\frac{c\sigma\sqrt{n}}{n} < V_n(c\sigma\sqrt{n}), \tag{63}$$

or $c\sigma\sqrt{n} < \beta_n$, which completes the proof.

4. Generalization

In order not to lengthen the paper, we state only one generalization whose proof follows very closely those given above.

THEOREM 3. *Let $\alpha > \frac{1}{2}$; then there exists a stopping variable $\tau \in T$ for which*

$$E\frac{s_\tau}{\tau^\alpha} = \sup_{t \in T_\infty} E\frac{s_t}{t^\alpha}. \tag{64}$$

For such τ we have

$$0 < E\frac{s_\tau}{\tau^\alpha} < \sigma\left(\sum_{i=1}^{\infty}\frac{1}{i^{2\alpha}}\right)^{1/2}. \tag{65}$$

Such a τ may be defined through a suitable increasing sequence of positive numbers $\beta_1(\alpha), \beta_2(\alpha), \cdots$ as follows: $\tau = n$ if n is the smallest positive integer for which $s_n \geq \beta_n(\alpha)$. The $\beta_n(\alpha)$ satisfy

$$c_2(\alpha) \leq \liminf_{n=\infty}\frac{\beta_n(\alpha)}{\sigma\sqrt{n}} \leq \limsup_{n=\infty}\frac{\beta_n(\alpha)}{\sigma\sqrt{n}} \leq c_1(\alpha) \tag{66}$$

with positive finite $c_1(\alpha)$ and $c_2(\alpha)$.

PROOF. As above for $\alpha = 1$ we obtain, similarly to (10),

$$E\frac{s_t^2}{(a+t)^{2\alpha}} < \sum_{i=1}^{\infty}\frac{\sigma^2}{(a+i)^{2\alpha}}, \tag{67}$$

from which we have in lieu of (16),

$$E\frac{s_t}{(a+t)^\alpha} < \frac{\sigma}{(2\alpha-1)^{1/2}a^{\alpha-1/2}}. \tag{68}$$

Lemmas 4, 5, and 6 hold with the same proof if we replace everywhere in the denominators a, $a + t, \cdots$, by a^α, $(a+t)^\alpha$, and so on. Similarly, lemma 7 remains valid when $a + t^*$ and $2a$ in (27) are replaced by $(a+t^*)^\alpha$ and $(2a)^\alpha$ respectively, and we have, instead of (28),

$$E\frac{1}{(a+t^*)} < \frac{1}{(2a)^\alpha} + \frac{2^\alpha - 1}{(2a)^\alpha}\cdot\frac{4\sigma^2 a}{b^2}. \tag{69}$$

As in the proof of lemma 8, we obtain from (68) and (69) an inequality similar to (34), which may be rewritten as $Q_\alpha\,(b/\sigma\sqrt{a}) < 0$ where Q_α is a polynomial of the second degree with leading coefficient 1, and the other coefficients negative (and depending on α). This shows, similarly to lemma 8, that the conditional expectation of s_t/t^α, given that $t \geq n$ and $s_n \geq c_1(\alpha)\sigma\sqrt{n}$, is maximized by putting $t = n$. Since lemma 9 remains valid, with the same proof, when s_n/n is replaced by s_n/n^α, we can apply lemma 10 and deduce the existence of $\tau \in T$ satisfying (64) as well as the last inequality of (66). The second inequality of

(65) follows from (68), and the first is proved exactly as the first inequality of (6). Finally, the first inequality of (66) follows again from considerations of asymptotic normality, similarly to the proof of (50).

5. Remarks

1. Various other generalizations and extensions are possible. Thus we can study in a similar fashion $E((b + s_t)^+)^\beta/(a + t)^\alpha$ provided $2\alpha > \max(1, \beta)$. Also more complicated, and less explicit, functions of t and s_t may be considered.

2. The left inequality of (6) cannot be improved (without considering other features of the distribution of the x_i besides the variance). This is seen trivially by remarking that a random variable with zero mean and given variance can have an arbitrary small positive bound. On the other hand, $\pi/\sqrt{6}$ is not the best possible constant in (6).

3. The inequality (10) of lemma 2 cannot be improved. Indeed, let $0 < p < 1$, $q = 1 - p$, and each x_i assume the values $\sigma\sqrt{q/p}$ and $-\sigma\sqrt{q/p}$ with probabilities p and q respectively. Then for $t \in T_\infty$ defined by: $t = i$ if i is the smallest integer for which $x_i > 0$ and $i < q/p$, and $t = \infty$ otherwise, we have

$$E\left(\frac{s_t}{a + t}\right)^2 = \sigma^2 \sum_{i=1}^{[q/p]} pq^{i-1} \frac{((q/p)^{1/2} - (i - 1)(p/q)^{1/2})^2}{(a + i)^2}. \tag{70}$$

But as $p \to 0$ the right side of (70) approaches the limit $\sigma^2 \sum_{i=1}^{\infty} (a + i)^{-2}$. However, (15) in lemma 3 can be improved as is seen by examining the conditions under which little is lost by the passage from (10) to (15), as well as those under which (10) gives a close to best estimate, and observing that it is impossible to satisfy both simultaneously. This establishes, in particular, the assertion about (6) in the preceding remark.

A similar remark applies to (68) and (67).

4. It is not difficult to improve the constants c_1 and c_2 in (49) and (50) in theorem 2. But we do not know how to obtain the best constants. It is very likely that the lim sup and lim inf in (49) and (50) must coincide; that is, that $\lim_{n=\infty} \beta_n/\sqrt{n}$ must exist (though it may depend on aspects of the distribution of the x_i other than the variance). But we cannot prove this in general. The existence of this limit seems to be a most interesting problem connected with the structure of the optimal stopping rule.

5. Since the sequence $s_1, s_2, \cdots$ is Markovian, it is possible to describe every optimal stopping variable τ by a sequence of "absorbing" sets B_n of real numbers as follows: $\tau = n$ if n is the smallest positive integer for which $s_n \in B_n$.

It follows from lemma 5 that it is natural to take the sets B_n as half-lines: $B_n = \{u: u \geq b_n\}$. Theorem 2 gives such a description of the regular stopping variable τ_0, but more is true. If $\tau = n$ when n is the smallest integer for which $s_n \geq b_n$ describes an optimal stopping variable (in the sense of (5)), then the b_n necessarily satisfy (49) and (50) (with b_n in place of β_n). Indeed, it follows from the considerations of the beginning of section 3 that τ can be optimal only if

$P(\tau_0 = n, s_n \in I_n) = 0$ for all n, where I_n is the open interval with endpoints β_n and b_n. It can be easily seen that τ_0 is not bounded. If the random variables x_i are such that $P(x_i \in I) > 0$ for every interval I whose length exceeds some given number, it can be easily inferred that the sequence $b_n - \beta_n (n = 1, 2, \cdots)$ is bounded, and thus the b_n satisfy (49) and (50). The general case requires more elaborate arguments.

6. The assumption that the random variables x_i have zero mean is, of course, unimportant and can be dropped. We did, however, make essential use of the fact that they have finite variance. It seems possible to replace this condition by a weaker one, such as assuming the finiteness of an absolute moment of order greater than one, but this has not yet been done. It would also be interesting to relax the conditions of independence and identical distribution of the x_i.

7. We assumed that $\mathcal{B}_n$ is the σ-field generated by $x_1, \cdots, x_n$. All our considerations remain valid if we assume instead that $\mathcal{B}_n$ is the σ-field generated by $x_1, \cdots, x_n$ and $\mathcal{C}$, where $\mathcal{C}$ is a subfield of $\mathcal{B}$ independent of those generated by the x_n. This device makes it possible to treat certain randomized stopping variables.

8. Our method can easily be adapted to treat stochastic processes with a continuous time parameter. This may, however, necessitate a slight reformulation of the problem. Consider, for example, the standard Brownian motion process $s(h)$, $0 \leq h < \infty$. Denoting by $\mathcal{B}(h)$ the σ-field generated by $s(h')$, $0 \leq h' \leq h$, a nonnegative random variable t is called a stopping variable if $\{t \leq h\} \in \mathcal{B}(h)$ for all $0 \leq h < \infty$. Now, $\sup Es(t)/t$ taking over all stopping variables is ∞, and indeed if we define τ to be the smallest $h > 0$ for which $s(h) \geq \sqrt{h \log^+ 1/h}$, then we find $Es(\tau)/\tau = \infty$ (the almost sure continuity of $s(h)$ and the law of the iterated logarithm imply that τ is a stopping variable). If we wish to consider problems for which $\sup Es(t)/t$ is finite we must modify the original problem somewhat. We might, for example, confine our attention to stopping variables t satisfying $P(t \geq \delta) = 1$ for a given $\delta > 0$, or consider $s(t)/(a + t)$ instead of $s(t)/t$. Then the supremum of the expectations would be finite, and there would exist a stopping variable τ for which this supremum is achieved and, moreover, a result similar to theorem 2 would hold.

REFERENCES

[1] Y. S. Chow and Herbert Robbins, "On optimal stopping rules," *Z. Wahrscheinlichkeitstheorie und Verw. Gebeite*, Vol. 2 (1963), pp. 33–49.

[2] ———, "On optimal stopping rules for s_n/n," *Illinois J. Math.*, Vol. 9 (1965), pp. 444–454.

ON DISCRETE EVASION GAMES WITH A TWO-MOVE INFORMATION LAG

THOMAS S. FERGUSON
UNIVERSITY OF CALIFORNIA, LOS ANGELES

1. Introduction

This paper deals with extensions of the following game. Although there are other interpretations of this game, we use the traditional one of a ship trying to evade a bomber. This problem is sometimes called the bomber-battleship problem.

A ship is constrained to travel on the integer lattice of the real line. In one time unit, he may move one unit distance to the right or one unit distance to the left. He must move each time; he is not allowed to stay still. A bomber with exactly one bomb flies overhead and wants to drop the bomb on the ship. He may drop the bomb on any point he desires, but it takes two time units for the bomb to fall. He knows that at the end of two units time the ship can only be at one of three places: exactly where it is when he lets go of the bomb, two steps to the right, or two steps to the left. There is no use in dropping the bomb at any but these three points. The ship starts at the origin, and he does not know when or where the bomb is dropped until it hits. The bomber may observe the movements of the ship for as long as he likes before dropping the bomb. He wins one unit from the ship if the bomb hits the ship; otherwise, there is no payoff.

This game-theoretic problem, suggested by Rufus Isaacs, was first solved by Dubins in 1953 and published in [2]. It was solved independently by Isaacs and Karlin [4] using a different method. Further results were obtained by Isaacs [3] and by Karlin [5]. A general theory of games with information lag was studied by Scarf and Shapley [7].

In papers [2] and [4], it is shown that this game has a value $v_1 = (3 - \sqrt{5})/2 = .382\ldots$, that the ship has an optimal strategy which is explicitly described, and that the bomber does not have an optimal strategy (merely ϵ-optimal ones). In other words, there is a strategy for the ship such that no matter what strategy the bomber uses, the ship will not be hit with probability greater than v_1, and, for every $\epsilon > 0$, there is a strategy for the bomber such that no matter what strategy the ship uses, the ship will be hit with probability at least $v_1 - \epsilon$.

A strategy for the ship is a rule telling him at each step with what probability he should go to the right as a function of all his past moves. Such a strategy is said to be *Markov* if this probability depends on the past moves only through

Partially supported by National Science Foundation Grant GP-1606.

the preceding step. If this probability is also independent of time, the strategy is said to be *stationary Markov*. If the ship uses a (stationary) Markov strategy, the sequence of choices of left or right forms a (stationary) Markov chain. In the above problem, the ship has an optimal strategy which is stationary Markov and invariant under interchange of left and right. This strategy is as follows. At time zero, go right or left, each with probability $\frac{1}{2}$. At all future times, go in the same direction as the last move with probability $(\sqrt{5} - 1)/2 = .618\ldots$, and go in the opposite direction with probability $(3 - \sqrt{5})/2 = .382\ldots$.

This strategy insures that the ship will not be hit with probability greater than $v_1 = (3 - \sqrt{5})/2$, as is easily checked. To show that v_1 is in fact the value of the game is a more difficult problem. Dubins solved it by showing that the game has a value and that v_1 is the upper value of the game. Isaacs and Karlin solved it by dealing with the fundamental functional equation of this game. A third method is presented below in which an ϵ-optimal strategy for the bomber is explicitly exhibited.

The main objective of this paper is to give an extension of this result, where the ship is allowed to travel on a more general graph. Except for some related problems treated by Blackwell [1] and Matula [6], I know of no other extensions of this problem which have been solved. In section 5, some unsolved problems, possibly more direct generalizations of the main problem, are stated.

2. Graphs

By a graph, we mean a pair (V, E) where V is a set of points or vertices, and E is a set of *unordered* pairs of points of V. The elements of E are called paths or edges. In the following two sections, we treat the extension of the problem mentioned in the first section to the situation where the ship is constrained to travel on such a graph from one vertex to another vertex along one of the available edges in one time unit. It is immaterial whether or not there is more than one edge joining two vertices; for simplicity, we assume that to any pair of vertices there is at most one edge joining them. However, it is important that there does not exist an edge which joins one point to itself. One may take it that the definition of a graph excludes such a possibility. We prefer to list this restriction explicitly with two other restrictions which we place on graphs.

We define a *restricted n-graph* to be a graph which satisfies the following three conditions.

(i) *There does not exist an edge which joins a vertex to itself.*

(ii) *There are no four-sided figures.* In other words, there do not exist four distinct points A, B, C, and D such that $\{A, B\}$, $\{B, C\}$, $\{C, D\}$, and $\{A, D\}$ are edges.

(iii) *There are exactly $n + 1$ distinct vertices joined to each vertex by an edge.* It is assumed that n is a positive integer.

It should be noted that three-sided figures are allowed, but that (ii) requires that no two distinct three-sided figures have a common edge.

The lattice of integers on the real line is a restricted 1-graph. Other examples when $n = 1$ may easily be constructed for any finite number k of vertices in V, provided $k \geq 3$ and $k \neq 4$.

The hexagonal lattice in the plane provides an example of a restricted 2-graph (figure 1). Another example of a restricted 2-graph is provided by the vertices and edges of a dodecahedron, the regular polyhedron with twelve faces made of regular pentagons, with twenty vertices and with thirty edges.

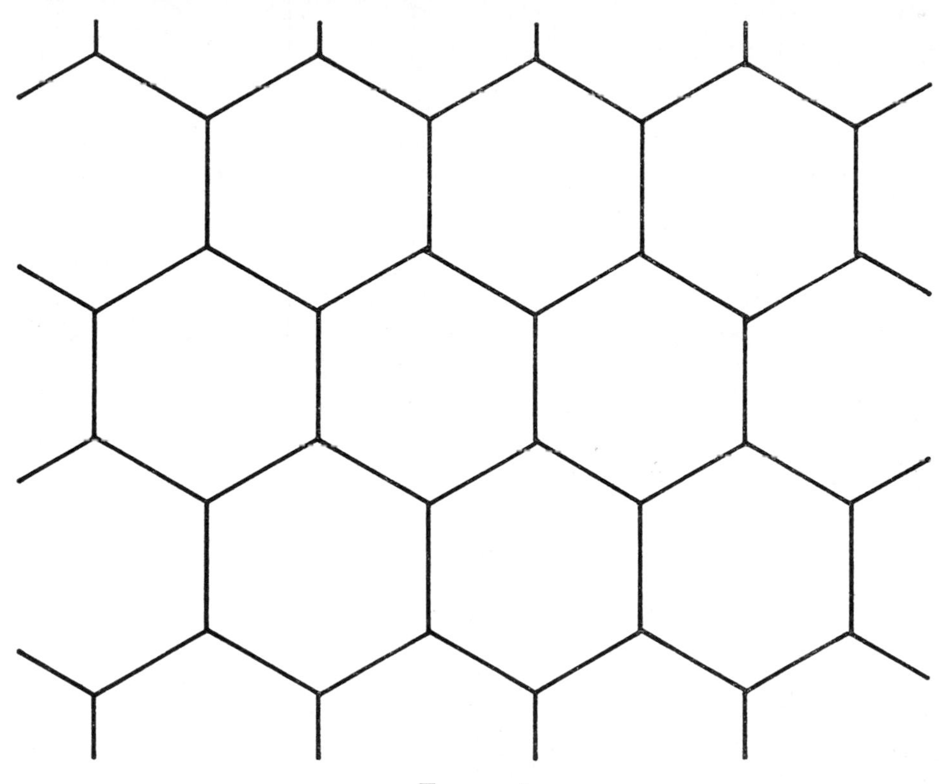

FIGURE 1

For arbitrary n, analogues of the hexagonal lattice in the plane can be described in n-dimensional space to provide examples of restricted n-graphs. Finally, figure 2 provides an example of a regular graph in the plane which is a restricted 3-graph.

3. An optimal strategy for the ship

In this section, we define a class of symmetric stationary Markov strategies for a ship traveling on a restricted n-graph. The minimax strategy within this class is then derived. That this strategy is in fact optimal within the class of all

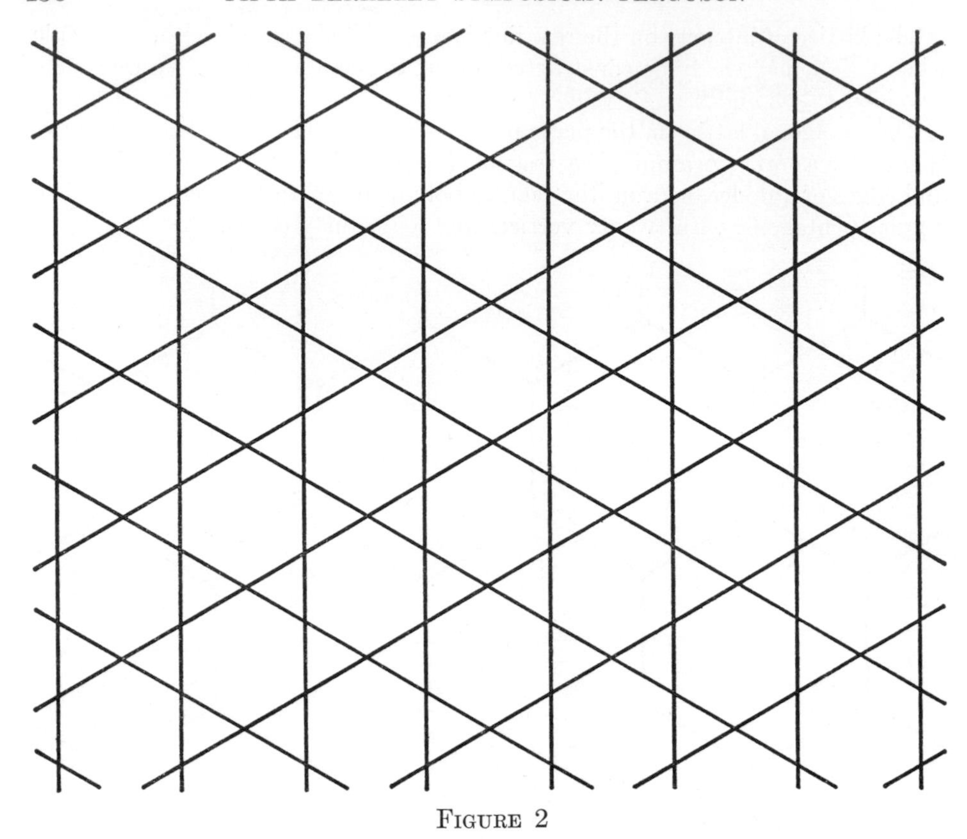

FIGURE 2

strategies is shown in the following section by exhibiting an ϵ-optimal strategy for the bomber.

Consider the following class of stationary Markov strategies for the ship. The first step is chosen at random among the $n + 1$ points one step away, each point having probability $1/(n + 1)$. From then on, the probabilities depend only on the point just vacated, the probability of a return to that point being $1 - np$ where $0 \leq p \leq 1/n$, and the probability of a trip to any of the other n available points being p. We shall find that value of p which minimizes the maximum probability of being at any point two steps ahead among this class of strategies.

At the very start, there is probability $p/(n + 1)$ of being at each of the $n^2 + n$ points two steps away from the starting point (exclusive of the starting point). The restriction (ii) on the graphs is used here to imply that these $n^2 + n$ points are distinct. The probability of returning to the starting point in two steps is $1 - np$.

At any time later, the probability of advancing to any of the n^2 points two steps away not passing through the point just vacated is p^2 for each point. The probability of retreating to any of the n points two steps away passing through the point just vacated is $p(1 - np)$ for each point. The probability of returning to the point occupied after two steps is $1 - np$.

To find the minimum of the maximum of these three functions, p^2, $p(1 - np)$, and $(1 - np)$, we note first that $p(1 - np) \leq (1 - np)$ for all p. Then, since p^2 is increasing in p and $1 - np$ is decreasing in p, the minimax strategy is found as that value of p for which $p^2 = 1 - np$. This equation has two roots $p = (-n \pm \sqrt{n^2 + 4})/2$, of which one

$$p = (\sqrt{n^2 + 4} - n)/2 \tag{1}$$

is positive. The minimax value, denoted by v_n is easily found to be

$$v_n = p^2 = 1 - np = (n^2 + 2 - n\sqrt{n^2 + 4})/2. \tag{2}$$

We must check that the probabilities of being at the various points after the very first steps does not exceed v_n. That is, we must show $p/(n + 1) \leq v_n$ and $1 - np \leq v_n$, where p and v_n satisfy (1) and (2). The inequality $1 - np \leq v_n$ is obviously satisfied with equality from (2). The inequality $p/(n + 1) \leq v_n$ is equivalent to $(n + 1)p \geq 1$ or to $(n + 1)\sqrt{n^2 + 4} \geq n^2 + n + 2$, which is easily checked by squaring both sides.

When $n = 1$, we find that $p = (\sqrt{5} - 1)/2$ and $v_1 = (3 - \sqrt{5})/2$, the solution given by Dubins. When $n = 2$, we find $p = \sqrt{2} - 1 = .414\ldots$ and $v_2 = 3 - 2\sqrt{2} = .172\ldots$. Thus, on the hexagonal lattice of figure 1, the optimal policy of the ship is to return to the point just vacated with probability .172 . . . and to move to one of the other two points with probability .414 . . . each.

4. An ϵ-optimal strategy for the bomber

Let $w < v_n = (n^2 + 2 - n\sqrt{n^2 + 4})/2$. We will describe a strategy for the bomber which guarantees him that he will hit the ship with probability at least w.

Let the initial position of the ship be called point A. At time zero, drop the bomb on point A with some small probability p_0, the exact value of which will be determined later. The first step of the ship carries him to some point which we shall call B. At time one, drop the bomb on point B and each of the n^2 points two steps away from B not passing through A with equal probabilities p_1 each. (The p_i are unconditional probabilities, *not* conditional probabilities, given that the bomb has not been dropped.)

If the next step of the ship is to a point other than A, he is bound on step three to go to a vertex hit with probability p_1. Hence, the conditional probability that he will be hit, given that the bomb is dropped at times zero or one, is $p_1/(p_0 + (n^2 + 1)p_1)$. We require of p_0 and p_1 that

$$w = p_1/(p_0 + (n^2 + 1)p_1). \tag{3}$$

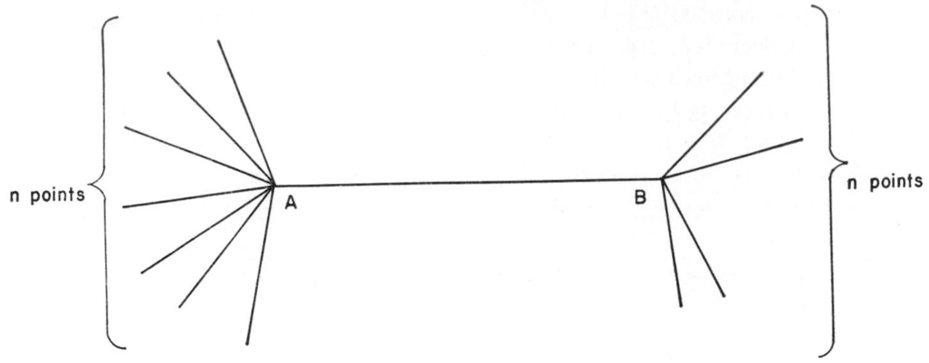

FIGURE 3

Thus, if the ship does not return to point A at step two, he will be hit with conditional probability w, given that the bomb was dropped at times zero or one. If, in this case, the bomb was not dropped, we may start over. Eventually (with probability one), the bomb will be dropped, and we will have probability w of hitting the ship.

Suppose now that the ship returns to A after two steps. He immediately gets hit with probability p_0, or with conditional probability $p_0/(p_0 + (n^2 + 1)p_1)$, given that the bomb was dropped at times zero or one. If this conditional probability is at least w, we are finished. Ordinarily, however, it will be impossible to choose p_0 and p_1 satisfying (3) and $p_0/(p_0 + (n^2 + 1)p_1) \geq w$. In such a case, we bomb A and the n^2 points two steps away from A not passing through B with equal probabilities p_2 each. If the ship goes to a point other than B at step three, he is bound at step four to go to a vertex hit with probability p_2. The overall conditional probability that he will be hit is $(p_0 + p_2)/(p_0 + (n^2 + 1)(p_1 + p_2))$. We require

$$w = (p_0 + p_2)/(p_0 + (n^2 + 1)(p_1 + p_2)). \tag{4}$$

Thus, if the ship does not return to B at step three, he will be hit with conditional probability w, given the bomb has been dropped.

If he does return to B at step three, he gets hit with probability p_1, so that if $(p_0 + p_1)/(p_0 + (n^2 + 1)(p_1 + p_2)) \geq w$, we are finished. But again, it is likely that equalities (3) and (4) imply $(p_0 + p_1)/(p_0 + (n^2 + 1)(p_1 + p_2)) < w$, so that in order to obtain probability w of hitting the ship, we again bomb A and the n^2 points two steps distant not going through B with equal probabilities p_3 each. We continue in this manner hoping that after some k steps we will have not only

$$\frac{p_1}{p_0 + (n^2 + 1)p_1} = \frac{p_0 + p_2}{p_0 + (n^2 + 1)(p_1 + p_2)} = \cdots = \frac{p_0 + p_1 + \cdots + p_{k-2} + p_k}{p_0 + (n^2 + 1)(p_1 + \cdots + p_k)} = w, \tag{5}$$

but also

$$\frac{p_0 + p_1 + \cdots + p_{k-2} + p_{k-1}}{p_0 + (n^2 + 1)(p_1 + \cdots + p_k)} \geq w. \tag{6}$$

Inequality (6) together with the last equality of (5) imply that $p_{k-1} \geq p_k$. If w were not less than v_n, then equalities (5) would imply, as we shall see, that the p_i are strictly increasing, so that there would not exist a k such that (6) is satisfied. But all we need to show is that if $w < v_n$, then (5) and (6) may be satisfied.

In the equations

$$\begin{aligned} w &= \left(\sum_0^j p_i - p_{j-1}\right) \Big/ \left(p_0 + (n^2 + 1) \sum_1^j p_i\right) \\ &= \left(\sum_0^{j+1} p_i - p_j\right) \Big/ \left(p_0 + (n^2 + 1) \sum_1^{j+1} p_i\right), \end{aligned} \tag{7}$$

the ratio of the difference of the numerators to the difference of the denominators is also w:

$$w = (p_{j+1} - p_j + p_{j-1})/((n^2 + 1)p_{j+1}) \qquad \text{for} \quad j = 1, \cdots, k - 1. \tag{8}$$

In another form,

$$p_{j+1} = \alpha(p_j - p_{j-1}) \qquad \text{for} \quad j = 1, \cdots, k - 1 \tag{9}$$

where

$$\alpha = (1 - (n^2 + 1)w)^{-1}. \tag{10}$$

The difference equation (9) is to be solved subject to the boundary condition (3), rewritten as

$$p_1 = (\alpha - 1)p_0/(n^2 + 1). \tag{11}$$

Since the equations (5) and (6) involve only ratios of the p_i, we may arbitrarily set $p_0 = 1$ for the purposes of the computation, and later norm the p_i so that $p_0 + (n^2 + 1)(p_1 + \cdots + p_k) = 1$. Thus we add the boundary condition,

$$p_0 = 1. \tag{12}$$

The general solution of (9) is

$$p_j = C_1\left(\frac{\alpha + \sqrt{\alpha^2 - 4\alpha}}{2}\right)^j + C_2\left(\frac{\alpha - \sqrt{\alpha^2 - 4\alpha}}{2}\right)^j \tag{13}$$

where C_1 and C_2 are arbitrary constants. Boundary condition (12) implies that $C_1 + C_2 = 1$. Boundary condition (11) implies

(14)

$$(\alpha - 1)p_0/(n^2 + 1) = C_1((\alpha + \sqrt{\alpha^2 - 4\alpha})/2) + (1 - C_1)((\alpha - \sqrt{\alpha^2 - 4\alpha})/2),$$

or equivalently,

$$2C_1 = 1 - \frac{(n^2 - 1)\alpha + 2}{(n^2 + 1)\sqrt{\alpha^2 - 4\alpha}}. \tag{15}$$

We want to show that all quantities involved are real, provided that $w < v_n$ and w is sufficiently close to v_n. This amounts to showing that $\alpha > 4$. But $\alpha > 4$ if and only if $\frac{3}{4} < (n^2 + 1)w < 1$. In other words, we must show $\frac{3}{4} < (n^2 + 1)v_n \leq 1$. But it is easily checked that $(n^2 + 1)v_n$ is an increasing function of $n \geq 1$, that $(n^2 + 1)v_n \to 1$ as $n \to \infty$, and that $(n^2 + 1)v_n$ for $n = 1$ is $3 - \sqrt{5}$, which is between $\frac{3}{4}$ and 1.

The question remaining is whether or not (6) can be satisfied for some k. As noticed, this is equivalent to asking if for some k, $p_k \leq p_{k-1}$. Since $\alpha > 4$, this question is answered affirmatively provided $C_1 < 0$, as can be seen from equations (9) and (13). But from equation (15), $C_1 < 0$ if and only if (when $\alpha > 4$)

$$n^2\alpha^2 - n^2(n^2 + 3)\alpha - 1 < 0 \tag{16}$$

or, equivalently, if and only if

$$w^2 - (n^2 + 2)w + 1 > 0 \tag{17}$$

using (10). This is satisfied if $w < v_n$ as is easily seen.

In summary, for a given w satisfying $\frac{3}{4} < (n^2 + 1)w$ and $w < v_n$, compute α from (10) and C_1 from (15). You will find that $\alpha > 4$ and $C_1 < 0$. Let $C_2 = 1 - C_1$ and compute p_j from (13). Let k be the first integer for which $p_k \leq p_{k-1}$. Normalize these numbers by dividing by a constant so that $p_0 + (n^2 + 1) \sum_1^k p_i = 1$. With probability $(n^2 + 1)p_j$, the bomb is dropped at time j, provided the ship has alternated between the points A and B up to that time, and then the target is chosen at random among the point where the ship is and the n^2 points two steps away from that point not reversing direction, each of the $n^2 + 1$ points being equally likely. If, at any time, the ship goes to a point other than A or B, and the bomb has not been dropped, the whole procedure is repeated starting at the new point. Eventually with probability one, the bomb will be dropped giving probability at least w of hitting the ship.

5. Unsolved problems

There are (at least) three natural extensions of the problem mentioned in the first section which the present methods do not treat and which are still unsolved problems.

Problem 1. Give the ship the option of staying still. Thus at the end of two moves, he may be at one of five positions: where he is now, one or two steps to the left, or one or two steps to the right. The restricted n-graph could be modified to contain this problem by removing condition (i).

Problem 2. Consider the original problem on the square lattice on the plane. Such a graph does not satisfy condition (ii) of the restricted 3-graph.

Problem 3. Assume that the bomb takes three moves to fall. This is known as the three-move lag problem.

For each of these problems, it is known that the game has a value, that the ship has an optimal strategy, and that the bomber does not have an optimal strategy. The reason that these three problems seem more difficult than those treated in this paper is that those treated in this paper have optimal stationary Markov strategies for the ship, whereas for the above three problems it is conjectured that no stationary Markov strategy is optimal.

The outstanding unsolved problem in my opinion is number three, the three-move lag problem on the integer lattice on the real line. In this problem it is known that the optimal strategy of the ship is not Markov. But since the bomb takes three moves to fall, one would expect the ship to use 2-dependent strategies, those strategies which depend on the past only through the two previous positions. The general stationary 2-dependent strategies invariant under interchange of right and left may be described by two probabilities q_0 and q_1, where q_0 represents the probability of continuing in the same direction as the last move, given that the last two moves were in the same direction, and where q_1 represents the probability of continuing in the same direction as the last move, given that the last two moves were in opposite directions. For stationary Markov strategies $q_0 = q_1$. The best that can be done with these stationary Markov strategies is to choose $q_0 = q_1 = \frac{2}{3}$, and this ensures that the ship will not be hit with probability greater than $(\frac{2}{3})^3 = \frac{8}{27} \sim .296 \ldots$. I claim that the best that can be done with these stationary 2-dependent strategies is to choose $q_0 = (3 - \sqrt{3})/2 \sim .634 \ldots$ and $q_1 = \sqrt{3} - 1 \sim .732 \ldots$, and this ensures that the ship will not be hit with probability greater than

$$3(3\sqrt{3} - 5)/2 \sim .294 \ldots . \tag{18}$$

It is unknown whether or not this last strategy is optimal. In fact, it has been conjectured that no strategy with finite memory (that is, a strategy which depends only on the last m moves for some finite integer m) is optimal for the three-move lag problem.

Blackwell [1] treats a class of problems on the prediction of sequences, very similar to those treated in this paper. Matula [6], in extending some of these results, gives an example of a problem with a four-move lag in which he finds an optimal strategy without finite memory for the sequence chooser (the ship), and gives convincing arguments that no optimal strategy for the sequence chooser has finite memory. Such a state of affairs could hold for the three-move lag problem above. A lower bound on the value, v, of the three-move lag problem has been obtained by investigating strategies of the bomber similar to the ϵ-optimal strategies found in the previous section for the two-move lag case. Using methods found there and analyses twice as complicated, it may be shown that $v \geq \frac{23}{81} = .284 \ldots$. Combined with the upper bound of (18), we have $.284 \ldots \leq v \leq .294 \ldots$.

REFERENCES

[1] D. Blackwell, "The prediction of sequences," RAND Memorandum 1570 (1955).

[2] L. Dubins, "A discrete evasion game," *Contributions to the Theory of Games*, Vol. 3, Annals of Mathematics Studies No. 39 (1957), pp. 231–255.

[3] R. Isaacs, "The problem of aiming and evasion," *Naval Res. Logist. Quart.*, Vol. 2 (1956). pp. 47–67.
[4] R. Isaacs and S. Karlin, "A game of aiming and evasion," RAND Memorandum 1316 (1954).
[5] S. Karlin, "An infinite move game with a lag," *Contributions to the Theory of Games*, Vol. 3, Annals of Mathematics Studies No. 39 (1957), pp. 257–272.
[6] D. Matula, "1100 and other embedded sequence games," Mimeographed notes, Berkeley, University of California, 1964.
[7] H. E. Scarf and L. S. Shapley, "Games with partial information," *Contributions to the Theory of Games*, Vol. 3, Annals of Mathematics Studies No. 39 (1957), pp. 213–229.

TWO-ACTION COMPOUND DECISION PROBLEMS

M. V. JOHNS, JR.
STANFORD UNIVERSITY

1. Introduction and summary

The compound decision problem considered here consists of a sequence of component problems, in each of which one of two possible actions must be selected. The loss structure is the same for each component decision problem. Each component problem involves independent identically distributed observations whose common distribution function is unknown but belongs to some specified parametric or nonparametric family of distributions (for example, the family of all Poisson distributions with parameter λ bounded above by some finite number B). This family remains fixed for all component problems. It is assumed that, at the time a decision is made in any particular component problem, the available information includes the data obtained in all previous component decision problems in the sequence.

Compound decision problems of this type arise in situations where routine testing and evaluation programs are in operation. For example, in routine lot by lot acceptance sampling for quality control purposes, each lot of items is sampled, and the lot is either accepted or rejected on the basis of the observations obtained. Another example arises in routine medical diagnosis where a decision between two alternative treatments must be made for each of a continuing sequence of patients on the basis of results obtained from a diagnostic test performed on each patient. In either of these examples records of all past observations could certainly be accumulated.

In the compound decision problem as formulated here, no relationships whatever are assumed to exist among the distributions governing the observations associated with different component decision problems (aside from the requirement that all these distributions are members of a specified general family). A strictly "objective" approach to this situation appears, at first glance, to require that each component problem be treated in isolation with the decision for each problem being based on the observations obtained for that problem alone. It has been known for some time, however, that for certain types of compound decision problems, substantially better performance in terms of average risk incurred for a number of component problems may be obtained by using "compound decision procedures" which make explicit use at each

This research supported in part under Office of Naval Research Contract Nonr-225(53) (NR 042-002).

stage of the seemingly irrelevant data from previous component problems. A number of authors have investigated this aspect of compound decision problems, notably Robbins [5], Hannan and Robbins [1], Samuel [8], [9], Hannan and Van Ryzin [2], Van Ryzin [11], and Swain [10]. These references are cited chronologically to indicate stages in the evolution of the subject and are not exhaustive. In the earlier papers [5], [1], and [8] the space of "states of nature," that is, the family of distribution functions governing the observations, is assumed to be finite, so that these models are not suitable for most applications. In these papers, and in [9] as well, the main results are concerned only with the convergence to zero of the difference between the average risk and a certain "optimal" goal (discussed in detail below) as the number of component problems becomes large. In two of the more recent papers ([2], [11]) the finite state model has been retained, but stronger results involving bounds on the deviations of the average risk from the desired goal and rates of convergence to "optimality" are obtained. The papers of Samuel [9] and Swain [10] deal with standard (infinite state) estimation problems with squared error loss, and their results are therefore immediately relevant to applications. In all of these papers except [10] the "optimal" goal asymptotically achieved by the average risk is defined in essentially the same way. For each n, the average risk for the first n component problems is compared to the Bayes optimal risk one could achieve for a single component problem if the parameter of interest had a known a priori distribution equal to the empirical distribution of the parameter values associated with the first n component problems. This criterion does not, however, represent the best that can be achieved by compound decision procedures, and in fact, a variety of more stringent criteria may be defined which take into account empirical dependencies of various orders which may occur in the sequence of parameter values. At the suggestion of the present author, these more stringent criteria were considered by Swain in [10] and were shown to be asymptotically achievable for the compound estimation problem. Swain also obtains bounds and rates of convergence for some cases.

The object of the present paper is to find bounds for the deviations of the average risk from various optimal goals for the two-action compound decision problem. Attention is confined to certain classes of loss functions and compound decision procedures, and to the case of discrete-valued observations. Both parametric and nonparametric models are treated and the convergence of the bounds to zero is shown to be ratewise sharp. In order to state these results explicitly, the problem must be presented more formally.

The compound decision problem consists of a sequence of component problems where the j-th component problem has the following structure:

(a) The distribution governing the observations is denoted by F_j and is a member of a specified family $\mathfrak{F}$ of distribution functions, each assigning probability one to a fixed denumerable set of numbers $x_1, x_2, \cdots$.

(b) The statistician obtains k independent observations with common distribution function F_j. The observations are denoted by the vector

$$X_j = (X_{1,j}, X_{2,j}, \cdots, X_{k,j}).$$

(c) For the parametric case the parameter of interest determines F_j completely and is denoted by λ_j. For the nonparametric case, $\lambda_j = Eh(X_{1,j})$, where $h(\cdot)$ is a specified function.

(d) On the basis of the observations the statistician selects one of two actions and incurs loss $L_a(\lambda_j)$, $a = 1, 2$, if action a is selected.

A typical compound decision rule for the j-th component problem is represented by $\Delta_j(x)$, where $E\Delta_j(x)$ is the probability of taking action one if $X_j = x$. For each value of the vector x, $\Delta_j(x)$ is a random variable depending on the mutually independent random vectors $X_1, X_2, \cdots, X_{j-1}$. The risk for the j-th problem is given by

$$r_j = (L_1(\lambda_j) - L_2(\lambda_j))E\Delta_j(X_j) + L_2(\lambda_j). \tag{1.1}$$

Letting $p_j(x)$ be the probability that $X_j = x$, and

$$\alpha_j(x) = (L_1(\lambda_j) - L_2(\lambda_j))p_j(x), \tag{1.2}$$

the average risk for the first n component problems is given by

$$\begin{aligned} \bar{r}_n &= \frac{1}{n}\sum_{j=1}^{n} r_j \\ &= \frac{1}{n}\sum_{j=1}^{n}\sum_{x}(L_1(\lambda_j) - L_2(\lambda_j))E\{\Delta_j(x)|X_j = x\}p_j(x) + \frac{1}{n}\sum_{j=1}^{n} L_2(\lambda_j) \\ &= \frac{1}{n}\sum_{j=1}^{n}\sum_{x}\alpha_j(x)E\Delta_j(x) + \frac{1}{n}\sum_{j=1}^{n} L_2(\lambda_j). \end{aligned} \tag{1.3}$$

The "classical" goal that one attempts to achieve asymptotically, is defined by considering a *hypothetical* Bayesian version of a typical component problem. Suppose that for such a problem it is known that the sampling distribution F is chosen randomly according to the discrete a priori probability measure on $\mathfrak{F}$ which assigns probability n^{-1} to each element of the set $\{F_1, F_2, \cdots, F_n\}$ of sampling distributions arising in the first n component problems. If one uses the decision rule $\delta(x)$ (based only on the observations obtained for the single component problem under consideration), where $\delta(x)$ is the probability of taking action one when x is observed, the risk incurred is

$$\rho_n = \frac{1}{n}\sum_{j=1}^{n}\sum_{x}\alpha_j(x)\delta(x) + \frac{1}{n}\sum_{j=1}^{n} L_2(\lambda_j). \tag{1.4}$$

Letting

$$m_j(x) = \sum_{i=1}^{j}\alpha_i(x), \qquad j = 1, 2, \cdots, \tag{1.5}$$

it is easily seen that the Bayes optimal decision rule is given by

$$\delta^*(x) = \begin{cases} 1, & m_n(x) < 0, \\ 0, & m_n(x) \geq 0, \end{cases} \tag{1.6}$$

and the optimal Bayes risk is

$$\rho_n^* = \frac{1}{n} \sum_x m_n(x)^- + \frac{1}{n} \sum_{j=1}^{n} L_2(\lambda_j), \tag{1.7}$$

where $m_n(x)^-$ indicates the negative part of $m_n(x)$.

The object is to discover compound decision procedures having the property that the resulting average risks $\bar{r}_n$ satisfy

$$|\rho_n^* - \bar{r}_n| < b(n), \qquad \text{for all } n, \tag{1.8}$$

where $b(n) \to 0$, as $n \to 0$, and where $b(n)$ is independent of the particular sequence $F_1, F_2, \cdots$, occurring. Theorem 1 of section 2 gives conditions under which a class of compound decision procedures will satisfy (1.8) with $b(n) = Kn^{-1/2}$, for a certain positive constant K independent of the sequence of F_j's. It is also noted that $n^{-1/2}$ is the best possible rate of convergence for this class of procedures. Typically, of course, neither $\bar{r}_n$ nor ρ_n^* will themselves converge to limits.

In section 3, specific compound decision procedures satisfying the conditions of theorem 1 are presented for certain parametric cases (Poisson, negative binomial, and so on) involving families of sampling distributions of exponential type. The nonparametric case is also discussed and procedures satisfying theorem 1 are given. A very simple loss structure is used throughout. In fact, it is assumed that

$$L_1(\lambda) - L_2(\lambda) = c(\lambda - b), \tag{1.9}$$

where b, c are specified constants. It is also assumed that $L_1(\lambda)$ and $L_2(\lambda)$ are bounded on any bounded interval of λ's. The particular loss functions

$$L_1(\lambda) = \begin{cases} 0, & \lambda < b, \\ c(\lambda - b), & \lambda \geq b, \end{cases} \tag{1.10}$$

$$L_2(\lambda) = \begin{cases} c(b - \lambda), & \lambda < b, \\ 0, & \lambda \geq b, \end{cases} \tag{1.11}$$

where $c > 0$, clearly satisfy (1.9), and are quite reasonable for many two-action problems of the one-sided hypothesis testing type. The arguments presented extend almost without change to the case where $L_1(\lambda) - L_2(\lambda)$ is any specified polynomial in λ. All of the compound decision procedures considered here are based on the construction of consistent unbiased estimates for each x of the quantities $m_j(x)$, $j = 1, 2, \cdots$, defined by (1.5). Action one is then chosen in the j-th component problem if and only if the estimate of $m_{j-1}(X_j)$ is negative.

The compound decision problem is closely related to the "empirical Bayes" problem where an actual unknown a priori distribution is assumed to exist. The empirical Bayes problem corresponding to the compound decision problem considered here is discussed in the nonparametric case by the present author in [3], and in the parametric case by Robbins [6] and Samuel [7]. With the exception of the necessity for a certain amount of auxiliary randomization, the compound decision procedures exhibited in section 3 are essentially the same as those suggested for the corresponding empirical Bayes problems.

The "classical" goal for compound decision problems described above may be generalized to produce a sequence of more stringent goals by extending the definition of the hypothetical Bayes decision problem. Instead of assuming that the present sampling distribution F is selected by a uniform a priori measure over $F_1, F_2, \cdots, F_n$, one may assume that the vector $(\tilde{F}_1, \tilde{F}_2, \cdots, \tilde{F}_t)$ of sampling distributions corresponding to the $t - 1$ most recent component problems and the present problem respectively, is a random vector with a discrete a priori probability measure on the t-fold product $\mathfrak{F} \times \mathfrak{F} \times \cdots \times \mathfrak{F}$, which assigns probability $(n - t + 1)^{-1}$ to each of the vectors

$$(1.12) \qquad (F_{j-t+1}, F_{j-t}, \cdots, F_j),$$

$j = t, t + 1, \cdots, n$. The optimal Bayes decision rule for such a problem must involve the observations obtained in the $t - 1$ most recent component problems as well as the present one. If the resulting Bayes risk is denoted by $\rho^*_{t,n}$, it is intuitively plausible that this quantity should be decreasing in t since advantage is taken of possible empirical dependencies of higher order as t is increased. Theorem 2 of section 4 shows that for each $t \geq 1$,

$$(1.13) \qquad \rho^*_{t+1,n} < \rho^*_{t,n} + \xi_n,$$

where $\xi_n = 0(n^{-1})$. For "most" sequences of F_j's one would expect $\rho^*_{t+1,n}$ to be significantly smaller than $\rho^*_{t,n}$ when t is small, since "most" sequences will exhibit substantial empirical dependencies of small order. In section 4 certain "t-fold" compound decision procedures are considered and the attainment of the goal $\rho^*_{t,n}$ is discussed. Illustrations of specific t-fold compound decision procedures are given for the problems considered in the "classical" case in section 3.

Some suggestions for further generalizations are given in section 5.

2. General results

In this section we assume the existence for each x of an estimator $\hat{\alpha}(x)$, which for any element F of $\mathfrak{F}$ is an unbiased estimator of $\alpha(x) = (L_1(\lambda) - L_2(\lambda))p(x)$, where λ is the parameter value and $p(x)$ the probability mass function associated with F. The estimator $\hat{\alpha}(x)$, which may be randomized, must depend only on observations having F as their common c.d.f., and is assumed to have a finite third absolute moment for each x. For each x, let $\sigma^2(x) = \text{var}(\hat{\alpha}(x))$ and $\gamma^3(x) = E|\hat{\alpha}(x) - \alpha(x)|^3$.

We now introduce two conditions which impose certain restrictions on $\mathfrak{F}$ and $\hat{\alpha}$.

CONDITION 1. *There exists a finite number B and a function $p_0(x)$ such that* (a) $\sum_x p_0^{1/2}(x) < \infty$, *and for each element of $\mathfrak{F}$ the corresponding λ and $p(x)$ satisfy* (b) $|\lambda| < B$, *and* (c) $p(x) \leq p_0(x)$ *for all x.*

CONDITION 2. *There exists a finite number $C > 0$ and a positive function $\epsilon(x) < 1$ such that* (a) $\sum_x \epsilon(x) < C$, (b) $\sum_x p_0(x)\epsilon^{-3}(x) < C$, *and for each element of $\mathfrak{F}$ and each x,* (c) $\epsilon^2(x) \leq \sigma^2(x) < C(\epsilon^2(x) + p_0(x))$, *and* (d) $\gamma^3(x) < C$.

For any sequence $F_1, F_2, \cdots,$ of elements of $\mathfrak{F}$ and for each x, let $\hat{\alpha}_j(x)$, $\sigma_j^2(x)$, and $\gamma_j^3(x)$ represent $\hat{\alpha}(x)$, $\sigma^2(x)$, and $\gamma^3(x)$ respectively for the sampling distributions F_j, $j = 1, 2, \cdots$. It is apparent that for fixed x, the sequence $\hat{\alpha}_j(x)$, $j = 1, 2, \cdots$, is a sequence of independent random variables, provided that any randomization involved is performed independently for each j. For each x and for $j = 1, 2, \cdots$, let

$$S_j(x) = \sum_{i=1}^{j} \hat{\alpha}_i(x). \tag{2.1}$$

We observe that $ES_j(x) = m_j(x)$, and denote the variance of $S_j(x)$ by $s_j^2(x) = \sum_{i=1}^{j} \sigma_i^2(x)$. The compound decision procedure to be evaluated is given for $j > 1$ by

$$\Delta_j(x) = \begin{cases} 1, & S_{j-1}(x) < 0, \\ 0, & S_{j-1}(x) \geq 0. \end{cases} \tag{2.2}$$

The decision rule $\Delta_1(x)$ for the first component problem may be arbitrary. We now state and prove the following theorem.

THEOREM 1. *If conditions* 1 *and* 2 *are satisfied, then there exists a finite constant K such that the average risk for the compound decision procedure* (2.2) *satisfies*

$$|\bar{r}_n - \rho_n^*| < Kn^{-1/2}, \tag{2.3}$$

for all n, for every sequence of elements of $\mathfrak{F}$.

PROOF. Recalling (1.3), (1.7), and (2.2) we have

$$n|\bar{r}_n - \rho_n^*| < \sum_x \left| \sum_{j=2}^{n} \alpha_j(x) P\{S_{j-1}(x) < 0\} + \xi(x) - m_n(x)^- \right|, \tag{2.4}$$

where $\xi(x)$ represents the contribution to the risk due to the arbitrary decision rule $\Delta_1(x)$ used in the first component problem. Since by condition 1 and (1.9) $\sum_x |\xi(x)|$ is bounded, it will be ignored in the subsequent argument. We now consider an arbitrary fixed value of x and suppress this value whenever it appears as the argument of a previously defined function. Letting $\Phi(\cdot)$ represent the c.d.f. of a standard normal random variable, we know by the Berry-Esseen theorem (see, for example, [4], p. 288) that there exists a constant C_0 such that

$$\begin{aligned} &\left| \sum_{j=2}^{n} \alpha_j P\{S_{j-1} < 0\} - \sum_{j=2}^{n} \alpha_j \Phi\left(-\frac{m_{j-1}}{s_{j-1}}\right) \right| \\ &\qquad \leq \sum_{j=2}^{n} |\alpha_j| \left| P\left\{\frac{S_{j-1} - m_{j-1}}{s_{j-1}} < -\frac{m_{j-1}}{s_{j-1}}\right\} - \Phi\left(-\frac{m_{j-1}}{s_{j-1}}\right) \right| \\ &\qquad \leq C_0 \sum_{j=2}^{n} \frac{|\alpha_j|}{s_{j-1}^3} \left| \sum_{i=1}^{j-1} \gamma_i^3 \right| = R_1^{(n)}. \end{aligned} \tag{2.5}$$

We seek a bound on the second sum on the left-hand side of (2.4) under the assumption that $m_n^- = 0$, that is, $m_n \geq 0$. For any particular sequence $F_1, F_2, \cdots$, such that $m_n \geq 0$, for $y > 1$, let

$$m(y) = m_{j-1} + \alpha_j(y - j + 1), \qquad j - 1 < y \leq j, \tag{2.6}$$

$$s(y) = \begin{cases} s_{j-1}, & j-1 < y \le j - j^{-2}, \\ s_{j-1} + j^2(s_j - s_{j-1})(y - j + j^{-2}), & j - j^{-2} < y < j, \end{cases} \tag{2.7}$$

for $j = 2, 3, \cdots$. Thus, we have

$$\int_{j-1}^{j} m'(y)\Phi\left(-\frac{m(y)}{s(y)}\right) dy = \alpha_j \int_{j-1}^{j} \Phi\left(-\frac{m(y)}{s(y)}\right) dy. \tag{2.8}$$

Also, since $\Phi(\cdot)$ is monotone and bounded by one, and $m(y)/s(y)$ is monotone on the interval $(j-1, j-j^{-2}]$ for each $j > 1$, we have

$$\left|\int_{j-1}^{j} \Phi\left(-\frac{m(y)}{s(y)}\right) dy - \Phi\left(-\frac{m_{j-1}}{s_{j-1}}\right)\right| \le \left|\Phi\left(-\frac{m_j}{s_{j-1}}\right) - \Phi\left(-\frac{m_{j-1}}{s_{j-1}}\right)\right| + 2j^{-2}. \tag{2.9}$$

Hence, letting $\varphi(\cdot) = \Phi'(\cdot)$,

$$\begin{aligned} &\left|\sum_{j=2}^{n} \alpha_j \Phi\left(-\frac{m_{j-1}}{s_{j-1}}\right) - \int_1^n m'(y)\Phi\left(-\frac{m(y)}{s(y)}\right) dy\right| \\ &\quad = \left|\sum_{j=2}^{n} \alpha_j \left\{\Phi\left(-\frac{m_{j-1}}{s_{j-1}}\right) - \int_{j-1}^{j} \Phi\left(-\frac{m(y)}{s(y)}\right) dy\right\}\right| \\ &\quad \le \sum_{j=2}^{n} |\alpha_j| \left|\Phi\left(-\frac{m_j}{s_{j-1}}\right) - \Phi\left(-\frac{m_{j-1}}{s_{j-1}}\right)\right| + 2\sum_{j=2}^{n} |\alpha_j| j^{-2} \\ &\quad \le \sum_{j=2}^{n} |\alpha_j| \left|\Phi\left(\frac{|\alpha_j|}{2s_{j-1}}\right) - \Phi\left(-\frac{|\alpha_j|}{2s_{j-1}}\right)\right| + 2\sum_{j=2}^{n} |\alpha_j| j^{-2} \\ &\quad \le \varphi(0) \sum_{j=2}^{n} \frac{\alpha_j^2}{s_{j-1}} + 2\sum_{j=2}^{n} |\alpha_j| j^{-2} = R_2^{(n)}. \end{aligned} \tag{2.10}$$

We must now bound the integral appearing on the left-hand side of (2.10) uniformly in all functions $m(y)$ and $s(y)$ corresponding to sequences $F_1, F_2, \cdots$, such that $m(n) \ge 0$. Let $h(y) = (m(y)/s(y))$ so that $m'(y) = s'(y)h(y) + s(y)h'(y)$ (except at the points $y = j-1,\ j - j^{-2},\ j = 2, 3, \cdots$, where $m'(y)$ is not defined). Let

$$\begin{aligned} I &= \int_1^n m'(y)\Phi\left(-\frac{m(y)}{s(y)}\right) dy \\ &= \int_1^n s'(y)h(y)\Phi(-h(y))\, dy + \int_1^n s(y)h'(y)\Phi(-h(y))\, dy. \end{aligned} \tag{2.11}$$

Integrating the first expression by parts and integrating the resulting expression by parts again, we have

(2.12)

$$\begin{aligned} I &= s(n)h(n)\Phi(-h(n)) - s(1)h(1)\Phi(-h(1)) + \int_1^n s(y)h(y)h'(y)\varphi(-h(y))\, dy \\ &= s(n)h(n)\Phi(-h(n)) - s(n)\varphi(-h(n)) + \int_1^n s'(y)\varphi(-h(y))\, dy \\ &\qquad + s(1)\varphi(-h(1)) - s(1)h(1)\Phi(-h(1)). \end{aligned}$$

Now, observing that $\max_{Z>0} Z\Phi(-Z) = C_1\Phi(-C_1)$, where $\Phi(-C_1) = C_1\varphi(C_1)$, we have

$$|I| \leq (s(n) + s(1))(C_1^2\varphi(C_1) + 2\varphi(0)) = R_3^{(n)}. \tag{2.13}$$

Combining (2.5), (2.10), and (2.13) we see that for any fixed x, the summand on the right-hand side of (2.4) is bounded by $R_1^{(n)} + R_2^{(n)} + R_3^{(n)}$ for any case where $m_n(x) \geq 0$. The same result holds when $m_n(x) < 0$, since then

$$\sum_{j=2}^{n} \alpha_j P\{S_{j-1} < 0\} - m_n^- = \sum_{j=1}^{n} \alpha_j(P\{S_{j-1} < 0\} - 1) = -\sum_{j=1}^{n} \alpha_j P\{S_{j-1} \geq 0\}, \tag{2.14}$$

and essentially the same argument applies.

We now reintroduce the suppressed variable x and undertake to demonstrate that the quantity $\sum_x (R_1^{(n)}(x) + R_2^{(n)}(x) + R_3^{(n)}(x))$ is bounded by $Kn^{1/2}$ where K is independent of the sequence $F_1, F_2, \cdots$. Letting $C_2 = c(B + |b|)$ and recalling (1.2), we see that by condition 1 (b) and (c), $|\alpha_j(x)| < C_2p_j(x) < C_2p_0(x)$, for each x and j. Thus, referring to (2.5), we have by conditions 2 (b), (c), and (d)

$$\sum_x R_1^{(n)}(x) \leq C_0C_2C \sum_x \frac{p_0(x)}{\epsilon^3(x)} \sum_{j=2}^{n} (j-1)^{-1/2} \leq 2C_0C_2Cn^{1/2}. \tag{2.15}$$

Similarly, referring to (2.10), we have

$$\begin{aligned} \sum_x R_2^{(n)}(x) &\leq \varphi(0)C_2^2 \sum_x \frac{p_0(x)}{\epsilon(x)} \sum_{j=2}^{n} (j-1)^{-1/2} + 2C_2 \sum_x p_j(x) \sum_{j=2}^{n} j^{-2} \\ &\leq 2\varphi(0)C_2^2Cn^{1/2} + 2C_2C_3, \end{aligned} \tag{2.16}$$

where $C_3 = \sum_{j=2}^{\infty} j^{-2}$. For $R_{(n)}^3(x)$ given by (2.13), we note that for each s, $s(n) = s_n(x)$, so that by condition 2 (c)

$$s_n(x) \leq C^{1/2}n^{1/2}(\epsilon^2(x) + p_0(x))^{1/2}. \tag{2.17}$$

Hence by conditions 1 (a) and 2 (a)

$$\sum_x s_n(x) \leq C^{1/2}n^{1/2} \sum_x (\epsilon(x) + p_0^{1/2}(x)) \leq C^{1/2}(C + B_0)n^{1/2}, \tag{2.18}$$

where $B_0 = \sum_x p_0^{1/2}(x)$. This completes the proof of the theorem.

REMARK 1. The result of theorem 1 is ratewise sharp since the conditions of the theorem do not, for example, exclude sequences $F_1, F_2, \cdots, F_n$ such that $\lambda_j = b$ (that is, $\alpha_j(x) = 0$) for $j < n - n^{1/2}$, and $\lambda_j = b_0 > b$ (that is, $\alpha_j(x) = c(b_0 - b)p_j(x)$) for $n - n^{1/2} \leq j \leq n$. For such sequences the contribution of the terms $\sum_{j=1}^{n} \alpha_j(x)P\{S_{j-1}(x) < 0\}$ appearing in (2.4) will typically be of the order of $n^{1/2}$ and positive for each x. Many sequences having this property may be constructed, and such sequences can occur in both the parametric and nonparametric applications discussed in the next section. The constant K appearing in the statement of theorem 1 is defined implicitly in the proof and the value so determined is not "best" in any sense.

REMARK 2. The maximization of the integral I, defined by (2.11), over the class of all bounded continuous differentiable functions $m(y)$, may be viewed as a classical variational problem whose solution would yield valuable insight concerning "least favorable" sequences $F_1, F_2, \cdots$. Unfortunately, the variational problem is singular and cannot be solved by standard methods.

3. Applications

A. *The parametric case.* The parametric families for which estimators $\hat{\alpha}_j(x)$ satisfying the conditions of theorem 1 can be constructed are essentially those for which the compound estimation problem is tractable (see, for example, [9]).

The first example, which includes the Poisson and negative binomial families as special cases, is the exponential family with probability mass function

$$p(x) = g(x)\beta(\lambda)\lambda^x, \qquad \text{for } x = 0, 1, \cdots, \tag{3.1}$$

where $g(x) > 0$ and $g(x)/g(x+1)$ is bounded for all $x \geq 0$. The family $\mathfrak{F}$ consists of all distributions having probability mass functions of this form for a given $g(x)$ with $0 \leq \lambda < B$, where B and $B_1 > B$ are chosen so that $\sum_x g(x)B_1^x < \infty$. For this example, we confine attention to situations where a single observation is obtained for each component problem, that is, $X_j = X_{1,j}$. This observation may be regarded as the value assumed by a sufficient statistic perhaps based on a larger number of observations.

For each x and j let

$$\hat{\alpha}_j(x) = \begin{cases} \dfrac{cg(x)}{g(x+1)} + Z_j(x), & \text{for } X_j = x + 1, \\ -cb + Z_j(x), & \text{for } X_j = x, \\ Z_j(x), & \text{otherwise,} \end{cases} \tag{3.2}$$

where for each x, $Z_1(x), Z_2(x), \cdots$, is a sequence of independent random variables independent of the X_j's, such that $EZ_j(x) = 0$, $EZ_j{}^2(x) = \epsilon^2(x)$, and the third absolute moments of the $Z_j(x)$'s are bounded uniformly in x and j. The significance of the $Z_j(x)$'s which represent auxiliary randomization is discussed in remark 3 below. It is evident that for each x and j, $E\hat{\alpha}_j(x) = c(\lambda_j - b)p_j(x) = \alpha_j(x)$, $\epsilon^2(x) < \sigma_j^2(x) < C(\epsilon^2(x) + p_j(x))$, and $\gamma_j^3(x) < C$, for some suitably chosen C. Letting $p_0(x) = g(x)B^x/g(0)$, and noting that $g(0) \leq \sum_x g(x)\lambda^x = \beta^{-1}(\lambda)$, we see that for each x, $p_0(x) \geq p(x)$ for all elements of $\mathfrak{F}$ and $\sum_x p_0(x)^{1/2} < \infty$ since $\sum_x g(x)B_1^x < \infty$ for $B_1 > B$. Condition 1 and conditions 2 (c) and (d) are therefore satisfied by estimators of the form (3.2). To show that theorem 1 holds for these estimators, it remains to exhibit $Z_j(x)$'s satisfying conditions 2 (a) and (b) with $p_0(x)$ as defined above. For fixed $\delta > 0$, let

$$Z_j(x) = \begin{cases} (x+1)^{-(1+\delta)}, & \text{with probability } = \frac{1}{2}, \\ -(x+1)^{-(1+\delta)}, & \text{with probability } = \frac{1}{2}. \end{cases} \tag{3.3}$$

Then $EZ_j(x)^2 = \epsilon^2(x) = (x+1)^{-2(1+\delta)}$ and condition 2 (a) is satisfied. Since $\sum_x g(x)B_1^x$ converges for $B_1 > B$, it follows that $\sum_x (x+1)^{3(1+\delta)}p_0(x)$ converges and condition 2 (b) is satisfied.

Operationally, only one randomization need be performed at each stage since for fixed j, the $Z_j(x)$'s need not be independent for different x's and may be computed on the basis of the outcome of the same randomization experiment.

A second parametric example involves the family of distributions with probability mass functions of the form

$$p(x) = g(x)\beta(\lambda)\left(\frac{\lambda}{a_1+\lambda}\right)^x, \qquad x = 0, 1, \cdots; \lambda \geq 0, \tag{3.4}$$

where a_1 is a specified positive constant,

$$g(x) = \frac{a_1(a_1+1)\cdots(a_1+(x-1))}{x!}, \qquad x = 1, 2, \cdots, \tag{3.5}$$

$g(0) = 1$, and

$$\beta(\lambda) = \left(\frac{a_1}{a_1+\lambda}\right)^{a_1}. \tag{3.6}$$

This family possesses the interesting property that $EX = \lambda$ for all λ. These distributions are actually reparameterizations of negative binomial distributions. For each x and j let

$$\hat{\alpha}_j(x) = \begin{cases} \dfrac{ca_1g(x)}{g(x+t)} + Z_j(x), & X_j = x+t,\ t = 1, 2, \cdots, \\ -cb + Z_j(x), & X_j = x, \\ Z_j(x), & \text{otherwise}, \end{cases} \tag{3.7}$$

where the $Z_j(x)$'s are defined as in the previous example. Again $E\hat{\alpha}_j(x) = \alpha_j(x)$, and under the same conditions on the λ's as in the previous example, and with an analogous definition of $p_0(x)$, it is easily verified that theorem 1 holds for this example also.

The important case of the binomial distribution is treated below as a special case of the nonparametric problem.

B. *The nonparametric case.* We now consider the situation where the probability mass functions $p(x)$ corresponding to elements of $\mathfrak{F}$ are not assumed to have a known functional form and are not necessarily in a one-to-one relationship with the values of λ. For this case, $\lambda = Eh(X)$, where $h(\cdot)$ is a specified function and X is a typical observation having probability mass function $p(x)$. Thus, for instance, λ might be EX as in the second parametric example above. Other possibilities are $\lambda = E(X-t)^2$, or $\lambda = P\{X \leq t\}$ for some specified t.

Since so little is assumed about the probability structure of the problem, it is not surprising that the goal which is attainable in this case is slightly less stringent than that achieved in the parametric case. Specifically, if k observations are obtained for each component problem, the procedure discussed below will satisfy the conditions of theorem 1 with ρ_n^* interpreted as the optimal risk for a hypothetical Bayes problem involving only $k-1$ observations. Thus, one observation is sacrificed in the interests of generality or as the price of ignorance.

For the case of k observations ($k \geq 2$), $p(x) = p'(x_1)p'(x_2) \cdots p'(x_k)$, where $x = (x_1, x_2, \cdots, x_k)$ and $p'(\cdot)$ is the probability mass function for a single observation. Letting $x' = (x_1, x_2, \cdots, x_{k-1})$ and recalling (1.2) and (1.3), we see that if a compound decision rule $\Delta_j(x')$ based on x' is used, then the expression for $\bar{r}_n$ remains unchanged except that x is replaced by x' throughout. We therefore seek a suitable estimate of $\alpha_j(x')$. Let $y(x') = (y_1, y_2, \cdots, y_{k-1})$, where $y_1 \leq y_2 \leq \cdots \leq y_{k-1}$ are the ordered values of the components of x'. For $t = 1, 2, \cdots, k$, and all j, let $X_j^{(t)} = (X_{1,j}, X_{2,j}, \cdots, X_{t-1,j}, X_{t+1,j}, \cdots, X_{j,k})$. Finally, for each j and x' let

$$\hat{\alpha}_j(x') = \begin{cases} c(h(X_{t,j}) - b)M(x') + Z_j(x'), & \text{for } y(X_j^{(t)}) = y(x') \text{ and } t = 1, 2, \cdots, k, \\ Z_j(x'), & \text{otherwise} \end{cases} \tag{3.8}$$

with $M(x') = (m_1!m_2! \cdots m_{k-1}!)/k!$, where m_i is the number of components of x' having the i-th smallest distinct value. Even though it is possible for $y(X_j^{(t)})$ to equal $y(x')$ for more than one value of t, $\hat{\alpha}_j(x')$ is still well defined since $X_{t,j}$ will have the same value for each such case.

If $EZ_j(x') = 0$, it is evident that $E\hat{\alpha}_j(x') = \alpha_j(x')$. If we assume that $E|h(X)|^3 < C < \infty$, for any single observation X with probability mass function corresponding to an element of $\mathfrak{F}$, and if we assume the existence of a function $p_0'(\cdot)$ dominating each $p'(\cdot)$ corresponding to an element of $\mathfrak{F}$ and satisfying $\sum_{x_1} p_0'(x_1)^{1/2} < \infty$, we see that condition 1 is satisfied with x' replacing x. The choice of the randomizing $Z_j(x')$'s so that condition 2 is satisfied depends on the particular denumerable set of values which the observations may assume. If this set is the set of integers, then letting

$$Z_j(x') = \begin{cases} -\prod_{i=1}^{k-1} |x_i + \tfrac{1}{2}|^{-(1+\delta)}, & \text{with probability} = \tfrac{1}{2}, \\ \prod_{i=1}^{k-1} |x_i + \tfrac{1}{2}|^{-(1+\delta)}, & \text{with probability} = \tfrac{1}{2}, \end{cases} \tag{3.9}$$

for some $\delta > 0$, we see that condition 2 is satisfied with x replaced by x' provided $\sum_{x_1} |x_1|^{3(1+\delta)} p_0'(x_1) < \infty$. Under such circumstances, the result of theorem 1 holds with the interpretation of ρ_n^* given above. It should be noted that the case of the binomial distribution is included in this framework if we allow only the values zero and one for each individual observation, and set $h(x) = x$ so that $\lambda = p'(1) = 1 - p'(0)$. This case is not really "nonparametric" since the value of λ determines the distribution of the observations.

REMARK 3. If the λ's are bounded away from zero in the two parametric examples discussed in part A of this section, it is easily verified that condition 2, and hence theorem 1, holds without the introduction of the randomizing $Z_j(x)$'s.

The author knows of no examples within the context of the present paper (parametric or nonparametric) for which randomization can be demonstrated to be necessary for the result of theorem 1, provided the conditions unrelated to randomization are satisfied. It is conjectured that such randomization is *not* essential, although because of the form of the Berry-Esseen bound, it is required for the method of proof used here.

4. The *t*-dependent case

In this section criteria based on generalizations of ρ_n^* are introduced.

Consider a hypothetical Bayes decision problem in which one of two actions is chosen on the basis of k-dimensional vectors of observations $X_1, X_2, \cdots, X_t$, having a random joint probability mass function $\tilde{p}(x_1, x_2, \cdots, x_t) = \tilde{p}_1(x_1)\tilde{p}_2(x_2) \cdots \tilde{p}_t(x_t)$, where the $\tilde{p}_i(\cdot)$'s are random functions whose structure is described below. Note that x_i now stands for a k-dimensional vector and not a real component as was the case heretofore.

Now suppose that the vector of random probability mass functions $(\tilde{p}_1(\cdot), \tilde{p}_2(\cdot), \cdots, \tilde{p}_t(\cdot))$ corresponds to the random vector of sampling distributions $(\tilde{F}_1, \tilde{F}_2, \cdots, \tilde{F}_t)$, chosen according to the discrete a priori probability measure on the t-fold product space $\mathfrak{F} \times \mathfrak{F} \times \cdots \times \mathfrak{F}$ which assigns probability $(n - t + 1)^{-1}$ to each of the vectors $(F_{j-t+1}, F_{j-t+2}, \cdots, F_j), j = t, t + 1, \cdots, n$. Assuming that the losses depend only on the value of the parameter λ_t associated with $\tilde{p}_t(\cdot)$, the risk incurred if the arbitrary decision rule $\delta(x_1, x_2, \cdots, x_t)$ is used, is given by

$$\rho_{t,n} = \frac{1}{n-t+1} \sum_{(x_1,x_2,\cdots,x_t)} \delta(x_1, x_2, \cdots, x_t) \sum_{j=t}^{n} \alpha_{t,j}(x_1, x_2, \cdots, x_t) + \frac{1}{n-t+1} \sum_{j=t}^{n} L_2(\lambda_j), \tag{4.1}$$

where, for $j \geq t$

$$\alpha_{t,j}(x_1, x_2, \cdots, x_t) = (L_1(\lambda_j) - L_2(\lambda_j))p_{j-t+1}(x_1)p_{j-t+2}(x_2) \cdots p_j(x_t). \tag{4.2}$$

Letting

$$m_{t,j}(x_1, x_2, \cdots, x_t) = \sum_{i=t}^{j} \alpha_{t,i}(x_1, x_2, \cdots, x_t), \tag{4.3}$$

for $j \geq t$, the optimal Bayes risk is clearly given by

$$\rho_{t,n}^* = \frac{1}{n-t+1} \sum_{(x_1,x_2,\cdots,x_t)} m_{t,n}(x_1, x_2, \cdots, x_t)^- + \frac{1}{n-t+1} \sum_{j=t}^{n} L_2(\lambda_j), \tag{4.4}$$

and is achieved by the decision rule

$$\delta^*(x_1, x_2, \cdots, x_t) = \begin{cases} 1, & m_{t,n}(x_1, x_2, \cdots, x_t) < 0, \\ 0, & \text{otherwise.} \end{cases} \tag{4.5}$$

If the sequence $\tilde{p}_1(\cdot), \tilde{p}_2(\cdot), \cdots,$ were a function-valued stochastic process with known probability structure involving dependencies of order $t + 1$, one would expect the Bayes risk based on $t + 1$ vectors of observations to be smaller, in general, than that based on only t vectors of observations. In the present case, the hypothetical a priori probability measure changes as t changes, but an analogous result holds as is shown by the following elementary theorem.

THEOREM 2. *If* $|L_i(\lambda)| < K_0 < \infty$, $i = 1, 2$, *for all* λ*'s corresponding to elements of* $\mathfrak{F}$, *then there exists a finite number* K_1 *such that, for*

$$\rho^*_{t+1,n} \leq \rho^*_{t,n} + K_1(n - t)^{-1} \tag{4.6}$$

for any fixed t *and* $n > t$, *for every sequence of elements of* $\mathfrak{F}$.

PROOF. The proof is based on the elementary fact that for any $b_1, b_2, \cdots, b_n$, $(\sum_{j=1}^n b_j)^- \geq \sum_{j=1}^n b_j^-$. Thus

(4.7)

$$\begin{aligned}
&\rho^*_{t+1,n} - \rho^*_{t,n} \\
&\leq \frac{1}{n-t} \sum_{(x_1,\cdots,x_{t+1})} \left(\sum_{j=t+1}^{n} \alpha_{t+1,j}(x_1, \cdots, x_{t+1}) \right)^- \\
&- \frac{1}{n-t} \sum_{(x_1,\cdots,x_t)} \left(\sum_{j=t+1}^{n} \alpha_{t,j}(x_1, \cdots, x_t) + \alpha_{t,t}(x_1, \cdots, x_t) \right)^- + 2K_0(n-t)^{-1} \\
&\leq \frac{1}{n-t} \sum_{(x_2,\cdots,x_{t+1})} \left(\sum_{j=t+1}^{n} \sum_{x_1} \alpha_{t+1,j}(x_1, \cdots, x_{t+1}) \right)^- \\
&- \frac{1}{n-t} \sum_{(x_1,\cdots,x_t)} \left\{ \left(\sum_{j=t+1}^{n} \alpha_{t,j}(x_1, \cdots, x_t) \right)^- - \alpha_{t,t}(x_1, \cdots, x_t)^- \right\} + 2K_0(n-t)^{-1} \\
&\leq 4K_0(n-t)^{-1},
\end{aligned}$$

since $\sum_{x_1} \alpha_{t+1,j}(x_1, \cdots, x_{t+1}) = \alpha_{t,j}(x_2, \cdots, x_{t+1})$. This establishes the desired result.

As was remarked in section 1, it is to be expected that many sequences of sampling distributions will exhibit regularities that are equivalent to empirical dependencies. Such sequences will tend to yield values of $\rho^*_{t+1,n}$ substantially smaller than those for $\rho^*_{t,n}$, especially when t is small.

We now consider the use of compound decision rules of the form

$$\Delta_{t,j}(x_{j-t+1}, \cdots, x_j)$$

for the j-th component problem for $j \geq t$. It is understood that $\Delta_{t,j}(\cdot)$ may depend on observations obtained for component problems prior to that with index $j - t + 1$, and for $i \leq j < t$, $\Delta_{t,j}(\cdot)$ is arbitrary. Letting $x^*_t = (x_1, x_2, \cdots, x_t)$ for notational simplicity, the average risk for the t-th to the n-th component problems then becomes

$$\bar{r}_{t,n} = \frac{1}{n-t+1} \sum_{j=t}^{n} \sum_{x^*_t} \alpha_{t,j}(x^*_t) E\Delta_{t,j}(x^*_t) + \frac{1}{n-t+1} \sum_{j=t}^{n} L_2(\lambda_j). \tag{4.8}$$

For $j \geq 2t$ we assume that there exist estimators $\hat{\alpha}_{t,j}(\cdot)$ of $\alpha_{t,j}(\cdot)$ which are unbiased and which depend on the vectors of observations $X_1, X_2, \cdots, X_{j-t}$. Let

$$S_{t,j}(x^*_t) = \sum_{i=2t}^{j} \hat{\alpha}_{t,i}(x^*_t), \tag{4.9}$$

for $j \geq 2t$. For $j \geq 2t$ we consider compound decision rules of the form

$$\Delta_{t,j}(x_t^*) = \begin{cases} 1, & S_{t,j-t}(x_t^*) < 0, \\ 0, & \text{otherwise}. \end{cases} \tag{4.10}$$

For $t < j < 2t$, $\Delta_{t,j}(\cdot)$ may be arbitrary.

The problem as formulated thus far appears to be essentially the same as that considered in section 2. However, an additional difficulty arises from the fact that, for all cases of interest, the sequence $\hat{\alpha}_{t,2t}(\cdot), \hat{\alpha}_{t,2t+1}(\cdot), \cdots$, is a t-dependent sequence of random functions. That is, $\hat{\alpha}_{t,j}(\cdot)$ and $\hat{\alpha}_{t,j'}(\cdot)$ are independent only if $|j - j'| > t$.

The author has been able to show that if compound decision rules of the form (4.10) are used, then there exist an $\epsilon > 0$ and a finite K such that for all n

$$|\bar{r}_{t,n} - \rho_{t,n}^*| < Kn^{-\epsilon}, \tag{4.11}$$

for all sequences $F_1, F_2, \cdots$. The conditions for this result to hold are straightforward generalizations to the t-dependent case of conditions 1 and 2. The proof of (4.11), which is rather complex, will not be reproduced here since the author is convinced that, in fact, (4.11) holds with $\epsilon = \frac{1}{2}$. A "proof" of this conjecture has been produced which requires a suitable version of the Berry-Esseen theorem for t-dependent random variables. Unfortunately, no such theorem seems to be available.

The parametric and nonparametric estimators of the α's given in section 3 are readily adaptable to the t-dependent case. This is illustrated by considering the simplest parametric case, that is, the case of the geometric distribution. For this case a single observation having probability mass function $p_j(x) = \lambda_j^x(1 - \lambda_j)$, $x = 0, 1, \cdots$, is obtained for the j-th component problem. Thus, recalling that $x_t^* = (x_1, x_2, \cdots, x_t)$,

$$\alpha_{t,j}(x_t^*) = c(\lambda_j - b)\lambda_{j-t+1}^{x_1}\lambda_{j-t+2}^{x_2} \cdots \lambda_j^{x_t}(1 - \lambda_{j-t+1}) \cdots (1 - \lambda_j). \tag{4.12}$$

For $j \geq 2t$ let

$$\hat{\alpha}_{t,j}(x_t^*) = \begin{cases} c + Z_j(x_t^*), & X_j = x_t + 1, X_{j-1} = x_{t-1}, \cdots, X_{j-t-1} = x_1, \\ -cb + Z_j(x_t^*), & X_j = x_t, X_{j-1} = x_{t-1}, \cdots, X_{j-t+1} = x_1, \\ Z_j(x_t^*), & \text{otherwise}, \end{cases} \tag{4.13}$$

where for some $\delta > 0$,

$$Z_j(x_t^*) = \begin{cases} -\prod_{i=1}^{t} (x_i + 1)^{-(1+\delta)}, & \text{with probability } = \frac{1}{2}, \\ \prod_{i=1}^{t} (x_i + 1)^{-(1+\delta)}, & \text{with probability } = \frac{1}{2}. \end{cases} \tag{4.14}$$

If we restrict the possible values of λ to $0 \leq \lambda < B < 1$, then (4.11) holds for the compound decision rule (4.10) based on these $\hat{\alpha}_{t,j}$'s. The other parametric and nonparametric cases are disposed of in a similar fashion.

REMARK 4. Since the t-dependent case involves the "matching" of t vectors of observations with sequences of t consecutive past observation vectors, it is clear that, if t is much greater than one, the number of component problems must be quite large before good results can be expected. This consideration, together with the fact that the improvement in $\rho^*_{t+1,n}$ compared with $\rho^*_{t,n}$ tends to be greatest when t is small, indicates that in most cases one should use values of t on the order of one, two, or three.

5. Conclusion

As is customary in papers in this area, we take note of the fact that when the number of component problems is small, the procedures suggested will be relatively ineffective. Thus, as a practical matter, it is necessary to provide some means of orderly transition from "classical" decision procedures to compound decision procedures as the number of component problems increases.

Hopefully, the results of the present paper can be generalized in at least two directions. First, it would be very desirable to find similar results for finite action problems with more than two possible actions. Often such formulations conform more closely to real situations. Furthermore, greater flexibility in the choice of the loss structure can be obtained even under the restriction that the pairwise differences in the loss functions be linear in the parameter of interest.

A second important generalization would be the extension of the present methods to cases involving continuous random variables. Some such results are obtained for both the parametric and nonparametric compound estimation problems in [9] and [10]. It is conjectured that, for sufficiently sophisticated methods, bounds of order arbitrarily close to $n^{-1/2}$ on the difference between the average risk and the appropriate goal can be obtained in the continuous case.

REFERENCES

[1] J. F. Hannan and H. Robbins, "Asymptotic solutions of the compound decision problem for two completely specified distributions," *Ann. Math. Statist.*, Vol. 26 (1955), pp. 37–51.

[2] J. F. Hannan and J. R. Van Ryzin, "Rate of convergence in the compound decision problem for two completely specified distributions," *Ann. Math. Statist.*, Vol. 36 (1965), pp. 1743–1752.

[3] M. V. Johns, "Nonparametric empirical Bayes procedures," *Ann. Math. Statist.*, Vol. 28 (1957), pp. 649–669.

[4] M. Loève, *Probability Theory*, Princeton, Van Nostrand, 1960 (2d ed.).

[5] H. Robbins, "Asymptotic subminimax solution of compound decision problems," *Proceedings of the Second Berkeley Symposium on Mathematical Statistics and Probability*, Berkeley and Los Angeles, University of California Press, 1951, pp. 131–148.

[6] ———, "The empirical Bayes approach to testing statistical hypotheses," *Rev. Inst. Internat. Statist.*, Vol. 31 (1963), pp. 195–208.

[7] E. Samuel, "An empirical Bayes approach to the testing of certain parametric hypotheses," *Ann. Math. Statist.*, Vol. 34 (1963), pp. 1370–1385.

[8] ———, "Convergence of the losses of certain decision rules for the sequential compound decision problem," *Ann. Math. Statist.*, Vol. 35 (1964), pp. 1606–1621.

[9] ———, "Sequential compound estimators," *Ann. Math. Statist.*, Vol. 36 (1965), pp. 879–889.

[10] D. D. SWAIN, "Bounds and rates of convergence for the extended compound estimation problem," Statistics Department, Stanford University Technical Report, 1965.
[11] J. R. VAN RYZIN, "The sequential compound decision problem with $m \times n$ finite loss matrix," Argonne National Laboratory, Applied Mathematics Division Technical Memorandum No. 54, 1965.

HORIZON IN DYNAMIC PROGRAMS

JERZY ŁOŚ
INSTITUTE OF MATHEMATICS OF THE POLISH ACADEMY OF SCIENCES

1. Introduction

The theory of dynamical programs deals with undertaking decisions in time. Usually we have a functional over a set of sequences (or functions), and the task consists in finding a minimum of this functional. The components of the sequences (or the value of the functions—when time is considered to be continuous) represent the decisions, which are to be carried out at the appropriate point of time. As the solution—minimizing the functional—we get a sequence of decisions, which tells us what to do at all future times.

This is a considerable simplification of problems we face in applications. Usually in applications we are not interested in all sequences of decisions, but indeed, we are interested in the particular one which we must carry out at the present stage. However, the functional to be minimized is not completely known to us. This means that many data are needed to define a functional. These data will occur in time, finally allowing selection of one functional from a family of many possible functionals. But when making the first decision, we do not know which one will finally be selected.

In several cases, to compute the optimal first step decision, we do not need all the data of the functional, but only a part of them; for instance, those which will occur up to a specific point of time h in the future. Such a point is called the horizon of the problem. This is the point up to which one has to know the future in order to compute the optimal decision at the present stage.

The idea of horizon goes back to Modigliani, who in [6] and [7] defined it in an intuitive manner. But the ideas of Modigliani were not worked out to a precise form, and therefore, the term "horizon," which may be found in many papers concerned with dynamical programs, is used with various meanings.

In this paper we present a rigorous definition of the notion of horizon. An auxiliary notion is that of a dynamical parameter, which serves to express the information concerning data of the functional occurring in time.

There are two groups of problems basic to the theory of horizon. One of them deals with the properties of solutions computed with the help of a given horizon ("horizonal solutions"); the other one is concerned with the existence of the horizon in specific cases. Since this paper has an introductory character, both groups of problems are represented here, but by weak theorems only.

Stronger results may be obtained by additional assumptions on the families of problems concerned.

Notations. Throughout this paper we shall use a standard notation, with a few exceptions, which will be mentioned here.

Usually a lower case letter, like x or ζ, denotes infinite sequence:

$$x = \langle x_1, x_2, \cdots \rangle, \qquad \zeta = \langle \zeta_1, \zeta_2, \cdots \rangle. \tag{1}$$

By $x|k$ we denote the finite sequence $\langle x_1, x_2, \cdots, x_k \rangle$ of k first coordinates of x. By $k|x$ we denote the infinite sequence $\langle x_{k+1}, x_{k+2}, \cdots \rangle$. If A_n is a function of n variables, then $A_n(\overline{x}|k, x|n-k) = A_n(\overline{x}_1, \cdots, \overline{x}_k, x_1, \cdots, x_{n-k})$. The same notation applies to functional A over sequences x:

$$A(\overline{x}|k, x) = A(\overline{x}_1, \cdots, \overline{x}_k, x_1, x_2, \cdots). \tag{2}$$

The symbol R^+ denotes the set of nonnegative real numbers and $+\infty$. The symbol $\chi(\ \ )$ denotes the characteristic function of the relation in parentheses. For instance,

$$\chi(0 < \alpha) = \begin{cases} 1, & \text{if } 0 < \alpha, \\ 0, & \text{if } 0 \geq \alpha. \end{cases} \tag{3}$$

2. Simple dynamic programming problems

A simple dynamic programming problem (d.p.p.) is defined by two sequences: $X_1, X_2, \cdots, A_1, A_2, \cdots$. The first one is a sequence of sets, the second one is a sequence of functions A_n: $X_1 \times \cdots \times X_n \to R^+$. By the "problem" we mean the problem of finding a minimum of the function A: $X \to R^+$, where $X = X_1 \times X_2 \cdots$ and $A(x) = \sum_{i=1}^{\infty} A_i(x|i)$. A d.p.p. is denoted (X_n, A_n).

EXAMPLE 2.1. The problem lies in finding a minimum of the function $C(x) = \sum_{i=1}^{\infty} c_i \cdot x_i$ for x's satisfying $x_i \geq 0$ and $\sum_{i=0}^{n} x_i \geq \sum_{i=1}^{n} d_i$. Here c_i (cost coefficients), d_i (demands), and x_0 (initial stock) are nonnegative numbers. One can assume that $\sum_{i=1}^{\infty} c_i \cdot d_i < +\infty$. To convert this problem into a d.p.p. we will set $X_n = R^+$ and

$$A_n(x_1, \cdots, x_n) = \begin{cases} c_n \cdot x_n, & \text{if } \sum_{i=0}^{k} x_i \geq \sum_{i=1}^{k} d_i \text{ for } k = 1, \cdots, n; \\ +\infty, & \text{in the opposite case.} \end{cases} \tag{4}$$

EXAMPLE 2.2 (Modigliani, Hohn [7]). Let us consider the function

$$C(x) = \sum_{i=1}^{\infty} \beta^{i-1} \left[c(x_i) + \alpha \left(\sum_{j=0}^{i-1} x_j - \sum_{j=1}^{i-1} d_j \right) \right], \tag{5}$$

where c is a convex, monotone-increasing function, positive for $x_i > 0$ (cost function), x_0 (initial stock), α (storage cost) are nonnegative numbers, and where $0 \leq \beta \leq 1$ (discount factor) and $d_j \geq 0$ (demands). To transform the problem of finding a minimum of C over the set of x's satisfying $x_i \geq 0$ and $\sum_{i=0}^{n} x_i \geq \sum_{i=1}^{n} d_i$ into a d.p.p., we set $X_n = R^+$ and

$$(6) \qquad A_n(x_1, \cdots, x_n) = \begin{cases} \beta^{n-1}\left[c(x_n) + \alpha\left(\sum_{j=0}^{n-1} x_j - \sum_{j=0}^{n-1} d_j\right)\right], \\ \text{if } \sum_{j=0}^{k} x_j \geq \sum_{j=1}^{k} d_j, \text{ for } k = 1, \cdots, n, \\ +\infty, \text{ in the opposite case.} \end{cases}$$

EXAMPLE 2.3 (Bellman, Glicksberg, Gross [3]). We are given two positive numbers c (cost coefficient) and α (cost of increasing rate of production). For $x_0 \geq 0$, $(i = 1, 2, \cdots)$ we define

$$(7) \qquad A_n(x_{n-1}, x_n) = \begin{cases} a(x_{n-1}, x_n), & \text{if } x_n \geq d_n, \\ +\infty, & \text{if } x_n < d_n \end{cases}$$

where $a(x_{n-1}, x_n) = c \cdot x_n + \alpha(x_n - x_{n-1})\chi(x_n > x_{n-1})$. The sequence of sets $X_n = R^+$ and the sequence of functions A_n define a d.p.p. which is called the problem of production planning without storage.

EXAMPLE 2.4 (Wagner, Whitin [8]). Let us define for the nonnegative numbers x_0, s_i, m_i, d_i,

$$(8) \qquad A_n(x_1, \cdots, x_n) = \begin{cases} s_n\chi(0 < x_n) + m_n\left(\sum_{j=0}^{n-1} x_j - \sum_{j=1}^{n-1} d_j\right), \\ \text{if } \sum_{j=0}^{k} x_j \geq \sum_{j=1}^{k} d_j \text{ for } k = 1, \cdots, n; \\ +\infty, \text{ in the opposite case.} \end{cases}$$

The sequence of function A_n together with the sequence of sets $X_n = R^+$ form a d.p.p.

EXAMPLE 2.5 (Blackwell [4]). We are given two finite sets S (states) and A (actions), and moreover, two real functions $r\colon S \times A \to R^+$ and $p\colon S \times A \times S \to R^+$; the latter, $p(s'; \text{if } a, s)$, is a probability distribution in s'.

For every n, let X_n be the set of functions $x_n\colon S \to A$. Given an s_0 in S and $x = \langle x_1, x_2, \cdots \rangle$ in $X = X_1 \times X_2 \times \cdots$, we define

$$(9) \qquad p_0(s; \text{if } s_0, x|0) = \begin{cases} 1, & \text{if } s = s_0, \\ 0, & \text{if } s \neq s_0, \end{cases}$$

$$(10) \qquad p_{n+1}(s; \text{if } s_0, x|n+1) = \sum_{s' \in S} p(s'; \text{if } x_1(s_0), s_0) \cdot p_n(s; \text{if } s', (1|x) \mid n)$$

where $1|x$ denotes $\langle x_2, x_3, \cdots \rangle$ (and therefore, $(1|x) \mid n = \langle x_2, \cdots, x_{n+1} \rangle$).

We define

$$(11) \qquad A_n(x|n) = \beta^{n-1} \sum_{s \in S} r(s, x_n(s)) p_{n-1}(s; \text{if } s_0, x|n-1).$$

This gives us a simple d.p.p. composed of the sequences $X_1 = X_2 = \cdots$ and $A_1, A_2, \cdots$.

3. Families of d.p.p. The dynamic parameter

If to every element ζ of a set P corresponds a simple d.p.p., $(X_n, A_n(\cdot\,; \zeta))$, $n = 1, 2, \cdots$, then we have a family of d.p.p. over the set of parameters P.

If the set P is a subset of a product $Z = Z_1 \times Z_2 \times \cdots$, and for every ζ in P and every n, $A_n(\cdot\,;\zeta)$ do not depend on the entire sequence ζ but only upon their first n coordinates $\zeta|n$, then we will call the family $(X_n, A_n(\cdot\,;\zeta|n))$, $\zeta \in P$, a family with a dynamic parameter.

EXAMPLE 3.1. In this example, fixing x_0 and c_i's, the sequence of d_i's is a dynamic parameter. The function A_n, in fact, does depend on the first n of the d_i's, and so we can write $A_n(x_1, \cdots, x_n; d_1, \cdots, d_n) = A_n(x|n; d|n)$. The set of parameters P is in this case the whole product $R^+ \times R^+ \times \cdots$, but for various reasons, it may be restricted to its subset.

By fixing only x_0 and assuming not only d_i's but also c_i's as being variable, we obtain a larger family with d_i's and c_i's occurring as a dynamic parameter. Strictly speaking, in order to conform to the definition, we have to accept as a dynamic parameter the sequence of pairs

$$\zeta = \langle\langle c_1, d_1\rangle, \langle c_2, d_2\rangle, \cdots\rangle. \tag{12}$$

By varying x_0 we can change the considered d.p.p.'s. But x_0 is not a dynamic parameter.

EXAMPLE 3.2. In this example α, β, x_0 and d_i's are parameters. Only d_i's may be considered as a dynamic parameter. If instead of the discount factor β we adopt varying factors β_i, then the sequence of β_i's may be also considered as a dynamic parameter. This may have some meaning when studying discount fluctuations on a market.

Note that the choice of the dynamic parameter depends on the problem we plan to study.

EXAMPLE 3.3. Here, the numbers c, α, x_0 and all d_i's are parameters, but only d_i's form a dynamic parameter. Following the definition of A_n which we accepted, these functions depend only on the two last coordinates of $x|n$ and on the last coordinate of $d|n$. It neither affects the definition of d.p.p. nor that of the dynamic parameter. Defining functions $\overline{A}_n$ as

$$\overline{A}_n(x_1, \cdots, x_n) = \begin{cases} a(x_{n-1}, x_n), & \text{if } x_k \geq d_k, \text{ for } k = 1, \cdots, n, \\ +\infty, & \text{in the opposite case,} \end{cases} \tag{13}$$

we obtain another d.p.p. These d.p.p.'s considered as a family with a dynamic parameter $d = \langle d_1, d_2, \cdots\rangle$ no longer have the property mentioned earlier. All functions $\overline{A}_n$ depend essentially on $x|n$ and $d|n$. In spite of the identity $A(x) = \sum_{n=1}^{\infty} A_n(x|n) = \overline{A}(x) = \sum_{n=1}^{\infty} \overline{A}_n(x|n)$ (for a fixed d), both d.p.p. (X_n, A_n) and $(X_n, \overline{A}_n)$ must be considered as different d.p.p.'s, because generally $A_n \neq \overline{A}_n$.

EXAMPLE 3.4. All sequences s_i, m_i, and d_i may be considered as a dynamic parameter. The number x_0 is not one.

EXAMPLE 3.5. The element s_0 is the only varying factor in the A_n's. It cannot be considered as a dynamic parameter.

4. Truncated and partially completed d.p.p.'s. The initial parameter

Let us define for a given d.p.p. (X_n, A_n),

$$A_n(x|n; 1) = A_n(x|n) \quad \text{and} \quad A_n(x|n; 0) = 0. \tag{14}$$

The family $(X_n, A_n(\cdot\,; t))$, $t \in T_0$, where T_0 is the set of all sequences t with $t_1 = t_2 = \cdots = t_N = 1$, $t_{N+1} = t_{N+2} = \cdots = 0$ for some natural number N, is called the family of truncated d.p.p.'s of (X_n, A_n).

For a d.p.p. (X_n, A_n), $\bar{x}$ in $X = X_1 \times X_2 \times \cdots$ and natural number k, we define the d.p.p. partially completed by $\bar{x}|k$ as the d.p.p. $(X_n, \bar{A}_n)$ with $\bar{X}_n = X_{n+k}$ and $\bar{A}_n(x|n) = A_{n+k}(\bar{x}|k, x|n)$ for every x in $X_{k+1} \times X_{k+2} \times \cdots$.

Let us consider a family of d.p.p.'s $(X_n, A_n(\cdot\,; \zeta|n))$, $\zeta \in P$ with the (only) dynamic parameter ζ. For a given $\bar{x}$ in $X = X_1 \times X_2 \times \cdots$, $\bar{\zeta}$ in P and a natural number k, we define the family partially completed by $\bar{x}|k$ and $\bar{\zeta}|k$ as the family of all d.p.p.'s $(X_n, A_n(\cdot\,; \zeta|n))$ with $\zeta \in P$ and $\zeta|k = \bar{\zeta}|k$, partially completed by $\bar{x}|k$. This is again a family with a dynamic parameter ξ which runs over the set $P(\bar{\zeta}|k)$ of all $\xi = \langle \xi_1, \xi_2, \cdots \rangle$ such that $\langle \bar{\zeta}_1, \cdots, \bar{\zeta}_k, \xi_1, \xi_2, \cdots \rangle$ belongs to P.

Let us now assume that the family $(X_n, A_n(\cdot\,; \zeta|n, x_0))$, $\zeta \in P$, $x_0 \in X_0$ is such that $X_1 = X_2 = X_3 = \cdots$ and $P(\bar{\zeta}|k) = P$ for every $\bar{\zeta}$ in P and natural k.

In this case the partial completion of a subfamily $(X_n, A_n(\cdot\,; \zeta|n, x_0))$, $\zeta \in P$, with a fixed x_0, by $\bar{x}|k$ and $\bar{\zeta}|k$, results in a family defined onto the same sets X_n and with the same dynamic parameter $\bar{\zeta}$ in P. It may happen that this family is one of the subfamilies of the whole family, only with another nondynamic parameter x_0'. If this occurs for every $\bar{x}|k$, $\bar{\zeta}|k$ and x_0 in X_0, then the parameter x_0 is called the initial parameter of the whole family.

EXAMPLES 4.1–4.4. In all these examples x_0 is an initial parameter. Let us consider, for instance, example 1. If the initial stock x_0 demands $\bar{d}_1, \cdots, \bar{d}_k$ and productions $\bar{x}_1, \cdots, \bar{x}_k$ are given for the first k periods, then at the $k + 1$-period the initial stock is $x_0' = x_0 + \sum_{i=1}^{k} \bar{x}_i - \sum_{i=1}^{k} \bar{d}_i$, and this is the only influence of the past on the coming periods. In order to cover the case when $\bar{x}_1, \cdots, \bar{x}_k$ is not feasible for $\bar{d}_1, \cdots, \bar{d}_k$ (for instance, $x_0' < 0$), we may assume that x_0 takes nonnegative values and -1.

EXAMPLE 4.5. In this example s_0 is not an initial parameter, even in the case when $\beta = 1$. If we complete the program by $\langle \bar{x}_1, \cdots, \bar{x}_k \rangle = \bar{x}|k$, starting with a given s_0 in S, then the initial s in the partially completed program is known to us only through the distribution $p_k(s; \text{if } s_0, \bar{x}|k)$.

In order to have an initial parameter in our family of problems, we may extend the set of parameters S to the set Π of (unconditional) distribution over S. Then for a given π in Π we will have

$$\bar{p}_n(s; \text{if } \pi, x|n) = \sum_{s_0 \in S} p_n(s; \text{if } s_0, x|n)\pi(s_0). \tag{15}$$

Now for a given π and $\bar{x}|k$, the parameter π of the partially completed problem will be $\bar{p}_k(\cdot\,; \text{if } \pi, \bar{x}|k)$ which is in Π.

The parameter π may be considered as an initial parameter also in the case when $\beta < 1$, provided that we agree to consider two d.p.p.'s which differ only by a positive coefficient (that is, $A_n = \alpha A_n'$, for all n, with $\alpha > 0$) as equal. (This remark also applies in example 2, where the discount factor β is introduced.)

Finally, let us note that the family considered in this example has no proper dynamic parameter. In order to fit it to the definition we may always introduce a dynamic parameter P consisting of a constant sequence. Such a P fulfills the requirement of the definition.

5. The horizon and horizonal solutions

From this point on we will assume that all d.p.p.'s with which we are concerned attain a minimum at some point of the product of their sets.

Now let us fix a family $(X_n, A_n(\cdot\,; \zeta|n))$, $\zeta \in P \subset Z = Z_1 \times Z_2 \times \cdots$. The function connected with a ζ in P shall be denoted by $A(x; \zeta) = \sum_{n=1}^{\infty} A_n(x|n; \zeta|n)$. By $v(\zeta)$ we shall denote the value of the d.p.p. with the parameter ζ: $v(\zeta) = \min_{x \in X} A(x; \zeta)$. A d.p.p. with $v(\zeta) < +\infty$ is called convergent.

We define a relation between a natural number h and an element $\bar{\zeta}$ in Z (but not necessarily in P) in the following way: h is the horizon for $\bar{\zeta}$ if there exists an element x_1^* in X_1, such that for every ζ in P and such that $\zeta|h = \bar{\zeta}|h$, there exists an x in X with $A(x, \zeta) = v(\zeta)$ and $x_1 = x_1^*$.

Roughly speaking, h is a horizon for $\bar{\zeta}$ if there exists a "first step decision" x_1^*, which may be extended to the minimal solution of every d.p.p. of the family concerned, provided the dynamic parameter ζ of that program agrees with $\bar{\zeta}$ in h first coordinates.

If h is a horizon for $\bar{\zeta}$, then the element x_1^*, which the definition asserts to exist, is called the horizonal element for $\bar{\zeta}$.

We should point out that the notion of horizon depends on the family of d.p.p.'s under consideration. The correct way of expressing the defined relation is the following: "h is the horizon for $\bar{\zeta}$ in the family. . . ." The notion of a horizonal element is strongly dependent on h. If h is a horizon for $\bar{\zeta}$, then every $h_1 > h$ is also a horizon for $\bar{\zeta}$. But an element x_1^* which satisfies the definition for h_1 does not necessarily satisfy it for h.

A sequence x^* in X is called a horizonal solution for $\bar{\zeta}$, iff for every k, x_{k+1}^* is a horizontal element for $\langle \bar{\zeta}_{k+1}, \bar{\zeta}_{k+2}, \cdots \rangle$ in the family of d.p.p.'s partially completed by $x^*|k$ and $\bar{\zeta}|k$.

A very important lemma on the horizonal solutions is the following.

LEMMA 5.1. *If x^* is a horizonal solution for $\bar{\zeta}$, then for every k there exists an h_k such that, for every ζ in P satisfying $\zeta|k + h_k = \bar{\zeta}|k + h_k$, there exists an x in X with $A(x; \zeta) = v(\bar{\zeta})$ and $x|k + 1 = x^*|k + 1$.*

The proof of this lemma is by induction on k, and we will not give it here.

We should note that a $\bar{\zeta}$ for which there exists a horizonal solution cannot be completely arbitrary. In order to have the family partially completed by $x^*|k$ and $\bar{\zeta}|k$, a $\zeta^{(k)}$ in P with $\zeta^{(k)}|k = \bar{\zeta}|k$ is needed. In this case we say that $\bar{\zeta}$ is a limit of the sequence $\zeta^{(k)}$, and we write $\zeta^{(k)} \to \bar{\zeta}$. In order to have a horizonal solution, $\bar{\zeta}$ has to be a limit of parameters in P.

We say that a simple d.p.p. has a horizon if the constant sequence $t = \langle 1, 1, \cdots \rangle$ has a horizon in the family of truncated d.p.p.'s of the given d.p.p. In the same way, we say that x^* is a horizonal solution of a simple d.p.p. This means that it is a horizonal solution for $t = \langle 1, 1, \cdots \rangle$ in the family of truncated d.p.p.'s.

EXAMPLE 5.1. In this example, h is a horizon for $\langle\langle \bar{c}_1, \bar{d}_1\rangle, \langle \bar{c}_2, \bar{d}_2\rangle, \cdots\rangle$ in the family with an initial parameter x_0, iff $\bar{c}_h \leq \bar{c}_1$ and $\bar{c}_k > \bar{c}_1$, for $k = 2, \cdots, h-1$. Then if $\sum_{i=1}^{h-1} \bar{d}_i \leq x_0$, $x_1^* = 0$ is the horizonal element and if $\sum_{i=1}^{h-1} \bar{d}_i > x_0$, then $x_1^* = \sum_{i=1}^{h-1} \bar{d}_i - x_0$ is the horizonal element.

If, for example, $\liminf \bar{c}_i = 0$ (which is a reasonable assumption, as usually the cost $\bar{c}_i$ will be the real costs reduced by a discount factor), then the horizonal solution exists.

EXAMPLE 5.2. This is the classical example for the horizonal solution. If we allow the parameters $d = \langle d_1, d_2, \cdots\rangle$ to run through the set of all nonnegative sequences, then there obviously is no horizon for any sequence. But if we restrict d to a set P of uniformly bounded sequences (that is, $d_i \leq M$, for all i and d in P), then in that family with the arbitrary initial parameter there exists a horizon for every sequence in P. Since the whole family is a family with an initial parameter, then for every $\bar{\zeta}$ in P a horizonal solution for $\bar{\zeta}$ exists. Some generalizations of this theorem have recently been proved (see section 6).

EXAMPLE 5.3. It may be easily shown that for a given $d = \langle d_1, d_2, \cdots\rangle$, h is a horizon for d in the family with an initial parameter x_0, if $d_h \geq x_0$ and $d_k < x_0$ for $k = 1, \cdots, h-1$. A better result states that independently of x_0 the natural number h' with $(\alpha/c) < h' \leq (\alpha/c) + 1$ is a horizon for every d. The first horizon h, as a function of x_0 and d, is neither defined everywhere, nor bounded on the set where it is defined. The second one is defined everywhere and bounded, but it may happen that $h < h'$. Hence, the horizon h' is not always the shortest one.

Following the method presented in Arrow, Karlin [2], it may be shown that there exists a horizon even in the continuous case with a convex cost function (see section 6).

EXAMPLE 5.4. There is no horizon for all sequences of the dynamic parameters. (This fact was established by A. Brauner.) It is proved by showing that if $m_i = 2$, $d_i = 1$, and $s_i = 3 + a_i$, where $0 < a_i < a_{i+1} < 1$, then the minimal solution of the d.p.p. truncated on N is either $\langle 2, 0, 2, 0, \cdots, 2, 0\rangle$ if N is even, or $\langle 1, 2, 0, 2, 0, \cdots, 2, 0\rangle$ if N is odd.

It can be shown that a d.p.p. with all dynamic parameters constant, namely $s_i = s_1 > 0$, $m_i = m_1 > 0$, $d_i = d_1 > 0$, has a horizon in the family of its truncated programs.

EXAMPLE 5.5. Not every d.p.p. with $\beta < 1$, belonging to the family presented in section 1, has a horizon in the family of its truncated d.p.p.'s. The example is the following. The set S contains four states: s_0, s_1, s_2, s_3. All actions a in A lead from one state to another in a deterministic manner (that is, $p(s'; \text{if } a, s) = 1$ or 0). We are given actions leading from s_0, s_1, s_2 to every other state, but s_3 is an absorbing state. This means that every action leads from s_3 to s_3 only. For transitions $s_0 \rightarrow s_1$, $s_0 \rightarrow s_2$, $s_1 \rightarrow s_2$, $s_2 \rightarrow s_1$, the loss r is equal to 1. For transition $s_2 \rightarrow s_3$ there is no loss, that is, r is equal to 0. All other transitions have the loss $r > 1$; in particular, the transition $s_3 \rightarrow s_3$ has a loss which is large in comparison to β, let us say $2/\beta$.

If we start with s_0 and the program is infinite, then the best we can do is to go to either s_1 or s_2, and then to change at every step from s_1 to s_2 and from s_2 to s_1. Proceeding this way we incur a minimal loss equal to $v = \sum_{i=1}^{\infty} \beta^{i-1}$. If the program is finite, let us say of the length N, then the best policy, when starting with s_0, is to go in $N - 1$ first steps through transitions with loss 1 and then to finish with the transition $s_2 \rightarrow s_3$. The total loss in such a case will be $v_N = \sum_{i=1}^{N-1} \beta^{i-1}$, and it is the minimal one. But to achieve this we must make in the first step the transition $s_0 \rightarrow s_2$, if N is even, and the transition $s_0 \rightarrow s_1$, if N is odd. This shows that there is no horizon for the infinite problem in the family of its truncated d.p.p.'s.

In spite of the nonexistence of the horizon for some d.p.p.'s in our family, we can show that in some cases the horizon does exist.

We want to recall that

$$A_n(x|n; s_0) = \beta^{n-1} \sum_{s \in S} r(s, x_n(s))p_{n-1}(s; \text{if } s_0, x|n - 1). \tag{16}$$

Let us form the family of truncated d.p.p.'s. By $t^{(N)}$ we shall denote the sequence with $t_n^{(N)} = 1$ or 0, according to $n \leq N$ or $n > N$. Then the functions of truncated problems are

$$A_n(x|n; t_n^{(N)}, s_0) = t_n^{(N)} \cdot A_n(x|n; s_0) \tag{17}$$

and

$$A(x; t^{(N)}, s_0) = \sum_{n=1}^{N} A_n(x|n; s_0). \tag{18}$$

It follows from the inductive definition of the conditional distribution p_n that

(i) $A(x; t^{(N)}, s_0) = r(s_0, x_1(s_0)) + \beta \sum_{s \in S} p(s; \text{if } x_1(s_0), s_0) \cdot A(1|x, t^{(N-1)}, s)$.

Starting with this formula it can easily be proved by induction that

(ii) For every N, there exists an $x^{(N)}$ such that $A(x^{(N)}; t^{(N)}, s_0) = v(t^{(N)}, s_0)$, for every s_0 in S.

Another easy preparatory lemma is the following.

(iii) For every s_0 in S, $v(t^{(N)}, s_0) \rightarrow v(s_0)$, when $N \rightarrow \infty$.

We associate with each function $f\colon S \rightarrow A$ (then f is in X_1) and every function $\phi\colon S \rightarrow R^+$, a function $L(f, \phi)\colon S \rightarrow R^+$, which is defined by

$$L(f, \phi)(s_0) = r(s_0, f(s_0)) + \beta \sum_{s \in S} p(s; \text{if } f(s_0), s_0)\phi(s). \tag{19}$$

Then we have

(iv) $A(x; t^{(N)}, s_0) = L(x_1, A(1|x; t^{(N-1)}, \cdot))(s_0)$.

Let us call an x_1^* in X_1 a minimal element with respect to the function $\phi\colon S \rightarrow R^+$, iff $L(f, \phi)(s_0) \geq L(x_1^*, \phi)(s_0)$ for all f in X_1 and s_0 in S.

(v) For an x_1^* in X_1, to have an extension $x^{(N)}$ (that is, $x_1^{(N)} = x_1^*$) with $A(x^{(N)}; t^{(N)}, s_0) = v(t^{(N)}, s_0)$ for all s_0 in S, it is necessary and sufficient to be a minimal element with respect to $v(t^{(N-1)}, \cdot)$.

It is easy to show that if x_1^* fulfills the condition, then for every $x^{(N-1)}$ which minimizes the problem of the length $N - 1$, $\langle x_1^*, x_1^{(N-1)}, x_2^{(N-1)}, \cdots \rangle$ minimize the problem of the length of N.

On the other hand, if $x^{(N)}$ minimizes the problem of the length of N and $x_1^* = x_1^{(N)}$, then, as $A(1|x^{(N)}, t^{(N-1)}, s) \geq v(t^{(N-1)}, s)$ for every s and L is monotonic, we have

$$\begin{aligned}(20)\qquad L(x_1^*, A(1|x^{(N)}, t^{(N-1)}, \cdot))(s_0) &= A(x^{(N)}, t^{(N)}, s_0) \\ &\geq L(x_1^*, v(t^{(N-1)}, \cdot))(s_0) = v(t^{(N)}, s_0) \qquad \text{for every} \quad s_0.\end{aligned}$$

It follows that $L(x_1^*, v(t^{(N-1)}, \cdot))(s_0) \leq L(f, v(t^{(N-1)}, \cdot))(s_0)$ for every s_0 and f, which means that x_1^* is a minimal element with respect to $v(t^{(N-1)}, \cdot)$.

Now let us remark that since $L(f, \phi)$ is continuous in ϕ, then

(vi) if f is not minimal with respect to ϕ, then there exists a neighborhood U of ϕ such that f is not minimal with respect to every ψ in U.

It follows from (vi) that,

(vii) there exists a neighborhood U of $v(\cdot)$ such that if x_1^* is minimal with respect to a given ψ in U, then x_1^* is minimal with respect to $v(\cdot)$.

Now we can prove the following theorem.

THEOREM 5.1. *If the minimal element x_1^* with respect to $v(\cdot)$ is unique, then there exists a horizon h for the infinite problem, and the horizonal element is x_1^*.*

Let U be the neighborhood as described in (vii) and h a number such that, if $N \geq h$, then $v(t^{(N)}, \cdot)$ belongs to U. It follows from the construction that the minimal element x_1^* with respect to $v(\cdot)$ is minimal with respect to $v(t^{(N)}, \cdot)$, and it follows from (v) that it can be extended to an $x^{(N)}$ with $A(x^{(N)}, t^{(N)}, s_0) = v(t^{(N)}, s_0)$ for all s_0 in S. This proves the theorem.

6. Optimal properties of horizonal solutions

One of the most important problems of the theory of the horizon is to establish when a horizonal solution is a minimal one. It is not always minimal, but the theorem presented in this section will cover some important cases when it is so.

As in the preceding section, we will fix a family $(X_n, A_n(\cdot\,; \zeta|n))$, $\zeta \in P \subset Z = Z_1 \times Z_2 \times \cdots$ and will use the notation $A(x; \zeta) = \sum_{n=1}^{\infty} A_n(x|n; \zeta|n)$, $v(\zeta) = \min_{x \in X} A(x; \zeta)$ and $\zeta^{(n)} \to \zeta$.

THEOREM 6.1. *If $\zeta^{(n)} \in P$, $\zeta^{(n)} \to \zeta$, $v(\zeta^{(n)}) \leq M$ and x^* is a horizonal solution for $\bar{\zeta}$, then $A(x^*; \bar{\zeta}) \leq M$.*

PROOF. Since $\zeta^{(n)} \to \bar{\zeta}$ and x^* is a horizonal solution for $\bar{\zeta}$, then, by the lemma in section 5, for every k we may find a number n and $x^{(n)}$ such that $\zeta^{(n)}|k = \bar{\zeta}|k$, $A(x^{(n)}, \zeta^{(n)}) = v(\zeta^{(n)})$ and $x^{(n)}|k = x^*|k$. Hence,

$$\begin{aligned}(21)\qquad \sum_{i=1}^{k} A_i(x^*|i; \bar{\zeta}|i) &= \sum_{i=1}^{k} A_i(x^{(n)}|i; \zeta^{(n)}|i) \\ &\leq A(x^{(n)}; \zeta^{(n)}) = v(\zeta^{(n)}) \leq M,\end{aligned}$$

which proves the theorem.

THEOREM 6.2. *If $\zeta^{(n)} \in P$, $\zeta^{(n)} \to \bar{\zeta}$, $v(\zeta^{(n)}) \leq v(\bar{\zeta})$ and x^* is a horizonal solution for $\bar{\zeta}$, then $A(x^*; \bar{\zeta}) = v(\bar{\zeta})$.*

PROOF. By applying theorem 6.1 with $M = v(\bar{\zeta})$, we obtain $A(x^*; \bar{\zeta}) \leq v(\bar{\zeta})$.

THEOREM 6.3. *If $\bar{\zeta} \in P$ and x^* is a horizonal solution for $\bar{\zeta}$, then $A(x^*; \bar{\zeta}) = v(\bar{\zeta})$.*

PROOF. Since $\bar{\zeta} = \zeta^{(n)} \to \bar{\zeta}$, then this is the corollary of theorem 6.2.

Following theorem 6.3, a horizonal solution for a d.p.p. in the family with respect to which the horizonal solution has been constructed is a minimal one. This theorem was proved by Maria W. Łoś in 1962.

THEOREM 6.4. *If x^* is a horizonal solution of a simple d.p.p. (that is, in the family of its truncated problems), then it is a minimal solution.*

PROOF. If $t^{(N)}$ is the sequence with $t_n^{(N)} = 1$ for $n \leq N$ and $t_n^{(N)} = 0$ for $n > N$, then $t^{(N)} \to t = \langle 1, 1, \cdots \rangle$. Obviously, $v(t^{(N)}) \leq v(t) = v$. Therefore, this theorem follows from theorem 6.2.

EXAMPLE 5.1. Let us suppose we are given a family of d.p.p.'s with the parameter s_0 in S as described in section 2 and later studied in section 5. Moreover, let us assume that there is a horizon in this family. In order to have an initial parameter, we extend S to the set Π of all distributions over S, and we consider the functions A_n with average distributions $\bar{p}_n(s; \text{if } \pi, x|n)$, as shown in section 4.

It is easy to check that extending S to Π does not affect the existence of the horizon, and moreover, that both the horizon and the horizonal element may be chosen independently of the parameter π.

Since the family being considered has a horizon and an initial parameter, then there exists a horizonal solution for every simple d.p.p. in this family. As the horizonal element does not depend on π and, going step by step, the same horizonal element may be used, then the horizonal solution will be a sequence x with $x_1 = x_2 = x_3 = \cdots$. Such a solution is called a stationary solution.

It follows from theorem 4 that this horizonal stationary solution is a minimal one.

In the paper by Blackwell [4] it is shown that every family of d.p.p.'s as studied here has a stationary minimal solution. It may be shown by easy examples that not every stationary minimal solution is a horizonal one, even in the case when a horizonal solution does exist.

7. Horizon for d.p.p.'s with continuous time

By studying dynamical programming problems with continuous time, the theory of the horizon changes in several respects. Without going into detail we shall show by two examples how in these cases the notion may be applied. Both examples are indeed continuous versions of formerly presented examples.

We are given a nonnegative function c, defined for $x \geq 0$ and such that $c'(x) > 0$, $c''(x) \geq 0$. Furthermore, we are given two nonnegative constants α and x_0. The problem lies in minimizing the functional (see Arrow, Karlin [1])

$$(22) \qquad \mathfrak{J}_0^T(x, \zeta) = \int_0^T z(x(t)) + \alpha\left[x_0 + \int_0^t (x(\tau) - \zeta(\tau))\, d\tau\right] dt$$

over the set of all nonnegative functions x, continuous and differentiable for all but a finite number of points, and such that

$$(23) \qquad y(x, \zeta, t) = x_0 + \int_0^t (x(\tau) - \zeta(\tau))\, d\tau \geq 0 \qquad \text{for} \quad 0 \leq t \leq T.$$

We shall assume that the function ζ—which is indeed the dynamical parameter of the problem—belongs to a set Z_0 of nonnegative, continuous, and differentiable functions. This set will be more exactly specified later.

We say that the function $\bar{\zeta}$ has a horizon H for $T_0 > 0$ in the above-defined family with parameters in Z_0, iff the following is true.

There exists a function $\bar{x}$ such that for every $T \geq H$ and every ζ in Z_0 with $\zeta(t) - \bar{\zeta}(t)$, for $0 \leq t \leq H$, there exists a function x^* with $y(x^*, \zeta, t) \geq 0$ for $0 \leq t \leq T$, $\mathfrak{J}_0^T(x^*, \zeta) = \min_{y(x,\zeta,t) \geq 0} \mathfrak{J}_0^T(x, \zeta)$ and $x^*(t) = \bar{x}(t)$ for $0 \leq t \leq T_0$.

This is certainly not true if we do not restrict Z_0 to be a uniformly bounded set of functions. But if we do restrict Z_0 to be the set of functions with $0 \leq m \leq x(t) \leq M < \infty$, then, as is shown by Blikle [5], the above statement is true for $H = T_0 + (1/\alpha)[c'(M) - c'(m)]$.

Now let c be a nondecreasing and nonnegative differentiable function, and let ψ be a decreasing positive continuous function. Finally, let α be a positive constant. Let us suppose we are interested in minimizing the functional

$$(24) \qquad \mathfrak{F}_0^T(x) = \int_0^T [c(x(t)) + \alpha \cdot x'(t)\chi(0 < x'(t))]\psi(t)\, dt$$

over the set of continuous nonnegative functions x, differentiable in all but a finite number of points t, and such that $x(t) \geq \zeta(t)$, for all t.

Here again ζ—the dynamical parameter—is a function which is assumed to belong to a set Z_0 of continuous differentiable and nonnegative functions.

Independent of what the set Z_0 is assumed to be, it may be shown, following methods given in Arrow, Karlin [2], that there exists a horizon for every ζ in Z_0. In particular we have the following theorem.

THEOREM 7.1. *For a given* $\bar{\zeta}$ *and* $T_0 > 0$, *there exists a function* $\bar{x}$ *such that, for every* ζ *in* Z_0 *with* $\zeta(t) = \bar{\zeta}(t)$ *for* $0 \leq t \leq T_0 + \max_{0 \leq t \leq T_0} \alpha/c'(\bar{\zeta}(t))$, *there exists a function* x^* *with* $x^*(t) \geq \zeta(t)$, $0 \leq t \leq T$, $\mathfrak{F}_0^T(x^*) = \min_{x(t) \geq \zeta(t)} \mathfrak{F}_0^T(x)$ *and* $x^*(t) = \bar{x}(t)$, *for* $0 \leq t \leq T_0$.

This theorem may be stated briefly, as follows.

In every family with parameters in Z_0, $H = T_0 + \max_{0 \leq t \leq T_0} \alpha/c'(\bar{\zeta}(t))$ *is a horizon for an arbitrary function* $\bar{\zeta}$.

Neither of the theorems we have given in this section is stated in its strongest form. In both cases the defined horizon is not the shortest one for a given parameter $\bar{\zeta}$. For the sake of simplicity we have taken their weaker form, since the aim in presenting them was only to give an example of horizons in dynamical programming problems with continuous time.

REFERENCES

[1] K. J. ARROW and S. KARLIN, "Production over time with increasing marginal costs," *Studies in the Mathematical Theory of Inventory and Production*, Stanford, Stanford University Press, 1958, pp. 61–69.

[2] ———, "Production planning without storage," *Studies in the Mathematical Theory of Inventory and Production*, Stanford, Stanford University Press, 1958, pp. 85–91.

[3] R. BELLMAN, I. GLICKSBERG, and O. GROSS, "The theory of dynamic programming as applied to a smoothing problem," *J. Soc. Indust. Appl. Math.*, Vol. 2 (1954), pp. 82–88.

[4] D. BLACKWELL, "Discrete dynamic programming," *Ann. Math. Statist.*, Vol. 33 (1962), pp. 719–725.

[5] A. J. BLIKLE, and J. ŁOŚ, "On the notion of horizon in dynamic programs with continuous time," to appear.

[6] F. MODIGLIANI, "The measurement of expectations," abstract of a paper presented at the Boston Meetings, December 1952, *Econometrica*, Vol. 20 (1952), pp. 481–482.

[7] F. MODIGLIANI and F. HOHN, "Production planning over time and the nature of expectation and planning horizon," *Econometrica*, Vol. 23 (1955), pp. 46–66.

[8] H. M. WAGNER and T. M. WHITIN, "Dynamic version of the economic lot size model," *Management Sci.*, Vol. 5 (1959), pp. 89–96.

INFORMATION SCIENCE AND ITS CONNECTION WITH STATISTICS

TOSIO KITAGAWA
KYUSHU UNIVERSITY

1. Introduction

The purpose of this paper is to give an exposition of the general scope of information science and to make clear its connection with statistics. It is expected that this will be helpful in finding directions of active development of the latter. For this purpose we must develop several preliminary considerations in order to secure the understanding of the readers on the reasons why the author wishes to present the topics on such an occasion. By information science we mean a newly organized branch of science which has at least two characteristic aspects. The first aspect is shown by the fact that it is presently drafting a form of blue-print of its own future development. It is indeed a science planned in its scope and thus is sharply distinct from the natural growth which has been experienced in a majority of areas of pure sciences. The second characteristic aspect comes from its construction in which constituents range over a vast ensemble of individual sciences, and it is apparent that information science is an amalgamated science whose constituent branches have their respective scientific principles. In short, information science is a planned, consolidated, and integrated science having several different branches, and because of these two characteristic aspects, its methodology ought to be explained in some detail.

We shall explain our points of view on these two characteristic aspects, both from our experience and from somewhat more logical considerations. In giving our explanation we have to refer to various topics which will range over physics, biology, electronics, and so on, none of which seems at first sight to be an appropriate topic of the *Symposium on Mathematical Statistics and Probability*. Moreover, we do not intend to give a report on established results of some scientific area, but are merely trying to give our ideas on how to organize a new field of science. In spite of these two unusual circumstances, we intend to refer to a topic which has a definite connection with the development of statistics in the coming days.

2. The road leading to information science

The author has been working in the area of successive processes of statistical inference and control [37]–[45] for the fifteen years since 1950. The image of information science which we shall expound in the following sections is indeed, for the author, one of the natural directions of extension of the research areas which can be imagined from such developments.

We have explained in our previous papers [41], [45] the connection between our statistical theory and cybernetics in the sense of Wiener [78]. The former is concerned with successive processes of statistical inference and control, whereas the latter is concerned with communication and control in men, animals, and machines. In spite of the differences between the scopes of these two topics, their logical aspects have a deeper connection than might be imagined at first appearance. The intimate connection comes from the following circumstances.

(1) The theory of communication and control discussed in cybernetics is based on a mathematical apparatus which is nothing but that of stochastic processes. Indeed the science of communication so far developed in cybernetics by Wiener is essentially statistical science ([78], p. 10).

(2) The theory of successive processes of inference and control is concerned with an automatically controlled sequence of statistical procedures, as we have explained in detail in one of our recent papers [45]. Indeed, the theory is a cybernetical formulation of statistical inference and control whose logical aspects were explained in another paper of the author [41].

These two circumstances will be sufficient to explain the connection between statistical theory and cybernetics. Now it is also important to refer to the differences between them. The two crucial differences which have a certain connection with the birth of information science are the following.

(a) Cybernetics is a prototype of integrated science, whereas the theory of successive processes of statistical inference and control belongs to the area of statistics, although its fields of application cover various areas of science and technology.

(b) The theory of successive processes of statistical inference and control deals with learning from experience in a somewhat broader sense than that developed in the first 1948 edition of *Cybernetics*.

The road which has led us to a certain formulation of information science was suggested by our struggle to reach a higher level at which these two differences (a) and (b) can be "aufgehoben."

The crucial problems confronting us in this connection are the following:

(i) whether or not a methodological science can be developed in an efficient way without some systematic plan of connections with substantive sciences;

(ii) whether or not either cybernetics or the theory of successive processes of statistical inference and control can efficiently contain the theory of learning processes and self-organization in their respective frames.

Keeping these problems in mind, we shall enter in the following section into the more fundamental discussion regarding the topics of three categories of sciences. Indeed, in the area of statistics a long sequence of disputes on whether this science is methodological or substantive has never been settled. It is one of the byproducts of this paper that we can and shall be in a position to give a discussion, from a broad point of view, on this problem which has stood for three centuries.

3. Three categories of sciences

The following current classification of sciences into three categories has been adopted very frequently, and it seems to us to be at least convenient to start with the following description.

Category 1. *Object Science.* The sciences of this category have their respective assigned objects of research but no assigned methods of research. Their scientific information as a whole cannot be expected to be organized in a deductive form by one methodological procedure. On the other hand, they are always concerned with the same object. In this sense scientific information on objective sciences can be mostly descriptive.

Category 2. *Elementary Science.* These sciences have their respective assigned objects of research, and moreover, they have a certain unified methodology of attack on their assigned objects of research. Their scientific information admits an intention of research workers and scientists to organize them, at least partly, in a deductive form. It is one of their objectives to search for principles of how to organize, as far as possible, at each stage of its development, their information in a deductive form.

Category 3. *Methodological Science.* This type of science has no assigned object of research, but it has an organized information pattern which can be applied to certain sciences belonging to the first and the second categories. Its job may sometimes be to abstract a common feature of various methodologies developed in the first and the second categories, but its main function is to develop a unified methodology on its own systematic foundation with an intention to establish scientific achievements of its own with consequences applicable to other categories of sciences.

Cybernetics in the original sense of Wiener ([78], 1st ed.) is manifestly intended to be a science of the third category, and it has been understood as such by a majority of people. Statistics is understood by some statisticians, including most of the mathematical statisticians, to be a science of the third category, whereas other people working in substantive sciences, including economics, still insist that the science of statistics cannot be understood merely from its methodological aspects and that statistics cannot exist in itself apart from coexistence with sciences of the first category. Mathematics can be recognized to be one of the purest and most extreme cases of the third category. Logic is manifestly the methodology to be valid in every science. However, it so happens that one can dispute whether or not logic should be considered a science. Some may insist that logic is not an individual science, but rather a super entity without which no theoretical scientific reasoning can be secure. This viewpoint is indeed intimately connected with the definition of science in which experience must be accumulated through observation and experimentation. It is our purpose in this paper to collect these fundamental methodologies within broadly defined science families and then to compare them with each other in order to make clear the distinctions among them. This is in itself the principle of collection and differentiation which can be considered a fundamental method of logic. By such a method, we shall be able to make a systematic approach to analyzing learning processes. Indeed, our spectrum contains logic, mathematics, statistics, and cybernetics belonging to the third category of science in its broad sense.

4. Methodological aspects of statistics

In a previous paper of the author [41], logical aspects of successive processes of statistical inference and control were discussed in some detail. The three main logical aspects are concerned with (i) objectivity, (ii) subjectivity, and (iii) practices. Each of these consists of two fundamental principles, and hence the total logical aspects are concerned with six fundamental principles:

(1) Objectivity
 (a) principle of probabilistic scheme,
 (b) principle of deterministic scheme,
(2) Subjectivity
 (c) principle of strategy,
 (d) principle of valuation,
(3) Practices
 (e) principle of efficiency,
 (f) principle of successiveness.

The background upon which such a system of six principles was proposed is manifestly influenced by the particular studies of the author on successive processes of statistical inference and control, and cannot be claimed to cover the whole statistical activity of human beings and human society. Nevertheless, it is the constant pursuit of the author to adhere to the characteristic aspects of successive processes of statistical inference and control and then to endeavor a generalization of various ideas established in specific restricted areas to more general aspects of different areas in statistical research and statistical activity. In this sense data analysis was discussed in one of the papers of the author [41]. According to the author [41], data analysis does not belong to either descriptive statistics or statistical inference theory.

The decision function approach has its validity in some restricted area, but it cannot cover all the aspects of successive learning processes unless it should become equipped with modification procedures for cost functions, and the methods of pattern recognition under which it is formulated. In discussing data analysis in statistics the author [41] placed an emphasis on its learning procedures, and it is the consequence of these discussions that some fundamental aspects of learning processes in data analysis have several common features with those under which successive processes of statistical inference and control are discussed.

In 1963 a paper of the author [45] discusses the fundamental aspects of automatically controlled sequences of statistical procedures and their implications for statistical programming to be used in automatic digital computers and for statistics in its general sense. This series of papers [41], [44], and [45] since 1960 is entirely concerned with an elaboration and a generalization of the ideas developed in the series published in the Fifties. Therefore, we are always conscious of the fact that our logical considerations have been prepared to give a rearrangement of what we have obtained in our specific approaches and to give a bird's-eye view of the whole domain so as to reach a better understanding of what area we have covered and how we have obtained our results, in an expectation that further developments can be searched for leading to a still better understanding of what we have done. Having reached the standpoint of our 1963 paper, we are now very conscious of our distance from the classical theory of R. A. Fisher [24], [25], and from that of Neyman-Pearson [56], [57]. So much has been left undeveloped in an area which we pointed out to be worthy of cultivation. It is, however, the present opinion of the author that the undeveloped area of statistical science can be more realistically scheduled for development only after establishing an understanding of the science of information which should be deeper than anything we have hitherto achieved.

5. Methodological aspects of cybernetics (1)

A more direct and deeper understanding of the science of information and control may be expected to be more easily obtained in the realm of cybernetics than in that of statistics. It is true that the logical principles enunciated in the previous section were derived from our studies and reconsiderations of specific approaches on successive processes of statistical inference and control, but we believed that they were identical to the logical principles to be expected in the formulation of cybernetics, as we have explicitly emphasized in section 4 of our paper [41]. In short, our approach may be termed the "cybernetical formulation of statistical procedures." With this affinity and similarity in mind, it is still true that cybernetics has at least one more aspect which statistics neither covers nor is responsible for. This is the fact that cybernetics attempts to be a theory of communication and control in real entities, that is, in men, animals, and machines. That is to say, cybernetics has the responsibility of explaining and discussing the real phenomena in real existence. Therefore, it is quite natural to expect to find there some techniques, some ideas, and some achievements which we do not have in statistics but which can be carried over to the formulation of statistics, particularly in reference to learning procedures. It has turned out, however, that statistics has not benefitted much from cybernetics, at least in its early stage. This has left some feeling of disappointment toward a science on which high hopes had once been placed for help, or at least guidance in developing understanding of learning processes in statistics.

The reason why cybernetics has not shown achievement commensurate with the expectations of the Fifties derives from its methodology.

Fruitfulness of a science belonging to the third category can be expected only to the degree that it cooperates with the other sciences belonging to the first and second categories.

In the search for effective progress in the area of statistics we have now reached a notion of information science from which, we believe, we can obtain aid and guidance. At the same time, statistics is one of the indispensable constituent elements in a consolidation of various scientific methods in information science. Statistics has its reason for existence, but it can flourish and further develop by maintaining the intimate connection with information science which we are now going to explain.

6. Structural aspects and constituents of information science

Information science aims at being a unified scientific approach integrating various phenomena connected with information; that is, production of information, transmission of information, transformation of information, storage of information, deduction of information, pattern recognition due to information, retrieval of information, various operational uses of information, and so on. These information phenomena can be found in biological existence, social lives of human beings, as well as in machines manufactured by human beings. A science of information which treats information phenomena dealing with common aspects irrespective of particular details of real existences is naturally one of the sciences belonging to the third category in the sense of section 3. By this assertion we deny the idea that any science of the third category is an abstract theory which can be treated by symbols and notations such as in mathematics and in some of mathematical statistics. On the contrary, we are now proposing a science of information which has a certain structure, consisting of the following five indispensable research branches.

Branch 1. Physical information phenomena.
Branch 2. Theoretical formulation of information phenomena.
Branch 3. Information system analysis.
Branch 4. Information phenomena in biological existence.
Branch 5. Artificial realization of information phenomena.

Branch 1. This branch contains the following research divisions.

1.1 Metal and magnetic elements.
1.2 Semiconductor elements.
1.3 Cryogenic elements.
1.4 Optical information elements.
1.5 Fluid information elements.
1.6 Elastic wave information elements.
1.7 Chemical information elements.
1.8 Dielectric information elements.

Branch 2.

2.1 Logic.

2.2 Theory of recognition.
2.3 Theory of linguistics.
2.4 Theory of information networks.
2.5 Self-organizing systems.
2.6 Learning theory.
2.7 Theory of information transmission.
2.8 Information theory.
2.9 Mathematical programming.
2.10 Statistical theory.
Branch 3.
3.1 Information processing systems.
3.2 Computation systems.
3.3 Information control.
3.4 System analysis.
3.5 Operational analysis.
3.6 Man-machine systems.
Branch 4.
4.1 Neurophysiology.
4.2 Integration of the central nervous system.
4.3 Sensory information processing.
4.4 Transmission of genetic information.
4.5 Exhibition of genetic information.
4.6 Adjustment of genetic information.
Branch 5.
5.1 Information circuits.
5.2 Information transmission apparata (transducers).
5.3 Information transformation apparata.
5.4 Recognition apparata.
5.5 Language apparata.
5.6 Thinking apparata.
5.7 Adaptation apparata.
5.8 Education apparata.

Branch 1 is devoted to research on physical materials and phenomena associated with these materials, which have or will have some connection with information phenomena. Results and achievements obtained in Branch 1 have been and will be the basis of developments in Branch 5. Realization of information phenomena by virtue of cooperation between Branches 1 and 5 have been and will be a stimulus to a new theory of information in Branch 2, and will provide powerful experimental apparata to research in the area of Branch 4. An introduction of new information apparata will involve new problems regarding information systems to be discussed in Branch 3, and sometimes even revolutionary effects will be experienced as we have already seen in the case of electronic computers.

Branch 1 is responsible for the creation of new materials as carriers of information phenomena and as the basis for information apparata through which biological information phenomena can be investigated and by means of which new mathematical models can be introduced.

Branch 1 will contain various researches on various matters and on physical and chemical phenomena, including solid matter, fluid, elastic waves, optical, and chemical phenomena.

The role of Branch 2 is to prepare various theoretical models through which information phenomena can be described and analyzed. Its research domain will cover theoretical investigations on control, learning, self-organization, statistics, and programming, and hence it should also contain fundamental research on recognition, foundations of language and of mathematics. It is also to be noticed that the theory to be developed in this branch is responsible for giving theoretical foundation of automation, naturally with the cooperation of the other branches of information science.

Remarkable advances have been recently experienced in this area, and revolutionary effects are now anticipated in epidemiology, logic, linguistics, and psychology. More direct influences have already been observed in mathematical theories regarding statistics, computation, and control.

In view of recent remarkable progress, it may be almost impossible to overestimate the tremendous and far-reaching effects of information science in the coming ten years on social and biological sciences. On the other hand, it is frankly admitted that the social and biological sciences that are currently somewhat less developed are so because of the lack of development of information science in these areas, in spite of the evident facts that information plays the crucial role in their formulations. These are important problems which will be facing us in the near future.

The objects of research in Branch 3 are various information systems with information apparata which are to be invented and investigated in Branch 5. Human societies cannot continue to keep their various activities in sound condition without adequate systems of information. For instance, we can mention here production control, inventory control, economic planning, and so on. Among these information systems there can be found a set of fundamental information systems such that any information system is a composition of some of these fundamental ones. In the present state of our social and academic activities we can mention the following set of fundamental information systems: (i) statistical data processing, (ii) computation systems, (iii) documentation systems, and (iv) information control systems. Branch 3 also contains research on systems analysis, operations research, and man-machine systems as its essential constituent research techniques. Researches and problems in this area of Branch 3 will have direct and strong influence upon mathematics and statistics. It will also prepare various powerful techniques for the development of economic science and management science.

Branch 4 is responsible for investigations regarding information phenomena encountered in the biological sciences. There are various biological phenomena to be investigated in the light of information science, that is to say, through uses of theoretical information models and with the aid of information science techniques. On the other hand, results established regarding biological information phenomena will give rise to many problems to be discussed in Branch 2, and hence strong stimuli for development of theories in information science can be expected to come from Branch 4. Without deep investigation of information systems in the biological realm some characteristic aspects of biological phenomena cannot be made wholly clear, however far their material and energy aspects may have been investigated. In this sense Branch 4 will constitute one of the fundamental divisions upon which essential progress in biological science will depend.

Branch 5 is devoted to the development of artificial information apparata. At the present stage the central problems in this area are concerned with artificial intelligence, including various functions such as pattern recognition, translation of languages, information retrieval, adaptive control, learning procedures, and even thinking, in a certain sense. In such investigations Branch 5 will depend on the investigations of all other branches, and conversely, the achievements obtained in Branch 5 will provide powerful tools for investigations in other branches.

In short, each of the five branches constitutes an indispensable research department in information science, and it will give scientific or technical aid to every other branch directly. Hence, all these five branches are connected through mutual aid and cooperation. With any lack of such aid or cooperation, a science of information cannot be expected to realize its potential of sound and efficient development. With this idea in mind, let us make a series of surveys of various well-established sciences which are closely connected with information science. We shall do this in the next few sections.

7. Imbedding in and correlation with information science

Let us now turn to examining the actual character of various sciences which are, each in its own way, closely connected with the notion of information science described in the previous section. We shall then find two different situations: (i) imbedding and (ii) nonimbedding and correlation.

(1) There are many examples of sciences which are imbedded in the framework of information science. By an imbedding we mean here that the constituents of the science belong to one of the five branches enumerated in the previous section.

(2) There are again many examples of sciences, each of which has constituents belonging to one of the five branches but also has an area that is not covered by any branch of information science. These are the cases of nonimbedding and correlation.

We shall see that cybernetics and bionics can be recognized as sciences imbedded in the framework of the five branches, and so are examples of situation

(1). On the other hand, there are many other examples illustrating situation (2). Let us mention here a few examples: behavioral sciences, mathematical sciences, medical electronics, human engineering, and so on. The following few sections are devoted to analyzing these situations in some detail, with the idea in mind that such analysis can be useful in clarifying the role of information science in motivating scientific research of various kinds. These analyses will involve an objective observation on the present state of each individual scientific activity, and then there will follow naturally the task of classification of sciences. But this is not all that is to be done. We are rather seeking a strategy for promoting scientific research. We are not only a geographer who wants to prepare a map for everyone's use, but we are rather a mountain climber aiming at reaching the top of a mountain or a navigator going on the ocean with the intention of searching out a new continent. In this light there is an essential element that comes into our discussion. This is the subjective attitude of research workers regarding methodologies and motivations in their particular research areas. Each individual research worker in a team of research workers, and even a group of research workers itself, has a certain standpoint according to which a particular emphasis will be placed upon its activities. An objective description of research activities by a systematic analysis of mutual relationships has given us the structure of information science in terms of the five branches listed above. This much corresponds to the making of a map. But each research worker, and each group of research workers, is rather a traveler having his own intentions. Thus there is something more than just the map involved. This real picture will involve us in deeper considerations in discussing the two situations mentioned above.

8. Methodological aspects of cybernetics (2)

Cybernetics was introduced by N. Wiener in 1948, as a science of communication and control in man and machine. We understand cybernetics to be one of the prototypes of information science. In the earliest development of cybernetics its scientific emphasis was placed on the study of the principal aspects of communication and control by means of mathematical models based on stochastic processes and their transformation by linear (and later by nonlinear) filters. In the terminology of information science as we have set it forth above, cybernetics' main task was the establishment of Branch 2 to give an integrated picture of communication and control and their applications to Branch 4 and other branches of science and technology, including sociology and linguistics, computers, and automatic machinery.

In the later development of cybernetics there were many features different from those observed in early stages. For instance, deeper considerations on the foundations of cybernetics appeared in the French literature, while economic system analysis was stressed by the Russian school. Some of these investigations may be said to belong to either Branch 2 or Branch 3. Also it should be observed that Wiener and his collaborators were, from the earliest stages in the history of cybernetics, keenly interested in artificial intelligence machines; such investigations belong to Branch 5 of our information science. In the latest period of development various new features have emerged, as described in several publications of Wiener [79]–[83]. Thus, as regards the theoretical features of cybernetics, their scope has become large enough to cover the greater part of the constituent categories of Branch 2 detailed above and bio-cybernetics in particular [82] and [83] has shown a remarkable growth into many research areas.

In spite of these recent developments, the earlier characteristic aspects of cybernetics seem to be still most prominent in its present activity, for the reason that no systematic connection between Branch 1 and the others has been explored. This is why cybernetics is to be distinguished from bionics, and also why cybernetics cannot be identified with the whole range of information science.

9. Methodological aspects of bionics

Bionics—the name was coined by J. Steel in 1960—is one of the most interesting fields of scientific technology. Its main purpose is to realize simulations of biological systems by means of artificial apparata, and hence to invent a new field of engineering (cf. [14], [53]). There are many biological functions which engineers can probably bring to bear to solve their own problems. Particularly in the field of biological information phenomena we have many examples: sensory organs, pattern recognition, learning processes of animals, self-organizing mechanisms, and so on.

Now let us consider the five branches of information science. In its first appearance bionics contains only the two Branches 4 and 5, and its main job is to cultivate Branch 5 in the light of knowledge obtained in Branch 4. However, a more careful observation convinces us immediately of the fact that Branches 1, 2, and 3 are also indispensable in promoting its main job. Branch 4 is concerned with biological phenomena, but information obtained there is, in the first place, described in terms of physical and chemical concepts in order to secure the possibility of realizing corresponding information phenomena through artificial apparata. For this reason bionics must depend heavily on the development of Branch 1. Second, the main interest in bionics should be focused upon information phenomena in biological existence, and therefore the phenomena of chief concern in bionics should be described in the terminology of Branch 1 and explained in the light of various theories in Branch 1. Branch 3 is also indispensable in bionics. Indeed, without system analysis of information systems there can be no artificial intelligence that can be recognized as corresponding to biological information organs.

We have emphasized that the results coming out of Branch 4 will benefit Branch 5, through the cooperation of Branches 1, 2, and 3. However, the cooperation among these branches is by no means to be one-sided. For example, research in Branch 2 will gain from results in Branch 4, which can reveal many biological phenomena that will admit new directions for research and inspire new theory and fresh model building.

In this way we see that bionics has at least five branches in common with information science, although not all of the constituent fields of each branch are contained in bionics.

According to these observations, bionics, in terms of its objective structure, may be said to be imbedded in information science. However, we imagine that some researchers in bionics may not be completely satisfied with this identification. In addition to any such classification work there is need of deep understanding of the subjective attitude of scientific research workers. The subjective attitude common to most research workers in bionics may be said to be characterized by a conviction regarding the fruitfulness of results to be expected in the research direction "from biology to engineering," as a guiding principle. In this connection we want to make here the following three points which seem to be important.

(1) An emphasis on some subjective guiding principle in establishing a science and/or a technology can be expected to be fruitful in exploring a new domain of research, and sometimes it proves to be a powerful working hypothesis.

(2) At the same time it is necessary to have an objective picture of just where a researcher is working and of the complex of paths along which he can progress from there.

(3) In the actual course of development of research the principles of practice described in section 2 are also of considerable importance.

In short, there are three principal aspects of importance to be considered in establishing and developing a new field of science: (1) the subjective guiding principle, (2) the objective map, and (3) the practical procedure.

An original initiator starts with (1), and his successors tend to be overly concerned with (3). After many years of struggle, with many successes and many failures, a science will have attained its reason for existence. Then there will appear an author who will supply (2) in a systematic fashion. At this stage there arises the danger of losing the strong impulse of fostering further developments without an appeal to some guiding principle. In a period of science and technology which can open a new era for mankind, we should be conscious of the laws governing birth, growth, and death of a science, and we should have a "medical plan" which is effective throughout the entire life of a science. Bionics is a new field of science and technology, and it should be worthwhile to analyze its methodology from the standpoint of information science; this we have tried to do.

10. Methodological aspects of the mathematical sciences

The terminology of mathematics began, a few years ago, to show up increasingly in various fields of the natural and social sciences, thus giving evidence of extensive new achievements in the application of mathematical methods (cf.

[85]). There had been a long continuation of rather unfruitful and tedious disputes as to whether and how one could distinguish between pure and applied mathematics. During approximately a quarter of the century, from 1925 to 1960, remarkable progress has been made in areas where mathematical methods play an essential part. Econometrics in economics, biometrics in biology, and psychometrics in psychology have the common feature of emphasizing the role and effectiveness of mathematical and statistical methods in these substantive sciences. Econometrics and psychometrics established themselves and their respective domains in the realm of science during these years, whereas biometrics which had its birth in the first quarter of the century, and which also had experienced a somewhat slow if steady growth in that earlier period.

Moreover, various new topics, such as the design of experiments, decision procedures, amount of information, learning processes, computer of logic, systems analysis, and programming of operations, have become objects of mathematical studies. These extensions of the scope of mathematical study were remarkable in two ways. In the first place, they had not been introduced through any internal growth mechanism of the traditional fields of mathematics. In the past history of mathematics we have a number of examples showing the internal growth mechanism in traditional mathematics. For example, the birth of non-Euclidean geometry was stimulated by deep studies upon the foundation of Euclidean geometry and was realized as a consequence of the establishment of a new system of axioms. The growth and development of so-called abstract algebra in the German school in the early Twenties were likewise not influenced by any other science but came about through the accumulation of research in the area of algebra itself. Now, a sharp contrast can be seen between such spontaneous developments due to the internal mechanism of mathematics in the first quarter of the century and new developments which have been much influenced by contacts with other fields of scientific and technological research. These latter had stimulated new mathematical investigations and have led us to create a set of new fields of mathematics.

Indeed, these all belong to the topics under Branches 2 and 3 in our formulation of information science. We believe that this correspondence between new fields of mathematical science and the topics in Branches 2 and 3 of information science has a profound implication for the mathematician in regard to direction of research. First of all, this coincidence shows that new fields in mathematical science are concerned with model formulations and with the study of models of information phenomena in our broad sense. Second, this coincidence involves the question of whether possibly the ideas of the mathematical sciences can be used as tools in exploring further in present and future areas of purely mathematical study.

The distinction between mathematics and mathematical sciences can be of use to mathematicians, because the notion of mathematical sciences will make them conscious of the origins of their mathematical problems and of the need for contact with the substantive sciences, which provide them with fruitful problems. It is also of use to people working in the sciences and in technology, because it will remind them of how much they owe to mathematical methodology for their studies. Nevertheless, there does exist the fundamental question of whether the notion of mathematical sciences is in itself sufficient to motivate new fields of mathematical study.

The essentially new topics which the present mathematical sciences have to explore all belong to Branch 2 of information science, and their vigorous development can be anticipated from the cooperation of the various branches of information science, through a network of interrelations connecting the different scientific and technological lines of research. Without this background of systematic cooperation, we cannot expect to have a satisfactory presentation of realistic and essential problems for study in the mathematical sciences.

11. Methodological aspects of the behavioral sciences

Behavior may be understood as an action or a sequence or system or pattern of actions, which are induced by a decision, or by a sequence or system or pattern of decisions. Since such decisions are more or less based upon three uses of information—(a) storage of information, (b) pattern recognition, and (c) operational use of information—it is not surprising that the behavioral sciences have an intimate connection with information science. However, it is important to look into the logical aspects of the behavioral sciences in order to find out exactly how they and information science are mutually correlated and, moreover, in order to be able to find a sound procedure for establishing a mode of cooperation that may be beneficial to both of them.

The behavioral sciences were born in the realm of the social sciences, but their methodologies are quite different from those that are traditional in the social sciences. The behavioral sciences deal mainly with human behavior as studied in psychology, economics, sociology, and politics. More specifically, the behavioral sciences are mainly concerned with the psychological, economic, social, and political behavior of individual human beings and of certain groups and organizations. It seems to us, however, that a theory in behavioral sciences can be properly developed only after setting down the principles of objectivity, subjectivity, and practice. For instance, economic behavior is human behavior with respect to economic affairs in economic systems, as we see it in our usual experience. But a theory of economic behavior can exist and can have scientific value only after three logical aspects of objectivity, subjectivity, and practice have been duly established, according to which there can be constructed an adequate model for discussing specific behavior. For example, the game theoretic approach to economic behavior by J. von Neumann and O. Morgenstern [72] is a case in which certain types of economic behavior within certain specified economic situations are described according to their formulations under what we have called three logical aspects: objective world, subjective attitude, and practice. It should be remarked at the same time that the economic man specified by the game-theory model due to von Neumann and Morgenstern is merely one possibility; he by no means exhausts all possibilities.

With regard to politics, sociology, and psychology, the situations are more or less the same as those of economics as just exemplified. For instance, in politics, political behavior can be the subject of the behavioral sciences only after we have answered the following questions:

(1) How do we define the objective world of politics? What are its principal model schemes?

(2) How do we define objectively the subjective principles of political man?

(3) How do we define objectively the principles of practice of political men?

Any theory in the behavioral sciences should be developed by preparing and working with a model that answers to all three logical aspects. We neither think that such a model has already been set in any area of the behavioral sciences, nor imagine that such models will be easily obtained within the next few years. All we can admit at the present moment is that such models have been proposed and adopted tentatively in some areas of the behavioral sciences with some successes in giving somewhat reasonable explanation and predictions of particular phenomena.

In addition to such general observations on the present states of the behavioral sciences, the more fundamental problems concerning the following issues should receive our attention:

(a) Under what conditions can model building be useful as a theoretical tool in exploring a given area of our research in the behavioral sciences?

(b) What are the features which characterize a particular branch of the behavioral sciences?

(c) What procedures can be introduced by which to improve a model in the behavioral sciences in the light of accumulating information obtained by experience and/or by experiments (if it be possible)?

(d) Under what conditions are experiments possible in a branch of the behavioral sciences?

As far as we are aware, these four fundamental problems have not been duly discussed by behavioral scientists; it has rather been characteristic of the attitudes of these scientists during these twenty years that tremendous adventures have been pursued, but without any deep considerations or reflections upon their logical and scientific basis. We do not think the present situation can continue forever, but we do think that at least these four fundamental problems should have been carefully discussed and should have been resolved, leading us to a unified understanding of the logical and scientific basis of the behavioral sciences before their development became accelerated.

Having laid out explicitly these fundamental problems in behavioral sciences, let us concentrate our discussion on problem (b) which must be presented in this section. Behavior may, in the simplest case, be a response to a stimulus, but in general it should be recognized as being closely connected with information concerning our objective world, our pattern of subjective attitudes, and our strategic principles of practice. Learning processes, adaptation, and self-organization are the proper topics of Branch 2 of information science, and a deep theory in the behavioral sciences can scarcely be expected to develop unless such topics have been adequately deliberated beforehand. Information systems such as data processing systems, documentation systems, computer systems, and statistical systems are also among the fundamental factors specifying the objective world in which economic, social, political, and psychological behavior takes place. In this sense Branch 3 of information science is one of the basic foundation stones upon which any branch of the behavioral sciences is to be built.

On the other hand, the behavioral sciences deal with human behavior which is neither entirely nor uniquely determined by information on the objective world. As we have said, it also depends on the subjective attitudes of the human being faced with certain objective worlds in which game-theoretic approaches, for instance, may have some validity.

In short, whereas the behavioral sciences require some branches of information science as indispensable constituents, they also require other principles which are not included in information science. But it is extremely important to remark here that information science should be developed much further before any appreciable amount of encouragement can be given for the progress of the behavioral sciences.

This remark is based on three reasons. The first is that any deep scientific achievement in Branches 2 and 3 can only be secured by mutual cooperation with the other three branches of information science, that is, with the whole realm of information science. The second reason is that a theory in the behavioral sciences cannot be very substantial without some specification of the objective world in which the pertinent human behavior takes place and such specification of the objective world can be more or less described by its information systems, which may be sometimes quite primitive, and rather loose, and sometimes quite advanced and well organized. The third reason is that human behavior should be analyzed with reference to the individual's storage of information and to his pattern of recognition, and these can only be suitably discussed in Branch 2 of information science.

On the other hand, we will not be fair if we do not point out also the possibility of indebtedness of information science to behavioral sciences. This possibility derives basically from the very fundamental fact that information, in our understanding, cannot exist without connection with control in its broad sense. Now, it is true that there does exist a long distance between the two notions, control and behavior, at the present stage of our understanding. But it should be observed also that it is the remarkable trend of the last quarter of this century that the distance between these two notions has been becoming shorter and shorter. The recent topics in the discussion of control problems are adaptive control, control in learning processes, and automation; and these will be recognized as taking into consideration the human behavior of learning and adaptation. On the other hand, all the branches of behavioral sciences seem to have had the remarkable common tendency to strive for model building, each within its particular theoretical framework; this should have some connection with Branches 2 and 3 of the information science. The shorter their distance becomes, the clearer are their common features. But this is not all that can be expected. Having found a clearer picture of the features common to these two notions, it will be easier to discover also the distinctions between them and also to seek for a scientific tool to add to the theory of control in order to better analyze human behavior in psychology, economics, sociology, and politics.

In closing this section it should be remarked that the expression "behavioral science" can and should be coined after, and because of, the establishment of common scientific principles which would be valid in every one of various branches of what is currently called "the behavioral sciences."

12. Library and documentation science

Before entering into the main topics of the present paper a few remarks should be given here regarding library and documentation science which, we assert to be a subdivision of information science as we have defined it above. We do not believe that very lengthy arguments are needed to establish the assertion we have just made. On the other hand, it is extremely important to remark that library and documentation science is not to be investigated independently of information science. Holding to the view that library and documentation science is an important subdivision of information science, we can look forward to keeping this science in constant and close touch with every new development in the branches of information science. For instance, information apparata, such as translation machines, learning machines, and teaching machines, can be quite helpful in solving technical problems encountered in library and documentation science and technology. The systems analysis of library activity is one of the most important topics to be investigated within the general framework of information systems; indeed, here is a very promising field as yet undeveloped research topics where operations research can be successfully applied. Fundamental research in the mathematical theory of linguistics will find immediate application in the field of library and documentation science. At the same time, the science will provide us with many interesting problems in practical information retrieval and mechanical translations, which are to be discussed and solved in Branches 2, 3, and 5 of information science. We may remark that library and documentation activities are now so prevalent in human activity that they are suitable subjects for statistical investigation and surveys which are easily designed and performed. Indeed this division of information science is a promising area for the application systems analysis and information apparata with the purpose of gathering large amounts of data to be used for further improvement, not only of libraries but also of information service centers in general.

13. Characteristic aspects of some of the recent developments in statistics

It is the purpose of this section and the following ones to pin-point some of the characteristic aspects of recent developments in statistics and to show the connections between statistics and information science which we have already mentioned. It is not our intention to cover the whole of recent developments in statistics; we are going to look only into some of the principal ones, which we believe to be important as indicating exceptional new directions of research in statistics, and show their close connection with developments in the various branches of information science. It is also to be remarked that we shall restrict ourselves to observations on tbe most recent trends in statistics, of the last five years. Thus we shall talk neither about the statistical decision function approach [74] nor about the successive process approach of statistical inference and control ([38]–[40]), both of which had been developed essentially in the ten years up to 1960. We shall talk about some of the very recent remarkable advances that have already been realized and some of the new frontiers of research that have been proposed, confining ourselves, moreover, to topics which have features distinguishing them from developments that took place during the Fifties. That is to say, we are going to point out some evolutionary directions which may be said to have been already programmed for the future development of statistics. It is true that (i) some of these evolutionary strains have been there during the whole process of the development in statistics, or that, in the case of others, (ii) their origins can be traced to some point in the history of statistics, or, for still others, that (iii) they can be understood most clearly from the standpoint of successive processes of statistical inference and control. Nevertheless, they should be recognized as evolutionary directions of statistics rather than outgrowths of statistics.

Let us just briefly note what we mean by an evolutionary direction in statistics in contrast to an outgrowth of statistics. By an instance of evolution in statistics we mean (i) the fact that some recent advances in statistics came about by treating problems which seemed to be neither authentic problems of statistics nor traditional cases for the application of statistics, and (ii) the recognition that such an advance has involved the reformation of the framework of informative patterns which have been traditionally associated with statistics and by which statistical information has been gathered and processed.

Now, what are characteristic aspects of recent advances in statistics which are to be recognized as evolutionary? Before answering this question, let us first give here a list of some topics whose descriptions may help us toward finding an answer.

Our list is claimed neither to be the unique solution to the present problem nor to be the best one; it is merely the product of our own experience in statistics and of our evaluation and interpretation of the work of some contemporary researchers in various fields employing statistical approaches. The list reads as follows:

(1) statistical programming,
(2) data analysis,
(3) statistical control processes,
(4) total quality control system,
(5) operations research and integrated product quality control.

Each of the following five sections is devoted to discussing one of these five topics. There then follows discussion on the formulation of statistics which can suitably cover the characteristic new developments enunciated by the list. It will be shown why and to what extent modern statistics will have deep connections with the information science in the days to come.

14. Statistical programming and automatically controlled sequences of statistical procedures

The principles of statistical analysis using large electronic computers have become one of the most important topics in statistics, and several contributions have been worked out by Terry [70], Yates [86], [87], Cooper [17], and M. G. Kendall and P. Wegner [36]. Among others, Terry [70] explained the role of the statistical programmer in the following two sentences:

(i) "The statistical programmer does not know a priori the exact analytic path that his data must follow."

(ii) "The statistician may very well prefer to let the data speak for itself and suggest the appropriate transformation to exclude from consideration on measurement deemed discordant, or to replace such measurements by derived measurements."

In realizing these principles and the role of the statistical programmer, Terry [70] suggested the broad aspects of adequate statistical programming:

(iii) "Now, with the advent of the disc file, which has the effect of increasing the storage capability of the computer to the order of two million measurements or more, we believe that it will be possible to store in this ancillary device many different statistical strategies, computational techniques, and statistical decision rules as well as large blocks of data."

(iv) "Then, by writing a program of the analytical strategy to be employed, we could permit the data to call in the appropriate analytical techniques and rules, and thus produce a much more effective final analysis."

Let us now pick up two particular topics in statistical programming in order to observe the characteristic aspects of statistical programming. The first topic is concerned with screening and validation procedures which are particularly important in the logic of statistical approaches. The second topic is a review of comprehensive programming systems developed recently by several statisticians.

Regarding the first topic, many experts on census and large scale sample surveys have been keenly aware of different types of errors occurring in the case of large-scale sample surveys.

Deming [20] gave a detailed listing and description of the different types of errors which should be taken into consideration, both in designing and analyzing sample surveys. Mahalanobis [49] gave the classification of different kinds of errors into three types, and "revealed the great importance of controlling and eliminating as far as possible the mistakes which occurred at the stage of the field survey." The interpenetrating sample procedure was introduced by him as one way of doing this.

Yates [87] referred to the general problem of preliminary editing of data before analysis and enunciated the uses of electronic computers in the following sentence. "Once appropriately instructed a computer will perform any required tests on each item of data as it is read in, and can draw attention to anomalies, reject suspicious items, or even in some cases make the appropriate correction." Several papers and memoranda have been written by various authors with particular reference to screening and validation problems. We can mention (i) preliminary assessment, by Reed [62], (ii) autostat, by Douglas and Mitchell [21], (iii) treatment of spotty data, by Tukey [71], and (iv) analysis of residuals by, Anscombe-Tukey [2]. Cooper [17] has published the first set of rigorous procedures for validating and controlling the presentation of data to a computer.

Several comprehensive programming systems have been prepared by various statisticians and various institutions.

Yates [87] pointed out several important aspects of the use of computers in research, saying that: "In research statistics the analysis must in fact proceed step by step, the exact nature of the next step being determined after examination of the results of the previous step. This presents considerable problems, since the results at each step must be stored (clearly on magnetic tape, if available), and indexed in such a manner that the required item can be specified as data for the program performing the next step."

The MUSP prepared by statisticians at Harvard University is said to consist of a set of 19 subprograms which can be called in by a program in a special-purpose language specially designed for MUSP. The sequential operation of subroutines is directed by a control program called MUSP Control Program which "accepts as input the symbolic specification of the problem to be solved in terms of a sequence of subroutine names and parameter values, checks the specification for obvious errors such as missing parameters, translates the specification into a machine-oriented representation, and then executes the resulting set of the specifications interpretively" (M. G. Kendall and Wegner [36]).

We have quoted several important sentences of various statisticians in connection with statistical programming which has become an important task for statisticians during the past five years. It is our opinion that the problems of statistical programming presented here will lead us to fundamental reconsiderations of statistical analysis and statistical information, as we shall make clear in section 18, from a standpoint involving the introduction of learning processes in statistics.

15. Data analysis and automatically controlled sequences of statistical procedures

One of the most valuable contributions to the foundations of statistics is the thorough consideration of the various aspects of data analysis given by Tukey [71]. Data analysis has always been considered to be a collection of fragmentary techniques, and its importance has not been properly recognized in the history of statistics. Some scholars have habitually classified statistics into two divisions, namely, (i) descriptive statistics and (ii) statistical inference theory. This classification is currently adopted by a majority of statisticians. In the first place, speaking from the theoretical point of view, the domain of application of descriptive statistics has been understood to be sharply distinguished from that of statistical inference because the latter divisions are exclusively concerned with random samples from a hypothetical population, while the former does not rely upon the notions of population and sample. In the second place, the classification has had real significance on practical grounds because each of two divisions has had its individual domain of application distinct from the other.

It is one of the remarkable trends to be observed in statistical activity today that both of these reasons, theoretical and practical, are losing their force to maintain the classification of statistics into descriptive statistics and inference theory. For instance, there are indications that the tasks of official statistics are becoming more and more analytic, and that the gap between the two divisions of statistics is now becoming much narrower than it has heretofore been. Illustrative examples in technological research problems in engineering industries as well as examples in large scale sample surveys and in designed experiments on biological phenomena can easily be given to indicate the diminishing distance between these two presumed divisions in statistics. Indeed, on one hand, we are now faced with the need for handling masses of data, while on the other hand, we are equipped with the feasibility of analytical studies in virtue of the use of high-speed electronic computers having rich memories.

We stand in need of developing a methodology for handling large masses of data and for conducting analytical studies of such data. We believe this circumstance is the real basis for data analysis becoming so serious a problem in recent years.

It will be worthwhile to discuss the following assertions by Tukey [71] in this connection:

(1) "We need to give up the vain hope that data analysis can be founded upon a logico-deductive system like Euclidean plane geometry." Data analysis is intrinsically an empirical science (section 46, p. 63).

(2) "Some may feel that if data analysis cannot be a logico-deductive system, it inevitably falls to the state of a crass technology. With them I cannot agree" (section 46, p. 63). "There will be also the hallmarks of stimulating science: intellectual adventure, demanding calls upon insights, and a need to find out" how things really are "by investigation and the confirmation of insights with experience" (section 46, p. 63).

(3) "I look forward to the automation of as many standardizable statistical procedures as possible" (section 17, p. 22). He mentioned three arguments in favor of his views.

16. Statistical control processes

Statistical control processes are understood to be control processes based upon probabilistic formulations and/or control processes which require statistical information during their course of evolution.

In his formulation of cybernetics N. Wiener [78] emphasized both of these characteristics of statistical control processes. In the first place, the theory of stochastic processes was developed from the study of Brownian motion in the effort to build a mathematical apparatus for the design of controls. Second, by the use of stochastic processes Wiener aimed to define the essential aspects of the semi-exact science that became the domain of cybernetics.

These two aspects of cybernetics are present in every statistical control process, and are the features that distinguish such a process from a deterministic one. Most deterministic approaches to control processes could be made more realistic by adding stochastic elements in place of some of the deterministic elements, the latter being usually simplified or idealized pictures of the former. However, this task of introducing stochastic elements is not necessarily as easily done as one might imagine. In fact, the actual situation is that deeper analyses of mechanisms have been performed in the realm of deterministic control processes because of the possibility of appealing to well-developed techniques of mathematical analysis, for example, differential equations and the calculus of variations, and more recently the method of dynamic programming. The works by Pontryagin [61] and his school, and those by Kalman and others, are far more precise than those obtained in the area of statistical control processes. A systematic approach to control processes has been developed by Bellman [7] through his original method of appealing to dynamic programming. Now let us mention here some of the formulations of statistical control processes:

(a) stochastic approximation by Robbins [64] and his successors,

(b) evolutionary operations programs and adaptive process controls by Box [16] and his school,

(c) successive processes of statistical control by the present author [38] and [40].

Other examples can be cited but these will suffice to enable us to compare the characteristic features of statistical approaches to control processes with those of deterministic approaches. Now what are the essential differences between these two? Our answer to this question is this. In statistical approaches (a) the principle of objectivity, (b) the principle of subjectivity and, (c) the principle of practice can be and are clearly formulated in full awareness of our information pattern, the accuracy and precision of our information, the objective of control, the cost of control, the sequential or successive approaches of control procedures.

In short, statistical control processes are discussed in the realm of information science, whereas most deterministic formulations, if not all, are mainly or exclusively concerned with idealized situations which assume, mostly if not always, persons who have complete information on the subject matter and whose controls are completely exact or completely under command. Due to the fact that statistical approaches to control processes have their basis in consideration of information patterns, amounts of information, accuracy and precision of controls, and strategic principles of control procedures, it is obvious that statistical control processes can take from the branches of information science any results and techniques that can be beneficial to their theories. In particular, the recent advances in learning processes and adaptive controls are now challenging us to generalize the present scheme of statistical control processes to the point that these results, obtained mainly in deterministic schemes, can be combined with probabilistic formulations.

17. Statistical quality control (SQC) and total quality control (TQC)

Statistical quality control was introduced by W. A. Shewhart [67] in the first quarter of this century, and has made a great contribution to the development of statistical methods as well as to industrial engineering, where it has been successfully applied. The Shewhart principles of statistical quality control were rather precisely described in his monograph [67], and gave us the basis upon which statistical techniques could be invented and applied. During the past forty years, in which extensive experience with quality control has accumulated and much success has been enjoyed throughout the world and in all industries, there have appeared some new or revised viewpoints regarding the principles underlying quality control. It seems worthwhile to observe, among other things, how and why the original idea of statistical quality control has been transformed into the new idea of total quality control. The object of this section is to show the characteristic aspects of these two different ideas on quality control, particularly with reference to the notions and the systems of information control attached to each of them. It is our opinion that by doing this we can get a deeper understanding of how intimately quality control problems are connected with information science.

In the original Shewhart principles of statistical quality control, great emphasis was placed on the consequences of the applications of statistical methods to quality control. It seems to be rather exact to say that the essence of the Shewhart principles of quality control is the application of statistical methods. The applicability of statistical methods to engineering production was essentially based on the approximate admissibility of introducing into actual production circumstances a probability field in terms of which to interpret the fluctuation of product quality during production. This situation could be assumed to exist when specification and standardization were attained at a certain level of control of the effects on products. It seems fair enough to say that without statistical methods no systematic or scientific approach to quality control would have been introduced.

However, the accumulation of experience in various industries has occasioned reflection on the meaning and the implication of each word in the expression "statistical quality control." In the first place, it was argued that quality should be interpreted broadly enough to cover various stages of planning, production, inspection, marketing, and management. Instead of interpreting quality as simply the quality of the manufactured product, there were strong arguments prevailing in a wide range of industries in favor of generalized interpretation of the quality which would comprise, along with "quality" in the narrower sense, the completion date for production, cost, and afterservices.

The second point of issue was concerned with the meaning of "control." Control chart procedures were one of the fundamental tools in statistical quality control, and they worked quite well in various fields of industrial engineering at certain levels of control of production, before the stage when automation was introduced. Now the introduction of automatic control in some of these fields has produced a change in the fundamental aspects of production and sometimes also a restriction on the use of statistical quality control in that interruption of production by human action was required for controlling the process. At the same time, the combined development of large scale electronic computers and of theories and apparatus for control led to the introduction of new and more advanced features of control, including adaptive control, optimizing control, control with learning, and so on. It is now almost self-evident that quality control would not be effective in controlling quality, either in the narrow or in the broad sense, unless the meanings and the procedures of control were broad enough to cover these new notions of control. In fact, the more automation proceeds in an industry, the more urgently needed it is to generalize the notion of control beyond its original formulation in the SQC of Shewhart.

The third point of issue was concerned with the statistical methodology of statistical quality control. In view of the generalizations of the two fundamental notions, quality and control, it has become a matter of course to encounter needs for broader methodologies than those which sufficed under the narrow sense definitions. Thus, having the completion date as one quality of production in the generalized sense, we naturally require various techniques that may be available for production and inventory control, such as those belonging to the areas of linear programming, dynamic programming, queueing theory, scheduling and flow techniques, and so on.

In a word, after the elapse of only three years following the introduction of Shewhart's principles of statistical quality control (SQC) in Japanese industries, we experienced the need for a generalization of his principles with respect to each of S, Q, and C. We may conveniently use the symbols S^*, Q^*, C^*, to indicate the generalizations of the corresponding original notions, S, Q, and C.

Let us summarize here what we have explained, in the following abbreviated form;

$$
\begin{aligned}
(1) \qquad SQC &= \text{Statistical Control of Quality} \\
&= SCQ \to SCQ^* \to SC^*Q^* \to S^*C^*Q^* \\
&= M^*C^*Q^*, \\
&= M^*Q^*C^*
\end{aligned}
$$

where we have put, for the moment,

$$
(2) \qquad Q^* = \begin{Bmatrix} Q\text{: Quality in the strict sense} \\ C\text{: Cost of production} \\ D\text{: Date of production} \\ S\text{: Afterservices and care} \end{Bmatrix},
$$

$$
(3) \qquad C^* = \begin{Bmatrix} \text{Adjustment control} \\ \text{Control by analysis} \\ \text{Adaptive control} \\ \text{Control by learning} \end{Bmatrix},
$$

$$
\begin{aligned}
(4) \qquad S^* &= \begin{Bmatrix} \text{Statistical method} \\ \text{Operations research} \\ \text{Management Science techniques} \end{Bmatrix}, \\
&= M^*.
\end{aligned}
$$

The transformation of SQC into $M^*Q^*C^*$ was necessary in most of industrial engineering, and it was realized in a majority of actual cases by involving the cooperation of all persons in a given company. In the consequence, quality control has been concerned with every job in the company. Quality control must be performed in every job in the company by every person in the company. This understanding of the quality control movement implied the transformation of SQC into $M^*Q^*C^*$, and it was indeed the practical basis on which some remarkable benefits were brought about. Nevertheless, this understanding could not have been sufficient as a theoretical basis for the further development of quality control, unless it had been supplemented with stricter observations of real situations.

In this connection the idea of total quality control is worthy of discussion. The terminology "total quality control" has been used with somewhat different meanings by different authors in the field of quality control. In order to avoid misunderstanding, we shall set down certain characteristic aspects of total quality control according to our understanding of the term, which we believe is close to that of Juran [34] and Feigenbaum [22], [23], but which is not identical with either.

The characteristic aspects of total quality control are embodied in the following two properties:

(1) a total valuation system is introduced in terms of which every activity company can be evaluated;

(2) a total information system is introduced which makes information available for controlling every activity in the company.

These two aspects are characteristic of a total quality control system, they may be only approximately fulfilled in any actual company. By having a common unit of evaluation, the merit or demerit of each activity of the company can be evaluated, and the manager can give an exact indication for the job of any particular person. By having a total information system, the company can secure its self-regulation.

Here again we have come to the realm of information science, which can provide us with the effective organization for advancing total quality control.

18. Operations research and integrated control of product quality

Operations research [1] is concerned with operations that are applied to man-machine systems in which controls are required to attain the desired objectives. For instance, production lines in a plant, or railway systems, entail operations in this sense. Department stores are concerned with various kinds of selling operations; the wood and pulp industry has to do with operations that include transportation and tree planting. The development of a city implies a large assembly of operations, including the building of roads, houses, and electrical network, a water system, and a transportation system for its inhabitants.

Operations research is understood to be the scientific approach to operations. An operation research team endeavors to develop suitable theories for predicting the responses to operations indicated by a manager and performed by an organization; and also to introduce measuring procedures so that actual performances and states of the system can be measured, and to set up information systems through which such information can be transmitted to managers.

In short, operations research gives to a manager both a theoretical tool and a tool of observation by means of which the manager can determine suitable management objectives and can then put into effect operations that will be effective in attaining these objectives.

Now the essential role of an operations research team consists in its working as one constituent element in a closed cycle of a feedback system of information. An operations research team can and must be ready to improve proposed theoretical tools in the light of data accumulated, and it must take constant responsibility for checking whether the system is working well and for informing the manager of any significant change in order that the manager may make the necessary decisions to improve the situation and bring to bear suitable operations on the system before it turns out to cause any serious loss.

This understanding of the role of the operations research team will suffice to make clear the following two observations.

The first is concerned with the relationship between operations research and information science. Operations research is not to be viewed in terms of individual final solutions to restricted types of operational problems, but rather as being imbedded in the framework of an information system through which constant checking and improvement can and should be performed. In consequence, the following assertions can be deduced.

(1) Operations research should be organized and carried out in such a way that it is working routinely.

(2) Operations research should be recognized as a learning process in which experience can be accumulated and improvement of operations realized as a result of these experiences.

(3) Operations research should be systematized in such a way that total optimalization as well as suboptimation should be brought about or at least searched for in the handling of the various problems of management in its broadest sense.

The second observation is this. In terms of what we have said in our first observation, it is now quite clear that the notion of integrated control of product quality as advocated by various quality control engineers in Europe corresponds exactly to the role of operations research as we have described it. The remarkable differences between total quality control (TQC) and integrated control of product quality (ICPQ) arise from the following three factors.

(1) The notion of product is broader in ICPQ than in TQC. In ICPQ the building of a library, a hospital, a railway, or a harbor is considered as the creation of a product, or, more properly, a superproduct, to which quality control can and should be applied.

(2) The objective of control must be broad enough to cover social welfare and is not to be confined to individual manufacturing companies.

(3) The methodologies of controlling products are not confined to statistical methods, but should rather be concerned with operations in the sense of operations research as just explained.

In summarizing these two observations, we can say that operations research can work as a routine job in management and production, just as SQC has been working in industry, leading to the notion of ICPQ, and that this is possible by imbedding operations research in the information science scheme associated with management and production. If this assertion is granted, then the duties and the legal status of operations research workers in industry can be formulated and agreed on. These deliberations show how and why the information science approach can serve in essential ways in the present age of industrialization.

19. Automatically controlled sequences of statistical procedures

The purpose of the present section is to show to what extent one of the recent new developments in statistics, namely, automatically controlled sequences of statistical procedures, abbreviated by ACSSP, is connected with the various developments explained in the last few sections, and then to give a preparation for the introduction of the following section in which we shall summarize the relationship between statistics and information science.

(1) One of the characteristic features of recent developments can be seen in the fact that the notion of automatically controlled sequences of statistical procedures (ACSSP) is realistic and totally applicable in statistical activity. In our two papers [44] and [45] we have explained the needs and the usefulness

of the notion of ACSSP. Automatically controlled sequences of statistical procedures are defined with reference to computer programs. They have turned out to be applicable in virtue of the development of computers, which are now indispensable information apparata in the sense of information science. Statistical programming is nothing but the setting up and programming of ACSSP into a computer in the terminology of that computer's particular program language. Indeed, programming implies automatically controlled sequences of procedures. In characterizing programming as statistical programming, we need deep and thorough understanding of the implications of statistical methodology.

The concepts of statistical programming emphasized by Terry [70], Yates [87], and others, are extremely important for a clear distinction between mathematical and statistical programming.

In a previous paper [45] we have also given our own idea on how to define a statistical procedure. This runs as follows.

(1) A statistical procedure is a procedure that deals with a set of data under the following two specifications:

(a) composite regularities are described in terms of probability fields, as specified in subsection 6.1 of [45];

(b) quantitative evidence is obtained through the application of two principles for acquisition of data, namely, randomization and transformation, as explained in subsection 6.2 of [45].

Evidence obtained from statistical procedures is a collection of inductive determinations based on a set of rejection rules for statistical hypotheses, in the sense explained in subsection 6.3.

The third terminology of automatic data analysis, that of Tukey [71], is essentially of the same content as ACSSP.

Indeed, automatic data analysis is performed according to a family of ACSSP's which is programmed into a computer. The usefulness of automatic data analysis is explained very clearly by Tukey [71], and his assertion holds true for the usefulness of statistical programming and the ACSSP family.

In short, automation of statistical analysis is one of the characteristic features in various developments of statistics.

(2) The second characteristic feature of recent developments in statistics lies in the introduction of various formulations of learning processes into the domain of statistics.

We have so far explained three uses of information obtained from data with reference to a tentatively specified model. The tentativeness of specified models is rather common in actual statistical analysis. And, in fact, it has been an essential job of statisticians to decide what course of analysis can and should be chosen in view of the data. The difference between classical statistical analysis and more recent statistical analysis does not lie in the recognition of need for learning by experience. Both of them have been based on the understanding of this need. The difference between the two comes in the matter of the learning processes being or not being formulated in terms of programs in computers.

In recent work in statistics, some learning processes have been formulated to the point that ACSSP can be used for their analysis. For example, statistical procedures of TE type—that is, estimations after preliminary test(s) of significance—are one of the simplest kinds of learning processes that are being formulated in the framework of the ACSSP. In general, an ACSSP pertains to a successive approach in which a feedback principle is adopted in order to adjust the course of procedures in view of future data. One can quote many examples showing the recognition of processes as learning processes and their introduction into statistical inference and control. These examples are given in a series of papers of the author dealing with successive processes of statistical inference and control.

(3) The third characteristic feature of recent developments in statistics is the emphasis of a fundamental fact which has not been duly recognized by some schools of mathematical statisticians. This is the fact that statistics as a whole is an empirical science and hence is not a logico-deductive system at each stage of its development. This is by no means contradictory to the assertion that a theory of statistics should be a logico-deductive system. Indeed, any one of the ACSSP approaches whose importance in recent developments in statistics has been recognized provides us with a logico-deductive system that can be used for determining operating characteristics. Nevertheless, we must always reserve the freedom to enlarge and revise any ACSSP family in view of data we have gathered and of the experience we have accumulated in applying ACSSP families in various domains of statistical problems. Statistics has no limited areas of application—the development of science, technology, and industry, as well as that of the social sciences, has repeatedly introduced new problems which had not previously been treated at earlier stages in the growth of statistics.

For example, quality control had not been a topic of statistics until the first ten years of this century. But we should duly recognize that statistical quality control has had much influence on the development of statistical method.

In view of the various influences on the development of statistics that have come from outside, it should be the attitude of statisticians to keep open the doors of statistics in order that such influences may always be accepted, but at the same time to organize the structure of statistics as systematically as possible at each stage of its development. What we have established should be programmed in a computer as ACSSP, and we should always be searching for procedures that may be useful in particular types of statistical problems. It is contradictory to our fundamental understanding of statistics as an empirical science to restrict the problems of statistics to within a certain prescribed range. In this connection Tukey [71] asserts the importance of statistics in cultivating new fields of data analysis by pointing out "the needs for collecting the results of actual experiences with specific data-analytic techniques," and by recommending "free use of ad hoc and informal procedures in seeking indications" (Tukey [71], section 45, p. 62).

(4) The fourth characteristic aspect of the recent developments we referred

to in the previous sections is the intimate connection between statistical inference and statistical control. In the formulation of statistical decision functions, decisions and actions are closely connected, mainly for the reason that decisions are recognized as operational uses of information. Due to our position on the uses of information in statistics, we do not think that operational uses of information should be the sole purpose for which we reduce our data; two other uses, storage of information and pattern recognition, should also be taken into consideration. With this general understanding of the three uses of information in statistics, it may be remarked that ACSSP makes particularly easy the connection of inference with control. Although in both of the papers [44] and [45] we have referred exclusively to automatically controlled sequences of statistical inference, it can be readily seen that statistical procedures can be enunciated both with inferences and with control. This can be seen by observing that controls can be understood as transformations of one objective world into another, usually by changing a set of parameter values into another set or by introducing some more drastic change of the objective world to another one which is still a member of the class of objective worlds within our concern.

20. Statistics and information science

The object of this section is to summarize all the foregoing considerations and observations in order to show the relationship between certain recent developments in statistics and those in information science, and also to point out some underdeveloped areas in statistics which it would seem to be worthwhile to cultivate in view of the relationship between these two sciences.

For this purpose we shall prepare ourselves with three different points of view. The first point of view is that of the three logical principles of statistics enunciated in the preceding section: (1) objectivity, (2) subjectivity, and (3) practice.

The second point of view arises from the three uses of information: (a) storage of information, (b) pattern recognition, and (c) operational uses of information.

Finally, the third point of view refers to the five branches of information science: (i) information elements, (ii) information theory, (iii) information systems, (iv) biological information, and (v) information apparata.

The first of these three points of view will be found to give a primary classification, while the two other standpoints combine to give subsidiary categories under the headings assigned by the primary classification.

[1] *Objectivity.* This category is concerned with the question of how to formulate models of the objective world which is to be dealt with in statistics. Broadly speaking, there are two fundamental schemes for formulating models of the objective world, namely, chaos and cosmos. These two principal features can be observed in the theoretical schemes of information science by examining the subject matter and scope of all constituent research fields in Branch 1 of information science. At the same time it should be added that one of the characteristic features of certain advances in statistics during the last five years is the

increasing need for elaboration and/or generalization of cosmos formulations. Indeed, it is in the area of information science that deeper consideration concerning logic, pattern recognition, information networks, and systems have appeared, raising the notion of cosmos into greater prominence than ever before. These new developments in information science will have great influence on the general formulation of statistics. In particular, they are deeply concerned with pattern recognition in statistical approaches, a subject that has not been fully discussed in the traditional area of statistics.

Here a close cooperation between the two sciences, statistics and information science, can be usefully initiated. In fact, information science owes its chaos formulation of the objective world to the field of statistics. On the other hand, statistics can and must learn much from information science, where the notion of cosmos has been studied in various connections as explained just above. In particular, the statistician can and must maintain intimate contact with the methods and achievements in the fields of communication theory, control theory, mathematical linguistics, and the logic of computers, all of which belong to the subject matter of Branch 1 of information science.

[2] *Subjectivity.* This category is concerned with the principles upon which is based our choice subjective attitude. In statistics our subjective attitudes and performances should be coordinated and integrated under two principles: (i) the principle of strategy and (ii) the principle of valuation. We have explained the implications of these two principles, so that they are seen to be valid not only in decision function approaches in statistics but also in the formulation of controls. It is to be remarked here that no theoretical formulation of information science can work well without reference to these two principles of subjectivity.

This category is indeed indispensable in a formulation of information science. A piece of information cannot have meaning without reference to some subjective existence, which conducts itself according to a certain set of subjectivity principles. In this sense the two principles valid over a wide range of statistical theories are common to theoretical formulations under Branch 1 of information science. As far as these principles can be formulated objectively, the category of subjectivity can be applied to the formulation of information science. In this sense information science can expect to gain from the formulation of statistics. Here, in speaking of statistics we are not confining ourselves to just decision function approaches. As we have explained in our paper [41], the Fisherian spirit of scientific inference ([24], [25]) can be understood within the framework we have explained in the previous section. In order to establish a theory in statistics we must present as many automatically controlled sequences of statistical procedures as possible, which can be dealt with from the standpoint of the two principles of subjectivity. In this connection emphasis should be placed upon every aspect of the three uses of information (cf. [43]), including (a) storage of information, (b) pattern recognition, and (c) operational uses of information. In virtue of such a systematic review of statistical principles, we shall be conscious of the need for introducing and developing the two principles, strategy and valuation, in the case of each of the two information uses (a) and (b). Indeed, the two principles have been applied mostly, if not exclusively, to the information use (c), as we observe it in Wald's theory of statistical decision functions [74].

Although we do not intend to give any extended discussion of the application of these two principles to the first two categories of uses of information use, it should be pointed out that strategy and valuation regarding storage and pattern recognition are crucial subjects in both information science and statistics. The problems suggest an as yet undeveloped area of statistics, in which some new systematic approaches should be tried. This assertion holds also for Branch 2 of information science.

[3] *Practice.* Once the subjective aspects of our strategies and valuations have been settled upon for our objective world, various principles of practices will enter to play the role of providing us with practical guides in dealing with real circumstances. In this connection two principles seem to be essential: (i) the principle of efficiency and (ii) the principle of successiveness, or the principle of feedback. As we have explained in our previous paper [41], the characteristic features of statistical analysis can be explained in terms of these two principles. Now, as we have already explained, in view of the recent developments in information science, the need is urgent for a generalization of the principle of efficiency as applied in various statistical problems, including the three uses of information. Thus, the operating characteristic approach of ACSSP has to refer not only to operational uses of information but also to storage and pattern recognition. Now, how can we define an efficiency of storage of information? How can we define an efficiency of pattern recognition? These have not been duly discussed in statistics. But we are now coming to the stage where we cannot help but be conscious of the present lack of any adequate generalized notion of efficiency which can be applied to such generalized problems in statistics. We are now only in a position to be able to suggest the importance of the problems before us. We should not forget that we are faced with these problems. In this connection we have already discussed the purposes and the implications of systems analysis applied to ACSSP.

An ACSSP can be considered as a system that is decomposable into a set of component subsystems, each of which is also an ACSSP, while it can also be considered as a component subsystem of a more complex system which is also an ACSSP (cf. [45]).

Systems analysis is concerned with several types of analysis, such as (i) component analysis, (ii) composition analysis, (iii) stability analysis, (iv) flexibility analysis and (v) reliability analysis. We have explained the implication of each of these five types of analysis in our previous paper [45]. Equipped with all these methods of analysis applicable to an ACSSP, we have a rather broad view of the various possible functions of an ACSSP, not only with respect to operational uses of information, but also partly with respect to storage and pattern recognition, as, for example, (i) component analysis and (ii) composition analysis.

There has been a great deal of progess in the domain of information science bearing on the principle of successiveness. It is concerned with learning processes and self-organizing systems. On the other hand, one of the mainstreams of development in statistics can be evaluated as development along the lines of learning processes.

Data in themselves are indeed a bulk accumulation of information, and any statistical understanding obtained from data should have a certain integration pattern of statistical recognition.

An integration pattern of statistical recognition can be defined with reference to an information summary scheme and an information evaluation scheme. An information summary scheme is a recognition pattern according to which information obtained from data is stored and arranged. An information evaluation scheme is a set of utility functions associated with an information summary scheme, according to which the utility of information stored and arranged in the information summary scheme is measured. In our previous paper we gave a detailed analysis of the Wald decision function approach [74] in statistics with reference to these two schemes and discussed the reasons why some parts of statistics are not completely satisfactory with respect to this formulation. In order to achieve more freedom and a higher degree of flexibility, whereby both uses of information, storage and pattern recognition, can be admitted in the schemes of statistics, we have suggested in our previous paper [45] the formulation as follows.

(i) An ACSSP system can be decomposed into a set of subsystems each of which has an integration pattern admitting a risk function.

(ii) A coordination principle is set according to which all these subsystems are coordinated into a single whole. (This principle does not necessarily refer to the corresponding category in the utilitarian viewpoint, which is common in information evaluation schemes of subsystems.)

(iii) A coordination is not necessarily geared to any integration pattern having a risk function evaluation scheme, and hence, the whole may be decomposed.

(iv) A coordination is to be flexible enough to learn from experience so as to be able to introduce a tentative integration pattern over the whole as data accumulates. Such a tentative integration is always subject to replacement by another more suitable one as more information is gathered.

In short, a decomposable coordination of subsystems has been suggested in our previous paper [45]. We believe this will serve as a framework covering the three aspects of information use, and at the same time will admit learning processes, which are essential in the development of statistics.

A few words may be in order to clarify the relationship between statistics and the three particular branches of information science, information elements, biological information, and information apparata. We have already shown the intimate connections between statistics and the other two branches of information science, information theory and information systems.

Briefly, statistics has no direct connection with these three branches, but it has an important indirect connection with each of them. An understanding of this is, we believe, indispensable to the development of statistics in the coming years, because it provides us with a systematic way of maintaining contact with the achievements in information science.

21. Summary and comments

This paper consists of twenty-one sections, which can be broadly divided into three parts. In the first part of the first five sections, we explain the impetus and the reasons for the need to establish an information science. In the second part, consisting of the middle seven sections, from section 6 and section 12, a description of information science is given, and its connections with some neighboring sciences, such as cybernetics, bionics, the mathematical sciences, the behavioral sciences, library and documentation sciences, are illustrated in order to give a better understanding of the scope and role of information science. In the third part, consisting of the last nine sections, some of the characteristic aspects of recent statistical problems are described, which show up implications of information science for the development of statistics. We can summarize our points of view in the following assertions.

(1) There is an urgent need to develop an information science which will satisfy the following specific requirements:

(i) it is a planned area of consolidated (integrated) scientific approaches which must start with the five indispensable branches set forth in section 6;

(ii) its theoretical formulation has a scope which is broad enough to cover the three principal logical aspects: (a) objectivity, (b) subjectivity, (c) practices, as explained in section 20.

(2) Statisticians and mathematicians should have intimate contact with such an information science for the following two reasons:

(i) statisticians and mathematicians should be aware of the crucial problems in information science to whose solution they can contribute,

(ii) statisticians and mathematicians should be aware of essential achievements in the many branches of information science; these will in turn serve them in their selections of problems and thereby further their research.

(3) In particular, some of the crucial problems for formulations in mathematics and statistics are now coming to be common topics of discussion among scientists and engineers working in information science, and statisticians and mathematicians will have much to gain from new developments in information science.

In the conclusion of the paper the author takes the occasion to give several remarks on the contemporary works and views of other authors who deal with what we call information science.

REMARK 1. The scope and role of information science as we describe it may be said to be almost identical with those which the Russian scientists call "kibernetiki" (cybernetics). From the publications [46], [48], and [75] I have the impression that "kibernetiki" has larger scope and role than the cybernetics of Wiener ([78] through [83]), as in the case with our information science. The reason we adopted the terminology of "information science" rather than "cybernetics" is that we think it better to retain the latter term in the original meaning given it by the inventor and his collaborators. The substantial difference that may exist between kibernetiki and our information science is that we propose to construct our science in accordance with the requirements just explained. This position of ours may be explained as intending to keep our research within the area of science.

REMARK 2. In the later stage of the development of his cybernetics, Wiener emphasized that an important part of cybernetics in the future would lie in the field of organization (cf. [80], p. 12). He also referred to a problem of economic planning with the assertion that such a problem is of a statistical character, and therefore of an information character (cf. [80], p. 12). In view of these affirmations, it might be thought that cybernetics in the sense of Wiener has been growing to the point where there is no substantial difference of coverage between it and our information science. However, it is to be noted that he did not seem inclined to include certain econometric and psychometric research within the area of cybernetics. This is one of the crucial differences between our information science and cybernetics. Our position is to refer to three logical aspects, which include subjectivity and practice as well as objectivity. It is our view that no scientific theory of organization can be built without reference to a set of logical aspects including subjectivity and practice. The author is in complete agreement with the view of Ashby [3] when he suggests that the recognition of the limitation implied by the law of requisite variety may prove useful by ensuring that our scientific strategies in the case of a complex system shall be new strategies genuinely adapted to the special peculiarities of that system. He then suggested that the strategies appropriate to complex systems might be those already becoming well known under the title "operational research." However, it seems to us that the logical aspects of operations research can be reduced to those which, as we have already explained, are more fundamental in formulating the methodology of science.

REMARK 3. In recent years several symposia ([14], [15], [19], [51], [53], [68], [78], [82]) have been held in the areas of artificial intelligence, perception, bionics, and biocybernetics, and many works have been published concerning pattern recognition, learning processes, and self-organization; among these quite a few are executed from the standpoint of statistical approach (cf. [8], [14], [15], [18], [27], [28], [33], [35], [51] through [58], [59], [63], [68], [73], [76] through [78], [81] through [84]). In particular, decision function approaches have been applied to pattern recognition and learning processes by several authors (cf. [27], [28], [66]). The full range of statistical approaches is not covered by the Wald formulation of statistical decision functions, as we have explained in section 20, as well as in [37], (3), [47], [43], [44], and [45]. There are few works which are clearly statistical approaches to given problems but which cannot claim to come under the decision function approach (cf. [35], [58]). Furthermore, there do exist several recent statistical works searching for different possibilities (cf. [33], [59], [76], [77], [84]), and these seem to us to be more or less conscious of the problems which we have explained in section 20. There are also many stimulating pieces of research on mathematical models of learning and on neurophysiological models ([51], [52], [53] through [55], [58], [65]). We look forward to mutual assistance and cooperation among these areas with the consequence that there may be a more definitive realization of our ideas suggested in section 20.

REMARK 4. In section 18, we emphasized that any ultimate formulation of control theory should be given within the framework of information science. In spite of all the quite excellent work on control theory, we have to agree with A. T. Fuller ([29], p. 291) when he points out that at present there exists no theory of control which allows simultaneously for the noisiness, variability, and partial nonlinearity and the jelly-like behavior of plants. It must also be recognized that it is not possible to measure all of the components of a state vector and that its dynamic equation may not be known by the controller. In this connection the recent work of Bellman and his colleagues ([9], [11], [12], and [15]) seem to be of importance; it is devoted to the identification problem and the determination of unmeasurable state variables on the basis of the observations of a process.

REMARK 5. Our standpoint given in section 18 does seem to us to be almost in agreement with the views of Beer [4], [5], and [6].

REMARK 6. The terminology of information science(s) has sometimes been used in a sense different from ours. Tayler [69] gave a clear description of what he means by information science and it is broad enough to cover what we call library and documentation science. His definition is quite suitable for bringing together various areas which are essentially in need of librarians and documentators. From our explanations in section 12 it can be seen that all the areas suggested by Tayler are imbedded within our framework.

REMARK 7. M. C. Goodall [31] points out that the theory of cognitive systems deals with inductive inference from a new point of view, that of constructibility; that constructibility brings to light a basic paradox which is analogous to the situation that developed in logic with the discovery of Russel's [9] paradox. Bellman and Kalaba [8] have suggested the construction of a hierarchy of processes whereby to introduce precision into the concepts of "intelligence" and "thinking." Their standpoint can be understood in terms of ours because we have arrived at our information science by starting from the actual problems of statistics and step-by-step generalizing the notions which have been actually needed in this research. Although we have been most heavily concerned here with the abstract notions, one of the characteristic properties of our research strategy is to always have a basis which is adequately realistic and whose analytical methods promise ever new development.

This paper is dedicated to the celebration of the seventieth birthday of Professor Jerzy Neyman with congratulations on his scientific achievements, which are monumental in the history of statistics, and on his successful efforts in fostering international cooperation among statisticians and mathematicians.

REFERENCES

[1] R. L. Ackoff, *Progress in Operations Research*, Vol. 1, Publications in Operations Research, New York, Wiley, 1961.

[2] J. J. Anscombe and J. W. Tukey, "The criterium of transformation," unpublished manuscript, 1954.

[3] W. R. Ashby, "Requisite variety and its implications for the control of complex systems," *Cybernetica*, Vol. 1 (1958), pp. 83–99.

[4] S. Beer, "Operational research and cybernetics," *Proceedings of the 1st International Congress on Cybernetics*, Namur, Paris (1958), pp. 29–45.

[5] ———, "A technical consideration of the cybernetic analogue for planning and programming," *Proceedings of the 1st International Congress on Cybernetics*, Namur, Paris (1958), pp. 515–534.

[6] ———, "The impact of cybernetics on the concept of industrial organization," *Proceedings of the 1st International Congress on Cybernetics*, Namur, Paris (1958), pp. 535–554.

[7] R. Bellman, *Adaptive Control Processes, A Guided Tour*, Princeton, Princeton University Press, 1961.

[8] R. Bellman and R. Kalaba, "Dynamic programming, intelligent machines, and self-organizing systems," *RAND Memo* 3175-PR (1962).

[9] R. Bellman, H. Kagiwada, and E. Kalaba, "Quasilinearization, system identification, and prediction," *RAND Memo* 3812-PR (1963).

[10] R. Bellman, R. Kalaba, and R. Sridhar, "Adaptive control via quasilinearization and differential approximation," *RAND Memo* 3928-PR (1965).

[11] R. Bellman, B. Gluss, and R. Roth, "On the identification of systems and the unscrambling of data: some problems suggested by neuro-physiology," *RAND Memo* 4266-PR (1964).

[12] ———, "Segmental differential approximation and the 'black box' problem," *RAND Memo*, 4269-PR (1964).

[13] R. Bellman and B. Gluss, "Adaptive segmental differential approximation," *RAND Memo* 4314-PR (1964).

[14] E. E. Bernard and K. R. Kare, *Biological Prototypes and Synthetic Systems*, Vol. 1, Second Bionics Symposium, sponsored by Cornell University and the General Electric Company, Advanced Electronics Center, New York, Plenum Press, 1962.

[15] H. Billing, "Lernende Automaten," Bericht über die Fachtagung der Nachrichtentechnischen gesellschaft im DVE (NTG), Fachausschluss 6, *Informationsverarbeitung*, München, 1961.

[16] G. E. P. Box, "Evolutionary operation: a method for increasing industrial productivity," *Appl. Statist.*, Vol. 6 (1957), pp. 3–23.

[17] B. E. Cooper, "Designing the data presentation of statistical program for the experimentalist," *The 34th Session of the International Statistical Institute at Ottawa,* Vol. 40 (1963), pp. 567–587.
[18] P. W. Cooper, "The hyperplane in pattern recognition," *Cybernetica,* Vol. 5 (1962), pp. 215–238.
[19] L. Couffignal, "Le symposium de Zurich et les concepts de base de la cybernetique," *Cybernetica,* Vol. 1 (1958), pp. 15–31.
[20] W. E. Deming, *Some Theory of Sampling,* New York, Wiley, 1950.
[21] A. S. Douglas and A. J. Mitchell, "AUTOSTAT: A language for statistical data processing," *Comput. J.,* Vol. 3 (1960), pp. 61–66.
[22] A. V. Feigenbaum, *Quality Control, Principles, Practices, and Administration,* New York, McGraw-Hill, 1961.
[23] ———, *Total Quality Control—Engineering and Management,* New York, McGraw-Hill, 1961.
[24] R. A. Fisher, "Statistical method and scientific induction," *J. Roy. Statist. Soc. Ser. B,* Vol. 17 (1955), pp. 69–78.
[25] ———, *Statistical Methods and Scientific Inferences,* London, Oliver and Boyd, 1959 (2d rev. ed.).
[26] J. D. N. de Fremery, personal communication.
[27] K. S. Fu, "A sequential decision model for optimum networks," *Biological Prototypes and Synthetic Systems,* Vol. 1, edited by E. E. Bernard and M. R. Kare, New York, Plenum Press, 1962.
[28] ———, "A statistical approach to the design of intelligent machines—pattern recognition and learning," *Cybernetica,* Vol. 5 (1962), pp. 88–102.
[29] A. T. Fuller, "Directions of research in control," *Automatics,* Vol. 1 (1963), pp. 289–296.
[30] G. Fürst, "Changing tasks in official statistics," *The 34th Session of the International Statistical Institute at Ottawa,* Vol. 15 (1963), pp. 196–206.
[31] F. H. George, *The Brain as a Computer,* New York, Pergamon Press, 1962.
[32] M. C. Goodall, "Induction and logical types," *Biological Prototypes and Synthetic Systems,* Vol. 1, edited by E. E. Bernard and M. R. Kare, New York, Plenum Press, 1962.
[33] K. Jankowski, "The dynamic investigation of stochastic sequences," *Cybernetica,* Vol. 5 (1962), pp. 155–197.
[34] J. M. Juran, L. A. Seder, and L. M. Gryna, *Quality Control Handbook,* New York, McGraw-Hill, 1961 (2d ed.).
[35] L. Kanal, "Evaluation of a class of pattern-recognition networks," *Biological Prototypes and Synthetic Systems,* edited by E. E. Bernard and M. R. Kare, New York, Plenum Press, 1962.
[36] M. G. Kendall and P. Wegner, "An introduction to statistical programming," *The Review of the International Statistical Institute at Ottawa,* Vol. 31 (1963), pp. 305–360.
[37] T. Kitagawa, "Successive process of statistical inferences," *Mem. Fac. Sci. Kyushu Univ. Ser. A,* Vol. 5 (1950), pp. 139–180; Vol. 6 (1951), pp. 54–95; Vol. 6 (1952), pp. 131–155; Vol. 7 (1953), pp. 81–105; Vol. 8 (1953), pp. 1–29; *Bull. Math. Statist.,* Vol. 5 (1951), pp. 35–50.
[38] ———, "Successive process of statistical controls," *Mem. Fac. Sci. Kyushu Univ. Ser. A,* Vol. 7 (1952), pp. 13–26; Vol. 13 (1959), pp. 1–16; Vol. 14 (1960), pp. 1–33.
[39] ———, "Successive process of statistical inferences applied to linear regression analysis and its specializations to response surface analysis," *Bull. Math. Statist.,* Vol. 8 (1959), pp. 80–114.
[40] ———, "A mathematical formulation for evolutionary operation programs," *Mem. Fac. Sci. Kyushu Univ. Ser. A,* Vol. 15 (1961), pp. 21–71.
[41] ———, "The logical aspects of successive processes of statistical inferences and controls," *Bull. Inst. Internat. Statist.,* Vol. 38 (1961), pp. 151–164.

[42] ———, "Estimation after preliminary test of significance," *Univ. California Publ. Statist.*, Vol. 3 (1963), pp. 147–186.

[43] ———, "The relativistic logic of mutual specification in statistics," *Mem. Fac. Sci. Kyushu Univ. Ser. A*, Vol. 17 (1963), pp. 76–105.

[44] ———, "Automatically controlled sequence of statistical procedure in data analysis," *Mem. Fac. Sci. Kyushu Univ. Ser. A*, Vol. 17 (1963), pp. 106–129.

[45] ———, "Automatically controlled sequence of statistical procedures," *Bernoulli, Bayes, Laplace, Anniversary Volume*, Berlin, Springer-Verlag (1965), pp. 146–178.

[46] L. J. Krieger, "Soviet philososphy, science and cybernetics," *RAND Memo* 3619-PR (1963).

[47] R. Levien and M. E. Maron, "Cybernetics and its development in the Soviet Union," *RAND Memo* 4156-PR (1964).

[48] A. A. Lyapunov, "General problems of cybernetics," *Problems of Cybernetics, I*, New York, Pergamon Press, 1960, pp. 1–21.

[49] P. C. Mahalanobis, "On large-scale sample surveys," *Philos. Trans. Roy. Soc. London Ser. B*, Vol. 231 (1944), pp. 329–451.

[50] ———, "Some aspects of the design of sampling surveys," *Sankhyā*, Vol. 12 (1952), pp. 1–7.

[51] M. Minsky and O. G. Selfridge, "Learning in random nets," *Information Theory*, Fourth London Symposium, edited by Colin Cherry, London, Butterworth, 1961.

[52] M. E. Maron, "Mechanisms underlying predicative behavior for an intelligent machine," *RAND Memo* 3011-PR (1962).

[53] G. A. Muses, "Aspects of the theory of artificial intelligence," *Proceedings of the First International Symposium on Biosimulation, Lacarno 1960*, New York, Plenum Press, 1962.

[54] A. Newell, "Learning, generality and problem-solving," *RAND Memo* 3285-PR (1962).

[55] ———, "Some problems of basic organization in problem-solving problems," *RAND Memo* 3283-PR (1962).

[56] J. Neyman, "On the problem of the most efficient tests of statistical hypotheses," *Philos. Trans. Roy. Soc. Ser. A*, Vol. 23 (1933), pp. 289–337.

[57] J. Neyman and E. S. Pearson, "The testing of statistical hypotheses in relation to probabilities a priori," *Proc. Cambridge Philos. Soc.*, Vol. 29 (1938), pp. 492–510.

[58] S. Papert, "Some mathematical models of learning," *Information Theory*, Fourth London Symposium, edited by Colin Cherry, London, Butterworth, 1961.

[59] G. Pask and H. von Foerster, "A predicative model for self-organizing systems," *Cybernetica*, Vol. 3 (1960), pp. 258–300; Part II, *Cybernetica*, Vol. 3 (1961), pp. 20–55.

[60] K. Pearson, *The Grammar of Science*, Oxford University Press, 1896.

[61] L. S. Pontryagin, *The Mathematical Theory of Optimal Processes*, New York, Prentice Hall, 1962. (Translated from the Russian.)

[62] S. J. Reed, "Screening rules," M.S. thesis, Rutgers University, 1959.

[63] K. H. Reid, "Design studies of conditional probability computers," *Biological Prototypes and Synthetic Systems*, Vol. 1, edited by E. E. Bernard and M. R. Kare, New York, Plenum Press, 1962.

[64] H. Robbins and S. Monroe, "A stochastic approximation method," *Ann. Math. Statist.*, Vol. 22 (1951), pp. 400–407.

[65] L. Rosenblatt, *Principles of Neuro-Dynamics; Perception and the Theory of Brain Mechanisms*, Washington, 1962.

[66] G. S. Sebestyen, *Decision-Making Processes in Pattern Recognition*, New York, Macmillan, 1962.

[67] W. A. Shewhart, "Statistical method from the viewpoint of quality control," Washington, D. C. (1939).

[68] K. Steinbuch and S. W. Wagner, "Neuere Ergebnisse der Kybernetik," Bericht über die Tagung Karlsruhe 1963 der Deutschen Arbeitsgemschaft Kybernetik, Munchen, 1964.

[69] R. S. TAYLER, "The information sciences," *Library Journal*, Vol. 88 (1965), pp. 4161–4163.

[70] M. E. TERRY, "The principles of statistical analysis using large electronic computers," *The 34th Session of the International Statistical Institute at Ottawa*, Vol. 40 (1963), pp. 547–552.

[71] J. W. TUKEY, "The future of data analysis," *Ann. Math. Statist.*, Vol. 33 (1962), pp. 1–68.

[72] J. VON NEUMANN and O. MORGENSTERN, *Theory of Games and Economic Behavior*, Princeton, Princeton University Press, 1953.

[73] R. D. TURNER, "First-order experimental concept formation," *Biological Prototypes and Synthetic Systems*, Vol. 1, edited by E. E. Bernard and M. R. Kare, New York, Plenum Press, 1962.

[74] A. WALD, *Statistical Decision Functions*, New York, Wiley, 1950.

[75] W. H. WARE and W. B. HOLLAND, "Soviet cybernetics technology: I Soviet cybernetics, 1959–1962," *RAND Memo* 3675-PR (1963).

[76] S. WATANABE, "Information theoretical analysis of multivariate correlation," *IBM J. Res. Develop.*, Vol. 4 (1960), pp. 66–82.

[77] ———, "A note on the formulation of concept and of association by information-theoretical correlation analysis," *Information and Control*, Vol. 4 (1961), pp. 291–296.

[78] N. WIENER, *Cybernetics or Control and Communication in the Animal and the Machine*, New York, Wiley, 1948 (1st ed.); New York, M.I.T. Press, 1961 (2d ed.).

[79] ———, "The mathematics of self-organizing system," *Recent Development in Information and Decision Processes*, edited by R. E. Machol and P. Gray, New York, Macmillan, 1962, pp. 1–21.

[80] ———, "My connection with cybernetics, its origins and its future," *Cybernetics*, Vol. 1 (1958), pp. 1–14.

[81] ———, *God and Golem, Inc., A Comment on Certain Points where Cybernetics Impinges on Religion*, New York, M.I.T. Press, 1964.

[82] N. WIENER and J. P. SCHADE, "Nerve, brain and memory models," *Progress in Brain Research*, Vol. 2, Amsterdam/London/New York, Elsevier, 1963.

[83] ———, *Progress in Biocybernetics*, Vol. 1, Amsterdam/London/New York, Elsevier, 1964.

[84] R. A. WIESEN and E. H. SHUFORD, "Bayes strategies as adaptive behavior," *Biological Prototypes and Synthetic Systems*, Vol. 1, edited by E. E. Bernard and M. R. Kare, New York, Plenum Press, 1962.

[85] S. S. WILKS, "New fields and organizations in the mathematical sciences," *The Math. Teacher*, Vol. 51 (1958), pp. 2–9.

[86] F. YATES and H. R. SIMPSON, "A general program for the analysis of surveys," *Comput. J.*, Vol. 3 (1960), pp. 136–140.

[87] L. YATES, "Computer in research-promise and performance," The Presidential Address, *Comput. J.*, Vol. 4 (1961), pp. 273–279.

ON SOME BASIC PROBLEMS OF STATISTICS FROM THE POINT OF VIEW OF INFORMATION THEORY

ALFRÉD RÉNYI
MATHEMATICAL INSTITUTE OF THE HUNGARIAN ACADEMY OF SCIENCES

1. Introduction

Many problems of mathematical statistics consist in that one has to extract from certain observations the information which is needed. In other words, one has to separate the relevant information from the irrelevant. For instance, it is a generally accepted view that if for a parameter there exists a sufficient statistic, its sufficiency means that it contains all the information which is present in the sample and is relevant for determining the parameter. Although this is generally admitted, it is not usual to go a step further and ask: *how much information is contained in a statistic (sufficient or not) concerning a parameter?* (According to the author's knowledge, the first to consider this question was D. V. Lindley [1].) In view of the success of information theory in other fields (especially in the theory of information-transmission), which success was achieved by attributing a numerical measure to amounts of information, this question is a very natural one. It seems to the author that the reason why this question is usually not asked in current statistical practice is that a meaningful answer to this question can be given only if one accepts the Bayesian point of view; that is, if one considers the unknown parameter as a random variable and attributes to it a prior distribution. As a matter of fact, the amount of information in a random variable concerning another random variable is a well-defined concept of information theory, whereas the amount of information in a random variable concerning a constant is always zero. The amount of information in a random variable ξ concerning another random variable θ is equal to the average decrease of uncertainty (entropy) concerning θ which results if ξ is observed. In order to measure this decrease of uncertainty, our prior knowledge about θ has to be taken into account.

If we have some prior knowledge about θ, this causes no difficulty. If we have no prior knowledge about θ, except that we know the set of its possible values, from the point of view of information theory it seems to be natural to attribute to θ that prior distribution on the admissible set of values which has the largest entropy, that is which corresponds to maximal uncertainty. Even if the parameter is in reality a constant, if its value is unknown to us, I do not find any logical fault in attributing to θ a prior distribution, if this is needed to compare different statistics and to choose that which is the best for our purposes. Usually the choice depends only weakly on the prior distribution.

In this paper we do not want to go into the philosophical aspects of the question—however interesting they may be—as our aim is to deal with certain purely mathematical problems which arise when one tries to apply the concepts of information theory to the mentioned statistical problems.

The problem which will be discussed in what follows is: *how can one decide whether or not a certain sequence of observations contains all the information which is needed* (for example, to find the true value of the parameter)? In general, if a statistician is confronted with a concrete problem, his first task is to decide whether or not the required information is fully present. If the answer is positive, then the second step consists in trying to find an appropriate decision procedure, for instance one which is optimal in some respect. However, if the answer to the first question is negative, then it is futile to take the second step, as even the "best" decision procedure will not yield the required information. In this case the statistician has to look for some other source of information, that is, make some other observations.

In the present paper we shall discuss from this point of view the following problem. Let us be given an infinite sequence $\{\xi_n\}$, $(n = 1, 2, \cdots)$ of observations. We suppose that the distributions of the random variables ξ_n, $(n = 1, 2, \cdots)$ depend on a parameter θ, whose set of possible values is *finite*. We suppose further that for each fixed value of θ the random variables ξ_n, $(n = 1, 2, \cdots)$ are independent, but in general do not have the same distribution. We shall be especially interested in the case where the amount of information on θ contained in the observation ξ_n decreases when n increases. Such problems are often encountered. Let us imagine for instance that an event $\mathcal{E}$ happened at time $t = 0$, and ξ_n is the value of some quantity connected with the aftereffects of this event measured by some instrument at time $t = n$. Usually as time passes, the aftereffects of the event $\mathcal{E}$ become weaker and weaker, and thus as n increases, ξ_n gives gradually less and less information on the event $\mathcal{E}$.

In section 2 we shall consider the amount of information on θ which is still missing after having observed the values $\xi_1, \xi_2, \cdots, \xi_n$ and compare it with the error of the "standard" decision, consisting in deciding always in favor of the hypothesis which has the largest posterior probability.

In section 3 we give an upper bound for the amount of missing information (theorem 2) and give a necessary and sufficient condition for the convergence to 0 of this quantity for n tending to $+\infty$.

In section 4 we shall discuss some special cases, some of which have been discussed in previous papers of the author (see [2], [3], [4]), and also others which are presented here for the first time. In section 5 we compare some of our results with a theorem of S. Kakutani [5].

For the sake of brevity we deal in detail only with the case where θ may have only two values. The generalization to the case when the set of possible values of θ is an arbitrary *finite* set is quite straightforward and presents no difficulty. It should be added, however, that our results cannot be generalized immediately to the case when the set of possible values of θ is infinite. This case presents some peculiar difficulties. We hope to return to this question in another paper.

2. An inequality between the amount of missing information and the error of the standard decision

Let $\{\xi_n\}$, $(n = 1, 2, \cdots)$ be a sequence of random variables. Let us suppose that the distribution of ξ_n depends on a parameter θ, which may take on two different values θ_0 and θ_1. We suppose that the random variables ξ_n are independent under the condition $\theta = \theta_0$, as well as under the condition $\theta = \theta_1$. We suppose further (this restriction is made only to simplify notations) that all the distributions in question are absolutely continuous. Let $f_n(x)$ and $g_n(x)$ denote the density functions of ξ_n under the conditions $\theta = \theta_0$ and $\theta = \theta_1$, respectively. As regards θ we suppose that it is a random variable, taking on the values θ_0 and θ_1 with the corresponding positive probabilities W_0 and W_1, $(W_0 + W_1 = 1)$.

These suppositions can be formulated as follows. Let $[\Omega, \mathcal{A}, P] = S$ be a probability space. Let us be given a partition of Ω into two $\mathcal{A}$-measurable sets Ω_0 and $\Omega_1 = \Omega - \Omega_0$, with $P(\Omega_0) = W_0$, $P(\Omega_1) = W_1 = 1 - W_0$. Let S_0 and S_1 denote the probability spaces $S_0 = [\Omega, \mathcal{A}, P_0]$ and $S_1 = [\Omega, \mathcal{A}, P_1]$ where $P_0(A) = (P(A\Omega_0)/P(\Omega_0))$ and $P_1(A) = (P(A\Omega_1)/P(\Omega_1))$ for $A \in \mathcal{A}$. Let $\xi_n = \xi_n(\omega)$, $(\omega \in \Omega)$, $(n = 1, 2, \cdots)$ be a sequence of $\mathcal{A}$-measurable real functions. Then we can consider the ξ_n as random variables on the probability space S as well as on the probability spaces S_0 and S_1. We suppose that the random variables ξ_n are independent on S_0 as well as on S_1. Note that on the probability space S the random variables ξ_n are usually not independent.

Let us suppose now that we observe the values of all the variables ξ_n, and we want to decide on this basis whether $\theta = \theta_0$ or $\theta = \theta_1$. In other words, we suppose that the sequence of values $\xi_n(\omega^*)$, $(n = 1, 2, \cdots)$ is given for a single unknown $\omega^* \in \Omega$, and want to decide, whether $\omega^* \in \Omega_0$ or $\omega^* \in \Omega_1$. We especially want to find out under what conditions on the density functions $f_n(x)$ and $g_n(x)$ can a correct decision be made for almost all $\omega^* \in \Omega$ (with respect to the measure P)?

Let us denote by ζ_n the random n-dimensional vector with components $(\xi_1, \xi_2, \cdots, \xi_n)$. Let I_n denote the amount of information contained in ζ_n concerning θ. Then we have

$$I_n = H(\theta) - E(H(\theta|\zeta_n)), \tag{2.1}$$

where

$$H(\theta) = W_0 \log \frac{1}{W_0} + W_1 \log \frac{1}{W_1} \tag{2.2}$$

is the entropy of the random variable θ, and $H(\theta|\zeta_n)$ is the conditional entropy of θ given ζ_n, that is

$$(2.3)\qquad H(\theta|\zeta_n) = P(\theta = \theta_0|\zeta_n) \log \frac{1}{P(\theta = \theta_0|\zeta_n)} + P(\theta = \theta_1|\zeta_n) \log \frac{1}{P(\theta = \theta_1|\zeta_n)},$$

and $E(H(\theta|\zeta_n))$ denotes the expectation of the random variable $H(\theta|\zeta_n)$. Here and in what follows log always denotes logarithm with base 2.

According to our supposition and the Bayes' theorem, one has

$$(2.4)\qquad P(\theta = \theta_0|\zeta_n) = \frac{W_0 f_1(\xi_1) f_2(\xi_2) \cdots f_n(\xi_n)}{W_0 f_1(\xi_1) \cdots f_n(\xi_n) + W_1 g_1(\xi_1) \cdots g_n(\xi_n)}$$

and similarly,

$$(2.5)\qquad P(\theta = \theta_1|\zeta_n) = \frac{W_1 g_1(\xi_1) g_2(\xi_2) \cdots g_n(\xi_n)}{W_0 f_1(\xi_1) \cdots f_n(\xi_n) + W_1 g_1(\xi_1) \cdots g_n(\xi_n)}.$$

For the sake of brevity, we introduce the notations $x^{(n)} = (x_1, \cdots, x_n)$, $\varphi_n(x^{(n)}) = f_1(x_1) \cdots f_n(x_n)$, and $\psi_n(x^{(n)}) = g_1(x_1) \cdots g_n(x_n)$.

With these notations, letting $\chi_n(x^{(n)}) = W_0\varphi_n(x^{(n)}) + W_1\psi_n(x^{(n)})$ we have

$$(2.6)\qquad E(H(\theta|\zeta_n)) = \int_{X_n} \left[W_0\varphi_n(x^{(n)}) \log \frac{\chi_n(x^{(n)})}{W_0\varphi_n(x^{(n)})} + W_1\psi_n(x^{(n)}) \log \frac{\chi_n(x^{(n)})}{W_1\psi_n(x^{(n)})} \right] dx^{(n)}$$

where X_n is the n-dimensional Euclidean space and $dx^{(n)}$ stands for $dx_1 dx_2 \cdots dx_n$.

The quantity $E(H(\theta|\zeta_n))$ may be interpreted as *the amount of missing information on θ after observing ζ_n*.

It is easy to see that I_n is nondecreasing for $n = 1, 2, \cdots$ and $I_n \leq H(\theta)$. Thus $\lim_{n\to+\infty} I_n = I^*$ always exists. If $I^* = H(\theta)$, we shall say that *the sequence of observations $\{\xi_n\}$, $(n = 1, 2, \cdots)$ gives us full information on θ*, whereas in the case $I^* < H(\theta)$ we shall say that the observations $\{\xi_n\}$ do not give full information on θ.

Clearly, the most natural decision after having observed ζ_n is to accept θ_0 if $P(\theta_0|\zeta_n) > P(\theta_1|\zeta_n)$ and to accept θ_1 if $P(\theta_1|\zeta_n) > P(\theta_0|\zeta_n)$, and if $P(\theta_0|\zeta_n) = P(\theta_1|\zeta_n)$ to make a random choice between θ_0 and θ_1 with probabilities W_0 and W_1. We shall call this the *standard decision*. Let us define the random variable $\Delta_n = \Delta_n(\xi_n)$ as follows:

$$(2.7)\qquad \Delta_n = \begin{cases} \theta_0 & \text{if the standard decision means acceptance of } \theta_0, \\ \theta_1 & \text{if the standard decision means acceptance of } \theta_1. \end{cases}$$

The *error ϵ_n of the standard decision* after taking n observations is defined as the probability of the standard decision being false. We have clearly

$$(2.8)\qquad \epsilon_n = P(\Delta_n \neq \theta) = W_0 P(\Delta_n = \theta_1|\theta = \theta_0) + W_1 P(\Delta_n = \theta_0|\theta = \theta_1),$$

where $P(A|B)$ denotes the conditional probability of the event A under condition B. Obviously, $\Delta_n = \theta_0$ if $(\varphi_n(\zeta_n)/\psi_n(\zeta_n)) > (W_1/W_0)$ and $\Delta_n = \theta_1$ if $(\varphi_n(\zeta_n)/\psi_n(\zeta_n)) < (W_1/W_0)$. Thus

$$(2.9)\qquad \epsilon_n = W_0 \int_{A_n} \varphi_n(x^{(n)})\, dx^{(n)} + W_1 \int_{B_n} \psi_n(x^{(n)})\, dx^{(n)}$$

where

$$A_n = \left\{\frac{\psi_n(x^{(n)})}{\varphi_n(x^{(n)})} \geq \frac{W_0}{W_1}\right\} \quad \text{and} \quad B_n = \left\{\frac{\psi_n(x^{(n)})}{\varphi_n(x^{(n)})} < \frac{W_0}{W_1}\right\}. \tag{2.10}$$

It is easy to prove the following.

THEOREM 1. *One has*

$$\frac{\epsilon_n}{2} \leq H(\theta) - I_n \leq h(\epsilon_n) \tag{2.11}$$

where

$$h(p) = p \log \frac{1}{p} + (1 - p) \log \frac{1}{1 - p}, \qquad (0 \leq p \leq 1). \tag{2.12}$$

PROOF OF THEOREM 1. One has clearly

$$E(H(\theta|\zeta_n)) = W_0 E(H(\theta|\zeta_n)|\theta = \theta_0) + W_1 E(H(\theta|\zeta_n)|\theta = \theta_1); \tag{2.13}$$

$E(\eta|B)$ denotes the conditional expectation of η under condition B. Now evidently

(2.14)

$$W_1 E(H(\theta|\zeta_n)|\theta = \theta_1) \geq \int_{A_n} \frac{\varphi_n(x^{(n)}) W_0}{\chi_n(x^{(n)})} \log \left(1 + \frac{W_1 \psi_n(x^{(n)})}{W_0 \varphi_n(x^{(n)})}\right) W_1 \psi_n(x^{(n)})\, dx^{(n)}$$

and thus, as on the set defined by $(\psi_n(x^{(n)})/\varphi_n(x^{(n)})) \geq (W_0/W_1)$ one has $(W_1\psi_n(x^{(n)})/\varphi_n(x^{(n)})) \geq \frac{1}{2}$, it follows that

$$W_1 E(H\theta|\zeta_n)|\theta = \theta_1) \geq \tfrac{1}{2} \int_{A_n} W_0 \varphi_n(x^{(n)})\, dx^{(n)} = \frac{W_0}{2} P(\Delta = \theta_1|\theta = \theta_0). \tag{2.15}$$

Similarly we obtain

$$W_0 E(H(\theta|\zeta_n)|\theta = \theta_0) \geq \frac{W_1}{2} P(\Delta = \theta_0|\theta = \theta_1). \tag{2.16}$$

Thus it follows that

$$E(H(\theta|\zeta_n)) \geq \frac{\epsilon_n}{2}. \tag{2.17}$$

This proves the lower inequality in (2.11).

This proof was given earlier in [2]; it is reproduced here for the convenience of the reader. To prove the upper inequality in (2.11) we apply the following integral form of Jensen's inequality:

$$\frac{\int_A h(a(x))k(x)\, dx}{\int_A k(x)\, dx} \leq h\left(\frac{\int_A a(x)k(x)\, dx}{\int_A k(x)\, dx}\right), \tag{2.18}$$

valid for every nonnegative function $k(x)$ and every concave function $h(x)$; we apply (2.18) to the concave function $h(x)$ defined by (2.12) and the domain A in X_n defined by $(\varphi_n(x^{(n)})W_0/\psi_n(x^{(n)})W_1) \geq 1$ and for the complementary domain $\bar{A}$. Thus we obtain

$$E(H(\theta|\zeta_n)) \leq \alpha h(p) + \beta h(q) \tag{2.19}$$

where

$$\alpha = \int_A \chi_n(x^{(n)})\, dx^{(n)}, \qquad \beta = \int_{\bar{A}} \chi_n(x^{(n)})\, dx^{(n)} = 1 - \alpha,$$

(2.20)

$$p = \frac{W_1 \int_A \psi_n(x^{(n)})\, dx^{(n)}}{\alpha} \quad \text{and} \quad q = \frac{W_0 \int_{\bar{A}} \varphi_n(x^{(n)})\, dx^{(n)}}{\beta}.$$

Again applying Jensen's inequality in the form

$$\alpha h(p) + \beta h(q) \leq h(\alpha p + \beta q), \tag{2.21}$$

valid for any concave function h and for $\alpha \geq 0$, $\beta \geq 0$, $\alpha + \beta = 1$, since $\alpha p + \beta q = \epsilon_n$, it follows that

$$E(H(\theta|\zeta_n)) \leq h(\epsilon_n) = \epsilon_n \log \frac{1}{\epsilon_n} + (1 - \epsilon_n) \log \frac{1}{1 - \epsilon_n}. \tag{2.22}$$

Thus theorem 1 is proved.

We obtain easily from theorem 1 the following corollary.

COROLLARY. *The error of the standard decision tends to zero for $n \to \infty$ if and only if the amount of missing information tends to zero, that is,* $\lim_{n\to+\infty} \epsilon_n = 0$ *if and only if* $\lim_{n\to\infty} I_n = H(\theta)$.

It should be added that by a slight modification of the usual proof of the Neyman-Pearson fundamental lemma one can prove (see [3]) that the error of any decision function is at least as large as that of the standard decision. Thus if $\lim_{n\to\infty} I_n < H(\theta)$, that is, if the amount of information in the first n observations does not tend to the total amount of information needed, then there cannot exist a decision procedure whose error tends to zero, whereas if $\lim_{n\to\infty} I_n = H(\theta)$, there certainly exists such a procedure, namely the standard decision.

3. An upper bound for the amount of missing information and a criterion for obtaining full information

We prove now the following theorem.

THEOREM 2. *Let us write*

$$\lambda_k = \int_{-\infty}^{+\infty} \sqrt{f_k(x) g_k(x)}\, dx, \qquad (k = 1, 2, \cdots). \tag{3.1}$$

Then the following inequality holds:

$$0 \leq H(\theta) - I_n \leq B\sqrt{W_0 W_1} \prod_{k=1}^{n} \lambda_k \tag{3.2}$$

where $B > 0$ is an absolute constant.

PROOF OF THEOREM 2. The function $(h(x)/\sqrt{x})$ where $h(x)$ is defined by (2.12) is clearly continuous in the closed interval $0 \leq x \leq 1$. Let

$$C = \max_{0 \leq x \leq 1} \frac{h(x)}{\sqrt{x}}. \tag{3.3}$$

Since $h(x) = h(1 - x)$, we also have

$$C = \max_{0 \le x \le 1} \frac{h(x)}{\sqrt{1-x}}. \tag{3.4}$$

It follows that

$$H(\theta|\zeta_n) \le C\sqrt{P(\theta = \theta_1|\zeta_n)} \tag{3.5}$$

and also that

$$H(\theta|\zeta_n) \le C\sqrt{P(\theta = \theta_0|\zeta_n)}. \tag{3.6}$$

Thus we have, in view of (2.4) and (2.5),

$$H(\theta|\zeta_n) \le C\sqrt{\frac{W_0}{W_1}} \prod_{k=1}^{n} \sqrt{\frac{f_k(\xi_k)}{g_k(\xi_k)}} \tag{3.7}$$

and

$$H(\theta|\zeta_n) \le C\sqrt{\frac{W_1}{W_0}} \prod_{k=1}^{n} \sqrt{\frac{g_k(\xi_k)}{f_k(\xi_k)}}. \tag{3.8}$$

From (3.8) we obtain

$$W_0 E(H(\theta|\zeta_n)|\theta = \theta_0) \le C\sqrt{W_0 W_1} \prod_{k=1}^{n} \lambda_k, \tag{3.9}$$

and from (3.7) we obtain

$$W_1 E(H(\theta|\zeta_n)|\theta = \theta_1) \le C\sqrt{W_0 W_1} \prod_{k=1}^{n} \lambda_k. \tag{3.10}$$

Adding (3.9) and (3.10) we get

$$E(H(\theta|\zeta_n)) \le 2C\sqrt{W_0 W_1} \prod_{k=1}^{n} \lambda_k, \tag{3.11}$$

which proves theorem 2 with $B = 2C$.

Note that according to the Cauchy-Schwarz inequality, $0 \le \lambda_k \le 1$ and $\lambda_k = 1$ if and only if $f_k(x) = g_k(x)$ almost everywhere. Further, $\lambda_k = 0$ if and only if the intersection of the sets on which $f_k(x) > 0$ and $g_k(x) > 0$ is of Lebesgue measure zero. In this case, of course, the observation of ξ_k alone is sufficient with probability 1 to decide whether $\theta = \theta_0$ or $\theta = \theta_1$.

We shall now prove the following.

THEOREM 3. *One has*

$$\prod_{k=1}^{n} \lambda_k \le \sqrt{\frac{\epsilon_n}{W_0 W_1}} \tag{3.12}$$

where λ_k *is defined by* (3.1), *and* ϵ_n *is the error of the standard decision.*

PROOF OF THEOREM 3. Clearly,

$$\prod_{k=1}^{n} \lambda_k = \int_{X_n} \sqrt{\varphi_n(x^{(n)})\psi_n(x^{(n)})}\, dx^{(n)}. \tag{3.13}$$

Let us denote again by A the subset of X_n on which $(W_0\varphi_n(x^{(n)})/W_1\psi_n(x^{(n)}) \ge 1$ and put $\bar{A} = X_n - A$. Taking into account that φ_n is a density function, the Cauchy-Schwarz inequality gives

$$\int_A \sqrt{\varphi_n(x^{(n)})\psi_n(x^{(n)})}\, dx^{(n)} \le \left(\int_A \psi_n(x^{(n)})\, dx^{(n)}\right)^{1/2}. \tag{3.14}$$

Similarly we obtain

$$\int_{\bar{A}} \sqrt{\varphi_n(x^{(n)})\psi_n(x^{(n)})}\, dx^{(n)} \leq \left(\int_{\bar{A}} \varphi_n(x^{(n)})\, dx^{(n)}\right)^{1/2}. \tag{3.15}$$

Thus using again the Cauchy inequality we obtain

$$\prod_{k=1}^{n} \lambda_k \leq \left(\frac{1}{W_0} + \frac{1}{W_1}\right)^{1/2} \sqrt{\epsilon_n} = \sqrt{\frac{\epsilon_n}{W_0 W_1}}. \tag{3.16}$$

This proves theorem 3.

Now we can prove the following theorem.

THEOREM 4. *If* $\lambda_k > 0$ *for* $k = 1, 2, \cdots$, *the sequence of observations* ξ_n $(n = 1, 2, \cdots)$ *contains full information on* θ *if and only if the series*

$$\sum_{k=1}^{\infty} (1 - \lambda_k) \tag{3.17}$$

is divergent.

As regards the connection of theorem 4 with a theorem of Kakutani, see section 4.)

PROOF OF THEOREM 4. Since $1 - x \leq e^{-x}$, if the series $\sum_{k=1}^{\infty} (1 - \lambda_k)$ is divergent, one has $\lim_{n\to\infty} \prod_{k=1}^{n} \lambda_k = 0$, and thus by theorem 2 it follows that

$$\lim_{n\to+\infty} I_n = H(\theta). \tag{3.18}$$

This proves the "if" part of the theorem. On the other hand, using the inequality $1 - x \geq e^{-(x/1-x)}$, $(0 \leq x \leq 1)$, we obtain

$$\prod_{k=1}^{n} \lambda_k \geq \exp\left\{-\sum_{k=1}^{n}\left(\frac{1-\lambda_k}{\lambda_k}\right)\right\}. \tag{3.19}$$

Now if $\sum_{k=1}^{\infty} (1 - \lambda_k)$ is convergent, then $\lim_{k\to\infty} \lambda_k = 1$, and since by assumption $\lambda_k > 0$ for $k = 1, 2, \cdots$, it follows that the sequence λ_k has a positive lower bound: $\lambda_k \geq c > 0$ for $k = 1, 2, \cdots$. It follows that the series $\sum_{k=1}^{\infty} (1 - \lambda_k/\lambda_k)$ is also convergent, and thus $\prod_{k=1}^{n} \lambda_k$ has a positive lower bound. By theorem 3 this implies that ϵ_n has a positive lower bound. Therefore, by theorem 1 the sequence $H(\theta) - I_n$ has a positive lower bound too. This proves the "only if" part of theorem 4.

THEOREM 5. *A sequence of statistics* $\alpha_n = \alpha_n(\xi_1, \cdots, \xi_n)$ *converging in probability to the true value of the parameter (real-valued) can exist only if the sequence* ξ_n $(n = 1, 2, \cdots)$ *contains full information with respect to* θ, *that is if* (3.18) *holds. Conversely if* (3.18) *holds, there exists a sequence of statistics* α_n *which converges with probability* 1 *to* θ.

PROOF OF THEOREM 5. If $\lim_{n\to\infty} P(|\alpha_n - \theta| > \epsilon) = 0$ for every $\epsilon > 0$, then we can construct the following decision function:

$$\begin{cases} \text{if } |\alpha_n - \theta_1| < |\alpha_n - \theta_0|, & \text{accept } \theta_1; \\ \text{if } |\alpha_n - \theta_0| < |\alpha_n - \theta_1|, & \text{accept } \theta_0; \\ \text{if } |\alpha_n - \theta_0| = |\alpha_n - \theta_1|, & \text{choose at random between } \theta_0 \text{ and} \\ & \theta_1 \text{ with probabilities } W_0 \text{ and } W_1. \end{cases} \tag{3.20}$$

Let us suppose that $\theta_0 < \theta_1$. Clearly,

$$P(|\alpha_n - \theta_0| \le |\alpha_n - \theta_1| \,|\theta = \theta_1) = P\left(\alpha_n \le \frac{\theta_0 + \theta_1}{2}\Big|\theta = \theta_1\right), \tag{3.21}$$

and thus by assumption,

$$\lim_{n\to\infty} P(|\alpha_n - \theta_0| \le |\alpha_n - \theta_1| \,|\theta = \theta_1) = 0. \tag{3.22}$$

Similarly,

$$\lim_{n\to\infty} P(|\alpha_n - \theta_1| \le |\alpha_n - \theta_0| \,|\theta = \theta_0) = 0. \tag{3.23}$$

Thus the error of the decision in question tends to zero for $n \to \infty$. A fortiori, the error of the standard decision tends to zero which implies by theorem 1 that (3.18) holds. This proves the "only if" part of theorem 5. Conversely, if (3.18) holds and θ_0 and θ_1 are different real numbers, let us choose a sequence n_r such that the series $\sum_{r=1}^{\infty} \epsilon_{n_r}$ is convergent. Then by the Borel-Cantelli lemma the standard decisions Δ_{n_r} tend with probability 1 to θ. More exactly, $\Delta_{n_r} = \theta$ for all but a finite number of values of r. Now if we put $\alpha_n = \Delta_{n_r}$ for $n_r \le n < n_{r+1}$, then the sequence of statistics $\alpha_n = \alpha_n(\xi_1, \cdots, \xi_n)$ tends with probability 1 to θ.

If the variables ξ_n have a discrete distribution, all our results remain valid; only the quantity λ_k has to be defined accordingly. If ξ_n can take the values $a_1, \cdots, a_l, \cdots$ and if

$$P(\xi_k = a_l|\theta = \theta_0) = p_{k,l}, \tag{3.24}$$

whereas

$$P(\xi_k = a_l|\theta = \theta_1) = q_{k,l}, \tag{3.25}$$

write

$$\lambda_k = \sum_{l=1}^{\infty} \sqrt{p_{k,l}q_{k,l}}. \tag{3.26}$$

All our results remain valid for this case.

4. Some examples

EXAMPLE 1. Let us suppose that $f_k(x) = f(x)$ and $g_k(x) = g(x)$, $(k = 1, 2, \cdots)$; that is, the random variables ξ_k are identically distributed under condition $\theta = \theta_i$, $(i = 1, 2)$. Suppose further that

$$1 > \lambda = \int_{-\infty}^{+\infty} \sqrt{f(x)g(x)}\, dx > 0. \tag{4.1}$$

Then it follows from theorem 2 that

$$H(\theta) - I_n \le B\sqrt{W_0W_1}\lambda^n, \tag{4.2}$$

and thus the missing information tends exponentially to zero for $n \to \infty$. This question was treated in [2] and [3]. In [4] we have been concerned with the special case where the random variables ξ_n take only the values 0 and 1, and where one has

$$P(\xi_n = 1|\theta = \theta_j) = p_j, \qquad P(\xi_n = 0|\theta = \theta_j) = q_j = 1 - p_j \tag{4.3}$$

for $j = 1, 2$. We have shown that the smallest value of λ for which (4.2) holds is $\lambda = 2^{-d(p_1,p_2)}$ where

$$d(p_1, p_2) = \rho \log \frac{\rho}{p_1} + (1 - \rho) \log \frac{1-\rho}{1-p_1} = \rho \log \frac{\rho}{p_2} + (1 - \rho) \log \frac{1-\rho}{1-p_2} \tag{4.4}$$

with

$$\rho = \frac{\log (q_1/q_2)}{\log (p_2 q_1/p_1 q_2)}. \tag{4.5}$$

In the special case where $p_1 = q_2 = p$, $q_1 = p_2 = q = 1 - p$, one has simply $\lambda = 2\sqrt{pq}$.

EXAMPLE 2. Let ξ_n be normally distributed with variance S_n^2 and with mean $m = m_0$ or $m = m_1 \neq m_0$ according to whether $\theta = \theta_0$ or $\theta = \theta_1$. We want to find the true value of m. Clearly

$$\lambda_k = \exp\left(-\frac{(m_0 - m_1)^2}{8S_k^2}\right). \tag{4.6}$$

It follows that we have full information on m if and only if the series $\sum_{k=1}^{\infty} (1/S_k^2)$ is divergent. Note that the statistic

$$\eta_n = \frac{\sum_{k=1}^{n} (\xi_k/S_k^2)}{\sum_{k=1}^{n} (1/S_k^2)}, \tag{4.7}$$

being the unbiased linear estimate of m with the least variance, is normally distributed with mean m and variance $(\sum_{k=1}^{n} (1/S_k^2))^{-1}$. If $\sum_{k=1}^{\infty} (1/S_k^2) = +\infty$, then η_n converges in probability to m and a suitably chosen subsequence η_{n_r}, such that $\sum_{r=1}^{\infty} (\sum_{k=1}^{n_r} (1/S_k^2))^{-1} < \infty$, converges with probability 1 to m.

EXAMPLE 3. Let an urn contain $a + b$ balls ($a > 0$, $b > 0$, $a \neq b$) of which either a are red and b white, or conversely, b are red and a white. Suppose that both cases have the prior probability $\frac{1}{2}$. We draw a ball from the urn, notice its color and put it back, and add, independently of the color of the ball drawn, $c_1 \geq 1$ red balls. After mixing the balls we draw again a ball, notice its color and put it back, adding $c_2 \geq 1$ red balls.

Let us continue this process indefinitely so that after the n-th step $c_n \geq 1$ new red balls are added. Can we determine with probability 1 the original composition of the urn?

Let us put $\xi_n = 1$ if the ball drawn at the n-th step is white and $\xi_n = 0$ if it is red. Let θ denote the number of white balls contained originally in the urn; thus $\theta_0 = b$ and $\theta_1 = a$. In this case

$$p_{k,1} = P(\xi_k = 1|\theta = \theta_0) = \frac{b}{a + b + \sum_{i=1}^{k-1} c_i} \tag{4.8}$$

and

$$q_{k,1} = P(\xi_k = 1 | \theta = \theta_1) = \frac{a}{a + b + \sum_{i=1}^{k-1} c_i}. \tag{4.9}$$

Putting $N_k = a + b + \sum_{i=1}^{k-1} c_i$, one has

$$\lambda_k = \frac{\sqrt{ab}}{N_k} + \sqrt{\left(1 - \frac{a}{N_k}\right)\left(1 - \frac{b}{N_k}\right)}, \tag{4.10}$$

and thus

$$\lambda_k = 1 - \frac{\frac{a+b}{2} - \sqrt{ab}}{N_k} + 0\left(\frac{1}{N_k^2}\right). \tag{4.11}$$

Therefore, $\sum_{k=1}^{\infty} (1 - \lambda_k)$ is divergent or convergent according to whether $\sum_{k=1}^{\infty} (1/N_k)$is divergent or convergent. Thus $\sum_{k=1}^{\infty} (1/N_k) = +\infty$, for example if $c_n = 1$ for $n = 1, 2, \cdots$, one can find out the original composition of the urn with probability 1, whereas if $\sum_{k=1}^{\infty} (1/N_k) < \infty$ (for instance if $(c_n = n)$ this is impossible.

EXAMPLE 4. Suppose that ξ_n has under the condition that $\theta = \theta_i$ a Poisson distribution with mean value $\alpha_i \delta_n$, $(i = 1, 2)$ where $\alpha_1 > 0$, $\alpha_2 > 0$, $\alpha_1 \neq \alpha_2$ and $\delta_n > 0$, $\lim_{n\to\infty} \delta_n = 0$. In this case

$$\lambda_k = \exp\left(-\left(\frac{\alpha_1 + \alpha_2}{2} - \sqrt{\alpha_1\alpha_2}\right)\delta_k\right). \tag{4.12}$$

Clearly, $\sum_{k=1}^{\infty} (1 - \lambda_k)$ is divergent if and only if $\sum_{k=1}^{\infty} \delta_k$ is divergent. Thus, for instance, if $\delta_k = (\gamma/k)$, $(k = 1, 2, \cdots; \gamma > 0)$, it is possible to decide with probability 1 whether $\theta = \theta_1$ or $\theta = \theta_2$, but if $\delta_k = (\gamma/k^2)$, this is not possible.

EXAMPLE 5. Let ξ_n have a binomial distribution of fixed order N and parameter $\delta_n\theta_i$ under the condition that $\theta = \theta_i$, $(i = 0, 1)$ where $\theta_0 > 0$, $\theta_1 > 0$, $\theta_0 \neq \theta_1$. Then we can easily see that the observations ξ_n, $(n = 1, 2, \cdots)$ contain full information on θ if and only if $\sum_{k=1}^{\infty} \delta_n = +\infty$.

EXAMPLE 6. Let η_n have the density function $f(x)$ where $f(x)$ is everywhere positive and has a continuous derivative such that $\int_{-\infty}^{+\infty} (f'^2(v)/f(v))\, dv < +\infty$. Suppose that $\xi_n = m_0 + c_n\eta_n$ if $\theta = \theta_0$ and $\xi_n = m_1 + c_n\eta_n$ if $\theta = \theta_1$ where $m_0 \neq m_1$ and $c_n \to \infty$. Writing $d = m_1 - m_0$, we have

$$1 - \lambda_k = \tfrac{1}{2} \int_{-\infty}^{+\infty} \left(\sqrt{f\left(u + \frac{d}{c_n}\right)} - \sqrt{f(u)}\right)^2 du. \tag{4.13}$$

Write $p(u) = \sqrt{f(u)}$; then we have

$$1 - \lambda_k = \tfrac{1}{2} \int_{-\infty}^{+\infty} \left(\int_u^{u+(d/c_n)} p'(v)\, dv\right)^2 du. \tag{4.14}$$

Thus

$$1 - \lambda_k \leq \frac{d^2}{8c_k^2} \int_{-\infty}^{+\infty} \frac{f'^2(v)}{f(v)}\, dv. \tag{4.15}$$

It follows that if the series $\sum_{k=1}^{\infty} (1/c_k^2) < +\infty$, then $\sum_{k=1}^{\infty} 1 - \lambda_k < +\infty$, and we do not get full information on θ.

On the other hand, the divergence of the series $\sum (1/c_k{}^2)$ ensures that the sequence of observations $\{\xi_n\}$ does contain full information on θ. As a matter of fact, since $p'(x)$ is continuous, there exists an interval (a, b) in which $p'(x)$ does not change sign and $|p'(x)| \geq \delta > 0$. Letting $h_n = d/c_n$, it follows that

$$1 - \lambda_n \geq \delta^2(b - a - h_n)h_n^2. \tag{4.16}$$

Thus if $\sum_{n=1}^{\infty} (1/c_n^2) = +\infty$, then $\sum_{n=1}^{\infty} (1 - \lambda_n) = +\infty$.

5. Further remarks

S. Kakutani [5] has proved the following.

THEOREM K. *Let $(\Omega_n, \mathcal{A}_n)$ be measurable spaces $(n = 1, 2, \cdots)$. Let Ω be the product space $\prod_{n=1}^{\infty} \Omega_n$. Let μ_n and ν_n be two equivalent probability measures on Ω_n, and let μ and ν denote the product measures $\mu = \prod_{n=1}^{\infty} \mu_n$ and $\nu = \prod_{n=1}^{\infty} \nu_n$ on Ω. Then the measures μ and ν are either equivalent or orthogonal according to whether the infinite product $\prod_{n=1}^{\infty} \rho(\mu_n, \nu_n)$ is convergent or divergent. Here $\rho(\mu_n, \nu_n)$ denotes the Hellinger distance of the measures μ_n and ν_n; that is, if m is any measure such that μ_n and ν_n are both absolutely continuous with respect to m,*

$$\rho(\mu_n, \nu_n) = \int_{\Omega_n} \sqrt{\frac{d\mu_n}{dm} \cdot \frac{d\nu_n}{dm}}\, dm \tag{5.1}$$

where $(d\mu_n/dm)$ and $(d\nu_n/dm)$ are Radon-Nikodym derivatives.

(The value of ρ is clearly independent of the choice of m.)

Obviously our results are closely connected with the above mentioned theorem of S. Kakutani. (My thanks are due to Professor K. Jacobs for calling my attention to this fact.)

As a matter of fact, according to Kakutani's theorem, if μ_n and ν_n denote the measures with density $f_n(x)$ and $g_n(x)$ with respect to the Lebesgue measure on the real line, then according to whether $\theta = \theta_0$ or $\theta = \theta_1$, one obtains in the sequence-space $x = (\xi_1, \xi_2, \cdots, \xi_n, \cdots)$ the product measure μ or ν. Further, our λ_k is equal to the Hellinger distance ρ_k of μ_k and ν_k. It follows that $\mu \sim \nu$ or $\mu \perp \nu$ according to whether $\prod_{k=1}^{\infty} \lambda_k > 0$ or $\prod_{k=1}^{\infty} \lambda_k = 0$.

If $\prod_{k=1}^{\infty} \lambda_k = 0$, that is, if $\mu \perp \nu$, then there exists a measurable set A in X such that $\mu(A) = 1$ and $\nu(A) = 0$. In this case given the infinite sequence $\{\xi_n\}$ one can clearly decide with probability 1 whether $\theta = \theta_0$ or $\theta = \theta_1$: if the infinite sequence $(\xi_1, \cdots, \xi_n, \cdots) \in A$ we decide for $\theta = \theta_0$, and in the opposite case for $\theta = \theta_1$. Now one can obviously find a sequence of sets A_n such that A_n is a cylinder set of X, its base belonging to the finite product set $\prod_{k=1}^{n} \Omega_k$, such that $\mu(A_n) \to 1$ and $\nu(A_n) \to 0$ for $n \to +\infty$. Thus if after observing $\xi_1, \cdots, \xi_n$ we decide for θ_0 or θ_1 according to whether or not the point $(\xi_1, \cdots, \xi_n)$ belongs to the base of A_n, the error of our decision tends to 0 for $n \to \infty$.

Conversely, it is easy to see that such a sequence of sets cannot exist if $\mu \sim \nu$.

Thus, in view of the corollary of theorem 1, theorem 4 can be deduced from theorem K. Note, however, that our direct approach is not only more elementary than this one via Kakutani's theorem, it also gives somewhat more, as it shows not only that $\epsilon_n \to 0$ if and only if $\lim_{n\to\infty} \prod_{k=1}^{n} \lambda_k = 0$, but also gives estimates between these quantities for finite values of n.

REFERENCES

[1] D. V. Lindley, "On a measure of the information provided by an experiment," *Ann. Math. Statist.*, Vol. 27 (1956), pp. 986–1005.

[2] A. Rényi, "On the amount of information concerning an unknown parameter in a sequence of observations," *Publ. Math. Inst. Hungar. Acad. Sci.*, Vol. 9 (1964), pp. 617–624.

[3] ———, "On the amount of missing information and the Neyman-Pearson lemma," *Festschrift for J. Neyman*, Wiley, 1966, pp. 281–288.

[4] ———, "On the amount of information in a frequency-count," *The 35th Session of the International Statistical Institute at Beograd*, 1965, pp. 1–8.

[5] S. Kakutani, "On equivalence of infinite product measures," *Ann. of Math.*, Vol. 49 (1948), pp. 214–226.

APPROXIMATIONS IN INFORMATION THEORY

MILLU ROSENBLATT-ROTH
BUCHAREST UNIVERSITY

1. Introduction

In the papers [8]–[11], [14] the author studied stochastic processes and channels, stationary, or nonstationary, with discrete time and arbitrary sets of states. In these papers, for regular processes and channels, two basic theorems of Shannon type [15] are proved for the case in which the states of the process and the channel input states are discrete and the output states arbitrary.

In this study, the essential role of the differential entropy of probability fields, processes, and channels appears. Obviously, if the sets of states are discrete, instead of the differential entropy, the correspondent entropy appears. Here we study the problem of approximation of processes with continuous sets of states by discrete processes and also of channels with continuous input-sets by channels with discrete input-sets.

In this study an essential role is played by the concept of ϵ-entropy of a set, of a probability field, of a channel, and of a complex source-channel. We may observe also the role played by differential entropy in the approximation problem. The constructions used here in the approximation problems are such that the essential properties of the given object are preserved.

2. The differential entropy of probability fields

Let us consider the measure space (X, S, μ) where X is a set of elements x and S a σ-algebra of subsets of X and μ a measure in S. Over X let us consider a probability field A, defined by the probability density $p(x)$ with respect to μ. By M we denote the expectation.

DEFINITION 2.1. *The value* $h(A) = -M \log p(x)$ *is the differential entropy of* A *with respect to* μ.

Obviously, $h(A)$ exists only if $M |\log p(x)| < +\infty$, and from $|h(A)| < M |\log p(x)|$ it follows that in this case it is finite.

Let (X, S, μ), (Y, Σ, ν) be measure spaces, $\pi(x, y)$ the probability density of some field C over their product, A the field defined by the probability density $p(x)$ induced by $\pi(x, y)$ in X, and $q_x(y) = \pi(x, y)/p(x)$ the conditional probability density of some probability field B_x over Y. We denote $C = AB$ (the union).

DEFINITION 2.2. *The conditional differential entropy of B with respect to A, for a given measure ν is* $h_A(B) = h(B|A) = Mh(B_x)$.

THEOREM 2.1. *If from $h(A)$, $h_A(B)$, $h(AB)$, two exist, then the third of them exists also and* $h(AB) = h(A) + h_A(B)$.

The proof is analogous with that of the corresponding theorem of the entropy case, using Fubini's theorem.

Different properties of the differential entropy of a field may be found in [8]–[11], [14]. In another publication by the same author will be given an axiomatic approach to the differential entropy of probability fields.

3. The approximation of probability fields

3.1. *The ϵ-entropy of a set.* Let us consider the measure space (X, S, μ), separable with respect to the distance $\rho(x, y)$, $x \in X$, $Z \in S$.

DEFINITION 3.1.1. *The sequence θ of measurable sets $Z_i \in S$, $(1 \leq i \leq n)$ is a* cover *of X if* (a) *these sets are nonoverlapping, and* (b) *X is their sum.*

DEFINITION 3.1.2. *The sequence θ_ϵ of measurable sets $Z_i^\epsilon \in S$, $(1 \leq i \leq n)$ is an* ϵ-cover *of X if it is a cover of X, and if $d(Z_i^\epsilon) \leq 2\epsilon$, $(1 \leq i \leq n)$, (d = the diameter).*

DEFINITION 3.1.3. *The space X is* centering *if in it, for every set $Z \subset X$ with $d(Z) = 2r$, there exists an element x, the center of Z, for which $\rho(x, y) \leq r$ for any $y \in Z$.* (See [6], p. 8.)

We may proceed as if every separable metric space were centering. Indeed, in [6] by means of the known theorem of Mazur-Banach ([1], chapter XI, section 8, theorem 10) and of theorem VI from ([6], section 1), it is proved that *every separable metric space X may be imbedded in a centering space X_** [16]. For totally bounded spaces let us denote by $N_\epsilon(X)$ the minimal number of elements in any ϵ-cover θ_ϵ.

DEFINITION 3.1.4. *The number $K'_\epsilon(X) = \log N_\epsilon(X)$ is the (minimal) ϵ-entropy of the set X.* (See [6], [16].)

DEFINITION 3.1.5. *The number* $K_\epsilon(X) = \log [N_\epsilon(X)/\mu(X)] = K'_\epsilon(X) - \log \mu(X)$ *is the normed (minimal) ϵ-entropy of the set X.*

3.2. *The discrete ϵ-entropy of a probability field.*

We shall use the following symbols:

(i) $D(X)$ will denote the totality of probability fields over (X, S); A and A' will be elements of $D(X)$;

(ii) if x and x' are elements of X, the probability density of A with respect to μ will be denoted by $p(x)$, and the corresponding conditional probability will be written as $p(x|x')$;

(iii) $D^0(X)$ will denote the totality of discrete fields with states $x_i \in X$, and $I(A, A') = h(A) - h(A|A')$.

DEFINITION 3.2.1. *If $Z_{x_i}^\epsilon$ is the sphere in X with center x_i and radius ϵ, let $W_\epsilon(AA')$ denote the property that for every state x_i of the discrete field $A' \in D^0(X)$ the condition*

$$P_{AA'}\{Z^{\epsilon}_{x_i}|x_i\} = 1 \tag{3.2.1}$$

is satisfied.

If the property $W_\epsilon(AA')$ is satisfied, the conditional field A_{x_i} possesses a set of states $Z_i \subset Z^{\epsilon}_{x_i}$; obviously, we may consider $Z_i \in \theta_\epsilon$, where θ_ϵ is any ϵ-cover of X.

DEFINITION 3.2.2. *The quantity $H_\epsilon(A) = \inf I(A, A')$, where the lower bound is considered for all pairs AA' for which the property $W_\epsilon(AA')$ is satisfied for the given A, is the discrete ϵ-entropy of the field A.*

THEOREM 3.2.1. *The discrete ϵ-entropy $H_\epsilon(A)$ is equal to $H(A) + K_\epsilon(X)$.*

PROOF. (a) From the definition of $H_\epsilon(A)$ it follows that $H_\epsilon(A) = h(A) - \sup h(A|A')$ where the upper bound is taken over all pairs AA' for which the property $W_\epsilon(AA')$ is satisfied for the given A; consequently, we must prove only that $\sup h(A|A') = -K_\epsilon(X)$. We shall prove that for any given probability field A' we may construct another probability field A^0 so that $h(A|A') \leq h(A|A^0) = -K_\epsilon(X)$.

(b) If θ_ϵ is any ϵ-cover of X, $Z_i \in \theta_\epsilon$ $(1 \leq i \leq n)$, and x_i the center of $Z_i (1 \leq i \leq n)$, let us consider the probability field A' with elementary events x_i and any arbitrarily determined probabilities $P(x_i)$, $(1 \leq i \leq n)$. Obviously, if $p(x|x_i)$ does not vanish only for $x \in Z_i$, the condition $W_\epsilon(AA')$ is satisfied. In this case $h(A|x_i) \leq \log \mu(Z_i)$; $h(A|A') \leq \sum_{i=1}^{n} P(x_i) \log \mu(Z_i)$.

(c) In the same conditions as above, if we consider $p(x|x_i) = 1/\mu(Z_i)$ for $x \in Z_i$ and zero in the rest, we define the field A'' so that $h(A|x_i) = \log \mu(Z_i)$;

$$h(A|A'') = \sum_{i=1}^{n} P(x_i) \log \mu(Z_i). \tag{3.2.2}$$

(d) If θ'_ϵ is any ϵ-cover of X, $Z'_i \in \theta'_\epsilon$, $(1 \leq i \leq n')$, with all elements Z'_i of the same μ-measure defined by

$$u = \sum_{i=1}^{n} P(x_i)\mu(Z_i) \tag{3.2.3}$$

and n' given by the entire part of $\mu(X)/u$, let us define the probability field A''' with elementary events x'_i, $(1 \leq i \leq n')$ the centers of Z'_i, and any arbitrarily determined probabilities $P(x'_i)$, $(1 \leq i \leq n')$. If we consider that $p(x|x'_i) = 1/u$ for $x \in Z'_i$ and zero in the rest, then obviously the condition $W_\epsilon(AA''')$ is satisfied and

$$h(A|x'_i) = \log u; \; h(A|A''') = \log u. \tag{3.2.4}$$

From the convexity of the function $\log x$ we obtain the inequality

$$\sum_{i=1}^{n} P(x_i) \log \mu(Z_i) \leq \log u \tag{3.2.5}$$

so that $h(A|A') \leq h(A|A'') \leq h(A|A''') \leq \log u \leq \log (\mu(X)/n')$.

(e) Let us consider any ϵ-cover θ^0_ϵ of X, $Z^0_i, \in \theta^0_\epsilon$, $(1 \leq i \leq n_0 = N_\epsilon(X))$, with $\mu(Z^0_i) = \mu(X)/n_0$, $(1 \leq i < n_0)$, x^0_i the centers of Z^0_i, $(1 \leq i \leq n_0)$, and $P(x^0_i)$ any arbitrarily given probabilities. We define the probability field A^0 by $p(x|x_i^0) = 1/\mu(Z_i^0)$ for $x \in Z_i^0$ and zero in the rest. Obviously,

$$h(A|A^0) = \sum_{i=1}^{n} P(x_i^0) \log \mu(Z_i^0) = \log \frac{\mu(X)}{n_0} = -K_\epsilon(X). \tag{3.2.6}$$

Because $n_0 < n'$, it follows that $h(A|A''') \leq h(A|A^0) = -K_\epsilon(X)$ so that

$$h(A|A') \leq h(A|A'') \leq h(A|A''') \leq h(A|A^0) = -K_\epsilon(X). \tag{3.2.7}$$

Consequently, the upper bound of $h(A|A')$ for all A' which satisfies the condition $W_\epsilon(AA')$ is equal to the upper bound of $h(A|A^0)$, that is to $-K_\epsilon(X)$, and our theorem is proved.

THEOREM 3.2.2. *Taking the upper bound for* $A \in D(X)$ *one has* $\sup H_\epsilon(A) = K'_\epsilon(X)$.

PROOF. Analogously, as for the entropy, it is easy to see that the upper bound of $h(A)$ for $A \in D(X)$ is $\log \mu(X)$; from theorem 3.2.1 it follows that

$$\sup H_\epsilon(A) = \log \mu(X) + K_\epsilon(X) = K'_\epsilon(X). \tag{3.2.8}$$

Let us suppose that $p(x)$ is a uniformly continuous function.

THEOREM 3.2.3. *For any* $\epsilon > 0$, *for a given probability field* $A \in D(X)$ *which possesses finite differential entropy* $h(A)$ *there exists a discrete probability field* $A_\epsilon \in D^0(X)$ *with states not depending on* A, *such that* (a) *the property* $W_\epsilon(AA_\epsilon)$ *is satisfied*, (b) $H(A_\epsilon) = H_\epsilon(A) + o(1)$, *and* (c) $I(A, A_\epsilon) = H(A_\epsilon) + o(1)$.

PROOF. Let us consider any ϵ-cover θ_ϵ^0 of X, $Z_i^0 \in \theta_\epsilon^0$, $(1 \leq i \leq n_0 = N_\epsilon(X))$, with $\mu(Z_i^0) = \mu(X)/n_0$. We define the field $A_\epsilon \in D^0(X)$ with the elementary events x_i^0 (the centers of Z_i^0) and $P_{A_\epsilon}(x_i^0) = P_A(Z_i^0) = p_i\mu(Z_i^0)$, where

$$p_i \in [\inf p(x), \sup p(x)], \tag{3.2.9}$$

and the lower and upper bounds are considered for $x \in Z_i^0$. We define the conditional probability field $(A|x_i^0)$ by means of $p_{A|A_\epsilon}(x|x_i^0) = 1/\mu(Z_i^0)$ for $x \in Z_i^0$ and zero in the rest. Obviously, $W_\epsilon(AA_\epsilon)$ is satisfied and $h(A|x_i^0) = h(A|A_\epsilon) = -K_\epsilon(X)$,

$$\begin{aligned} H(A_\epsilon) &= -\sum_{i=1}^{n_0} p_i\mu(Z_i^0) \log [p_i\mu(Z_i^0)] \\ &= -\sum_{i=1}^{n} (p_i \log p_i)\mu(Z_i^0) + K_\epsilon(X) \\ &= h(A) + K_\epsilon(X) + o(1). \end{aligned} \tag{3.2.10}$$

Using theorem 3.2.1, (b) follows and

$$\begin{aligned} I(AA_\epsilon) = h(A) - h(A|A_\epsilon) &= h(A) + K_\epsilon(X) = H_\epsilon(A) \\ &= H(A_\epsilon) + o(1). \end{aligned} \tag{3.2.11}$$

4. The differential entropy of stochastic processes

4.1. *Generalities.* Let us denote: (i) I = the set of all entire numbers; (ii) I^+ = the set of all natural numbers; (iii) $(X_\tau, S_\tau, \mu_\tau)$ = a measure space $(\tau \in I)$,

$x_\tau \in X_\tau$, $Z_\tau \in S_\tau$; (iv) $\alpha \subset I$ = a finite set of $|\alpha|$ numbers $\tau_i \in I$; (v) $(X^\alpha, S^\alpha, \mu^\alpha) = \times_{\tau \in \alpha} (X_\tau, S_\tau, \mu_\tau)$; (vi) $x^\alpha = (x_{\tau_1}, x_{\tau_2}, \cdots, x_{\tau|\alpha|}) = \{x_\tau, \tau \in \alpha\}$, $Z^\alpha \in S^\alpha$; (vii) $\alpha^* = I - \alpha$, $(X^{\alpha^*}, S^{\alpha^*}) = \times_{\tau \in \alpha^*} (X_\tau, S_\tau)$; (viii) $(X, S) = \times_{\tau \in I} (X_\tau, S_\tau)$, $x \in X$, $Z \in S$.

Let us consider that in the spaces X^α there exists a stochastic process A, that is, a consistent system of probability measures $P^\alpha(Z^\alpha)$, and let us denote by $P(Z)$ the extension of the measures $P^\alpha(Z^\alpha)$ in X.

We shall suppose that the measures P^α are μ^α-absolutely continuous, and let us denote by $\pi^\alpha(x^\alpha)$ the probability density; we also denote by $\pi^{\{\beta|\alpha\}}(x^\beta|x^\alpha)$ the probability density of $P^{\{\beta|\alpha\}}(Z^\beta|x^\alpha)$.

If $|\alpha| \cdot f^\alpha(x) = -\log \pi^\alpha(x^\alpha)$, it follows that $|\alpha| \cdot Mf^\alpha(x) = h(A^\alpha)$ with $A^\alpha = [\pi^\alpha(x^\alpha), X^\alpha, \mu^\alpha]$. If $\alpha_n = [t, t + n - 1]$, let us denote

$$(4.1.1)\qquad \begin{aligned} p^{\alpha_{n+1}}(x) &= \pi^{\{t+n|\alpha_n\}}(x_{t+n}|x^{\alpha_n}) = \pi^{\alpha_{n+1}}(x^{\alpha_{n+1}})/\pi^{\alpha_n}(x^{\alpha_n});\\ g^{\alpha_n}(x) &= -\log p^{\alpha_n}(x);\ \beta_m = [t + m, t + n - 1];\\ \pi^{\{\alpha_m|\beta_m\}}(x^{\alpha_m}|x^{\beta_m}) &= \frac{\pi^{\alpha_n}(x^{\alpha_n})}{\pi^{\beta_m}(x^{\beta_m})};\\ |\beta_m| \cdot \varphi_{t,n}^{(m)}(x) &= -\log \pi^{\{\alpha_m|\beta_m\}}(x^{\alpha_m}|x^{\beta_m}). \end{aligned}$$

It follows that

$$(4.1.2)\qquad |\beta_m| \cdot M\varphi_{t,n}^{(m)}(x) = h(A^{\alpha_m}|A^{\beta_m}).$$

Let us also denote

$$(4.1.3)\qquad \lambda_t^{(m)}(A) = \lim_{n\to\infty} M\varphi_{t,n}^{(m)}(x), \qquad (t \in I,\ m \in I^+)$$

if this limit exists and is finite.

DEFINITION 4.1.1. *The limit* $h_t(A) = \lim_{n\to\infty} n^{-1}h(A^{\alpha_n})$ *(if it exists) is the differential entropy of the process A at the instant t.*

In [14] are given different properties of $h_t(A)$.

4.2. *The entropy stability.*

DEFINITION 4.2.1. *The stochastic process A possesses* (a) *the weak,* (b) *the strong, and* (c) *in the norm the entropy stability property at the instant t, if* $f^{\alpha_n}(x)$ *converges to* $h_t(A)$, *respectively,* (a) *in probability,* (b) *almost everywhere, and* (c) *in the norm in the Banach space* L_1.

We shall denote these properties by $E_t^{(i)}(A)$, $(i = 1, 2, 3)$.

THEOREM 4.2.1. *In order that the stochastic process A possesses the property* $E_t^{(i)}(A)$, *it is necessary and sufficient that the sequence* $\{g^{\alpha_k}(x)\}$ *satisfies the law of large numbers, respectively in* (a) *the weak sense* $(i = 1)$, (b) *the strong sense* $(i = 2)$, *and* (c) *in the norm* $(i = 3)$.

The proof is the same as in ([14], theorem 1.2).

THEOREM 4.2.2. *If one of the properties* $E_t^{(i)}(A)$, $E_{t+m}^{(i)}(A)$, $(m \in I^+)$ *is satisfied, then in order that the other property be satisfied also, it is necessary and sufficient that the convergence of* $\varphi_{t,n}^{(m)}(x)$ *to* $\lambda_t^{(m)}(A)$ *holds* (a) *in the probability* $(i = 1)$, (b) *almost everywhere* $(i = 2)$, *and* (c) *in the norm* $(i = 3)$.

The proof is the same as in ([14], theorem 3.4.).

DEFINITION 4.2.2. *If the property* $E_t^{(i)}(A)$ *is satisfied for all* $t \in I$, *then the property* $E^{(i)}(A)$ *is satisfied for* $(i = 1, 2, 3)$.

Obviously, from theorem 4.2.2 we may immediately obtain the necessary and sufficient conditions for $E^{(i)}(A)$, as in ([14], theorem 3.5). If A possesses only discrete sets of states X_t and finite $H(A_t)$, $(t \in I)$ (for example, if X_t is finite, $(t \in I)$), from the property $E_{t_0}^{(1)}(A)$ for an arbitrary $t_0 \in I$, property $E^{(1)}$ follows ([7], theorem 3.2). In [14] are given different properties of $E_t^{(3)}(A)$.

DEFINITION 4.2.3. *If* $h_t(A)$ *exists and has the same finite value for all* $t \in I$ *and the property* $E^{(1)}(A)$ *is satisfied, then* A *is regular.*

DEFINITION 4.2.4. *If* A, B *are two stochastic processes,* $I_t(A, B) = \lim_{n\to\infty} n^{-1}\cdot I(A^{\alpha_n}, B^{\alpha_n})$ *(if it exists) is the common quantity of information of* A, B, *at the instant* t.

For a stationary A, we denote $g_n(x) = g^{\alpha_n}(x)$.

THEOREM 4.2.3. *In order that the stationary stochastic process* A *possess the property* $E^{(i)}(A)$, *it is necessary and sufficient that the sequence* $\{g_n(U^n x)\}$ (U *is the shift operator) verify the law of large numbers, respectively in* (a) *the weak sense* $(i = 1)$, (b) *the strong sense* $(i = 2)$, *and* (c) *in the norm* $(i = 3)$.

In ([14], theorem 3.9 and 3.10) are given different sufficient conditions for $E^{(i)}$ $(i = 1, 3)$. Analogous results may be obtained for $E^{(2)}$. (The particular case of discrete sets of states was studied in ([2], [3]).)

5. The approximation of stochastic processes

5.1. *Notations.* Let us consider the sequence of measure spaces $(X_\tau, S_\tau, \mu_\tau)$, separable for the respective distances $\rho_\tau(x_\tau, y_\tau)$, and let us retain all the notations in 4.1. Further, let $\rho_\alpha(x^\alpha, y^\alpha) = \max_{\tau\in\alpha} \rho_\tau(x_\tau, y_\tau)$; $\rho(x, y) = \sup_{\tau\in I} \rho_\tau(x_\tau, y_\tau)$; $D(X)$ be the totality of stochastic processes over (X, S); $\theta_{(\tau)\epsilon}$, θ_ϵ^α, θ_ϵ, be ϵ-covers of the spaces X_τ, X^α, X, respectively. Obviously, $\theta_\epsilon^\alpha = \times_{\tau\in\alpha} \theta_{(\tau)\epsilon}$, $\theta_\epsilon = \times_{\tau\in I} \theta_{(\tau)\epsilon}$; that is, if $i^\alpha = \{i_\tau, \tau \in \alpha\}$, $i = \{i_\tau, \tau \in I\}$, $Z^\alpha \in \theta_\epsilon^\alpha$, $Z \in \theta_\epsilon$, there exist i^α and i such that $Z^\alpha = Z_{i^\alpha} = \times_{\tau\in\alpha} Z_{i_\tau}$, $Z = Z_i = \times_{\tau\in I} Z_{i_\tau}$, $Z_{i_\tau} \in \theta_{(\tau)\epsilon}(\tau \in I)$.

Let $D^0(X)$ denote the totality of discrete stochastic processes with states in $Z_\tau(\tau \in I)$. If $Z^\epsilon(x_{i_\tau})$, $Z^\epsilon(x_{i^\alpha})$, $Z^\epsilon(x_i)$ are spheres in X_τ, X^α, X, respectively, with centers x_{i_τ}, x_{i^α}, x_i and radius ϵ (for the distances ρ_τ, ρ^α, ρ), obviously

$$Z^\epsilon(x_{i^\alpha}) = \mathop{\times}_{\tau\in\alpha} Z^\epsilon(x_{i_\tau}), \qquad Z^\epsilon(x_i) = \mathop{\times}_{\tau\in I} Z^\epsilon(x_{i_\tau}). \tag{5.1.1}$$

Let us denote by $W_\epsilon(AA')$ the property that for every sample x_i of the discrete process $A' \in D^0(X)$ the condition $P_{AA'}\{Z^\epsilon(x_i)|x_i\} = 1$ holds; that is, the property $W_\epsilon(A^\alpha A'^\alpha)$ is satisfied for any $\alpha \subset I$.

5.2. *The* ϵ*-entropy of a sequence of sets.*

LEMMA 5.2.1. *The normed* ϵ*-entropy* $K_\epsilon(X^\alpha)$ *is equal to* $\sum_{\tau\in\alpha} K_\epsilon(X_\tau)$.

The proof follows from the definition of the distance $\rho_\alpha(x^\alpha, y^\alpha)$.

DEFINITION 5.2.1. *The quantity* $K_{t,\epsilon}(X) = \lim_{n\to\infty} n^{-1}K_\epsilon(X^{\alpha_n})$, *if it exists, is the normed* ϵ*-entropy of the sequence of sets* $\{X_\tau\}$ *at the instant* t.

DEFINITION 5.2.2. *If $K_{t,\epsilon}(X)$ exists and has the same finite value for all $t \in I$, then the sequence $\{X_\tau\}$ is regular.*

5.3. *The discrete ϵ-entropy of a stochastic process.*

DEFINITION 5.3.1. *The quantity $H_{t,\epsilon}(A) = \lim_{n\to\infty} n^{-1}H_\epsilon(A^{\alpha_n})$, if it exists, is the discrete ϵ-entropy of the stochastic process A at the instant t.*

We shall suppose that $H_{t,\epsilon}(A)$, $h_t(A)$, $K_{t,\epsilon}(X)$ exist and are finite for a fixed $t \in I$.

THEOREM 5.3.1. *The discrete ϵ-entropy $H_{t,\epsilon}(A)$ is equal to $h_t(A) + K_{t,\epsilon}(X)$.*

The proof follows from theorem 3.2.1.

THEOREM 5.3.2. *Let us assume that the stochastic process A possesses finite differential entropy $h_t(A)$, the property $E_t^{(i)}(A)$, $(i = 1, 2, 3)$, and that the normed ϵ-entropy $K_{t,\epsilon}(X)$ exists and is finite.*

Then, for any $\epsilon > 0$, there exists a discrete stochastic process $A_\epsilon \in D^0(X)$ with states not depending on A, such that

(a) *the property $W_\epsilon(AA_\epsilon)$ is satisfied;*

(b) *$H_{t,\epsilon}(A)$ and $H_t(A_\epsilon)$ exist, are finite, and $H_t(A_\epsilon) = H_{t,\epsilon}(A) + o(1)$;*

(c) *$I_t(A, A_\epsilon) = H_t(A_\epsilon) + o(1)$;*

(d) *A_ϵ possesses the corresponding property $E_t^{(i)}$, $(i = 1, 2, 3)$;*

(e) *if $\{X_\tau\}$ is regular, from the regularity of A follows that of A_ϵ;*

(f) *from the stationarity of A follows that of A_ϵ.*

PROOF. (a) In every X_τ let us consider an ϵ-cover $\theta_{(\tau)\epsilon}$ with $Z^{i_\tau} \in \theta_{(\tau)\epsilon}$ such that $\mu_\tau(Z_{i_\tau}) = \mu_\tau(X_\tau)/N_\epsilon(X_\tau)$. From the definition of ρ_α it follows that in this manner is generated an ϵ-cover θ_ϵ^α with $Z_{i^\alpha}^\epsilon = \times_{\tau\in\alpha} Z_{i_\tau}^\epsilon$, and $\mu^\alpha(Z_{i^\alpha}^\epsilon) = \mu^\alpha(X^\alpha)/N_\epsilon(X^\alpha)$ for all i^α.

Let us denote by $x_{i_\tau}^\epsilon$ the center of $Z_{i_\tau}^\epsilon$ and by $x_{i^\alpha}^\epsilon = \{x_{i_\tau}^\epsilon, \tau \in \alpha\}$ the center of $Z_{i^\alpha}^\epsilon$. We define the probability field $A_\epsilon^\alpha \in D^0(X^\alpha)$ with the elementary events $x_{i^\alpha}^\epsilon$ and

$$(5.3.1) \qquad P_{A_\epsilon^\alpha}(x_{i^\alpha}^\epsilon) = P_{A^\alpha}(Z_{i^\alpha}^\epsilon) = p_{i^\alpha}^\epsilon \cdot \mu(Z_{i^\alpha}^\epsilon)$$

where $p_{i^\alpha}^\epsilon \in [\inf p^\alpha(x^\alpha), \sup p^\alpha(x^\alpha)]$, with the lower and upper bounds taken for $x^\alpha \in Z_{i^\alpha}^\epsilon$.

For any $\tau \in \alpha$, we define the union $A_\tau A_{(\tau)\epsilon}$ by means of the probability density

$$(5.3.2) \qquad p_{A_\tau|A_{(\tau)\epsilon}}(x_\tau|x_{i_\tau}^\epsilon) = N_\epsilon(X_\tau)/\mu_\tau(X_\tau)$$

when $x_\tau \in Z_{i_\tau}^\epsilon$ $(\tau \in \alpha)$ and by zero in the remainder, and the union $A^\alpha A_\epsilon^\alpha$ by means of the probability density

$$(5.3.3) \qquad p_{A_\alpha|A_\epsilon^\alpha}(x^\alpha|x_{i^\alpha}^\epsilon) = \prod_{\tau\in\alpha} p_{A_\tau|A_{(\tau)\epsilon}}(x_\tau|x_{i_\tau}^\epsilon) = \frac{N_\epsilon(X^\alpha)}{\mu^\alpha(X^\alpha)} = \prod_{\tau\in\alpha} \frac{N_\epsilon(X_\tau)}{\mu_\tau(X_\tau)}$$

for $x^\alpha \in Z_{i^\alpha}^\epsilon$, and by zero in the rest.

Obviously, the properties $W_\epsilon(A_\tau A_{(\tau)\epsilon})$, $W_\epsilon(A^\alpha A_\epsilon^\alpha)$, and $W_\epsilon(AA_\epsilon)$ are satisfied.

(b) We obtain immediately, as in theorem 3.2.3, that

$$(5.3.4) \qquad \begin{aligned} h(A^\alpha|x_{i^\alpha}^\epsilon) &= h(A|A_\epsilon^\alpha) = -K_\epsilon(X^\alpha), \\ H(A_\epsilon^\alpha) &= h(A^\alpha) + K_\epsilon(X^\alpha) + o(1), \\ I(A^\alpha, A_\epsilon^\alpha) &= H(A_\epsilon) + o(1). \end{aligned}$$

We obtain immediately the results (b) and (c) if we recall the definitions of $h_t(A)$, $H_t(A_\epsilon)$, $H_{t,\epsilon}(A)$, $I_t(A, A_\epsilon)$.

(c) Obviously for $x \in Z_{i\alpha}^\epsilon$,

$$\begin{aligned} (5.3.5) \qquad f_{A_\epsilon}^\alpha(x_{i\alpha}^\epsilon) &= -n^{-1} \cdot \log P_{A_\epsilon^\alpha}(x_{i\alpha}^\epsilon) = -n^{-1} \cdot \log p_{i\alpha}^\epsilon + n^{-1} K_\epsilon(X^\alpha) \\ &= -n^{-1} \log p^\alpha(x) + n^{-1} K_\epsilon(X^\alpha) + o(1) \\ &= -n^{-1} \log p^\alpha(x) + K_{t,\epsilon}(X) + o(1), \end{aligned}$$

and consequently,

$$(5.3.6) \qquad f_{A_\epsilon}^\alpha(x_{i\alpha}^\epsilon) - H_t(A_\epsilon) = f_A^\alpha(x) - h_t(A) + o(1).$$

Because P_{A_ϵ} is derived from P_A, from $E_t^{(i)}(A)$ follows $E_t^{(i)}(A_\epsilon)$, $(i = 1, 2, 3)$.

The results (e) and (f) follow immediately from the construction of the stochastic process A_ϵ.

6. The approximation of stochastic transition functions

6.1. *The metric space of stochastic transition functions*

Let us denote: (X, S), (X', S') two measurable spaces; $x \in X$, $x' \in X'$, $Z \in S$, $T \in S'$, $R(X', S')$ the totality of probability measures $P'(T)$ with the domain of definition (X', S'), $R(X, S, X', S')$ the totality of stochastic transition functions $P(x, T)$ with the domain of definition (X, S, X', S').

If P' and P_1' are elements of $R(X', S')$, let us denote by $\beta'(P', P_1')$ the total variation of $P' - P_1'$.

DEFINITION 6.1.1. *If P and P_1 are elements of $R(X, S, X', S')$ and if for a given $x \in X$ we denote by $P(x, \cdot)$, $P_1(x, \cdot)$ the corresponding measures, elements in $R(X', S')$, we define*

$$(6.1.1) \qquad \beta(P, P_1) = \sup_{x \in X} \beta'[P(x, \cdot), P_1(x, \cdot)] = \sup_{x \in X, T \in S'} |P(x, T) - P_1(x, T)|.$$

DEFINITION 6.1.2. (See [4].) *The ergodic coefficient of $P \in R(X, S, X', S')$ may be defined by*

$$(6.1.2) \qquad \alpha(P) = 1 - \sup_{x, x_1 \in X} \beta'[P(x, \cdot), P(x_1, \cdot)].$$

Obviously, $0 \leq \beta(P, P_1) \leq 1$.

DEFINITION 6.1.3. *Two stochastic transition functions $P, P_1 \in R(X, S, X', S')$ are mutually almost singular, if for each $\epsilon > 0$ there exist some elements $x_\epsilon \in X$, $T_{x_\epsilon} \in S'$ such that $P(x_\epsilon, T_{x_\epsilon}) < \epsilon$, $P_1(x_\epsilon, T_{x_\epsilon}) < \epsilon$ where * denotes the complement.*

LEMMA 6.1.1. (a) *In order that $\beta(P, P_1) = 0$, it is necessary and sufficient that* $P \equiv P_1$; (b) *in order that $\beta(P, P_1) = 1$, it is necessary and sufficient that P, P_1 be mutually almost singular.*

PROOF. The proof of (a) is obvious; therefore, we shall prove only (b).

Necessity. If $\beta(P, P_1) = 1$, for any $\epsilon > 0$ there exist some $x_\epsilon \in X$ and some $T_{x_\epsilon} \in S'$, such that

$$(6.1.3) \qquad 1 - \epsilon < |P(x_\epsilon, T_{x_\epsilon}) - P_1(x_\epsilon, T_{x_\epsilon})| < 1.$$

From the equality

$$(6.1.4) \qquad P(x, T) - P_1(x, T) = -[P(x, T^*) - P_1(x, T^*)]$$

it follows that we may limit ourselves to the case where

$$(6.1.5) \qquad P(x_\epsilon, T_{x_\epsilon}) - P_1(x_\epsilon, T_{x_\epsilon}) < 0,$$

so that

$$(6.1.6) \qquad 1 - \epsilon < 1 - \epsilon + P(x_\epsilon, T_{x_\epsilon}) < P_1(x_\epsilon, T_{x_\epsilon});$$
$$P(x_\epsilon, T_{x_\epsilon}) < P_1(x_\epsilon, T_{x_\epsilon}) - 1 + \epsilon < \epsilon$$

that is $P(x_\epsilon, T_{x_\epsilon}) < \epsilon$, $P_1(x_\epsilon, T^*_{x_\epsilon}) < \epsilon$.

Sufficiency. If P and P_1 are mutually almost singular, for any $\epsilon > 0$ there exist some $x_\epsilon \in X$, $T_{x_\epsilon} \in S'$ such that the inequalities in definition 6.1.3 are satisfied, and consequently,

$$(6.1.7) \qquad 1 - 2\epsilon < 1 - \epsilon - P(x_\epsilon, T_{x_\epsilon}) < P_1(x_\epsilon, T_{x_\epsilon}) - P(x_\epsilon, T_{x_\epsilon})$$
$$< P_1(x_\epsilon, T_{x_\epsilon}) < 1,$$

that is, $\beta(P, P_1) = 1$.

LEMMA 6.1.2. *If P and P_1 belong to $R(X, S, X', S')$, then $|\alpha(P) - \alpha(P_1)| \leq 2\beta(P, P_1)$.*

PROOF. Let us suppose that $\alpha(P) \leq \alpha(P_1)$. Obviously

$$(6.1.8) \qquad |P(x, T) - P(x_1, T)| \leq |P(x, T) - P_1(x, T)| + |P_1(x, T) - P_1(x_1, T)|$$
$$+ |P_1(x_1, T) - P(x_1, T)|.$$

Taking everywhere the upper bound for all $x \in X$, $x_1 \in X$, $T \in S'$, it follows immediately that $\alpha(P_1) - \alpha(P) \leq 2\beta(P, P_1)$, which proves the theorem.

THEOREM 6.1.1. *The space $R(X, S, X', S')$ is a complete metric space for the distance $\beta(P, P_1)$.*

PROOF. (a) *The function $\beta(P, P_1)$ is a distance.* The function β is symmetric, and in lemma 6.1.1 we have seen that from $\beta(P, P_1) = 0$ it follows that $P = P_1$. Let us consider $P_i(x, T) \in R(X, S, X', S')$, $(i = 1, 2, 3)$ and $|P_i(x, T) - P_j(x, T)| = u_{i,j}(x, T)$, $(i = 1,\ j = 2;\ i = 2,\ j = 3;\ i = 3,\ j = 1)$. From $u_{1,3} \leq u_{1,2} + u_{2,3}$, if we take everywhere the upper bound for $x \in X$, $T \in S'$, the triangular inequality follows for β.

(b) *The space $R(X, S, X', S')$ is complete.* Let $P_n(x, T)$, $(n \in I^+)$ be a β-fundamental sequence in this space, that is, $\beta(P_n, P_m) \to 0$, $(n, m \to \infty)$.

(b_1) From the definition of β it follows that the numerical sequence $P_n(x, T)$, $(n \in I)$ is fundamental for each pair of *fixed elements* $x \in X$, $T \in S'$, so that from the completeness of the real line there exists a limit $P(x, T)$ to which $P_n(x, T)$ converges as $n \to \infty$. From $P_n(x, T) \in R(X, S, X', S')$ it follows that $P(x, T) \in R(X, S, X', S')$.

(b_2) Because $P_n(x, T)$, $(n \in I)$ is a β-fundamental sequence, it follows that for any fixed $\epsilon > 0$ we may find a number $N = N(\epsilon)$ such that $\beta(P_n, P_m) < \epsilon$ for any $m, n \geq N(\epsilon)$, that is, $|P_n(x, T) - P_m(x, T), \leq \epsilon$ for all $x \in X$, $T \in S'$, and for all $m, n \geq N(\epsilon)$.

If m increases to infinity, from (b_1) it follows that $|P_n(x, T) - P(x, T)| \leq \epsilon$ for every fixed x, T with $n \geq N(\epsilon)$; that is, the convergence of P_n to P is uniform with respect to all $x \in X$, $T \in S'$ so that $\beta(P_n, P) \leq \epsilon$ for $n \geq N(\epsilon)$; that is, $\beta(P_n, P) \to 0$; in other words, the space $R(X, S, X', R')$ is complete.

THEOREM 6.1.2. *For β-convergence, the ergodic coefficient is continuous; that is, from $\beta(P_n, P) \to 0$ as $n \to \infty$ it follows that $\alpha(P_n) \to \alpha(P)$.*

The proof follows from lemma 6.1.2.

THEOREM 6.1.3. *The β-convergence is equivalent to convergence in distribution uniformly in $x \in X$, $T \in S'$.*

The proof follows from the definition of the distance β.

6.2. *The metric space of stochastic transition operators.* Let us consider some measurable space (X, S), and let us denote by V_X the Banach space of all real-valued generalized measures μ on the σ-algebra S, with norm $\|\mu\|$ one half of the total variation of μ. Obviously, for any probability measure μ, it follows that $\|\mu\| = \frac{1}{2}$.

Let us consider (see [4]) the subspace $L_X \subset V_X$ of all functions $\lambda \in V_X$ for which $\lambda(X) = 0$. In [4] it is proved that

$$\|\lambda\| = \sup_{Z \in S} |\lambda(Z)|. \tag{6.2.1}$$

If P, $P_1 \in R(X, S, X', S')$, it follows for any fixed $x \in X$ that $P(x, \cdot)$, $P_1(x, \cdot) \in V_{X'}$ and $\nu(x, \cdot) = P(x, \cdot) - P_1(x, \cdot) \in L_{X'}$ so that

$$\|\nu(x, \cdot)\| = \sup_{T \in S'} |\nu(x, T)|, \tag{6.2.2}$$

and consequently,

$$\beta(P, P_1) = \sup_{x \in X} \|\nu(x, \cdot)\|. \tag{6.2.3}$$

DEFINITION 6.2.1. *We define the stochastic transition operator Q which corresponds to the stochastic transition function $P(x, T)$ as a map from V_X to $V_{X'}$: $\mu' = Q\mu$, by means of the equality*

$$\mu'(T) = \int_X P(x, T)\, \mu(dx) \tag{6.2.4}$$

with $\mu \in V_X$, $\mu' \in V_{X'}$, $T \in S'$.

Obviously, Q is linear and continuous. If G_X is the subspace of all probability measures in V_X, it is obvious that Q maps G_X into $G_{X'}$ and its norm is one. If Q_1 corresponds to P_1 in the same manner as Q to P, let us consider the linear continuous operator $Q - Q_1$ which maps V_X into $L_{X'}$. We denote by $N(Q - Q_1)$ the norm of $Q - Q_1$, that is,

$$N(Q - Q_1) = \sup_{\mu \in V_X} \{\|(Q - Q_1)\mu\|/\|\mu\|\} = 2 \sup_{\mu \in V_X} \|(Q - Q_1)\mu\|. \tag{6.2.5}$$

LEMMA 6.2.1. *For any $\mu \in G_X$, if $\mu' = Q\mu \in G_{X'}$, $\mu_1' = Q_1\mu \in G_{X'}$, the inequality $\beta'(\mu', \mu_1') \leq \beta(P, P_1)$ is satisfied.*

PROOF. From the definition of μ', μ_1' it follows that for any $T' \in S'$,

$$(6.2.6) \qquad |\mu'(T) - \mu_1'(T)| \leq \int_X |P(x, T) - P_1(x, T)| \, \mu(dx) \leq \sup_{x \in X} |\nu(x, T)|$$

so that

$$(6.2.7) \qquad \beta'(\mu', \mu_1') = \sup_{T \in S'} |\mu'(T) - \mu_1'(T)| \leq \sup_{x \in X, T \in S'} |\nu(x, T)| = \beta(P, P_1).$$

THEOREM 6.2.1. *The norm $N(Q - Q_1)$ is equal to $2\beta(P, P_1)$.*

PROOF. (a) Let $e_x(Z)$, $(x \in X)$ denote the probability measure with $e_x(Z) = 1$, if $x \in Z$, and $e_x(Z) = 0$, if $x \in Z^*$, so that $e_x \in G_X$ for any fixed $x \in X$. If $e_x' = Q_{e_x}$, it follows that

$$(6.2.8) \qquad e_x'(T) = \int_X P(x_1, T) \, e_x(dx_1) = P(x, T).$$

Considering also the analogous relation corresponding to $e_{(1)x}' = Q_1 e_x$, it follows that

$$(6.2.9) \qquad [(Q - Q_1)e_x](T) = P(x, T) - P_1(x, T),$$

so that, using the definition of $N(Q - Q_1)$,

$$(6.2.10) \qquad N(Q - Q_1) \geq 2 \cdot \|Qe_x - Q_1 e_x\| = 2\|e_x' - e_{(1)x}'\| = 2 \cdot \|\nu(x, \cdot)\|$$

for any $x \in X$. This implies the inequality

$$(6.2.11) \qquad N(Q - Q_1) \geq 2 \cdot \sup_{x \in X} \|\nu(x, \cdot)\| = 2 \cdot \beta(P, P_1).$$

(b) If $\mu \in G_X$, then for any $T \in S'$,

$$(6.2.12) \qquad |[(Q - Q_1)\mu](T)| = \left|\int_X \nu(x, T) \, \mu(dx)\right| \leq \int_X \sup_{x \in X} |\nu(x, T)| \, \mu(dx) = \sup_{x \in X} |\nu(x, T)| \leq \beta(P, P_1).$$

Obviously $(Q - Q_1)\mu \in L_{X'}$. Let us suppose that X_+', X_-' are respectively the positive and negative sets of a Hahn decomposition of X' for the function $(Q - Q_1)\mu$. From the above inequality it follows in particular that $[(Q - Q_1)\mu] \, (X_+') \leq \beta(P, P_1)$. Consequently, it is easy to see that $[(Q - Q_1)\mu] \, (X_+') = -[(Q - Q_1)\mu] \, (X_-') = \|(Q - Q_1)\mu\|$ so that $\|(Q - Q_1)\mu\| \leq \beta(P, P_1)$ for any $\mu \in G_X$ and $N(Q - Q_1) \leq 2\beta(P, P_1)$, which proves our lemma. Obviously $N(Q - Q_1)$ is a distance in the metric space of all probability transition operators.

6.3. *Another expression of $\beta(P, P_1)$.* Let us consider the measurable space (X, S).

DEFINITION 6.3.1. *Between two measures, μ_1 and $\mu_2 \in V_X$, there exists the relation $\mu_1 \prec \mu_2$ if $\mu_1(Z) \leq \mu_2(Z)$ for any $Z \in S$.*

For $\rho \in V_X$ let us denote

$$(6.3.1) \qquad \sigma(\mu_1, \mu_2) = \sup_{\rho \;\; \mu_1, \rho \prec \mu_2} \rho(X).$$

A. N. Kolmogorov pointed out ([1], section 1) that $\sigma(\mu_1, \mu_2)$ may also be defined by

$$\sigma(\mu_1, \mu_2) = \inf \sum_{i=1}^{m} \min [\mu_1(Z_i), \mu_2(Z_i)] \tag{6.3.2}$$

where the lower bound is taken over all possible finite covers θ of X, and $Z_i \in \theta$, $(1 \leq i \leq m < \infty)$. It is known ([4], section 1) that $\|\mu_1 - \mu_2\| = 1 - \sigma(\mu_1, \mu_2)$. If $\mu_1 = P(x, \cdot)$, $\mu_2 = P_1(x, \cdot)$, we obtain $\|P(x, \cdot) - P_1(x, \cdot)\| = 1 - \sigma[P(x, \cdot), P_1(x, \cdot)]$ so that we obtain the following theorem.

THEOREM 6.3.1. *The following equalities hold:*

$$\begin{aligned} \beta(P, P_1) &= \sup_{x \in X} \|\nu(x, \cdot)\| \\ &= 1 - \inf_{x \in X} \sigma[P(x, \cdot), P_1(x, \cdot)]. \end{aligned} \tag{6.3.3}$$

If X, X' are denumerable sets with the states x_i, $(i \in I)$, then $P(x, T)$ and $P_1(x, T)$ are given by means of the stochastic matrices Q, Q_1 with elements $p_{k,m}$, $p_{k,m}^{(1)}$.

THEOREM 6.3.2. *The distance $\beta(P, P_1)$ is equal to*

$$\beta(P, P_1) = 1 - \inf_{1 \leq k < \infty} \sum_{m=1}^{\infty} \min (p_{k,m}, p_{k,m}^{(1)}). \tag{6.3.4}$$

PROOF. It is easy to see [4] that the expression given by A. N. Kolmogorov for $\sigma(\mu_1, \mu_2)$ does not change if we consider not only finite covers of X but also denumerable covers of it.

Let us observe that the sum in this expression cannot decrease if instead of the cover θ we consider another cover θ', finer than θ, that is, in which each set in θ is a sum of certain sets in θ'. Because the cover θ_0, each set of which contains only one element $Z_i = x_i$, is finer than any arbitrary cover θ, from the expression of σ it follows that

$$\sigma(\{p_{k,m}\}, \{p_{k,m}^{(1)}\}) = \sum_{m=1}^{\infty} \min (p_{k,m}, p_{k,m}^{(1)}). \tag{6.3.5}$$

From theorem 6.3.1, the desired result follows.

6.4. *The discrete case.* Here we shall prove theorem 6.2.1 using the definition of $\beta(P, P_1)$ from theorem 6.3.1.

Let us consider [4] the linear space F of those infinite dimensional vectors $q = \{q_i\}$, $(i \in I^+)$ for which the sum of the components vanishes and the sum of their absolute values converges. If $U = \{u_{i,j}\}$, $U_1 = \{u_{i,j}^{(1)}\}$ are some stochastic matrices and $q \in F$, then $Uq \in F$, $U_1 q \in F$. Let us define the norm of q by

$$\|q\| = \sum_{i=1}^{\infty} |q_i| = 2 \sum_{i=1}^{\infty} (q_i)^+ = -2 \cdot \sum_{i=1}^{\infty} (q_i)^- \tag{6.4.1}$$

where $(a)^+ = \max (a, 0)$, $(a)^- = \min (a, 0)$. Obviously, $(a + b)^+ \leq (a)^+ + (b)^+$. Let us denote

$$\beta(U, U_1) = 1 - \inf_{i} \sum_{j=1}^{\infty} \min (u_{i,j}, u_{i,j}^{(1)}). \tag{6.4.2}$$

THEOREM 6.4.1. *The norm $N(U - U_1)$ is equal to*

$$(6.4.3) \qquad N(U - U_1) = \sup_{q \in F} \frac{\|Uq - U_1 q\|}{\|q\|} = 2\beta(U, U_1).$$

PROOF. (a) First we shall prove that the number on the left side is not greater than the one on the right; for this it is sufficient to prove that for any $q \in F$, $\|Uq - U_1 q\| \leq 2\|q\| \cdot \beta(U, U_1)$.

(a_1) If q has only two nonvanishing components $q_{i_1} = \|q\|/2 = \lambda$, $q_{i_2} = -\lambda$, using the same method as in ([4], p. 372) and the notation

$$(6.4.4) \qquad \sum_{k=1}^{\infty}{}' = \sum_{l=i_1,i_2} \sum_{k=1}^{\infty},$$

we obtain the inequalities

$$\begin{aligned}
(6.4.5) \qquad & \|(U - U_1)q\| \\
&= 2 \sum_{k=1}^{\infty} [q_{i_1}(u_{i_1k} - u_{i_1k}^{(1)}) + q_{i_2}(u_{i_2k} - u_{i_2k}^{(1)})]^+ \\
&\leq 2 \sum_{k=1}^{\infty}{}' [q_l(u_{lk} - u_{lk}^{(1)})]^+ = \|q\| \sum_{k=1}^{\infty} [(u_{i_1k} - u_{i_1k}^{(1)})^+ + (u_{i_2k}^{(1)} - u_{i_2k})^+] \\
&= \|q\| \sum_{k=1}^{\infty} \{[u_{i_1k} - \min(u_{i_1k}, u_{i_1k}^{(1)})] + [u_{i_2k}^{(1)} - \min(u_{i_2k}, u_{i_2k}^{(1)})]\} \\
&= \|q\| \sum_{l=i_1,i_2} \left[1 - \sum_{k=1}^{\infty} \min(u_{lk}, u_{lk}^{(1)})\right] \leq 2\|q\| \cdot \beta(U, U_1).
\end{aligned}$$

(a_2) If q is any vector in F, it is easy to see that it may be represented as an absolute convergent sum

$$(6.4.6) \qquad q = \sum_{i=1}^{\infty} q^{(i)}$$

of vectors $q^{(i)} \in F$ in such a way that each vector $q^{(i)}$ has only two nonvanishing components, and also

$$(6.4.7) \qquad \|q\| = \sum_{i=1}^{\infty} \|q^{(i)}\|.$$

From (a_1) we obtain the relations

$$\begin{aligned}
(6.4.8) \qquad \|(U - U_1)q\| &\leq \sum_{i=1}^{\infty} \|(U - U_1)q^{(i)}\| \leq 2\beta(U, U_1) \cdot \sum_{i=1}^{\infty} \|q^{(i)}\| \\
&= 2\beta(U, U_1)\|q\|.
\end{aligned}$$

(b) We shall prove the inverse inequality.

(b_1) From the given definition of $\beta(U, U_1)$ it follows that for any $\epsilon > 0$ there exist two different numbers i_1, i_2 such that for $l = i_1, i_2$, the inequality

$$(6.4.9) \qquad \left|\sum_{j=1}^{\infty} \min(u_{l,j}, u_{l,j}^{(1)}) - [1 - \beta(U, U_1)]\right| < \epsilon$$

is satisfied.

The existence of one value i_1 with the indicated property follows from the definition of the lower bound; in the case where another value $i_2 \neq i_1$ with the indicated property does not exist, we may use the following method. Instead of the matrices U, U_1 with the states $\{x_i\}$, $(i = 1, 2, \cdots)$, we consider the matrices T, T_1 with the states $\{x_i\}$, $(i = 0, 1, 2, \cdots)$ where $t_{i,j} = u_{i,j}$, $t_{0,j} = u_{i_1,j}$, $(i, j = 1, 2, \cdots)$, $t_{i,0} = 0$ $(i = 0, 1, 2, \cdots)$, and analogously for T_1. Obviously, $\beta(U, U_1) = \beta(T, T_1)$; here $i_2 = 0$ and $i_1 \neq i_2 = 0$ have the desired property. Consequently, from the beginning we may suppose that U, U_1 possesses this property.

(b_2) For a fixed vector q which possesses only two components $q_{i_1} = \|q\|/2$, $q_{i_2} = -q_{i_1}$, from the inequalities in (a_1), using the inequality in (b), we obtain $\|(U - U_1)q\| \geq 2\|q\| [\beta(U, U_1) - \epsilon]$, or

$$N(U - U_1) \geq \frac{\|(U - U_1)q\|}{\|q\|} \geq 2\beta(U, U_1) - 2\epsilon, \tag{6.4.10}$$

and consequently, $N(U - U_1) \geq 2\beta(U, U_1)$.

THEOREM 6.4.2. *The following equalities hold:*

$$N(U - U_1) = 2 \cdot \beta(U, U_1) = \sup_i \sum_{k=1}^{\infty} |u_{i,k} - u_{i,k}^{(1)}|. \tag{6.4.11}$$

PROOF. We may observe that the next to last inequality in (b_2) shows that the upper bound in the last inequality in (b_2) is attained for vectors q which possess only two nonvanishing components. If F_1 is the totality of these vectors, it follows that

$$\begin{aligned} \|(U - U_1)q\| &= \|q\| \cdot \sum_{k=1}^{\infty} \{(u_{i_1,k} - u_{i_1,k}^{(1)})^+ + (u_{i_2,k}^{(1)} - u_{i_2,k})^+\} \\ &= \|q\| \cdot \sum_{k=1}^{\infty}{}' |u_{l,k} - u_{l,k}^{(1)}|, \end{aligned} \tag{6.4.12}$$

and consequently,

$$\begin{aligned} \sup_{q \in F_1} \frac{\|(U - U_1)q\|}{\|q\|} &= 2\beta(U, U_1) = \tfrac{1}{2} \cdot \sup_{i_1, i_2} \sum_{k=1}' |u_{l,k} - u_{l,k}^{(1)}| \\ &= \sup_{i \in I} \sum_{k=1} |u_{i,k} - u_{i,k}^{(1)}|. \end{aligned} \tag{6.4.13}$$

6.5. *The approximation theorems of stochastic transition functions.* Let us suppose that $P(x, T) \in R(X, S, X', S')$, that θ is a cover of X, and that x_i^0 is an arbitrarily fixed element in $Z_i \in \theta$. We define the stochastic transition function $P_1(x, T)$ equal to $P(x_i^0, T)$ for any $x \in Z_i \in \theta$.

LEMMA 6.5.1. *The distance $\beta(P, P_1)$ satisfies the inequality $\beta(P, P_1) \leq 1 - \alpha(P)$.*

PROOF. One can write

$$\begin{aligned} \beta(P, P_1) &= \sup |\nu(x, T)| = \sup |P(x, T) - P(x_i^0, T)| \\ &\leq \sup |P(x, T) - P(x_1, T)| \leq \sup |P(x, T) - P(x_1, T)| = 1 - \alpha(P) \end{aligned} \tag{6.5.1}$$

where the first upper bound is taken for $x \in X$, $T \in S'$, the second for $i \in I^+$,

$x \in Z_i$, $T \in \sum'$, the third for $i \in I$, $x \in Z_i$, $x_1 \in Z_i$, $T \in S'$, the fourth for x, $x_1 \in X$, $T \in S'$.

Let us consider that (X, S) is a separable metric space with the distance $\rho(x, x_1)$, and $P(x, T)$ is uniformly continuous in $x \in X$, uniform for all $T \in S'$. That is, for any $\delta > 0$ there exists a number $\epsilon = \epsilon(\delta) > 0$ such that

$$|P(x, T) - P(x_1, T)| < \delta \tag{6.5.2}$$

for all x, $x_1 \in X$ for which $\rho(x, x_1) < \epsilon$ and for all $T \in S'$. If θ_ϵ is an ϵ-cover of X, $Z_i^\epsilon \in \theta_\epsilon$, x_i^ϵ the center of Z_i^ϵ, $(i \in I)$, let us define the stochastic transition function $P_\epsilon(x, T)$ equal to $P(x_i^\epsilon, T)$ for $x \in Z_i^\epsilon$, $(i \in I)$, $T \in S'$.

THEOREM 6.5.1. *For any* $\delta > 0$, *there exists a number* $\epsilon = \epsilon(\delta)$ *such that* $\beta(P, P_\epsilon) < \delta$.

The *proof* follows from the first two equalities of the proof of lemma 6.5.1 letting $Z_i = Z_i^\epsilon$, $x_i^0 = x_i^\epsilon$, $P_1 = P_\epsilon$ if we observe that for all $T \in S'$, $x \in Z_i^\epsilon$, $i \in I^+$, the following inequality is satisfied:

$$|P(x, T) - P(x_i^\epsilon, T)| < \delta. \tag{6.5.3}$$

Here we shall study the simultaneous approximation of a probability field and of a stochastic transition function which transforms it.

Let us denote by $p(x, x')$ the conditional probability density of $P(x, T)$; if we consider also the probability distribution $P_A(Z)$ of the field A, then the conditional distribution $P_{A_{x'}}(Z|x')$ of the field $A_{x'} \in D(X)$ is completely defined for any $x' \in X'$. Let us denote by $p(x|x')$ the probability density of $A_{x'}$. By P_A and $P(x, T)$, a field $B \in D(X')$ is completely defined also.

Let us denote by $R^0(X, S, X', S')$ the totality of probability transition functions with domain of definition (X_1, S_1, X', S') where X_1 is any discrete subset of X.

THEOREM 6.5.2. *Let us consider* $\delta > 0$, $A \in D(X)$, $P(x, T) \in R(X, S, X', S')$ *uniformly continuous in* $x \in X$, *uniformly for* $T \in S'$.

There exists a number $\epsilon = \epsilon(\delta)$, *discrete probability fields* $A_\epsilon \in D^0(X)$, $(A_{x'})_\epsilon \in D^0(X)$, $(x' \in X')$, *and a discrete stochastic transition function* $P_\epsilon(x, T) \in R^0(X, S, X', S')$ *such that*

(a) *the properties* $W_\epsilon(AA_\epsilon)$, $W_\epsilon[A_{x'}(A_{x'})_\epsilon]$ *are satisfied,*

(b) *if* Q, Q_ϵ *are stochastic transition operators defined by* P, P_ϵ *respectively, and* $P_B = Q \cdot P_A$, $P_{B\epsilon} = Q \cdot P_{A\epsilon} = Q_\epsilon \cdot P_{A\epsilon}$, *then*

$$\beta'(P_B, P_{B\epsilon}) \leq \beta(P, P_\epsilon) < \delta, \tag{6.5.4}$$

(c) $I(A, B) = I(A_\epsilon, B_\epsilon) + o(1)$.

PROOF. (a) From theorem 3.2.3 it follows that for any $x' \in X'$ there is a discrete probability field $(A_{x'})_\epsilon$ such that the condition $W_\epsilon[A_{x'}(A_{x'})_\epsilon]$ is satisfied. Obviously, if θ_ϵ is an ϵ-cover, then the states of $(A_{x'})_\epsilon$ and those of A_ϵ are the centers x_i^ϵ of $Z_i^\epsilon \in \theta_\epsilon$. We may observe that x_i^ϵ does not depend on $x' \in X'$. Let us define the probability in $(A_{x'})_\epsilon$ by

$$P_{(A_{x'})_\epsilon}(x_i^\epsilon|x') = P_{A_{x'}}(Z_i^\epsilon|x') = \int_{Z^\epsilon} p(x|x')\, dx = p_i(x') \cdot \mu(Z_i^\epsilon) \tag{6.5.5}$$

where $p_i(x')$ is a number between the lower and the upper bounds of $p(x|x')$, for $x \in Z_i^\epsilon$. Let us also define the union $A_{x'}(A_{x'})_\epsilon$ by the density $p_{A|(A_{x'})_\epsilon}(x|x_i^\epsilon, x') = p_{A|A_\epsilon}(x|x_i^\epsilon)$ equal to $N_\epsilon(X)/\mu(X)$ for $x \in Z_i^\epsilon$ and zero in the rest, for any arbitrary $x' \in X'$. Consequently,

$$p_{A_{x'}(A_{x'})_\epsilon}(x, x_i^\epsilon|x') = p_{A|(A_{x'})_\epsilon}(x|x_i^\epsilon, x')\cdot P_{(A_{x'})_\epsilon}(x_i^\epsilon|x'), \tag{6.5.6}$$

which is equal to $p_i(x')$ for $x \in Z_i^\epsilon$ and to zero in the rest.

Obviously the condition $W_\epsilon[A_{x'}(A_{x'})_\epsilon]$ is satisfied. We also obtain

$$h(A|x_i^\epsilon, x') = h(A|(A_{x'})_\epsilon) = K_\epsilon(X). \tag{6.5.7}$$

If $\varphi(t) = t \log t$, then

$$\begin{aligned} H[(A_{x'})_\epsilon] &= -\sum_i \varphi[P_{(A_{x'})_\epsilon}(x_i^\epsilon|x')] = -\sum_i \varphi[P_{(A_{x'})_\epsilon}(Z_i^\epsilon|x')] \\ &= -\sum_i \varphi[p_i(x')\mu(Z_i^\epsilon)] = -\sum_i \varphi[p_i(x')]\cdot\mu(Z_i^\epsilon) \\ &\quad -\sum_i P_{A_x}(Z_i^\epsilon|x')\log\mu(Z_i^\epsilon) = h(A_{x'}) + K_\epsilon(X) + o(1). \end{aligned} \tag{6.5.8}$$

Let us consider the probability fields defined by

$$P_B(T) = \int_X P_A(dx)\,P(x, T), \qquad P_{B_\epsilon}(T) = \int_X P_A(dx)\,P_\epsilon(x, T). \tag{6.5.9}$$

(b) From lemma 6.2.1 and from theorem 6.5.1, it follows that for any $\delta > 0$ there exists a number $\epsilon = \epsilon(\delta)$ such that $\beta'(P_B, P_{B_\epsilon}) \le \beta(P, P_\epsilon) < \delta$, and consequently, for any $Z \in S$, $P_{B_\epsilon}(Z) = P_B(Z)(1 + o(1))$.

(c) Consequently,

$$\begin{aligned} H(A_\epsilon|B_\epsilon) &= M_{B_\epsilon}H[(A_{x'})_\epsilon] = M_B H[(A_{x'})_\epsilon](1 + o(1)) \\ &= h(A|B) + K_\epsilon(X) + o(1), \end{aligned} \tag{6.5.10}$$

and using theorems 3.2.3 (b) and 3.2.1, it follows that $I(A_\epsilon, B_\epsilon) = H(A_\epsilon) - H(A_\epsilon|B_\epsilon) = I(A, B) + o(1)$.

7. The stochastic complex source-channel

7.1. *The differential entropy of* $[A, \Delta]$. The stochastic channel Δ is defined by (a) the input-elements $x_\tau \in X_\tau$, $(\tau \in I)$; (b) the output-elements y_τ which form the measure space $(Y_\tau, V_\tau, \nu_\tau)$, $(\tau \in I)$, $(Y^{\alpha_n}, V^{\alpha_n}, \nu^{\alpha_n}) = X_{\tau\in\alpha_n}(Y_\tau, V_\tau, \nu_\tau)$; $(Y, V) = X_{\tau\in I}(Y_\tau, V_\tau)$, where $y^{\alpha_n} \in Y^{\alpha_n}$, $y \in Y$; (c) the transmission law which is defined by the probability density $\pi_{B|A}^{\alpha_n'}(y^{\alpha_n}|x)$ (with respect to the ν^{α_n}-measure) of the realization of the element $y^{\alpha_n} \in Y^{\alpha_n}$ in the time $\alpha_n = [t, t + n - 1]$ by the output of the channel, if it is known that by the input, $x \in X$ is entered.

In this manner, for any $t \in I$, $n \in I^+$, $T^{\alpha_n} \in V^{\alpha_n}$ the measure $P_{B|A}^{\alpha_n}(T^{\alpha_n}|x)$ is defined, and consequently, their extension $P_{B|A}(T|x)$ for $T \in V$, $x \in X$. Let us denote the channel defined in this manner by $\Delta = [X, P_{B|A}(\cdot|x), Y]$.

We shall use the ordinary concept of a nonanticipative channel with finite memory. A channel is stationary if $\pi^{\alpha'_n}_{B|A}(Uy^{\alpha_n}|Ux) = \pi^{\alpha_n}_{B|A}(y^{\alpha_n}|x)$ for any $y^{\alpha_n} \in Y^{\alpha_n}$, $x \in X$, $t \in I$, $n \in I^+$. Here $\alpha'_n = \{\tau, t+1 \leq \tau \leq t+n\}$, and U is the shift operator.

If X_τ is simultaneously the set of states of the input process A and of the input of the channel Δ at the instant $\tau (\tau \in I)$, we may consider the composite process AB with sets of states $X_\tau \times Y_\tau$ and also the output-process B with sets of states Y_τ. In this case, let us denote by $[A, \Delta]$ the complex of the input-process A and the channel Δ. We also denote

$$(7.1.1)\qquad \begin{aligned} g^{\alpha_n}_{A|B}(x, y) &= g^{\alpha_n}_{AB}(x, y) - g^{\alpha_n}_{B}(x); & f^{\alpha_n}_{A|B}(x, y) &= f^{\alpha_n}_{AB}(x, y) - f^{\alpha_n}_{B}(y), \\ G^{\alpha_n}_{AB}(x, y) &= g^{\alpha_n}_{A}(x) - g^{\alpha_n}_{A|B}(x, y); & F^{\alpha_n}_{AB}(x, y) &= f^{\alpha_n}_{A}(x) - f^{\alpha_n}_{A|B}(x, y). \end{aligned}$$

DEFINITION 7.1.1. *The differential entropy of the complex* $[A, \Delta]$ *is the quantity*

$$(7.1.2)\qquad h_t(A|B) = \lim_{n\to\infty} n^{-1} \cdot h(A^{\alpha_n}|B^{\alpha_n})$$

(if it exists). The rate of information transmission in the complex $[A, \Delta]$ *is the quantity*

$$(7.1.3)\qquad I_t(A, B) = \lim_{n\to\infty} n^{-1} \cdot I(A^{\alpha_n}, B^{\alpha_n})$$

(if it exists)

Different properties of these concepts are given in ([8]–[11], [14]).

7.2. *The entropy stability for the complex* $[A, \Delta]$.

DEFINITION 7.2.1. *The complex* $[A, \Delta]$ *possesses* (a) *the weak,* (b) *the strong, and* (c) *the norm entropy stability* (resp. *information stability*) *at the instant t if* $f^{\alpha_n}_{A|B}(x, y)$ (resp. $F^{\alpha_n}_{AB}(x, y)$) *converges to* $h_t(A|B)$ (resp. $I_t(A, B)$) *respectively* (a) *in probability,* (b) *almost everywhere,* (c) *in the norm in the Banach space* L_1.

We shall denote these properties by $E^{(i)}_t(A|B)$, $J^{(i)}_t(AB)$, $(i = 1, 2, 3)$.

THEOREM 7.2.1. *In order that the complex* $[A, \Delta]$ *possess the property* $E^{(i)}_t(A|B)$ (resp. $J^{(i)}_t(AB)$), *it is necessary and sufficient that the sequence* $\{g^{\alpha_n}_{A|B}(x, y)\}$ (resp. $\{G^{\alpha_n}_{AB}(x, y)\}$) *satisfy respectively* (a) *the weak* $(i = 1)$, *(b) the strong* $(i = 2)$, *and* (c) *the norm* $(i = 3)$ *law of large numbers.*

The proof is analogous to that of theorem 4.2.1.

DEFINITION 7.2.2. *If the property* $E^{(i)}_t(A|B)$ (resp. $J^{(i)}_t(AB)$), $(i = 1, 2, 3)$ *is satisfied for all* $t \in I$, *then the property* $E^{(i)}(A|B)$, $(J^{(i)}(AB))$ *is satisfied.*

As in the case of the processes (see 4.2), here also results may be obtained concerning the existence of the properties $E^{(i)}(A|B)$, $J^{(i)}(AB)$ and also concerning the stationary complexes $[A, \Delta]$, (see [8]–[11], [14]).

DEFINITION 7.2.3. *The regular set of sources* F_Δ *of the channel* Δ *is the totality of regular sources A with the same states* X_τ *as in the input of* Δ *at the same instant, for which* $I_t(A, B)$ *exists, is finite, does not depend on the time, and satisfies the property* $J^{(1)}(A, B)$.

DEFINITION 7.2.4. *The channel* Δ *is regular if it is nonanticipative and* F_Δ *is not void.*

DEFINITION 7.2.5. *The regular capacity of the channel* Δ *is* $C = \sup I(A, B)$ *where the upper bound is taken for* $A \in F_\Delta$.

8. The approximation of stochastic channels

8.1. *The metric space of stochastic channels.* Let us suppose that the nonanticipative channel Δ with finite memory m is given by means of the measure spaces $(X_\tau, S_\tau, \mu_\tau)$, $(Y_\tau, V_\tau, \nu_\tau)$, $(\tau \in I)$ and of the probability transition functions $P^{\alpha_n}(X^{\alpha'_n}, T^{\alpha_n}) \in R(X^{\alpha'_n}, S^{\alpha'_n}, Y^{\alpha_n}, V^{\alpha_n})$ where $\alpha_n = [t, t+n-1]$, $\alpha'_n = [t-m, t+n-1]$, $(t \in I, n \in I^+)$.

Let us denote by $R(X, S, Y, V)$ the totality of channels over (X, S, Y, V).

DEFINITION 8.1.1. *If $\Delta, \Delta_1 \in R(X, S, Y, V)$, and P^α, P_1^α are the corresponding probability transition functions $(\alpha \subset I)$, we define $\gamma(\Delta, \Delta_1) = \sup \beta(P^\alpha, P_1^\alpha)$, where the upper bound is taken over all $\alpha \subset I$.*

LEMMA 8.1.1. *If $\alpha \subset \alpha_1$, then $\beta(P^\alpha, P_1^\alpha) \leq \beta(P^{\alpha_1}, P_1^{\alpha_1})$.*

The proof follows immediately from the definition of the distance β.

THEOREM 8.1.1. *The space $R(X, S, Y, V)$ is a complete metric space with the distance $\gamma(\Delta, \Delta_1)$.*

PROOF. The function $\gamma(\Delta, \Delta_1)$ is a distance because $\beta(P^\alpha, P_1^\alpha)$ is a distance for any $\alpha \subset I$. The space $R(X, S, Y, V)$ is complete for the distance $\gamma(\Delta, \Delta_1)$, because $R(X^{\alpha'}, S^{\alpha'}, Y^\alpha, V^\alpha)$ is complete for the distance $\beta(P^\alpha, P_1^\alpha)$.

8.2. *The approximation of the system (A, Δ).* Let us consider a cover θ_τ of X_τ, $Z_{i_\tau} \in \theta_\tau$, and let x_{i_τ} be any arbitrarily given element in Z_{i_τ}, $(\tau \in I)$. In this manner is also determined a cover $\theta^\alpha = \times_{\tau \in \alpha} \theta_\tau$ in X^α such that $x_{i^\alpha} = \{x_{i_\tau}, \tau \in \alpha\} \in Z_{i^\alpha} \in \theta^\alpha$. For any given channel $\Delta \in R(X, S, Y, V)$ let us define another channel Δ_1 by means of the probability transition functions $P_1^\alpha(x^{\alpha'}, T^\alpha)$ equal to $P^\alpha(x_{i^{\alpha'}}, T^\alpha)$ for $x^\alpha \in Z_{i^\alpha}$, $\alpha' = [t-m, t+n-1]$, $\alpha = [t, t+n-1]$. If $a(P^\lambda)$ is the ergodic coefficient of P^λ, let us denote $\sigma(\Delta) = \inf a(P^\lambda)$ where the lower bound is taken for all $\lambda \subset I$.

LEMMA 8.2.1. *The distance $\gamma(\Delta, \Delta_1)$ is less than or equal to $1 - \sigma(\Delta)$.*

The proof follows from lemma 6.5.1.

With the hypotheses and notation of 5.1, let us suppose that $P^\alpha(x^{\alpha'}, T^\alpha)$ is uniformly continuous in $x^\alpha \in X^\alpha$, (uniformly for all $\alpha \subset I$, $T^\alpha \in V^\alpha$). Let us define the stochastic channel Δ_ϵ by means of the probability transition functions $P_\epsilon^\alpha(x^{\alpha'}, T^\alpha) = P^\alpha(x_{i^{\alpha'}}^\epsilon, T^\alpha)$ for $x^{\alpha'} \in Z_{i^{\alpha'}}^\epsilon$. From theorem 6.5.1 follows immediately theorem 8.2.1.

THEOREM 8.2.1. *For any $\delta > 0$ there exists a number $\epsilon = \epsilon(\delta)$ such that $\gamma(\Delta, \Delta_\epsilon) < \delta$. Let us denote: $R^0(X, S, Y, V)$ the totality of stochastic processes with domain of definition (X_1, S_1, Y, V) where X_1 is any discrete subset of X; $C_\epsilon = \sup_{A_\epsilon} I_t(A_\epsilon, B_\epsilon)$.*

THEOREM 8.2.2. *Let us consider* (1) *a stochastic process $A \in D(X)$, which possesses finite $h_t(A)$ and the property $E_t^{(i)}(A)$;* (2) *a stochastic channel $\Delta \in R(X, S, Y, V)$ which is defined by uniformly continuous $P^\alpha(x^{\alpha'}, T^\alpha)$ (uniformly in $\alpha \subset I$, $T^\alpha \in V^\alpha$), and possesses finite $h_t(A|B)$ and the property $E_t^{(i)}(A|B)$* (resp. *$I_t(A, B)$ and $J_t^{(i)}(A, B)$*).

For any given $\delta > 0$, we may determine a number $\epsilon = \epsilon(\delta)$ such that

(1) *there exists a discrete stochastic process $A_\epsilon \in D^0(X)$ with finite entropy*

(8.2.1) $$H_t(A_\epsilon) = h_t(A) + K_{t,\epsilon}(X) + o(1)$$

and the property $E_t^{(i)}(A_\epsilon)$, $(i = 1, 2, 3)$;

(2) *there exists a discrete stochastic channel* $\Delta_\epsilon \in R^0(X, S, Y, V)$, *such that if connected with* A_ϵ, *there exists finite*

(8.2.2) $$H_t(A|B) = h_t(A|B) + K_{t,\epsilon}(X) + o(1),$$

and the property $E_t^{(i)}(A|B)$, *or respectively*, $I_t(A_\epsilon, B_\epsilon)$, $J_t^{(i)}(A, B)$, $(i = 1, 2, 3)$;

(3) *the property* $W_\epsilon(AA_\epsilon)$ *is satisfied, and* $\gamma(\Delta, \Delta_\epsilon) < \delta$;

(4) *from the regularity of* $\{X_\tau\}$, Δ, *the same thing follows for* Δ_ϵ;

(5) *from the stationarity of* Δ *the same thing follows for* Δ_ϵ;

(6) $I_t(A, B) = I_t(A_\epsilon, B_\epsilon) + o(1)$;

(7) $C = C_\epsilon + o(1)$.

For the proof we may use the process A_ϵ constructed in theorem 5.3.2 and the channel Δ_ϵ constructed in theorem 8.2.1; the proof runs analogously to that of theorem 6.5.2.

9. The basic theorems of Shannon type

We shall suppose here that $\{X_\tau\}$ is regular.

THEOREM 9.1. *Let us consider* (1) *a regular channel* Δ *with uniformly continuous probability transition functions, with continuous input sets of states, with finite memory, and with finite regular capacity* C;

(2) *a regular process* $\mathring{A}$ *with continuous input sets of states and* $h(\mathring{A}) < C$.

For a given $\delta > 0$, *if we determine* $\epsilon = \epsilon(\delta)$, $\mathring{A}_\epsilon$, Δ_ϵ *as in theorems* 5.3.2, 8.2.2, *obviously* $W_\epsilon(\mathring{A}\mathring{A}_\epsilon)$ *is satisfied and* $\gamma(\Delta, \Delta_\epsilon) < \delta$. *If*

(9.1.1) $$H(\mathring{A}_\epsilon) = h(\mathring{A}) + K_\epsilon(\mathring{X}) < C + o(1),$$

then concerning the possibility of transmission of the production of the process A_ϵ *through the channel* Δ_ϵ *with the error probability not greater than a given* λ, *the first basic theorem of Shannon type is true* ([14], *p.* 243).

If $\mathring{X}_\tau$, the sets of states of $\mathring{A}$ are totally bounded, then $\mathring{A}_\epsilon$ has at each instant a finite number n_τ of states. Let

(9.1.2) $$\overline{K}_\epsilon(X) = \limsup_{n\to\infty} n^{-1} \sum_{k=0}^{n-1} \log \mathring{n}_{t+k}.$$

THEOREM 9.2. *Under the conditions of theorem* 9.1, *if the sets of states of* $\mathring{A}$ *are totally bounded and* $\overline{K}_\epsilon(X) < \infty$, *then concerning the possibility of the choice of a code such that the transmission rate in the system* $[\mathring{A}_\epsilon, \Delta_\epsilon]$ *is as close to* $H(\mathring{A}_\epsilon) = h(\mathring{A}) + K_\epsilon(\mathring{X}) + o(1)$ *as one wishes, the second basic theorem of Shannon type is true* ([14], *p.* 244).

The proofs of these two theorems follow from the fact that $\mathring{A}_\epsilon$, Δ_ϵ verifies the conditions of the basic theorems in ([14], pp. 243–244).

REFERENCES

[1] S. BANACH, *Théorie des Opérations Linéaires,* Warszawa, 1932.
[2] L. BREIMAN, "The individual ergodic theorem of information theory," *Ann. Math. Statist.,* Vol. 28 (1957), pp. 809–811.
[3] K. L. CHUNG, "A note on the ergodic theorem on information theory," *Ann. Math. Statist.,* Vol. 32 (1961), pp. 612–614.
[4] R. L. DOBRUSHIN, "The central limit theorem for non-homogeneous Markov chains," *Theor. Probability Appl.,* Vol. 1 (1956), pp. 72–89 and 365–425.
[5] A. I. KHINTCHIN, "On the basic theorems of information theory," *Uspehi Mat. Nauk,* Vol. 11 (1956), pp. 17–75.
[6] A. N. KOLMOGOROV and V. M. TIHOMIROV, "The ϵ-entropy and the ϵ-content of sets in the functional spaces," *Uspehi Mat. Nauk,* Vol. 14 (1959), pp. 3–86.
[7] A. N. KOLMOGOROV, "On some asymptotic characteristics of the totally bounded metric spaces," *Dokl. Akad. Nauk SSSR,* Vol. 108 (1956), pp. 385–388.
[8] M. ROSENBLATT-ROTH, "The concept of entropy in the probability theory and its applications in the theory of transmission through channels," *Transactions of the Third All-union Mathematical Congress,* Moscow, Vol. 2 (1956), pp. 132–133.
[9] ———, "The concept of entropy in the probability theory and its applications in the theory of transmission through channels," thesis, Moscow State University, 1956.
[10] ———, "The entropy of stochastic processes," *Dokl. Akad. Nauk SSSR,* Vol. 112 (1957), pp. 16–19.
[11] ———, "The theory of transmission of information through stochastic channels," *Dokl. Akad. Nauk SSSR,* Vol. 112 (1957), pp. 202–205.
[12] ———, "Normed ϵ-entropy and the transmission of the information of continuous sources through continuous channels," *Dokl. Akad. Nauk SSSR,* Vol. 130 (1960), pp. 265–268.
[13] ———, "Normed ϵ-entropy of sets and the theory of transmission of information," *Transactions of the Second Prague Conference on Information Theory, Statistical Decision Functions and Random Processes,* Prague, 1960, pp. 569–577.
[14] ———, "The concept of entropy in the probability theory and its applications in the theory of transmission by channels," *Theor. Probability Appl.,* Vol. 9 (1964), pp. 238–261.
[15] C. E. SHANNON, "A mathematical theory of communication," *Bell System Tech. J.,* Vol. 27 (1948), pp. 378–423 and 623–656.
[16] A. G. VITUSHKIN, "The absolute entropy of the metric spaces," *Dokl. Akad. Nauk SSSR,* Vol. 117 (1957), pp. 745–748.

APPROXIMATION WITH A FIDELITY CRITERION

J. WOLFOWITZ
CORNELL UNIVERSITY

1. Introduction

The present paper, as so many others in information theory, was stimulated by a paper of Shannon's [1]. The interesting theorem 1 below is due to him; the new result is theorem 2. We give a different proof of theorem 1. Actually this proof is not very new and is essentially the one used to prove theorem 1 of [3] (reproduced in [2] as theorem 3.2.1). The relation between the notion of "distortion" and that of "being generated" will be clear from this proof.

In the present paper we keep separate the ideas of approximating and coding. Then theorem 1 says essentially that, by embedding a certain number of sequences one can achieve a prescribed bound on the distortion, and theorem 2 says essentially that this cannot be done with fewer sequences. Shannon's results on coding are described in section 4. Some of his generalizations and additional suggestions for further generalizations are described in section 5.

It may perhaps be of interest to mention that theorem 4.9 of [4] is a special case of (4.3) below (the latter is theorem 1 of [1]). In fact, the probability of error defined in (4-65) of [4] is a special case of Shannon's distortion function ((2.1) below).

In [1], and in the present paper, the "source" digits (components of u below) are chance variables with a given (fixed) distribution. This is also true in the situation treated in theorem 4.9 of [4]. In the strong converse proved in [3], and in the others proved in [2], the messages are *not* stochastic and are chosen arbitrarily by the sender. If they should be chosen by a chance process their distribution can be arbitrary. The claims made in ([4], p. 219) on behalf of theorem 4.9 of [4] are therefore without the least basis in fact.

2. The approximating theorem

Consider the alphabets $M = \{m_1, \cdots, m_a\}$ and $Z = \{z_1, \cdots, z_b\}$. Let M^* (resp. Z^*) be the space of n-sequences (sequences of length n) in the M-alphabet (resp. the Z-alphabet). Let $\pi = (\pi_1, \cdots, \pi_a)$ be a probability a-vector which will be fixed in all that follows. When we speak of the probability distribution on M^*, we shall always mean the distribution implied by n independent chance variables with the common distribution π.

Research under contract with the Office of Naval Research.

Let d be a nonnegative function, called the "distortion" function, defined on $(M \times Z)$. Let $u_o = (x_1, \cdots, x_n)$ be any sequence in M^* and $v_o = (y_1, \cdots, y_n)$ be any sequence in Z^*. We define the distortion $d(u_o, v_o)$ between u_o and v_o by

$$d(u_o, v_o) = \frac{1}{n} \sum_{i=1}^{n} d(x_i, y_i). \tag{2.1}$$

Let D_o and D_{oo} be, respectively, the minimum and maximum values of d. Let D be a variable which temporarily takes values in the open interval (D_o, D_{oo}). For any value of D let $w(j|i|D)$, $i = 1, \cdots, a$; $j = 1, \cdots, b$, be nonnegative numbers (if they exist) such that $w(\cdot|i|D)$ is a probability b-vector with the following properties:

$$\sum_{i,j} \pi_i w(j|i|D)\, d(i, j) = L(w(\cdot|\cdot|D)) \text{ (say)} \leq D \tag{2.2}$$

and

$$R(w(\cdot|\cdot|D)) \text{ (say)} = \sum_{i,j} \pi_i w(j|i|D) \log \left(\frac{w(j|i|D)}{\sum_i \pi_i w(j|i|D)} \right) \leq \sum_{i,j} \pi_i w(j|i) \log \left(\frac{w(j|i)}{\sum_i \pi_i w(j|i)} \right), \tag{2.3}$$

where $w(\cdot|\cdot)$ is any channel probability function ((c.p.f.), that is, $w(j|i) \geq 0$, $i = 1, \cdots, a$; $j = 1, \cdots, b$, and $w(\cdot|i)$ is a probability b-vector) such that $L(w(\cdot|\cdot)) \leq D$. Henceforth we write

$$\pi'(D) = \left(\sum_i \pi_i w(1|i|D), \cdots, \sum_i \pi_i w(b|i|D)\right). \tag{2.4}$$

When we use π and $\pi'(D)$ to multiply matrices we shall consider them to be column vectors. Let $W(D)$ be the $(b \times a)$-matrix with element $w(j|i|D)$ in the j-th row and i-th column. Then

$$\pi'(D) = W(D)\pi. \tag{2.5}$$

To simplify the notation, we shall write $R(D)$ for $R(w(\cdot|\cdot|D))$. From the definition of $R(D)$ it is obvious that $R(D)$ is a monotonically nonincreasing function of D. Let $D_1 < D_2$ be any two values of D, and consider

$$w_o(\cdot|\cdot) = \tfrac{1}{2} w(\cdot|\cdot|D_1) + \tfrac{1}{2} w(\cdot|\cdot|D_2). \tag{2.6}$$

We have

$$L(w_o(\cdot|\cdot)) = \tfrac{1}{2} L(w(\cdot|\cdot|D_1)) + \tfrac{1}{2} L(w(\cdot|\cdot|D_2)) \tag{2.7}$$

and

$$R(w_o) \leq \tfrac{1}{2}(R(D_1) + R(D_2)). \tag{2.8}$$

Hence, $R(D)$ is a convex function of D, and hence, a (strictly) monotonically decreasing function of D.

The minimum value $D_{\min}$ of D, which we shall need to consider, can be found as follows: fix i; let j_o be such that

$$d(i, j_o) = \min_j d(i, j), \tag{2.9}$$

and let $w(j_o|i) = 1$. Then $D_{\min} = \sum_i \pi_i \min_j d(i, j)$. The maximum value $D_{\max}$ of D which we shall need to consider is the smallest value of D for which $R = 0$. If $R = 0$, then $w(j|i|D_{\max})$ is independent of i, say $w_o(j)$. Then

$$L(w(\cdot|\cdot|D_{\max})) = \min_{w_0} \sum_j w_o(j) \sum_i \pi_i \, d(i, j) \tag{2.10}$$
$$= D_{\max} = \min_j \sum_i \pi_i \, d(i, j).$$

Henceforth, D will be a variable with values in the open interval $(D_{\min}, D_{\max})$. What happens at the ends of the interval will be discussed separately later, or else will be obvious.

Let S be a set of n-sequences in Z^*. For any element u_o of M^*, let

$$d(u_o, S) = \min_{v_o \in S} d(u_o, v_o). \tag{2.11}$$

Let u be a chance sequence with values in M^* and the distribution already defined on M^*. For any set S, the expected value $E\, d(u, S)$ is thus defined.

THEOREM 1. (The approximating theorem.) *Let $\epsilon^* > 0$ be arbitrary. There exists a function $n_o(\epsilon^*)$ of ϵ^* such that, for $n > n_o(\epsilon^*)$, we have the following: for any $D(D_{\min} < D < D_{\max})$ there exists a set $S(D) \subset Z^*$ containing N elements such that*

$$E\, d(u, S(D)) < D + \epsilon^* \tag{2.12}$$

and

$$N \le \exp_2 \{nR(D)\}. \tag{2.13}$$

PROOF. We may assume that $D < D_{\max} - \epsilon^*$, or the theorem is trivially true. Let

$$\epsilon = \frac{\epsilon^*}{2(1 + D_{oo})}, \qquad D' = D + \frac{\epsilon^*}{2}, \tag{2.14}$$

and

$$h = \min_y \left[R(y) - R\left(y + \frac{\epsilon^*}{2}\right) \right] \tag{2.15}$$

where the minimum is taken over the range $D_{\min} \le y \le D_{\max} - \epsilon^*/2$. Hence, $h > 0$. Throughout the course of the present proof (and only then), write π', for short, in place of $\pi'(D') = W(D')\pi$. Let $w'(\cdot|\cdot|D')$ be defined by

$$w'(i|j|D') = \frac{\pi_i w(j|i|D')}{\pi_j'}, \qquad i = 1, \cdots, a; \quad j = 1, \cdots, b. \tag{2.16}$$

We define the chance variable (u, v) (u has already been defined) with values in $M^* \times Z^*$ and distribution determined by either of the following (which give the same result):

(i) the (marginal) distribution of u is as given above, and the conditional distribution of the k-th component of v, $(k = 1, \cdots, n)$, given $u = u_o$ and all the other components of v, is $w(\cdot|x_k|D')$, or

(ii) the (marginal) distribution of v is that of a sequence of independent chance variables with common distribution π', and the conditional distribution of the k-th component of u, $(k = 1, \cdots, n)$, given $v = v_o$ and all the other components of u, is $w'(\cdot|y_k|D')$.

Let $N(i|u_o)$ be the number of elements i in u_o, and similarly for v_o. Let $N(i, j|u_o, v_o)$ be the number of k, $k = 1, \cdots, n$, such that $x_k = i$ and $y_k = j$. We shall say that u_o is generated by v_o if

$$|N(i, j|u_o, v_o) - N(j|v_o)w'(i|j|D')| \leq \delta[N(j|v_o)w'(i|j|D'))(1 - w'(i|j|D'))]^{1/2} \tag{2.17}$$

for $i = 1, \cdots, a; j = 1, \cdots, b$. Here $\delta > 0$ is such that

$$P\{u \text{ is generated by } v_o|v = v_o\} > 1 - \frac{\epsilon}{4}. \tag{2.18}$$

(The symbol $P\{\quad\}$ denotes the probability of the relation in braces. The symbol $P\{A|B\}$ denotes the probability of A, conditional upon B.) We shall say that a sequence v_o in Z^* is a π'-sequence if

$$|N(j|v_o) - n\pi_j'| \leq z\sqrt{n\pi_j'(1 - \pi_j')}, \qquad j = 1, \cdots, b \tag{2.19}$$

where $z > 0$ is such that

$$P\{v \text{ is a } \pi'\text{-sequence}\} > 1 - \frac{\epsilon}{4}. \tag{2.20}$$

It follows from (2.2), (2.16), (2.17), and (2.19) that, for all n sufficiently large and any pair (u_o, v_o) such that v_o is a π'-sequence and u_o is generated by v_o, we have

$$d(u_o, v_o) < D' + \epsilon. \tag{2.21}$$

Let

$$\{(v_1, A_1), \cdots, (v_N, A_N)\} \tag{2.22}$$

be a code $(n, N, 1 - \epsilon/4)$ as follows:

(2.23) $v_1, \cdots, v_N$ are π'-sequences;

(2.24) A_i, $i = 1, \cdots, N$, consists of all n-sequences in M^* generated by v_i and not in $A_1 \cup \cdots \cup A_{i-1}$;

$$P\{u \in A_i|v = v_i\} \geq \frac{\epsilon}{4}, \qquad i = 1, \cdots, N; \tag{2.25}$$

(2.26) it is impossible to increase N and maintain (2.23)–(2.25).

As in ([2], (3.2.5)), we conclude that, when v_o is any π'-sequence not in the set $\{v_1, \cdots, v_N\}$, we have

$$P\{u \text{ is generated by } v_o \text{ and is in } A_1 \cup \cdots \cup A_N|v = v_o\} > 1 - \frac{\epsilon}{2}. \tag{2.27}$$

For, if (2.27) did not hold, we could increase N by adding to (2.22) the pair (v_o, A_o), where A_o is the set of sequences generated by v_o and not in $A_1 \cup \cdots \cup A_N$.

Now, to each A_i, add enough sequences generated by v_i so that, calling the enlarged set B_i,

$$P\{u \in B_i | v = v_i\} > 1 - \frac{\epsilon}{2}, \qquad i = 1, \cdots, N. \tag{2.28}$$

We conclude from (2.27) and (2.28) that, for *any* π'-sequence v_o we have

$$P\{u \in (B_1 \cup \cdots \cup B_N) | v = v_o\} > 1 - \frac{\epsilon}{2}. \tag{2.29}$$

From (2.20) and (2.29) we conclude that

$$P\{u \in (B_1 \cup \cdots \cup B_N)\} > 1 - \epsilon. \tag{2.30}$$

Let

$$S(D) = \{v_1, \cdots, v_N\}. \tag{2.31}$$

From (2.30), (2.21), and the fact that every sequence in B_i is generated by the π'-sequence v_i, we obtain that

$$E\, d(u, S(D)) < D' + \epsilon + \epsilon D_{oo} = D + \epsilon^*. \tag{2.32}$$

From ([2], lemma 3.3.1) we obtain that

$$N < \exp_2 \{n[R(D') + h]\} \leq \exp_2 \{nR(D)\} \tag{2.33}$$

for n sufficiently large. In the above argument, whenever n had to be sufficiently large, its lower bound could be made to depend only on ϵ^* and not on D or D'. (The lemma of [2] which we invoked is valid with constants which do not depend on the channel probability function.) The theorem is therefore proved.

It is obvious that we can replace ϵ^* by zero in (2.12) if we replace $R(D)$ by $(R(D) + \epsilon^*)$ in (2.13). From this, one can easily conclude what the theorem is when $D = D_{\min}$ or $D_{\max}$.

THEOREM 1′. *The set $S(D)$ whose existence is proved in theorem 1 may consist only of π'-sequences.*

This is a consequence of (2.23).

3. Converse of the approximating theorem

THEOREM 2. *Let $\epsilon^* > 0$ be arbitrary. There exists a function $n_{oo}(\epsilon^*)$ of ϵ^* such that, for $n > n_{oo}(\epsilon^*)$, we have the following: for any D ($D_{\min} < D < D_{\max}$), any set $S(D) \subset Z^*$ which contains N elements and satisfies*

$$E\, d(u, S(D)) \leq D, \tag{3.1}$$

must also satisfy

$$N > \exp_2 \{n[R(D) - \epsilon^*]\}. \tag{3.2}$$

PROOF. Let $\epsilon > 0$ be a number to be chosen later. Write $D + \epsilon = D^*$. We have

$$P\{d(u, S(D)) < D^*\} \geq \frac{\epsilon}{D^*} = 2\alpha \text{ (say)} \tag{3.3}$$

by (3.1). Define the set G' by

$$G' = \{u_o \in M^* | d(u_o, S(D)) < D^*\}. \tag{3.4}$$

Let u_o be any sequence in M^*. We shall say that u_o is a π-sequence if

$$|N(i|u_o) - n\pi_i| \leq z'\sqrt{n\pi_i(1 - \pi_i)}, \qquad i = 1, \cdots, a \tag{3.5}$$

where $z' > 0$ is such that

$$P\{u \text{ is a } \pi\text{-sequence}\} > 1 - \alpha. \tag{3.6}$$

Hence,

$$P\{u \in G\} > \alpha, \tag{3.7}$$

where G is the set of π-sequences which are members of G'. For n sufficiently large the number of sequences in G exceeds

$$\alpha \cdot \exp_2 \{n[H(\pi) - \epsilon]\}, \tag{3.8}$$

where $H(\pi) = -\sum_i \pi_i \log \pi_i$; this is proved exactly as in lemma 2.1.7 of [2]. The lower bound on n does not depend on D.

Let u_o be any sequence in G and v_o any sequence in $S(D)$ such that

$$d(u_o, v_o) < D^*. \tag{3.9}$$

Now

$$\begin{aligned} nd(u_o, v_o) &= \sum_{i,j} N(i, j|u_o, v_o)\, d(i, j) \\ &= \sum_{i,j} N(i|u_o) w(j|i|u_o, v_o)\, d(i, j) \end{aligned} \tag{3.10}$$

where

$$w(j|i|u_o, v_o) = \frac{N(i, j|u_o, v_o)}{N(i|u_o)}. \tag{3.11}$$

(We stop for a moment to dispose of the case $N(i|u_o) = 0$. If no component of π is zero, then for n sufficiently large this can never occur. If $\pi_i = 0$ let the probability vector $w(\cdot|i|u_o, v_o)$ be defined arbitrarily.) Let $W(u_o, v_o)$ be the $(b \times a)$-matrix whose (j, i)-th element is $w(j|i|u_o, v_o)$. Since u_o is a π-sequence, it follows from (3.9), (3.10), and the definition of R in (2.3), that

$$R(w(\cdot|\cdot|u_o, v_o)) > R(D^*) - \psi_1(\epsilon), \tag{3.12}$$

where $\psi_1(\epsilon) \to 0$ as $\epsilon \to 0$ and $n \to \infty$.

To each u_o in G we may assign some v_o which satisfies (3.9). Since the right member of (3.11) is the ratio of two integers, it follows that the number of possible matrices $W(u_o, v_o)$ is at most $n^{a(b+1)}$. Let $W = \{w(j|i)\}$ be any matrix obtained as in (3.11), and let B be the set of pairs (u_o, v_o) such that $w(j|i|u_o, v_o) = w(j|i)$, $i = 1, \cdots, a; j = 1, \cdots, b$. Let K be the set of different v_o which occur among the elements of B. Suppose (3.2) does not hold. Then the number N_o of elements in K satisfies

$$N_o \leq \exp_2 \{n[R(D) - \epsilon^*]\}. \tag{3.13}$$

Let v_o be any sequence in K. It follows from the definition of G that

$$|N(j|v_o) - n\varphi_j| < n\psi_2(\epsilon), \qquad j = 1, \cdots, b \tag{3.14}$$

where

$$\varphi = W\pi, \tag{3.15}$$

$\psi_2(\epsilon) \to 0$ as $\epsilon \to 0$ and $n \to \infty$, and $W = W(u_o, v_o)$ is a matrix which corresponds to any pair (u_o, v_o) whose second element is the present v_o. Let $w(j|i)$ be the element in the j-th row and i-th column of W. Define

$$w'(i|j) = \frac{\pi_i w(j|i)}{\varphi_j}, \qquad \begin{array}{l} i = 1, \cdots, a, \\ j = 1, \cdots, b. \end{array} \tag{3.16}$$

From (3.11), (3.16), and the fact that u_o is a π-sequence, we obtain

$$|N(i, j|u_o, v_o) - n\varphi_j w'(i|j)| < n\psi_3(\epsilon), \qquad \begin{array}{l} i = 1, \cdots, a, \\ j = 1, \cdots, b, \end{array} \tag{3.17}$$

where $\psi_3(\epsilon) \to 0$ as $\epsilon \to 0$ and $n \to \infty$. It follows from (3.14), (3.17), and ([2], lemma 2.1.6), that the number of pairs in B, *with the same* v_o, is less than

$$\exp_2 \{n[\sum_j \varphi_j H(w'(\cdot|j)) + \psi_4(\epsilon)]\} \tag{3.18}$$

where $\psi_4(\epsilon) \to 0$ as $\epsilon \to 0$ and $n \to \infty$. From (3.12) we obtain

$$\begin{aligned} \sum \varphi_j H(w'(\cdot|j)) &< \sum_j \pi_j'(D^*) H(w'(\cdot|j|D^*)) + \psi_1(\epsilon) \\ &= \sum_j \pi_j'(D) H(w'(\cdot|j|D)) + \psi_5(\epsilon), \end{aligned} \tag{3.19}$$

where $\psi_5(\epsilon) \to 0$ as $\epsilon \to 0$ and $n \to \infty$. The right member of (3.13) is equal to

$$\exp_2 \{n[H(\pi) - \sum_j \pi_j'(D) H(w'(\cdot|j|D)) - \epsilon^*]\}. \tag{3.20}$$

From (3.13), (3.20), and (3.18), we conclude that the number of different sequences in G is less than

$$\begin{aligned} \exp_2 \{n[H(\pi) &- \sum_j \pi_j'(D) H(w'(\cdot|j|D)) - \epsilon^* \\ &+ \max_W \sum_j \varphi_j H(w'(\cdot|j)) + \psi_4(\epsilon)] + a(b+1)\log_2 n\} \end{aligned} \tag{3.21}$$

which, by (3.19), is less than

$$\exp_2 \{n[H(\pi) + \psi_5(\epsilon) + \psi_4(\epsilon) - \epsilon^*] + a(b+1)\cdot\log_2 n\}. \tag{3.22}$$

From (3.8) and (3.22), we obtain

$$-n\epsilon + \log_2 \alpha < n(\psi_5(\epsilon) + \psi_4(\epsilon) - \epsilon^*) + a(b+1)\cdot\log n. \tag{3.23}$$

Now $\epsilon^* > 0$ is fixed. Let ϵ be sufficiently small and n sufficiently large. We obtain that (3.23) cannot hold. This contradiction and the fact that, whenever n had to be sufficiently large in the above proof, the lower bound on n did not depend on D, complete the proof of theorem 2.

4. Coding and approximating

Suppose that a discrete memoryless channel Γ of capacity C per letter is given. Each sequence in M^* is coded into an n'-sequence in the input alphabet of Γ, the latter is sent over Γ, and the then received n'-sequence in the output alphabet of Γ is decoded by the receiver into a sequence of Z^*. What is the expected value of the distortion d^* between the sequence in M^* and the one in Z^*? Essentially, the answer to this question has been given in [1].

Suppose that $D(D_{\min} < D < D_{\max})$ and an arbitrary $\epsilon > 0$ are given. If n and n' satisfy

$$\frac{n'C}{nR(D)} > 1 + \epsilon, \tag{4.1}$$

then, for all such n and n' greater than lower bounds which depend on D and ϵ, the expected value of d^* is less than $D + \epsilon$. To see this, one takes the set $S(D)$ embedded in Z^* according to theorem 1 ($\epsilon^* = \epsilon/2$). Let K_o be a code $(n', 2^{nR(D)}, \lambda)$ for channel Γ (that is, word length n', code length $2^{nR(D)}$, probability of error $\leq \lambda$). Because of (4.1), we can, by making n and n' sufficiently large, make λ as small as we wish. To each sequence in $S(D)$ we make correspond a transmitted (message) sequence of K_o in any manner, provided only that no transmitted sequence corresponds to more than one sequence in $S(D)$. Let $u_o \in M^*$ be any sequence. We code u_o into that transmitted sequence of K_o which corresponds to that (or any) sequence v_o in $S(D)$ such that

$$d(u_o, v_o) = d(u_o, S(D)). \tag{4.2}$$

After receiving the received sequence, the receiver decides which sequence was transmitted and then decodes the latter into its inverse in $S(D)$, if such an inverse exists. If it does not exist then he decodes into an arbitrary element of $S(D)$. Since λ can be made arbitrarily small the desired result is obvious.

(After this paper was completed, the author concluded that the problem described in this paragraph was considered in greater generality by Dobrushin [5]. It is extremely likely that the result attributed to Shannon in our paragraph above was also obtained by Dobrushin. (The verification of the latter's conditions is a formidable task.) Theorem 1 of our paper does not seem to be in [5] and seems, therefore, to be due to Shannon alone, as ascribed above. Our theorem 2 is not in [5] and, as stated in the introduction, is the new result of the present paper.)

Shannon ([1]) has proved the following nonasymptotic result:

$$E\,d^* \geq R^{-1}\left(\frac{n'}{n}\,C\right). \tag{4.3}$$

An intuitive explanation of this result is easy to give. According to theorem 2, we must embed approximately $\exp_2 \{nR(D)\}$ sequences in Z^* in order to attain $Ed(u, S(D)) \leq D$. Only $\exp_2 \{n'C\}$ sequences (approximately) can be sent over Γ and be distinguished from each other. Hence, operationally speaking, $S(D)$ acts as if it contained $\exp_2 \{n'C\}$ sequences. By theorem 2 the minimum distortion S which can be achieved with this many sequences in $S(D)$, satisfies $nR(s) = n'C$.

5. Generalizations

Theorems 1 and 2 can be generalized. For example, we used the fact that the distribution on M^* is that implied by independent, identically distributed chance variables in order to obtain (2.20) and (3.6). These will hold if the process on M^* is merely stationary and ergodic. Suppose now that the distortion function d is defined over the Cartesian product of g copies of M and g copies of Z, and, in place of (2.1), one defines

$$d(u_o, v_o) = \frac{1}{n - g + 1} \sum_{k=1}^{n-g+1} d(x_k, x_{k+1}, \cdots, x_{k+g-1}, y_k, \cdots, y_{k+g-1}). \tag{5.1}$$

Shannon [1] has treated both these generalizations. Finally, it is not necessary that the channel Γ be discrete memoryless. This case, too, has been treated in [1].

The reader will recognize that the distortion function of (5.1) corresponds to the discrete finite-memory channel of ([2], chapter 5), just as the distortion function of (2.1) corresponds to the discrete memoryless channel. This suggests that one could employ even "nonlocal" distortion measures which correspond to other channels, for example, that of ([2], section 6.6).

REFERENCES

[1] C. E. Shannon, "Coding theorems for a discrete source with a fidelity criterion," *Information and Decision Processes*, edited by Robert E. Machol, New York, McGraw-Hill, 1960.

[2] J. Wolfowitz, *Coding Theorems of Information Theory*, Berlin, Springer-Verlag, 1961 (1st ed.), or 1964 (2d ed.).

[3] ———, "The coding of messages subject to chance errors," *Illinois J. Math.*, Vol. 1 (1957), pp. 591–606.

[4] R. G. Gallager, "Information theory," *The Mathematics of Physics and Chemistry*, edited by H. Margenau and G. M. Murphy, Chapter IV, Vol. 2, Princeton, Van Nostrand, 1964.

[5] R. L. Dobrushin, "General formulation of the fundamental theorem of Shannon in information theory," *Uspehi Mat. Nauk.*, Vol. 14 (1959), pp. 3–104.

SOME CONTRIBUTIONS TO THE THEORY OF ORDER STATISTICS

PETER J. BICKEL
UNIVERSITY OF CALIFORNIA, BERKELEY

1. Introduction and summary

This paper arose from the problem of proving the asymptotic normality of linear combinations of order statistics which was first posed by Jung [9]. In the course of this investigation, several facts of general interest in the study of moments of order statistics, which either had not been stated or had not been proved in their most satisfactory form, were established. These are collected in theorems 2.1 and 2.2 of section 2. Briefly we show in theorem 2.1 that any two order statistics are positively correlated, and in theorem 2.2 we give necessary and sufficient conditions for the existence of moments of quantiles and the convergence of the suitably normalized moments to those of the appropriate normal distribution.

Section 3 contains an "invariance principle" for order statistics more elementary than the one given by Hájek [7] but requiring fewer regularity conditions and adequate for our purposes in section 4. In an as yet unpublished paper, J. L. Hodges and the author give another application of this principle in deriving the asymptotic distribution of an estimate of location in the one sample problem.

Section 4 contains the principal results of the paper. We consider linear combinations of order statistics which do not involve the extreme statistics to a more significant extent than the sample mean does. For this class of statistics we establish asymptotic normality and convergence of normalized moments to those of the appropriate Gaussian distribution.

2. Some properties of moments of order statistics

Let $X_1, \cdots, X_n$ be a sample from a population with distribution F and density f which is continuous and strictly positive on $\{x|0 < F(x) < 1\}$. Then $F^{-1}(t)$ is well-defined and continuous for $0 < t < 1$, and for those values of t we may define $\psi(t) = f[F^{-1}(t)]$. We denote by $Z_{1,n} < \cdots < Z_{n,n}$ the order statistics of the sample.

The following two theorems will be proved in this section.

THEOREM 2.1. *Suppose that* $E(Z_{i,n}^2) + E(Z_{k,n}^2) < \infty$. *Then*, $\text{cov}(Z_{i,n}, Z_{j,n}) \geq 0$.

THEOREM 2.2. *Suppose that* $\lim_{x\to\infty} |x|^\epsilon[1 - F(x) + F(-x)] = 0$ *for some* $\epsilon > 0$. *Then*,

Prepared with the partial support of the National Science Foundation Grant GP 2593.

(a) *for any natural number* $k \geq 0, 0 < \alpha < 1$, *there exists* $N(k, \alpha, \epsilon)$ *such that* $E(Z_{r,n}^k)$ *exists for* $\alpha n \leq r \leq (1 - \alpha)n$ *and* $n \geq N(k, \alpha, \epsilon)$. *Conversely, if* $E|Z_{r,n}^k| < \infty$ *for some* k, n, *then for some* $\epsilon > 0$, $\lim_{x \to \infty} |x|^\epsilon(1 + F(-x) - F(x)) = 0$.

(b) *Then* $E[Z_{r,n} - F^{-1}(r/(n+1))]^k = n^{-k/2}\sigma^k(p_n)\mu_k + o(n^{-k/2})$ *uniformly for* $\alpha n \leq r \leq (1 - \alpha)n$, n *sufficiently large, where* (i) $p_n = (r/n)$, (ii) $\sigma^2(p_n) = [(r/n)(1 - r/n)(\psi(r/n)]^{-2}$, *and* (iii) μ_k *is k-th central moment of the standard normal distribution.*

REMARK. Theorem 2.1, though useful and interesting, as we shall see in section 3, seems not to have appeared in the literature previously but was independently proved by Lehmann in a work, as yet unpublished, on positive dependance. Theorem 2.2(a) is trivial but seemed worth isolating. Theorem 2.2(b) has been proved in the literature, under assorted regularity conditions, by several authors, including Hotelling and Chu [3], Sen [11], [12], and Blom [2]. The last author obtains better estimates of the error than $o(n^{-k/2})$ under various conditions of differentiability and boundedness on F^{-1} and stipulations of the exact form of the tails of f. However, under the given minimal assumptions for $k = 1$, he shows that the error is $0(n^{-1/2})$ which is insufficient for our purposes.

To prove theorem 2.1 we require a lemma stated without proof in Tukey [13]. The elegant simplification of the author's original proof, which we present below, is due to Dr. S. S. Jogdeo.

LEMMA 2.1. *Let* X, Y *be random variables such that* $E(X^2) + E(Y^2) < \infty$ *and* $E(Y|X)$ *is a monotone increasing function of* X a.s.; *that is, there exists a monotone increasing function* $s(x)$, *such that* $s(X)$ *is a version of* $E(Y|X)$. *Then,* $\text{cov}(X, Y) \geq 0$.

PROOF. Let $s(x) = E(Y - E(Y)|x)$. Then since $s(x)$ is monotone increasing and $E(s(X)) = 0$, there exists a number c such that $s(x) \leq 0$ if $x \leq c$ and $s(x) \geq 0$ if $x \geq c$. But then, it is easily seen that

$$\begin{aligned}\text{cov}(X, Y) &= E\{XE[Y - E(Y)|X]\} \\ &= E[(X - c)s(X)] \geq 0. \end{aligned} \tag{2.1}$$

Q.E.D.

We now prove theorem 2.1. By lemma 2.1 it suffices to show that if $i < j$, $E(Z_{k,n}|Z_{i,n})$ is a continuous monotone increasing function of Z_{in}. It is well known that given $Z_{i,n}$, $Z_{j,n}$ is distributed as the $(j - i)$-th order statistic of a sample of $n - i$ from a population with density $f(x)/(1 - F(Z_{i,n}))$ for $x \geq Z_{i,n}$ and 0 otherwise. Then,

$$\begin{aligned} & E(Z_{k,n}|Z_{i,n}) \\ &= (j - i)\binom{n - i}{j - i}\int_0^1 F^{-1}[(1 - F(Z_{i,n}))t + F(Z_{i,n})]t^{j-i-1}(1 - t)^{n-j}\,dt, \end{aligned} \tag{2.2}$$

by a standard representation of the expected value of an order statistic. (See Wilks [14], p. 236). Monotonicity of $E(Z_{j,n}|Z_{i,n})$ now follows readily since $(1 - s)t + s$ is monotone in s for $0 \leq t < 1$. Left and right continuity of $E(Z_{j,n}|Z_{i,n})$ also is a consequence of (2.2), the continuity of F and F^{-1}, and the dominated convergence theorem. Theorem 2.1 is proved.

We now proceed to the proof of theorem 2.2(a). The given condition is equivalent to

$$\lim_{s\to 0} s^{1/\epsilon}F^{-1}(s) = 0 = \lim_{s\to 1}(1-s)^{1/\epsilon}F^{-1}(s). \tag{2.3}$$

Let j be the next largest natural number after $1/\epsilon$. Then $|F^{-1}(s)|^k \leq M^k[s(1-s)]^{-kj}$. Upon again applying the standard fact that $F(Z_{r,n})$ has a beta $(r, n-r+1)$ distribution we find that

$$\begin{aligned} E|Z_{r,n}|^k &= r\binom{n}{r}\int_0^1 |F^{-1}(s)|^k s^{r-1}(1-s)^{n-r}\,ds \\ &\leq M^k\binom{n}{r}\int_0^1 s^{r-kj-1}(1-s)^{n-r-kj}\,ds. \end{aligned} \tag{2.4}$$

Theorem 2.2(a), part 1, now follows upon taking $N(k, \alpha, \epsilon) = [kj/\alpha] + 1$ where $[x]$ is the greatest integer in x.

Conversely, if $E|Z_{n,r}|^\lambda < \infty$ for some $\lambda > 0$, then $\lim_{x\to\infty} x^\lambda P[|Z_{r,n}| \geq x] = 0$, which implies that

$$\lim_{x\to\infty} x^\lambda \int_x^\infty F^{r-1}(t)(1-F(t))^{n-r}\,dF(t) = 0. \tag{2.5}$$

Choose t_0 such that $F(t_0) > 0$. Then

$$\begin{aligned} \int_x^\infty F^{r-1}(t)(1-F(t))^{n-r}\,dF(t) &\geq F^{r-1}(t_0)\int_x^\infty (1-F(t))^{n-r}\,dF(t) \\ &= F^{r-1}(t_0)\,\frac{(1-F(x))^{n-r+1}}{(n-r+1)}. \end{aligned} \tag{2.6}$$

We conclude that

$$\lim_{x\to\infty} x^{\lambda/(n-r+1)}(1-F(x)) = 0, \tag{2.7}$$

and similarly,

$$\lim_{x\to\infty} x^{\lambda/(r+1)}F(-x) = 0. \tag{2.8}$$

Theorem 2.2(a) is proved.

The proof of 2.2(b) proceeds by a series of lemmas.

LEMMA 2.2. *Let $U_{1,n} < \cdots < U_{n,n}$ be the order statistics of a sample from the uniform distribution on* $[0, 1]$. *Let*

$$g_{n,k}(x) = k\binom{n}{k}x^{k-1}(1-x)^{n-k}, \qquad 0 \leq x \leq 1 \tag{2.9}$$

denote the density of $U_{k,n}$. Let

$$g^*_{n,k}(x) = n^{-1/2}g_{n,k}(xn^{-1/2} + k(n+1)^{-1}). \tag{2.10}$$

Then for every $\alpha > 0$ there exists $\tau(\alpha) > 0$, $M(\alpha) > 0$ such that $g^*_{n,k}(x) \leq M(\alpha)\exp - (\tau(\alpha)x^2)$ for $\alpha n \leq k \leq (1-\alpha)n$.

PROOF. By Stirling's approximation,

$$k\binom{n}{k} \leq Cn^{n+1/2}k^{-(k-1/2)}(n-k)^{-[(n-k)+1/2]}, \tag{2.11}$$

where C is independent of n, k. Let $p_n = k/n$. Then,

$$g^*_{n,k}(x) \leq Cp_n^{1/2}(1 - p_n)^{-1/2}p_n^{-k}(1 - p_n)^{n-k} \tag{2.12}$$

$$\left[n^{-1/2}x + \frac{np_n}{(n+1)}\right]^{k-1}\left[1 - \left(n^{-1/2}x + \frac{np_n}{(n+1)}\right)\right] n - k$$

for

$$-\frac{kn^{1/2}}{(n+1)} \leq x \leq n^{1/2}\left(1 - \frac{k}{(n+1)}\right). \tag{2.13}$$

Hence, after some simplification we obtain

$$\begin{aligned} &g^*_{n,k}(x) \\ &\leq C[p_n(1 - p_n)]^{-1/2}\left\{\left(1 + \frac{(x - \epsilon_n)}{p_n}n^{-1/2}\right)^{p_n - 1/n}\left(1 - \frac{(x - \epsilon_n)}{(1 - p_n)}n^{-1/2}\right)^{1-p_n}\right\}, \end{aligned} \tag{2.14}$$

where $\epsilon_n = n^{1/2}p_n(n + 1)^{-1}$ and $-k(n + 1)^{-1} \leq n^{-1/2}x \leq (1 - k(n + 1)^{-1})$.

Consider the function

$$q(y, \epsilon) = M^{-\epsilon}(1 + ya^{-1})^{a-\epsilon}(1 - yb^{-1})^b \exp \lambda y^2/2, \tag{2.15}$$

where

$$\begin{aligned} &0 \leq \lambda \leq \min\,[(a - \epsilon)/(a + b)^2, b/(a + b)^2] \leq 1, \\ &\qquad b \geq 0, \quad a \geq \epsilon \geq 0, \quad M \geq a/(a - \epsilon). \end{aligned} \tag{2.16}$$

Now,

$$\frac{\partial^2 \log q(y, \epsilon)}{\partial y^2} = \lambda - (a - \epsilon)(a + y)^{-2} - b(b + y)^{-2}, \tag{2.17}$$

and from the given restrictions on λ, it follows that for $-a \leq y \leq b$, $(\partial^2 q(y, \epsilon)/\partial y^2) \leq 0$. Moreover,

$$\frac{\partial \log q(y, \epsilon)}{\partial y} = \lambda y - (a - \epsilon)(a + y)^{-1} - b(b - y)^{-1}, \tag{2.18}$$

and from (2.18) we may see that,

$$\frac{\partial \log q(0, \epsilon)}{\partial y} \leq 0, \qquad \frac{\partial \log q(-\epsilon, \epsilon)}{\partial y} \geq 0, \tag{2.19}$$

since $\lambda \leq 1$.

Hence, $q(y, \epsilon)$ reaches its maximum, whatever be M, for $-\epsilon \leq y \leq 0$. We now show that for the given M, $q(y, \epsilon) \leq 1$, $-a \leq y \leq b$. Remark that $\log q(y, 0) \leq 0$ since $\log q(0, 0) = 0$, $\partial \log q(0, 0)/\partial y = 0$. But,

$$\frac{\partial \log q(y, \epsilon)}{\partial \epsilon} = -\log M - \log a^{-1}(a + y) \tag{2.20}$$

is ≤ 0 for M given, $-\epsilon \leq y \leq 0$, $\epsilon \geq 0$, and the inequality follows. We conclude that $(1 + ya^{-1})^{a-\epsilon}(1 - yb^{-1})^b \leq M^\epsilon \exp -\lambda y^2/2$ for $-a \leq y \leq b$, M and λ as given.

It follows from (2.18) and our preceding remarks that

$$g^*_{n,k}(x) \leq C[p_n(1 - p_n)]^{-1/2}M_n \exp - \lambda_n^2/2(x - \epsilon_n)^2 \tag{2.21}$$

for $-p_n \leq n^{-1/2}(x - \epsilon_n) \leq (1 - p_n)$ and $M_n = p_n(p_n - n^{-1})^{-1}$ and $\lambda_n = \min\{p_n - n^{-1}, (1 - p_n)\}$. But $-n^{1/2}p_n + \epsilon_n = -n^{1/2}k(n + 1)^{-1}$, and λ_n and M_n can be uniformly bounded away from 0 and ∞ since $\alpha \leq p_n \leq (1 - \alpha)$. The lemma is therefore proved since $g^*_{n,k}(x)$ vanishes off the given range.

REMARK. It is well known that $n^{-1/2}g_{n,k}(n^{-1/2}x + k(n + 1)^{-1})$, the density of $n^{1/2}[U_{k,n} - k(n + 1)^{-1}]$, converges to a normal density uniformly on compacts if $\alpha \leq p_n \leq (1 - \alpha)$. More precisely,

$$\sup_{\alpha < p_n < (1-\alpha)} |g^*_{n,k}(x) - [\tau(p_n)]^{-1}\varphi(x/\tau(p_n))| \tag{2.22}$$

converges to 0 uniformly on bounded intervals if $\tau^2(p_n) = p_n(1 - p_n)$, and $\varphi(x)$ is the normal density (see, for example, Wilks [13], p. 270). It now follows from our lemma that $E(U_{k,n} - k(n + 1)^{-1})^r = n^{-r/2}\tau^r(p_n)\mu^r + o(n^{-r/2})$ uniformly for $\alpha \leq p_n \leq (1 - \alpha)$ since we can, in particular, conclude that $n^{r/2}[U_{k,n} - k(n + 1)^{-1}]^r$ is uniformly integrable for k in the given range. We now prove lemma 2.3.

LEMMA 2.3. *Let F satisfy the general conditions of this section, and in addition, suppose that $f(x)$ is $\geq \lambda > 0$ for all x such that $0 < F(x) < 1$. Then, the conclusion of theorem 2.2(b) holds.*

PROOF. We remark first that the given conditions imply that $\{x|0 < F(x) < 1\}$ is an open interval, and hence X_1 is bounded and $E(Z^r_{k,n})$ exists for every k, r. Let $U_{k,n} = F(Z_{k,n})$. Then $U_{1,n} < \cdots < U_{n,n}$ are the order statistics of a sample from the uniform distribution on [0, 1]. Then, by the mean value theorem,

$$n^{1/2}[Z_{k,n} - F^{-1}(k(n + 1)^{-1})] = [\psi(U^*_{k,n})]^{-1}V_{k,n}, \tag{2.23}$$

where $U^*_{k,n}$ lies between $U_{k,n}$ and $k(n + 1)^{-1}$ and $V_{k,n} = n^{1/2}(U_{k,n} - k(n + 1)^{-1})$. Then,

$$\begin{aligned} &|E[n^{r/2}[Z_{k,n} - F^{-1}(k(n + 1)^{-1})]^r - \sigma^r(p_n)\mu_r| \\ &\leq \sup [|V_{k,n}| \leq A]|[\psi(U^*_{k,n})]^{-r} - [\psi(p_n)]^{-r}|E|V_{k,n}|^r \\ &+ [\psi(p_n)]^{-r}\left|\int_{[|V_{k,n}| \leq A]} V^r_{k,n}\, dP - [\tau(p_n)]^{-1}\right. \\ &\left.\int_{[|x| \leq A]} x^r\varphi\left(\frac{x}{\tau(p_n)}\right) dx\right| \\ &+ \frac{[\psi(p_n)]^{-r}}{\tau(p_n)} \int_{[|x| \leq A]} x^r\varphi\left(\frac{x}{\tau(p_n)}\right) dx \\ &+ \lambda^{-r}\int_{[|V_{k,n}| \geq A]} |V_{k,n}|^r\, dP. \end{aligned} \tag{2.24}$$

Since $|V_{k,n}| \leq A \Leftrightarrow |U_{k,n} - k(n + 1)^{-1}| \leq An^{-1/2}$ implies $|U^*_{k,n} - p_n| \leq An^{-1/2} + p_n n^{-1}$ and since ψ is uniformly continuous on the interval $[\beta, 1 - \beta]$ strictly contained in [0, 1], we may conclude, using lemma 2.2, that, as $n \to \infty$, the first two terms on the right-hand side of the inequality (2.30) go to 0 uniformly for $\alpha \leq p_n \leq (1 - \alpha)$. Again by lemma 2.2 the last term goes to 0 as

$A \to \infty$, uniformly in n, and the third term is evidently $o(1)$ as $A \to \infty$ uniformly for $\alpha \leq p_n \leq (1 - \alpha)$. The lemma follows.

We now prove theorem 2.2(b). Let $0 < \alpha - \delta$, and let $c = F^{-1}(\alpha - \delta)$, $d = F^{-1}(1 - (\alpha - \delta))$. Define

(i) $f_{c,d}(x) = f(x)$ for $c \leq x \leq d$,

(ii) $\phantom{f_{c,d}(x)} = f(c)$ for $c - (\alpha - \delta)[f(c)]^{-1} \leq x \leq c$,

(iii) $\phantom{f_{c,d}(x)} = f(d)$ for $d \leq x \leq d + (\alpha - \delta)[f(d)]^{-1}$.

Define $F_{c,d}$ to be the distribution with density $f_{c,d}$, $\psi_{c,d}$ to be the corresponding $f_{c,d}(F_{c,d}{}^{-1})$. Given our original sample, $X_1, \cdots, X_n$ generates a sample $\hat{X}_1, \cdots, \hat{X}_n$ from $f_{c,d}$ by defining $\hat{X}_i = X_i$ if $c \leq X_i \leq d$, $\hat{X}_i = T_1^i$ if $X_i < c$, $\hat{X}_i = T_2^i$ if $X_i > d$, where $\{T_1^i\}$, $\{T_2^i\}$, $1 \leq i \leq n$ are distributed independently of each other and the X_i's according to the uniform distribution on $(d - (\alpha - \delta)[f(c)]^{-1}, c)$ and $(d, d + (\alpha - \delta)[f(d)]^{-1})$ respectively. Let $\hat{Z}_{1,n} < \cdots < \hat{Z}_{n,n}$ denote the order statistics of $\{\hat{X}_i\}$, $1 \leq i \leq n$. Then, by lemma 2.3, $E(\hat{Z}_{k,n} - F^{-1}(k(n+1)^{-1}))^r = n^{-r/2}\sigma(p_n) + o(n^{-r/2})$ uniformly for $\alpha \leq p_n \leq (1 - \alpha)$, since for n sufficiently large

$$F_{c,d}^{-1}(k(n+1)^{-1}) = F^{-1}(k(n+1)^{-1}) \tag{2.25}$$

and $\psi_{c,d} = \psi$ if $\alpha \leq p_n \leq (1 - \alpha)$. Hence, to prove the theorem, it suffices to show that $n^{r/2}E|Z_{k,n} - \hat{Z}_{k,n}|^r \to 0$ uniformly for $\alpha \leq p_n \leq (1 - \alpha)$.

Suppose that $c < 0$, $d > 0$. The cases where c, d have the same sign may be dealt with similarly. Then

$$|Z_{k,n} - \hat{Z}_{k,n}| = |Z_{k,n} - \hat{Z}_{k,n}|(I[Z_{k,n} < c] + I[Z_{k,n} > d]) \tag{2.26}$$

where $I(A)$ is the indicator function of the event A. We may conclude that

$$\begin{aligned} n^{r/2}E|Z_{k,n} - \hat{Z}_{k,n}|^r &\leq n^{r/2}E[(|Z_{k,n}| + |c|)^r I[Z_{k,n} < c]] \\ &\quad + E[(|Z_{k,n}| + d + (\alpha + \delta)[f(d)]^{-1})^r I[Z_{k,n} > d]]\}. \end{aligned} \tag{2.27}$$

It therefore suffices to show

$$E(|n^{1/2}Z_{k,n}|^r I[Z_{k,n} < c]) \text{ and } E(|n^{1/2}Z_{k,n}|^r I[Z_{k,n} > d]) \to 0 \tag{2.28}$$

since it then follows that $|c|^r E(I[Z_{k,n} < c]) \to 0$. The other term behaves similarly.

By assumption there exists a natural number j such that $|F^{-1}(y)| \leq M[y(1 - y)]^{-j}$. Now,

$$\begin{aligned} E|n^{1/2}Z_{k,n}|^r I[Z_{k,n} < c] &= E(|n^{1/2}F^{-1}(U_{k,n})|^r I[U_{k,n} < \alpha - \delta]) \\ &\leq M^r n^{r/2} E(U_{k,n}{}^{-rj}(1 - U_{k,n})^{-rj} I[U_{k,n} < \alpha - \delta]) \\ &= M^r n^{r/2} \int_{[x < \alpha - \delta]} k \binom{n}{k} x^{k-rj-1}(1 - x)^{n-k-rj}\, dx. \end{aligned} \tag{2.29}$$

Without loss of generality, take r to be a natural number and choose n sufficiently large so that $(\alpha - \delta/2) \leq (k - rj)/(n - 2rj + 1)$ for all $k \geq \alpha n$. Then,

$$(2.30)\qquad n^{r/2}\int_{[x<\alpha-\delta]} k\binom{n}{k} x^{k-rj-1}(1-x)^{n-k-rj}\,dx$$
$$= n^{r/2}\frac{n(n-1)\cdots(n-2rj+1)}{(k-rj)\cdots(k-1)(n-k-rj+1)\cdots(n-k+1)}$$
$$\int_{[x<\alpha-\delta]}(k-rj)\binom{n-2rj}{k-rj}x^{k-rj-1}(1-x)^{(n-2rj)-(k-rj)}\,dx.$$

Now, $x < \alpha - \delta$ and $(k - rj) \geq (\alpha - \delta/2)(n - 2rj + 1)$ imply

$$(2.31)\qquad (n-2rj)^{1/2}(x-(k-rj)(n-2rj+1)^{-1}) < -(n-2rj)^{1/2}\delta/2.$$

Hence, the expression on the right of (2.30) is not larger than

$$(2.32)\qquad n^{r/2+2rj}\int_{[x<-(n-2rj)^{1/2}\delta/2]} g^{*}_{(n-2rj)(k-rj)}(x)\,dx$$
$$\leq Me^{-K(n-2rj)}(n-2rj)^{-1/2}n^{(r/2)+2rj},$$

where K, M depend only on α, by lemma 2.2 and the well-known approximation to the tail of the normal distribution (Feller [5], p. 166). The theorem is proved.

REMARK. The hypothesis that f be continuous and positive throughout on the carrier of F may obviously, if one is interested in the moments of a single percentile $Z_{[\alpha n]n}$, be weakened to f continuous in some neighborhood of $F^{-1}(\alpha)$. Our results thus contain the results of Hotelling and Chu [3] and Sen [11], [12]. Upon putting supplementary conditions on the local behavior of f, we may similarly obtain better estimates of the error term thus refining the results of Blom.

3. An invariance principle for the quantile function

We keep the general assumptions of section 2. Let us define a process on $[0, 1]$ by,

$$(3.1)\qquad Z_n(t) = n(Z^{*}_{k,n} - Z^{*}_{(k-1)n})t + Z^{*}_{k,n}(1-k) + kZ^{*}_{(k-1)n}$$

on $[(k-1)/n, k/n)$; $1 \leq k \leq n$, where $Z^{*}_{k,n} = Z_{k,n} - F^{-1}(k/(n+1))$ and $Z^{*}_{0,n} = 0$, $Z_n(1) = Z^{*}_{n,n}$.

Then, for every n, $Z_n(t)$ is a process with continuous sample functions. For each $0 < \alpha < \beta < 1$, there is a natural correspondence between $\{Z_n(t), \alpha \leq t \leq \beta\}$ and a probability P_n which belongs to the set $\mathcal{P}(C[\alpha, \beta])$ of all probability measures on the set of all continuous functions on $[\alpha, \beta]$ endowed with the uniform norm and the appropriate Borel field.

Let Q_n, Q be members of $\mathcal{P}(C[\alpha, \beta])$. Let $Q_n(t)$, $Q(t)$ be processes with continuous sample functions on $[\alpha, \beta]$, inducing the measures Q_n, Q on $C[\alpha, \beta]$. Then Q_n converges to Q in the sense of Prohorov if and only if for every h continuous and bounded on $C[\alpha, \beta]$, $\mathcal{L}[h(Q_n(t))] \rightarrow \mathcal{L}[h(Q(t))]$. Hájek [8] has shown that a necessary and sufficient condition for Prohorov convergence to a $Q \in \mathcal{P}[C[\alpha, \beta]]$ is,

(3.2) $$\lim_{\delta\to 0}\limsup_{n\to\infty} P[\sup_{s,t\in[\alpha,\beta],|t-s|<\delta}|Q_n(s) - Q_n(t)| \geq \epsilon] = 0$$

for every $\epsilon > 0$, and that,

(3.3) $$\mathfrak{L}[Q_n(s_1), \cdots, Q_n(s_k)] \to \mathfrak{L}[Q(s_1), \cdots, Q(s_k)]$$

for all $s_1, \cdots, s_k \in [\alpha, \beta]$.

For $Z_n(t)$ condition (3.2) is readily seen to be equivalent to

(3.4) $$\lim_{\delta\to 0}\limsup_{n\to\infty} P[\sup_{k,m\in[\alpha n,\beta n],|k-m|<\delta n}|Z^*_{k,n} - Z_{m,n}| \geq \epsilon] = 0.$$

After obtaining the main theorem of this section we were informed of an unpublished monograph of Hájek [7] in which a more general theorem than ours is stated under a regularity condition. Since in our simpler situation the regularity condition is unnecessary and our proof quite short, we felt it worthwhile to include theorem 3.1. Our original proof has been further simplified by a lemma ascribed to Rubin [7].

THEOREM 3.1. *Let $Z_n(t)$ be as above, $0 < \alpha < \beta < 1$. Then there exists a centered Gaussian process $Z(t)$ on $[\alpha, \beta]$ with continuous sample functions and covariance $s(1 - t)/\psi(s)\psi(t)$, $s \leq t$, such that $n^{1/2}Z_n(t)$ converges to $Z(t)$ in the sense of Prohorov on $[\alpha, \beta]$.*

PROOF. Let $U_n(t) = n(U^*_{k,n} - U^*_{(k-1)n})t + U^*_{k,n}(1 - k) + kU^*_{(k-1),n}$ on $[(k - 1)/n, k/n)$, $1 \leq k \leq n$, where $= U^*_0 = 0$, $U^*_{k,n}F(Z_{k,n}) - k/(n + 1)$, $U_n(1) = Z^*_{n,n}$. Let $V_n(t) = (U_{k,n} - U_{(k-1),n})^{-1}(U^*_{(k-1),n} - U^*_{k,n})t + (U_{k,n} - U_{(k-1),n})^{-1}$ $\{U_{(k-1),n}U^*_{k,n} - U_{k,n}U^*_{(k-1),n}\}$ on $[U_{(k-1),n}, U_{k,n})$, $1 \leq k \leq n + 1$, where $U_{(n+1),n} = 1$, $U^*_{(n+1),n} = 0$.

Now, $V_n(t)$ is essentially a version of the empirical cumulative, and Donsker [4] has shown that $n^{1/2}V_n(t)$ converges on $[0, 1]$, in the sense of Prohorov, to a Gaussian process $V(t)$ centered at 0 with continuous sample functions and covariance $s(1 - t)$ for $s \leq t$, a process known as the Brownian bridge.

From this follows (Rubin) lemma 3.1.

LEMMA 3.1. *The process $n^{1/2}U_n(t)$ converges in the sense of Prohorov on $[0, 1]$ to the Brownian bridge.*

PROOF. Clearly, (3.2) is satisfied in this case. To prove (3.1) remark that,

(3.5)
$$\begin{aligned} &P[\sup_{|s-t|<\delta} n^{1/2}|U_n(s) - U_n(t)| \geq \epsilon] \\ &\leq P[\sup_{|k-m|<2\delta n} n^{1/2}|U^*_{k,n} - U^*_{m,n}| \geq \epsilon] \qquad \text{for } n \geq \delta^{-1} \\ &\leq P[\sup_{|k-m|<2\delta n} n^{1/2}|U^*_{k,n} - U^*_{m,n}| \geq \epsilon, \max_{1\leq j\leq n}|U_{k,n} - k/(n + 1)| < \delta] \\ &\quad + P[\max_{1\leq j\leq n} |U_{k,n} - k/(n + 1)| \geq \delta]. \end{aligned}$$

Now $|k/n - m/n| < 2\delta$, $|U_{k,n} - k/(n + 1)| \leq \delta$, $|U_{m,n} - m/(n + 1)| \leq \delta$ implies $|U_{k,n} - U_{m,n}| < 5\delta$ for $n \geq \delta^{-1}$. We conclude that

(3.6)
$$\begin{aligned} &P[\sup_{|k-m|<2\delta n} n^{1/2}|U^*_{k,n} - U^*_{m,n}| \geq \epsilon, \max_{1\leq j\leq n} |U_{k,n} - k/(n + 1)| \leq \delta] \\ &\qquad \leq P[\sup_{|s-t|<5\delta} n^{1/2}|V_n(s) - V_n(t)| \geq \epsilon] \to 0. \end{aligned}$$

Therefore, to prove the lemma it suffices to show that

(3.7) $$P[\max_{1 \leq k \leq n} |U_{k,n} - k/(n+1)| \geq \epsilon] \to 0.$$

But,

(3.8) $$\begin{aligned} P[\max_{1 \leq k \leq n} |U_{k,n} - k/(n+1)| \geq \epsilon] &\leq \sum_{n\alpha \leq k \leq n(1-\alpha)} P[U_{k,n} - k/(n+1)| \geq \epsilon] \\ &+ P[\max_{1 < k < \alpha n} |U_{k,n} - k/(n+1)| > \epsilon] \\ &+ P[\max_{(1-\alpha)n \leq k \leq n} |U_{k,n} - k/(n+1)| \geq \epsilon]. \end{aligned}$$

Choose α such that $\alpha < \epsilon/2$. Then

(3.9) $$P[\max_{1 \leq k \leq \alpha n} |U_{k,n} - k/(n+1)| \geq \epsilon] \leq P[U_{[\alpha n]n} \geq \epsilon/2] \to 0.$$

Similarly, $P[\max_{(1-\alpha)n \leq k \leq n} |U_{k,n} - k/(n+1)| \geq \epsilon] \to 0$ and by lemma 2.2,

(3.10) $$\sum_{n\alpha \leq k \leq n(1-\alpha)} P[|U_{k,n} - k/(n+1)| \geq \epsilon] \leq 2nM(\alpha) \exp - Kn \to 0.$$

In fact, by the Borel-Cantelli lemma, $\max |U_{k,n} - k/(n+1)|$ converges almost surely to 0. Lemma 3.1 follows.

We require the following generalization to processes of a well-known theorem of Slutsky.

LEMMA 3.2. *Let $\{Q_n\}$ be a sequence of processes with continuous sample functions on $[\alpha, \beta]$ which converge in the sense of Prohorov to Q on $[\alpha, \beta]$. Let $\{V_n\}$ be a sequence of processes with continuous sample functions such that*

(3.11) $$P[\sup_{\alpha \leq t \leq \beta} |V_n(s) - b(s)| \geq \epsilon] \to 0$$

for every $\epsilon > 0$ and a fixed continuous function $b(s)$. Then,

(a) *the sequence $Q_n(s)V_n(s)$ converges in the sense of Prohorov to $b(s)Q(s)$;*

(b) *the sequence $Q_n(s) + V_n(s)$ converges to $b(s) + Q(s)$.*

PROOF. We prove (a); the proof of (b) is similar. It is clear that

(3.12) $$\mathcal{L}[V_n(s_1)Q_n(s_1), \cdots, V_n(s_k)Q_n(s_k)] \to \mathcal{L}[b(s_1)Q(s_1), \cdots, b(s_k)Q(s_k)]$$

for all $\alpha < s_i < \beta$, k finite, by the convergence of Q_n and V_n and the ordinary Slutsky theorem. It remains to show that V_nQ_n satisfies (3.2). Let $M_{1,n} = \sup_{\alpha \leq t \leq \beta} |Q_n(t)|$, $M_{2,n} = \sup_{\alpha \leq t \leq \beta} |V_n(t)|$. Then by an elementary inequality

(3.13) $$\begin{aligned} P[\sup_{s,t \in [\alpha,\beta], |t-s| < \delta} |Q_n(s)V_n(s) - Q_n(t)V_n(t)| \geq \epsilon] \\ \leq P[\sup_{s,t \in [\alpha,\beta], |t-s| < \delta} M_{1,n}|V_n(t) - V_n(s)| \geq \epsilon/2] \\ + P[\sup_{s,t \in [\alpha,\beta], |t-s| < \delta} M_{2,n}|Q_n(s) - Q_n(t)| \geq \epsilon/2]. \end{aligned}$$

But by the convergence of Q_n and V_n, there exists M_a such that $P[M_{1,n} \leq M_a] \geq 1 - a$ and $P[M_{2,n} \leq M_a] \geq 1 - a$ for all n. We conclude that

(3.14) $$\begin{aligned} \lim_{\delta \to 0} \limsup_n P[\sup_{s,t \in [\alpha,\beta], |t-s| \leq \delta} M_{1,n}|V_n(t) - V_n(s)| \geq \epsilon/2] \\ \leq \lim_{\delta} \limsup_n P[\sup_{s,t \in [\alpha,\beta], |t-s| < \delta} |V_n(t) - V_n(s)| > \epsilon/2M_a] + a. \end{aligned}$$

The lemma follows by applying a similar argument to the second term of (3.13).

Now, define $\tilde{Z}_n(t) = F^{-1}(U_n(t) + nt/(n+1)) - F^{-1}(nt/(n+1))$, $\alpha \leq t \leq \beta$.

By the mean value theorem,

$$n^{1/2}\tilde{Z}_n(t) = (\psi[Y_n(t)])^{-1}n^{1/2}U_n(t) \tag{3.15}$$

where $Y_n(t)$ lies between $nt/(n+1)$ and $U_n(t) + nt/(n+1)$. The process $[\psi(Y_n(t))]^{-1}$ necessarily possesses continuous sample functions on $[\alpha, \beta]$. From the convergence of $n^{1/2}U_n(t)$ it follows that

$$P[\sup_{0 \leq t \leq 1} |U_n(t)| \geq \epsilon] \to 0 \tag{3.16}$$

for every $\epsilon > 0$, and therefore, since $[\psi(x)]^{-1}$ is uniformly continuous for $0 < a < x < b < 1$, that

$$P[\sup_{\alpha \leq t \leq \beta} |[\psi(Y_n(t))]^{-1} - [\psi(t)]^{-1}| \geq \epsilon] \to 0 \tag{3.17}$$

for every $\epsilon > 0$. Hence by lemma 3.2 we obtain that $n^{1/2}\tilde{Z}_n(t)$ converges on $[\alpha, \beta]$ to $\hat{Z}(t)[\psi(t)]^{-1}$, a centered Gaussian process with the covariance structure given in the statement of theorem 3.1. Now, to show that (3.2) holds for $n^{1/2}Z_n(t)$, it suffices to check that (3.4) is satisfied. But $Z_n(k/n) = \tilde{Z}_n(k/n) + Z^*_{k,n}$. By the necessity of (3.2),

$$\begin{aligned} 0 &= \lim_{\delta \to 0} \limsup_{n} P[\sup_{s,t \in [\alpha,\beta], |t-s| < \delta} n^{1/2}|\tilde{Z}_n(s) - \tilde{Z}_n(t)| \geq \epsilon] \\ &\geq \lim_{\delta \to 0} \limsup_{n} P[\sup_{\alpha n < k,m < \beta n, |k-m| < \delta n} n^{1/2}|Z^*_{k,n} - Z^*_{m,n}| \geq \epsilon], \end{aligned} \tag{3.18}$$

and the theorem follows.

4. Convergence of linear systematic statistics

Let $\{a_{k,n}\}$ $1 \leq k \leq n$, $n \geq 1$ be a double sequence of constants. Form the statistic $T_n = \sum_{k=1}^{n} a_{k,n} Z_{k,n}$. Such quantities are known as systematic statistics and are of use in estimation and testing (cf. Jung [9] and Blom [2]). The convergence of moments of T_n and the asymptotic normality of T_n have been investigated by several writers, including Jung [9], Blom [2], Hájek [7], and more recently, Gastwirth, Chernoff, and Johns (private communication), [6], and Govindarajulu [15] under various regularity conditions.

Our conditions are somewhat simpler, though by no means inclusive. We are essentially able to deal with all systematic statistics which involve the extremal statistics to the same extent as the sample mean or less.

Let us define $M_n(t) = \sum_{k<nt} a_{k,n}$. Then $M_n(t)$ is of bounded variation and

$$T_n - \int_0^1 F^{-1}(nt/(n+1))\, dM_n(t) = \int_0^1 Z_n(t)\, dM_n(t). \tag{4.1}$$

We then have a modification and generalization of a theorem of Hájek [7].

THEOREM 4.1. *Under the general conditions of section 2, suppose that there exists $\alpha > 0$ such that $a_{k,n} = 0$ for $k \leq \alpha n$, $k \geq (1-\alpha)n$ for all $n \geq N$. Suppose that there exists $M(t)$ of bounded variation in $[\alpha, 1-\alpha]$ such that $M_n(t) \to M(t)$ on a dense set of t, $\alpha \leq t \leq (1-\alpha)$ and that $\overline{V}(M_n) \leq M' < \infty$ for all n where $\overline{V}(M_n)$ denotes the total variation of M_n. Then,*

$$\mathfrak{L}[n^{1/2}(T_n - \int_0^1 F^{-1}(t)\, dM_n(t))] \to N(0, \sigma^2(M, F)) \tag{4.2}$$

where N denotes the normal distribution and

$$\sigma^2(M, F) = 2 \int_0^1 \int_0^t s(1 - t)[\psi(s)\psi(t)]^{-1}\, dM(s)\, dM(t). \tag{4.3}$$

PROOF. We remark that since M is constant off $(\alpha, 1 - \alpha)$ and the integrand is bounded in that interval, by our assumptions $\sigma^2(M, F) < \infty$. To prove the theorem it suffices to show that

$$\mathfrak{L}\left(\int_0^1 n^{1/2} Z_n(t)\, dM_n(t)\right) \to \mathfrak{L}\left(\int_0^1 Z(t)\, dM(t)\right), \tag{1}$$

and that

$$\int_0^1 |F^{-1}(t) - F^{-1}(nt/(n + 1))| dM_n(t) = o(n^{-1/2}), \tag{2}$$

since by (4.1) it readily follows that $\int_0^1 Z(t)\, dM(t)$ has the desired distribution. By theorem 3.1 and a theorem of Prohorov ([10], p. 166), relation (1) holds if

$$\int_0^1 f(t)\, dM_n(t) \to \int_0^1 f(t)\, dM(t) \tag{4.4}$$

uniformly for equicontinuous, uniformly bounded (compact) sets of continuous functions f on $[\alpha, (1 - \alpha)]$. But this readily follows from our assumptions upon using the method of proof of Helly's theorem. Relation (2) follows trivially since

$$\left| F^{-1}\left(\frac{nt}{(n + 1)}\right) - F^{-1}(t) \right| \leq \frac{M''t}{(n + 1)} \tag{4.5}$$

for $t \in [\alpha, (1 - \alpha)]$ by the mean value theorem and continuity of $\psi(t)$. Theorem 4.1 is proved. The following corollaries are immediate.

COROLLARY 4.1. *If* $\overline{V}(M_n - M) = o(n^{-1/2})$, *then theorem* 4.1 *holds with* $\int_0^1 F^{-1}(t)\, dM_n(t)$ *replaced by* $\int_0^1 F^{-1}(t)\, dM(t)$.

COROLLARY 4.2. *If*

$$M_n(t) = n^{-1} \sum_{kn^{-1} < t} h(kn^{-1}), \tag{4.6}$$

that is, $a_{k,n} = n^{-1}h(kn^{-1})$, $h = 0$ *on* $[\alpha, (1 - \alpha)]^c$, *and* h *is continuously differentiable or, more generally, obeys a Lipschitz condition of order* $> \frac{1}{2}$ *on* $[\alpha, (1 - \alpha)]$, *then theorem* 4.1 *holds with* $\int_0^1 F^{-1}(t)\, dM_n(t)$ *replaced by* $\int_0^1 F^{-1}(t)h(t)\, dt$ *and* $M(t) = \int_0^t h(s)\, ds$.

PROOF. The condition is clearly sufficient to guarantee

$$\int_0^1 F^{-1}(t)\, dM_n(t) = n^{-1} \sum_{k=1}^{n} h(kn^{-1}) \tag{4.7}$$

to equal $\int_0^1 F^{-1}(t)h(t)\, dt + o(n^{-1/2})$.

REMARK. (1) In particular, corollary 4.1 applies if

$$a_{k,n} = \int_{(k-1)/n}^{k/n} h(t)\, dt + o(n^{-3/2}) \tag{4.8}$$

uniformly for $\alpha n \leq k \leq (1-\alpha)n$ for some function $h(t)$ in $L_1([\alpha, (1-\alpha)])$. This provides an alternative system of weights for the estimates considered by Jung.

(2) Corollary 4.2 establishes the asymptotic normality of the trimmed and Winsorized means of Tukey (see Bickel [1]).

THEOREM 4.2. *Under the conditions of theorem* 4.1 *if* $|x|^\epsilon[1 - F(x) + F(-x)]$ *tends to* 0 *as* $x \to \infty$ *for some* $\epsilon > 0$, $E(T_n^k)$ *exists eventually for every natural number* k *and*

$$n^{k/2}E(T_n - \int_0^1 F^{-1}(t)\, dM_n(t))^k \to \sigma^2(M, F)\mu_k, \qquad \text{as} \quad n \to \infty. \tag{4.9}$$

PROOF. By the linearity property of the expectation and (4.1),

$$\begin{aligned} E\left(T_n - \int_0^1 F^{-1}(t)\, dM_n(t)\right)^k \\ = \int_\alpha^{(1-\alpha)} \int_\alpha^{(1-\alpha)} E\left(\prod_{i=1}^k [Z_n(s_i) - F^{-1}(s_i)]\right) \prod_{i=1}^k dM_n(s_i). \end{aligned} \tag{4.10}$$

An easy extension of theorem 2.2 implies that

$$n^{k/2}E\left(\prod_{i=1}^k [Z_n(s_i) - F^{-1}(s_i)]\right) \to E\left[\prod_{i=1}^k Z(s_i)\right] \tag{4.11}$$

uniformly for $\alpha \leq s_i \leq (1-\alpha)$. We conclude that

$$E\left(T_n - \int_0^1 F^{-1}(t)\, dM_n(t)\right)^k \to E\left[\int_\alpha^{(1-\alpha)} Z(t)\, dM(t)\right]^k, \tag{4.12}$$

and the theorem is proved.

REMARK. (1) This establishes convergence of the variance for the trimmed and Winsorized means as stated in Bickel [1].

(2) Under the conditions of corollaries 4.1 or 4.2, $\int_0^1 F^{-1}(t)\, dM_n(t)$ may be replaced by $\int_0^1 F^{-1}(t)\, dM(t)$. We can now prove the following theorem.

THEOREM 4.3. *Suppose* $E(X_1^2) < \infty$. *Let* $M_n(t)$ *defined as before tend to* $M(t)$ *on a dense set in* $[0, 1]$, $\overline{V}(M_n) < \infty$ *on* $[0, 1]$. *Assume, furthermore, that for some* $\alpha > 0$, $|a_{k,n}| \leq M''n^{-1}$ *for all* $k \leq \alpha n$, $k \geq (1-\alpha)n$. *Then,*

$$\mathcal{L}[n^{1/2}(T_n - E(T_n))] \to N(0, \sigma^2(M, F)). \tag{4.13}$$

PROOF. We require first the following lemma.

LEMMA 4.1. *Let* X_n *be a sequence of random variables. Let* $Y_{m,n}$ *be another double sequence of random variables such that,*

(i) $\mathcal{L}(Y_{m,n}) \to \mathcal{L}(Y_m)$ *for each* m *as* $n \to \infty$,

(ii) $\mathcal{L}(Y_m) \to \mathcal{L}(Y)$ *as* $m \to \infty$,

(iii) $\limsup_m \limsup_n P[|X_n - Y_{m,n}| \geq \delta] = 0$,

for every $\delta > 0$. *Then,* $\mathcal{L}(X_n) \to \mathcal{L}(Y)$.

PROOF. First note that

$$(4.14)\quad \begin{aligned} &|P[Y_{m,n} < x] - P[X_n < x]| \\ &\quad \leq P[Y_{m,n} < x, X_n \geq x] + P[X_n < x, Y_{m,n} \geq x] \\ &\quad \leq P[x - \delta < Y_{m,n} < x, X_n \geq x] + P[X_n < x, x \leq Y_{m,n} < x + \delta] \\ &\qquad + 2P[|X_n - Y_{m,n}| \geq \delta] \\ &\quad \leq P[x - \delta < Y_{m,n} < x + \delta] + 2P[|X_n - Y_{m,n}| \geq \delta]. \end{aligned}$$

Let x, $x - \delta$, $x + \delta$ be points of continuity of $\mathfrak{L}(Y_m)$ for all m and of $\mathfrak{L}(Y)$ as well. Then,

$$(4.15)\quad \begin{aligned} &\limsup_n |P[Y_{m,n} < x] - P[X_n < x]| \\ &\qquad \leq P[x - \delta < Y_m < x + \delta] + 2 \limsup_n P[|X_n - Y_m| \geq \delta]. \end{aligned}$$

Now, take the lim sup as $m \to \infty$, reducing the second term to 0, the first to $P[x - \delta < Y < x + \delta]$, and finally the limit as $\delta \to 0$. It follows that

$$(4.16)\quad \lim_n P[X_n < x] = \lim_m \lim_n P[Y_{m,n} < x] = P[Y < x].$$

The lemma is proved.

By theorem 4.2, if $U_n(t)$ is the distribution function of the measure which assigns mass $1/n$ to i/n, $1 \leq i \leq n$, then

$$(4.17)\quad \operatorname{var} n^{1/2} \int_\beta^{(1-\beta)} Z_n(t)\, dU_n(t) \to 2 \int_\beta^{(1-\beta)} \int_\beta^t s(1 - t)[\psi(s)\psi(t)]^{-1}\, ds\, dt.$$

Let $\lambda_1 = F^{-1}(\beta)$, $\lambda_2 = F^{-1}(1 - \beta)$. Then,

$$(4.18)\quad 2 \int_\beta^{(1-\beta)} \int_\beta^t s(1 - t)[\psi(s)\psi(t)]^{-1}\, ds\, dt = 2 \int_{\lambda_1}^{\lambda_2} \int_{\lambda_1}^y F(x)(1 - F(y))\, dx\, dy$$

by a change of variable. The latter integral may readily be evaluated. Thus,

$$(4.19)\quad \begin{aligned} 2 \int_{\lambda_1}^{\lambda_2} \int_{\lambda_1}^y F(x)\, dx\, dy &= 2\lambda_2 \int_{\lambda_1}^{\lambda_2} F(x)\, dx - 2 \int_{\lambda_1}^{\lambda_2} y d\left(\int_{\lambda_1}^y F(x)\, dx\right) \\ &= \int_{\lambda_1}^{\lambda_2} y^2\, dF(y) + \beta\lambda_1^2 - (1 - \beta)\lambda_2^2 + 2\lambda_2 \int_{\lambda_1}^{\lambda_2} F(x)\, dx, \end{aligned}$$

and

$$(4.20)\quad 2 \int_{\lambda_1}^{\lambda_2} F(y) \left(\int_{\lambda_1}^y F(x)\, dx\right) dy = \int_{\lambda_1}^{\lambda_2} d\left(\int_{\lambda_1}^y F(x)\, dx\right)^2 = \left(\int_{\lambda_1}^{\lambda_2} F(x)\, dx\right)^2.$$

We obtain, therefore, after some simplification,

$$(4.21)\quad \begin{aligned} &2 \int_\beta^{(1-\beta)} \int_\beta^t s(1 - t)[\psi(s)\psi(t)]^{-1}\, ds\, dt \\ &\qquad = \int_{\lambda_1}^{\lambda_2} t^2\, dF(t) + \beta(\lambda_1^2 + \lambda_2^2) - \left(\int_{\lambda_1}^{\lambda_2} t\, dF(t) + \beta(\lambda_1 + \lambda_2)\right)^2. \end{aligned}$$

Since $E(X_1^2) < \infty$, it readily follows that as $\beta \to 0$,

$$(4.22)\quad \begin{aligned} &2 \int_\beta^{(1-\beta)} \int_\beta^y s(1 - t)[\psi(s)\psi(t)]^{-1}\, ds\, dt \to \operatorname{var} X_1 \\ &\qquad = 2 \int_0^1 \int_0^t s(1 - t)[\psi(s)\psi(t)]^{-1}\, ds\, dt. \end{aligned}$$

Let $\alpha \geq \beta_m \to 0$ and define

$$Y_{m,n} = \int_{\beta_m}^{(1-\beta_m)} [Z_n(t) - E(Z_n(t)] \, dM_n(t). \tag{4.23}$$

By theorem 4.2, $\mathcal{L}(Y_{m,n}) \to \mathcal{L}(Y_m)$ as $n \to \infty$, where Y_m is

$$N\left[0, \int_{\beta_m}^{(1-\beta_m)} s(1-t)[\psi(s)\psi(t)]^{-1} \, dM(s) \, dM(t)\right]. \tag{4.24}$$

Of course, $\mathcal{L}(Y_m) \to N(0, \sigma^2(M, F))$ as $m \to \infty$, where

$$\sigma^2(M, F) \leq (M'') \operatorname{var} X_1 + 2 \int_\alpha^{(1-\alpha)} \int_\alpha^t s(1-t)[\psi(s)\psi(t)]^{-1} \, dM(s) \, dM(t), \tag{4.25}$$

which is finite. Now

$$P[n^{1/2}|Y_{m,n} - (T_n - E(T_n))| \geq \epsilon] \leq \epsilon^{-2} n \operatorname{var}(Y_{m,n} - T_n) \tag{4.26}$$

by Tchebichev's inequality. But,

$$\begin{aligned} \operatorname{var}(Y_m - T_n) &= | \sum_{k,\ell \in [\beta_m n, (1-\beta_m)n]^c} a_{k,n} a_{\ell,n} \operatorname{cov}(Z_{k,n}, Z_{\ell,n}) \\ &\leq n^{-2}[M'']^2 \sum_{k,\ell \in [\beta_m n, (1-\beta_m)n]^c} \operatorname{cov}(Z_{k,n}, Z_{\ell,n}) \end{aligned} \tag{4.27}$$

by theorem 2.1. Now it follows that

$$n \operatorname{var}(Y_{m,n} - T_n) \leq M'' n \operatorname{var}\left(\overline{X} - \int_{\beta_m}^{(1-\beta_m)} Z_n(t) \, dU_n(t)\right) \tag{4.28}$$

where $\overline{X}$ is the sample mean. But again, by theorem 2.1,

$$\begin{aligned} n \operatorname{var}\left(\overline{X} - \int_{\beta_m}^{(1-\beta_m)} Z_n(t) \, dU_n(t)\right) \\ \leq n \operatorname{var} \overline{X} - n \operatorname{var} \int_{\beta_m}^{(1-\beta_m)} Z_n(t) \, dU_n(t). \end{aligned} \tag{4.29}$$

We conclude that,

$$\begin{aligned} \limsup_n P[n^{1/2}|Y_{m,n} - (T_n - E(T_n))| \geq \epsilon] \\ \leq \operatorname{var} X_1 - \int_{\beta_m}^{(1-\beta_m)} \int_{\beta_m}^t s(1-t)[\psi(s)\psi(t)]^{-1} \, ds \, dt. \end{aligned} \tag{4.30}$$

By our previous remarks we see that the requirements of lemma 4.1 are satisfied and the theorem is proved.

REMARK. Theorem 4.3 implies the asymptotic normality of the estimates considered by Jung.

The following corollary is immediate.

COROLLARY 4.2. *Under the conditions of theorem* 4.3, $n \operatorname{var} T_n \to \sigma^2(M, F)$.

In the general case we can only establish the following corollary.

COROLLARY 4.3. *Under the conditions of theorem* 4.3, *if* $n^{1/2}\overline{V}(M_n - M) \to 0$ *on* $[0, 1]$, *then,*

$$E(T_n) \to \int_0^1 F^{-1}(t) \, dM(t). \tag{4.31}$$

PROOF. Since clearly

$$E\left(\int_{\beta_m}^{(1-\beta_m)} Z_n(t)\, dM_n(t)\right) \to 0, \tag{4.32}$$

it suffices to show that lim $\sup_m$ lim $\sup_n |E(T_n - \hat{Y}_{m,n})| \to 0$ where

$$\hat{Y}_{m,n} = \int_{\beta_m}^{(1-\beta_m)} Z_n(t)\, dM_n(t) + \int_{\beta_m}^{(1-\beta_m)} F^{-1}(t)\, dM_n(t). \tag{4.33}$$

But,

$$|E(T_n - Y_{m,n})| \le n^{-1}M'' \sum_{k \in [\beta_m n, (1-\beta_m)n]^c} |E(Z_{k,n})|. \tag{4.34}$$

Define $R_n(t)$ to be the measure assigning mass $1/n$ to k/n if $E(Z_{k,n}) \ge 0$, $-(1/n)$ otherwise, $1 \le k \le n$. Then,

$$n^{-1} \sum_{\beta_m n \le k \le (1-\beta_m)n} |E(Z_{k,n})| = E \int_{\beta_m}^{(1-\beta_m)} Z_n(t)\, dR_n(t) + \int_{\beta_m}^{(1-\beta_m)} |F^{-1}(t)|\, dR_n(t) \to \int_{\beta_m}^{(1-\beta_m)} |F^{-1}(t)|\, dt. \tag{4.35}$$

Now, $n^{-1} \sum_{k=1}^n |E(Z_{k,n})| \le n^{-1} \sum_{k=1}^n E|Z_{k,n}| = E|X_1|$. The corollary follows from (4.34) and (4.35).

This result is, of course, unsatisfactory since it is precisely as an asymptotically normal estimate of $\int_0^1 F^{-1}(t)\, dM(t)$ that T_n is usually employed. Slightly less general but more satisfactory is corollary 4.4.

COROLLARY 4.4. *Under the conditions of theorem 4.3, if there exists A such that $f(x)$ is monotone for $|x| \ge A$, and $n^{1/2}\overline{V}(M_n - M) \to 0$ on $[0, 1]$, then*

$$n^{1/2}\left[E\left(T_n - \int_0^1 F^{-1}(t)\, dM(t)\right)\right] \to 0, \tag{4.36}$$

and hence $n^{1/2}(T_n - \int_0^1 F^{-1}(t)\, dM(t))$ has asymptotically an $N(0, \sigma^2(M, F))$ distribution.

PROOF. Let $0 < \beta_m + \delta < \min(F(-A), 1 - F(A), \alpha)$. Denote $F^{-1}(\beta_m + \delta)$ by λ_m. Let

$$\begin{aligned} f_m(x) &= f(x), \quad x < \lambda_m \\ &= f(\lambda_m), \qquad \lambda_m \le x < \lambda_m + (1 - \beta_m)[f(\lambda_m)]^{-1}. \end{aligned} \tag{4.37}$$

Define,

$$\begin{aligned} X_i(m) &= X_i, \qquad X_i < \lambda_m, \\ &= T_i, \qquad X_i > \lambda_m, \end{aligned} \tag{4.38}$$

where $\{T_i\}$ $1 \le 1 \le n$ is a sequence of random variables uniform on $(\lambda_m, \lambda_m + (1 - \beta_m)[f(\lambda_m)]^{-1})$ and independent of each other and of the X_i. Let $Z_{1,n}(m) < \cdots < Z_{n,n}(m)$ denote the order statistics of the $X_i(m)$. Then, clearly,

$$\sum_{k \le \beta_m n} a_{k,n}(Z_{k,n} - Z_{k,n}(m)) < nM''(|Z_{([\beta_m n]+1)n}| + K)I[Z_{([\beta_m n]+1)n} > \lambda_m] \tag{4.39}$$

where $K = \max(|\lambda_m|, |\lambda_m + (1 - \beta_m)[f(\lambda_m)]^{-1}|)$. We can show by arguments similar to those employed in the proof of theorem 2.2 that

(4.40) $$n^{1/2}(E[\sum_{k \leq \beta_m n} (Z_{k,n} - Z_{k,n}(m)]) \to 0$$

for every fixed m. Since

(4.41) $$n^{1/2}E\left(\int_{\beta_m}^{(1-\beta_m)} Z_n(t)\, dM_n(t)\right) = n^{1/2}\left(E(T_n) - \int_{\beta_m}^{(1-\beta_m)} F^{-1}(t)\, dM(t) + o(n^{-1/2})\right) \to 0,$$

by the remarks following theorem 4.2, we conclude from our previous remark that we need only show

(4.42) $$\limsup_m \limsup_n n^{1/2}\left|E\left(\sum_{k \leq \beta_m n} a_{k,n}Z_{k,n}(m) - \int_0^{\beta_m} F^{-1}(t)\, dM_n(t)\right)\right| = 0$$

and a similar proposition for the upper tail.

It clearly suffices to establish that

(4.43) $$n^{-1/2} \sum_{k \leq \beta_m n} |E(Z_{k,n}(m) - F^{-1}(k/(n+1))| \to 0.$$

Let F_m be the distribution of $X_1^{(m)}$. Then, F_m is either convex or concave depending on whether f_m is monotone increasing or decreasing for $x \leq A$. Hence, F_m^{-1} is concave or convex, and by Jensen's inequality,

(4.44) $$E(Z_{k,n}(m) - F_m^{-1}(k/(n+1)) = E[F_m^{-1}(U_{k,n})] - F_m^{-1}(E(U_{k,n}))$$

has the same sign for all k. But $F_m^{-1}(k/(n+1)) = F^{-1}(k/(n+1))$ for $k \leq \beta_m n$, and we conclude that,

(4.45) $$n^{-1/2} \sum_{k \leq \beta_m n} |E(Z_{k,n}(m)) - F^{-1}(k/(n+1))| = n^{-1/2}|E(\sum_{k \leq \beta_{m,n}} Z_{k,n}(m) - F^{-1}(k/(n+1))|.$$

Now $\limsup_m \limsup_n n^{-1/2}E[\sum_{k \leq \beta_m n} (Z_{k,n}(m) - F^{-1}(k/(n+1))] = 0$ readily follows from theorem 4.2, and the identity

(4.46) $$E[n^{-1} \sum_{k=1}^{n} Z_{k,n}(m)] = \int_0^1 F_m^{-1}(t)\, dt,$$ Q.E.D.

REMARK. The condition $E(X_1) < \infty$ clearly suffices for corollaries 4.3 and 4.4.

Jung [9] and Blom [2] have shown convergence of moments under various conditions. The condition of Jung in the case of convergence of the mean may be weakened to $a_{k,n} = n^{-1}\, a(k/n)$ when a has at least two derivatives bounded on $[0, 1]$. Corollary 4.2 then holds with the error being $0(n^{-1})$ rather than just $o(n^{-1/2})$ as shown in corollary 4.4.

This completes our present study of linear systematic statistics. Clearly, there are still many open questions. The restriction on $M(t)$ leaves statistics which involve the extremal order statistics in a more significant fashion than the mean undealt with. On the other hand, the restriction $E(X_1^2) < \infty$ seems too restrictive for systematic statistics involving the extremes to a lesser extent than the mean. Hájek [7], using his more general invariance principle, states a theorem which covers some situations we cannot deal with. Unfortunately his regularity conditions do not cover the mean itself.

The invariance principle of section 3, simple though it is, has other interesting applications. In a forthcoming paper J. L. Hodges and the author have applied it to determine the behavior of

$$\operatorname*{med}_{k \leq n} \tfrac{1}{2}[Z_{k,(2n)} + Z_{(2n-k+1),(2n)}], \tag{4.47}$$

an asymptotically nonnormal robust estimate of location, which is much easier to compute than the Hodges-Lehmann estimate $\text{med}_{i \leq j} (X_i + X_j)/2$.

Note added in proof. Results similar to theorem 2.2 (a) and (b) have appeared in W. Van Zwet, Convex Transformations of Random Variables, Thesis, Amsterdam, 1964. (In particular, 2.2(a) was noted and a stronger form of 2.2(b) proved under the assumption that f is continuously differentiable.)

REFERENCES

[1] P. J. Bickel, "On some robust estimates of location," *Ann. Math. Statist.*, Vol. 36 (1965), pp. 847–858.

[2] G. Blom, *Statistical Estimates and Transformed Beta Variables*, New York, Wiley; Stockholm, Almqvist and Wiksell, 1958.

[3] J. T. Chu and H. Hotelling, "Moments of the sample median," *Ann. Math. Statist.*, Vol. 26 (1955), pp. 593–606.

[4] M. D. Donsker, "Justification and extension of Doob's heuristic approach to the Kolmogorov-Smirnov theorems," *Ann. Math. Statist.*, Vol. 23 (1952), pp. 277–281.

[5] W. Feller, *Introduction to Probability Theory and Its Applications*, New York, Wiley, 1957 (2d ed.).

[6] H. Chernoff, J. T. Gastwirth, and M. V. Johns, "Linear combinations of functions of order statistics," Stanford University Technical Report, 1965.

[7] J. Hájek, Lectures delivered at IMS Summer Institute, Michigan State University, 1962, unpublished.

[8] ———, "Extension of the Kolmogorov-Smirnov test to regression alternatives," *Bernoulli-Bayes-Laplace Memorial Seminar*, Berlin, Springer Verlag, 1965.

[9] J. Jung, "On linear estimates defined by a continuous weight function," *Ark. Mat.*, Vol. 3 (1955), pp. 199–209.

[10] Y. Prohorov, "Convergence of random processes and limit theorems in probability theory," *Theor. Probability Appl.*, Vol. 1 (1956), pp. 157–214.

[11] P. K. Sen, "On the moments of the sample quantiles," *Calcutta Statist. Assoc. Bull.*, Vol. 9 (1959), pp. 1–20.

[12] ———, "Order statistics and their role in some problems of statistical inference," Ph.D. thesis, Calcutta University, 1961.

[13] J. W. Tukey, "A problem of Berkson and minimum variance orderly estimate," *Ann. Math. Statist.*, Vol. 29 (1958), pp. 588–592.

[14] S. S. Wilks, *Mathematical Statistics*, New York, Wiley, 1962.

[15] Z. Govindarajulu, "Asymptotic normality of linear functions of order statistics in one and multi samples," Case Institute of Technology, Air Force Report, 1965.

TOPICS IN RANK-ORDER STATISTICS

RALPH A. BRADLEY
FLORIDA STATE UNIVERSITY

This paper is dedicated to the memory of Dr. Frank Wilcoxon.

1. Introduction

Frank Wilcoxon joined the faculty of the Florida State University in 1960. He brought with him a number of ideas for further research in rank-order statistics and proceeded to develop them in association with colleagues and graduate students. This paper is largely expository and reports research based on his suggestions.

Two major topics are presented: sequential, two-sample, rank tests and multivariate, two-sample, rank procedures.

2. Sequential two-sample rank tests

2.1 *Preliminary remarks.* The two-sample, rank-sum test was introduced by Wilcoxon [29], [30]. Two populations, X- and Y-populations, are given with distribution functions,

$$P(X \leq u) = F(u), \qquad P(Y \leq u) = G(u), \tag{2.1}$$

X and Y being the random variables associated with the two populations. The basic null hypothesis tested is that

$$H_0\colon G(u) \equiv F(u), \tag{2.2}$$

usually with the assumed alternative of location change, $G(u - \theta) \equiv F(u)$.

Samples of independent observations of sizes m and n from X- and Y-populations respectively are taken and ranked in joint array. The sum of ranks, T for the X-sample or S for the Y-sample, is taken as the test statistic, and departures of the statistic from its mean under H_0, $\frac{1}{2}m(m + n + 1)$ or $\frac{1}{2}n(m + n + 1)$, are judged for significance. In order that ties in ranks between X- and Y-observations occur with probability zero, one may restrict F and G to be continuous.

Small-sample tables are available as are approximate, large-sample distributions for the rank sum under H_0. A recent extensive set of tables was developed by Wilcoxon, Katti, and Wilcox [31]. This table is divided into four sections corresponding to four levels of significance, 0.05, 0.025, 0.01, and 0.005 for a

This paper is based in part on research supported by the Army, Navy, and Air Force under an Office of Naval Research Contract, NONR-988(08), NR 042-004. Reproduction in whole or in part is permitted for any purpose of the United States Government.

one-sided test, and exact probabilities are given to four decimal places for the rank totals which bracket these significance levels. The critical rank totals are tabulated for sample sizes, $m, n = 3(1)50$.

Much has been written on the asymptotic properties of ranking procedures for large sample sizes. Properties of these procedures for small samples are difficult to investigate and depend on the forms of distribution functions like F and G, although empirical methods have been the bases for a limited number of such studies. Lehmann [15] suggested a class of alternatives to H_0,

$$H_a\colon G(u) \equiv F^k(u), \qquad k > 0. \tag{2.3}$$

For this class of alternatives, the small-sample power of the rank-sum test may be evaluated with the power dependent upon k but free of the form of $F(u)$. The basic sequential methods of Wald [27] and the class of alternatives of Lehmann permitted the development of two sequential, two-sample, grouped rank tests.

2.2 *Basic procedures.* Define a basic group of observations to consist of m X-observations and n Y-observations with ranking effected within the group in joint array as for the original Wilcoxon method. The sequential aspect of the experimentation consists of deciding at the end of each group of observations whether to continue experimentation by taking an additional group of observations or to terminate experimentation with a decision to accept the null hypothesis or with a decision to accept an alternative hypothesis. In this paper we restrict attention to H_0 of (2.2) and take the specific, one-sided, alternative among those of (2.3) to be

$$H_1\colon G(u) \equiv F^{k_1}(u), \qquad k_1 > 1. \tag{2.4}$$

This is the basic sequential system in [32]; the two-sided procedure is developed briefly in [33]. More generally, it is possible to consider a null hypothesis $H_0^*\colon G(u) \equiv F^{k_0}(u)$ and an alternative hypothesis

$$H_1^*\colon G(u) \equiv F^{k_1}(u), \qquad k_1 > k_0 > 0. \tag{2.5}$$

Consider the γ-th group of observations. Let $r_{1,\gamma}, \cdots, r_{m,\gamma}$ and $s_{1,\gamma}, \cdots, s_{n,\gamma}$ be the ranks assigned to X- and Y-observations respectively. The relevant information is retained if we consider only the Y-ranks and take $s_{1,\gamma} < \cdots < s_{n,\gamma}$. From [15] or [32] the probability that the Y-sample is given the indicated ranks under (2.3) is

$$P(s_{1,\gamma}, \cdots, s_{n,\gamma}|m, n, k) = \frac{k^n}{\binom{m+n}{n}} \prod_{j=1}^{n} \frac{\Gamma(s_{j,\gamma} + jk - j)\Gamma(s_{j+1,\gamma})}{\Gamma(s_{j+1,\gamma} + jk - j)\Gamma(s_{j,\gamma})} \tag{2.6}$$

where $s_{n+1,\gamma}$ is taken as $(m + n + 1)$.

For a sequential rank test of H_0 versus H_1 based on the actual configuration of ranks, the probability ratio for the γ-th group is

$$r_\gamma(m, n, k_1, 1) = \frac{k_1^n(m+n)!}{(s_{1,\gamma} - 1)!} \prod_{j=1}^{n} \frac{\Gamma(s_{j,\gamma} + jk_1 - j)}{\Gamma(s_{j+1,\gamma} + jk_1 - j)}, \tag{2.7}$$

the ratio of two probabilities of the form (2.6) with $k = k_1$ in the numerator probability and $k = 1$ in the denominator probability. If experimentation has proceeded to the end of t groups, the probability ratio for the complete experiment to that stage is, from an assumption of independence of groups of observations,

$$p_{1t}/p_{0t} = \prod_{\gamma=1}^{t} r_\gamma(m, n, k_1, 1), \tag{2.8}$$

the statistic required for the Wald sequential analysis. The procedure based on (2.8) has been called the configural rank test.

For a sequential rank test of H_0 versus H_1 based on rank sums, let

$$S_\gamma = \sum_{j=1}^{n} s_{j,\gamma} \quad \text{and} \quad T_\gamma = \sum_{i=1}^{m} r_{i,\gamma}. \tag{2.9}$$

The probability of S_γ may be formally expressed as

$$P\left(\sum_{j=1}^{n} s_{j,\gamma} = S_\gamma | m, n, k\right) = \sum P(s_{1,\gamma}, \cdots, s_{n,\gamma} | m, n, k) \tag{2.10}$$

where the summation in (2.10) is over the limits,

$$1 \leq s_{1,\gamma} < \cdots < s_{n,\gamma} \leq m + n, \qquad \sum_{j=1}^{n} s_{j,\gamma} = S_\gamma. \tag{2.11}$$

The γ-th group probability ratio corresponding to (2.7) is

$$R_\gamma(m, n, k_1, 1) = P\left(\sum_{j=1}^{n} s_{j,\gamma} = S_\gamma | m, n, k_1\right) / P\left(\sum_{j=1}^{n} s_{j,\gamma} = S_\gamma | m, n, 1\right), \tag{2.12}$$

and that corresponding to (2.8) is

$$P_{1t}/P_{0t} = \prod_{\gamma=1}^{t} R_\gamma(m, n, k_1, 1). \tag{2.13}$$

The procedure based on (2.13) has been called the rank-sum test.

In sequential analysis, α and β, the probabilities of Type I and Type II errors respectively, are used to form the constants,

$$A = (1 - \beta)/\alpha \quad \text{and} \quad B = \beta/(1 - \alpha). \tag{2.14}$$

Let P_{1t}/P_{0t} be the generic probability ratio at stage t. Then the decision process in logarithmic form is as follows.

(i) If $\ln B < \ln (P_{1t}/P_{0t}) < \ln A$, take another observation (another group of observations).

(ii) If $\ln (P_{1t}/P_{0t}) \leq \ln B$, terminate experimentation and accept H_0.

(iii) If $\ln (P_{1t}/P_{0t}) \geq \ln A$, terminate experimentation and accept H_1.

Substitution of (2.8) or (2.13) for P_{1t}/P_{0t} is made for the sequential rank tests.

Insight into the interpretation of the Lehmann model and the specification of k_1 may be obtained from the properties of the model. First, note that if X and Y are randomly selected from their respective populations,

$$p = P(X \leq Y) = k/(k + 1), \tag{2.15}$$

and conversely, $k = p/(1 - p)$. Second, when $G(u) \equiv F^k(u)$, $k > 1$, $G(u)$ is skewed to the right relative to $F(u)$, and if $F(u)$ is a standard normal distribution function, the mean of the Y-population is $\mu_y > 0$ and the variance is $\sigma_y^2 < 1$. Thus, associated with a value of k, we may consider also values of p and μ_y as defined. It is noted also that if k is an integer, $G(u)$ is the distribution function of the largest of k independent observations on X.

To facilitate use of the sequential, rank-sum test, tables [32] have been prepared for $m = n = 1(1)9$ giving values of T_γ, S_γ and corresponding values of

$$P_\gamma = P\left(\sum_{j=1}^{n} s_{j,\gamma} = S_\gamma | m, n, k_1\right) \tag{2.16}$$

defined in (2.10) and $\ln R_\gamma$ with R_γ defined in (2.12). The values of P_γ and $\ln R_\gamma$ are computed for $k_1 = 1.5, 2.33, 4$, and 9 with associated $p_1 = 0.6(0.1)0.9$ and $\mu_y = 0.282, 0.658, 1.029$, and 1.485. To facilitate use of the sequential, configural rank test, Wilcoxon has devised an ingenious algorithm for the computation of (2.6); its use is illustrated in [32]. The method consists of setting forth the ordered array of m X's and n Y's for the group and placing unity under each X and k_1 under each Y. Then cumulative totals from the left are obtained and the probability of (2.6) results from division of $k_1^n m!n!$ by the product of these cumulative totals.

2.3 *Examples.* The sequential rank tests were illustrated [32] for a screening experiment on chemical compounds for possible amelioration of the harmful effects of radiation. At each stage of the screening process ten laboratory animals were chosen, exposed to equal doses of radiation, divided randomly into equal size Control (X-sample) and Experimental (Y-sample) samples with the Experimental sample being subjected to injection of the chemical compound under study. The example is summarized in table II.1; although survival times are given in [32], only ranks representing orders of death are given here.

TABLE II.1

RANKS FOR CONTROL AND EXPERIMENTAL SAMPLES BY GROUPS IN A SCREENING EXPERIMENT TOGETHER WITH TEST STATISTICS

Group	X- and Y-ranks	T_γ	S_γ	$\ln r_\gamma$	$\ln (p_{1t}/p_{0t})$	$\ln R_\gamma$	$\ln (P_{1t}/P_{0t})$
1	1, 2, 3, 5, 8	19					
	4, 6, 7, 9, 10		36	1.511	1.511	1.403	1.403
2	1, 2, 5, 9, 10	27					
	3, 4, 6, 7, 8		28	0.182	1.693	−0.498	0.905
3	1, 2, 3, 7, 9	22					
	4, 5, 6, 8, 10		23	1.070	2.763*	0.669	1.574
4	1, 2, 3, 6, 7	19					
	4, 5, 8, 9, 10		36	—	—	1.403	2.977*

* The sequential process terminates with acceptance of H_1.

For this illustration, $m = n = 5$, $k_1 = 2.33$ $(p_1 = 0.7)$, $\alpha = 0.15$, $\beta = 0.05$, $\ln A = 1.85$, and $\ln B = -2.83$. Values of $\ln r_\gamma$, $\ln (p_{1t}/p_{0t})$, $\ln R_\gamma$, and ln

(P_{1t}/P_{0t}) are shown in table II.1. The Wilcoxon algorithm was used to compute $\ln r_\gamma$ and thence $\ln (p_{1t}/p_{0t})$, and the tables noted were used to record $\ln R_\gamma$ and thence $\ln (P_{1t}/P_{0t})$. Both the sequential, configural rank test and the sequential, rank-sum test led to rejection of H_0 and acceptance of H_1, the former with one less group of observations than the latter. Termination is judged by comparison of $\ln (p_{1t}/p_{0t})$ or $\ln (P_{1t}/P_{0t})$ with $\ln A$ and $\ln B$ as explained in the preceding subsection.

Numerous other possible applications exist in the medical-biological area and, indeed, many of these necessitate the use of within-group ranking. A second example is given in [33] and is based on visual ordering of severity of ulceration in rats.

The sequential ranking methods may be particularly useful in research wherein measurement is difficult but subjective ordering within groups of limited size is possible. Sequential triangle and duo-trio tests for the selection of expert taste panels have been discussed [2]; Kramer [13], [14] has extended the matching process to more samples. Sequential rank methods could be used for judge selection for tests where X- and Y-samples differ in and are to be ordered by basic taste characteristics.

Other applications may be noted in life testing. A specific case noted by the author involved a testing machine with a rotating cam which flexed six small springs under test. If two sets of three springs each, the sets differing in metal composition, were randomly allocated to test-machine positions, an appropriate sequential experiment could be devised and depend on order of failure of the springs.

2.4 *Properties.* Wald has provided means of evaluation of the average sample number (ASN) function and the power function of sequential tests at special values of the parameter under test. These formulas have been applied to the sequential rank tests [32] but do not adequately characterize the functions.

Monte Carlo studies were undertaken [3] to further evaluate properties of the sequential tests. These studies included empirical calculation of points on the ASN and power functions for the sequential rank-sum test with $\alpha = \beta = .05$, the calculations for each set of design parameters being based on 500 simulated sequential experiments. The studies were made for data generated in accordance with the Lehmann model and also for data generated from normal populations with unit variance and with means differing by μ_y corresponding to the parallel value of k in the Lehmann model.

It was found that the power function $\Phi(k, \mu_y)$ could be adequately represented by the probit model,

$$\Phi(k, \mu_y) = \frac{1}{\sqrt{2\pi}} \int_{-\infty}^{a+b\mu_y-5} e^{-w^2/2}\, dw. \tag{2.17}$$

Estimates of a and b for the various designs are given in table II.2; k and μ_y are related as noted above and as tabled in [32]. In general, it was found tha α and β were somewhat less than the nominal values of the designs. Note that $\hat{a}$ tends to

TABLE II.2

ESTIMATED PARAMETER VALUES FOR FITTED POWER FUNCTIONS FOR THE SEQUENTIAL RANK-SUM TEST

(Main entries are for the Lehmann model; values in parentheses are for the normal model.)

Design		$m = n$ 2	3	4	5
$k_1 = 2.33$	$\hat{a}$	3.89 (3.68)	3.44 (3.31)	3.16 (3.99)	3.51 (3.86)
$\alpha = \beta = .05$	$\hat{b}$	4.34 (4.25)	5.01 (4.80)	5.59 (3.79)	4.93 (3.84)
$k_1 = 4$	$\hat{a}$	2.59 (2.98)	3.16 (3.10)	2.63 (2.81)	2.71 (3.08)
$\alpha = \beta = .05$	$\hat{b}$	4.30 (3.43)	3.60 (3.43)	4.27 (3.70)	4.17 (3.40)
$k_1 = 9$	$\hat{a}$	2.83 (3.05)	2.56 (2.65)	2.40 (2.69)	2.42 (2.47)
$\alpha = \beta = .05$	$\hat{b}$	2.87 (2.30)	3.11 (2.69)	3.42 (2.75)	3.36 (2.97)

be larger for the normal model than for the Lehmann model, and $\hat{b}$ tends to be smaller indicating a larger Type I error and lower power for positive μ_y or k in excess of unity; $\hat{a}$ and $\hat{b}$ are estimates of a and b respectively in (2.17).

In table II.3, some typical values of the ASN functions are given for the sequential, rank-sum test with data generated from the Lehmann model and from the normal model. This table also contains information on the ASN function for a modified configural rank test to be discussed in the next subsection. Note that ASN values tend to be slightly higher for the studies on the normal model but are not excessively so.

The Monte Carlo studies suggest an element of robustness for the sequential rank-sum test and give confidence in its use when the Lehmann model may not be entirely appropriate.

Additional information from the Monte Carlo studies is given in [3]. Effects of truncation were considered as were the distributions of termination numbers.

2.5 *Modified sequential rank tests.* It appears intuitively that better sequential rank tests might be obtained when it is feasible to effect complete re-ranking of the totality of X- and Y-observations at each stage of the sequential process. Such a procedure has at least theoretical interest.

Suppose that X- and Y-observations are still taken in groups of m and n and that no group or block effects are present. Then, at the t-th stage of such a process, mt X-observations and nt Y-observations are ranked in joint array. If the Lehmann model is used again, a modified, configural rank test might be based on the probability ratio,

$$p^*_{1t}/p^*_{0t} = r(mt, nt, k_1, 1) \tag{2.18}$$

with r defined in (2.7). Similarly, a modified, rank-sum test might be based on

$$P^*_{1t}/P^*_{0t} = R(mt, nt, k_1, 1) \tag{2.19}$$

with R as in (2.12). Use of Wald bounds (2.14) may be tried. The case with

TABLE II.3

AVERAGE SAMPLE NUMBERS OF THE MODIFIED CONFIGURAL RANK TEST IN COMPARISON WITH THE GROUPED RANK-SUM TESTS

(Main entries for grouped rank-sum tests are for the Lehmann model; values in parentheses are for the normal model.)

Design	k	μ_y	ASN*: Modified Configural Rank Test	ASN*: Grouped Rank-Sum Tests $m = n = 2$	$m = n = 3$	$m = n = 4$	$m = n = 5$
$k_1 = 1.5$	1	0	54.93**	116.20	110.00	110.22	112.52
$\alpha = \beta = .05$	3	0.846	21.93	31.96 (35.31)	30.62 (35.21)	30.30 (34.78)	29.83 (33.05)
	5	1.163	16.66	22.84 (25.45)	21.61 (25.46)	21.84 (24.87)	21.66 (24.82)
$k_1 = 2.33$	1	0	19.33	29.16	30.47	29.39	28.61
$\alpha = \beta = .05$	3	0.846	14.78	21.36 (26.20)	20.68 (26.03)	21.15 (23.73)	20.71 (24.93)
	5	1.163	10.12	13.71 (16.86)	14.27 (16.82)	13.60 —	14.20 (16.87)
$k_1 = 4$	1	0	8.78	12.46	13.62	14.23	13.75
$\alpha = \beta = .05$	3	0.846	13.09	18.89 (20.74)	18.44 (20.78)	19.90 (23.93)	19.66 (23.24)
	5	1.163	8.64	11.68 (13.97)	11.37 (13.70)	12.58 (14.22)	12.24 (15.20)
	7	1.352	6.98	—	9.55 —	10.18 (12.46)	10.02 (12.82)
$k_1 = 9$	1	0	4.53	6.65	6.82	7.65	7.89
$\alpha = \beta = .05$	3	0.846	9.07	12.65 (12.68)	13.84 (15.20)	14.85 (15.92)	16.08 (15.41)
	5	1.163	7.29	9.54 (11.22)	11.60 (13.09)	10.81 (13.20)	12.07 (14.99)
	7	1.352	6.05	7.95 (9.22)	9.31 (11.54)	9.40 (10.78)	10.36 (12.89)
	9	1.485	5.65	7.05 (8.70)	8.45 (10.15)	7.70 (10.11)	8.61 (11.13)

* Average Sample Numbers are average numbers of observations from each population.

** Based on 30 simulated experiments only; all remaining entries based on 500 simulated experiments.

$m = n = 1$ assumes more interest now and is the case considered in the Monte Carlo studies noted below.

Savage and Savage [22] have demonstrated that the bounds (2.14) are appropriate for the modified, configural rank test but have not so demonstrated for the modified, rank-sum test. In the former case, they have shown the properties of finite termination and finite expected termination under H_0 and H_1. Savage and Sethuraman [23] are developing stronger termination results following somewhat the approach of Jackson and Bradley [12]. Hall, Wijsman, and Ghosh [11] discuss the problem as does Berk [1]. Further theoretical work is required to obtain information on ASN functions and power functions when the basic assumption of Wald's methods are not met. Much more extensive tables of $\ln R$ would be required to facilitate use of a modified, rank-sum test.

Limited Monte Carlo studies have been conducted under the Lehmann model for the modified, configural rank test, and values of the ASN functions are shown in table II.3. Values of the power function are very close to those for the grouped, sequential rank tests. The comparison in table II.3 is somewhat confounded in that the modified, configural rank test with $m = n = 1$ is compared with the grouped, rank-sum test. Appreciable reductions in ASN values are shown.

More complete discussion of the modified, sequential, rank tests is given in [4].

2.6 *Other research.* Milton [17] has computed probabilities of possible rank configurations for $1 \leq n \leq m \leq 7$ and $n = 1$, $m = 8(1)12$ for $F(u)$ and $G(u)$ normal with unit variances and differences in means, $\mu_y = 0(0.2)1, 1.5, 2, 3$. He has used these tables for various nonparametric power and efficiency computations and comparisons [18] and also to develop grouped, rank-sum and configural rank tests of the normal shift hypothesis [19].

For the sequential rank tests, Milton has tabled values of the group probability ratios comparable to (2.7) and (2.12). He has used Wald's formulas to evaluate ASN and power functions at the selected points for which such formulas are available. It is difficult to compare his results with the Monte Carlo studies [3], [4] because his values of μ_y under the alternative hypothesis do not match those corresponding with our values of k_1 or p_1 very well. Limited comparisons suggest that his ASN values are slightly lower than those of our Monte Carlo studies, but it appears that the Wald formulas may underestimate ASN values, at least for the Lehmann model (see [3] table 4).

Parent [20] has defined *sequential ranks* and applied the concept to the two-sample problem discussed above and also to the paired-sample problem leading to a type of sequential, signed rank procedure. The sequential rank of an observation X_t relative to the set $X_1, \cdots, X_t$ is k if X_t is the k-th smallest observation of the set; $X_1, \cdots, X_{t-1}$ are not given new ranks at stage t, and several of these X's could have received rank k also as they were observed in sequence. It is pointed out that the sequential ranks for $X_1, \cdots, X_t$ are uniquely determined, and moreover, that they uniquely determine the ordering of $X_1, \cdots, X_t$.

The use of sequential ranks provides some simplifications in modified sequential rank tests discussed above. However, sequential ranks are not independent for the Lehmann model unless $k = 1$, and they do not avoid the requirement for stronger basic theory of sequential analysis than that provided by Wald.

Signed sequential ranks are discussed by Parent also. Independence of the signed sequential ranks is demonstrated when the observations giving rise to them are independent and equally distributed from a population with cdf $F(u)$ satisfying the "symmetry" relationship,

$$F(-u) = F(0)[1 - F(u) + F(-u)], \qquad u > 0. \tag{2.20}$$

This condition is met by distributions of positive, negative, and symmetric-about-zero random variables, but for random variables taking both positive and negative values with median different from zero, the condition is rather restrictive, ruling out many common distributions. A sequential, signed rank test analogous to Wilcoxon's procedure is not developed; rather, a procedure to detect a change in distribution from $F(u)$ to some $G(u)$ at some stage in a sequence of observations is developed and applied to process control.

3. Multivariate two-sample rank tests

3.1 *The multivariate problem.* Consider two, p-variate populations with associated, column-vector variates, X and Y. Let $F(u)$ and $G(u)$ be the distribution functions as in (2.1), X and u now being vectors. The null hypothesis is expressed again as in (2.2), and alternatives specifying location change only are usually the ones of interest. Samples, $x_1, \cdots, x_m$ and $y_1, \cdots, y_n$ of independent, column-vector observations from X- and Y-populations respectively are taken.

The problem considered is basically the two-sample form of Hotelling's problem with the generalized Student ratio when $F(u)$ and $G(u)$ are multivariate normal with common dispersion matrix Σ and, under H_0, identical mean vectors. The well-known statistic used then is

$$T^2 = \frac{mn}{m+n} (\bar{x} - \bar{y})'S^{-1}(\bar{x} - \bar{y}) \tag{3.1}$$

where

$$\bar{x} = \frac{1}{m} \sum_{\alpha=1}^{m} x_\alpha, \qquad \bar{y} = \frac{1}{n} \sum_{\beta=1}^{n} y_\beta \tag{3.2}$$

and

$$S = \left[\sum_{\alpha=1}^{m} x_\alpha x'_\alpha - m\bar{x}\bar{x}' \sum_{\beta=1}^{n} y_\beta y'_\beta - n\bar{y}\bar{y}' \right] / (m + n - 2). \tag{3.3}$$

Given the multivariate normal populations, it is known that $(m + n - p - 1)T^2/(m + n - 2)p$ has the variance-ratio distribution with p and $m + n - p - 1$ degrees of freedom while, asymptotically with m and n as they become large in constant ratio, T^2 has the chi-square distribution with p degrees of freedom. These distributions are central under H_0 but noncentral under the location change alternative with noncentrality parameter,

$$\lambda^2 = \frac{mn}{m+n} \mu' \Sigma^{-1} \mu \tag{3.4}$$

where $\mu = \mu_x - \mu_y$, μ_x and μ_y, the mean vectors of X- and Y-populations. When the multivariate normal assumption is removed, little is known about the small-sample distribution of T^2 and nonparametric methods may be needed.

3.2 *Nonparametric procedures.* Wald and Wolfowitz [28] seem to have been first to consider multivariate, two-sample, randomization tests. (They considered the univariate case in some detail and indicated that extensions to the multivariate case were straightforward.) A modified statistic, proportional to

$$T_M^2 = \frac{mn}{m+n} (\bar{x} - \bar{y})' S_M^{-1} (\bar{x} - \bar{y}), \tag{3.5}$$

was used, where

(3.6)

$$S_M = \left[\sum_{\alpha=1}^{m} x_\alpha x_\alpha' + \sum_{\beta=1}^{n} y_\beta y_\beta' - (m+n)^{-1}(m\bar{x} + n\bar{y})(m\bar{x} + n\bar{y})' \right] / (m + n - 1).$$

The statistic T_M^2 is monotonically related to T^2 through the relationship,

$$T_M^2 = (m + n - 1)T^2/[(m + n - 2) + T^2]. \tag{3.7}$$

The randomization test of H_0 is conditional on the numerical values of the observation vectors. Let $z_1, \cdots, z_{m+n}$ constitute the complete set of vectors, $x_1, \cdots, x_m, y_1, \cdots, y_n$. Under H_0, the designation of a z-vector as an X- or Y-vector is taken as a matter of random labeling; each of the $\binom{m+n}{m}$ possible distinct assignments of m X-labels and n Y-labels is taken to be equally likely. For each labeling, T_M^2 is evaluated, and one of these values is the observed one, say, $T_{M,\text{obs.}}^2$. Let a test with significance level α be desired. Let η be the number of values of $T_M^2 \geq T_{M,\text{obs.}}^2$. If $\eta \leq \alpha \binom{m+n}{m}$, the observed value $T_{M,\text{obs.}}^2$ is taken to be significant and H_0 is rejected. Since S_M is invariant under the random labeling and S is not, T_M^2 is used for the test because it is considerably easier to compute or study than T^2. The test based on T_M^2 is equivalent to the similar test based on T^2 because of (3.7). It is seen that the randomization test in the multivariate case follows the same principles as in the univariate case.

Minor simplifications may be made. The constant multiplier in (3.5) may be dropped, and we may replace $(\bar{x} - \bar{y})$ by $(\sum_{\alpha=1}^{m} x_\alpha - m\bar{z})$ where $\bar{z} = \sum_{\gamma=1}^{m+n} z_\gamma/(m+n)$ and obtain a new statistic monotonically related to T^2 and T_M^2. Since $\bar{z}$ is fixed for given observation vectors, the latter substitution yields a statistic for which only the vector $\sum_{\alpha=1}^{m} x_\alpha$ changes from one labeling to the next. Even so, the randomization test is numerically difficult and tabling is not possible.

Wald and Wolfowitz show that the limiting distribution of T_M^2 is the central chi-square distribution with p degrees of freedom for the randomization test subject to mild restrictions on the sequence of vectors of real constants, $z_1, \cdots,$

z_{m+n}. It follows directly that T^2 has the same limiting distribution under H_0 for the randomization test. Thus, for moderate sizes of m and n, one might assume that use of the limiting distribution is adequate for applications as an approximation to the randomization test.

Bradley and Patel [5] in work in progress have considered moments of T_M^2 over the randomization distribution. We note only the first two moments here:

$$E(T_M^2) = p,$$

$$E(T_M^4) = \left(\frac{c_1}{c_0^2}\right) p(p+2) + \left(\frac{m+n}{mn}\right)^2 (c_0 - 6c_1) \sum_{\gamma=1}^{m+n} \lambda_{\gamma\gamma}^2 \tag{3.8}$$

$$\approx p(p+2) \qquad \text{for } m, n \text{ large},$$

where

$$c_0 = mn/(m+n)(m+n-1),$$

$$c_1 = m(m-1)n(n-1)/(m+n)(m+n-1)(m+n-2)(m+n-3), \tag{3.9}$$

$$\lambda_{\gamma\gamma} = (z_\gamma - \bar{z})' S_M^{-1} (z_\gamma - \bar{z}).$$

A basis for an approximation to the randomization distribution of T_M^2 is to fit a continuous density function of appropriate type to it. The statistic $T^2/(m+n-1)$ has the beta distribution on $(0, 1)$ with parameters $\frac{1}{2}p$ and $\frac{1}{2}(m+n-p-1)$ under normal theory. Suppose a beta distribution is the appropriate type and determine its unknown parameters $\frac{1}{2}\nu_1$ and $\frac{1}{2}\nu_2$ by the method of moments, two moments of $T_M^2/(m+n-1)$ being available from (3.8). Then

$$\nu_1 = \phi p, \qquad \nu_2 = \phi(m+n-p-1) \tag{3.10}$$

with

$$\phi = \frac{2}{m+n-1}\left[\left\{\frac{p(m+n-p-1)}{v}\right\} - 1\right] \tag{3.11}$$

where

$$v = \text{var}\,(T_M^2) = E(T_M^4) - [E(T_M^2)]^2 \tag{3.12}$$

from (3.8). The corresponding approximation to the randomization distribution of $(m+n-p-1)T^2/(m+n-2)p$ is the variance-ratio distribution with ν_1 and ν_2 degrees of freedom as computed from (3.10). Note that ϕ approaches unity with large m, n since $v \approx 2p$.

In the univariate problem, it is possible to replace the original observations with ranks. Then the randomization distribution depends only on m and n, and tables are available [31] as we have seen. Use of ranks in the multivariate problem leads to only slight simplifications. Suppose that observations on each variate are ranked separately as in the univariate problem, and let $r_1, \cdots, r_m, s_1, \cdots, s_n$ be the resulting vectors of ranks. The computation of S_M is simplified as all diagonal elements are known and equal, but the nondiagonal elements are proportional to the various rank correlations of the data. In different problems, for given m and n, different arrays of rank correlations will arise and tabling of the distribution of T_M^2 is not feasible. Bradley and Patel have considered the use of ranks as well as normal scores and have developed some large-sample theory associated with them.

Following the paper by Wald and Wolfowitz, much new theory of nonparametric and rank-order statistics was developed, but there was an interval wherein little more was done on the multivariate problem, an exception being the work of Lynch and Freund [16]. Recently, there has been more activity. In addition to [5], Chatterjee and Sen [6] discuss the bivariate problem with use of ranks and give some consideration to the nonnull distribution of T_M^2. Sen [24], [25], Sen and Govindarajulu [26], Govindarajulu [10], and Chatterjee and Sen [7] provide more general results including two-sample, multivariate problems, C-sample, multivariate problems, and limit theory. Robson [21] proposes a distance method with application to ecology.

Wilcoxon had a long-term interest in the multivariate generalization of the rank-sum test. He was seeking a procedure of relative simplicity and proposed two bivariate methods with that characteristic. Neither of these methods has an adequate theoretical base, but both have intuitive appeal. They are presented here in order to record his ideas and perhaps to stimulate further consideration of them.

3.3 *Wilcoxon's first bivariate method.* Consider m bivariate X-observations and n bivariate Y-observations with corresponding rank vectors $(r_{11}, r_{21}), \cdots,$ $(r_{1m}, r_{2m}), (s_{11}, s_{21}), \cdots, (s_{1n}, s_{2n})$. Let the sample mean vectors be $(\bar{r}_1, \bar{r}_2)$ and $(\bar{s}_1, \bar{s}_2)$, and let r be the pooled correlation coefficient calculated from the ranks,

(3.13)
$$r = \frac{\left[\sum_{\alpha=1}^{m} (r_{1\alpha} - \bar{r}_1)(r_{2\alpha} - \bar{r}_2) + \sum_{\beta=1}^{n} (s_{1\beta} - \bar{s}_1)(s_{2\beta} - \bar{s}_2)\right]}{\left\{\left[\sum_{\alpha=1}^{m} (r_{1\alpha} - \bar{r}_1)^2 + \sum_{\beta=1}^{n} (s_{1\beta} - \bar{s}_1)^2\right]\left[\sum_{\alpha=1}^{m} (r_{2\alpha} - \bar{r}_2)^2 + \sum_{\beta=1}^{n} (s_{2\beta} - \bar{s}_2)^2\right]\right\}^{1/2}}.$$

Wilcoxon computed Fisher's discriminant function from the ranks, the linear function that has the greatest variance between samples relative to the variance within samples. A quantity proportional to this discriminant function for an arbitrary point (t_1, t_2) is

$$z = t_1 + t_2 \tan \theta \tag{3.14}$$

where

$$\tan \theta = \frac{\Delta r_2 - r\, \Delta r_2}{\Delta r_1 - r\, \Delta r_1} \tag{3.15}$$

with

$$\Delta r_1 = \sum_{\alpha=1}^{m} r_{1\alpha} - \tfrac{1}{2} m(m + n + 1), \qquad \Delta r_2 = \sum_{\alpha=1}^{m} r_{2\alpha} - \tfrac{1}{2} m(m + n + 1). \tag{3.16}$$

Substitution of each rank vector $(r_{1\alpha}, r_{2\alpha})$ and $(s_{1\alpha}, s_{2\alpha})$ for (t_1, t_2) in (3.14) yields

$(m + n)$ values of z. The z's are then ranked and the ranks associated with the X- and Y-samples, the ranks being $R_1, \cdots, R_m, S_1, \cdots, S_n$. Wilcoxon again used the rank-sum statistic, but in the function

$$(3.17) \qquad \Delta^2 = 12\left[\sum_{\alpha=1}^{m} R_\alpha - \tfrac{1}{2}m(m+n+1)\right]^2 \Big/ mn(m+n+1).$$

The procedure is simple geometrically. The observation vectors are transformed to rank vectors as a scaling process, since first and second variates in an observation vector may not otherwise be commensurate. The rank vectors are plotted in the two-dimensional space, and a line with slope $\tan\theta$ of (3.15) is drawn, say, through the mean point $[\frac{1}{2}(m+n+1), \frac{1}{2}(m+n+1)]$. Each plotted rank-vector point is projected orthogonally onto the line, and the projection points are ranked along the line yielding the required ranks, $R_1, \cdots, R_m, S_1, \cdots, S_n$.

The distribution of Δ^2 is not known. It is formulated from the univariate problem in which Δ may be taken to be standard normal under H_0 for moderate sizes of m and n. In the bivariate problem, the line chosen for the final ranking gives a maximum or near maximum value for Δ^2, and univariate tables for the small-sample distribution of $\sum_{\alpha=1}^{m} R_\alpha$ are not satisfactory. Wilcoxon believed that Δ^2 had approximately a chi-square distribution with two, rather than one, degrees of freedom under H_0. This belief was based on empirical studies; his notebook contains many calculations of Δ^2 for various sets of data and some limited calculations of the randomization distribution of Δ^2 for special examples. Further study is needed to substantiate his belief.

3.4 *Wilcoxon's second bivariate method.* Again consider the bivariate rank vectors $(r_{1\alpha}, r_{2\alpha})$ and $(s_{1\beta}, s_{2\beta})$, $\alpha = 1, \cdots, m$; $\beta = 1, \cdots, n$, of subsection 3.3. Wilcoxon transformed the rank variates to yield new vectors, $(u_{1\alpha}, u_{2\alpha})$ and $(v_{1\beta}, v_{2\beta})$, wherein $u_{1\alpha} = r_{1\alpha} - r_{2\alpha}$, $u_{2\alpha} = r_{1\alpha} + r_{2\alpha}$; $v_{1\beta} = s_{1\beta} - s_{2\beta}$, $v_{2\beta} = s_{1\beta} + s_{2\beta}$. Note that a randomly selected vector (t_1, t_2) from the set of r- and s-vectors yields a correlation between t_1 and t_2, but that a randomly selected vector (w_1, w_2) from the set of u- and v-vectors has zero correlation between w_1 and w_2, a result that follows since t_1 and t_2 have equal variances.

Wilcoxon supposed that lack of correlation might justify an approximate procedure, properly valid when w_1 and w_2 are independent. Thus he suggested re-ranking of the variates in the u- and v-vectors leading to new rank vectors, say, $(R_{1\alpha}, R_{2\alpha})$ and $(S_{1\alpha}, S_{2\alpha})$. Then he computed $\sum_{\alpha=1}^{m} R_{1\alpha}$ and $\sum_{\alpha=1}^{m} R_{2\alpha}$, Δ_1^2 and Δ_2^2 from (3.17) by substituting the two new rank sums for $\sum_{\alpha=1}^{m} R_\alpha$ in that formula, and W^2, his proposed statistic where

$$(3.18) \qquad W^2 = \Delta_1^2 + \Delta_2^2.$$

On the basis of the assumed independence, he took W^2 to have the chi-square distribution with two degrees of freedom under H_0.

This method should be correct asymptotically with large m and n. It does not provide easy generalization beyond the bivariate case.

4. Other research

Wilcoxon had a third major interest in research in rank-order statistics. This was in regard to the distributions of ranges of rank totals in a two-way classification. His notes contain much preliminary study of the problem, and he worked with Dunn-Rankin [9] in the development of a dissertation on the topic in the School of Education. In addition, in research in progress, McDonald and Thompson have developed new results in this area at the Florida State University.

Daniel and Wilcoxon [8] have a paper on the design of factorial experiments scheduled for publication in *Technometrics* in the near future.

REFERENCES

[1] R. H. Berk, "Asymptotic properties of sequential probability ratio tests," Ph.D. thesis Harvard University, 1964.

[2] R. A. Bradley, "Some statistical methods in taste testing and quality evaluation," *Biometrics*, Vol. 9 (1953), pp. 22–38.

[3] R. A. Bradley, D. C. Martin, and F. Wilcoxon, "Sequential rank tests, I. Monte Carlo studies of the two-sample procedure," *Technometrics*, Vol. 7 (1965), pp. 463–483.

[4] R. A. Bradley, S. D. Merchant, and F. Wilcoxon, "Sequential rank tests, II. A modified two-sample procedure," *Technometrics*, Vol. 8 (1966).

[5] R. A. Bradley and K. M. Patel, "A randomization test for equality of means of two multivariate populations with common covariance matrix," *Ann. Math. Statist.*, Vol. 36 (1965), pp. 730–731. (Abstract.)

[6] S. K. Chatterjee and P. K. Sen, "Non-parametric tests for the bivariate two-sample location problem," *Calcutta Statist. Assoc. Bull.*, Vol. 13 (1964), pp. 18–58.

[7] ———, "Nonparametric tests for the multisample multivariate location problem," NSF Technical Report, University of California, Berkeley, 1965.

[8] C. Daniel and F. Wilcoxon, "2^{p-q} plans robust against linear and quadratic trends," *Technometrics*, Vol. 8 (1966), pp. 259–278.

[9] P. Dunn-Rankin, "The true probability distribution of the range of rank totals and its application to psychological scaling," Ed.D. thesis, Florida State University, Tallahassee, 1965.

[10] Z. Govindarajulu, "Asymptotic normality of a class of nonparametric test statistics," AFOSR Technical Report, Case Institute of Technology, Cleveland and University of California, Berkeley, 1965.

[11] W. J. Hall, R. A. Wijsman, and J. K. Ghosh, "The relationship between sufficiency and invariance with applications in sequential analysis," *Ann. Math. Statist.*, Vol. 32 (1965), pp. 575–614.

[12] J. E. Jackson and R. A. Bradley, "Sequential χ^2- and T^2-tests," *Ann. Math. Statist.*, Vol. 32 (1961), pp. 1063–1077.

[13] C. Y. Kramer, "A method of choosing judges for a sensory experiment," *Food Research*, Vol. 20 (1955), pp. 492–496.

[14] ———, "Additional tables for a method of choosing judges for a sensory experiment," *Food Research*, Vol. 21 (1956), pp. 598–600.

[15] E. L. Lehmann, "The power of rank tests," *Ann. Math. Statist.*, Vol. 24 (1953), pp. 23–43.

[16] L. Lynch and J. E. Freund, "On the analysis of paired ranked observations," ONR Technical Report No. 32, Virginia Polytechnic Institute, Blacksburg, 1957.

[17] R. C. Milton, "Rank order probabilities; two-sample normal shift alternatives," *Ann. Math. Statist.*, Vol. 36 (1965), pp. 1613–1614. (Abstract.)

[18] ———, "Power of two-sample nonparametric tests against the normal shift alternative," *Ann. Math. Statist.*, Vol. 36 (1965), p. 1614. (Abstract.)

[19] ———, "Sequential two-sample rank tests of the normal shift hypothesis," *Ann. Math. Statist.*, Vol. 36 (1965), p. 1614. (Abstract.)

[20] E. A. Parent, Jr., "Sequential ranking procedures," ONR Technical Report, Stanford University, Stanford, 1965.

[21] D. S. Robson, "A rank-sum test of whether two multivariate samples were drawn from the same population," ONR Technical Report No. 14, Cornell University, Ithaca, 1964.

[22] I. R. Savage and L. J. Savage, "Finite stopping time and finite expected stopping time," *J. Roy. Statist. Soc. Ser. B*, Vol. 27 (1965), pp. 284–289.

[23] I. R. Savage and J. Sethuraman, "Properties of the generalized SPRT based on ranks for Lehmann alternatives," ONR Technical Report No. M-84, Florida State University, Tallahassee and NSF Technical Report No. 7, Stanford University, Stanford, 1965.

[24] P. K. Sen, "On a class of multisample multivariate non-parametric tests," USPH Technical Report, University of North Carolina, Chapel Hill, 1965.

[25] ———, "On a class of two sample bivariate non-parametric tests," NSF Technical Report, University of California, Berkeley, 1965.

[26] P. K. Sen and Z. Govindarajulu, "On a class of C-sample weighted rank-sum tests for location and scale," NSF and AFOSR Technical Report, University of California, Berkeley, 1965.

[27] A. Wald, *Sequential Analysis*, New York, Wiley, 1947.

[28] A. Wald and J. Wolfowitz, "Statistical tests based on permutations of the observations," *Ann. Math. Statist.*, Vol. 15 (1944), pp. 358–372.

[29] F. Wilcoxon, "Individual comparisons by ranking methods," *Biometrics*, Vol. 1 (1945), pp. 80–83.

[30] ———, "Probability tables for individual comparisons by ranking methods," *Biometrics*, Vol. 3 (1947), pp. 119–122.

[31] F. Wilcoxon, S. K. Katti, and R. A. Wilcox, *Critical Values and Probability Levels for the Wilcoxon Rank Sum Test and the Wilcoxon Signed Rank Test*, Pearl River and Tallahassee, American Cyanamid Company and Florida State University, 1963.

[32] F. Wilcoxon, L. J. Rhodes, and R. A. Bradley, "Two sequential two-sample grouped rank tests with applications to screening experiments," *Biometrics*, Vol. 19 (1963), pp. 58–84.

[33] F. Wilcoxon and R. A. Bradley, "A note on the paper 'Two sequential two-sample grouped rank tests with application to screening experiments'," *Biometrics*, Vol. 20 (1964), pp. 892–895.

GENERALIZATIONS OF THEOREMS OF CHERNOFF AND SAVAGE ON THE ASYMPTOTIC NORMALITY OF TEST STATISTICS

Z. GOVINDARAJULU, L. LECAM, and M. RAGHAVACHARI
UNIVERSITY OF CALIFORNIA, BERKELEY

1. Introduction

The purpose of the present paper is to generalize the results obtained by Chernoff and Savage [5] on the asymptotic normality of a large class of two-sample nonparametric test statistics.

The assumptions made in [5] involve a certain function J which is assumed to possess two derivatives satisfying boundedness restrictions. However, certain test statistics, for instance those proposed by Ansari and Bradley [1] and Siegel and Tukey [15], do not satisfy the regularity conditions imposed by Chernoff and Savage. In particular, the *first* derivative of the appropriate function J fails to exist at certain points, so that the arguments of Chernoff and Savage are no longer directly applicable.

It will be shown here that the basic asymptotic normality result of [5] remains valid without any assumptions whatsoever or the existence of second derivatives. The assumption of existence of the first derivative is replaced by an assumption of absolute continuity. It should be noted that even this assumption is somewhat too stringent if one is willing to impose restrictions on the couple (F, G). However, the discussion of such possibilities remains beyond the purview of the present paper.

Section 2 of the paper gives a number of definitions which will be used throughout. Section 3 summarizes some properties of the set of functions J which will be used later. The main results are a lemma (lemma 2) on uniform square integrability and a continuity theorem (lemma 3) for the variances of the normal approximations to the distributions of the Chernoff-Savage statistics. Section 4 gives an account of convergence properties of empirical cumulative distributions and of their inverse functions.

The tails of the Chernoff-Savage statistics are bounded in section 5, and the main asymptotic normality theorem appears in section 6. Finally, natural extensions to the c-sample situation are provided in section 7.

Prepared with the partial support of the Mathematics Division of the Air Force Office of Scientific Research under Grant No. AF-AFOSR-741-65 and with the partial support of the U. S. Army Research Office (Durham) Grant DA-31-124-AROD-548.

2. Standing assumptions and notations

Let $X_1, X_2, \cdots, X_m$ and $Y_1, Y_2, \cdots, Y_n$ be random samples of sizes m and n drawn from populations with cumulative distribution functions (= c.d.f.) F and G respectively. Let $N = m + n$ and let $N\lambda_N = m$. *It will be assumed throughout that there is a* $\lambda_0 > 0$ *such that* $0 < \lambda_0 \leq \lambda_N \leq 1 - \lambda_0$ *and that the distribution functions* F *and* G *have no common discontinuities.*

The function $H = \lambda_N F + (1 - \lambda_N)G$ will be called the combined population cumulative distribution function. Let F_m and G_n be the empirical c.d.f.'s of the X's and Y's respectively. The function $H_N = \lambda_N F_m + (1 - \lambda_N)G_n$ is called the combined empirical c.d.f. It will be assumed that (F, G) and λ_N vary with m and n. However, to avoid an excess of indices the notation suppresses this fact. Another reason for this simplified notation is that the following theorems are 'uniform' and are valid whether the distributions are constant, tend to a limit, or vary rather arbitrarily with N.

Define

$$Z_{N,i} = \begin{cases} 1 & \text{if the } i\text{-th smallest in the combined sample is an } X, \\ 0 & \text{otherwise.} \end{cases} \tag{2.1}$$

Then, we will be concerned with statistics of the form

$$mT_N = \sum_{i=1}^{N} E_{N,i} Z_{N,i}, \tag{2.2}$$

where the $E_{N,i}$ are given constants. Many statistics occurring in nonparametric statistical inference can be reduced to the form (2.2). For examples the reader is referred to Chernoff and Savage [5]. We will, as Chernoff and Savage did, use the following representation:

$$T_N = \int_{-\infty}^{\infty} J_N\left(\frac{N}{N+1} H_N\right) dF_m(x). \tag{2.3}$$

The representations (2.2) and (2.3) are equivalent when $E_{N,i} = J_N(i/(N + 1))$. Although J_N need be defined at $1/(N + 1), 2/(N + 1), \cdots, N/(N + 1)$, we can conveniently extend its domain of definition to $(0, 1)$ by letting J_N be constant on $(i/(N + 1), (i + 1)/(N + 1))$, $(i = 0, 1, 2, \cdots, N)$. Our J_N is slightly different from that used by Chernoff and Savage [5]. In (2.3) they use $J_N(H_N)$. Consequently, their J_N need be defined at $1/N, 2/N, \cdots, N/N$, that is, in $(0, 1]$. Our main purpose in slightly changing the J_N function is to avoid asymmetry and eliminate the possibility that F_m gives mass at points where the argument of J_N is unity. The implication of this symmetry will be clear in the statements of the main theorems, in which one of the assumptions of [5] can be dispensed with. The problem of asymmetry had also been recognized by J. Pratt and I. R. Savage who informed one of the authors via personal communication.

It is easily verified that one could replace the assumption that F and G have no common discontinuities by the apparently stronger requirement that both F

and G be continuous. However, the more general case reduces immediately to the continuous one as we shall now show.

If F has a jump of size α at a point t, remove the point t from the real line and insert in its place a closed interval of length α. Distribute the probability mass α uniformly over this interval. The cumulative distribution G is kept constant over the inserted interval. Proceed similarly for the jumps of G. The new cumulative distributions F^* and G^* so obtained are continuous. For samples obtained from F^* and G^* the relative order relations between X's and Y's have the same probability distribution as if the samples were obtained from F and G.

If F and G had common discontinuities, ties would occur with positive probability. The definition of the variables $Z_{N,i}$ would no longer be complete. Thus, taking into account the possibility of "continuization" as performed above, the assumption of continuity of both F and G is equivalent to the assumption that ties between X's and Y's occur with probability equal to zero.

As a further reduction, let us show that there is in fact no loss of generality in assuming that the following assumption holds.

ASSUMPTION (A). *For each integer N the cumulative distributions (F, G) and the number λ_N are such that $H = \lambda_N F + (1 - \lambda_N) G$ is the cumulative $H(x) \equiv x$, $x \in [0, 1]$, of the uniform distribution and $0 < \lambda_0 \leq \lambda_N \leq 1 - \lambda_0$.*

To show this, note that if after removal of discontinuities, the function H^* remains constant over certain intervals, no observations will occur in these intervals. Thus, these intervals can be deleted from the line without affecting the order of the observations. This will leave us with a continuous strictly increasing cumulative distribution which can now be transformed to the uniform cumulative $H(x) \equiv x$ by a strictly increasing continuous transformation.

In view of this we shall assume throughout that assumption A holds and, if necessary, indicate the original distributions before transformation by $(\check{F}, \check{G})$ instead of (F, G).

When assumption A is satisfied the measure dF induced by F possesses a density φ with respect to the Lebesgue measure dH on $[0, 1]$. The inequality $0 \leq \varphi \leq \lambda_0^{-1}$ will play an important role in the sequel.

To describe a class of functions to which our asymptotic normality results will apply, it is convenient to introduce the following definition.

DEFINITION 1. *A function f, $f \geq 1$ defined on the interval $(0, 1)$ will be said to belong to the class $\mathfrak{U}_1$ (respectively $\mathfrak{U}_2$) if it is integrable (respectively square integrable) for the Lebesgue measure and if in addition there is some $\alpha \in (0, 1)$ such that f is monotone decreasing in $(0, \alpha]$ and monotone increasing in $[\alpha, 1)$.*

Let b denote a constant $0 < b < \infty$. Let f_0, f, and g be three nonnegative functions defined on $(0, 1)$. Assume that f_0 is Lebesgue integrable, and that $f \in \mathfrak{U}_1$ and that $g \in \mathfrak{U}_2$.

Consider functions J defined by integrals of the type $J(x) = \int_{1/2}^{x} J'(\xi)\, d\xi$. We shall say that J belongs to the class $\mathcal{S}_0$ if $|J'| \leq fg$, that $J \in \mathcal{S}$ if $|J'| \leq f_0 + fg$ and that $J \in \mathcal{S}_1$ if $J' = J_1' + J_2'$ with $|J_2'| \leq fg$ and $\int |J_1'(x)|\, dx \leq b$.

One could also introduce functions J which differ from the integrals

$\int_{1/2}^{x} J'(\xi)\, d\xi$ by a constant. However, this will not change the difference studied below. Thus the consideration of functions of the type $a + \int_{1/2}^{x} J'(\xi)\, d\xi$ is left to the care of the reader.

In the sequel the product fg will play essentially the same role as the function $f(x)g(x) = K[x(1-x)]^{-1+\delta}[x(1-x)]^{-(1/2)+\delta} = K[x(1-x)]^{-(3/2)+2\delta}$ of Chernoff and Savage. For this special choice of product fg one can also prove the following result. Let $\xi_{N,k}$ be the k-th order statistic in a sample of size N from the uniform distribution on $[0, 1]$. For any function J let $\bar{J}_N$ be the function defined on $(0, 1)$ as follows. If $y = (N+1)^{-1}k$, $k = 1, 2, \cdots, N$, let

$$\bar{J}_N(y) = EJ(\xi_{N,k}) = \int J(x)\beta_N(x, k)\, dx \tag{2.4}$$

where

$$\beta_N(x, k) = \frac{\Gamma(N+1)}{\Gamma(k)\Gamma(N+1-k)} x^{k-1}(1-x)^{N-k} \tag{2.5}$$

is the density of $\xi_{N,k}$.

Complete the definition of $\bar{J}_N$ by interpolating linearly between successive values $\{k/(N+1), (k+1)/(N+1)\}$ and leaving $\bar{J}_N$ constant below $1/(N+1)$ and above $N/(N+1)$.

LEMMA 1. *Assume that there exists a constant K and a δ, $0 < \delta < \frac{1}{2}$, such that $|J'(x)| \leq K[x(1-x)]^{-(3/2)+\delta}$. Then, there exists a constant K_1 and an N_0 such that $N \geq N_0$ implies*

$$|\bar{J}'_N(x)| \leq K_1[x(1-x)]^{-(3/2)+\delta}. \tag{2.6}$$

Furthermore, if $\{J'_\nu\}$ is a sequence such that J'_ν converges to J' in Lebesgue measure and $|J'_\nu| \leq K[x(1-x)]^{-(3/2)+\delta}$, for all ν, then $\bar{J}'_{\nu,N} - J'_\nu$ converges to zero in Lebesgue measure as $N \to \infty$ uniformly in the index ν.

PROOF. Decompose J' into its positive and negative parts and then separate each of these into two pieces, one of which vanishes on $(0, \frac{1}{2}]$ and the other on $(\frac{1}{2}, 1)$. If the results hold for each of these four parts separately, they will hold for J' itself. Owing to the symmetry of the situation, it will be sufficient to prove the result for a function J' such that $J' \leq 0$ and $J'(x) = 0$ for $x > \frac{1}{2}$ with the additional restriction $|J'| \leq x^{-(3/2)+\delta}$. Let $J(x) = \int_x^{\frac{1}{2}} |J'(u)|\, du$. The slope of the function $\bar{J}_N$ between two successive points $k/(N+1)$ and $(k+1)/(N+1)$ is given by the expression

$$\begin{aligned} s_N(k) &= (N+1) \int_0^1 J(x)[\beta_N(x, k) - \beta_N(x, k+1)]\, dx, \\ &= (N+1) \int_0^1 \left[J(x) - J\left(\frac{k}{N}\right)\right] [\beta_N(x, k) - \beta_N(x, k+1)]\, dx. \end{aligned} \tag{2.7}$$

In this expression the integrand is nonnegative, since J is decreasing. This implies in particular that the slope $s_N(k)$ is smaller than the slope obtainable from the function $J'(x) = -x^{-(3/2)+\delta}$. Thus

$$s_N(k) \leq \frac{2(N+1)}{1-2\delta} \int_0^1 x^{-(1/2)+\delta}[\beta_N(x, k) - \beta_N(x, k+1)]\, du. \tag{2.8}$$

This last expression is easily expressible in terms of gamma functions and the first result follows by direct computation for $k = 1$ and by application of Stirling's formula for $k > 1$.

To prove the second result note that the slope $s_N(k)$ can also be written in the form

$$s_N(k) = \int \frac{N[J(x) - J(k/N)]}{k - Nx} B_N(x, k)\, dx, \tag{2.9}$$

with $B_N(x, k)$ equal to the probability density

$$B_N(x, k) = \frac{N+1}{N} \frac{|k - Nx|^2}{k(1-x)} \beta_N(x, k). \tag{2.10}$$

For every $\epsilon > 0$ and every $y \in (0\ 1)$, if $(k/N) \to y$, then

$$\int_{|x-y| \geq \epsilon} \frac{N[J(x) - J(k/N)]}{k - Nx} B_N(x, k)\, dx \leq \int_{|x-y| \geq \epsilon} \frac{[x^{-(1/2)+\delta} - y^{-(1/2)+\delta}]}{y - x} B_N(x, k)\, dx. \tag{2.11}$$

This quantity tends to zero as $N \to \infty$. Thus it is sufficient to consider the behavior of the integral taken for $|x - y| < \epsilon$. For this purpose note first that when y is fixed, $y \in (0, 1)$ and $0 < y - \epsilon$, then the ratio $N[J(x) - J(y)][y - x]^{-1}$ remains bounded, independently of the choice of J, in the interval $[y - \epsilon, y + \epsilon]$. Therefore, taking for k_N the integer part of $(N + 1)y$, one can select a number $c < \infty$ and an N_0 such that

$$\int_{\sqrt{N}|x-y| \geq c} \frac{N[J(x) - J(k_N/N)]}{k_N - Nx} B_N(x, k_N)\, dx < \epsilon \tag{2.12}$$

for every J and every $N \geq N_0$.

Suppose then that the sequence $\{J'_\nu\}$ converges in measure to a limit J'. Taking a subsequence, if necessary, one can assume that $J'_\nu \to J'$ almost everywhere. In this case, for every $\alpha > 0$ there exist a compact subset S of the interval (0, 1) such that S^c has a Lebesgue measure inferior to α and such that the J'_ν are continuous when restricted to S and such that $J'_\nu(x)$ converges to $J'(x)$ uniformly for $x \in S$. Suppose that y is a point of density of the set S and consider the integrals

$$I_{N,\nu} = \int \left[\frac{N[J_\nu(x) - J_\nu(k_N/N)]}{k_N - Nx} - J'_\nu(y)\right] B_N(x, k_N)\, dx, \tag{2.13}$$

taken over the set $S_N = \{x\colon x \in S \text{ and } \sqrt{N}|x - y| \leq c\}$. A simple change of variable $x = y + \xi/\sqrt{N}$ will show immediately that $I_{N,\nu}$ converges to zero uniformly in ν. Furthermore, an analogous integral taken over the set $\{x\colon x \in S^c$ and $\sqrt{N}|x - y| \leq c\}$ must tend to zero, since the point y is assumed to be a point of density of S. Taking into account the fact that almost all points of S are points of density, the result follows.

One could also apply the same argument to functions J_N^* obtained by the formula

$$J_N^*(\xi) = \int J_N(x)\beta_N[x, (N+1)\xi]\, dx, \tag{2.14}$$

for all values of ξ such that $1 \leq (N+1)\xi \leq N$.

3. Properties of functions which belong to $\mathcal{S}$

In this section we shall assume that the functions f_0, f and g are fixed and derive certain boundedness and integrability properties for the elements of the corresponding set $\mathcal{S}$ of functions.

LEMMA 2. *There is a number* b_0 *such that* $\sup \{\int J^2(u)\, du; J \in \mathcal{S}\} < b_0$. *Furthermore, for every* $\epsilon > 0$ *there is a number* b *such that*

$$\int_{|J(u)|>b} J^2(u)\, du < \epsilon \tag{3.1}$$

for every $J \in \mathcal{S}$.

PROOF. If $J' \in \mathcal{S}$, so are its positive and negative parts. Thus, it is sufficient to prove the result assuming $J' \geq 0$. In addition, the part $J_1' = \min [f_0, J']$ contributes a bounded term to the indefinite integral J. Therefore, it is sufficient to prove the lemma assuming $0 \leq J' \leq fg$. Take α so small that both f and g are monotone decreasing in $(0, \alpha]$. For every $\xi \in (0, \alpha]$ one can write $\xi f(\xi) \leq \int_0^\xi f(u)\, du$ and $\xi g^2(\xi) \leq \int_0^\xi g^2(u)\, du$. Let $c^2(\alpha)$ be the number

$$c^2(\alpha) = \max\left\{\int_0^\alpha g^2(u)\, du, \left[\int_0^\alpha f(u)\, du\right]^2\right\}. \tag{3.2}$$

Let $\varphi(u) = \int_u^\alpha J'(\xi)\, d\xi$. One can write

$$\begin{aligned}\int_0^\alpha \varphi^2(u)\, du &\leq \int_0^\alpha \int_0^\alpha f(\xi)g(\xi)f(y)g(y) \min(\xi, y)\, d\xi\, dy \\ &= 2\int_0^\alpha \left\{\int_0^y f(\xi)g(\xi)\xi\, d\xi\right\} f(y)g(y)\, dy \\ &\leq 2\int_0^\alpha c^2(\alpha)\left\{\int_0^y \frac{1}{\sqrt{\xi}}\, d\xi\right\} f(y)g(y)\, dy \\ &\leq 4c^3(\alpha)\int_0^\alpha f(y)\, dy \leq 4c^4(\alpha).\end{aligned} \tag{3.3}$$

A similar argument applies to the interval $[1-\alpha, 1)$ for α sufficiently small. If a is the maximum of $f(x)g(x)$ for $x \in [\alpha, 1-\alpha]$, the term $\int_\alpha^{1-\alpha} J'(\xi)\, d\xi$ remains bounded by $\int f_0(\xi)\, d\xi + a$. Hence the result.

Assuming as usual that H is the cumulative distribution of the Lebesgue measure on $[0, 1]$, let φ be the density $\varphi = [dF/dH]$ and let $\psi = [dG/dH]$. By assumption, $\lambda_N\varphi + [1 - \lambda_N]\psi$ is identically unity on $[0, 1]$.

Let L and M be the functions defined by the equalities

$$L(x) = \int_{1/2}^{x} J'(\xi)\, dF(\xi) = \int_{1/2}^{x} J'(\xi)\varphi(\xi)\, d\xi, \tag{3.4}$$

and

$$M(x) = \int_{1/2}^{x} J'(\xi)\, dG(\xi) = \int_{1/2}^{x} J'(\xi)\psi(\xi)\, d\xi. \tag{3.5}$$

If the function J belongs to $\mathcal{S}$, then both L and M belong to the set $\lambda_0^{-1}\mathcal{S} = \{v\colon \lambda_0 v \in \mathcal{S}\}$. Therefore, the preceding lemma applies to L and M as well as to J.

The remainder of the present section is devoted to continuity theorems which are easily proved under the assumption $H(x) \equiv x$ for $x \in [0, 1]$. However, to make them more directly applicable they will be stated for distributions on the line. For this purpose let $\mathfrak{D}$ be the set of pairs $(\check{F}, \check{G})$ of distributions on the real line subject to the only restriction that $\check{F}$ and $\check{G}$ have no common discontinuities. One could topologize $\mathfrak{D}$ as usual by the requirement that $(\check{F}_\nu, \check{G}_\nu) \to (\check{F}, \check{G})$ if $\check{F}_\nu(x) \to \check{F}(x)$ and $\check{G}_\nu(x) \to \check{G}(x)$ at every point of continuity. This topology can also be induced by the BL-norm (for Bounded Lipschitz) defined as follows (see [8]). If P and Q are two finite signed measures on the line, then

$$\|P - Q\|_{BL} = \sup_h \left| \int h\, dP - \int h\, dQ \right| \tag{3.6}$$

where the supremum is taken over all functions h such that $|h| \le 1$ and $|h(x) - h(y)| \le |x - y|$.

The space $\mathcal{S}' = \{J'\colon |J'| \le f_0 + fg\}$ will be topologized by the topology of convergence in Lebesgue measure. This topology can be induced by the metric

$$\operatorname{dist}(J_1', J_2') = \int_0^1 \frac{|J_1'(x) - J_2'(x)|}{1 + |J_1'(x) - J_2'(x)|}\, dx. \tag{3.7}$$

To each pair $(\check{F}, \check{G}) \in \mathfrak{D}$ and $J' \in \mathcal{S}'$ and each $\lambda \in [\lambda_0, 1 - \lambda_0]$ corresponds a pair (L, M) of functions defined on the interval $[0, 1]$. This pair is obtained by first reducing $H = \lambda F + (1 - \lambda)G$ to be uniform on $[0, 1]$ as explained in the introduction and then defining $L(x) = \int_{1/2}^{x} J'(\xi)\, dF(\xi)$, and so forth.

Let us say that $L_\nu \to L$ if $\int |L_\nu(x) - L(x)|^2\, dx \to 0$ and if $\sup \{|L_\nu(x) - L(x)|;\ x \in S\} \to 0$ for every compact subset S of the open interval $(0, 1)$ and similarly for M.

Lemma 3. *The map which makes correspond to* $[(F, G), J', \lambda] \in \mathfrak{D} \times \mathcal{S}' \times [\lambda_0, 1 - \lambda_0]$ *the pair* (L, M) *is jointly continuous for the topologies defined above.*

Proof. Since the topologies in question are all metrizable, it is sufficient to show that whenever a sequence $\{((\check{F}_k, \check{G}_k), J_k', \lambda_k)\}$ converges to a limit $((\check{F}, \check{G}), J', \lambda)$, then the corresponding pairs (L_k, M_k) converge to the appropriate pair (L, M).

Let $\check{H}_k = \lambda_k \check{F}_k + (1 - \lambda_k)\check{G}_k$ and let (F_k, G_k) be the pair obtained by the process described in the introduction. Let ξ be a number $\xi \in (0, 1)$. Consider the graph Γ_k of $\check{H}_k$ augmented by inserting vertical lines at jumps. If the horizontal line at the ordinate ξ meets Γ_k at a point which is not a jump of $\check{F}_k$, then $F_k(\xi) =$

$\check{F}_k[\check{H}_k(x)]$ for any point x such that $\check{H}_k(x-0) \leq \xi \leq \check{H}_k(x+0)$. It follows that F_k converges to F and that G_k converges to G. Thus we can assume that $(\check{F}_k, \check{G}_k)$ has been replaced by (F_k, G_k). However, in this case $\varphi_k = dF_k/dH_k$ is a bounded measurable function, $0 \leq \varphi_k \leq \lambda_0^{-1}$. Therefore, convergence of F_k to F implies that

$$\int \varphi_k(x)v(x)\,dx \to \int \varphi(x)v(x)\,dx \tag{3.8}$$

for every integrable function v. This, in turn, implies that $\int \varphi_k(x)v_k(x)\,dx \to \int \varphi(x)v(x)\,dx$ whenever $\int |v_k(x) - v(x)|\,dx \to 0$. Therefore, $L_k(x)$ converges to $L(x)$ uniformly on every interval of values of x which is bounded away from zero and unity. The convergence in quadratic mean follows from this and from the uniform integrability asserted by lemma 2. This proves the desired result.

A simple consequence of lemmas 2 and 3 which will be used in section 6 is the following. Let B_N be the random variable

$$\begin{aligned} B_N &= \int J(x)\,d(F_m - F) - \int L(x)\,d[H_N(x) - H(x)] \\ &= (1-\lambda_N)\left\{\int M(x)\,d(F_m - F) - \int L(x)\,d[G_n(x) - G(x)]\right\}. \end{aligned} \tag{3.9}$$

This expression is equivalent to the formula

$$\begin{aligned} \sqrt{N}B_N &= \frac{1-\lambda_N}{\sqrt{\lambda_N}}\frac{1}{\sqrt{m}}\sum_{j=1}^{m}[M(X_j) - EM(X_j)] \\ &\quad + \sqrt{1-\lambda_N}\frac{1}{\sqrt{n}}\sum_{j=1}^{m}[L(Y_j) - EL(Y_j)], \end{aligned} \tag{3.10}$$

where the variables X_j and Y_j are all independent, and each X_j has distribution F, whereas each Y_j has distribution G. The variance of $\sqrt{N}B_N$ is given by

$$\begin{aligned} \sigma_N^2[F, G, J, \lambda_N] &= \frac{(1-\lambda_N)^2}{\lambda_N} \text{ variance } M(X_1) \\ &\quad + (1-\lambda_N) \text{ variance } L(Y_1). \end{aligned} \tag{3.11}$$

PROPOSITION 1. *Let P_N be the distribution of $\sqrt{N}B_N$ and let Q_N be the normal distribution which has variance $\sigma_N^2[F, G, J, \lambda_N]$ and expectation zero. For every $\epsilon > 0$ there exists an $N(\epsilon)$ such that $N \geq N(\epsilon)$ implies $\|P_N - Q_N\|_{BL} < \epsilon$ for every $J \in \mathcal{S}$, and every triple $[(F, G), \lambda_N]$.*

In addition, there is an $N(\epsilon, a)$ such that $N \geq N(\epsilon, a)$ and $\sigma_N^2[F, G, J, \lambda_N] \geq a$ implies $\sup_x |P_N\{(-\infty, x]\} - Q_N\{(-\infty, x]\}| < \epsilon$ for all $J \in \mathcal{S}$ and all triples $[(F, G), \lambda_N]$.

PROOF. The first statement follows immediately from the usual central limit theorem and the uniform integrability asserted by lemma 2. The second statement follows from the first by the simple procedure of considering $\sqrt{N}B_N/\sigma_N$ instead of $\sqrt{N}B_N$. Hence the result.

In this connection the following lemma is of some interest.

LEMMA 4. *The equality* $\sigma_N^2[F, G, J, \lambda_N] = 0$ *implies that* $J'(x)\varphi(x)\psi(x) = 0$ *almost everywhere on the interval* $(0, 1)$. *However the identity* $J'\varphi\psi \equiv 0$ *is not sufficient to imply* $\sigma_N^2 = 0$.

PROOF. If, for instance, variance $M(X) = 0$, then M is almost everywhere constant on the set $E = \{x\colon \varphi(x) > 0\}$. Therefore, the derivative $M'(x) = J'(x)\psi(x)$ must be equal to zero at all points of density of the set E. This implies the stated result.

Let $\mathfrak{D} \times \mathcal{S}' \times [\lambda_0, 1 - \lambda_0]$ be topologized by the product topology used in lemma 3. For every $a \geq 0$ the set of triples $((F, G), J', \lambda)$ such that $\sigma_N^2 > a$ is an open subset of $\mathfrak{D} \times \mathcal{S}' \times [\lambda_0, 1 - \lambda_0]$. This implies the following corollary.

COROLLARY. *Assume that* J *is not constant. If* $(\check{F}_0, \check{G}_0) \in \mathfrak{D}$ *is a pair such that* $\varphi_0(x)\psi_0(x) > 0$ *almost everywhere for some* $\lambda \in [\lambda_0, 1 - \lambda_0]$, *then there is an* $a > 0$ *and an open neighborhood of* $(\check{F}_0, \check{G}_0)$ *such that* $\sigma_N^2 > a$ *for every pair* $(\check{F}, \check{G})$ *in this neighborhood.*

PROOF. The condition $\varphi_0(x)\psi_0(x) > 0$ almost everywhere with respect to the Lebesgue measure is equivalent to the condition that the measures induced by $\check{F}_0$ and $\check{G}_0$ are mutually absolutely continuous. Thus it is independent of the choice of λ. Since σ_N^2 is continuous in λ, the values $\sigma_N^2[(F_0, G_0), J, \lambda]$ attain their minimum as λ varies in $[\lambda_0, 1 - \lambda_0]$. According to lemma 4, this minimum value is a positive number, say, $2a > 0$. For each $\lambda \in [\lambda_0, 1 - \lambda_0]$ let V_λ be a neighborhood of $(\check{F}_0, \check{G}_0)$ and let W_λ be a neighborhood of λ such that $\sigma_N^2 > a$ for $(\check{F}, \check{G}) \in V_\lambda$ and $\xi \in W_\lambda$. There is a finite system $\{W_{\lambda_j}\}$ which covers $[\lambda_0, 1 - \lambda_0]$. If $(\check{F}, \check{G}) \in \bigcap_j V_{\lambda_j}$, one has $\sigma_N^2 > a$ for every $\lambda \in [\lambda_0, 1 - \lambda_0]$. Hence the result.

More specifically, the following lemma holds.

LEMMA 5. *Assume that* J *is not constant. Let* $\{J_k\}$ *be a sequence such that* $J_k \in \mathcal{S}$ *and such that* $J_k' \to J'$ *in Lebesgue measure. Let* $\{(\check{F}_k, \check{G}_k)\}$ *be a sequence of pairs converging to a pair* $(\check{F}, \check{G})$ *at all points of continuity of the pair* $(\check{F}, \check{G})$. *Then if* $\check{F} = \check{G}$,

$$\frac{\lambda_N}{1 - \lambda_N} \sigma_N^2[\check{F}_k, \check{G}_k, J_k, \lambda_N] \tag{3.12}$$

converges uniformly in N to $\int_0^1 J^2(u)\, du - [\int_0^1 J(u)\, du]^2 > 0$.

PROOF. It is sufficient to apply lemma 3 and compute the limiting value of σ_N^2. This limit is equal to

$$\begin{aligned} \frac{(1 - \lambda_N)^2}{\lambda_N} \quad \text{variance} \quad M(X_1) + (1 - \lambda_N) \quad \text{variance} \quad L(Y_1) \\ = \left[\frac{(1 - \lambda_N)^2}{\lambda_N} + (1 - \lambda_N)\right] \quad \text{variance} \quad J(X_1), \end{aligned} \tag{3.13}$$

since $L \equiv M$, and since X_1 and Y_1 have the same distribution. The result follows.

4. Certain properties of empirical distribution functions

For this section we shall derive several inequalities and limit theorems which can be used to show that the higher order random terms occurring in theorem 4.1

tend to zero as N tends to infinity. The first results are inequalities on the tails of empirical distribution functions and a sharpened form of a theorem of Donsker [7]. For simplicity of notation the results are given for the uniform distribution [0, 1]. There is no difficulty in rewording them to apply to arbitrary continuous distributions.

A convenient tool in the derivation of these results is a replacement of binomial variables by Poisson variables which can be described as follows.

Let $\epsilon \in (0, 1)$. Let $\{u_j, j = 1, 2, \cdots\}$ be a sequence of independent random variables which are uniformly distributed on $(0, \epsilon)$. Let (r, s) be a pair of integer-valued random variables independent of the u_j. Assume that the joint distribution of (r, s) is such that marginally r has a binomial distribution, $B(m, \epsilon)$, corresponding to m trials with probability of success ϵ. Assume also that s has a Poisson distribution with expectation $m\epsilon$. Let U_m and V_m be the processes defined for $t \in (0, \epsilon)$ by taking $mU_m(t)$ equal to the number of u_j's such that $u_j \leq t$ and $j \leq r$ and taking $mV_m(t)$ equal to the number of u_j's such that $u_j \leq t$ and $j \leq s$.

LEMMA 6. *There is a joint distribution for the pair (r, s) such that*

$$P\{U_m(t) \equiv V_m(t) \text{ all } t \in (0, \epsilon)\} \geq 1 - 2\epsilon. \tag{4.1}$$

PROOF. It is sufficient to select a joint distribution for (r, s) such that $P[r \neq s] \leq 2\epsilon$. The possibility of such a selection results from a theorem of Prohorov [12].

Note that if F_m is the empirical cumulative obtained from m uniformly distributed independent variables on [0, 1], then the two processes $\{F_m(t); t \in (0, \epsilon)\}$ and $\{U_m(t); t \in (0, \epsilon)\}$ have identical distributions.

LEMMA 7. *Let g be a positive nonincreasing function defined on $(0, \epsilon)$. Then*

$$P\{\sup_{0<t<\epsilon} \sqrt{m}g(t)|U_m(t) - t| \geq 1\} \leq 2\epsilon + \int_0^\epsilon g^2(u)\, du. \tag{4.2}$$

PROOF. According to lemma 6, it is sufficient to show that

$$P\{\sup_{0<t<\epsilon} \sqrt{m}g(t)|V_m(t) - t| \geq 1\} \leq \int_0^\epsilon g^2(u)\, du. \tag{4.3}$$

This follows immediately from the remark that V_m has independent increments such that $EV_m(t) = t$ and $Em[V_m(t) - t]^2 = t$. The process $Z(t) = m|V_m(t) - t|^2$ is a semimartingale to which the Hájek-Rényi inequalities [10], or their generalization by Birnbaum and Marshall [3], can be applied. This gives the stated result.

After this was written, we became aware of results of B. Rosen [14] which give similar inequalities without using the Poisson approximation. Also, certain deeper results of D. M. Čibisov [6] could be used to obtain sharper inequalities.

Another result needed in the sequel is the following lemma.

LEMMA 8. *Let U_m be the empirical cumulative distribution obtained from m independent observations on the uniform distribution on* [0, 1]. *For every $\epsilon > 0$ there exists a $\beta > 0$ such that*

$$P\left\{\sup_t \frac{U_m(t)}{t} > \frac{1}{\beta}\right\} < \epsilon \tag{4.4}$$

and

$$P\{U_m(t) \geq \beta t \text{ for every } t \text{ such that } U_m(t) > 0\} > 1 - \epsilon. \tag{4.5}$$

PROOF. For $t > \delta > 0$ this follows, for instance, from Donsker's theorem, or from the Kolmogorov-Smirnov theorems. For t small one can again reduce the problem to an equivalent one concerning the standard Poisson processes. For the standard Poisson process the result is well known and easily verifiable.

Consider now two cumulative distribution functions F and G and two integers m and n. Let $N = m + n$ and let $\lambda_N = m/N$. Assume $0 < \lambda_0 < \lambda_N < 1 - \lambda_0 < 1$. Assume also that $H(t) = \lambda_N F(t) + (1 - \lambda_N)G(t)$ is identical to t for $t \in [0, 1]$. If $H_N = \lambda_N F_m + (1 - \lambda_N)G_n$ is the combined sample cumulative obtainable from m observations with distribution F and n observations with distribution G, one can obtain bounds on H_N from the bounds on the component cumulative distributions F_m and G_n. Further information can also be obtained as follows.

Let $\varphi = dF/dH$ be the density of F with respect to the Lebesgue measure on $[0, 1]$. Let S be the set $S = (0, \delta] \cup [1 - \delta, 1)$ with $0 < 2\delta < 1$. Classify points to be placed on the interval $(0, 1)$ in four categories, according to whether they are in S or S^c and according to whether they are labeled X or Y. For the pair (F_m, G_n) this gives a matrix $\nu = \{(\nu_{i,j}); i = 1, 2; j = 1, 2\}$ with $\nu_{1,1} + \nu_{2,1} = m$ and $\nu_{1,2} + \nu_{2,2} = n$. Let p be the probability S for F and let q be the probability of S for G. One can form another matrix ν^* such that $\nu^*_{1,1}$ and $\nu^*_{1,2}$ are independent Poisson variables with expectations $E\nu^*_{1,1} = E\nu_{1,1} = mp$ and $E\nu^*_{1,2} = E\nu_{1,2} = nq$. Taking $\nu^*_{2,1} = \nu_{2,1}$ and $\nu^*_{2,2} = \nu_{2,2}$, Prohorov's theorem insures the existence of a joint distribution such that $P[\nu \neq \nu^*] \leq 2(p + q)$.

Consider also another matrix $\tilde{\nu}$ whose distribution is given by a multinomial distribution with N trials and probabilities $p_{1,1} = \lambda_N p$ and $p_{1,2} = (1 - \lambda_N)q$ and $p_{2,1} = \lambda_N(1 - p)$, and finally $p_{2,2} = (1 - \lambda_N)(1 - q)$. One could find a joint distribution such that $P\{(\nu^*_{1,1}, \nu^*_{1,2}) \neq (\tilde{\nu}_{1,1}, \tilde{\nu}_{1,2})\} \leq 2(p + q)$. Therefore, one can find a joint distribution such that $P\{(\nu_{1,1}, \nu_{1,2}) \neq (\tilde{\nu}_{1,1}, \tilde{\nu}_{1,2})\} \leq 4(p + q)$. For such a joint distribution, one can construct the second row of the matrix $\tilde{\nu}$ by selecting $\tilde{\nu}_{2,1}$ from a binomial distribution with probability of success $[\lambda_N(1 - p)]\,[\lambda_N(1 - p) + (1 - \lambda_N)(1 - q)]^{-1}$ and $[N - (\tilde{\nu}_{1,1} + \tilde{\nu}_{1,2})]$ trials.

Another matrix ν' can be constructed with $(\nu'_{1,1}, \nu'_{1,2}) = (\tilde{\nu}_{1,1}, \tilde{\nu}_{1,2})$ and $\nu'_{1,1} + \nu'_{2,1} = m$ and $\nu'_{1,2} + \nu'_{2,2} = n$. Then $P[\nu \neq \nu'] \leq 4(p + q)$. Given the matrix $\tilde{\nu}$ one can place independently $\tilde{\nu}_{1,1}$ points in S and $\tilde{\nu}_{2,1}$ points in S^c, according to the distribution F. One can also place independently $\tilde{\nu}_{1,2}$ points in S and $\tilde{\nu}_{2,2}$ points in S^c, according to the distribution G. It is easily verified that the system so obtained has exactly the same distribution as the system of points obtainable by the following procedure. First select points $\{\xi_j, j = 1, 2, \cdots, N\}$ independently according to the Lebesgue measure on $[0, 1]$. Then, for each ξ_j and independently of the rest, with probability $\lambda_N \varphi(\xi_j)$ label it X, and with probability $1 - \lambda_N \varphi(\xi_j)$, label it Y.

Since the combined cumulative H_N ignores the distinction between X and Y, the above argument shows that, except for cases having total probability at most $4(p + q)$, the behavior of H_N on S will be the same as that of a sample cumulative from N independent uniformly distributed observations. To apply the preceding lemmas to the study of H_N, let us introduce the following notation. If v is a numerical function defined on $(0, 1)$ and g is an element of $\mathfrak{U}_2$, the g-norm of v is the number

$$\|v\|_g = \sup_t \{|v(t)g(t)|; t \in (0, 1)\}. \tag{4.6}$$

Let $B(g)$ be the set of functions v which are defined on $(0, 1)$ and have a finite g norm. Let $S = \{s_j; j = 0, 1, 2, \cdots, h\}$ be a finite subset of the interval $[0, 1]$ such that $0 = s_0 < s_1 < s_2 < \cdots < s_{h-1} < s_h = 1$. To such a set S associate a projection Π_S of $B(g)$ into itself by the requirements that $(\Pi_S v)(s_j) = v(s_j)$ for every $s_j \in S$ and that

$$\{[\Pi_S v](s) - (\Pi_S v)s_j\} = \frac{s - s_j}{s_{j+1} - s_j} [v(s_{j+1}) - v(s_j)] \tag{4.7}$$

for $s \in (s_j, s_{j+1})$.

Let $\{W_N(t); t \in [0, 1]\}$ be the process $W_N(t) = \sqrt{N}[H_N(t) - t]$.

PROPOSITION 2. *Let g be an element of $\mathfrak{U}_2$. For every $\epsilon > 0$ there is a $\delta > 0$ and an integer N_0 depending on ϵ and g only such that $N \geq N_0$ implies*

$$P\{\|W_N - \Pi_S W_N\|_g > \epsilon\} < \epsilon \tag{4.8}$$

for every pair (F, G) of continuous distribution functions and every set $S = \{s_j; j = 0, 1, 2, \cdots, k\}$; $0 = s_0 < s_1 < \cdots < s_{k-1} < s_k = 1$, such that $s_{j+1} - s_j < \delta$ for every $j = 0, 1, \cdots, k - 1$.

PROOF. Let α be a number so small that g becomes monotone in $(0, \alpha]$ and $[1 - \alpha, 1)$ and that

$$16 \left\{\alpha + \frac{1}{\epsilon^2} \int_0^\alpha g^2(\xi)\, d\xi + \frac{1}{\epsilon^2} \int_{1-\alpha}^1 g^2(\xi)\, d\xi\right\} < \epsilon. \tag{4.9}$$

Since $tg^2(t) \leq \int_0^t g^2(\xi)\, d\xi \leq \int_0^\alpha g^2(\xi)\, d\xi$ for $t \leq \alpha$, it follows from lemma 7 that

$$P\{\sup_t [g(t)|W_N(t) - tW_N(\alpha)|; t \in (0, \alpha]] > \epsilon\} \leq 2\alpha + \frac{4}{\epsilon^2} \int_0^\alpha g^2(\xi)\, d\xi. \tag{4.10}$$

A similar inequality holds for values of t belonging to $[1 - \alpha, 1)$. Therefore, it will be sufficient to prove the assertion for the process $\{W_N(t); t \in (\alpha, 1 - \alpha)\}$ and a function g which is bounded. The process W_N can be written

$$W_N(t) = \sqrt{\frac{m}{N}} \sqrt{m}[F_m(t) - F(t)] + \sqrt{\frac{n}{N}} \sqrt{n}[G_n(t) - G(t)]. \tag{4.11}$$

According to the argument of Donsker [5], there exist an N_0 and a $\delta > 0$ such that $m \geq \lambda_0 N_0$ and $\sup_j [F(s_j) - F(s_{j-1})] < \delta$ implies that if Ω_j' is the oscillation of $\sqrt{m}[F_m(t) - F(t)]$ in the interval $[s_{j-1}, s_j]$, then

$$\sum_j P\left[\Omega_j' > \frac{\epsilon}{2\|g\|}\right] < \frac{\epsilon}{4}. \tag{4.12}$$

The same result can be applied to $\sqrt{n}[G_n(t) - G(t)]$. Thus, if $[s_j - s_{j-1}] < \delta\lambda_0$ and $N \geq N_0$, and if Ω_j is the oscillation of W_N in the interval $[s_{j-1}, s_j]$, one can write $\sum_j P[\Omega_j > \epsilon/\|g\|] < \epsilon/2$. This implies the desired result.

COROLLARY. *For every $g \in \mathfrak{U}_2$ and every $\epsilon > 0$ there is an $N_0 < \infty$ and a finite set $\{v_j; j = 1, 2, \cdots, k\}$ of continuous functions defined on $[0, 1]$ such that*

$$P\{\inf_j \|W_N - v_j\|_g > \epsilon\} < \epsilon \tag{4.13}$$

for every $N \geq N_0$ and every pair (F, G) of distributions having no common discontinuities.

Consider the process Z_m defined by $Z_m(t) = \sqrt{m}\{F_m(t) - F(t)\}$, where F_m is the empirical cumulative from a sample of size m drawn from the distribution F. Assume as usual that $\lambda_N F + (1 - \lambda_N)G = H$ is the uniform distribution on $[0, 1]$ and that $0 < \lambda_0 \leq \lambda_N \leq 1 - \lambda_0 < 1$.

LEMMA 9. *For each integer m let K_m be a random process defined on the interval $[0, 1]$. Let $Z_m^*(x) = Z_m[K_m(x)]$. If $P\{\sup_t |K_m(t) - t| > \epsilon\} \to 0$ for every $\epsilon > 0$, then*

$$P^*\{\sup_x |Z_m^*(x) - Z_m(x)| > \epsilon\} \tag{4.14}$$

tends to zero for every $\epsilon > 0$.

PROOF. According to proposition 2, or according to Donsker's theorem, for every $\epsilon > 0$ there exists an $N(\epsilon) < \infty$ and a finite set of continuous functions $\{v_j: j = 1, 2, \cdots, k\}$ such that

$$P\{\inf_j \|Z_m - v_j\| > \epsilon/3\} < \epsilon \tag{4.15}$$

for every $m \geq N(\epsilon)$. Let $\gamma_j(z) = 1$ if the *first* index i such that $\|z - v_i\| \leq \epsilon/3$ is precisely equal to j. Let $\gamma_j(z) = 0$ otherwise. According to the above inequality, if $\overline{Z}_m = \sum_j \gamma_j(Z_m)v_j$, then $P\{\|\overline{Z}_m - Z_m\| > \epsilon/3\} < \epsilon$. Therefore, eliminating cases having probability at most ϵ, one can also write $\sup_x |\overline{Z}_m[K_m(x)] - Z_m[K_m(x)]| \leq \epsilon/3$.

Furthermore, there exists a δ such that $|s - t| < \delta$ implies $|v_j(t) - v_j(s)| < \epsilon/3$ for every $j = 1, 2, \cdots, k$. Therefore, if $P\{\sup_t |K_m(t) - t| \geq \delta\} < \epsilon$, one can write $\|\overline{Z}_m[K_m] - \overline{Z}_m\| < \epsilon/3$, except in cases having probability at most 2ϵ. The result follows.

The preceding lemma 9 can be used under the following circumstances. Let H_N be the combined empirical cumulative. Let K_N be the function defined by $K_N(x) = \inf\{t: H_N(t) \geq x\}$. Assume as usual that $H(u) = u$ for $u \in [0, 1]$. Since $\|H_N - H\| \to 0$ in probability, the difference $\sup_x |K_N(x) - x|$ must also tend to zero in probability. It follows that $\|Z_m[K_N] - Z_m\|$ tends to zero in probability.

For values of x of the type $x = j/N$, the variable $Z_m[K_N(x)]$ is simply equal to $\sqrt{m}[F_m(\xi_j) - F(\xi_j)]$ where ξ_j is the order statistic of rank j in the combined sample. In other words, the number $mF_m[K_N(j/N)]$ is the number of X_j's whose rank is inferior or equal to j.

For the next proposition, it is convenient to introduce the space $\mathfrak{M}$ of all finite signed measures on the interval (0, 1) and their indefinite integrals. If $\mu \in \mathfrak{M}$, let $J_\mu(x) = \mu\{(0, x]\}$ and let $\|\mu\|$ be the total mass of μ. The functions J_μ are simply those functions of bounded variation on [0, 1] which are right continuous and vanish at zero.

PROPOSITION 3. *For every $\epsilon > 0$ there exists an $N(\epsilon)$ such that $N \geq N(\epsilon)$ implies*

$$P\left\{\sqrt{N}\left|\int [J_\mu(H_N) - J_\mu(H)]\, d(F_m - F)\right| > \epsilon\|\mu\|\right\} < \epsilon \tag{4.16}$$

for every $\mu \in \mathfrak{L}$ and every pair (F, G) of distribution functions having no common discontinuities.

REMARK. In the above proposition one could replace H_N by $(N/(N + 1))H_N$, since $\sqrt{N} \int [J_\mu((N/(N + 1))H) - J_\mu(H)]\, dF$ is of order $1/\sqrt{N}$.

PROOF. The integral

$$I_m = \sqrt{m} \int J_\mu(H_N)\, d(F_m - F) = \int J_\mu[H_N(x)]\, dZ_m(x) \tag{4.17}$$

can also be written

$$I_m = \int \{Z_m(1) - Z_m[K_N(\xi) - 0]\}\, \mu(d\xi). \tag{4.18}$$

Therefore,

$$\begin{aligned} &\left|\sqrt{m} \int [J_\mu(H_N) - J_\mu(H)]\, d(F_m - F)\right| \\ &\qquad = \left|\int \{Z_m(\xi - 0) - Z_m[K_N(\xi) - 0]\}\, \mu(d\xi)\right. \\ &\qquad \leq \|\mu\| \sup_x |Z_m(x) - Z_m[K_N(x)]|. \end{aligned} \tag{4.19}$$

This implies the desired result by application of lemma 9. Another result which may be useful in the investigation of the Chernoff-Savage statistics (but will not be needed for our purposes) is a theorem relative to the behavior of the quantile function K_N defined on the interval [0, 1] by the formula

$$K_N(u) = \inf \{x\colon H_N(x) \geq u\}. \tag{4.20}$$

For the present purposes the assumptions that $H(x) \equiv x$ for $x \in (0, 1)$ and that $H = \lambda_N F + (1 - \lambda_N)G$ with $0 < \lambda_0 \leq \lambda_N \leq 1 - \lambda_0$ are rather important. Consider, under these conditions, the process Z_N defined by $Z_N(u) = \sqrt{N}[K_N(u) - u]$, $u \in [0, 1]$.

Since the process $\sqrt{N}(H_N - H)$ is asymptotically Gaussian, it follows from the equivalence $\{K_N(u) > x\} \Leftrightarrow \{H_N(x) < u\}$ that for any finite set $0 = u_0 < u_1 < \cdots < u_{r-1} < u_r = 1$ the distribution of the vector $\{Z_N(u_j); j = 1, 2, \cdots, r\}$ is also asymptotically normal, with the same covariance function

$$C_N(u, v) = \frac{m}{N} F(u)[1 - F(v)] + \frac{n}{N} G(u)[1 - G(v)] \tag{4.21}$$

for $u \leq v$ as the process $\sqrt{N}(H_N - H)$ itself.

The following proposition strengthens this result. Let S be a finite set $S = \{u_j; j = 0, 1, 2, \cdots, r\}$, with $0 = u_0 < u_1 < \cdots < u_r = 1$. Let $Z_{N,S}(u)$ be defined by $Z_{N,S}(u) = Z_N(u)$ if $u \in S$. If u is between two consecutive points u_j and u_{j+1} of S, define $Z_{N,S}(u)$ by linear interpolation.

PROPOSITION 4. *For every $\epsilon > 0$ there exists a finite set S and an integer $N(\epsilon)$ such that $N \geq N(\epsilon)$ implies*

$$P\{\|Z_{N,S} - Z_N\| > \epsilon\} < \epsilon \tag{4.22}$$

for every pair (F, G).

PROOF. Let ϵ be a positive number $0 < \epsilon < 1$ and let b and r be positive integers. Select b such that $2P\{\sqrt{N}\|H_N - H\| > b\} < \epsilon$. For any function h defined on $[0, 1]$ let Πh be the function obtained by taking $(\Pi h)(j/r) = h(j/r)$ for $j = 0, 1, 2, \cdots, r$ and interpolating linearly between successive values. One can find a number r and an integer N_0 such that $N \geq N_0$ implies $2P\{\sqrt{N}\|\Pi H_N - H_N\| \geq \epsilon(1 - \epsilon)\} < \epsilon$ and $2r[b(\epsilon) + 1]^2 < \epsilon\sqrt{N}$.

Let $H_N^* = H_N$ if $\sqrt{N}\|H_N - H\| < b$ and $\sqrt{N}\|\Pi H_N - H_N\| < \epsilon(1 - \epsilon)$. Let $H_N^* = H$ otherwise. It follows that $N \geq N_0$ implies $P\{H_N^* \neq H_N\} < \epsilon$. Furthermore, $\sqrt{N}\|\Pi H_N^* - H_N^*\| < \epsilon(1 - \epsilon)$ and $\sqrt{N}\|H_N^* - H\| < b$ without exception. The second inequality implies that the segments of lines which compose ΠH_N^* have slopes which differ from unity by no more than $\alpha < [2rb(\epsilon)]/\sqrt{N} < \epsilon$. This implies in particular that ΠH_N^* is increasing. Further, let K_N^* be the function related to H_N^* by the equation $K_N^*(u) = \inf\{x: H_N^*(x) \geq u\}$ and let $\tilde{K}_N$ be the corresponding function relative to $\tilde{H}_N = \Pi H_N^*$.

Since the slope of ΠH_N^* is always larger than $1 - \epsilon$, the inequality $\sqrt{N}\|\Pi H_N^* - H_N^*\| < \epsilon(1 - \epsilon)$ implies $\sqrt{N}\|\tilde{K}_N - \tilde{K}_N^*\| < \epsilon$. Thus, it will be sufficient to prove the result for $\tilde{K}_N$, instead of K_N^* or K_N. Interpolate $\tilde{K}_N$ just as before. That is, let $\bar{K}_N(u) = \tilde{K}_N(u)$ if $ru = j$, $j = 0, 1, 2, \cdots, r$ and interpolate linearly between these values. Suppose that $\sqrt{N}[\tilde{H}_N(j/r) - H(j/r)] = z_1 > 0$ and that $\sqrt{N}[\tilde{H}_N((j + 1)/r) - H((j + 1)/r)] = z_2 < 0$. Since the slopes of the segments composing $\tilde{H}_N$ are between $1 - \alpha$ and $1 + \alpha$, this implies $\tilde{K}_N(j/r) \geq j/r - (1 - \alpha)z_1/\sqrt{N}$ and $\tilde{K}_N((j + 1)/N) \leq (j + 1)/r + (1 + \alpha)|z_2|/\sqrt{N}$. It follows from this that

$$\begin{aligned} \sqrt{N}\left\{\bar{K}_N\left[\frac{j}{r} + \frac{z_1}{\sqrt{N}}\right] - \frac{j}{r}\right\} \tag{4.23} \\ &\leq r\left\{(1 + \alpha)\frac{|z_2|}{\sqrt{N}} + (1 - \alpha)\frac{z_1}{\sqrt{N}} + \frac{1}{r}\right\} z_1 - (1 - \alpha)z_1 \\ &\leq z_1\left\{\alpha + (1 + \alpha)r\left[\frac{z_1 + |z_2|}{\sqrt{N}}\right]\right\}. \end{aligned}$$

Since $|z_i| \leq b$ and since $2rb^2 < \epsilon\sqrt{N}$, this is smaller than $b\alpha + (1 + \alpha)\epsilon < 3\epsilon$. The desired result is an immediate consequence of these inequalities.

REMARK. The result of proposition (4) is well known for the case where $F = G$. In fact, one can reverse the chain of arguments leading to the proof of proposition (4) to obtain a simple and rather elementary proof of Donsker's theorem. For the case where H is not the uniform distribution and for applications, see [2] and [9].

A consequence of proposition (4) is the following result.

COROLLARY. *There exist joint distributions for pairs of processes* (Z'_N, W_N) *such that*

(i) *the distribution of* Z'_N *is the same as that of* $\sqrt{N}[K_N - K]$;

(ii) *the process* W_N *is Gaussian, with mean zero and covariance*

$$EW_N(s)W_N(t) = \frac{m}{N} F(s)[1 - F(t)] + \frac{n}{N} G(s)[1 - G(t)] \tag{4.24}$$

for $s \leq t$;

(iii) *for every* $\epsilon > 0$ *there is an* $N(\epsilon)$ *such that* $P\{\|Z'_N - W_N\| > \epsilon\} < \epsilon$ *if* $N \geq N(\epsilon)$.

This follows immediately from proposition (4) and a theorem of V. Strassen [16].

Propositions (3) and (4) can be used to investigate the asymptotic properties of statistics of the Chernoff-Savage type as follows.

Let $\mathfrak{M}$ be the space of finite signed measures on the interval [0, 1]. For each $\mu \in M$ let $J_\mu = \mu\{(0, x]\}$. Let T_N be the expression

$$T_N = \sqrt{N}\left\{\int J_\mu(H_N)\, dF_m - \int J_\mu(H)\, dF\right\}. \tag{4.25}$$

Introduce the pair of stochastic processes (Z_N, Z^*_N) by the equalities

$$Z_N = \sqrt{N}[K_N - K], \qquad Z^*_N(\xi) = \sqrt{m}[F_m(\xi - 0) - F(\xi)]. \tag{4.26}$$

Let T^*_N be the expression

$$\begin{aligned} T^*_N &= \sqrt{N} \int [J_\mu(H_N) - J_\mu(H)]\, dF + \sqrt{N} \int J_\mu(H)\, d(F_m - F) \\ &= \sqrt{N} \int \left[F(\xi) - F\left[\xi + \frac{1}{\sqrt{N}} Z_N(\xi)\right]\right] \mu(d\xi) - \sqrt{\frac{N}{m}} \int Z^*_N(\xi)\mu(d\xi). \end{aligned} \tag{4.27}$$

According to proposition 3, the difference $T^*_N - T_N$ converges to zero in probability, uniformly for $\|\mu\|$ bounded.

According to propositions 2 and 4, for $\epsilon > 0$, both Z_N and Z^*_N admit linear interpolations of bounded rank which differ from them by less than ϵ, except in cases of probability ϵ. Therefore, one could find a suitable probability space and pairs (W_N, W^*_N) of Gaussian processes with appropriate covariances such that for every $\epsilon > 0$,

$$P\{\|Z_N - W_N\| + \|Z_N^* - W_N^*\| > \epsilon\} < \epsilon \tag{4.28}$$

for $N \geq N(\epsilon)$ and for every pair (F, G). The functions

$$T_N^*(W, W^*) = \sqrt{N} \int \left[F(\xi) - F\left(\xi + \frac{W(\xi)}{\sqrt{N}}\right)\right] \mu(d\xi) - \sqrt{\frac{N}{m}} \int W^*(\xi)\mu(d\xi) \tag{4.29}$$

satisfies a Lipschitz condition

$$|T_N^*(u, u^*) - T_N^*(v, v^*)| \leq [\|u - v\| + \|u^* - v^*\|] \frac{1}{\lambda_0} \|\mu\|, \tag{4.30}$$

since $\lambda_N F$ has a derivative bounded by unity and since $\lambda_0 \leq \lambda_N$. As a consequence, one can state the following corollary.

PROPOSITION 5. *Let T_N, W_N, and W_N^* be the objects defined above. Let P_N be the distribution of T_N and let Q_N be the distribution of*

$$\sqrt{N} \int \left\{F(\xi) - F\left[\xi + \frac{W_N}{\sqrt{N}}\right]\right\} \mu(d\xi) - \sqrt{\frac{N}{m}} \int W_N^*(\xi)\mu(d\xi). \tag{4.31}$$

For every $\epsilon > 0$ there exists an $N(\epsilon)$ such that $N \geq N(\epsilon)$ implies $\|P_N - Q_N\|_{BL} \leq \epsilon\|\mu\|$ for every $\mu \in \mathfrak{M}$ and every pair (F, G).

REMARK. The above proposition remains valid if T_N is modified by replacing H_N by $(N/(N + 1))H_N$. It is easily checked that such a replacement amounts to a slight modification of the measure μ and the introduction of terms which are at most of order $[\|\mu\|/\sqrt{N}]$.

5. Bounds for the tails of the Chernoff-Savage statistics

For the purposes of the present section, let f and g be the functions defining the set $\mathcal{S}$ and let $\mathcal{S}_0$ be the set of indefinite integrals of the type $J(x) = \int_{1/2}^{x} J'(\xi)\, d\xi$ with $|J'| \leq fg$. If τ is a number such that $0 < 2\tau < 1$, let

$$\Delta_N^*(J, \tau) = \sqrt{N} \int_A \left\{J\left(\frac{N}{N+1} H_N\right) - J(H)\right\} dF_m, \tag{5.1}$$

where the integral is taken over the set $A = (0, \tau] \cup [1 - \tau, 1)$.

PROPOSITION 6. *For every $\epsilon > 0$ there exists a number τ_0 such that*

$$P\{\sup [|\Delta_N^*(J, \tau)|; 0 < \tau \leq \tau_0, J \in \mathcal{S}_0] > \epsilon\} < \epsilon \tag{5.2}$$

for every N and every pair (F, G).

PROOF. If one reverses the order of the observations by changing x to $1 - x$, the part of $\Delta_N^*(J, \tau)$ arising from the integral over $[1 - \tau, 1)$ is transformed into a similar integral, for a different function J, over the interval $(0, \tau]$. Thus, it will be sufficient to bound the part relative to $(0, \tau]$. Let δ be a number $0 < 8\delta < \lambda_0^{-1}\epsilon$ such that both f and g are monotone decreasing in the interval

$(0, \delta]$. One can assume throughout that $\tau \leq \delta$. In this case, replacing binomial distributions by Poisson distributions as explained in section 4, one can bound instead of Δ_N^* the simpler expression

$$S_N^*(J, \tau) = \frac{\sqrt{N}}{m} \sum_{\xi_i \leq \tau} \left| J\left(\frac{i}{N+1}\right) - J(\xi_i) \right| Z_{N,i} \tag{5.3}$$

where the ξ_i are the order statistics from a sample of size N taken from the uniform distribution. The $Z_{N,i}$ are independently selected with conditional probability of being equal to unity given by

$$P[Z_{N,i} = 1 | \xi_1, \xi_2, \cdots, \xi_N] = 1 - P[Z_{N,i} = 0 | \xi_1, \cdots, \xi_N] = \lambda_N \varphi(\xi_i). \tag{5.4}$$

Instead of using the above representation, one can also introduce independent random variables $\{U_j\}$, $j = 1, 2, \cdots, N$ which are uniformly distributed on $[0, 1]$ and their ranks R_j. The whole system $\{(U_j, R_j); j = 1, 2, \cdots, N\}$ will be denoted by the letter W. Taking this possibility into account and the fact that $\lambda_N \geq \lambda_0$, it will be sufficient to bound

$$S_N(J, \tau) = \frac{1}{\sqrt{N}} \sum_{U_i \leq \tau} \left| J\left(\frac{R_i}{N+1}\right) - J(U_i) \right|. \tag{5.5}$$

According to lemma 8, there is a number $\beta > 0$ such that

$$P\left\{\beta U_i \leq \frac{R_i}{N+1} \leq \frac{1}{\beta} U_i \text{ for all } i\right\} > 1 - \epsilon/4. \tag{5.6}$$

Let f_β be the function $f_\beta(x) = f(\beta x)$ defined for $x \in (0, \delta]$. Define g_β similarly. Once β is chosen, lemma 7 implies the existence of a number c_1 such that

$$P\left\{\sup_i \left[\sqrt{N} g_\beta(U_i) \left|\frac{R_i}{N} - U_i\right|; U_i \leq \delta\right] \geq c_1\right\} < \epsilon/4. \tag{5.7}$$

In addition, if $\beta(N+1)U_i \leq R_i \leq \beta^{-1}(N+1)U_i$, one can write

$$\max_i \left\{\frac{1}{\sqrt{N}} g_\beta(U_i) \frac{R_i}{N+1}\right\} \leq \frac{1}{\beta\sqrt{N}} \max_i U_i g_\beta(U_i). \tag{5.8}$$

If the maximum is restricted to those values i such that $U_i \leq \tau$, this last term is not larger than

$$\frac{1}{\beta\sqrt{N}} \sup_{x \leq \tau} x g_\beta(x) = \frac{1}{\beta\sqrt{N}} \sup_{x \leq \tau} x g(\beta x). \tag{5.9}$$

Since $xg(x) \to 0$ as $x \to 0$, there is a τ_1 and a c such that $P[W \in \mathcal{R}] > 1 - \epsilon$, if $\mathcal{R}$ is the set of systems $W = \{(U_i, R_i)\}$ which satisfy

(i) $(N+1)\beta U_i \leq R_i \leq (N+1)\beta^{-1} U_i$

and

(ii) $\sup_i \left\{\sqrt{N} g_\beta(U_i) \left|\frac{R_i}{N+1} - U_i\right|; U_i \leq \tau_1\right\} \leq c.$

However, if $W \in \mathcal{R}$, then $|J'(u)| \leq f_\beta(U_i) g_\beta(U_i)$ for every point u belonging to the interval between U_i and $R_i/(N+1)$.

It follows that

$$S_N(J, \tau) = \frac{1}{\sqrt{N}} \sum_{U_i \leq \tau} \left| J\left(\frac{R_i}{N+1}\right) - J(U_i) \right| \leq \frac{c}{N} \sum_{U_i \leq \tau} f_\beta(U_i). \tag{5.10}$$

The integral $\int_0^\tau f_\beta(u)\, du$ is equal to $(1/\beta) \int_0^{\tau/\beta} f(v)\, dv$. Therefore there exists a $\tau_0 \leq \tau_1$ such that $c \int_0^{\tau_0} f_\beta(u)\, du \leq \epsilon^2$. The desired result follows by application of Markov's inequality.

The quantity which appears in the study of Chernoff-Savage statistics is not exactly equal to $\Delta_N^*(J, \tau)$ but to

$$\Delta_N(J, \tau) = \sqrt{N} \left\{ \int_A J\left(\frac{N}{N+1} H_N\right) dF_m - \int_A J(H)\, dF \right\}, \tag{5.11}$$

where A is again equal to $(0, \tau] \cup [1 - \tau, 1)$. Clearly,

$$\Delta_N = \Delta_N^* + \sqrt{N} \int_A J(H)\, d(F_m - F). \tag{5.12}$$

The difference term $\Delta_N^* - \Delta_N$ is a normalized sum with expectation zero and a variance bounded by expressions of the type $4\lambda_0^{-1} c^4(\tau)$ where $c(\tau)$ is a function described in the proof of lemma 2. Thus for every $\epsilon > 0$ there is a τ_0 such that $P\{|\Delta_N^* - \Delta_N| > \epsilon\} < \epsilon$ for $J \in \mathcal{S}_0$ and $\tau \leq \tau_0$. In other words, the following corollary holds.

COROLLARY. *Let A be the set $A = (0, \tau] \cup [1 - \tau, 1)$ and let*

$$\Delta_N(J, \tau) = \sqrt{N} \left| \int_A J\left(\frac{N}{N+1} H_N\right) dF_m - \int_A J(H)\, dF \right|. \tag{5.13}$$

For every $\epsilon > 0$ there is a number $\tau_0 > 0$ such that $\tau < \tau_0$ and $J \in \mathcal{S}_0$ implies

$$P\{|\Delta_N(J, \tau)| > \epsilon\} < \epsilon \tag{5.14}$$

for every N and every pair (F, G).

For some purposes it is convenient to eliminate a few terms in the tails of the Chernoff-Savage statistics. In this connection, let us mention the following easy result. Suppose that $|J|$ is monotone decreasing in the interval $(0, \delta]$. Let k be an integer and let $y = k/(N + 1)$. Then

$$\begin{aligned} \frac{\sqrt{N}}{m} \left| \sum_{i=1}^{k} J\left(\frac{i}{N+1}\right) \right| &\leq \frac{\sqrt{N}}{m} (N+1) \int_0^y |J(x)|\, dx \\ &\leq \frac{N+1}{m} \sqrt{\frac{Nk}{N+1}} \left\{ \int_0^y J^2(x)\, dx \right\}^{1/2}. \end{aligned} \tag{5.15}$$

Therefore, whenever J stays in a family of monotone functions which are uniformly square integrable the sum $(\sqrt{N}/m) \sum_{i=1}^{k} |J(i/(N + 1))|$ tends to zero for each fixed k as $N \to \infty$. It follows that for each fixed $y \in (0, \infty)$ the terms

$$\sqrt{N} \int_{Nx \leq y} \left| J\left(\frac{N}{N+1} H_N(x)\right) \right| dF_m(x) \tag{5.16}$$

tend to zero in probability as $N \to \infty$.

6. The asymptotic behavior of the Chernoff-Savage statistics

For the purposes of the present section it is convenient to consider the set $\mathcal{S}_1 \supset \mathcal{S}$ of functions J which are indefinite integrals of functions J' such that $J' = J_1' + J_2'$ with $|J_2'| \leq fg$ and $\int |J_1'(x)|\,dx \leq b$. The set $\mathcal{S}_1$ will be topologized as follows. A sequence $\{J_k\}$ converges to J if $\int_A |J_k'(x) - J'(x)|\,dx \to 0$ for every interval $A = [\tau, 1 - \tau]$ with $0 < 2\tau < 1$. A subset S of $\mathcal{S}_1$ is called relatively compact if every sequence $\{J_k\} \subset S$ admits a convergent subsequence.

Triples $\{F, G, \lambda\}$ such that $H = \lambda F + (1 - \lambda)G$ is the uniform distribution on $[0, 1]$ will be topologized by requiring that when $[F_\nu, G_\nu, \lambda_\nu] \to [F, G, \lambda]$, the densities $\lambda_\nu \varphi_\nu = \lambda_\nu (dF_\nu)/dH$ converges in measure. Of course, it is still assumed that $0 < \lambda_0 \leq \lambda \leq 1 - \lambda_0$.

Consider the functions L and M defined in section 3 by $L(x) = \int^x J'(\xi)\,dF(\xi)$ and $M(x) = \int^x J'(\xi)\,dG(\xi)$. Let σ_N^2 be the variance $\sigma_N^2[(F, G), J, \lambda_N]$ introduced in section 3.

THEOREM 1. *Let J be an element of $\mathcal{S}_1$ and let*

$$T_N = \sqrt{N}\left\{\int J\left[\frac{N}{N+1} H_N\right] dF_m - \int J(H)\,dF\right\}. \tag{6.1}$$

Let P_N be the distribution of T_N and let Q_N be the normal distribution which has expectation zero and variance $\sigma_N^2[(F, G), J, \lambda_N)]$. If S is a relatively compact subset of $\mathcal{S}_1$, then for every $\epsilon > 0$ there is an $N(\epsilon)$ such that $N \geq N(\epsilon)$ implies $\|P_N - Q_N\|_{BL} < \epsilon$ for every $J \in S$ and every triple $\{(F, G), \lambda_N\}$. Similarly, if $\mathfrak{F}$ is a relatively compact set of triples $[(F, G), \lambda]$, then for every $\epsilon > 0$ there exists an $N(\epsilon)$ such that $N \geq N(\epsilon)$ implies $\|P_N - Q_N\|_{BL} < \epsilon$ for every $J \in \mathcal{S}$ and every triple $\{(F, G), \lambda\} \in \mathfrak{F}$.

On sets such that σ_N^2 stays bounded away from zero the bounded Lipschitz norm $\|P_N - Q_N\|_{BL}$ may be replaced by the Kolmogorov vertical distance.

PROOF. Suppose $J' = J_1' + J_2'$ with $|J_2'| \leq fg$. According to proposition 6, for any given $\epsilon > 0$ there is a number $\tau > 0$ and an $N_0(\epsilon)$ such that if A is the set $(0, \tau] \cup [1 - \tau, 1)$ and if T_N^τ is the expression

$$T_N^\tau = \sqrt{N}\left\{\int_A J_2\left[\frac{N}{N+1} H_N\right] dF_m - \int_A J_2(H)\,dF\right\}, \tag{6.2}$$

then $P\{|T_N^\tau| > \epsilon\} < \epsilon$ for every J_2' and every triple $\{(F, G), \lambda\}$. Since the function $|J_2'|$ is integrable and bounded by $\sup\{f(x)g(x), \tau \leq x \leq 1 - \tau\}$ on the interval $[\tau, 1 - \tau]$, it will be sufficient to prove the theorem for integrable functions J' such that $\int |J'(x)|\,dx \leq b$.

In this case, according to proposition 3 or proposition 5, one can replace the variable T_N by $T_N^* = \sqrt{N}B_N + R_N$ with

$$\sqrt{N}B_N = \sqrt{N}\int J(H)\,d(F_m - F) + \sqrt{N}\int [H_N - H]J'(H)\,dF, \tag{6.3}$$

$$R_N = \sqrt{N}\int [J(H_N) - J(H) - (H_N - H)J'(H)]\,dF. \tag{6.4}$$

For every $\epsilon > 0$ there is an $N_1(\epsilon)$ such that $N \geq N_1(\epsilon)$ implies $P\{|T_N^* - T_N| > \epsilon\} < \epsilon$, whatever may be J and whatever may be F, G, and λ_N. Since $\sqrt{N}B_N$ is precisely the term introduced in section 3, the results claimed in the statement of the theorem will depend on the evaluation of appropriate bounds for R_N.

First note that given $\epsilon > 0$ there is a $c < \infty$ such that $P\{\sqrt{N}\|H_N - H\| \geq c\} < \epsilon$. Let $\tilde{H}_N = H_N$ if $\sqrt{N}\|H_N - H\| < c$ and let $\tilde{H}_N = H$ otherwise. Let

$$\tilde{R}_{N,1} = \sqrt{N} \int [J(\tilde{H}_N) - J(H)]\, dF \tag{6.5}$$

and let

$$\tilde{R}_{N,2} = \int \sqrt{N}[\tilde{H}_N - H]J'(H)\, dF. \tag{6.6}$$

If $\varphi = dF/dH$ and if $\tilde{K}_N(x) = \inf \{t\colon \tilde{H}_N(t) \geq x\}$, one can also write $\tilde{R}_{N,1}$ and $\tilde{R}_{N,2}$ in the form

$$\tilde{R}_{N,1} = \sqrt{N} \int \{F(\xi) - F[\tilde{K}_N(\xi)]\} J'(\xi)\, d\xi, \tag{6.7}$$

$$\tilde{R}_{N,2} = \int \sqrt{N}[\tilde{H}_N(\xi) - H(\xi)]J'(\xi)\varphi(\xi)\, d\xi. \tag{6.8}$$

Since $0 \leq \lambda_0\varphi \leq 1$ and since $\sqrt{N}\|\tilde{H}_N - H\| \leq c$ implies $\sqrt{N}\|\tilde{K}_N(\xi) - \xi\| \leq c$, both $\tilde{R}_{N,1}$ and $\tilde{R}_{N,2}$ satisfy Lipschitz conditions in J' for the norm $\|J'\| = \int |J'(s)|\, dx$. If $\tilde{R}_N[F, J'] = \tilde{R}_{N,1} - \tilde{R}_{N,2}$, this implies $|\tilde{R}_N(F, J')| \leq (2c/\lambda_0)\|J'\|$. Therefore, the first statement of the theorem, with uniformity of the convergence on compact subsets of $\mathcal{S}_1$ will follow if we show that for a fixed J' the term $\tilde{R}_N[F, J']$ converges to zero uniformly in F.

Let $\rho_N(x)$ be the ratio

$$\rho_N(x) = \sup_{|\xi| \leq c} \sqrt{N}\left| J\left[x + \frac{\xi}{\sqrt{N}}\right] - J(x) - \frac{\xi}{\sqrt{N}} J'(x)\right|. \tag{6.9}$$

By definition of the derivative, this converges to zero for almost every x. In addition, $\int \rho_N(x)\, dx$ converges to zero. However,

$$|\tilde{R}_N[F, J']| \leq \int \rho_N(x)\, dF(x) \leq \frac{1}{\lambda_0} \int \rho_N(x)\, dx. \tag{6.10}$$

The first statement follows.

For the second statement, note that if $J \in \mathcal{S}$ and if the part of J' carried by the set $(0, \tau] \cup [1 - \tau, 1)$ has been removed, the remaining part of J' is smaller than a certain integrable function $\omega = a + f_0$, with a equal to sup $\{f(x)g(x); \tau \leq x \leq 1 - \tau\}$. The term $\tilde{R}_{N,2}$ can be written

$$\tilde{R}_{N,2}(\varphi) = \int \sqrt{N}[\tilde{H}_N(\xi) - H(\xi)] \left[\frac{J'(\xi)}{\omega(\xi)}\right] \omega(\xi)\, \varphi(\xi) d\xi. \tag{6.11}$$

This satisfies a Lipschitz condition,

$$|\tilde{R}_{N,2}(\varphi)| \leq c\|\varphi\|, \tag{6.12}$$

for the norm $\|\varphi\| = \int \omega(\xi)|\varphi(\xi)|\, d\xi$. Similarly,

$$\tilde{R}_{N,1}(\varphi) = \int \sqrt{N}[J(\tilde{H}_N) - J(H)]\varphi(\xi)\, d\xi \tag{6.13}$$

satisfies the condition

$$\tilde{R}_{N,1}(\varphi) \leq \int \gamma_N(\xi)|\varphi(\xi)|\, d\xi \tag{6.14}$$

with

$$\gamma_N(\xi) = \sup_{|x| \leq c} \sqrt{N}\left|\Omega\left[\xi + \frac{x}{\sqrt{N}}\right] - \Omega(\xi)\right|, \tag{6.15}$$

$$\Omega(x) = \int_0^x \omega(\xi)\, d\xi. \tag{6.16}$$

Suppose then that $\varphi_\nu \to \varphi$ in measure; then

$$\int \omega(\xi)|\varphi_\nu(\xi) - \varphi(\xi)|\, d\xi \to 0. \tag{6.17}$$

Furthermore,

$$\int \gamma_N(\xi)|\varphi_\nu(\xi) - \varphi(\xi)|\, d\xi \to 0 \tag{6.18}$$

uniformly in N since the functions γ_N are uniformly integrable. The result follows by the usual argument. This completes the proof of the theorem.

REMARK 1. The convergence is *not* uniform on the set of systems $\{(F, G), J, \lambda\}$ such that $J \in \mathcal{S}$. In fact, suppose that $J'_N(x)$ is equal to -1 for $2k/2^N < x \leq (2k+1)/2^N$ and to $+1$ for $(2k+1)/2^N < x \leq (2k+2)/2^N$, $k = 0, 1, 2, \cdots, 2^N$. Then $|J_N| \leq 2^{-N}$. Suppose that $2\lambda_N = 1$ and that F_N has a density φ_N equal to 2 for $2k/2^N < x \leq (2k+1)/2^N$ and to zero otherwise. Then the function L_N differs little from $L(x) = -x$ and M_N differs little from $M(x) = +x$. The expression $|T_N|$ is smaller than $2^{-N}\sqrt{N}$, but $4\sigma^2$ is approximately equal to unity.

In the paper of Chernoff and Savage, it is assumed that the second derivative J'' satisfies a restriction of the type $|J''(x)| \leq K[x(1-x)]^{-(5/2)+\delta}$. This implies in particular that the available family $\{J'\}$ is relatively compact for uniform convergence on the compact intervals of $(0, 1)$. Thus theorem 1 asserts uniformity of the convergence on that class.

REMARK 2. One particular case in which the uniformity asserted in the theorem may be of interest is the following.

Suppose that for each N the distribution $\check{F}$ of the original observations labeled X is given by a distribution function Ψ which admits a density. Suppose also that the distribution $\check{G}$ of the variables Y is given by $\Psi\{(x - \theta)/\beta\}$ with $\beta > 0$. In this case if $\theta_\nu \to \theta_0$ and $\beta_\nu \to \beta_0 \neq 0$, the density of $\Psi[(x - \theta_\nu)/\beta_\nu]$ converges in measure to that of $\Psi[(x - \theta_0)/\beta_0]$. It follows that the corresponding distributions (F, G) reduced to the interval $[0, 1]$ converge in the same sense. The convergence $\|P_N - Q_N\|_{BL} \to 0$ asserted in the theorem is therefore uniform for $J \in \mathcal{S}$, $\lambda_N \in [\lambda_0, 1 - \lambda_0]$ and every bounded set of values $[\theta, \log \beta]$. In addition, if for the integer N the value of (θ, β) is (θ_N, β_N) and $\theta_N \to 0$ and $\beta_N \to 1$, then the Kolmogorov distance $|P_N - Q_N|$ also tends to zero uniformly for any set $S \subset \mathcal{S}$ such that $J \in S$ implies $\int J^2(u)\,du - [\int J(u)\,du]^2 \geq \alpha > 0$. This follows immediately from lemma 5.

In many cases the functions J are obtained using expectations of suitable order statistics. In this connection the following theorem may be of interest.

For each integer N let J'_N be a nonnegative function such that $0 \leq J'_N(x) \leq K[x(1-x)]^{-(3/2)+\delta}$ for some fixed $K < \infty$ and some fixed δ, $0 < 2\delta < 1$. Let J_N be an integral of J'_N and let $\bar{J}_N(i/(N+1))$ be the expected value $E[J_N(\xi_{N,i})]$ where $\xi_{N,i}$ is the i-th smallest order statistic in a sample of size N from the uniform distribution on $[0, 1]$. Complete the definition of $\bar{J}_N$ by linear interpolation between successive integers i and by leaving $\bar{J}_N$ constant below $(1/(N+1))$ and above $(N/(N+1))$.

THEOREM 2. *Suppose that the relation* $|J'_N(x)| \leq K[x(1-x)]^{-(3/2)+\delta}$ *is satisfied and that* J'_N *converges in Lebesgue measure to a limit* J'. *Let* T_N *be the expression*

$$T_N = \sqrt{N}\left\{\int \bar{J}_N\left(\frac{N}{N+1}H_N\right)dF_m - \int J_N(H)\,dF\right\}. \tag{6.19}$$

Let P_N *be the distribution of* T_N *and let* Q_N *be the normal approximation of theorem* 1. *Then* $\|P_N - Q_N\|_{BL}$ *converges to zero, uniformly in* $[(F, G), \lambda]$ *as* $N \to \infty$.

PROOF. Let $T_N^* = \sqrt{N}\{\int \bar{J}_N(N/(N+1)H_N)\,dF_m - \int \bar{J}_N(H)\,dF\}$. According to lemma 1, the function $\bar{J}'_N$ satisfies the conditions of theorem 1, with $f = K[x(1-x)]^{-1+(\delta/2)}$ and $g^2 = [x(1-x)]^{-(1-\delta)}$. In addition, J'_N converges in measure to a limit J'; hence, $\bar{J}'_N$ converges to J' according to lemma 1. It follows from this that theorem 2 would be proved if T_N was replaced by T_N^*. To complete the proof, it will be sufficient to bound the difference $T_N - T_N^* = \sqrt{N}\{\int [\bar{J}_N(H) - J_N(H)]\,dF$. However, this is smaller than $\lambda_N^{-1}\int \sqrt{N}|\bar{J}_N(x) - J_N(x)|\,dx$.

Since, in this last integral, the terms in the absolute value sign are linear functions of J'_N and since J'_N can be split into positive and negative parts and each of these into parts supported by $[0, \frac{1}{2}]$ and $[\frac{1}{2}, 1]$ respectively, it will be sufficient to prove the result under the assumption $J_N(x) = \int_x^{\frac{1}{2}} h_N(\xi)\,d\xi$, $0 \leq h_N(x) \leq x^{-(3/2)+\delta}$, $h_N(x) = 0$ for $x > \frac{1}{2}$. The parts relative to the interval $[\frac{1}{2}, 1]$ are handled by changing x into $1 - x$.

With this definition, note that

$$\sqrt{N}\int_0^x J_N(\xi)\,d\xi \leq K\sqrt{N}x^{-(1/2)+\delta}, \tag{6.20}$$

and that a similar inequality holds for $\bar{J}_N$. Thus it will be sufficient to prove that $\sqrt{N}\int_{\epsilon_N}^1 |\bar{J}_N(x) - J_N(x)|\,dx$ tends to zero for $N\epsilon_N = N^{\delta/2}$.

We shall proceed by showing that $\sqrt{N}|\bar{J}_N(x) - J_N(x)|$ stays bounded by an integrable function and that $\sqrt{N}|\bar{J}_N(x) - J_N(x)| \to 0$ in measure. For the first part it is convenient to divide the range $[\epsilon_N, 1]$ into two parts $[\epsilon_N, \tau]$ and $(\tau, 1]$ with τ fixed but $\tau < 1$.

Consider first the part relative to the interval (ϵ_N, τ). Let $\beta_N(x, k)$ be the density of the k-th order statistics from a uniform sample of size N, and let

$$\beta_N(x, k) = \frac{\Gamma(N+1)}{\Gamma(k)\Gamma(N+1-k)} x^{k-1}(1-x)^{N-k} \tag{6.21}$$

even for noninteger values of the symbols N and k. Let $\xi = (k/(N+1))$ and let $I_1(\xi)$ be the integral

$$I_1(\xi) = \int_0^{\xi/4} J_N(x)\beta_N(x, k)\, dx. \tag{6.22}$$

Since $0 \leq J_N(x) \leq x^{-\alpha}$, $\alpha = \frac{1}{2} - \delta$, one can bound $I_1(\xi)$ by the integral

$$I_2(\xi) = \int_0^{\xi/4} \frac{\Gamma(N+1)\Gamma(k-\alpha)}{\Gamma(k)\Gamma[N+1-\alpha]} \beta_{N-\alpha}[x, k-\alpha]\, dx. \tag{6.23}$$

Consider also the function

$$\gamma_N(x, k-\alpha) = \frac{(N-\alpha)^{k-\alpha}}{\Gamma(k-\alpha)} x^{k-\alpha-1}e^{-(N-\alpha)x}. \tag{6.24}$$

A simple application of Stirling's formula shows that

$$\beta_N[x, k-\alpha] \leq c\left[1 - \frac{\alpha}{N}\right]^{1/2}\left[1 - \frac{k}{N}\right]^{-1/2} \gamma_N[x, k-\alpha] \tag{6.25}$$

for a certain constant c and for all values of N and k such that $N - k - \alpha \geq 1$.

In addition, $\int_0^{\xi/4} \gamma_N[x, k-\alpha]\, dx$ can be bounded by Markov's inequality as follows. If $\mu = (k-\alpha)/(N-\alpha)$, then, for $s > 0$,

$$\int e^{-2s[N-\alpha][x-\mu/2]}\gamma_N[x, k-\alpha]\, dx = \left[\frac{e^s}{1+2s}\right]^{k-\alpha}. \tag{6.26}$$

Therefore, $\int_0^{\mu/2} \gamma_N[x, k-\alpha]\, dx \leq \rho^{k-\alpha}$, with $\rho = \inf \{(e^s/(1+2s)), s \geq 0\} < 1$. Since $\xi/4 \leq \mu/2$ for $N \geq N_0$ and $k \geq N_0^{\delta/2}$, this implies

$$\begin{aligned} I_2(\xi) &\leq c\rho^{k-\alpha}\left(1 - \frac{\alpha}{N}\right)^{1/2}\left(1 - \frac{k}{N}\right)^{-1/2} \frac{\Gamma(N+1)\Gamma(k-\alpha)}{\Gamma(k)\Gamma(N+1-\alpha)} \\ &\leq c'\rho^k\left(\frac{N-\alpha}{k-\alpha-1}\right)^{\alpha}. \end{aligned} \tag{6.27}$$

Therefore, given $\epsilon > 0$, there exists an $N_1(\epsilon)$ such that $N \geq N_1$ and $k \geq N^{\delta/2}$ implies $\sqrt{N}I_2(\xi) \leq \epsilon\xi^{(1/2)-\delta}$ for $\epsilon_N < \xi \leq \tau$. Thus this term will become negligible. Consider the term

$$\begin{aligned} I_3(\xi) &= \sqrt{N}\int_{\xi/4}^1 [J_N(x) - J_N(\xi)]\beta_N(x, k)\, dx \\ &= \int_{\xi/4}^1 \frac{J_N(x) - J_N(\xi)}{|x-\xi|} \sqrt{N}|x-\xi|\beta_N[x, (N+1)\xi]\, dx. \end{aligned} \tag{6.28}$$

Note that

$$\sqrt{N}\int |x-\xi|\beta_N[x, (N+1)\xi]\, dx \leq \sqrt{\frac{N}{N+2}}\sqrt{\xi(1-\xi)}. \tag{6.29}$$

Furthermore, in I_3 the differential ratio involving J_N is, by assumption, smaller than the same ratio involving the function $\Omega(x) = x^{-1/2+\delta}$. For the latter function, the maximum value of the ratio is obtained at $x = \xi/4$ giving

$$I_3(\xi) \leq (\tfrac{1}{4}\xi)^{-(3/2)+\delta}\sqrt{\xi(1-\xi)} = c_1\sqrt{1-\xi}\,\xi^{-(1/2)+\delta}. \tag{6.30}$$

This is an integrable function.

To show that $I_3(\xi)$ does in fact tend to zero for almost all ξ, if J'_N converges to a function J', it is sufficient to repeat an argument similar to the argument of lemma 1. Note also that $J^*_N(\xi) = \int J_N(x)\beta_N[x, (N+1)\xi]\,dx$ provides a decreasing interpolation of the function $\overline{J}_N$. From this one concludes that $\sqrt{N}|J^*_N(\xi) - J_N(\xi)| \to 0$ for almost every ξ, and therefore $\sqrt{N}|\overline{J}_N(\xi) - J_N(\xi)| \to 0$ for almost every ξ. Furthermore, for $N \geq N_0$ one has $\sqrt{N}|\overline{J}_N(\xi) - J_N(\xi)| \leq c_2\xi^{-1/2+\delta}$ for every $\xi \geq \epsilon_N$. The result follows.

Remark. The preceding theorem 2 corresponds to theorem 2 of Chernoff and Savage. As a particular application, let us mention the following corollary.

Corollary. *Let k be a fixed integer and let a_j, $j = 1, 2, \cdots, k$ be bounded constants. Let*

$$J_N\left(\frac{i}{N+1}\right) = \sum_{j=1}^{k} a_j E(\xi^j_{N,i}) \tag{6.31}$$

where $\xi^j_{N,i}$ is the i-th order statistic in a sample of size N from a population whose cumulative distribution is the inverse of a function S_j.

If $|(dS^j(x)/dx)| \leq K[x(1-x)]^{-(3/2)+\delta}$ for $j = 1, 2, \cdots, k$, then the functions J_N satisfy the conditions of theorem 1.

Proof. This follows from the linearity of the transformation $J \rightsquigarrow \overline{J}_N$ used to define the functions which occur in theorem 2.

For the case where $a_1 = 0$, $a_2 = 1$, $k = 2$, and $S = \Phi^{-1}$ the resultant test statistic is the one considered by Capon [4] and Klotz [11].

7. The c-sample case

In this section we shall extend the results of section 6 to c-sample situations. Without additional assumptions Puri [13] had extended the Chernoff-Savage results to c-sample cases. Our theorems 3 and 4 are direct extensions of Puri's lemma 5.1 and theorem 6.1.

Let $X_{j,k}$, $k = 1, 2, \cdots, n_j$ be a random sample from a population having a continuous cumulative distribution function $F^{(j)}$. Assume that the c-samples obtained for $j = 1, 2, \cdots, c$ are independent. Let $N = \sum n_j$ and let $\lambda_j = n_j/N$. Assume that there is a $\lambda_0 > 0$ such that $\lambda_0 \leq \lambda_j \leq 1 - \lambda_0$ for every $j = 1, 2, \cdots, c$ and every N. Let $H = \sum_j \lambda_j F^{(j)}$ and $H_N = \sum_j \lambda_j F^{(j)}_{n_j}$ be respectively the combined cumulative and the combined empirical cumulative based on the samples $\{X_{j,k}\}$.

Let $T_{N,j} = n_j^{-1}\sum_{i=1}^{N} E_{N,i,j}Z_{N,i,j}$, $j = 1, 2, \cdots, c$, where $Z_{N,i,j} = 1$ if the i-th smallest observation from the combined sample of size N belongs to j-th sample and where $Z_{N,i,j}$ is equal to zero otherwise.

The $E_{N,i,j}$ are given constants. Following the notation of section 2, one can represent $T_{N,j}$ in an integral form

$$T_{N,j} = \int_{-\infty}^{+\infty} J_{N,j}\left[\frac{N}{N+1} H_N\right] dF_{n_j}^{(j)}(x). \tag{7.1}$$

For simplicity we have assumed that the functions $F^{(j)}$ are continuous and state a result analogous to theorem 1 in a form similar to the form of theorem 1 of Chernoff and Savage, for the original distribution functions.

THEOREM 3. *Assume that for all* $j = 1, 2, \cdots, c$ *the following conditions hold:*

(1) $J_j(H) = \lim J_{N,j}(H)$ *exists for* $0 < H < 1$, *and this limit is not a constant and it is absolutely continuous on* $(0, 1)$;

(2) $\int_{0<H_N\leq 1} \left\{J_{N,j}\left(\frac{N}{N+1} H_N\right) - J_j\left(\frac{N}{N+1} H_N\right)\right\} dF_{n_j}^{(j)}(x) = o_p(N^{-1/2})$.

(3) $\left|\frac{dJ_j(x)}{dx}\right| \leq K[x(1-x)]^{-(3/2)+\delta}$ with $0 < 2\delta < 1$.

Let $\mu_{N,j} = \int_{-\infty}^{+\infty} J_j[H(x)]\, dF^{(j)}(x)$ *and*

$$\begin{aligned} \sigma_{N,j}^2 = \sum_{i\neq j} 2\lambda_i \iint_{-\infty<x<y<+\infty} F^{(i)}(x)[1 - F^{(i)}(y)]J_j'[H(x)]J_j'[H(y)] \\ \cdot dF^{(j)}(x)dF^{(j)}(y) \\ + \frac{2}{\lambda_j} \iint_{x<y} F^{(j)}(x)[1 - F^{(j)}(y)]J_j'[H(x)]J_j'[H(y)] \\ \cdot d[H(x) - \lambda_j F^{(j)}(x)]\, d[H(y) - \lambda_j F^{(j)}(y)]. \end{aligned} \tag{7.2}$$

Then, if $\liminf_{N\to\infty} \sigma_{N,j}^2 > 0$, *one has*

$$\lim_{N\to\infty} P\left\{\sqrt{N}[T_{N,j} - \mu_{N,j}] < t\sigma_{N,j}\right\} = \frac{1}{\sqrt{2\pi}} \int_{-\infty}^{t} e^{-x^2/2}\, dx. \tag{7.3}$$

PROOF. Conditions (1) and (2) of the theorem imply that one may consider instead of $(T_{N,j} - \mu_{N,j})\sqrt{N}$ the expression

$$\tilde{T}_{N,j} = \sqrt{N}\left\{\int J_j\left[\frac{N}{N+1} H_N\right] dF_{n_j}^{(j)} - \int J_j(H)\, dF^{(j)}\right\}, \tag{7.4}$$

which is similar to the expression covered by theorem 1. One can proceed exactly as in theorem 1, along the following sequence of steps. First one can show that an integral of the type

$$\Delta_N[J_j, \tau] = \sqrt{N}\left\{\int_A J_j\left[\frac{N}{N+1} H_N\right] dF_{n_j}^{(j)} - \int_A J_j(H)\, dF^{(j)}\right\} \tag{7.5}$$

with $A = (0, \tau] \cup [1 - \tau, 1)$ can be made small by selecting τ small enough. Second, removing an appropriate term from J_j' (on the set A), one is left with functions J'_j which are bounded, and one shows, by an argument similar to that of proposition 3, that terms of the type

$$\sqrt{N} \int \left[J_j \left[\frac{N}{N+1} H_N \right] - J_j(H) \right] d[F^{(j)}_{n_j} - F^{(j)}] \tag{7.6}$$

can also be neglected. This replaces our $\tilde{T}_{N,j}$ by a term of the type

$$T^*_{N,j} = \sqrt{N} B_{N,j} + R_{N,j} \tag{7.7}$$

with

$$\sqrt{N} B_{N,j} = \sqrt{N} \int J_j(H)\, d(F^{(j)}_{n_j} - F^j) + \sqrt{N} \int [H_N - H] J'_j(H)\, dF^{(j)}, \tag{7.8}$$

and

$$R_{N,j} = \sqrt{N} \int [J_j(H_N) - J_j(H) - (H_N - H) J'_j(H)]\, dF. \tag{7.9}$$

An argument similar to that of theorem 1 shows that $R_{N,j}$ also tends to zero. Thus, the only nonnegligible term left is the term $\sqrt{N} B_{N,j}$, which can be written as sums of independent variables. The result is then obtainable by appropriate algebra and the central limit theorem.

To proceed through these steps, the necessary tools are the appropriate versions of lemmas 8 and 9 and propositions 2 and 3. Both lemmas 8 and 9 are proved there by substituting to binomial variables appropriate Poisson variables. This is still possible here. The difference in probabilities will be less than $2 \sum_{j=1}^{c} p_j$ where p_j is the probability attached to a set $(0, \epsilon] \cup [1 - \epsilon, 1)$ by the measure $F^{(j)}$ (reduced to the interval (0, 1) as before). Thus, in the tails, $\sqrt{N}\,[H_N - H]$ and H_N will still behave essentially as if they were obtained by taking N observations from the uniform distribution. Proposition 2 involves an interpolation which is feasible simply because none of the $\sqrt{N}[F^{(j)}_{n_j} - F^{(j)}]$ oscillates much on intervals which have small probability for the parent distribution. Since the derivatives $[dF^{(j)}/dH] = \varphi_j$ are still bounded, this is possible. Proposition 3 depends only on the behavior of $\sqrt{N}\,\|H_N - H\|$ and $\sqrt{N}\,\|F^{(j)}_{n_j} - F^{(j)}\|$. Hence it is still valid here. One could also extend proposition 4 to the present case, since its proof depends only on the validity of proposition 2 and on the fact that the interpolation formula of proposition 2, when applied to H_N itself, gives functions whose slope is arbitrarily close to unity. However, proposition 4 is not even needed for the proof of theorem 1.

Since in the present case one may have to consider the joint distribution of the statistics $T_{N,j}$, $j = 1, 2, \cdots, c$, it appears proper to mention that the random vector

$$T_N = \sqrt{N}\{(T_{N,1} - \mu_{N,1}), (T_{N,2} - \mu_{N,2}), \cdots, (T_{N,c} - \mu_{N,c})\} \tag{7.10}$$

has a joint limiting normal distribution, provided that the relevant covariances converge. This is the purpose of the following theorem.

THEOREM 4. *Let assumptions* (1), (2), *and* (3) *of theorem* 3 *be satisfied, and let* T_N *be the random vector* $T_N = \{\sqrt{N}(T_{N,j} - \mu_{N,j});\ j = 1, 2, \cdots, c\}$. *Let* P_N *be the distribution of* T_N *and let* Q_N *be the normal distribution which has expectation zero and covariance matrix* Γ_N. *Then* $\|P_N - Q_N\|_{BL} \to 0$ *provided* Γ_N *be given by* $\Gamma_N = ((\sigma_{N,i,j}))$ *with* $\sigma^2_{N,j}$ *equal to the quantity used in theorem* 3 *and*

$$
\begin{aligned}
(7.11)\qquad \sigma_{N,i,j} = \sum_{\substack{k=1\\ k\neq i,j}}^{c} \lambda_k \iint_{x<y} & F^{(k)}(x)[1 - F^{(k)}(y)]J_i'(H(x))J_j'(H(y)) \\
& \cdot dF^{(i)}(x)\, dF^{(j)}(y) \\
+ \iint_{x<y} & F^{(k)}(x)[1 - F^{(k)}(y)]J_i'(H(y))J_j'(H(x)) \\
& \cdot dF^{(i)}(y)\, dF^{(j)}(x) \\
- \iint_{x<y} & F^{(i)}(x)[1 - F^{(i)}(y)]J_i'(H(y))J_j'(H(x)) \\
& \cdot dF^{(j)}(x)\, d[H(y) - \lambda_i F^{(i)}(y)] \\
- \iint_{x<y} & F^{(i)}(x)[1 - F^{(i)}(y)]J_i'(H(x))J_j'(H(y)) \\
& \cdot dF^{(j)}(y)\, d[H(x) - \lambda_i F^{(i)}(x)] \\
- \iint_{x<y} & F^{(j)}(x)[1 - F^{(j)}(y)]J_i'(H(x))J_j'(H(y)) \\
& \cdot dF^{(i)}(x)\, d[H(y) - \lambda_j F^{(j)}(y)] \\
- \iint_{x<y} & F^{(j)}(x)[1 - F^{(j)}(y)]J_i'(H(y))J_j'(H(x)) \\
& \cdot dF^{(i)}(y)\, d[H(x) - \lambda_j F^{(j)}(x)].
\end{aligned}
$$

PROOF. The argument of theorem 3 shows that each term $\sqrt{N}\,[T_{N,j} - \mu_{N,j}]$ is asymptotically equivalent to a term $\sqrt{N}B_{N,j}$ defined by

$$
(7.12)\qquad \sqrt{N}B_{N,j} = \sqrt{N}\int J_j(H)\, d(F_{n_j} - F) + \sqrt{N}\int (H_N - H)J_j'(H)\, dF.
$$

Let L_j be the function $L_j(x) = \int_{1/2}^{x} J_j'(\xi)\, dF^j(\xi)$. Integrate by parts and separate the components of $[H_N - H]$. This gives

$$
\begin{aligned}
(7.13)\qquad \sqrt{N}B_{N,j} &= \sqrt{N}\int [J_j - \lambda_j L_j]\, d[F^{(j)}_{n_j} - F^{(j)}] \\
&\quad - \sqrt{N}\sum_{k\neq j}\int \lambda_k L_j\, d(F^{(k)}_{n_k}).
\end{aligned}
$$

This can also be written in the form

$$
\begin{aligned}
(7.14)\qquad \sqrt{N}B_{N,j} &= \frac{1}{\sqrt{\lambda_j}}\frac{1}{\sqrt{n_j}}\sum_{\nu}^{n_j} [J_j[X_{j,\nu}] - EJ_j(X_{j,\nu})] \\
&\quad - \sum_{k}\sqrt{\lambda_k}\sum_{\nu=1}^{n_k}\{L_j[X_{k,\nu}] - EL_j[X_{k,\nu}]\}.
\end{aligned}
$$

The central limit theorem applies to the sums, giving the result of theorem 4 upon evaluation of the covariance matrix. The explicit form given in the statement of the theorem is obtained by writing $\sqrt{N}B_{N,j}$ in still another form as follows.

Let $W_{N,j}$ be the process $W_{N,j} = \sqrt{n_j}\,[F_{n_j}^{(j)} - F^{(j)}]$ which has expectation zero and a covariance function

$$C_j(s, t) = F^{(j)}(s)[1 - F^{(j)}(t)] \tag{7.15}$$

for $s \leq t$. Then, the expression $\sqrt{N}B_{N,j}$ becomes

$$\sqrt{N}B_{N,j} = \sum_k \sqrt{\lambda_k} \int W_{N,k}(x)J_j'[H(x)]\,dH(x) - \frac{1}{\sqrt{\lambda_j}} \int W_{N,j}(x)J_j'[H(x)]\,dH(x). \tag{7.16}$$

The formula is obtained by taking the expectation of the product $\sqrt{N}B_{N,i}\sqrt{N}B_{N,j}$ in this form and using the fact that $EW_{N,k}(x)W_{N,\nu}(y) = 0$ for $k \neq \nu$, since then the processes $W_{N,k}$ and $W_{N,\nu}$ are independent.

Let us also note the following result.

THEOREM 5. *The conclusion of theorems* 3 *and* 4 *is still valid if the assumption* (3) *of theorem* 3 *is replaced by the condition that for each value of* $j = 1, 2, \cdots, c$ *the function* J_j' *is of the form* $J_{j,1}' + J_{j,2}'$ *with* $\int |J_{j,1}'(x)|\,dx < \infty$ *and* $|J_{j,2}'| \leq fg$ *with* $f \in \mathfrak{U}_1$ *and* $g \in \mathfrak{U}_2$. *The uniformity statements of theorem* 1 *extend to this case.*

PROOF. The proof is the same as that of theorem 3, with uniformity of the convergence depending on Lipschitz-type conditions as detailed in the proof of theorem 1.

Finally, let us note that the remark concerning location and scale families which follows theorem 1 is still applicable here and that theorem 2 provides a class of functions J_N which satisfy the assumptions (1) and (2) of theorem 3. The arguments which lead to this last statement do not in any way depend on the observations but only on properties of order statistics of the uniform distribution.

◇ ◇ ◇ ◇ ◇

ACKNOWLEDGMENTS

This paper was prepared during the Spring 1965 while the three authors were gathered at the University of California, Berkeley. The brief mention of functions of bounded variation (in proposition 5) was suggested by a remark of H. Chernoff.

In preparing the present version, one of the authors had the benefit of several conversations with J. Hájek to whom we owe several suggestions and several references. In addition, Hájek has informed one of us that he obtained recently a bound on the variances of Chernoff-Savage statistics which would allow drastic simplifications in the proofs of theorems such as theorem 1 and may also lead to extensions of its domain of validity.

REFERENCES

[1] A. R. Ansari and R. A. Bradley, "Rank-sum tests for dispersions," *Ann. Math. Statist.*, Vol. 31 (1960), pp. 1174–1189.

[2] P. Bickel, "Some contributions to the theory of order statistics," *Proceedings of the Fifth Berkeley Symposium on Mathematical Statistics and Probability*, University of California Press, Berkeley and Los Angeles, 1966, Vol. I, pp. 575–591.

[3] Z. W. Birnbaum and A. W. Marshall, "Some multivariate Chebyshev inequalities with extensions to continuous parameter processes," *Ann. Math. Statist.*, Vol. 32 (1961), pp. 687–703.

[4] Jack Capon, "Asymptotic efficiency of certain locally most powerful rank tests," *Ann. Math. Statist.*, Vol. 32 (1961), pp. 88–100.

[5] Hermann Chernoff and I. R. Savage, "Asymptotic normality and efficiency of certain non-parametric test statistics," *Ann. Math. Statist.*, Vol. 29 (1958), pp. 972–994.

[6] D. M. Čibisov, "Some theorems on the limiting behavior of an empirical distribution functions," *Trudy Mat. Inst. Steklov.*, Vol. 71 (1964), pp. 104–112.

[7] Monroe D. Donsker, "Justification and extension of Doob's heuristic approach to the Kolmogorov-Smirnov theorems," *Ann. Math. Statist.*, Vol. 23 (1952), pp. 277–281.

[8] R. M. Dudley, "Convergence of Baire measures," to appear in *Studia Math.*

[9] J. Hájek, "Extension of the Kolmogorov-Smirnov test to regression alternatives," *Bernoulli-Bayes-Laplace Anniversary Volume*, Berlin-Heidelberg-New York, Springer Verlag, 1965.

[10] J. Hájek and A. Rényi, "Generalization of an inequality of Kolmogorov," *Acta Math. Acad. Sci. Hungar.*, Vol. 6 (1955), pp. 281–283.

[11] Jerome Klotz, "Nonparametric tests for scale," *Ann. Math. Statist.*, Vol. 33 (1962), pp. 498–512.

[12] Yu. V. Prohorov, "Asymptotic behavior of the binomial distribution," *Uspehi. Mat. Nauk SSSR*, Vol. 8 (1953), pp. 135–142.

[13] Madan Lal Puri, "Asymptotic efficiency of a class of c-sample tests," *Ann. Math. Statist.*, Vol. 35 (1964), pp. 102–121.

[14] Bengt Rosen, "Limit theorems for sampling from finite populations," *Ark. Mat.*, Vol. 5 (1965), pp. 383–424.

[15] Sidney Siegel and John W. Tukey, "A nonparametric sum of ranks procedure for relative spread in unpaired samples," *J. Amer. Statist. Assoc.*, Vol. 55 (1960), pp. 429–444. (Errata, *ibid.* (1961), p. 1005.)

[16] V. Strassen, "The existence of probability measures with given marginals," *Ann. Math. Statist.*, Vol. 36 (1965), pp. 423–439.

ON A CLASS OF TWO-SAMPLE BIVARIATE NONPARAMETRIC TESTS

PRANAB KUMAR SEN
UNIVERSITY OF CALCUTTA, UNIVERSITY OF CALIFORNIA, BERKELEY

1. Introduction

The object of the present investigation is to propose and study a general class of nonparametric tests for the various types of problems that may usually arise in the case of two independent samples with bivariate observations. For this purpose, the concept of permutation tests has been used in the formulation of a class of tests based on appropriate generalized U-statistics, and the theory of permutation distribution of such generalized U-statistics has been developed further.

The advent of the theory of nonparametric methods in multivariate analyses may be regarded to be still in a more or less rudimentary stage, and only a few nonparametric contenders of some standard parametric multivariate procedures are available in the literature. The up-to-date development of distribution-free techniques in this field of research relates specifically to the problem of location in the single, as well as multisample case, and the problem of independence in the single sample case. In this study, I have confined myself to the multisample case only.

The earliest work on this line is the permutation test based on Hotelling's T^2-statistic, proposed and studied by Wald and Wolfowitz [25], in as early as 1944. This test is, however, a strictly value-permutation test and is subject to the usual limitations of this type of tests. Following this, there is a gap of nearly twenty years, during which practically no nonparametric test has evolved in this field. However, very recently, some attention has been paid to the development of nonparametric methods in multivariate multisample analyses.

Some genuine distribution-free tests for location in the bivariate two-sample, as well as p-variate c-sample ($p, c \geq 2$), case have been proposed and studied by Chatterjee and Sen [2], [4]. On the other hand, some tests for the same problem, which are only asymptotically distribution-free, have been considered by Bhapkar [1]. Chatterjee and Sen [3] have also considered some exact distribution-free tests for the two-sample bivariate association problem, and some of these tests have been extended to the c-sample case by Sen [22].

However, all these tests are based on specific forms of test criteria and relate specifically to the problem of location and association. No attempt has yet been

Prepared with the partial support of National Science Foundation, Grant GP-2593 at the Statistical Laboratory, University of California, Berkeley.

made to develop a general method of constructing suitable nonparametric tests for the different types of problems that may usually arise in the multivariate multisample case.

The scope of the present investigation has been confined only to the bivariate two-sample case, whereas the more complicated case with p-variates and c-samples $(p, c \geq 2)$ is intended to be considered separately. A class of permutation tests based on appropriate generalized U-statistics has been studied here, and these may be used to test the null hypothesis of permutation invariance against various types of admissible alternatives. The present study not only extends the scope of nonparametric tests to a more varied type of problems in the bivariate two-sample case, but also extends the theory of asymptotic permutation distribution to a more general class of statistics.

The literature on permutation tests relates mostly to the linear permutation statistics of the type considered by Wald and Wolfowitz [25], Noether [19], Hoeffding [11], [12], Dwass [5], [6], Motoo [17], Hájek [8], among others. The class of generalized U-statistics, considered here, is more general than the usual linear permutation statistics. Finally, in the univariate case, a class of multisample permutation tests based on appropriate U-statistics, has been proposed and studied by the present author [21], [22], and the present investigation also extends these findings in the multivariate case.

2. Preliminary notions

Let $X = (X^{(1)}, X^{(2)})$ be a vector-valued random variable, and let the first sample be composed of n_1 independent and identically distributed bivariate random variables (i.i.d.b.r.v.) $X_1, \cdots, X_{n_1}$, distributed according to the bivariate distribution function (cdf) $F_1(x)$, where $x = (x^{(1)}, x^{(2)})$. Similarly, let $Y_1, \cdots, Y_{n_2}$ be n_2 i.i.d.b.r.v., constituting the second sample drawn independently from another distribution with a bivariate cdf $F_2(x)$. Also, let Ω be the set of all pairs of nondegenerate bivariate cdf's, and it is assumed that

$$(F_1, F_2) \in \Omega. \tag{2.1}$$

It may be noted that Ω is the set of all possible types of pairs of bivariate cdf's, and it includes the family of pairs of continuous, or absolutely continuous cdf's, as subsets. Afterwards, some mild restrictions will have to be imposed on Ω, and these will be stated as and when necessary. Let $\mathcal{W}_0$ be a subset of points $(F_1, F_2) \in \Omega$, for which $F_1(x) \equiv F_2(x)$. Our problem is then to test the null hypothesis

$$H_0\colon (F_1, F_2) \in \mathcal{W}_0, \tag{2.2}$$

against various types of admissible alternatives. Since, under the null hypothesis (2.2), the joint distribution of the $N = n_1 + n_2$ observations of the combined sample, remains invariant under any permutation of the coordinated variables, the hypothesis (2.2) may also be termed the hypothesis of permutation invariance.

Now, the alternative hypotheses often relate to differences of location, scale, association pattern, or of some measurable characteristics of the two cdf's (F_1, F_2). In this context, I therefore introduce the role of *estimable parameters* or *regular functionals* (cf. Hoeffding [10], Lehmann [15]), which may be readily employed in the specification of a variety of alternative hypotheses. Also, in the bivariate case, we usually require a vector-valued regular functional to specify completely the alternative hypotheses. Thus, let

$$\boldsymbol{\theta}(F_1, F_2) = (\theta_1(F_1, F_2), \cdots, \theta_p(F_1, F_2)), \qquad p \geq 1; \tag{2.3}$$

be a vector-valued regular functional of the two cdf's (F_1, F_2), and (2.3) is assumed to be estimable, so that $\boldsymbol{\theta}(F_1, F_2)$ exists for all $(F_1, F_2) \in \Omega$.

Now to induce the nonparametric structure of the hypothesis (2.2), it is further assumed that

$$\boldsymbol{\theta}(F_1, F_2) = \boldsymbol{\theta}^0 = (\theta_1^0, \cdots, \theta_p^0) \qquad \text{for} \quad (F_1, F_2) \in \mathcal{W}_0, \tag{2.4}$$

where $\boldsymbol{\theta}^0$ is a vector with known elements.

Now let $\mathcal{W}_\theta$ be a subset of Ω for which $\boldsymbol{\theta}(F_1, F_2) = \boldsymbol{\theta}^0$. Obviously then, $\mathcal{W}_0 \subset \mathcal{W}_\theta$. We are now interested in the set of alternatives

$$H_\theta \colon (F_1, F_2) \in \Omega - \mathcal{W}_\theta \subset \Omega - \mathcal{W}_0, \tag{2.5}$$

that is, $\boldsymbol{\theta}(F_1, F_2) \neq \boldsymbol{\theta}^0$. Since $\boldsymbol{\theta}(F_1, F_2)$ is assumed to be estimable, there exists a vector-valued *kernel* of it, which is denoted by

$$\boldsymbol{\phi} = (\phi_i(X_{\alpha_1}, \cdots, X_{\alpha_{m_{i1}}}, Y_{\beta_1}, \cdots, Y_{\beta_{m_{i2}}}), i = 1, \cdots, p), \tag{2.6}$$

where ϕ_i is symmetric in its first m_{i1} arguments and also in its last m_{i2} arguments, though the roles of these two sets may not be symmetric, and where m_{i1}, m_{i2} are positive integers, for $i = 1, \cdots, p$. The *degree* of $\boldsymbol{\phi}$ is then denoted by

$$\mathbf{m} = \begin{pmatrix} m_{11}, & \cdots, & m_{p1} \\ m_{12}, & \cdots, & m_{p2} \end{pmatrix}. \tag{2.7}$$

It is further assumed that $\phi_1, \cdots, \phi_p$ are all linearly independent. Then, the generalized U-statistic corresponding to $\boldsymbol{\phi}$ is given by

$$\mathbf{U}_N = (U_{N1}, \cdots, U_{Np}), \qquad N = n_1 + n_2, \tag{2.8}$$

where

$$U_{Ni} = \binom{n_i}{m_{i1}}^{-1} \binom{n_2}{m_{i2}}^{-1} \sum_{S_i} \phi_i(X_{\alpha_1}, \cdots, X_{\alpha_{m_{i1}}}, Y_{\beta_1}, \cdots, Y_{\beta_{m_{i2}}}), \tag{2.9}$$

the summation S_i being extended over all possible

$$1 \leq \alpha_1 < \cdots < \alpha_{m_{i1}} \leq n_1; \quad 1 \leq \beta_1 < \cdots < \beta_{m_{i2}} \leq n_2, \quad \text{for} \quad i = 1, \cdots, p. \tag{2.10}$$

It is well known that under certain conditions on Ω (cf. Fraser [7], p. 142), $\mathbf{U}_N$ has uniformly the minimum concentration ellipsoid (as well as minimum risk with any convex loss function) among all unbiased estimators of $\boldsymbol{\theta}(F_1, F_2)$. Even when these conditions on Ω do not hold, the U-statistic corresponding to any unbiased estimator of $\boldsymbol{\theta}(F_1, F_2)$ has a concentration ellipsoid which cannot be larger than that of the estimator itself. Thus, it seems reasonable to base a test for H_0 in (2.2) against the set of alternatives in (2.5) on the values of $\mathbf{U}_N$, and the same has been accomplished here through a permutation approach.

We pool the two samples together into a combined sample of size $N = n_1 + n_2$ and denote these N (paired) observations by

$$\mathbf{Z}_N = (Z_1, \cdots, Z_N), \qquad Z_i = (Z_i^{(1)}, Z_i^{(2)}), \qquad i = 1, \cdots, N, \tag{2.11}$$

where conventionally we let

$$\begin{aligned} Z_i &= X_i, & i &= 1, \cdots, n_1, \\ Z_i &= Y_{i-n_1}, & i &= n_1 + 1, \cdots, N. \end{aligned} \tag{2.12}$$

In what follows, $\mathbf{Z}_N$ will be called the *collection vector*, as it is a collection of N random paired observations. Then, under the null hypothesis (2.2), $\mathbf{Z}_N$ is composed of N i.i.d.b.r.v., and hence, the joint distribution of $\mathbf{Z}_N$ is symmetric in its N arguments. Consequently, under (2.2) and given the collection matrix (2.11), all possible permutations of the coordinates of $\mathbf{Z}_N$ are equally likely, each such permutation having the same conditional probability $1/N!$.

Hence, all possible partitioning of the N variables into two subsets of n_1 and n_2 respectively are equally likely (conditionally), each having the same permutation probability $\binom{N}{n_1}^{-1}$. Since this probability is independent of $\mathbf{Z}_N$ as well as of $(F_1, F_2) \in \mathcal{W}_0$, we may readily use this to formulate various tests based on $\mathbf{U}_N$. Naturally, such a test is strictly distribution-free under the null hypothesis (2.2).

Now the formulation of the critical function $I(\mathbf{U}_N)$ depends evidently on the permutation distribution of $\mathbf{U}_N$. Consequently, we will study first some properties of the permutation distribution of $\mathbf{U}_N$ and later, with the aid of these, proceed further to consider $I(\mathbf{U}_N)$ and its various properties.

3. Permutation distribution of $\mathbf{U}_N$

Let us define first

$$\phi_i^*(Z_{\alpha_1}, \cdots, Z_{\alpha_{m_{i1}+m_{i2}}}) = \frac{1}{(m_{i1} + m_{i2})!} \sum_{S_i^*} \phi_i(Z_{\alpha_1}, \cdots, Z_{\alpha_{m_{i1}+m_{i2}}}), \tag{3.1}$$

where the summation S_i^* extends over all possible $(m_{i1} + m_{i2})!$ permutation of the variables

$$Z_{\alpha_1}, \cdots, Z_{\alpha_{m_{i1}+m_{i2}}} \tag{3.2}$$

in the ordered position of $\phi_i(\cdots)$, for $i = 1, \cdots, p$. Thus,

$$\boldsymbol{\phi}^* = (\phi_1^*, \cdots, \phi_p^*) \tag{3.3}$$

is the symmetric form of $\boldsymbol{\phi}$.

Then extending the idea of Sen [21], we say that $\mathbf{U}_N$ is a type A generalized U-statistic, if $\boldsymbol{\phi}^*$ is nonstochastic for all $Z_{\alpha_1}, \cdots, Z_{\alpha_{m_{i1}+m_{i2}}}$.

In this paper, we will be concerned with type A generalized U-statistics only. It may also be noted that they include as a special case the differences of individual sample U-statistics (cf. Sen [21]). Further, we will also assume that for the given class of generalized U-statistics, the dispersion matrix of

$$N^{1/2}\{\mathbf{U}_N - \boldsymbol{\theta}(F_1, F_2)\} \tag{3.4}$$

(where $\mathbf{U}_N$ and $\boldsymbol{\theta}(F_1, F_2)$ are defined in (2.8) and (2.3) respectively), has asymptotically a positive definite limit (as $N \to \infty$), for all $(F_1, F_2) \in \mathcal{W}_0$ (this limit may, of course, depend on the particular $(F_1, F_2) \in \mathcal{W}_0$).

Now the condition of nonstochasticness of $\boldsymbol{\phi}^*$, along with (2.4) and (3.1), implies that

$$\boldsymbol{\phi}^* = \boldsymbol{\theta}^0 \qquad \text{for all} \quad (F_1, F_2) \in \Omega \quad \text{and all} \quad \mathbf{Z}_N. \tag{3.5}$$

Let us now write

$$U_{Ni}(\mathbf{Z}_N) = \binom{N}{m_{i1} + m_{i2}}^{-1} \sum_{C_i} \phi_i^*(Z_{\alpha_1}, \cdots, Z_{\alpha_{m_{i1}+m_{i2}}}), \tag{3.6}$$

where the summation C_i extends over all possible

$$1 \leq \alpha_1 < \cdots < \alpha_{m_{i1}+m_{i2}} \leq N, \qquad \text{for} \quad i = 1, \cdots, p, \tag{3.7}$$

and let

$$\mathbf{U}_N(\mathbf{Z}_N) = (U_{N1}(\mathbf{Z}_N), \cdots, U_{Np}(\mathbf{Z}_N)). \tag{3.8}$$

Also, let $\mathcal{P}(\mathbf{Z}_N)$ denote the permutation probability distribution generated by the $N!$ permutations of the coordinates of $\mathbf{Z}_N$. It is then readily seen that

$$E\{\mathbf{U}_N | \mathcal{P}(\mathbf{Z}_N)\} = \mathbf{U}_N(\mathbf{Z}_N), \tag{3.9}$$

and hence, from (3.5), (3.6), (3.8) and (3.9), we obtain

$$E\{\mathbf{U}_N | \mathcal{P}\mathbf{Z}_N)\} = \boldsymbol{\theta}^0, \qquad \text{for all} \quad \mathbf{Z}_N \quad \text{and} \quad (F_1, F_2) \in \Omega. \tag{3.10}$$

For the time being, let us assume that

$$E\{\boldsymbol{\phi}' \cdot \boldsymbol{\phi} | \mathcal{P}(Z_N)\} < \infty. \tag{3.11}$$

and later, we will establish certain conditions under which (3.11) holds. Then, let the covariance of ϕ_i and ϕ_j (with respect to $\mathcal{P}(\mathbf{Z}_N)$), when c of the X_α's and d of the Y_β's are common between the two sets of X_α's and the two sets of Y_β's, be denoted by

$$\zeta_{cd}^{(i,j)}(\mathbf{Z}_N), \qquad \begin{matrix} 0 \leq c \leq \min(m_{i1}, m_{j1}), \\ 0 \leq d \leq \min(m_{i2}, m_{j2}), \end{matrix} \qquad \text{for} \quad i, j = 1, \cdots, p. \tag{3.12}$$

It may be noted then that all these quantities are random variables, as they depend on the random collection matrix $\mathbf{Z}_N$. Also, let

$$\zeta_{cd}^{(i,j)}(F), \qquad \begin{matrix} 0 \leq c \leq \min(m_{i1}, m_{j1}), \\ 0 \leq d \leq \min(m_{i2}, m_{j2}), \end{matrix} \tag{3.13}$$

be the unconditional covariance of ϕ_i and ϕ_j, when $\{X_\alpha\}$, $\{Y_\beta\}$ are i.i.d.b.r.v. distributed according to the cdf $F(x)$, and when c of the X_α's and d of the Y_β's are common between the two sets of X_α's and two sets of Y_β's, for $i = 1, \cdots, p$. Then, it follows by simple algebraic manipulations that

$$(3.14)\qquad \begin{aligned}\sigma_{ij}(\mathbf{Z}_N) &= \operatorname{cov}\{U_{Ni}, U_{Nj}|\mathcal{P}(\mathbf{Z}_N)\}\\ &= \binom{n}{m_{j1}}^{-1}\binom{n_2}{m_{j2}}^{-1}\sum_{c=0}^{m_{j1}}\sum_{d=0}^{m_{j2}}\binom{m_{i1}}{c}\binom{m_{i2}}{d}\\ &\quad\times\binom{n_1-m_{i1}}{m_{j1}-c}\binom{n_2-m_{i2}}{m_{j2}-d}\zeta_{cd}^{(i,j)}(\mathbf{Z}_N),\end{aligned}$$

for all $i, j = 1, \cdots, p$, and

$$(3.15)\qquad \begin{aligned}\sigma_{ij}(F) &= \operatorname{cov}\{U_{Ni}, U_{Nj}|F_1 \equiv F_2 \equiv F\}\\ &= \binom{n_1}{m_{j1}}^{-1}\binom{n_2}{m_{j2}}^{-1}\sum_{c=0}^{m_{j1}}\sum_{d=0}^{m_{j2}}\binom{m_{i1}}{c}\binom{m_{i2}}{d}\\ &\quad\times\binom{n_1-m_{i1}}{m_{j1}-c}\binom{n_2-m_{i2}}{m_{j2}-d}\zeta_{cd}^{(i,j)}(F),\end{aligned}$$

for all $i, j = 1, \cdots, p$.

Also let

$$(3.16)\qquad \begin{aligned}\boldsymbol{\Sigma}(\mathbf{Z}_N) &= ((\sigma_{ij}(\mathbf{Z}_N)))_{i,j=1,\cdots,p},\\ \boldsymbol{\Sigma}(F) &= ((\sigma_{ij}(F)))_{i,j=1,\cdots,p}.\end{aligned}$$

Then, we have the following theorems.

THEOREM 3.1. *For any real estimable* $\boldsymbol{\theta}(F_1, F_2)$ *and for type A generalized U-statistics,*

$$(3.17)\qquad \zeta_{00}^{(i,j)}(\mathbf{Z}_N) = 0 \qquad \text{for all}\quad i, j = 1, \cdots, p \quad \text{and all}\quad \mathbf{Z}_N.$$

The proof follows more or less on the same line as in Sen ([21], lemma 2.1), and hence is omitted.

THEOREM 3.2. *If* (3.11) *holds and* $(F_1, F_2) \in \mathcal{W}_0$, *then*

$$(3.18)\qquad \zeta_{cd}^{(i,j)}(\mathbf{Z}_N) \underset{\text{a.s.}}{\longrightarrow} \zeta_{cd}^{(i,j)}(F),$$

for all $0 \le c \le \min(m_{i1}, m_{j1})$, $0 \le d \le \min(m_{i2}, m_{j2})$, $i, j = 1, \cdots, p$ *(where* $F_1 \equiv F_2 \equiv F$*). Further, if* $\boldsymbol{\phi}$ *has finite fourth-order moments and if for the distribution F, the associated order statistic is complete, then* $\boldsymbol{\Sigma}(\mathbf{Z}_N)$ *has uniformly (for all* $(F_1, F_2) \in \mathcal{W}_0$*) the minimum concentration ellipsoid (as well as minimum risk with any convex loss function) among all unbiased estimators of* $\boldsymbol{\Sigma}(F)$.

PROOF. Let us write

$$(3.19)\qquad \begin{aligned}&g_{cd}^{(i,j)}(Z_{\alpha_1}, \cdots, Z_{\alpha_{m_{i1}+m_{i2}+m_{j1}+m_{j2}-c-d}})\\ &\qquad = \frac{c!(m_{i1}-c)!(m_{j1}-c)!d!(m_{i2}-d)!(m_{j2}-d)!}{(m_{i1}+m_{i2}+m_{j1}+m_{j2}-c-d)!}\\ &\qquad\quad \times \sum{}^{*}\phi_i(Z_{\alpha_1}, \cdots, Z_{\alpha_{m_{i1}+m_{i2}}})\phi_j(Z_{\beta_1}, \cdots, Z_{\beta_{m_{j1}+m_{j2}}}),\end{aligned}$$

where

(i) $\alpha_\ell = \beta_\ell$ for $\ell = 1, \cdots, c$;

(ii) $\alpha_{m_{i1}+\ell} = \beta_{m_{j1}+\ell}$ for $\ell = 1, \cdots, d$;

(iii) and $\alpha_i \neq \beta_j$ for any other (i, j),

(iv) and where the summation $\sum^*$ extends over every possible choice of α's and β's from $\alpha_1, \cdots, \alpha_{m_{i1}+m_{i2}+m_{j1}+m_{j2}-c-d}$.

It is readily seen that

$$\zeta_{cd}^{(i,j)}(\mathbf{Z}_N) = \binom{N}{m_{i1}+m_{i2}+m_{j1}+m_{j2}-c-d}^{-1} \times \sum g_{cd}^{(i,j)}(Z_{\alpha_1}, \cdots, Z_{\alpha_{m_{i1}+m_{i2}+m_{j1}+m_{j2}-c-d}}) - U_{Ni}(\mathbf{Z}_N)U_{Nj}(\mathbf{Z}_N). \tag{3.20}$$

Thus, $\zeta_{cd}^{(i,j)}(\mathbf{Z}_N) + U_{Ni}(\mathbf{Z}_N)U_{Nj}(\mathbf{Z}_N)$ is again a U-statistic of the N observations $\mathbf{Z}_N$, and hence, using the property of almost sure convergence of U-statistics (cf. Hoeffding [13]), it readily follows that if $E|g_{cd}^{(i,j)}| < \infty$ (which is implied by (3.11)),

$$\zeta_{cd}^{(i,j)}(\mathbf{Z}_N) + U_{Ni}(\mathbf{Z}_N)U_{Nj}(\mathbf{Z}_N) \underset{\text{a.s.}}{\longrightarrow} \zeta_{cd}^{(i,j)}(F) + \theta_i^0\theta_j^0. \tag{3.21}$$

Further, from (3.4), (3.9), and (3.10), we have $\mathbf{U}_N(\mathbf{Z}_N) = \boldsymbol{\theta}^0$, for all $\mathbf{Z}_N$, and hence, from (3.21), we obtain

$$\zeta_{cd}^{(i,j)}(\mathbf{Z}_N) \underset{\text{a.s.}}{\longrightarrow} \zeta_{cd}^{(i,j)}(F), \tag{3.22}$$

for all $0 \leq c \leq \min(m_{i1}, m_{j1})$; $0 \leq d \leq \min(m_{i2}, m_{j2})$, $i, j = 1, \cdots, p$. Consequently, from (3.14), (3.15), and (3.16), we find that

$$N\boldsymbol{\Sigma}(\mathbf{Z}_N) \underset{\text{a.s.}}{\longrightarrow} N\boldsymbol{\Sigma}(F) \qquad \text{for all} \quad (F_1, F_2) \in \mathcal{W}_0. \tag{3.23}$$

Again if $F_1 \equiv F_2 \equiv F$, and if for F, the associated order statistic is complete, it follows from a well-known theorem on U-statistics in the vector case (cf. Fraser [7], p. 142) that a vector-valued U-statistic has uniformly a minimum concentration ellipsoid among all unbiased estimators (having finite second moments) of the same parameter vector. Thus, from (3.20), we obtain after a few algebraic manipulations that $\zeta_{cd}^{(i,j)}(\mathbf{Z}_N)$'s jointly have uniformly the minimum concentration ellipsoid among all unbiased estimators of $\zeta_{cd}^{(i,j)}(F)$'s.

Hence, from (3.14) and (3.15), we directly get that if $\boldsymbol{\phi}$ has finite fourth moments, $\boldsymbol{\Sigma}(\mathbf{Z}_N)$ has uniformly the minimum concentration ellipsoid among all unbiased estimators of $\boldsymbol{\Sigma}(F)$. Hence the theorem.

Let us now consider the properties of $\boldsymbol{\Sigma}(\mathbf{Z}_N)$ when $(F_1, F_2) \notin \mathcal{W}_0$. In this case, no small sample property can be properly studied, and we shall consider here some asymptotic results. The term 'asymptotic' is used in the sense that $N \to \infty$ subject to

$$n_1/N \to \lambda : 0 < \lambda < 1. \tag{3.24}$$

Also, let us define

$$\begin{aligned} \bar{F}_N(x) &= (n_1/N)F_1(x) + (n_2/N)F_2(x), \\ \bar{F}(x) &= \lambda F_1(x) + (1-\lambda)F_2(x). \end{aligned} \tag{3.25}$$

Finally, by virtue of our assumption regarding $\mathbf{U}_N$, we have for $F_1 \equiv F_2 \equiv F$, $(F_1, F_2) \in \mathcal{W}_0$,

$$\lim_{N=\infty} N\boldsymbol{\Sigma}(F) = \boldsymbol{\Gamma}(F), \tag{3.26}$$

where $\boldsymbol{\Gamma}(F)$ is positive definite and finite for all $(F, F) \in \mathcal{W}_0$.

THEOREM 3.3. *If $\boldsymbol{\phi}$ has finite fourth-order moments for all $(F_1, F_2) \in \Omega$, and if* (3.24) *holds,*

$$\zeta_{cd}^{(i,j)}(Z_N) \xrightarrow{P} \zeta_{cd}^{(i,j)}(\overline{F}), \tag{3.27}$$

for all $0 \leq c \leq \min(m_{i1}, m_{j1})$; $0 \leq d \leq \min(m_{i2}, m_{j2})$, $i, j = 1, \cdots, p$; where $\overline{F}$ has been defined in (3.25). *Further, if* (3.26) *holds for both $\mathbf{U}_N$ and $N\boldsymbol{\Sigma}(\mathbf{Z}_N)$, and the associate order statistic is complete, then $N\boldsymbol{\Sigma}(\mathbf{Z}_N)$ is asymptotically the minimum concentration ellipsoid estimator of $\boldsymbol{\Gamma}(\overline{F})$, for all $(F_1, F_2) \in \Omega$.*

PROOF. In an earlier paper [23], it has been shown by the present author that a pooled sample U-statistic converges in probability to the associated regular functional of the cdf $\overline{F}$, (defined in (3.24)), when F_1, F_2 are not identical. From this result, it readily follows that if $\boldsymbol{\phi}$ has finite fourth-order moments for all $(F_1, F_2) \in \Omega$ and (3.24) holds, then

$$\zeta_{cd}^{(i,j)}(\mathbf{Z}_N) \xrightarrow{P} \zeta_{cd}^{(i,j)}(\overline{F}), \tag{3.28}$$

for all c, d, i, and j. Hence, the first part of the theorem.

In the same paper, it has also been shown that if the variance of a pooled sample U-statistic multiplied by the pooled sample size has a nonzero finite asymptotic limit (under $F_1 \equiv F_2 \equiv \overline{F}$), then the pooled sample U-statistic will asymptotically be the minimum variance unbiased estimate of the regular functional of the cdf $\overline{F}$, for all $(F_1, F_2) \in \Omega$.

The same result can be extended in a more or less straightforward manner to vector-valued U-statistics, and the asymptotic minimum variance unbiasedness can then be generalized to asymptotic minimum concentration ellipsoid unbiasedness. In our case, $N\boldsymbol{\Sigma}(\overline{F})$, by virtue of (3.26), is asymptotically equal to $\boldsymbol{\Gamma}(\overline{F})$, which is positive definite for all $(F_1, F_2) \in \Omega$. Hence, it follows from (3.14), (3.15), (3.16), and (3.26) that for all $(F_1, F_2) \in \Omega$,

$$N\boldsymbol{\Sigma}(\mathbf{Z}_N) \xrightarrow{P} \boldsymbol{\Gamma}(\overline{F}). \tag{3.29}$$

Finally, it follows from (3.14) and (3.16) that $N\boldsymbol{\Sigma}(\mathbf{Z}_N)$ is a $p \times p$ matrix, whose elements are linear functions of a set of pooled sample U-statistics. Since, these U-statistics are also assumed to satisfy a condition similar to (3.26), we readily get from (3.14), (3.15), and the discussion made above that $N\boldsymbol{\Sigma}(\mathbf{Z}_N)$ is asymptotically the minimum concentration ellipsoid unbiased estimator of $\boldsymbol{\Gamma}(\overline{F})$, for all $(F_1, F_2) \in \Omega$. Hence the theorem.

Thus the permutation covariance matrix $\boldsymbol{\Sigma}(\mathbf{Z}_N)$ possesses some convergence properties in both the situations when $(F_1, F_2) \in \mathcal{W}_0$ and $(F_1, F_2) \notin \mathcal{W}_0$.

THEOREM 3.4. *If $(F_1, F_2) \in \mathcal{W}_0$ and $\boldsymbol{\phi}$ has finite moments of the order $2 + \delta$ $(\delta > 0)$, or if $(F_1, F_2) \in \Omega - \mathcal{W}_0$ and $\boldsymbol{\phi}$ has finite fourth-order moments, then subject to* (3.26),

$$\mathcal{L}(N^{1/2}[\mathbf{U}_N - \boldsymbol{\theta}^0]) \xrightarrow{P} \mathbf{N}(\mathbf{0}, \boldsymbol{\Gamma}(\bar{F})), \tag{3.30}$$

where $\mathcal{L}$ stands for the convergence in distribution generated by the permutation probability function $\mathcal{P}(\mathbf{Z}_N)$, $\mathbf{N}$ for the p-variate normal distribution, and $\mathbf{0}$ for a null p-vector. Further, with respect to the same permutation probability measure

$$\mathcal{L}([\mathbf{U}_N - \boldsymbol{\theta}^0](\boldsymbol{\Sigma}(\mathbf{Z}_N))^{-1}[\mathbf{U}_N - \boldsymbol{\theta}^0]') \xrightarrow{P} \chi_p^2, \tag{3.31}$$

where χ_p^2 has the chi square distribution with p degrees of freedom.

PROOF. Let us define for each $i(= 1, \cdots, p)$,

$$\begin{aligned} \phi_{i(10)}(Z_\alpha) &= E\{\phi(Z_\alpha, Z_{\alpha_2}, \cdots, Z_{\alpha m_{i1}+m_{i2}}) | \mathcal{P}(\mathbf{Z}_N)\} \\ &= (N - 1_{P m_{i1}+m_{i2}-1})^{-1} \cdot \sum_{S_{i\alpha}^*} \phi(Z_\alpha, Z_{\alpha_2}, \cdots, Z_{\alpha m_{i1}+m_{i2}}), \end{aligned} \tag{3.32}$$

where the summation $S_{i\alpha}^*$ extends over all possible $\alpha_2 \neq \cdots \neq \alpha_{m_{i1}+m_{i2}} = 1, \cdots, N \ (\neq \alpha)$, and

$$\begin{aligned} \phi_{i(01)}(Z_\alpha) &= E\{\phi(Z_{\alpha_2}, \cdots, Z_{\alpha m_{i1}+m_{i2}}, Z_\alpha) | \mathcal{P}(\mathbf{Z}_N)\} \\ &= (N - 1_{P m_{i1}+m_{i2}-1})^{-1} \sum_{S_{i\alpha}^*} \phi(Z_{\alpha_2}, \cdots, Z_{\alpha m_{i1}+m_{i2}}, Z_\alpha). \end{aligned} \tag{3.33}$$

Also, let

$$\begin{aligned} V_{Ni} = (m_{i1}/n_1) &\sum_{j=1}^{n_1} \{\phi_{i(10)}(X_j) - \theta_i^0\} \\ &+ (m_{i2}/n_2) \sum_{j=1}^{n_2} \{\phi_{i(01)}(Y_j) - \theta_i^0\}, \qquad i = 1, \cdots, p, \end{aligned} \tag{3.34}$$

$$\mathbf{V}_N = (V_{N1}, \cdots, V_{Np}). \tag{3.35}$$

It then follows from the results of Nandi and Sen [18] and of Sen [21], with direct extension to the vector case, that if ϕ has finite second-order moments, then with respect to the permutation probability measure $\mathcal{P}(\mathbf{Z}_N)$,

$$N^{1/2}\{[\mathbf{U}_N - \boldsymbol{\theta}^0] - \mathbf{V}_N\} \xrightarrow{P} 0. \tag{3.36}$$

We will now show that, under (3.26), $N^{1/2}\mathbf{V}_N$ has a permutation distribution which is asymptotically a p-variate normal one. For this it is sufficient to show that if $\boldsymbol{\delta} = (\delta_1, \cdots, \delta_p)$ is any real nonnull vector, then $N^{1/2}(\boldsymbol{\delta}\mathbf{V}_N')$ has asymptotically a normal permutation distribution. If we now write

$$\begin{aligned} g_N(Z_\alpha|\boldsymbol{\delta}) = \sum_{i=1}^{p} \delta_i\{&m_{i1}[\phi_{i(10)}(Z_\alpha) - \theta_i^0] \\ &- \frac{n_1}{n_2} m_{i2}[\phi_{i(01)}(Z_\alpha) - \theta_i^0]\}, \qquad \alpha = 1, \cdots, N; \end{aligned} \tag{3.37}$$

then using (3.9), (3.10), (3.37), (3.38), and (3.39), we have, after some essentially simple steps,

$$\boldsymbol{\delta}\mathbf{V}_N' = \sum_{\alpha=1}^{N} C_{N\alpha} g_N(Z_\alpha|\boldsymbol{\delta}), \tag{3.38}$$

where

$$n_1 C_{N\alpha} = \begin{cases} 1, & \text{if } Z_\alpha \text{ belongs to the first sample,} \\ 0, & \text{otherwise; for } \alpha = 1, \cdots, N. \end{cases} \tag{3.39}$$

Now, we apply Wald-Wolfowitz-Noether-Hoeffding-Dwass-Motoo-Hájek theorem on the asymptotic permutation distribution of linear permutation statistics to our particular case of $\boldsymbol{\delta}\mathbf{V}'_N$, defined in (3.38). For this, it appears to be sufficient to show that $\{C_{N\alpha}\}$ satisfies the condition

$$\frac{(1/N)\sum_{\alpha=1}^{N}(C_{N\alpha}-1/N)^r}{\left\{(1/N)\sum_{\alpha=1}^{N}(C_{N\alpha}-1/N)^2\right\}^{r/2}} = 0(1) \qquad \text{for} \quad r = 3, 4, \cdots; \tag{3.40}$$

and $\{g_n(Z_\alpha|\boldsymbol{\delta})\}$ satisfies in probability, the condition

$$\lim_{N=\infty} \frac{\sum_{\alpha=1}^{N}|g_N(Z_\alpha|\boldsymbol{\delta})|^r}{\left\{\sum_{\alpha=1}^{N}|g_N(Z_\alpha|\boldsymbol{\delta})|^2\right\}^{r/2}} = 0, \qquad \text{for some} \quad r > 2. \tag{3.41}$$

Since n_1 of the $C_{N\alpha}$'s are equal to $1/n_1$ and the rest equal to 0, it is easily seen that if (3.24) holds, (3.40) also holds. Further, if $\boldsymbol{\phi}$ has finite moments of order $2 + \delta(\delta > 0)$, it is then readily seen that $g_N(Z_\alpha|\boldsymbol{\delta})$ has also a finite moment of order $2 + \delta$, uniformly in N. Proceeding, then, precisely on the same line as in Sen ([21], (3.6), (3.7)), we get that

$$\frac{1}{N}\sum_{\alpha=1}^{N}|g_N(Z_\alpha|\boldsymbol{\delta})|^r = 0_p(1), \tag{3.42}$$

for any given $r > 2$. Further, extending the results of Sen ([21], (2.23)) to the vector case in a more or less straightforward manner, it can be shown by following the lines of Nandi and Sen ([18], (3.10)) and using the results of Sen [23], that under (3.24),

$$\frac{1}{N}\sum_{\alpha=1}^{N}[g_N(Z_\alpha|\boldsymbol{\delta})]^2 \xrightarrow{P} \frac{1}{1-\lambda}\sum_{i=1}^{p}\sum_{j=1}^{p}\delta_i\delta_j\{m_{i1}m_{j1}\zeta_{10}^{(i,j)}(\overline{F}) + m_{i2}m_{j2}\zeta_{01}^{(i,j)}(\overline{F})\}, \tag{3.43}$$

provided either $(F_1, F_2) \in \mathcal{W}_0$ and $\boldsymbol{\phi}$ has finite second-order moments or $(F_1, F_2) \in \Omega - \mathcal{W}_0$ and $\boldsymbol{\phi}$ has finite fourth-order moments. Since, by (3.26), $\boldsymbol{\Gamma}(\overline{F})$ is positive definite, we get from (3.43) that the right-hand side of it is essentially positive for any nonnull $\boldsymbol{\delta}$. Consequently, from (3.42) and (3.43) we get that (3.41) holds, in probability. Hence, $N^{1/2}(\boldsymbol{\delta}\mathbf{V}'_N)$ has asymptotically, in probability, a normal distribution with mean zero and a finite variance for all nonnull $\boldsymbol{\delta}$. Thus, $N^{1/2}\mathbf{V}_N$ has asymptotically, in probability, a p-variate normal permutation distribution. The first part of the theorem then follows readily from (3.36) and the preceding two theorems.

To prove the second part of the theorem, we note that by virtue of theorems 3.2 and 3.3, under the conditions stated in the theorem,

$$N\boldsymbol{\Sigma}(\mathbf{Z}_N) \xrightarrow{P} \boldsymbol{\Gamma}(\overline{F}), \tag{3.44}$$

for all $(F_1, F_2) \in \Omega$, and hence,

$$\begin{aligned}[\mathbf{U}_N - \boldsymbol{\theta}^0](\boldsymbol{\Sigma}(\mathbf{Z}_N))^{-1}[\mathbf{U}_N - \boldsymbol{\theta}^0]' \\ \mathbf{P}\{N^{1/2}[\mathbf{U}_N - \boldsymbol{\theta}^0]\}(\boldsymbol{\Gamma}(\overline{F}))^{-1}\{N^{1/2}[\mathbf{U}_N - \boldsymbol{\theta}^0]'\} = S_N \text{ (say)}.\end{aligned} \tag{3.45}$$

Now by a well-known theorem (cf. Sverdrup [24]) on the limiting distribution of a continuous function of random variables, and from the distribution theory of quadratic forms of multinormal distributions, it follows from the first part of the theorem that S_N has asymptotically a chi square distribution with p degrees of freedom. Consequently, we get from (3.45) that

$$\mathcal{L}([\mathbf{U}_N - \boldsymbol{\theta}^0](\boldsymbol{\Sigma}(\mathbf{Z}_N))^{-1}[\mathbf{U}_N - \boldsymbol{\theta}^0]') \xrightarrow{P} \chi^2_p. \tag{3.46}$$

Hence, the theorem.

With these theorems, we will now proceed to consider our desired class of permutation tests.

4. The permutation test procedure

In the preceding two sections, the rationality of using $\mathbf{U}_N$ in the formulation of the tests as well as some properties of the permutation distribution of $\mathbf{U}_N$ have been discussed. Now, we are in a position to construct a suitable test function $I(\mathbf{U}_N)$ for testing the null hypothesis (2.2) against the set of alternatives (2.5). Since $I(\mathbf{U}_N)$ associates with each $\mathbf{U}_N$ a probability of rejecting H_0 in (2.2), and as this probability is determined by the permutation distribution function of $\mathbf{U}_N$ (conditioned on $\mathbf{Z}_N$), it follows readily that $I(\mathbf{U}_N)$ possesses the property of S-structure of tests (cf. Lehmann and Stein [16]). Consequently, it is a strictly distribution-free test.

Now $\mathbf{U}_N$ assumes values on a p-dimensional lattice, and conditioned on a given $\mathbf{Z}_N$, the number of points on this lattice is equal to $\binom{N}{n_1}^p$; though $\mathbf{U}_N$ can assume only $\binom{N}{n_1}$ values out of these, and at the remaining $\binom{N}{n_1}^p - \binom{N}{n_1}$ points, the permutation probability is zero. The permutation center of gravity of these mass points on the p-dimensional lattice is the point $\boldsymbol{\theta}^0$, and if (2.2) actually holds, then the permutation distribution will have a dense cluster around $\boldsymbol{\theta}^0$. Thus, we are to demarcate a set of points of this lattice, which will constitute the critical region. In small samples, all possible $\binom{N}{n_1}$ partitionings may be considered, and this set of points may be isolated. However, this procedure becomes prohibitively laborious as the sample sizes increase. In large samples, we are thus faced with the problem of using some suitable function of $\mathbf{U}_N$ as the test statistic and in approximating the permutation distribution of this statistics by some simple form.

Now using the usual concept of distance in the multivariate normal distribution, and noting that $\mathbf{U}_N$ has asymptotically, in probability, a multi-normal permutation distribution, it seems quite reasonable to propose the following test statistic,

$$T_N = [\mathbf{U}_N - \boldsymbol{\theta}^0](\boldsymbol{\Sigma}(\mathbf{Z}_N))^{-1}[\mathbf{U}_N - \boldsymbol{\theta}^0]', \tag{4.1}$$

and to reject (2.2) for large values of T_N. Thus, we consider the following test function:

$$\begin{aligned} I(\mathbf{U}_N) &= 1, && \text{if } T_N > T_{N,\epsilon}(\mathbf{Z}_N), \\ I(\mathbf{U}_N) &= a_\epsilon(\mathbf{Z}_N), && \text{if } T_N = T_{N,\epsilon}(\mathbf{Z}_N), \\ I(\mathbf{U}_N) &= 0, && \text{if } T_N < T_{N,\epsilon}(\mathbf{Z}_N), \end{aligned} \tag{4.2}$$

where $T_{N,\epsilon}(\mathbf{Z}_N)$ and $a_\epsilon(\mathbf{Z}_N)$ are so chosen that

$$E\{I(\mathbf{U}_N)|\mathcal{P}(\mathbf{Z}_N)\} = \epsilon, \qquad 0 < \epsilon < 1; \tag{4.3}$$

ϵ being the given level of significance. It then readily follows that

$$E\{I(\mathbf{U}_N)|(F_1, F_2) \in \mathcal{W}_0\} = \epsilon, \tag{4.4}$$

so that the test (4.2) has exactly the size ϵ. In small samples, the values of $T_{N\epsilon}(\mathbf{Z}_N)$ and $a_\epsilon(\mathbf{Z}_N)$ can be found out using the permutation distribution of $\mathbf{U}_N$, whereas in large samples, we have by virtue of theorem 3.4 that as N increases, subject to (3.24),

$$a_\epsilon(\mathbf{Z}_N) \xrightarrow{P} 0 \quad \text{and} \quad T_{N,\epsilon}(\mathbf{Z}_N) \xrightarrow{P} \chi^2_{p,\epsilon}, \tag{4.5}$$

where $\chi^2_{p,\epsilon}$ is the $100(1 - \epsilon)\%$ point of a χ^2 distribution with p degrees of freedom. Hence, asymptotically, the test (4.2) reduces to

$$\begin{aligned} I(\mathbf{U}_N) &= 1, \quad \text{if } T_N \geq \chi^2_{p,\epsilon}, \\ I(\mathbf{U}_N) &= 0, \quad \text{if } T_N < \chi^2_{p,\epsilon}. \end{aligned} \tag{4.6}$$

Equation (4.6) will be termed the asymptotic permutation test and (4.2), the exact permutation test.

THEOREM 4.1. *If $\boldsymbol{\phi}$ has finite fourth-order moments for all $(F_1, F_2) \in \Omega$ and if (3.26) holds, then the permutation test $I(\mathbf{U}_N)$ in (4.2) or (4.6) is consistent against the set of alternatives $H:(F_1, F_2) \in \Omega - \mathcal{W}_\theta$.*

PROOF. It follows from the well-known properties of generalized U-statistics that for $(F_1, F_2) \in \Omega - \mathcal{W}_\theta$,

$$\mathbf{U}_N - \boldsymbol{\theta}^0 \xrightarrow{P} \boldsymbol{\theta}(F_1, F_2) - \boldsymbol{\theta}^0 = \boldsymbol{\xi}(F_1, F_2), \tag{4.7}$$

where $\boldsymbol{\xi}(F_1, F_2)$ is nonnull for all $(F_1, F_2) \in \Omega - \mathcal{W}_\theta$. Also, it follows from our theorem 3.3 that under the stated conditions

$$N\boldsymbol{\Sigma}(\mathbf{Z}_N) \xrightarrow{P} \boldsymbol{\Gamma}(\overline{F}), \tag{4.8}$$

where $\boldsymbol{\Gamma}(\overline{F})$ is positive definite. Consequently, from (4.1), (4.7), and (4.8), we obtain for $(F_1, F_2) \in \Omega - \mathcal{W}_\theta$,

$$T_N/N \xrightarrow{P} \xi(F_1, F_2)(\boldsymbol{\Gamma}(\overline{F}))^{-1}\xi'(F_1, F_2) > 0, \tag{4.9}$$

and thus,

$$\lim_{N=\infty} P\{T_N \geq \chi^2_{p,\epsilon} | (F_1, F_2) \in \Omega - \mathcal{W}_\theta\} = 1. \tag{4.10}$$

Hence the theorem.

Let now $N\boldsymbol{\Sigma}(F)$ be defined as in (3.16), and let $N\hat{\boldsymbol{\Sigma}}(F)$ be any consistent estimate of $N\boldsymbol{\Sigma}(F)$. Then, we consider a statistic T_N^* of the form

$$T_N^* = [\mathbf{U}_N - \boldsymbol{\theta}^0](\hat{\boldsymbol{\Sigma}}(F))^{-1}[\mathbf{U}_N - \boldsymbol{\theta}^0]'. \tag{4.11}$$

It may be noted that T_N^* can be easily shown to have asymptotically, under H_0 in (2.2), a χ^2 distribution with p degrees of freedom. Consequently, an asymptotically distribution-free test for H_0 in (2.2) may be based on T_N^*, using the following test function:

$$\begin{aligned} &\text{if} \quad T_N^* \geq \chi^2_{p,\epsilon}, \quad \text{reject } H_0 \text{ in (2.2)}, \\ &\text{if} \quad T_N^* < \chi^2_{p,\xi}, \quad \text{accept } H_0 \text{ in (2.2)}. \end{aligned} \tag{4.12}$$

This type of test has been proposed by Bhapkar [1] for the location problem only. We will term this test an asymptotic unconditional test. It then follows from our results in the preceding two sections that under H_0 in (2.2),

$$T_N \overset{P}{\sim} T_N^*. \tag{4.13}$$

In the next section we will consider some further relations between T_N and T_N^*, and here we only note that the consistency of T_N^*-test follows similarly as in theorem 4.1.

5. Asymptotic power properties of the tests

For studying the asymptotic power properties of the test (4.2), (4.6), and (4.12), we require to study the asymptotic nonnull distribution of T_N and T_N^*, defined in (4.1) and (4.11) respectively. First, these have to be considered for some sequence of alternative specifications for which the power will lie in the open interval (0, 1), and second, we are to consider the unconditional distributions of T_N and T_N^*, as the same will be required to study the power.

Thus, we assume that the two cdf's $F_1(x)$ and $F_2(x)$ are replaced by two sequences $\{F_{1N}(x)\}$ and $\{F_{2N}(x)\}$ of cdf's, each converging to a common cdf $F(x)$ as $N \to \infty$, in such a manner that

$$H_N\colon \boldsymbol{\theta}(F_{1N}, F_{2N}) = \boldsymbol{\theta}^0 + N^{-1/2}\boldsymbol{\lambda}, \tag{5.1}$$

where $\boldsymbol{\lambda}$ is a p-vector with finite elements, and it is assumed to be nonnull. Then we have the following results.

THEOREM 5.1. *Under the sequence of alternatives* $\{H_N\}$,

$$\mathcal{L}(T_N) \to \chi^2_{p,\Delta}, \tag{5.2}$$

where $\chi^2_{p,\Delta}$ *has the noncentral* χ^2 *distribution with* p *degrees of freedom and the noncentrality parameter*

$$\Delta = \boldsymbol{\lambda}(\boldsymbol{\Gamma}(F))^{-1}\boldsymbol{\lambda}', \tag{5.3}$$

provided (3.24) *and* (3.26) *hold, and* $\boldsymbol{\phi}$ *has finite fourth-order moments for all* $(F_1, F_2) \in \Omega$.

PROOF. It is well known (cf. Fraser [7]) that whatever be $(F_1, F_2) \in \Omega$, under the stated regularity conditions $N^{1/2}\{\mathbf{U}_N - \boldsymbol{\theta}(F_{1N}, F_{2N})\}$ has asymptotically a multinormal distribution with a nonsingular dispersion matrix, as (3.26) holds. Using now the results of Sen [23], it can be readily seen that under $\{H_N\}$, the dispersion matrix of $N^{1/2}\{\mathbf{U}_N - \boldsymbol{\theta}(F_{1N}, F_{2N})\}$ converges to $\boldsymbol{\Gamma}(F)$, defined in (3.26). Consequently, it follows that

$$[N^{1/2}\{\mathbf{U}_N - \boldsymbol{\theta}^0\}(\boldsymbol{\Gamma}(F))^{-1}N^{1/2}\{\mathbf{U}_N - \boldsymbol{\theta}^0\}'] = S_N^* \text{ (say)}, \tag{5.4}$$

has asymptotically a noncentral χ^2 distribution with p degrees of freedom and the noncentrality parameter Δ, defined in (5.3). It also follows from theorem 3.3 and condition (5.1) that under $\{H_N\}$,

$$|N\boldsymbol{\Sigma}(\mathbf{Z}_N) - \boldsymbol{\Gamma}(\overline{F}_N)| \xrightarrow{P} 0; \qquad \overline{F}_N = \frac{n_1}{N}F_{1N} + \frac{n_2}{N}F_{2N}; \tag{5.5}$$

and from (5.1) we obtain, using the results of Sen [23], that under $\{H_N\}$ and (3.24)

$$\boldsymbol{\Gamma}(\overline{F}_N) \to \boldsymbol{\Gamma}(F) \quad \text{as} \quad N \to \infty. \tag{5.6}$$

Consequently, we get from (4.1), (5.4), (5.5), and (5.6) that

$$T_N \overset{P}{\sim} S_N^*, \qquad \mathcal{L}(T_N) \xrightarrow{P} \mathcal{L}(S_N^*) \to \chi^2_{p,\Delta}. \tag{5.7}$$

Hence the theorem.

THEOREM 5.2. *Under* $\{H_N\}$, $T_N \overset{P}{\sim} T_N^*$.

PROOF. Since $N\hat{\boldsymbol{\Sigma}}(F)$ estimates $N\boldsymbol{\Sigma}(F)$, and as under $\{H_N\}$ the dispersion matrix of $N^{1/2}\{\mathbf{U}_N - \boldsymbol{\theta}(F_{1N}, F_{2N})\}$ converges to $\boldsymbol{\Gamma}(F)$, which is also the limiting form of $N\boldsymbol{\Sigma}(F)$, it follows from (5.4) and a well-known convergence theorem that T_N^* is asymptotically equivalent to S_N^*, for the sequence of alternatives $\{H_N\}$.

Hence, from (5.7), we obtain under $\{H_N\}$,

$$T_N \overset{P}{\sim} T_N^* \sim S_N^*. \tag{5.8}$$

Hence the theorem.

Thus, it follows from a well-known result by Hoeffding ([12], p. 172) that the permutation test based on T_N and the asymptotically distribution-free test based on T_N^* are asymptotically power equivalent for the sequence of alternatives $\{H_N\}$.

The asymptotic power efficiency of the test based on T_N with respect to any other rival test can only be properly studied and made independent of $\boldsymbol{\lambda}$ in $\{H_N\}$, if the other test criterion has also (under $\{H_N\}$) a noncentral χ^2 distribution with the same degrees of freedom and the two noncentrality parameters are proportional for all $\boldsymbol{\lambda}$. Usually, these two tests may have noncentral χ^2 distributions with the same degrees of freedom, but their noncentrality parameters are not generally proportional (for all $\boldsymbol{\lambda}$). Thus, in general, the power efficiency depends on $\boldsymbol{\lambda}$, and in such a case, either one has to show that for all $\boldsymbol{\lambda}$, one of the two noncentrality parameters is at least as large as the other, or, one has to compute the supremum and infimum of the ratio of the two noncentrality parameters (with respect to $\boldsymbol{\lambda}$) and study the bounds for the asymptotic efficiency. The usual concept of Pitman efficiency is, generally, not adaptable in the multivariate case.

6. Illustrations and applications

Now, we will consider the two-sample bivariate location, scale, and association problem and study suitable nonparametric tests based on our results in the preceding sections. Let us first consider the location problem.

Here, let Ω be the set of all pairs of bivariate distributions, which are nondegenerate in the sense that the grade correlation of either of the cdf's is bounded away from ± 1. These two cdf's may be continuous or they may also be purely discrete distributions. Let us then define (with the same notations as in earlier sections)

$$(6.1)\qquad \begin{aligned} \theta_i(F_1, F_2) &= P\{X^{(i)} < Y^{(i)}\} + \tfrac{1}{2}P\{X^{(i)} = Y^{(i)}\}, \qquad i = 1, 2; \\ \boldsymbol{\theta}(F_1, F_2) &= (\theta_1(F_1, F_2), \theta_2(F_1, F_2)). \end{aligned}$$

Then for $(F_1, F_2) \in \mathcal{W}_0$, $\boldsymbol{\theta}(F_1, F_2) = (\frac{1}{2}, \frac{1}{2})$, whereas if for at least one of the two variates, the first sample observations are stochastically larger or smaller than the second sample observations, $\boldsymbol{\theta}(F_1, F_2) \neq (\frac{1}{2}, \frac{1}{2})$. Moreover, if we let

$$(6.2)\qquad F_2(\mathbf{x}) = F_1(\mathbf{x} + \boldsymbol{\delta}), \qquad \text{where } \boldsymbol{\delta} = (\delta_1, \delta_2),$$

then for nonnull $\boldsymbol{\delta}$, it is easily shown that $\boldsymbol{\theta}(F_1, F_2) \neq (\frac{1}{2}, \frac{1}{2})$.

Thus, for the location problem, we may use a permutation test based on the individual variate Wilcoxon-Mann-Whitney statistics, being compounded together as in (4.1). This follows more or less on the same line as in Chatterjee and Sen [2], with further generalizations to cover the case of discrete bivariate distributions too. Thus, we let

$$(6.3)\qquad \begin{aligned} \mathbf{U}_N &= (U_{N1}, U_{N2}), \\ U_{Ni} &= \frac{1}{n_1 n_2} \sum_{\alpha=1}^{n_1} \sum_{\beta=1}^{n_2} \phi(X_\alpha^{(i)}, Y_\beta^{(i)}), \qquad i = 1, 2; \end{aligned}$$

where

$$(6.4)\qquad \begin{aligned} \phi(a, b) &= 1 \quad \text{if} \quad a < b, \\ \phi(a, b) &= \tfrac{1}{2} \quad \text{if} \quad a = b, \\ \phi(a, b) &= 0 \quad \text{if} \quad a > b. \end{aligned}$$

Also among the N values of $Z_\alpha^{(i)}$, N_{ij} are equal to

$$Z_{\alpha_j}^{(i)}, j = 1, \cdots, k_{iN}, \qquad i = 1, 2, \tag{6.5}$$

where $N_{ij} \geq 0$. It is then easily shown that

$$V\{U_{Ni}|\mathcal{P}(\mathbf{Z}_N)\} = \frac{N^2}{n_1 n_2 (N-1)} \left\{\tfrac{1}{12}\left[1 - \sum_{j=1}^{k_{iN}} (N_{ij}/N)^3\right]\right\}, \qquad i = 1, 2; \tag{6.6}$$

and the permutation covariance of U_{N1}, U_{N2} is the rank correlation between $\{Z_\alpha^{(1)}, Z_\alpha^{(2)}; \alpha = 1, \cdots, N\}$, when both the sets of observations contain ties, and the expression for the same is available in Kendall ([14], p. 38). Once these are obtained, we can define T_N as in (4.1) and proceed similarly as in Chatterjee and Sen [2]. This test thus generalizes Putter's [20] Wilcoxon test to the bivariate case, and also Chatterjee and Sen's [2] test to the more general case of any pair of nondegenerate bivariate cdf's.

Let us next consider the scale problem. Extending the idea of Lehmann [15] to the bivariate case, let us define

$$\begin{aligned} \theta_i(F_1, F_2) &= P\{|X_\alpha^{(i)} - X_\beta^{(i)}| < |Y_\gamma^{(i)} - Y_\delta^{(i)}|\} \\ &\quad + \tfrac{1}{2}P\{|X_\alpha^{(i)} - X_\beta^{(i)}| = |Y_\gamma^{(i)} - Y_\delta^{(i)}|\}, \qquad i = 1, 2; \\ \boldsymbol{\theta}(F_1, F_2) &= (\theta_1(F_1, F_2), \theta_2(F_1, F_2)). \end{aligned} \tag{6.7}$$

Here also, for $(F_1, F_2) \in \mathcal{W}_0$, $\boldsymbol{\theta}(F_1, F_2) = (\frac{1}{2}, \frac{1}{2})$, while for any heterogeneity of scales, $\boldsymbol{\theta}(F_1, F_2) \neq (\frac{1}{2}, \frac{1}{2})$. Thus, if we define

$$\begin{aligned} \phi(a, b; c, d) &= 1, \quad \text{if} \quad |a - b| < |c - d|, \\ \phi(a, b; c, d) &= \tfrac{1}{2}, \quad \text{if} \quad |a - b| = |c - d|, \\ \phi(a, b; c, d) &= 0, \quad \text{if} \quad |a - b| > |c - d|, \end{aligned} \tag{6.8}$$

and write

$$\begin{aligned} U_{Ni} &= \binom{n_1}{2}^{-1}\binom{n_2}{2}^{-1} \sum \phi(X_\alpha^{(i)}, X_\beta^{(i)}; Y_\gamma^{(i)}, Y_\delta^{(i)}), \qquad i = 1, 2; \\ \mathbf{U}_N &= (U_{N1}, U_{N2}); \end{aligned} \tag{6.9}$$

an appropriate permutation test may be based on $\mathbf{U}_N$. Let us also define

$$g_N(Z_\alpha^{(i)}) = \frac{2}{(N-1)(N-2)(N-3)} \sum_{S_\alpha} \phi(Z_\alpha^{(i)}, Z_\beta^{(i)}, Z_\gamma^{(i)}, Z_\delta^{(i)}), \tag{6.10}$$

where the summation S_α extends over all possible choices of distinct β, γ, δ which are not equal to α, and where $\alpha = 1, \cdots, N$, $i = 1, 2$. Finally, let

$$\alpha_{Nij} = \frac{1}{N} \sum_{\alpha=1}^{N} g_N(Z_\alpha^{(i)}) g_N(Z_\alpha^{(j)}) - \tfrac{1}{4}, \qquad i, j = 1, 2. \tag{6.11}$$

It is then easily shown (cf. Sen [21] for the univariate case) that neglecting terms of the order N^{-2},

$$\operatorname{cov}(U_{Ni}U_{NJ}|\mathcal{P}(\mathbf{Z}_N)) = \frac{4N}{n_1 n_2} \alpha_{Nij} + 0(N^{-2}), \qquad \text{for} \quad i, j = 1, 2. \tag{6.12}$$

Thus, neglecting terms of the order N^{-2}, we may construct T_N as in (4.1) and proceed as in section 4. This test is thus an extension of Lehmann's [15] test to the bivariate as well as discrete type of cdf's case.

Finally, let us consider the association problem. Extending the notion of Hoeffding [9] to cover also the case of bivariate discrete distributions, let us define

$$\theta(F_1, F_2) = \theta(F_1) - \theta(F_2), \tag{6.13}$$

where

$$\begin{aligned}\theta(F_1) &= P\{\text{sign }(X_\alpha^{(1)} - X_\beta^{(1)}) \text{ sign }(X_\alpha^{(2)} - X_\beta^{(2)}) > 0\} \\ &\quad + \tfrac{1}{2}P\{(X_\alpha^{(1)} - X_\beta^{(1)})(X_\alpha^{(2)} - X_\beta^{(2)}) = 0\},\end{aligned} \tag{6.14}$$

and $\theta(F_2)$ is defined precisely on the same line with Y_α and Y_β. It may be noted that $\theta(F)$ may be treated as the probability of concordance in the general case, and it is related with another well-known measure of correlation, namely rank correlation τ (cf. Kendall [14]) by means of the simple relation

$$\tau(F) = 2\theta(F) - 1. \tag{6.15}$$

Thus for $(F_1, F_2) \in \mathcal{W}_0$, $\theta(F_1, F_2) = 0$, and it also implies that $\tau(F_1) = \tau(F_2)$.

Now, if we define for two vectors **a** and **b**

$$\begin{aligned}\phi(\mathbf{a}, \mathbf{b}) &= 1 \quad \text{if} \quad (a^{(1)} - b^{(1)})(a^{(2)} - b^{(2)}) > 0, \\ \phi(\mathbf{a}, \mathbf{b}) &= \tfrac{1}{2} \quad \text{if} \quad (a^{(1)} - b^{(1)})(a^{(2)} - b^{(2)}) = 0, \\ \phi(\mathbf{a}, \mathbf{b}) &= 0 \quad \text{if} \quad (a^{(1)} - b^{(1)})(a^{(2)} - b^{(2)}) < 0,\end{aligned} \tag{6.16}$$

and

$$U_N = \binom{n_1}{2}^{-1} \sum \phi(X_\alpha, X_\beta) - \binom{n_2}{2}^{-1} \sum \phi(Y_\alpha, Y_\beta), \tag{6.17}$$

then the permutation test is based on U_N.

This test has been considered by Chatterjee and Sen [3] in the case of continuous cdf's, while Sen [22] has also extended the test to the c-sample as well as discrete case. Hence, this is not considered in detail.

In this paper, we have not considered the asymptotic power efficiency aspect of the tests, discussed above. It may be noted that for bivariate continuous cdf's, Chatterjee and Sen [2] have considered the asymptotic power efficiency of their location test with respect to Hotelling's T^2-test. The asymptotic power efficiency of the association test is also under investigation. The details of this aspect of the tests is being kept pending for a further investigation.

REFERENCES

[1] V. P. Bhapkar, "Some non-parametric tests for the multivariate several sample location problem," *Inst. Statist. Univ. North Carolina*, Mimeo. Ser. No. 415 (1965).

[2] S. K. Chatterjee and P. K. Sen, "Non-parametric tests for the bivariate two sample location problem," *Calcutta Statist. Assoc. Bull.*, Vol. 13 (1964), pp. 18–58.

[3] ———, "Some non-parametric tests for the two sample bivariate association problem," *Calcutta Statist. Assoc. Bull.*, Vol. 14 (1965), pp. 14–35.

[4] ———, "Non-parametric tests for the multisample multivariate location problem," *Ann. Math. Statist.*, submitted.

[5] M. DWASS, "On the asymptotic theory of certain rank-order statistics," *Ann. Math. Statist.*, Vol. 24 (1953), pp. 303–306.

[6] ———, "On the asymptotic normality of some statistics used in non-parametric tests," *Ann. Math. Statist.*, Vol. 26 (1955), pp. 334–339.

[7] D. A. S. FRASER, *Nonparametric Methods in Statistics*, New York, Wiley, 1957.

[8] J. HÁJEK, "Some extensions of the Wald-Wolfowitz-Noether theorem," *Ann. Math. Statist.*, Vol. 32 (1961), pp. 506–523.

[9] W. HOEFFDING, "On the distribution of the rank correlation when the variates are not independent," *Biometrika*, Vol. 34 (1947), pp. 183–196.

[10] ———, "A class of statistics with asymptotically normal distributions," *Ann. Math. Statist.*, Vol. 19 (1948), pp. 293–325.

[11] ———, "A combinatorial central limit theorem," *Ann. Math. Statist.*, Vol. 22 (1951), pp. 558–566.

[12] ———, "The large sample power of tests based on permutation of observations," *Ann. Math. Statist.*, Vol. 23 (1952), pp. 169–192.

[13] ———, "The strong law of large numbers for *U*-statistics," *Inst. Statist. Univ. North Carolina*, Mimeo. Ser. No. 302 (1962).

[14] M. G. Kendall, *Rank Correlation Methods*, New York, Hafner, 1955.

[15] E. L. LEHMANN, "Consistency and unbiasedness of certain non-parametric tests," *Ann. Math. Statist.*, Vol. 22 (1957), pp. 165–179.

[16] E. L. LEHMANN and C. STEIN, "On the theory of some non-parametric hypotheses," *Ann. Math. Statist.*, Vol. 20 (1949), pp. 28–45.

[17] M. MOTOO, "On Hoeffding's combinatorial central limit theorem," *Ann. Inst. Statist. Math.*, Vol. 8 (1957), pp. 145–154.

[18] H. NANDI and P. K. SEN, "On the properties of *U*-statistics when the observations are not independent. Part two: Unbiased estimation of the parameters of a finite population," *Calcutta Statist. Assoc. Bull.*, Vol. 12 (1963), pp. 124–148.

[19] G. E. Noether, "On a theorem by Wald and Wolfowitz," *Ann. Math. Statist.*, Vol. 20 (1949), pp. 455–458.

[20] J. PUTTER, "The treatment of ties in some non-parametric tests," *Ann. Math. Statist.*, Vol. 26 (1955), pp. 368–386.

[21] P. K. SEN, "On some permutation tests based on *U*-statistics," *Calcutta Statist. Assoc. Bull.*, Vol. 14 (1965), pp. 106–126.

[22] ———, "On some multisample permutation tests based on a class of *U*-statistics," *J. Amer. Statist. Assoc.*, submitted.

[23] ———, "*U*-statistics and combination of independent estimates of regular functionals," *Ann. Math. Statist.*, submitted.

[24] E. SVERDRUP, "The limit distribution of a continuous function of random variables," *Skand. Aktuarietidskr.*, Vol. 35 (1952), pp. 1–10.

[25] A. WALD and J. WOLFOWITZ, "Statistical tests based on permutation of observations," *Ann. Math. Statist.*, Vol. 15 (1944), pp. 368–372.

ON SOME QUESTIONS CONNECTED WITH TWO-SAMPLE TESTS OF SMIRNOV TYPE

I. VINCZE
MATHEMATICAL INSTITUTE OF THE HUNGARIAN ACADEMY OF SCIENCES

1. Introduction

1.1. In the following we shall consider some questions concerning the comparison of two samples. The test around which our investigations will center is the Kolmogorov-Smirnov two-sample test, restricted always to the case of equal sample sizes.

In the first part we shall treat the power function for certain alternatives and make some remarks on the efficiency of the test considered in the case of small samples. In the second part some remarks will be given on distributions and limiting distributions occurring in connection with the treated problems. The investigations given here are closely connected with the author's work presented at the Fourth Berkeley Symposium.

1.2. For diminishing the difference in efficiency between parametric and non-parametric tests, the author has in his papers [9], [11] proposed the use of a pair of statistics instead of one statistic. In consequence of the Neyman-Pearson lemma, this results, for given alternatives, in a better test than the one based on either single test statistic. We apply the maximum deviation of the two empirical distribution functions as the first statistic, which ensures the asymptotic consistency of the test. Then we can add to this for several types of alternatives a suitable corresponding pair, for example, the first maximum index, the number of intersections, the Galton statistic, and so on. In order to examine the increase in the efficiency of the two-sample Smirnov test, we shall treat the situation in the case of a special alternative, for which the computation is relatively easy.

In our treatment we make use of the power functions of the original test and of the two-statistic test as well. The power function can be constructed easily in case of a (continuous) alternative containing piecewise linear parts. With such alternatives we can approximate any given alternative. Following Z. W. Birnbaum [1], these kinds of alternatives (for instance, the maximum and minimum alternatives) were treated by many authors in the one-sample case. As we shall see in section 1, this power function can be easily obtained in the two-sample case for all tests for which the distribution of the test statistic under null hypothesis is known; the idea used is the extension of the method used by

Z. W. Birnbaum in his mentioned paper. Nevertheless, this is simple enough for further considerations only when the number of linear sections of the distribution function is small. Further, we need the power function for the more complicated case of two-statistics tests, so we shall work in this first occasion with a very simple alternative knowing that our aim is to discover what we can expect at all from the use of pairs of statistics. We shall consider the power for moderate or small sample sizes which occur very often in the applications. Our formulae give the possibility of carrying out the program of D. Chapman [2] which was done for the one-sample case.

2. The power of two-sample tests for piecewise linear alternatives

2.1. *Notations.* Let ξ and η be random variables with continuous distribution functions $F(x)$ and $G(x)$ resp. for which the null hypothesis H_0: $G(x)$ and $F(x)$ are equally and uniformly distributed in interval (0, 1). We shall treat piecewise linear alternatives depending on vectors:

$$z = (z_0, z_1, \cdots, z_r), \qquad (0 = z_0 < z_1 < z_2 < \cdots < z_r = 1), \tag{2.1}$$

$$g = (g_1, g_2, \cdots, g_r), \tag{2.2}$$

$$(g_i \geq 0, i = 1, 2, \cdots, r) \quad \text{with} \quad \sum_{i=1}^{r} g_i(z_i - z_{i-j}) = 1:$$

$$H_1^{(r)}: \begin{cases} F(x) \text{ as in } H_0, \\ G'(x) = g(x) = \begin{cases} g_i & \text{if } z_{i-1} < x < z_i, \ i = 1, 2, \cdots, r, \\ 0 & \text{otherwise.} \end{cases} \end{cases} \tag{2.3}$$

In this case $\int_{-\infty}^{\infty} g(x)\, dx = \int_0^1 g(x)\, dx = 1.$

Let further $\xi_1, \xi_2, \cdots, \xi_n$ and $\eta_1, \eta_2, \cdots, \eta_n$ be independent observations on ξ and η resp. We denote the elements of the ordered samples by ξ_i^* and η_i^* resp., and the union of the two samples in order of magnitude by

$$\tau_1^* < \tau_2^* < \cdots < \tau_{2n}^*. \tag{2.4}$$

Let us define further the random variables for $i = 1, 2, \cdots, 2n$,

$$\vartheta_i = \begin{cases} +1, & \text{if } \tau_i^* = \xi_j \\ -1, & \text{if } \tau_i^* = \eta_\ell, \end{cases} \tag{2.5}$$

for some j and ℓ. With the usual notation $s_0 = 0$, $s_i = \vartheta_1 + \vartheta_2 + \cdots + \vartheta_i$, $(i = 1, 2, \cdots, 2n)$, $s_{2n} = 0$. The points (i, s_i) in the plane give the path of a random walk starting at the origin and returning after $2n$ steps to the point $(2n, 0)$.

In the present paper we shall consider the following statistics:

$$D_{n,n}^+ = \max_{(x)} (F_n(x) - G_n(x)) = \frac{1}{n} \max_{(i)} s_i, \tag{2.6}$$

$$D_{n,n} = \max_{(x)} |F_n(x) - G_n(x)| = \frac{1}{n} \max_{(i)} |s_i|, \tag{2.7}$$

$$(2.8) \qquad R^+_{n,n} = \min\left\{\frac{i}{n}: \quad s_i = nD^+_{n,n}\right\}$$

(2.9) $\Lambda_{n,n}$ = *the number of intersection points in the above mentioned random path, that is, the number of i's for which* $s_i = 0$ *and* $s_{i-1}s_{i+1} = -1$ *occurs, adding the point* $(2n, 0)$.

As to the numerical determination of the values of these statistics we mention the following. As it was pointed out in [3], $nD^+_{n,n}$ agrees with that index, for which in the translation scheme

$$(2.10) \qquad \begin{matrix} \xi^*_1, \cdots, \xi^*_\kappa, \xi^*_{\kappa+1}, \cdots, \xi^*_{\kappa+s}, \cdots, \xi^*_n \\ \eta^*_1, \cdots, \eta^*_s, \cdots, \eta^*_{n-\kappa}, \cdots, \eta^*_n \end{matrix}$$

it *first* occurs that each $\xi^*_{\kappa+i}$ exceeds the corresponding η^*_i. Translation in the opposite direction leads to $D^-_{n,n}$ and in this way to $D_{n,n}$. Further $nR^+_{n,n} = \kappa + 2s$ where $s = \min\{i: \xi^*_{\kappa+i} < \eta^*_{i+1}\}$, which can be seen easily.

Having the two ordered samples, let us define the random variables

$$(2.11) \qquad \epsilon_i = \begin{cases} +1, & \text{if} \quad \xi^*_i > \eta^*_i, \\ -1, & \text{if} \quad \xi^*_i < \eta^*_i. \end{cases}$$

Then $\Lambda_{n,n} - 1$ equals the number of changes of sign in the sequence $\epsilon_1, \epsilon_2, \cdots, \epsilon_n$ (see [12]).

2.2. *The power function of the two-sample Smirnov test under the hypothesis* H_0 *against* $H_1^{(r)}$. As is known the α-size critical region for one-sided alternatives $F(x) = G(x)$—this case will be treated in more detail later in this article—is determined by the relation

$$(2.12) \qquad P\left(D^+_{n,n} \geq \frac{k}{n} |H_0\right) = \frac{\binom{2n}{n-k}}{\binom{2n}{n}} = \alpha$$

with $k = k(\alpha, n)$.

Let us denote by the vectors $\boldsymbol{\nu} = (\nu_1, \nu_2, \cdots, \nu_r)$ and $\boldsymbol{\mu} = (\mu_1, \mu_2, \cdots, \mu_r)$ the events that out of the n ξ_i's exactly ν_j and (independently) out of the n η_i's exactly μ_j are contained in the intervals (z_{j-1}, z_j) $j = 1, 2, \cdots, r$. The probabilities of these events are clearly

$$(2.13) \qquad \begin{aligned} P(\boldsymbol{\nu}|H_0) &= P(\boldsymbol{\nu}|H_1) = n! \prod_{j=1}^{r} \frac{(z_j - z_{j-1})^{\nu_j}}{\nu_j!}, \\ P(\boldsymbol{\mu}|H_0) &= n! \prod_{j=1}^{r} \frac{(z_j - z_{j-1})^{\mu_j}}{\mu_j!}, \end{aligned}$$

$$(2.14) \qquad P(\boldsymbol{\mu}|H_1^{(r)}) = n! \prod_{j=1}^{r} \frac{[(z_j - z_{j-1})g_j]^{\mu_j}}{\mu_j!};$$

further, for $i = 0, 1$,

$$(2.15) \qquad P(\boldsymbol{\nu}, \boldsymbol{\mu}|H_i) = P(\boldsymbol{\nu}|H_i)P(\boldsymbol{\mu}|H_i).$$

Denoting now for $j = 1, 2, \cdots, r$ by n_j and m_j the partial sums $\nu_1 + \nu_2 +$

$\cdots + \nu_j$ and $\mu_1 + \mu_2 + \cdots + \mu_j$ resp., ($n_0 = m_0 = 0$), then the following holds.

THEOREM 2.1. *For the α-size one-sided two-sample Smirnov test the power function under $H_1^{(r)}$ against H_0 is given by*

$$W_n(H_1^{(r)}, \alpha) = 1 - (n!)^2 \sum_{(\nu)}^* \sum_{(\mu)}^* \prod_{j=1}^{r} \times \left(1 - \frac{\binom{\nu_j + \mu_j}{\mu_j + k - n_{j-1} + m_{j-1}}}{\binom{\nu_j + \mu_j}{\mu_j}}\right) \frac{(z_j - z_{j-1})^{\nu_j + \mu_j} g_j^{\mu_j}}{\nu_j! \mu_j!} \tag{2.16}$$

where $\sum_{(\nu)}^*$ and $\sum_{(\mu)}^*$ denote r-summation for all possible $\boldsymbol{\nu}$ and $\boldsymbol{\mu}$ vectors with $0 \leq \nu_j$, $\mu_j \leq n$ and $\sum_{j=1}^r \nu_j = \sum_{j=1}^r \mu_j = n$ with the further restriction $n_j - m_j < k$, $j = 1, 2, \cdots, r$.

Before outlining the proof we make some remarks.

(a) The expression of the power function seems to be suitable for asymptotic considerations or computational work in general for small values of r. For r large and for small intervals (z_{i-1}, z_i) we are interested in the mutual order of sample elements inside the intervals only for very peculiar alternatives. In these cases the problem can be reduced into the consideration of the null hypothesis

$$H_0^1\colon\ P(\xi \in (z_{i-1}, z_i)) = P(\eta \in (z_{i-1}, z_i)) \qquad \text{for } i = 1, 2, \cdots, r. \tag{2.17}$$

This means that our problem is the comparison of two multinomial distributions, which is treated recently by Hoeffding [6].

(b) If n remains finite and r tends to infinity, then we come essentially to the evaluation of the probabilities of each different array of the sample elements, arrays which belong to the critical region. This kind of expression was given by Hoeffding [5] and considered by Lehmann [7].

(c) In the two-sided case the combinatorial quantity just after the product sign is to be replaced by the expression corresponding to the two-barrier case

$$\frac{1}{\binom{\nu_j + \mu_j}{\mu_j}} \sum_{\gamma=-\infty}^{\infty} \left[\binom{\nu_j + \mu_j}{\nu_j + 2\gamma k} - \binom{\nu_j + \mu_j}{\nu_j + n_{j-1} - m_{j-1} + (2\gamma + 1)k}\right], \tag{2.18}$$

where for the summation the restriction $|n_j - m_j| < k$, $j = 1, 2, \cdots, r$ is to be made.

The proof is the consequence of some simple arguments which are used often for similar purposes and which are the following.

LEMMA 2.1. *Let ξ' be a continuous random variable in the interval* $(0, 1)$ *whose density function is constant in a subinterval* (a, b). *Then ξ' is uniformly distributed in* (a, b) *under the condition* $\{a < \xi' < b\}$.

A consequence of this is the following lemma.

LEMMA 2.2. *Let ξ' and η' be continuous random variables in the interval* $(0, 1)$, *the density functions of which are* (*not necessarily equal*) *constants in a subinterval* (a, b). *Let $\xi_1', \xi_2', \cdots, \xi_\nu'$ and $\eta_1', \eta_2', \cdots, \eta_\mu'$ be independent observations on ξ' and*

η' resp. falling in the interval (a, b). Under this condition, all possible arrays of the mentioned $\nu + \mu$ sample elements have the common probability $\binom{\nu+\mu}{\mu}^{-1}$.

A consequence of this lemma is that in each interval the conditional probability that $s_i < k$ given $\{\boldsymbol{\mu}, \boldsymbol{\nu}\}$, can be calculated under H_0.

Turning now to the determination of the probability

$$1 - W_n(H_1^{(r)}, \alpha) = P\left(D_{n,n}^+ < \frac{k}{n} | H_1^{(r)}\right), \tag{2.19}$$

we calculate this conditionally given $\{\boldsymbol{\mu}, \boldsymbol{\nu}\}$ and multiply this by $P(\boldsymbol{\nu}, \boldsymbol{\mu}|H_1^{(r)})$; then we have to sum over all $\boldsymbol{\nu}$ and $\boldsymbol{\mu}$. But under the mentioned condition the random walk falls into parts with division points $(i_j = n_j + m_j,\ s_{i_j} = n_j - m_j)$, $(j = 1, 2, \cdots, r)$ and according to the Markov property of the random walk, the relation

$$P\left(D_{n,n}^+ < \frac{k}{n} \mid \boldsymbol{\mu}, \boldsymbol{\nu}, H_1^{(r)}\right) = \prod_{j=1}^{r} P(s_i < k, \text{ for } n_{j-1} + m_{j-1} < i < n_j + m_j|\boldsymbol{\mu}, \boldsymbol{\nu}, H_1^{(r)}) \tag{2.20}$$

holds. Taking into account the elementary formula

$$P(s_i < k, \text{ for } n_{j-1} + m_{j-1} < i < n_j + m_j|\boldsymbol{\mu}, \boldsymbol{\nu}, H_1^{(r)}) = 1 - \frac{\binom{\nu_j + \mu_j}{\mu_j + k - n_{j-1} + m_{j-1}}}{\binom{\nu_j + \mu_j}{\mu_j}}, \tag{2.21}$$

and the relations (2.13), (2.14), and (2.15), we come to theorem 2.1.

2.3. *A special case.* For our comparative considerations we shall treat the following very simple alternative

$$H_1^{(2)}: \begin{cases} F(x) & \text{uniformly distributed in } (1, 1), \\ G'(x) = g(x) = \begin{cases} 0 & \text{if } 0 \le x < z, \\ \dfrac{1}{1-z} & \text{if } z \le x \le 1. \end{cases} \end{cases} \tag{2.22}$$

For this kind of alternative the first index of the maximum will occur with higher probability for smaller values than in the case of the null hypothesis. This alternative being simple enough, I have chosen it for a first comparison of tests based on 1 and on 2 statistics respectively.

Making use of the notation $\nu_1 = \nu$, then $\nu_2 = n - \nu$, and further, knowing that $P(\mu_1 = 0, \mu_2 = n|H_1^{(2)}) = 1$, we obtain for the power function

$$W_n(H_1^{(2)}, \alpha) = 1 - \sum_{\nu=0}^{k-1} \binom{n}{\nu} z^\nu (1 - z)^{n-\nu} \left(1 - \frac{\binom{2n - \nu}{n - \nu + k}}{\binom{2n - \nu}{n - \nu}}\right). \tag{2.23}$$

This simple case shows that in determining the power, the normal approximation of the binomial terms is not suitable, since $\nu < k(\alpha, n) \sim y_\alpha\sqrt{2n}$ when $n \to \infty$. A similar remark was made by J. Rosenblatt [8].

After some modification we may obtain the following form:

$$W_n(H_1^{(2)}, \alpha) = 1 - B_{1-z}(n - k + 1, k) + \frac{\alpha}{(1 - z)^k} B_{1-z}(n + 1, k), \tag{2.24}$$

where B is the beta-function

$$B_z(p, q) = \frac{\int_0^z t^{p-1}(1 - t)^{q-1}\, dt}{\int_0^1 t^{p-1}(1 - t)^{q-1}\, dt}. \tag{2.25}$$

This is suitable for immediate computation of the power against this simple kind of alternative. Now we give some numerical values for small samples using (n, k) pairs for which α is near 10% and $z = \max_{(x)} (F(x) - G(x)) = 0.1$; 0.2; 0.3. The following table gives the errors of the second kind for these values of z.

TABLE I

n	k	α	$z = 0.1$	$z = 0.2$	$z = 0.3$
20	7	0.0873	0.8177	0.6163	0.3363
30	8	0.1197	0.7230	0.4039	0.1164
40	9	0.1331	0.6587	0.2640	0.0377
50	11	0.7190	0.7190	0.2700	0.0200

2.4. *The power function of the test based on a pair of statistics.* Let us consider the two statistics $D_{n,n}^+$ and $R_{n,n}^+$ for the decision between H_0 and $H_1^{(2)}$, defined in 2.3. We shall introduce the random variable $S_{n,n}^+ = R_{n,n}^+ - D_{n,n}^+$. As $nD_{n,n}^+$ and $nR_{n,n}^+$ are of the same parity, $S_{n,n}^+$ is always even. Let us use the following notations:

$$P\left(D_{n,n}^+ = \frac{k}{n}, S_{n,n}^+ = \frac{s}{n} \mid H_0\right) = P_{k,s}^{(n)} \tag{2.26}$$

and

$$P\left(D_{n,n}^+ = \frac{k}{n}, S_{n,n}^+ = \frac{s}{n} \mid H_1\right) = Q_{k,s}^{(n)}. \tag{2.27}$$

Denoting the best critical region of the α-size test (restricting ourselves to the (k, s) plane) by $\mathcal{K}_\alpha$, this is defined with the aid of a suitable constant c_α, and can be written in the form

$$\mathcal{K}_\alpha = \left\{(s, k):\ \frac{Q_{k,s}^{(n)}}{P_{k,s}^{(n)}} > c_\alpha\right\} \tag{2.28}$$

and

$$\sum_{(k,s) \in \mathcal{K}_\alpha} P_{k,s}^{(n)} = \alpha. \tag{2.29}$$

The power function is

$$W_n(H_1^{(2)}, \alpha) = \sum_{(k,s) \in \mathcal{K}_\alpha} Q_{k,s}^{(n)}. \tag{2.30}$$

We need the probabilities $P_{k,s}^{(n)}$ and $Q_{k,s}^{(n)}$.

As it was proved in [10] the following relations are valid:

$$P_{0,s}^{(n)} = \frac{1}{2(2s-1)(n-s+1)} \frac{\binom{2s}{s}\binom{2n-2s}{n-s}}{\binom{2n}{n}}, \qquad s = 1, 2, \cdots, n, \tag{2.31}$$

$$P_{k,s}^{(n)} = \frac{k(k+1)}{(k+2s)(n-s+1)} \frac{\binom{k+2s}{s}\binom{-k+2n-2s}{n-s}}{\binom{2n}{n}}, \tag{2.32}$$

$$s = 0, 1, 2, \cdots, n-k.$$

We turn now to the determination of the probabilities $Q_{k,s}^{(n)}$. The number of ξ_i's in the interval $(0, z)$ may be $\nu = 0, 1, 2, \cdots, k$. As was mentioned, $\mu_1 = 0$ with probability 1. For $k = 0$, ν must be 0 and $P(\nu) = P(\nu = 0, n - \nu = n) = (1-z)^n$. In this case it can be seen immediately that

$$Q_{0,s}^{(n)} = P_{0,s}^{(n)}(1-z)^n, \qquad s = 1, 2, \cdots, n. \tag{2.33}$$

The case $s = 0$ (that is, $R_{n,n}^+ = D_{n,n}^+$) can happen only when $\nu = k$. In this case we have the second part of the path starting with probability 1 from the point $(k, s_k = k)$ and ending at $(2n, 0)$. The probability that this path will never reach the height $s_i = k + 1$ multiplied by $P(\nu) = P(\nu = k) = \binom{n}{k} z^k(1-z)^{n-k}$ gives the required probability

$$Q_{k,0}^{(n)} = \frac{k+1}{2n-k+1} \frac{\binom{2n-k+1}{n+1}}{\binom{2n-k}{n}} \binom{n}{k} z^k (1-z)^{n-k} \tag{2.34}$$

$$= \frac{k+1}{n+1} \frac{\binom{2n-k}{n}}{\binom{2n}{n}} \binom{2n}{k} z^k (1-z)^{n-k}, \qquad k = 1, 2, \cdots, n.$$

At the end, for $s > 0$, $k > 0$, we can construct the power function as given in 2.2 and evaluate the joint probabilities for the maximum and first maximum index in the case of the several ν's.

The resulting formula is

$$(2.35) \qquad Q_{k,s}^{(n)} = \frac{k+1}{n-s+1} \frac{\binom{-k+2n-2s}{n-s}}{\binom{2n}{n}} \sum_{\nu=0}^{k-1} \frac{k-\nu}{n-\nu} \binom{k-\nu+2s}{s} \binom{2n}{\nu} z^{\nu}(1-z)^{n-\nu}.$$

Using the above formulas for numerical calculation we can compare the second kind of errors for the use of $D_{n,n}^{+}$ alone and of the pair $(D_{n,n}^{+}, R_{n,n}^{+})$ respectively. The results are tabulated in table II. (The computation was carried out on a GIER electronic computer. I am indebted to A. Békéssy and G. Tusnády for their kind help in accomplishing these calculations.)

TABLE II

n	Error of first kind $D_{n,n}^{+}$	Error of first kind $(D_{n,n}^{+}, R_{n,n}^{+})$	Error of second kind, if $\Delta = z = 0, 1$: $D_{n,n}^{+}$	Error of second kind, if $\Delta = z = 0, 1$: $(D_{n,n}^{+}, R_{n,n}^{+})$	Error of second kind, if $\Delta = z = 0, 2$: $D_{n,n}^{+}$	Error of second kind, if $\Delta = z = 0, 2$: $(D_{n,n}^{+}, R_{n,n}^{+})$
10	0.0739	0.0839	0.8581	0.8071	0.7532	0.6198
	0.2005	0.2038	0.6815	0.5849	0.5215	0.3683
30	0.0675	0.0675	0.8263	0.6149	0.5573	0.2568
		0.0893		0.5636		
	0.1197	0.1197	0.7230	0.5036	0.4039	0.1703
50	0.0562	0.0562	0.8017	0.5105	0.3812	0.1179
	0.0893	0.0893	0.7190	0.4299	0.2700	0.0780
	0.1362	0.1362	0.6133	0.3488	0.1733	0.0473

3. The maximum and the number of intersections

3.1. *The nonconsistency of the number of intersections.* In our paper with E. Csáki [3] we considered the statistic $\Lambda_{n,n}$ (see 2.1), that is, the number of intersections. As a test statistic this has the following advantages. As we mentioned in 1.1, its value can be determined very easily. The distribution of $\Lambda_{n,n}$ under H_0 is very simple too:

$$(3.1) \qquad P(\Lambda_{n,n} = \ell | H_0) = \frac{2\ell}{n} \frac{\binom{2n}{n-\ell}}{\binom{2n}{n}}, \qquad \ell = 1, 2, \cdots, n.$$

In addition to these, the test based on $\Lambda_{n,n}$ has the same properties for one-sided and two-sided alternatives.

We conjectured that this statistic is consistent against all continuous alternatives. The grounds for this conjecture were the following: if $\max_{(x)} (F(x) - G(x)) = \Delta > 0$, then with probability greater than zero the statistic $nD_{n,n}^{+}$ will take values of order of magnitude n. Now the following theorem is valid.

THEOREM 3.1 (E. Csáki and I. Vincze). *Let* $\vartheta_1, \vartheta_2, \cdots$ *be independent random variables with* $P(\vartheta_i = 1) = P(\vartheta_i = -1) = 1/2$. *Let* $S_0 = 0$, $S_i = \vartheta_1 + \vartheta_2 + \cdots + \vartheta_i$, $i = 1, 2, \cdots$, *and let us define* λ_n *as the number of* i*'s for which* $S_i = 0$, $S_{i-1}S_{i+1} = -1$, $1 < i < n$. *Then the following relation holds with* $0 < c \leq 1$,

$$\lim_{n\to\infty} P(\lambda_n = \ell | S_n \sim cn) = \frac{2c}{1+c}\left(\frac{1-c}{1+c}\right)^{\ell}, \qquad \ell = 0, 1, 2, \cdots. \tag{3.2}$$

Further if $\psi(n) \to \infty$ *and* $\psi(n)/n^{1/2} < 1$, *then*

$$\lim_{n\to\infty} P\left(\lambda_n < \frac{n^{1/2}}{\psi(n)} y | S_n \sim c\psi(n)n^{1/2}\right) = 1 - e^{-2cy}, \qquad y > 0. \tag{3.3}$$

Now from this argument we would think that for n large enough the case $F(x) \equiv G(x)$ will lead to the greatest number of intersections (in probability), and thus the critical region will be the small values of $\Lambda_{n,n}$.

But E. M. Sarhan (unpublished) has given an example that $\Lambda_{n,n}$ is not consistent against the following alternative in (0, 1) for $z > 0$,

$$F(x) \equiv x, \qquad 0 \leq x \leq 1 \tag{3.4}$$

$$G(x) = \begin{cases} x, & \text{if } 0 \leq x < z, \\ \dfrac{x^2 + z}{1 + z}, & \text{if } z \leq x \leq 1. \end{cases} \tag{3.5}$$

On the other hand, he showed that the test based on $\Lambda_{n,n}$, using for the decision between H_0 and $H_1^{(2)}$ defined in 2.3, is more efficient than the one-sided Kolmogorov-Smirnov test. This way $\Lambda_{n,n}$ as a test statistic—by itself or in addition to the Smirnov statistic—seems not to be without interest.

3.2. *Joint distribution of the maximum deviation and the number of intersections.* In our paper [4] with E. Csáki the generating function of $D_{n,n}$ and $\Lambda_{n,n}$ is determined under H_0, which is the following

$$\sum_{n=1}^{\infty} \binom{2n}{n} P\left(D_{n,n} < \frac{k}{n}, \Lambda_{n,n} = \ell | H_0\right) z^n = 2\left(\frac{w - w^k}{1 - w^{k+1}}\right)^{\ell}, \qquad k, \ell = 1, 2, \cdots \tag{3.6}$$

where

$$w = \frac{1 - \sqrt{1 - 4z}}{1 + \sqrt{1 + 4z}}. \tag{3.7}$$

We can obtain without any difficulty the probabilities by series-expansion; these are the following:

$$P\left(D_{n,n} < \frac{k}{n}, \Lambda_{n,n} = \ell | H_0\right) = \frac{1}{\binom{2n}{n}} \sum_{i=0}^{\ell} \sum_{j=0}^{\infty} (-1)^{\ell} \binom{\ell}{i} \binom{\ell + j - 1}{j} \times \binom{2n}{n + i(k-1) + j(k+1) + \ell} \frac{i(k-1) + j(k+1) + \ell}{n}. \tag{3.8}$$

This formula is not suitable for the determination of the limiting distribution as $n \to \infty$, because in the sum each term tends to infinity. The way of solving the problem was the evaluation out of the integral form

$$\lim_{n\to\infty} \frac{\sqrt{2n}}{\binom{2n}{n}} \frac{2}{2\pi i} \oint \left(\frac{w - w^k}{1 - w^{k+1}}\right)^{l} \frac{dz}{z^{n+1}}, \tag{3.9}$$

where the integration path is a small circle around the origin. This was kindly done by N. G. de Bruijn, which we give in the following theorem.

THEOREM 3.2 (N. G. de Bruijn). *If* $x > 0$, $y > 0$, *then*

$$\lim_{n\to\infty} P\left(\sqrt{\frac{n}{2}}\, D_{n,n} < y,\, x \leq \frac{1}{\sqrt{2n}} \Lambda_{n,n} < x + \Delta x | H_0\right) \tag{3.10}$$
$$= \frac{2}{i\sqrt{2\pi}} \int_{1-i\infty}^{1+i\infty} \exp\left\{-2xu \frac{e^{uy} + e^{-uy}}{e^{uy} - e^{-uy}} + \tfrac{1}{2}u^2\right\} u\, du\, \Delta x + \sigma(\Delta x)$$

hold.

To the proofs of theorems 3.1 and 3.2 and a detailed treatment of the questions of the joint distribution law, we should like to return later.

REFERENCES

[1] Z. W. BIRNBAUM, "On the power of a one sided test of fit for continuous probability functions," *Ann. Math. Statist.*, Vol. 24, pp. 484–489.

[2] D. G. CHAPMAN, "A comparative study of several one-sided goodness-of-fit tests," *Ann. Math. Statist.*, Vol. 29 (1958), pp. 655–674.

[3] E. CSÁKI and I. VINCZE, "On some problems connected with the Galton-test," *Publ. Math. Inst. Hungar. Acad. Sci.*, Vol. 6 (1961), pp. 97–109.

[4] ———, "Two joint distribution laws in the theory of order statistics," *Mathematica (Cluj)*, Vol. 5 (1963), pp. 27–37.

[5] W. HOEFFDING, " 'Optimum' nonparametric tests," *Proceedings of the Second Berkeley Symposium on Mathematical Statistics and Probability*, Berkeley and Los Angeles, University of California Press, 1951, pp. 83–92.

[6] ———, "Asymptotically optimal tests for multinomial distributions," *Ann. Math. Statist.*, Vol. 36 (1965), pp. 369–401.

[7] E. L. LEHMANN, "The power of rank tests," *Ann. Math. Statist.*, Vol. 24 (1953), pp. 23–42.

[8] J. ROSENBLATT, "Some modified Kolmogorov-Smirnov tests of approximate hypotheses and their properties," *Ann. Math. Statist.*, Vol. 33 (1962), pp. 513–524.

[9] I. VINCZE, "Einige zweidimensionale Verteilungs und Grenzverteilungssätze in der Theorie der geordneten Stichproben," *Publ. Math. Inst. Hungar. Acad. Sci.*, Vol. 2 (1957), pp. 183–209.

[10] ———, "On some joint distributions and joint limiting distributions in the theory of order statistics," *Publ. Math. Inst. Hungar. Acad. Sci.*, Vol. 4 (1959), pp. 29–47.

[11] ———, "On two-sample tests based on order statistics," *Proceedings of the Fourth Berkeley Symposium on Mathematical Statistics and Probability*, Berkeley and Los Angeles, University of California Press, 1961, Vol. 1, pp. 695–705.

[12] ———, "Über einige Verteilungssätze in der Theorie der geordneten Stichproben," *Abh. Deutsch. Akad. Wiss. Berlin*, Vol. 4 (1964), pp. 123–126.